Improving Sustainable Viticulture and Winemaking Practices

Improving Sustainable Viticulture and Winemaking Practices

Edited by

J. Miguel Costa
LEAF — Centre Linking Landscape, Environment, Agriculture and Food, Instituto Superior de Agronomia, University of Lisbon, Lisbon, Portugal

Sofia Catarino
LEAF — Centre Linking Landscape, Environment, Agriculture and Food, Instituto Superior de Agronomia, University of Lisbon, Lisbon, Portugal

José M. Escalona
Agro-Environmental and Water Economics Institute (INAGEA-UIB) Resarch Group of Plant Biology under Mediterranean Conditions (PLATMED), Department of Biology, University of Balearic Islands, Palma, Spain

Piergiorgio Comuzzo
Department of Agricultural, Food, Environmental and Animal Sciences, University of Udine, Udine, Italy

Academic Press is an imprint of Elsevier
125 London Wall, London EC2Y 5AS, United Kingdom
525 B Street, Suite 1650, San Diego, CA 92101, United States
50 Hampshire Street, 5th Floor, Cambridge, MA 02139, United States
The Boulevard, Langford Lane, Kidlington, Oxford OX5 1GB, United Kingdom

Copyright © 2022 Elsevier Inc. All rights reserved.

No part of this publication may be reproduced or transmitted in any form or by any means, electronic or mechanical, including photocopying, recording, or any information storage and retrieval system, without permission in writing from the publisher. Details on how to seek permission, further information about the Publisher's permissions policies and our arrangements with organizations such as the Copyright Clearance Center and the Copyright Licensing Agency, can be found at our website: www.elsevier.com/permissions.

This book and the individual contributions contained in it are protected under copyright by the Publisher (other than as may be noted herein).

Notices

Knowledge and best practice in this field are constantly changing. As new research and experience broaden our understanding, changes in research methods, professional practices, or medical treatment may become necessary.

Practitioners and researchers must always rely on their own experience and knowledge in evaluating and using any information, methods, compounds, or experiments described herein. In using such information or methods they should be mindful of their own safety and the safety of others, including parties for whom they have a professional responsibility.

To the fullest extent of the law, neither the Publisher nor the authors, contributors, or editors, assume any liability for any injury and/or damage to persons or property as a matter of products liability, negligence or otherwise, or from any use or operation of any methods, products, instructions, or ideas contained in the material herein.

ISBN: 978-0-323-85150-3

For information on all Academic Press publications visit our website at https://www.elsevier.com/books-and-journals

Publisher: Nikki P. Levy
Acquisitions Editor: Megan R. Ball
Editorial Project Manager: Jai Marie Jose
Production Project Manager: Kumar Anbazhagan
Cover Designer: Mark Rogers

Typeset by TNQ Technologies

Contents

Contributors xvii
Acknowledgment to the external reviewers xxv
About the cover xxvii

CHAPTER 1 Achieving a more sustainable wine supply chain—Environmental and socioeconomic issues of the industry 1
J. Miguel Costa, Sofia Catarino, José M. Escalona and Piergiorgio Comuzzo

   1.1 Sustainability concept and issues 1
   1.2 The state of the wine industry—short overview 3
       1.2.1 The wine industry worldwide 3
       1.2.2 Risks and concerns of the modern wine industry 5
   1.3 Sustainability issues in wine industry 6
       1.3.1 Vineyard issues 6
       1.3.2 Winemaking issues 8
       1.3.3 Supply chain issues 11
   1.4 Legislation, standards, and certification of the wine sector—focus on the EU 12
       1.4.1 Legislation issues for sustainable soil and water management 12
       1.4.2 Wine quality certification issues: origin, quality, and socioenvironmentally sound 13
   1.5 Future prospects 14
       References 16
       Further reading 24

CHAPTER 2 Exploiting genetic diversity to improve environmental sustainability of Mediterranean vineyards 25
Josefina Bota, Rosa Arroyo-Garcia, Ignacio Tortosa and Hipólito Medrano

   2.1 Introduction 25
   2.2 Origin of cultivated grapevine and actual grapevine diversity 26
   2.3 Intercultivar variability in the physiological response to water stress 28
   2.4 Intracultivar variability in the physiological response to changing environments 30
   2.5 Rootstocks selection for better performance under semiarid conditions 31
   2.6 Progress in genomics tools and new breeding technologies 34
   2.7 Concluding remarks 35
       Acknowledgments 35
       References 35

CHAPTER 3 Optimizing conservation and evaluation of intravarietal grapevine diversity **45**
Elsa Gonçalves and Antero Martins

3.1 Introduction 45
3.2 Grapevine methodology for conservation, evaluation, and selection within a variety 47
    3.2.1 Representative sampling of intravarietal diversity 47
    3.2.2 Conservation of intravarietal diversity..47
    3.2.3 Evaluation of intravarietal diversity and polyclonal selection 49
    3.2.4 Establishment of multienvironmental trials for clonal selection......51
3.3 Advances in the methods for evaluation of genetic intravarietal grapevine diversity...53
3.4 Practical applications in Portugal..59
3.5 Concluding remarks 62
    Acknowledgments..62
    References..62

CHAPTER 4 Phenotyping for drought tolerance in grapevine populations: the challenge of heterogeneous field conditions **65**
Aude Coupel-Ledru, Eric Lebon, Jean-Pascal Goutouly, Angélique Christophe, Pilar Gago, Charlotte Brault, Patrice This, Agnès Doligez and Thierry Simonneau

4.1 Introduction 65
4.2 Phenotyping large populations in the field: the challenge of soil heterogeneity...66
    4.2.1 Variations in soil characteristics hinder drought tolerance studies ...66
    4.2.2 Statistical methods to handle soil heterogeneity and spatial variations..67
    4.2.3 Phenotyping for plant performance under water deficit: which traits for high-throughput measurements in the field?.............................67
4.3 Detection of genetic variability for water-use efficiency in field conditions: a case study 68
    4.3.1 Experimental setup..68
    4.3.2 Contrasted soil water scenarios established over the two years 69
    4.3.3 Spatial distribution of predawn leaf water potential within the field ...71
    4.3.4 Relationship between carbon isotope composition ($\delta^{13}C$) and predawn leaf water potential..72
    4.3.5 $\delta^{13}C$ correction procedure and effect of irrigation regimes on $\delta^{13}C$ measured on the whole progeny..73
    4.3.6 Genetic variability of $\delta^{13}C$ and QTL detection................................75

    4.4 Main outcomes 78
        4.4.1 Taking into account spatial heterogeneity of soil water deficit within blocks improves statistical power for QTL detection 78
        4.4.2 QTL detection under field conditions reveals new genomic regions as compared to those obtained on potted plants in phenotyping platforms with controlled conditions 78
        4.4.3 Minimal predawn leaf water potential in control plots was the best predictor of $\delta^{13}C$ measured in must 79
    4.5 Conclusions 79
        Acknowledgments 80
        References 80

CHAPTER 5 Soil management in sustainable viticultural systems: an agroecological evaluation... 85
Johanna Döring, Matthias Friedel, Maximilian Hendgen, Manfred Stoll and Randolf Kauer

    5.1 Introduction 85
    5.2 Sustainable management systems and their properties toward the avoidance of soil threats and the provision of soil ES 88
        5.2.1 Avoidance of soil compaction 88
        5.2.2 Erosion control 90
        5.2.3 Water quality and supply 90
        5.2.4 Avoidance of contamination for habitat provision 91
        5.2.5 Biodiversity conservation 92
        5.2.6 Biomass production 94
        5.2.7 Climate regulation 95
    5.3 Implications for future soil management of vineyards 96
        Acknowledgments 97
        References 97

CHAPTER 6 Vineyard water balance and use 105
Ignacio Buesa, Pascual Romero-Azorín, José M. Escalona and Diego S. Intrigliolo

    6.1 The water balance concept: from the single leaf to the whole vineyard 105
    6.2 Grapevine water status assessment: from soil to atmosphere 108
        6.2.1 Main indicators of soil-plant-atmosphere water status 109
    6.3 Vineyard water needs: crop coefficients in relation to vegetative development (LAI) and reproductive cycle. Crop stress coefficients 111
    6.4 Water-saving strategies and irrigation scheduling 113
    6.5 Use of nonconventional water for irrigation: wastewater and saline water. Effects on vine performance and grape composition 115

		6.5.1 Effects on vine performance and grape composition	117
	6.6	Concluding remarks	118
		Acknowledgments	118
		References	118

CHAPTER 7   Modern approaches to precision and digital viticulture ... 125
Sigfredo Fuentes and Jorge Gago

7.1	Introduction	125
7.2	Remote sensing for vineyard management	127
7.3	Artificial intelligence and remote sensing	132
	7.3.1 Computer vision	132
	7.3.2 Machine and deep learning in viticulture and winemaking	133
7.4	Conclusion	139
	References	139

CHAPTER 8   Novel technologies and Decision Support Systems to optimize pesticide use in vineyards ... 147
Cristina C.R. Carlos and Maria do Carmo M. Val

8.1	Introduction	147
8.2	Disease management	150
8.3	Pest management	153
8.4	Concluding remarks	159
	References	161

CHAPTER 9   Processed kaolin particles film, an environment friendly and climate change mitigation strategy tool for Mediterranean vineyards ... 165
Lia-Tânia Dinis, Tommaso Frioni, Sara Bernardo, Carlos Correia and José Moutinho-Pereira

9.1	Introduction	165
9.2	Climate change effects	166
	9.2.1 Phenology	166
	9.2.2 Physiology	166
	9.2.3 Leaf metabolites	167
	9.2.4 Yield and berry quality attributes	168
9.3	Kaolin case: short-term adaptation strategy	168
	9.3.1 Kaolin characterization	168
	9.3.2 Reflection of radiation excess and reduction of organ temperature	169
	9.3.3 Kaolin effects on vine water status and photosynthetic activity	171
	9.3.4 Impact on leaf metabolism	176
	9.3.5 Impact on berries and wine	177
9.4	Kaolin impacts: pros and cons	178
	9.4.1 For the environment	178
	9.4.2 Costs	178

	9.5	Concluding remarks and prospects...178
		Acknowledgments...180
		References..180

CHAPTER 10 Wine quality production and sustainability 187
Pierre-Louis Teissedre, Sofia Catarino and Piergiorgio Comuzzo

	10.1	Introduction..187
	10.2	Existing systems and initiatives at winery level..188
	10.3	Principal aspects to consider for a sustainable wine production......................189
		10.3.1 Carbon dioxide reuse...189
		10.3.2 Water management and saving 191
		10.3.3 Renewable energy 192
		10.3.4 Good practices in Oenology and winemaking process 192
		10.3.5 Functional biodiversity..193
		10.3.6 Management and use of by-products in Oenology............................194
	10.4	Concluding remarks..195
		References 196

CHAPTER 11 Water management toward regenerative wineries 201
Margarida Oliveira, Artur Saraiva, Milena Lambri, Joel Rochard, Rita Fragoso, Elia Romanini, Pedro Hipólito, Capri Ettore and Elizabeth Duarte

	11.1	Introduction..201
	11.2	Environmental impacts 202
	11.3	Regenerative wineries..203
		11.3.1 The water cycle 203
		11.3.2 Strategies toward regenerative wineries..206
	11.4	Case studies 208
		11.4.1 Case study—Portugal..208
		11.4.2 Case study—France...211
		11.4.3 Case study—Italy...212
		11.4.4 General overview and future challenges..213
	11.5	Conclusions..215
		References 215

CHAPTER 12 Energy use and management in the winery 221
Matia Mainardis and Rino Gubiani

	12.1	Introduction..221
	12.2	Energy audit in wineries..222
	12.3	Energy consumption in the winery 223
	12.4	Methodologies for reduction of energy demand..226

	12.5	Renewable energy utilization...228
		12.5.1 Anaerobic digestion..228
		12.5.2 Thermochemical conversion processes...................................229
		12.5.3 Solar systems...230
	12.6	Energy consumption and optimization in wineries: some case studies.............231
		12.6.1 Energy audit of an Italian winery..231
		12.6.2 TESLA research project..233
		12.6.3 Energy assessment related to wineries located in Veneto (Italy)....233
	12.7	Concluding remarks..236
		References..236

CHAPTER 13 Microbiological control of wine production: new tools for new challenges ... 239
M. Carmen Portillo and Albert Mas

	13.1	Introduction...239
	13.2	New tools..240
		13.2.1 "Omics" technologies: genomics, metagenomics, transcriptomics, metatranscriptomics, proteomics, and metabolomics........................240
		13.2.2 Genome editing: CRISPR/Cas9...244
	13.3	New challenges..246
		13.3.1 Grape microbiome and its control..246
		13.3.2 Reduction of SO_2 use...246
		13.3.3 Spontaneous versus inoculated fermentations.........................248
		13.3.4 The search for new strains..251
	13.4	Concluding remarks..251
		Acknowledgments..252
		References..252
		Further reading..258

CHAPTER 14 Sustainable use of wood in wine spirit production ... 259
*Sara Canas, Ilda Caldeira, Tiago A. Fernandes, Ofélia Anjos,
António Pedro Belchior and Sofia Catarino*

	14.1	Introduction...259
	14.2	The aged wine spirit and its production process.......................................260
		14.2.1 Wine spirit definition...260
		14.2.2 Technological process of aged wine spirit production.................261
		14.2.3 Main production regions worldwide..262
		14.2.4 Regulations...262
	14.3	The aging stage..264
		14.3.1 Main physicochemical phenomena and determining factors.........264
		14.3.2 The wood...265

	14.3.3 The aging technology	268
	14.3.4 How to assure a more sustainable aging using wooden barrels?	268
	14.3.5 Innovative technologies for wine spirit's aging	270
14.4	Concluding remarks	273
	Acknowledgments	274
	References	274
	Further reading	279

CHAPTER 15 Innovative processes for the extraction of bioactive compounds from winery wastes and by-products ... 281
Gianpiero Pataro, Daniele Carullo and Giovanna Ferrari

15.1	Introduction	281
15.2	Extraction technologies for bioactive compounds	284
15.3	Innovative extraction methods	285
	15.3.1 Electrotechnologies	285
	15.3.2 Ultrasound-assisted extraction	293
	15.3.3 Microwave-assisted extraction	294
	15.3.4 Supercritical fluids extraction	296
	15.3.5 Subcritical fluids extraction	297
15.4	Concluding remarks	297
	References	298

CHAPTER 16 The role of pressure-driven membrane processes on the recovery of value-added compounds and valorization of lees and wastewaters in the wine industry ... 305
Alexandre Giacobbo, Andréa Moura Bernardes and Maria Norberta de Pinho

16.1	Introduction	305
16.2	Value-added compounds found in wastewaters and by-products generated in wine industries	307
	16.2.1 Phenolic compounds	312
	16.2.2 Polysaccharides	313
16.3	General aspects about the recovery of value-added compounds from agro-industrial by-products and wastewaters	314
16.4	General aspects over pressure-driven membrane processes	315
16.5	PDMP in the recovery of polysaccharides and phenolic compounds	316
	16.5.1 Processing lees and winery wastewater	316
	16.5.2 Processing extracts from other winemaking by-products	319
16.6	Concluding remarks	320
	References	321

CHAPTER 17 Sustainable approach to quality control of grape and wine 327
Piergiorgio Comuzzo, Andrea Natolino and Emilio Celotti

17.1 Introduction and principles of green chemistry 327
17.2 Green Analytical Chemistry 328
17.3 Greening of analytical procedures 328
    17.3.1 Sampling 328
    17.3.2 Analytical methods and instruments 329
    17.3.3 Solvents and reagents 329
17.4 Sustainable grape analysis and quality control 331
    17.4.1 Laboratory methods 331
    17.4.2 On-field monitoring 331
    17.4.3 Grape quality control at delivery 334
17.5 Sustainable wine analysis and quality control 336
    17.5.1 Sustainability issues in winery labs 337
    17.5.2 Automation in winery labs 339
    17.5.3 Sustainability issues in service labs, public labs and research laboratories 341
17.6 Concluding remarks 342
    References 343

CHAPTER 18 Life cycle methods and experiences of environmental sustainability assessments in the wine sector 351
Almudena Hospido, Beatriz Rivela and Cristina Gazulla

18.1 The wine supply chain: from land to table 351
18.2 Life cycle—based studies on the wine sector: a review 353
    18.2.1 Lessons learnt from two decades of LCA application 353
    18.2.2 From a methodological framework point of view 357
18.3 Environmental product declarations in the wine sector 360
    18.3.1 Landscape of environmental labels on wine 360
    18.3.2 Environmental product declaration and related product category rules 362
    18.3.3 The Product Environmental Footprint process and its implications to the wine sector 363
18.4 Sustainability challenges in the wine sector from a life cycle perspective: circularity and methodological developments 366
    References 366

CHAPTER 19 Wine packaging and related sustainability issues 371
Fátima Poças, José António Couto and Timothy Alun Hogg

19.1 Introduction 371
19.2 Packaging systems used for wine 373
    19.2.1 Glass bottle 373
    19.2.2 Bag-in-box 374
    19.2.3 Laminated multimaterial boxes...375
    19.2.4 Metal (aluminum) can 375
    19.2.5 Plastic bottles..376
    19.2.6 Barrels and kegs 376
    19.2.7 Paperboard 377
    19.2.8 Cork as closure for wine bottles 377
19.3 LCA and environmental assessments for different packaging systems 379
    19.3.1 Carbon footprint and water footprint...379
    19.3.2 Comparison between different packages 380
    19.3.3 Comparison between single use and refillable glass bottle.....................382
    19.3.4 Impact of recycling of glass bottles...383
    19.3.5 Impact of lightweighting 384
    19.3.6 International distribution 384
19.4 Consumer perceptions of sustainable packaging options for wine 385
19.5 Concluding remarks..387
References 387
Further reading 390

CHAPTER 20 Standards and indicators to assess sustainability: the relevance of metrics and inventories 391
Ana Marta-Costa, Ana Trigo, J. Miguel Costa and Rui Fragoso

20.1 Introduction 391
20.2 Sustainability assessment: major approaches and methodologies....................392
    20.2.1 Conceptual theories 392
    20.2.2 Sustainability assessment tools: from simple indicators to complex frameworks 392
20.3 Indicators and metrics applied to grapes and wine production 393
    20.3.1 Environmental dimension and natural resources...................................395
    20.3.2 Social dimension and equity..395
    20.3.3 Economic dimension and efficiency..398
20.4 Sustainability assessment essay for winegrowing systems: a case study for the Douro's wine producing region...401
    20.4.1 Context, problem, and aims 401
    20.4.2 Research design and methodology...401
    20.4.3 Results and discussion 405
20.5 Future trends 407

20.6	Concluding remarks	408
	Acknowledgments	408
	References	408
	Further reading	413

CHAPTER 21 The guardianship of Aotearoa, New Zealand's grape and wine industry 415

Victoria Raw, Sophie Badland, Meagan Littlejohn, Marcus Pickens and Lily Stuart

21.1	Introduction	415
21.2	NZ's Māori heritage	416
	21.2.1 Māori values and principles	417
	21.2.2 Kaitiakitanga	418
	21.2.3 Māori and the wine industry	418
21.3	New Zealand Winegrowers	419
	21.3.1 Sustainable Winegrowing New Zealand	419
	21.3.2 Organic viticulture and wineries	420
	21.3.3 NZW's sustainability policy	421
	21.3.4 Grape and wine research	421
	21.3.5 Other NZW initiatives	423
	21.3.6 Sustainability guardians	423
21.4	Corporate social responsibility	424
	21.4.1 People, community, and culture	424
	21.4.2 Research	425
	21.4.3 Corporate environmental guardianship	425
21.5	Biosecurity	427
21.6	Natural disaster management	430
	21.6.1 2016. Kaikoura earthquake	430
	21.6.2 COVID-19	431
21.7	Filling the gap	431
21.8	Regional winegrower associations	433
	21.8.1 Role	434
	21.8.2 Wine Marlborough	434
	21.8.3 Wine Marlborough's activities	434
21.9	Conclusion	435
	Acknowledgments	436
	References	436
	Further reading	440

CHAPTER 22 Sustainable viticulture and behavioral issues: insights from VINOVERT project 441
Alexandra Seabra Pinto, Stéphanie Pérès, Yann Raineau, Isabel Rodrigo and Eric Giraud-Héraud

- **22.1** Introduction 441
- **22.2** VINOVERT—an innovative project 443
- **22.3** Consumers preferences for sustainable practices measured by experimental auctions 445
  - 22.3.1 Experimental auction carried out with the partnership of Portuguese wine companies 446
- **22.4** The behavioral hypothesis in viticulture validated by nudges 452
  - 22.4.1 Experience carried out with the partnership of Portuguese wine-growing cooperative 452
- **22.5** Concluding remarks: VINOVERT project insights 456
- References 457

CHAPTER 23 Interactive innovation is a key factor influencing the sustainability of value chains in the wine sector 461
José Muñoz-Rojas, María Rivera Méndez, José Francisco Ferragolo da Veiga, João Luis Barroso, Teresa Pinto-Correia and Åke Thidell

- **23.1** Introduction 461
  - 23.1.1 Interactive innovation and food value chain sustainability 461
  - 23.1.2 Eliciting interactive innovations influencing sustainability in the wine sector 463
  - 23.1.3 Wine value chains in Portugal and Alentejo 464
- **23.2** The Wines of Alentejo Sustainability Program: background and implementation 467
  - 23.2.1 History and outreach 467
  - 23.2.2 Sustainability quantitative assessments 468
  - 23.2.3 How have sustainability assessment results evolved? 473
- **23.3** Assessing WASP's interactive innovation toward enhanced sustainability 475
  - 23.3.1 Methods and stages 475
  - 23.3.2 Results of WASP's interactive innovation assessment 477
- **23.4** Final reflections and conclusions 479
- Acknowledgments 480
- References 480
- Further reading 483

CHAPTER 24 European wine policy framework—The path toward sustainability 485
João Onofre

  24.1 Introduction 485
  24.2 Environmental aspects of wine production 486
  24.3 Technical solutions to the challenges 488
  24.4 Wine production and climate change 489
  24.5 Markets and consumers expectations 490
  24.6 EU policy framework toward increased sustainability of the wine sector 492
      24.6.1 CAP reform proposals 492
      24.6.2 Green deal/farm to fork strategy 493
      24.6.3 Research and innovation in the wine sector 496
  24.7 Concluding remarks 497
      References 497

Index 501

Contributors

Ofélia Anjos
Instituto Politécnico de Castelo Branco, Quinta da Senhora de Mércules, Castelo Branco, Portugal; CEF — Centro de Estudos Florestais, Instituto Superior de Agronomia, Universidade de Lisboa, Lisboa, Portugal; Centro de Biotecnologia de Plantas da Beira Interior, Castelo Branco, Portugal

Rosa Arroyo-Garcia
Centro de Biotecnología y Genómica de Plantas (CBGP-INIA), Madrid, Spain

Sophie Badland
New Zealand Winegrowers Inc., Auckland, New Zealand

João Luis Barroso
Programa de Sustentabilidade dos Vinhos do Alentejo, Comissão Vitivinícola Regional Alentejana (CVRA), Évora, Portugal

António Pedro Belchior
Instituto Nacional de Investigação Agrária e Veterinária, Dois Portos, Portugal

Andréa Moura Bernardes
Post-Graduation Program in Mining, Metallurgical and Materials Engineering, (PPGE3M), Federal University of Rio Grande do Sul (UFRGS), Agronomia, Porto Alegre, Rio Grande do Sul, Brazil

Sara Bernardo
Centre for the Research and Technology of Agro-Environmental and Biological Sciences (CITAB), University of Trás-os-Montes e Alto Douro, Vila Real, Portugal

Josefina Bota
Research Group on Plant Biology Under Mediterranean Conditions, Departament de Biologia, Universitat de les Illes Balears (UIB) - Agro-Environmental and Water Economics Institute (INAGEA), Palma, Illes Balears, Spain

Charlotte Brault
AGAP Institut, University of Montpellier, CIRAD, INRAE, Institut Agro, Montpellier, France; Institut Français de la Vigne et du Vin, Montpellier, France; UMT Geno-Vigne®, IFV-INRAE-Institut Agro, Montpellier, France

Ignacio Buesa
University of the Balearic Islands (UIB), Research Group of Plant Biology under Mediterranean Conditions, Palma, Spain

Ilda Caldeira
Instituto Nacional de Investigação Agrária e Veterinária, Dois Portos, Portugal;
MED — Mediterranean Institute for Agriculture, Environment and Development, Instituto de formação avançada, Universidade de Évora, Évora, Portugal

Sara Canas
Instituto Nacional de Investigação Agrária e Veterinária, Dois Portos, Portugal;
MED — Mediterranean Institute for Agriculture, Environment and Development, Instituto de formação avançada, Universidade de Évora, Évora, Portugal

Cristina C.R. Carlos
ADVID — Associação para o Desenvolvimento da Viticultura Duriense, Edifício Centro de Excelência daVinha e do Vinho, Vila Real, Portugal; Centre for the Research and Technology of Agro-Environmental and Biological Sciences, CITAB, Universidade de Trás-os-Montes e Alto Douro, Vila Real, Portugal

Daniele Carullo
Department of Industrial Engineering, University of Salerno, Fisciano, Salerno, Italy

Sofia Catarino
LEAF — Centre Linking Landscape, Environment, Agriculture and Food Research Center, Instituto Superior de Agronomia, University of Lisbon, Lisbon, Portugal; CEFEMA — Center of Physics and Engineering of Advanced Materials, Instituto Superior Técnico, Universidade de Lisboa, Lisboa, Portugal

Emilio Celotti
Department of Agricultural Food, Environmental and Animal Science, University of Udine, Udine, Italy

Angélique Christophe
LEPSE, University of Montpellier, INRAE, Institut Agro, Montpellier, France

Piergiorgio Comuzzo
Department of Agricultural Food, Environmental and Animal Science, University of Udine, Udine, Italy

Carlos Correia
Centre for the Research and Technology of Agro-Environmental and Biological Sciences (CITAB), University of Trás-os-Montes e Alto Douro, Vila Real, Portugal

J. Miguel Costa
LEAF — Centre Linking Landscape, Environment, Agriculture and Food Research Center, Instituto Superior de Agronomia, University of Lisbon, Lisbon, Portugal

Aude Coupel-Ledru
LEPSE, University of Montpellier, INRAE, Institut Agro, Montpellier, France

José António Couto
Universidade Católica Portuguesa, CBQF - Centro de Biotecnologia e Química Fina, Laboratório Associado, Escola Superior de Biotecnologia, Rua Diogo Botelho, Porto, Portugal

Maria Norberta de Pinho
Center of Physics and Engineering of Advanced Materials, CeFEMA, Instituto Superior Técnico, University of Lisbon, Lisbon, Portugal; Chemical Engineering Department, Instituto Superior Técnico, University of Lisbon, Lisbon, Portugal

Lia-Tânia Dinis
Centre for the Research and Technology of Agro-Environmental and Biological Sciences (CITAB), University of Trás-os-Montes e Alto Douro, Vila Real, Portugal

Agnès Doligez
AGAP Institut, University of Montpellier, CIRAD, INRAE, Institut Agro, Montpellier, France; UMT Geno-Vigne®, IFV-INRAE-Institut Agro, Montpellier, France

Johanna Döring
Hochschule Geisenheim University, Department of General and Organic Viticulture, Geisenheim, Germany

Elizabeth Duarte
LEAF - Linking Landscape, Environment, Agriculture and Food, TERRA, Instituto Superior de Agronomia, Universidade de Lisboa, Lisboa, Portugal

José M. Escalona
Agro-Environmental and Water Economics Institute, University of the Balearic Islands, Palma, Spain

Capri Ettore
DiSTAS, Department for Sustainable Food Process, Università Cattolica del Sacro Cuore, Piacenza, Italy

Tiago A. Fernandes
CQE — Centro de Química Estrutural, Associação do Instituto Superior Técnico para a Investigação e Desenvolvimento (IST-ID), Universidade de Lisboa, Lisboa, Portugal; DCeT — Departamento de Ciências e Tecnologia, Universidade Aberta, Lisboa, Portugal

José Francisco Ferragolo da Veiga
MED-Mediterranean Institute for Agriculture, Environment and Development, Universidade de Évora, Évora, Portugal

Giovanna Ferrari
Department of Industrial Engineering, University of Salerno, Fisciano, Salerno, Italy; ProdAl Scarl — University of Salerno, Fisciano, Salerno, Italy

Rita Fragoso
LEAF - Linking Landscape, Environment, Agriculture and Food, TERRA, Instituto Superior de Agronomia, Universidade de Lisboa, Lisboa, Portugal

Rui Fragoso
CEFAGE — Center for Advanced Studies in Management and Economics, University of Évora, Évora, Portugal

Matthias Friedel
Hochschule Geisenheim University, Department of General and Organic Viticulture, Geisenheim, Germany

Tommaso Frioni
Department of Sustainable Crop Production, Università Cattolica del Sacro Cuore, Piacenza, Italy

Sigfredo Fuentes
Digital Agriculture, Food and Wine Group, School of Agriculture and Food, Faculty of Veterinary and Agricultural Sciences, University of Melbourne, Parkville, VIC, Australia

Jorge Gago
Plant Biology Research Under Mediterranean Conditions Group, University of the Balearic Islands, Palma, Spain

Pilar Gago
Misión Biológica de Galicia, Consejo Superior de Investigaciones Científicas, Pontevedra, Spain

Cristina Gazulla
Elisava Barcelona School of Design and Engineering, Barcelona, Spain

Alexandre Giacobbo
Post-Graduation Program in Mining, Metallurgical and Materials Engineering, (PPGE3M), Federal University of Rio Grande do Sul (UFRGS), Agronomia, Porto Alegre, Rio Grande do Sul, Brazil; Center of Physics and Engineering of Advanced Materials, CeFEMA, Instituto Superior Técnico, University of Lisbon, Lisbon, Portugal

Eric Giraud-Héraud
Univ. Bordeaux, INRAE, BSE, UMR CNRS 6060, USC INRAE 1441, ISVV, Villenave d'Ornon, France

Elsa Gonçalves
LEAF- Linking Landscape, Environment, Agriculture and Food, Instituto Superior de Agronomia, Universidade de Lisboa, Lisboa, Portugal

Jean-Pascal Goutouly
EGFV, Bordeaux Sciences Agro, INRAE, University of Bordeaux, ISVV, Villenave d'Ornon, France

Rino Gubiani
Department of Agricultural, Food, Environmental and Animal Sciences, University of Udine, Udine, Italy

Maximilian Hendgen
Hochschule Geisenheim University, Department of Soil Science and Plant Nutrition, Geisenheim, Germany

Pedro Hipólito
Instituto Superior de Agronomia, Universidade de Lisboa, Lisboa, Portugal; HUVA, SA — Herdade da Mingorra, Beja, Portugal

Timothy Alun Hogg
Universidade Católica Portuguesa, CBQF - Centro de Biotecnologia e Química Fina, Laboratório Associado, Escola Superior de Biotecnologia, Rua Diogo Botelho, Porto, Portugal

Almudena Hospido
CRETUS, Department of Chemical Engineering, Universidade de Santiago de Compostela, Santiago de Compostela, Spain

Diego S. Intrigliolo
Spanish National Research Council (CSIC), Desertification Research Center (CIDE), CSIC-UV-GV, Valencia, Spain

Randolf Kauer
Hochschule Geisenheim University, Department of General and Organic Viticulture, Geisenheim, Germany

Milena Lambri
DiSTAS, Department for Sustainable Food Process, Università Cattolica del Sacro Cuore, Piacenza, Italy

Eric Lebon
LEPSE, University of Montpellier, INRAE, Institut Agro, Montpellier, France

Meagan Littlejohn
New Zealand Winegrowers Inc., Auckland, New Zealand

Matia Mainardis
Department Polytechnic of Engineering and Architecture, University of Udine, Udine, Italy

Ana Marta-Costa
CETRAD — Centre for Transdisciplinary Development Studies, University of Trás-os-Montes e Alto Douro (UTAD), Vila Real, Portugal

Antero Martins
LEAF- Linking Landscape, Environment, Agriculture and Food, Instituto Superior de Agronomia, Universidade de Lisboa, Lisboa, Portugal

Albert Mas
Department de Bioquímica i Biotecnologia, Facultat d'Enologia, Universitat Rovira i Virgili, Tarragona, Spain

Hipólito Medrano
Research Group on Plant Biology Under Mediterranean Conditions, Departament de Biologia, Universitat de les Illes Balears (UIB) - Agro-Environmental and Water Economics Institute (INAGEA), Palma, Illes Balears, Spain

José Moutinho-Pereira
Centre for the Research and Technology of Agro-Environmental and Biological Sciences (CITAB), University of Trás-os-Montes e Alto Douro, Vila Real, Portugal

José Muñoz-Rojas
MED-Mediterranean Institute for Agriculture, Environment and Development, Universidade de Évora, Évora, Portugal

Andrea Natolino
Department of Agricultural Food, Environmental and Animal Science, University of Udine, Udine, Italy

Margarida Oliveira
LEAF - Linking Landscape, Environment, Agriculture and Food, TERRA, Instituto Superior de Agronomia, Universidade de Lisboa, Lisboa, Portugal; ESAS, UIIPS, CIEQV, Instituto Politécnico de Santarém, Quinta do Galinheiro, Santarém, Portugal

João Onofre
Head of Unit for Wines, Spirits and Horticultural Crops, European Commission, Brussels, Belgium

Gianpiero Pataro
Department of Industrial Engineering, University of Salerno, Fisciano, Salerno, Italy

Stéphanie Pérès
Univ. Bordeaux, Bordeaux Sciences Agro, BSE, UMR 6060, ISVV, Pessac, France

Marcus Pickens
Wine Marlborough, Blenheim, New Zealand

Teresa Pinto-Correia
MED-Mediterranean Institute for Agriculture, Environment and Development, Universidade de Évora, Évora, Portugal

Fátima Poças
Universidade Católica Portuguesa, CBQF - Centro de Biotecnologia e Química Fina, Laboratório Associado, Escola Superior de Biotecnologia, Rua Diogo Botelho, Porto, Portugal; Universidade Católica Portuguesa, CINATE, Escola Superior de Biotecnologia, Porto, Portugal

M. Carmen Portillo
Department de Bioquímica i Biotecnologia, Facultat d'Enologia, Universitat Rovira i Virgili, Tarragona, Spain

Yann Raineau
Univ. Bordeaux, Conseil Régional de Nouvelle-Aquitaine, BSE, UMR 6060, ISVV, Pessac, France

Victoria Raw
The New Zealand Institute for Plant and Food Research Limited, Blenheim, New Zealand

Beatriz Rivela
inViable Life Cycle Thinking, Madrid, Spain

María Rivera Méndez
MED-Mediterranean Institute for Agriculture, Environment and Development, Universidade de Évora, Évora, Portugal

Joel Rochard
VitisPlanet, Bouilly, France

Isabel Rodrigo
Instituto Superior de Agronomia, Universidade de Lisboa, Lisboa, Portugal

Elia Romanini
DiSTAS, Department for Sustainable Food Process, Università Cattolica del Sacro Cuore, Piacenza, Italy

Pascual Romero-Azorín
Instituto Murciano de Investigación y Desarrollo Agrario y Alimentario (IMIDA), Unidad "Fertirriego y Calidad Hortofrutícola", Unidad Asociada de I+D+I al CSIC, La Alberca, Spain

Artur Saraiva
LEAF - Linking Landscape, Environment, Agriculture and Food, TERRA, Instituto Superior de Agronomia, Universidade de Lisboa, Lisboa, Portugal; ESAS, UIIPS, CIEQV, Instituto Politécnico de Santarém, Quinta do Galinheiro, Santarém, Portugal

Alexandra Seabra Pinto
Instituto Nacional de Investigação Agrária e Veterinária, Oeiras, Portugal

Thierry Simonneau
LEPSE, University of Montpellier, INRAE, Institut Agro, Montpellier, France

Manfred Stoll
Hochschule Geisenheim University, Department of General and Organic Viticulture, Geisenheim, Germany

Lily Stuart
The New Zealand Institute for Plant and Food Research Limited, Blenheim, New Zealand

Pierre-Louis Teissedre
Université de Bordeaux, ISVV, Unité de Recherche Œnologie EA 4577, USC 1366 INRA, Bordeaux INP, Villenave d'Ornon, France

Åke Thidell
University of Lund, International Institute for Industrial Environmental Economics (IIIEE), Lund, Sweden

Patrice This
AGAP Institut, University of Montpellier, CIRAD, INRAE, Institut Agro, Montpellier, France; UMT Geno-Vigne®, IFV-INRAE-Institut Agro, Montpellier, France

Ignacio Tortosa
Research Group on Plant Biology Under Mediterranean Conditions, Departament de Biologia, Universitat de les Illes Balears (UIB) - Agro-Environmental and Water Economics Institute (INAGEA), Palma, Illes Balears, Spain

Ana Trigo
CoLAB Vines&Wines, Association for the Development of Viticulture in the Douro Region, Vila Real, Portugal

Maria do Carmo M. Val
ADVID — Associação para o Desenvolvimento da Viticultura Duriense, Edifício Centro de Excelência daVinha e do Vinho, Vila Real, Portugal

Acknowledgment to the external reviewers

Special appreciation for our expert colleagues who contributed by reviewing the materials for completeness and accuracy:

Alexandra Seabra Pinto (*Instituto Nacional de Investigação Agrária e Veterinária, Portugal*)
Alfredo Cassano (*Istituto per la Tecnologia delle Membrane, Consiglio Nazionale delle Ricerche, Italy*)
Almudena Hospido (*Universidade de Santiago de Compostela, Spain*)
António Augusto Areosa Martins (*Faculdade de Engenharia, Universidade do Porto, Portugal*)
Cassandra Collins (*University of Adelaide, Australia*)
Dênis Cunha (*Universidade de Viçosa, Brazil*)
Eric Giraud-Héraud (*INRAE-GREThA/Université de Bordeaux, France*)
Eliemar Campostrini (*Universidade Estadual do Norte Fluminense, Northern Rio de Janeiro State University, Brazil*)
Fernando Alves (*Symington Family Estates, Portugal*)
Giulio Malorgio (*Università di Bologna, Italy*)
Ian Vázquez-Rowe (*Pontificia Universidad Católica del Perú, Perú*)
Karen Titulaer (*Villa Maria Estate, Auckland, New Zealand*)
Kees Van Leeuwen (*Institut National de la Recherche Agronomique, Bordeaux, France*)
Laura Torres (*Universidade de Trás os Montes e Alto Douro, Portugal*)
Luigi Bavaresco (*Facoltà di Scienze Agrarie, Alimentari e ambientali; Università Cattolica del Sacro Cuore, Italy*)
Jorge Ricardo da Silva (*Instituto Superior de Agronomia, Universidade de Lisboa, Portugal*)
Josefina Bota (*Universitat de les Illes Balears, Palma, Spain*)
Maria Paz Diago (*Universidad de La Rioja, Spain*)
Maria Jose Frutos Fernández (*Universitas Miguel Hernández, Spain*)
Pedro Elez-Martínez (*Universitat de Lleida, Spain*)
Rolando Chamy (*Pontificia Universidad Católica de Valparaiso, Chile*)
Rosanna Tofalo (*Università Degli Studi di Teramo, Italy*)
Samuel Ortega-Farias (*Universidad de Talca, Chile*)
Stamatina Kallithraka (*Agricultural University of Athens, Greece*)
Xosé Antón Rodrígues (*Universidade de Santiago de Compostela, Spain*)
Vicente Sotés Ruiz (*Universidad Politécnica de Madrid/International Organisation of Vine and Wine (OIV), Spain*)
Victoria Raw (*NZ Institute for Plant and Food Research Limited, New Zealand*)
Victor de Freitas (*Faculdade de Ciências, Universidade do Porto, Portugal*)
Vittorino Novello (*Università degli Studi di Torino, Italy*)

About the cover

Improving Sustainable Practices in Viticulture and Enology involves scientific progress, technical innovation, and modernization along the wine supply chain. Scientific advances joint with digital technologies offer innovative solutions to viticulture and winemaking (e.g., via sensors, data management, communication tools) either to monitor processes, control quality, support decision, or information transfer along the supply chain. Digitalization of the wine industry has an increasing role and can be a major asset to support future solutions to face risks and implement more sustainable practices in the wine sector.

The Editors

CHAPTER 1

Achieving a more sustainable wine supply chain—Environmental and socioeconomic issues of the industry

J. Miguel Costa[1], Sofia Catarino[1], José M. Escalona[2] and Piergiorgio Comuzzo[3]

[1]*LEAF – Centre Linking Landscape, Environment, Agriculture and Food Research Center, Instituto Superior de Agronomia, University of Lisbon, Lisbon, Portugal;* [2]*Agro-Environmental and Water Economics Institute, University of the Balearic Islands, Palma, Spain;* [3]*Department of Agricultural Food, Environmental and Animal Science, University of Udine, Udine, Italy*

1.1 Sustainability concept and issues

Sustainability and sustainable development are important topics for all fields of the economy and society. In 1987, the United Nations Brundtland Commission defined sustainability as "meeting the needs of the present without compromising the ability of future generations to meet their own needs" (Brundtland, 1987).

Regarding the wine sector, the concept of sustainability should integrate economics, ecology, and community dimensions related to grape production and winemaking operations. According to the International Organization of Vine and Wine (OIV), a sustainable grape and wine industry is a "global strategy on the scale of the grape production and processing systems, incorporating at the same time the economic sustainability of structures and territories, producing quality products, considering requirements of precision in sustainable viticulture, risks to the environment, products safety and consumer health and valuing of heritage, historical, cultural, ecological, and landscape aspects" (OIV, 2004).

The wine industry relies heavily on manual labor inputs. Therefore, the social component of sustainability is highly relevant. The well-being and quality of life, the educational and work conditions, social benefits, and ethics must be accounted (Forbes et al., 2020; Martucci et al., 2019; Taylor, 2017).

The OIV reported a resolution on general principles of sustainable viticulture and winemaking in which it is stated that "Companies will have to consider the impact of their activities on the socioeconomic context and their involvement in the socioeconomic development of the territories (or areas)" (OIV, 2016). This resolution deals with working conditions, integration into the regional/local socioeconomic and cultural environment, consumer safety and health, and it emphasizes the need to ensure workers' health and safety, continuous training and the stability of the workforce (OIV, 2016).

Chapter 1 Achieving a more sustainable wine supply chain

The topic of "Sustainability in viticulture and winemaking" has received increased attention by the OIV and by academy research in recent decades (Gerling, 2015). This can be seen by the increasing number of publications combining topics such as sustainability, viticulture, wine production, and climate change in the last 10 years' time (2010–2020) (Fig. 1.1). A total of 223 articles (199 original research and 24 review articles) were published under the theme "Sustainability & Viticulture" whereas the topic of "Sustainability & Wine" was covered by 1766 papers (1496 research, 270 review

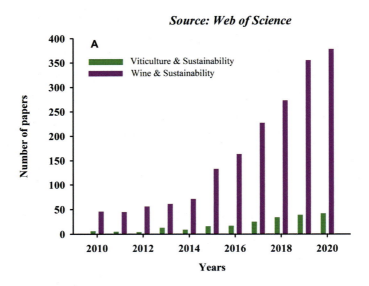

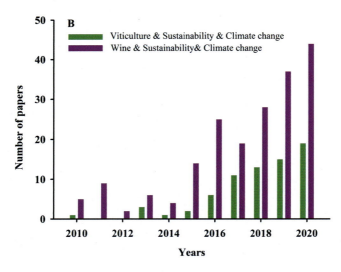

FIGURE 1.1

Number of ISI papers (original research articles, reviews) published in the last 10 years (2010–2020) based on the Web of Science database (WoS, 2021) on December 2021, and concerning the research topics "Viticulture & Wine & Sustainability" (A) and "Viticulture & Wine & Sustainability" related to climate change (B).

articles) (WoS, 2021). For the same 10-year period, the Google Scholar database provided 16,700 references on "Sustainability & Viticulture" and 50,200 focused on "Sustainability & Wine," with a marked annual increase (Google Scholar, 2021).

1.2 The state of the wine industry—short overview
1.2.1 The wine industry worldwide

Wine is a global and diversified product. Wine production has changed markedly over the last 50 years, in terms of the cultivated area, production value, market and consumer trends (OIV, 2021a; Pomarici, 2016). More than 40% of wine consumed at a global scale is imported from other countries (Anderson et al., 2017) and the European Union is the leading global exporter with a 70% share of all exports (OIV, 2021a).

The global surface area of vineyards was 7.3 million ha in 2020 and included the area devoted to the production of wine and juices, table grape and raisins, as well as young vines not yet in production (OIV, 2021a). About 55% of the total relates to grapes for wine production. In 2020, the global wine production was 260 million hL, largely based in countries such as Italy, Spain, France and USA (OIV, 2021a) (Fig. 1.2).

The global cultivated area reached a high point in late 1970s, averaging 10.2 million ha between 1975 and 1980 (WSET Alumni, 2015). Since then, the area has decreased due to a drop in the

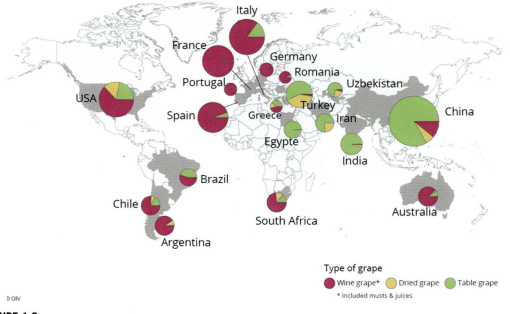

FIGURE 1.2

Distribution of the major production countries of grapes for wine, dried grapes, and table grape, according to the OIV's statistical report for the year 2019, on world grape and wine production (OIV, 2019).

production of the "Old World" countries such as France, Spain, Italy, and Portugal. In the 1980s, the European vineyard area represented 70% of the global area (WSET Alumni, 2015), but in 2019, it was only about 50% of the total global area (OIV, 2020a). This came about from the reforms implemented by European Union (EU) members between 2008 and 2011 to reduce the EU's wine production surplus (WSET Alumni, 2015), and the expansion elsewhere such as in Chile, Australia, South Africa, New Zealand, and China. The global wine trade increased by about 80% in volume between 2000 and 2020, and the trade value doubled in the same period (Fig. 1.3).

The case of China's growth is quite remarkable. The fast expansion of Chinese viticulture sector started with the "Reform and Opening up" policy in 1978 (Li & Bardají, 2017). The country became a global leader in table grape production. In fact, only 10% of the vineyard area in China (a total of 785,000 ha in 2020) is devoted to wine production estimated in 6.6 million hL in 2020 (OIV, 2021a). Shandong Province has the oldest history of winemaking but Hebei, Shanxi, Shaanxi, Ningxia (autonomous region), and Jilin are other important wine producing provinces (USDA, 2018). China maintains its position as a large wine importer, but counterfeit wine can be a problem for exporters to China (USDA, 2018). Chinese consumer trends tend for lower priced wines, mild taste profiles, and convenience (USDA, 2018).

The UK has also expanded fast its vineyard area (196 ha in 1975 to 3500 ha in 2019) (WineGB, 2021), which contributed positively to the increase of sparkling wine production globally (20 million

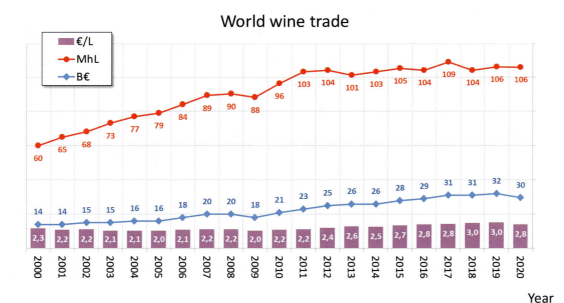

FIGURE 1.3

Wine worldwide trade in volume (millions of hectoliters, MhL) and Value (billions of Euros, B€), and price per liter (€/L), since the year 2000 until 2020.

Based on OIV. (2021a). State of the world vitivinicultural sector in 2020. https://www.oiv.int/public/medias/7909/oiv-state-of-the-world-vitivinicultural-sector-in-2020.pdf.

hL in 2018), with an overall increase of more than 57% since 2002 (OIV, 2020b). Wine production also expanded in India, but the area and production remain minor (about 2500 ha and 17,5 million L) while imports keep increasing (USDA, 2019).

Old World countries have been losing market share to New World players who have increased their scale, quality, and branding expertise. Countries like Chile and New Zealand increased their competitive performance suggesting that some wine suppliers are responding better to market needs (Anderson et al., 2017; Pomarici, 2016). After the global economic crisis in 2008, global wine consumption stabilized around 240–245 million hL since 2009 with a reduction in 2020 (234 million hL), less 3% than in 2019 (OIV, 2021a). The US market remained the largest wine consumer worldwide, with a total of 33 million hL in 2020 followed by France (24.7 million hL) and Italy (24.5.4 million hL) (OIV, 2021a).

COVID-19 had an impact on the wine sector. Tourism worldwide in 2020 and first half of 2021 shut down altogether with the stop of the HORECA channel that is made up of establishments and companies (hotels, restaurants, and catering) that prepare and serve food, meals, and drinks, as a rule, to be consumed on premise. The full reopening of businesses in 2021 did not occur and that remains still uncertain for 2022, although the global vaccination of people will bring improved trading opportunities.

The COVID-19 pandemic affected market and distribution channels and modified consumption patterns, prices and turnover, margins and profits of the sector, although with variation among producing countries (CEEV, 2021; IWSR, 2020). Wine consumers were encouraged to adopt contactless retail experiences and internet sales emerged as a complementary channel for wineries (IWSR, 2020; Nielsen, 2020; NZ Foreign Affairs & Trade, 2020; Recarte, 2020). COVID-19 also influenced training activities worldwide as well as trade relations between countries (e.g., the imposition of trade restrictions and tariffs on Australian wines exported to China in 2021). Governments were called to support the sector (e.g., in the EU funds were allocated to encourage the distillation of wine and green harvest) (Recarte, 2020).

1.2.2 Risks and concerns of the modern wine industry

The socioeconomic dimension of the wine industry goes far beyond vineyards and wineries. The sector covers a wide range of activities and subsectors including nursery stock, agrochemical products, machinery and equipment, cooperages, packaging, wine closure systems, enological products and technologies, quality control laboratories, certification and training, logistics, marketing and wine tourism (CEEV, 2016).

In parallel with sustainability aims, the wine industry must also handle challenges and risks posed by climate change. Sustainability and climate change are closely related because more adverse/extreme climate conditions will have a negative impact on yield, berry composition, vine's health, and on final wine quality, which ultimately, affects companies' performance and job security (Botha, 2020; Field et al., 2020; Santillán et al., 2020). Climate change has a negative impact on the availability and conservation of natural resources such as water and soil, and more erratic and extreme climate conditions increase unpredictability and risks (Costa et al., 2016; Hofmann et al., 2021; Lorenzo et al., 2021; Santillán et al., 2020).

1.3 Sustainability issues in wine industry
1.3.1 Vineyard issues

Modern vineyards must accommodate more sustainable practices and strategies to mitigate the impacts of climate change on berry yield and composition. Longer and dryer periods, warmer soil and air temperature conditions will be risky for phenology and for both grape yield, quality (Costa et al., 2016; Santos et al., 2020; Schultz & Jones, 2010). More frequent heat waves (Lorenzo et al., 2021) and increased water use pose more pressure on the already scarce water resources as irrigation is the main and easiest mitigating tool available in dry areas worldwide (Hayman et al., 2012). Nevertheless, irrigation is also used in cooler climates (e.g., New Zealand or Tasmania-Australia). Consequently, and due to some relaxation of irrigation restrictions in typically rainfed regions, irrigated viticulture has expanded in Southern Europe (Costa et al., 2016; Gambetta et al., 2020). In Spain, for example, the irrigated area increased from 2% of the total area in the 1950s to 40% of the total at present that represents 380.000 ha (Ayuda et al., 2020; MAPA, 2019) and in southern Portugal almost all newly planted vineyards are irrigated at present (Costa et al., 2016). Water scarcity has affected major producing regions worldwide namely in USA (e.g., California) (Folger, 2017; Gonsier et al., 2015), Australia (Coulter et al., 2019), and South Africa (Botha, 2020). In South Africa, the three-year drought period (2015–2018) forced the sector to reduce water consumption by 50% or more, with penalties for those not implementing it (Botha, 2020).

Scarcer water resources enabled there to be a more efficient use of water. The increasing use of sensors and data gathering on plant-soil-atmosphere relations will improve irrigation strategies, e.g., deficit irrigation (Romero et al., 2020; OIV, 2021c) and support soil and crop management (See Chapters 6 and 7). This will help to minimize inefficient water use and decrease wine water foot print (Costa et al., 2016; Ene et al., 2013; OIV, 2021c) (See Chapters 11 and 18). More adapted rootstocks and genotypes and clones will help to use water and nutrients more efficiently and mitigate climate change effects (Gambetta et al., 2020; Ollat et al., 2015; Simonneau et al., 2017; Tortosa et al., 2019). The grapevine is a species with large genetic variability and many autochthon varieties and/or clones remain uncharacterized regarding their tolerance to drought and heat (See Chapters 2 and 3). Knowledge of grapevine stress ecophysiology and grapevine stress responses has increased in recent years (Bota et al., 2016; Chaves et al., 2010; Simonneau et al., 2017), but an improved understanding of drought and heat stress responses is still needed to better predict adaptation to future climate scenarios (Gambetta et al., 2020) (See Chapters 4, 6, and 9). In the same way, improved knowledge on grapevine responses to flooding is relevant for some wine regions (e.g., in China) (Zhu et al., 2018).

Heat stress physiology at both leaf and berry levels should be evaluated especially with regards to the combined effect of high air and soil temperatures on leaf and berry metabolic responses, vine's water relations, and thermal regulation. The role of extreme soil/root zone temperatures on grapevine morphology and physiology in dry and warm areas requires more attention (Costa et al., 2016; Field et al., 2020). Relocation of vineyards to higher latitudes and altitudes has been suggested as solution (Van Leeuwen & Darriet, 2016), but this option may not be always feasible nor sustainable in both economic and social terms (See Chapter 24).

Sustainable water and energy use emerged as a critical corporate social responsibility issues for leading businesses including those in the wine industry. Many companies have already quantified and disclosed their water use embracing an end-to-end supply chain perspective (Costa et al., 2016;

Ene et al., 2013; Villanueva-Rey et al., 2018) (See Chapters 11 and 19). The use of unconventional water sources (saline and reclaimed water, treated wasted waters) has been suggested for dry regions (Mirás-Avalos & Intrigliolo, 2017) (See Chapter 6), but more studies on the long-term effect(s) of using alternative water sources on soil and plants are needed because it can result in higher soil salinity if poorly managed (Phogat et al., 2020).

Soil health and conservation are other crucial topics when dealing with a more sustainable viticulture (Lazcano et al., 2020) (See Chapter 5). The UN's Sustainable Development Goals considers sustainable soil use the basis of sustainable agriculture (Hou et al., 2020). Soil degradation can derive from practices that cause soil erosion, compaction, fertility loss, and pollution (e.g., by biocides, fertilizers) (Cataldo et al., 2020; Lazcano et al., 2020). Unfortunately, unsustainable practices in vineyards can contribute to soil loss (García-Ruiz et al., 2010). Soil erosion and loss of organic matter will decrease soil moisture retention and fertility endangering the economic sustainability of agricultural production (Hou et al., 2020; Novara et al., 2018). Improved understanding of the soil health concept can support and enable strategies for more sustainable soil management in vineyards. Correct fertilizer application based on soil analyses to support microzoning, adequate soil management practices (use of herbicide vs. mechanical weed control), or novel soil management practices (e.g., use of mycorrhiza) should be considered, and require more research for further commercial application (See Chapter 5).

Irrigation is not the solely tool to mitigate the effects of climate change. Complementary canopy management strategies should be also considered namely those modifying vine's sink/source ratio (leaf area/fruit ratio) or shading canopy strategies (ADVICLIM, 2016; Haymann et al., 2012; Zheng et al., 2017).

Pests and diseases cause major yield and quality losses in viticulture and reduce vine longevity (EIP-Agri, 2019) (See Chapter 8). The viticulture sector is an important user of pesticides in Europe and at global scale, but data are not easily available (EIP-Agri, 2019). Fungicides account for the largest share of pesticide treatments in most vineyards and the risks of soil and ground water pollution are high (Suciu et al., 2020). Therefore, the sector has been showing a progressive commitment to reduce the use of plant protection products (PPPs) (EIP-Agri, 2019). The restrictions imposed on the use of synthetic herbicides are good examples of reduced reliance of these agrochemicals (Silvia et al., 2020). Adoption of Integrated Pest Management (IPM) and biological control measures based on dynamic application methods that account for the variation on canopy leaf area, phenology, climate conditions, and modeling, and decision support systems are being implemented (Pérez-Expósito et al., 2017) (See Chapter 8).

Staff training is required to improve handling and use of PPPs. Moreover, increasing restrictions by the EU imposed to specific active substances are forcing changes in the viticulture sector. This will reduce the available solutions for growers to fight against pests and diseases. In parallel, an improved access to statistics on the use of PPPs, environmental monitoring, and environmental risk indicators are crucial to optimize the use of PPPs and to minimize their impact on the environment (ECA, 2020; Silva et al., 2018) (See Chapters 8 and 24).

The impact of PPPs on the environment, biodiversity, and on human health must be accounted to improve management practices (EU Commission, 2017). In all EU wine regions and after the introduction of the EU Directive on Sustainable use of Pesticides (Directive 2009/128/EC), that promotes the use of IPM practices (EIP-Agri, 2019), several leading wine companies have been converting large areas to organic vineyards (See Chapter 24). Also, organic and biodynamic wine production has been expanding mainly motivated by environmental issues (Castellini et al., 2017; FIBL & IFOAM, 2019; Meissner et al., 2019). The total organic vineyard area worldwide is expected to reach 550,000 ha by the year 2022 (SUDVINBIO, 2019), Like in Europe, the USA's market of organic agricultural products keeps growing and the wine sector follows the same trend (FIBL & IFOAM, 2019). However, there is still the need to better characterize consumer trends concerning this wine segment (Boncinelli et al., 2021).

The quality of the planting material is another critical issue for successful vineyard establishment, production, and longevity. Clean and healthy nursery stock material is crucial for a more sustainable viticulture. Low quality planting material can induce greater expenditure with biocides and/or fertilizers. The use of low quality and uncertified clean material also favors spreading of viruses, pests, and diseases on a global scale (Grohs, 2017; Waite et al., 2015), as many nursery businesses are exporting their products.

Regarding the high labor input in vineyards, mechanization is increasingly important to overcome the problem of labor shortage and reduce costs (Strub et al., 2021). In parallel, several socioeconomic issues such as the well-being and quality of life of workers, salaries and health, education and training, social benefits and ethics must be properly accounted by the sector (Forbes et al., 2020; Martucci et al., 2019; Taylor, 2017). Quantifying the social impacts associated with wine production is a complex task because it involves different actors and stakeholders as well as different production stages (Martucci et al., 2019). Nevertheless, certification programs must consider tools such as the Social Life Cycle Assessment (S-LCA) methodology (Martucci et al., 2019; Verones et al., 2017) (See Chapters 20 and 23). Certification focused on social issues is already adopted by some leading companies worldwide (e.g., B-Corp certification).

1.3.2 Winemaking issues

Similarly, to viticulture, winemaking is influenced by climate change and also requires more sustainable practices. Excessively high air temperatures negatively influence berry composition (e.g., higher sugars content, lower organic acids, higher pH, lower anthocyanins content, and lower total phenolic index) which results in higher alcoholic strength and modified wine sensory profile (Ollat et al., 2016; Schultz & Jones, 2010; Van Leeuwen & Darriet, 2016). These problems can be partly solved by correcting acidity or alcoholic strength, as well as by different enological strategies. The use of selected microorganisms for alcoholic fermentation, the processing of grape must by reverse osmosis to decrease sugar content prior to alcoholic fermentation, the modulation of alcohol content by physical processes such as membrane processes, spinning cone column, and distillation under vacuum are examples (Gonçalves et al., 2013; Mira et al., 2017; Schmitt & Christmann, 2019). The addition of water to musts is an authorized practice in some countries such as Australia or United States. Recently, and related to the circular economy approach, the concept of regenerative wineries is gaining relevance. Regenerative practices in wineries guarantee efficient use of resources but also regenerate resources by recovering value from grape and wine by-products and wastewater (See Chapters 10, 11, and 15).

Although water use or the water footprint in wine production largely relate to irrigation, water use and wastewater management are other priorities for modern wineries (Christ & Burritt, 2013; Matos & Pirra, 2020). More efficient water use without compromising hygiene is very important

(See Chapter 11) and large variation in water consumption in wineries (1–14 L water per L of wine) still occurs due to winery size, used technologies, and wine type (red *vs.* white) (Matos & Pirra, 2020; Oliveira et al., 2019; Teissèdre, 2018).

Several strategies can be implemented to reduce water consumption (See Chapters 10 and 11), including eco-designed buildings that enable the collection of rainwater and storage in tanks to be used during dry period (Boulton, 2019; Teissèdre, 2018). Efficient wastewater management is very important to minimize environmental impact. Winery wastewater relates mostly to cleaning activities that involve production of polluting residues such as filtration earths and fining agents as well as high organic load (Matos & Pirra, 2020; Oliveira & Duarte, 2016; Oliveira et al., 2019; Rinaldi et al., 2016). Devices connected by pipelines (e.g., wine bottling machines) are often cleaned with cleaning-in-place methods, and consume large amounts of water and chemical detergents that are not always biologically degradable (Englezos et al., 2019; Navarro et al., 2017; Sheridan et al., 2011). Ozone oxidation technologies are potential environmental alternatives to sterilize wood barrels (Guzzon et al., 2017) or cleaning-in-place of bottling machines (Englezos et al., 2019).

The global dimension of the wine industry and its massive scale of production make it a major energy user and greenhouse gas emitter (Navarro et al., 2017; Vela et al., 2017). Future competitiveness and sustainability of the sector depends therefore on a more efficient use of energy (Vela et al., 2017) (See Chapter 12). In the EU, energy use in wine production is estimated in about 1750 million kWh/year (500 million kWh/year in France, 500 million kWh/year in Italy, 400 million kWh/year in Spain, and 75 million kWh/year in Portugal) and using electricity as main energy source (over 90%) (Fuentes-Pila & García, 2014).

Energy reduction can be achieved by optimizing building characteristics. An excellent example of integrated winery energy management is the *"Jess S. Jackson" Sustainable Winery Building*, opened in 2013 at the University of California (UCD), Davis, USA. The winery is considered the "the first self-sustainable, zero-carbon teaching and research facility in the world". The winery has a system to capture CO_2, a solar-powered cooling equipment, reverse osmosis installations to recover wastewater and rainwater (to be used in cleaning), a solar cogeneration system combining photovoltaic panels and heat collection systems to deliver electricity and hot water, as well as a hydrogen-gas generator and a hydrogen fuel cell for energy production (wineserver.ucdavis.edu, 2020). The building is also able to passively cool down and heat up by itself as needed due to natural ventilation and thermal mass, and it allows energy generation via the use of a roof photovoltaic system.

Another project to reduce energy use by the sector is the EU Project TESLA—*Transferring energy save laid on agroindustry—IEE-12-324758, 2013–2016,* which encourages best available practices to evaluate the energy situation and promote energy savings by the agro-food industry, including wineries (Gubiani et al., 2019). The TESLA Project provided manuals and guides to improve energy use efficiency (EU Commission, 2020a; Fuentes-Pila & García, 2014). In Australia, shifting of peak to off-peak electricity helped wineries to save on electricity costs (Nordestgaard, 2013).

Major critical energy-consuming operations in wineries include temperature control and management (Malvoni et al., 2017). Therefore, global warming makes temperature control in wineries a major challenge for winery managers. In warm regions such as in Southern Europe, harvests frequently occur at mid-end of August, when air temperatures are 40°C, which increases energy requirements to cool down berries/musts and winery facilities. Nigh-time harvesting reduces berry temperature before processing and allows energy savings, but despite being a viable option when using mechanical harvesting, it increases labor costs, due to additional night working shifts in the winery. The cooling of grapes, musts, and wine, the temperature control during alcoholic and malolactic fermentation, and during storage in vat or barrel aging, cold stabilization, wine bottling (especially for sparkling wines), and product storage in warehouses are among the most energy demanding processes (Malvoni et al., 2017). Reduction of energy use during fermentation has been achieved by using specific microbiological strains (Giovenzana et al., 2016). In addition, suitable insulation of the fermentation vessels to reduce thermal losses, and the use of photovoltaic systems reduced annual energy consumption for cooling, from 11% to 21% and up to 41%, respectively, in a winery (Malvoni et al., 2017). In Spain, the use of solar energy in wineries enabled a 4%−36% reduction in the electricity costs (Gómez-Lorente et al., 2017).

Closed systems based on recycling of waste and effluents can help to save energy in winemaking. Solid waste such as glass, cardboard, paper, aluminum, steel, cork, and plastic are produced by the wine sector every year. This demands improved quantification. Glass bottles, for example, can contribute up to 40%−90% of total impact of bottling and packaging on the carbon footprint of wine and alternative bottling/packaging solutions are on demand (Ferrara & De Feo, 2018) (See Chapters 18 and 19). In turn, lees or other grape and wine by-products and effluents (stems, grape marc, must residues, and sludge from wastewater treatment plants) can be used as base material for clean energy production via anaerobic digestion, all of which lead to biogas and methane production (Da Ros et al., 2014; Guerini Filho et al., 2018) and/or used as soil remediates after proper treatment (Da Ros et al., 2014). Grape and wine by-products can be used to extract high value-added and bioactive compounds (e.g., polyphenols and antioxidant molecules, ethanol, tartaric acid, lignocelluloses, proteins, and polysaccharides) with potential use by the pharmaceutical and cosmetic industries (Ahmad et al., 2020; Barba et al., 2016; Gómez-Brandón et al., 2019). These compounds can be obtained via green nonconventional extraction methods (e.g., pulsed electric fields, high voltage electrical discharges, ohmic heating, microwaves, ultrasounds, and supercritical fluid extraction) (Barba et al., 2016), or by pressure-driven membrane processes (Giacobbo et al., 2017). The recent finding on the positive value of cork powder waste as a wine fining agent contributes to a more efficient and sustainable wine industry (Filipe-Ribeiro et al., 2018).

The use of cork stoppers by the wine industry also influences the environmental impact and the sustainability of the wine industry. Cork has unique physical properties such as its high flexibility, elasticity, compressibility, recovery, and good tightness to liquids (Silva et al., 2011) which makes it an excellent and renewable closure for wine bottles. By using cork, which is a renewable resource, the wine industry supports conservation of Mediterranean cork forests (e.g., the Portuguese "Montado" located in dry areas of South Portugal). According to the UNESCO, cork is crucial for economic development of Mediterranean countries as well as to avoid desertification (UNESCO, 2020).

Sustainable winemaking may suggest minimal intervention. Viticulture practices influence microbial diversity on grapes and later on the grape must and alcoholic fermentation (Capozzi et al., 2015; Mas, 2018). The use of starter cultures, either yeast or bacteria, is increasingly questioned, or excluded, just like it happens with "natural wines" or "biodynamic wines" (Guzzon et al., 2019). A major issue for microbiological control is about a more restrictive use of additives, e.g., sulfur dioxide (SO_2). To face global warming impact, microbial biodiversity of yeasts (nonconventional yeast species, in particular, non-*Saccharomyces* strains) and bacteria in wine production requires further investigation to reduce wine's alcohol content, while not altering sensory properties, which may occur with certain changes in the chemical composition of wine (e.g., higher sugar level, lower acidity, and higher pH) (Ollat et al., 2016). This scenario demands innovative approaches for microbiological control in winemaking (Mas, 2018) (See Chapter 13). By reducing the use of SO_2, this may also improve workers' health and safety as well as that of the consumer as it is known to be an irritant and the cause of allergic reactions (OIV, 2021b; Ribéreau-Gayon et al., 2006). Different techniques, or combination of techniques, can reduce SO_2 concentrations in wine (Comuzzo & Zironi, 2013) (See Chapter 13).

Another controversial wine additives issue is around the use of dimethyl dicarbonate (DMDC). Added to bottled wine, this additive eliminates microorganisms (especially yeasts), allowing for a reduction in the use of SO_2. The advantage of using DMDC is that it quickly and completely hydrolyzed at wine pH (3−4), forming small amounts of methanol and CO_2 (200 mg DMDC/L add 98 mg methanol/L to wine) (EU Commission, 2001). For this reason, some winemakers and suppliers consider DMDC a sustainable alternative to SO_2. However, other producers do not consider DMDC a sustainable additive due to its potential toxicity.

The use of wood in wine production and in wine-derived beverages (e.g., through distillation), during fermentation and aging, raises additional questions regarding sustainability (See Chapter 14). Aging is traditionally done by storing wine in wood barrels, but it has economic and environmental drawbacks. The availability of oak and chestnut wood is limited while the demand for these types of barrels increases (Martínez-Gil et al., 2018). The sector needs to use wood more efficiently and find alternative aging techniques (Canas et al., 2009), or promote the reuse of the barrels, from the aging of one beverage to another (Russell, 2003), e.g., the reuse of Port or Madeira wine barrels in whisky production (See Chapter 14).

Just like vineyards, wineries have an important socioeconomic impact and responsibility (Forbes et al., 2020; Taylor, 2017). Corporate social responsibility must become part of the wine industry by disclosing its environmental and social impacts (Ene et al., 2013; Villanueva-Rey et al., 2018) (See Chapters 20 and 23).

1.3.3 Supply chain issues
1.3.3.1 General aspects
A supply chain is a sequence of activities involved in the life cycle of a product, from the moment the product is designed and conceived to the moment it is consumed. The global dimension of the wine industry increases complexity of the wine supply chain, due to its logistics, transportation, and storage (Taylor, 2017). Wine logistics involve large amounts of products needing storage and inventory management to improve efficiency and reduce overhead costs (Taylor, 2017). Wine transportation and logistics have a major environmental impact (e.g., carbon footprint) (Hierlam, 2020). Some countries are increasing restrictions on imports of bottled wine and promoting imports of bulk wine. However, this may benefit larger producers at expenses of the smaller ones. Therefore, reducing weight of glass bottles or using alternative packaging is on demand (See Chapter 19).

Online marketing and e-commerce have expanded during the COVID-19 pandemics (IWSR, 2020). Online sales cannot be ignored especially when seen in relation to the increasing number of internet users for multiple purposes (education, culture, leisure, and commerce). New challenges may emerge from these new sales and marketing modes when considering the variability in the legislation on alcohol use/consumption existing among different countries or regions.

1.3.3.2 Metrics and analytical tools to assess supply chain performance

Data are crucial to support decision-making, auditing, and governance. Robust statistics are essential to support decision-making policies and successful planning and implementation of policies, e.g., more sustainable use of water resources (Gruère et al., 2020) or of pesticides (ECA, 2020). Unfortunately, statistics and data collection and management are often looked as an additional cost and a burden rather than a valuable tool (UNEP, 2014). Lack of proper monitoring, data analysis, and inventories decreases efficiency, increases costs and environmental impact. Detailed metrics on environmental, social, and financial issues can help implement benchmarking as well as more efficient management of inputs in vineyards (Costa et al., 2016) and wineries (Matos & Pirra, 2020) (See Chapters 11, 18, and 20). The context and specificities of each business should be accounted as well. The lack of primary data from the sector and other enterprises limits their evaluation and comparison as well as assessment of weaknesses and critical points. In fact, it is still quite common to find companies with no control mechanisms to store and provide these data (Barbosa et al., 2018). Vineyard and winery data collection should be thus promoted at local and regional levels (See Chapter 20).

Several types of metrics are used to assess performance of the supply chain. Life cycle thinking is a key concept to ensure a transition toward more sustainable production and consumption (Notarnicola et al., 2017). As a result, life cycle assessment (LCA) methodology has been increasingly applied to the agri-food sector (including the wine industry), to quantify environmental impacts and to support decision-making (Ferrara & De Feo, 2018; Renaud-Gentié et al., 2018; Taylor, 2017) (See Chapters 18 and 20). In the case of the wine sector, the available LCA studies are mainly in viticulture, winemaking, packaging, logistics, and retail (Ferrara & De Feo, 2018; Renaud-Gentié et al., 2018; Taylor, 2017). Data gathering, data availability, and data quality become even more relevant in light of the increasing use of "big data" analyses and data science approaches in the grape and wine sector.

1.4 Legislation, standards, and certification of the wine sector—focus on the EU

1.4.1 Legislation issues for sustainable soil and water management

Legislation influences government policies and their implementation. In relation to water issues, governments belonging to the Organization for Economic Co-operation and Development (OECD) and G20 countries have made commitments to improve agriculture and water policies (e.g., OECD Council "Recommendation on Water," the G20 Action Plan "Toward food and water security: Fostering sustainability, advancing innovation"). These commitments involve political actions on policies, investment, and research to help improve sustainability of water use in food and agricultural production (Gruère et al., 2020) and minimize water footprint of food and beverages. Under the EU Water Framework Directive dated from October 23, 2000, EU Member States are required to ensure adequate protection of water resources, but the EU Commission and some stakeholders are dissatisfied with its implementation so far, in particular with the use of exemptions to the environmental objectives (Boeuf et al., 2016). For example, water prices charged should reflect the full costs (e.g., operation and maintenance costs, capital costs, environmental and resource costs), but full recovery is not required and exemptions are possible in marginal areas or on grounds of social welfare.

1.4 Legislation, standards, and certification of the wine sector—focus on the EU

Other relevant EU directives related to water, soil, or circular economy were implemented in the EU (EU Commission, 2020b). The "Green Deal" strategy (December 2020) is a recent EU initiative focused on environmental protection and sustainability and it is part of the EU's *"Commission Work Program 2020"* (EU Commission, 2020c) (See Chapter 24). The proposal of the EU Commission for an EU Climate Law provides a framework to fulfill the global adaptation goal established in Article 7 of the Paris Agreement, in 2015. According to the EU regulation No 2018/1981, since January 1, 2019, the use of copper in agriculture has been severely restricted, which also covers organic farming practices (European Union, 2018) (See Chapter 24). Other relevant legislation related to residue processing and recycling highlight the variation observed among member countries (Spigno et al., 2017).

The economic impact of new policies on businesses should be evaluated by legislators, especially the impact on small businesses because less capitalized. Short- and long-term government incentives can be required to support investments of smaller companies. In parallel, education and training programs, and collaborative networks should support knowledge transfer between those involved, in order to improve workers' skills while promoting innovation (See Chapter 23). The EU, for example, has been keen to support technology transfer skills and peer-to-peer learning as a means to promote innovation in the agricultural sector (e.g., EU-NEFERTITI project).

1.4.2 Wine quality certification issues: origin, quality, and socioenvironmentally sound

Certification for sustainability increases transparency and brings value to the wine supply chain. Nowadays the concept of sustainability is no longer perceived by companies and governments as only referring to a minimization of their environmental impact, but it is a tool to mitigate and correct poor practices harming the welfare of employees and the community as a whole (Martucci et al., 2019).

The existing certification approaches followed in different countries worldwide show the commitment of the industry for sustainability issues, but there is large variation between them (Moscovici & Reed, 2018). Such variation can limit transferability and possibly generate confusion among consumers (Gerling, 2015). Certification also involves a financial cost that can delay its implementation by smaller businesses. The lack of transparency of certified information and limited cooperation between certification bodies can limit the development of certification programs (Moscovici & Reed, 2018).

Wine provenance and circulation are important for the wine industry. To boost wine quality, promote good practices and minimize fraud, wine authenticity was addressed by a regulatory "wine of origin" system in many countries. Geographical origin of wine represents value-added information and is a guarantee of quality and authenticity. Therefore, novel analytical methodologies to promote good production practices, guarantee product identity and its added value, as well as help to minimize counterfeit wine (e.g., in terms of geographic origin) are essential for the OIV. New analytical tools to ensure traceability and wine authenticity aim to discourage fraud, improve the carbon footprint and sustainability of a wine. Several studies on soil-related markers, namely analyzing for Strontium (the $^{87}Sr/^{86}Sr$ isotopic ratio) on wine from different regions worldwide, showed to be an effective method to determine a wine's provenance due to its unique marker (Catarino, 2021; Catarino et al., 2019; Epova et al., 2019).

1.5 Future prospects

The future of the wine sector will encompass changes in the short and long term to respond to the challenges posed by sustainability issues, by climate change, and by more competitive and diversified markets (Fig. 1.4).

Improved adaptation by the wine sector to climate change requires a better understanding of grapevine responses to the environment (e.g., climate extremes), alternative soil and novel management strategies. These responses are required to guarantee consistent yield and berry composition under more extreme climate conditions while ensuring conservation of soil, landscape, and water resources. Labor shortages puts increasing pressure on the wine sector to optimize mechanization of practices which demand investments in research and innovation, such as in robotics and in the use of artificial intelligence. In addition, wineries will also experience technological developments to minimize the use of energy, water, and additives and to decrease waste production by building more efficient and intelligent infrastructures as well as by undertaking new and more sustainable practices. Carbon and water footprint issues will become increasingly important especially in the more established consumer markets. Recycling and circularity will to be implemented in the wine supply chain as a means to minimize waste production and improve efficiency while reducing environmental impacts and maintaining a sustainable economic performance. Governmental support may be needed to partly support modernization costs especially for smaller businesses namely via grants/subsidies or low interest loans or lower taxes.

Increased efficiency of the wine supply chain depends on efficient logistics and marketing (Fig. 1.4). Digitalization can make logistics and marketing more efficient (Spadoni et al., 2019). The larger companies (vineyards and wineries) are often the first to adopt digital tools based on the Internet of Things, but modern smaller companies are eager to add this in their management practices (Deloitte, 2019). The increasing availability of data will require improved data governance. Data-driven agrifood businesses require new business models including data sharing platforms (See Chapter 7) although this entails public concern due to privacy and transparency issues. More uniform and robust metrics and sustainability indicators and standards are still required to support alternative frameworks to monitor sustainability performance (See Chapter 23). In addition, large databases will enable larger benchmarking programs.

Future technologies, it is hoped, will tend to be more affordable and more user friendly while being robust (Brunel et al., 2021). Low cost sensing and the introduction of digital methods along the supply chain will bring about gains in efficiency and cost reduction. Indeed this digitalization of the wine industry has an increasing role, and can be an asset, in the development and implementation of more sustainable practices in the wine sector".

The move of the wine industry toward a more organic and integrated production system will encompass more preventive rather than reactive actions and strategies, and will be based on more robust data to support decision-making. The digital transformation of the sector will require more training and education, especially for the small businesses. Upskilling and professional development are crucial to guarantee greater flexibility and competiveness at different levels of the supply chain (García-Alcaraz et al., 2017; Pomarici et al., 2021). Many sustainability programs already consider education as a major component for technological advances (Moscovici & Reed, 2018) and more skilled human resources can encourage flexibility and innovation in the wine industry (Garcia-Alcaraz, 2017).

1.5 Future prospects 15

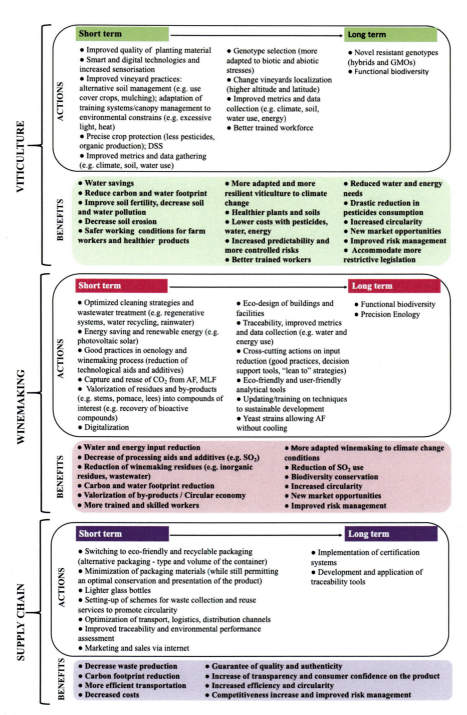

FIGURE 1.4

Summary of the major strategies (short, mid, and long term) and available tools to promote sustainability in the wine's supply chain, from viticulture to the final consumer.

R&D is important to promote sustainability of modern wine industry, namely in terms of crop selection and adaptation to stress, mechanization, use of sensors as means to improve water and energy savings and minimize environmental impacts. R&D is crucial to develop new sustainability evaluation frameworks for the wine industry namely to evaluate its environmental foot print as well as its social impact and the degree of social responsibility (Fig. 1.4). Research on novel management approaches (e.g., lean production) will favor efficiency and help to optimize returns on investments and make companies more resilient (Hill & Hathaway, 2016).

In conclusion, the wine sector faces multiple challenges that are largely related to sustainability and climate change issues (Fig. 1.1). The impact of COVID-19 on the global economy should not prevent the on-going ecological, social, and technological transitions and nino adaptation measures undertaken so far to respond to the risks posed by climate change. The wine industry must continue to implement more sustainable practices while guaranteeing profitability and increased social responsibility.

References

ADVICLIM. (2016). *Adapting viticulture to climate change guidance manual to support winegrowers' decision-making* Accessed on July 2021 https://www.adviclim.eu/wp-content/uploads/2019/06/B1-Guidance-Manual.pdf.

Ahmad, B., Yadav, V., Yadav, A., Ur Rahman, M., Yuan, W. Z., Li, Z., & Wang, X. (2020). Integrated biorefinery approach to valorize winery waste: A review from waste to energy perspectives. *Science of the Total Environment, 719*, 137315. https://doi.org/10.1016/j.scitotenv.2020.137315

Anderson, K., Nelgen, S., & Pinilla, V. (2017). *Global wine markets, 1860−2016: A statistical compendium* (p. 584). Adelaide, SA: University of Adelaide.

Ayuda, M. I., Esteban, E., Martín-Retortillo, M., & Pinilla, V. (2020). The blue water footprint of the Spanish wine industry: 1935−2015. *Water, 12*(7), 1872. https://doi.org/10.3390/w12071872

Barba, F. J., Zhu, Z., Koubaa, M., Sant'Ana, A., & Orlien, V. (2016). Green alternative methods for the extraction of antioxidant bioactive compounds from winery wastes and by-products: A review. *Trends in Food Science & Technology, 49*. https://doi.org/10.1016/j.tifs.2016.01.006

Barbosa, F. S., Scavarda, A. J., Sellitto, M. A., & Lopes Marques, D. I. (2018). Sustainability in the winemaking industry: An Analysis of Southern Brazilian companies based on a literature review. *Journal of Cleaner Production, 192*, 80−87.

Boeuf, B., Fritsch, O., & Martin-Ortega, J. (2016). Undermining European environmental policy goals? The EU water framework directive and the politics of exemptions. *Water, 8*, 388. https://doi.org/10.3390/w8090388

Boncinelli, F., Dominici, A., Gerini, F., & Marone, E. (2021). Insights into organic wine consumption: Behaviour, segmentation and attribute non-attendance. *Agricultural and Food Economics, 9*, 7. https://doi.org/10.1186/s40100-021-00176-6

Bota, J., Tomás, M., Flexas, J., Medrano, H., & Escalona, J. M. (2016). Differences among grapevine cultivars in their stomatal behavior and water use efficiency under progressive water stress. *Agricultural Water Management, 164*, 91−99.

Botha, E. (2020). An industry under pressure: The influence of economic, technological and environmental pressures on the social sustainability of the South African wine industry. In S. Forbes, T. De Silva, & A. Gilinsky (Eds.), *Social sustainability in the global wine industry — Concepts and cases* (pp. 15−25). Cham: Palgrave Macmillan.

Boulton, R. (2019). The design solutions for a self-sustainable zero-carbon winery. In *Wineries of the future solutions for the wine industry conference climate change leadership*. Porto. 6 March 2019 https://climatechange-porto.com/wp-content/uploads/2019/03/Session-4-Wineries-of-the-Future-Roger-Boultton-CCL2019.pdf.

References

Brundtland, G. (1987). *Report of the world commission on environment and development: Our common future.* United Nations General Assembly document A/42/427.

Brunel, G., Moinard, S., Ducanchez, A., Crestey, T., Pichon, L., & Tisseyre, B. (2021). Empirical mapping for evaluating an LPWAN (LoRa) wireless network sensor prior to installation in a vineyard. *OENO One, 55*(2), 301–313. https://doi.org/10.20870/oeno-one.2021.55.2.3102

Canas, S., Caldeira, I., & Belchior, A. P. (2009). Comparison of alternative systems for the ageing of wine brandy. Wood shape and botanical species effects. *Ciência e Técnica Vitivinícola, 24*, 90–99.

Capozzi, V., Garofalo, C., Chiriatti, M. A., Grieco, F., & Spano, F. (2015). Microbial terroir and food innovation: The case of yeast biodiversity in wine. *Microbiological Research, 181*, 75–83.

Castellini, A., Mauracher, C., & Troiano, S. (2017). An overview of the biodynamic wine sector. *International Journal of Wine Research, 9*, 1–11. https://doi.org/10.2147/IJWR.S69126

Cataldo, E., Salvi, L., Sbraci, S., Storchi, P., & Mattii, G. B. (2020). Sustainable viticulture: Effects of soil management in *Vitis vinifera*. *Agronomy, 10*(12). https://doi.org/10.3390/agronomy10121949, 1949.

Catarino, S. (2021). Strontium isotopic signatures for authenticity and wine geographical assessment. In *OENOVITI international 10th symposium "challenges in viticulture and oenology: Wine appellations, authenticity and innovation"* (pp. 88–96). Villenave d'Ornon, France: OENOVITI International network.

Catarino, S., Castro, F. P., Brazão, J., Moreira, L., Pereira, L., Fernandes, J. R., Eiras-Dias, J. E., Graça, A., & Martins-Lopes, P. (2019). 87Sr/86Sr isotopic ratios in vineyard soils and varietal wines from Douro Valley. *Bio Web of Conferences, 12*, 02031. https://doi.org/10.1051/bioconf/20191202031

CEEV. (2016). *European wine: A solid pillar of the European union economy.* Comité Européen des Entreprises Vins. https://www.ceev.eu/images/documents/press_releases/2016/Brochure_CEEV-High_resolution.pdf.

CEEV. (2021). *COVID-19 crisis and the EU wine sector: CEEV impact analysis & 2nd wine package*, 19 March 2021 https://www.ceev.eu/wp-content/uploads/CEEV-COVID-19-brief-and-2nd-Wine-Package-20210319.pdf Accessed on July 2021.

Chaves, M. M., Zarrouk, O., Francisco, R., Costa, J. M., Santos, T., Regalado, A. P., Rodrigues, M. L., & Lopes, C. M. (2010). Grapevine under deficit irrigation: Hints from physiological and molecular data. *Annals of Botany, 105*, 661–676.

Christ, K. L., & Burritt, R. L. (2013). Critical environmental concerns in wine production: An integrative review. *Journal of Cleaner Production, 53*, 232–242.

Comuzzo, P., & Zironi, R. (2013). Biotechnological strategies for controlling wine oxidation. *Food Engineering Reviews, 5*, 217–229.

Costa, J. M., Vaz, M., Escalona, J., Egipto, R., Lopes, C., Medrano, H., & Chaves, M. M. (2016). Modern viticulture in southern Europe: Vulnerabilities and strategies for adaptation to water scarcity. *Agricultural Water Management, 164*, 5–18.

Coulter, A., Cowey, G., Essling, M., Hoare, T., Holdstock, M., Longbottom, M., Simos, C., & Johnson, D. (2019). *Vintage 2019 – Observations from the AWRI helpdesk* Accessed on April 2021 https://www.awri.com.au/wp-content/uploads/2019/09/s2111.pdf.

Da Ros, C., Cavinato, C., Pavan, P., & Bolzonella, D. (2014). Winery waste recycling through anaerobic co-digestion with waste activated sludge. *Waste Management, 34*, 2028–2035.

Deloitte. (2019). *Growing smarter wine industry benchmarking and insights 2018.* New Zealand https://www2.deloitte.com/content/dam/Deloitte/nz/Documents/wine/nz_en_2018_%20wine_industry_benchmarking_and_insights_final.pdf Accessed on June 2021.

ECA. (2020). *Special report sustainable use of plant protection products: Limited progress in measuring and reducing risks.* European Court of Auditors. https://www.eca.europa.eu/Lists/ECADocuments/SR20_05/SR_Pesticides_EN.pdf.

EIP-Agri. (March 2019). *Focus group diseases and pests in viticulture – final report.* https://ec.europa.eu/eip/agriculture/sites/agri-eip/files/eip-agri_fg_diseases_and_pests_in_viticulture_final_report_2019_en.pdf.

Ene, S. A., Teodosiu, C., Robu, B., & Volf, I. (2013). Water footprint assessment in the winemaking industry: A case study for a Romanian medium size production plant. *Journal of Cleaner Production, 43*, 122−135.

Englezos, V., Rantsiou, K., Cravero, F., Torchio, F., Giacosa, S., Segade, S. R., Gai, G., Dogliani, E., Gerbi, V., Cocolin, L., & Rolle, L. (2019). Minimizing the environmental impact of cleaning in winemaking industry by using ozone for cleaning-in-place (CIP) of wine bottling machine. *Journal of Cleaner Production, 233*, 582−589.

Epova, E. N., Bérail, S., Séby, F., Vacchina, V., Bareille, G., Médina, B., Sarthou, L., & Donard, O. F. X. (2019). Strontium elemental and isotopic signatures of Bordeaux wines for authenticity and geographical origin assessment. *Food Chemistry, 294*, 35−45.

EU Commission. (2001). *Opinion of the Scientific Committee on Food on the use of dimethyl dicarbonate (DMDC) in wines*. SCF/CS/ADD/CONS/43 Final.

EU Commission. (2017). *Biodiversity protection in viticulture in Europe*. https://ec.europa.eu/environment/biodiversity/business/assets/pdf/Factsheet_eng_2017.pdf.

EU Commission. (2020a). *TESLA - Transferring energy save laid on agroindustry* Accessed on September 2020 https://ec.europa.eu/energy/intelligent/projects/en/projects/tesla.

EU Commission. (2020b). *Farm to fork strategy − for a fair, healthy and environmentally-friendly food system EU circular economy action plan. A new circular economy action plan for a cleaner and more competitive Europe.* https://ec.europa.eu/environment/circular-economy/.

EU Commission. (2020c). *The European Green Deal: Delivering the European Commission's ambitions to decouple resource use from economic growth*. https://www.oneplanetnetwork.org/european-green-deal-delivering-european-commissions-ambitions-decouple-resource-use-economic-growth.

European Union. (2018). Commission Implementing Regulation (EU) 2018/1981 of 13 December 2018 renewing the approval of the active substances copper compounds, as candidates for substitution, in accordance with Regulation (EC) No 1107/2009 of the European Parliament and of the Council concerning the placing of plant protection products on the market, and amending the Annex to Commission Implementing Regulation (EU) No 540/2011. *Official Journal of the European Union L, 317*, 16, 14.12.2018.

Ferrara, C., & De Feo, G. (2018). Life cycle assessment application to the wine sector: A critical review. *Sustainability, 10*, 395.

FIBL & IFOAM. (2019). *The world of organic agriculture: Statistics and emerging trends*. Acessed on July 2021 https://ciaorganico.net/documypublic/486_2020-organic-world-2019.pdf.

Field, S. K., Smith, J. P., Morrison, E., Neil Emery, R., & Holzapfel, B. (2020). Soil temperature prior to veraison alters grapevine carbon partitioning, xylem sap hormones, and fruit set. *American Journal of Enology and Viticulture, 71*, 52−61. https://doi.org/10.5344/ajev.2019.19038

Filipe-Ribeiro, L., Cosme, F., & Nunes, F. M. (2018). A simple method to improve cork powder waste adsorption properties: Valorization as a new sustainable wine fining agent. *ACS Sustainable Chemistry & Engineering, 7*(1), 1105−1112.

Folger, P. (2017). *Drought in the United States: Causes and current understanding*. https://fas.org/sgp/crs/misc/R43407.pdf.

Forbes, S. L., De Silva, T., & Gilinsky, A., Jr. (Eds.). (2020). *Social sustainability in the global wine industry*. Berlin: Springer Books. Springer, number 978-3-030-30413-3, June.

Fuentes-Pila, J., & García, J. L. (2014). *Handbook: Efficient wineries TESLA project deliverable D.6.6*. European Commission. http://teslaproject.chil.org/download-doc/62556.

Gambetta, G. A., Herrera, J. C., Dayer, S., Feng, Q., Hochberg, U., & Castellarin, S. D. (2020). The physiology of drought stress in grapevine: Towards an integrative definition of drought tolerance. *Journal of Experimental Botany, 71*, 4658−4676.

Garcia-Alcaraz, J. L., Maldonado-Macias, A. A., Hernandez-Arellano, J. L., Blanco-Fernandez, J., Jimenez-Macias, E., & SaenzDiez Muro, J. C. (2017). The impact of human resources on the agility, flexibility and performance of wine supply chains. *Agricultural Economics, 63*, 175–184.

García-Ruiz, J. M. (2010). The effects of land uses on soil erosion in Spain: A review. *Catena* (Vol. 81,(1), 1–11. https://doi.org/10.1016/j.catena.2010.01.001

Gerling, C. (2015). *Environmentally sustainable viticulture, practices and practicality.* In Chris Gerling (Ed.) (p. 399). CRS Press, ISBN 978-1-77188-112-8. Apple Academic Press 2015.

Giacobbo, A., Meneguzzi, A., Bernardes, A. M., & de Pinho, M. N. (2017). Pressure-driven membrane processes for the recovery of antioxidant compounds from winery effluents. *Journal of Cleaner Production, 155*, 172–178.

Giovenzana, V., Beghi, R., Vagnoli, P., Iacono, F., Guidetti, R., & Nardi, T. (2016). Evaluation of energy saving using a new yeast combined with temperature management in sparkling base wine fermentation. *American Journal of Enology and Viticulture, 67*, 308–314. https://doi.org/10.5344/ajev.2016.15115

Gómez-Brandón, M., Lores, M., Insam, H., & Domínguez, J. (2019). Strategies for recycling and valorization of grape marc. *Critical Reviews in Biotechnology, 39*, 437–450.

Gómez-Lorente, D., Rabaza, O., Aznar-Dols, F., & Mercado-Vargas, M. (2017). Economic and environmental study of wineries powered by Grid-connected photovoltaic systems in Spain. *Energies, 10*(2), 222. https://doi.org/10.3390/en10020222

Gonsier, S. (2015). *Trouble in paradise: How the California drought is affecting vineyards and the wine industry.* Theses 2015-Present. 2 https://fordham.bepress.com/environ_2015/2.

Gonçalves, F., Ribeiro, R., Neves, L., Lemperle, T., Lança, M., Ricardo da Silva, J., & Laureano, O. (2013). Alcohol reduction in wine by nanofiltration. Some comparisons with reverse osmosis technique. In *Proceedings of 1st international symposium of OENOVITI international network* (pp. 64–67). Villenave d'Ornon: ISVV.

Google Scholar (2021). https://scholar.google.com/

Grohs, D., Almança, M., Fajardo, T., Halleen, F., & Miele, F. (2017). Advances in propagation of grapevine in the world. *Revista Brasileira de Fruticultura, 39*(4). https://doi.org/10.1590/0100-29452017760

Gruère, G., Shigemitsu, M., & Crawford, S. (2020). *Agriculture and water policy changes: Stocktaking and alignment with OECD and G20 recommendations.* OECD food, agriculture and fisheries papers, No. 144. Paris: OECD Publishing. https://doi.org/10.1787/f35e64af-en

Gubiani, R., Pergher, G., & Mainardis, M. (2019). The winery in a perspective of sustainability: The parameters to be measured and their reliability. In *IEEE international workshop on metrology for agriculture and forestry (MetroAgriFor), 2019* (pp. 328–332). https://doi.org/10.1109/MetroAgriFor.2019.8909221

Guerini Filho, M., Lumi, M., Hasan, C., Marder, M., Leite, L. C. S., & Konrad, O. (2018). Energy recovery from wine sector wastes: A study about the biogas generation potential in a vineyard from Rio Grande do Sul, Brazil. *Sustainable Energy Technologies and Assessments, 29*, 44–49.

Guzzon, R., Bernard, M., Barnaba, C., Bertoldi, D., Pixner, K., & Larcher, R. (2017). The impact of different barrel sanitation approaches on the spoilage microflora and phenols composition of wine. *Journal of Food Science and Technology, 54*, 810–821.

Guzzon, R., Malacarne, M., Larcher, R., Franciosi, E., & Toffanin, A. (2019). The impact of grape processing and carbonic maceration on the microbiota of early stages of winemaking. *Applied Microbiology, 128*, 209–224.

Hayman, P., Longbottom, M., McCarthy, M., & Thomas, D. (2012). *Managing vines during heatwaves.* Wine Australia Factsheet. https://www.wineaustralia.com/getmedia/90cf20af-1579-462d-b06e35f343cbe129/201201_Managing-vines-duringheatwaves.pdf.

Hierlam, K. (2020). *Understanding the carbon footprint of the wine industry.* The Australian Wine Research Institute (AWRI) Accessed on July 2021 https://www.portoprotocol.com/wp-content/uploads/2021/07/Porto-Protocol-2021_Kieran.pdf.

Hill, M., & Hathaway, S. (2016). *Case study: The adoption of lean production in Australian wineries. Using the "Market, Message and Means of Communication" framework to design an extension strategy.* The State of Victoria Department of Economic Development, Jobs, Transport and Resources. https://www.wineaustralia.com/getmedia/c2cc36f6-3186-487f-a63d-9ba873671bf6/Adoption-of-RD-Lean-production-case-study.

Hofmann, M., Volosciuk, C., Dubrovský, M., Maraun, D., & Schultz, H. (2021). *Downscaling of climate change scenarios for a high resolution, site— specific assessment of drought stress risk for two viticultural regions with heterogeneous landscapes Earth System Dynamics.* https://doi.org/10.5194/esd-2021-9. Preprint).

Hou, D., Bolan, N. S., Tsang, D. C., Kirkham, M. B., & O'Connor, D. (2020). Sustainable soil use and management: An interdisciplinary and systematic approach. *Science of Total Environment, 2020*, 138961. https://doi.org/10.1016/j.scitotenv.2020.138961

IWSR. (2020). *Beverage alcohol in 2020 performs better than expected - IWSR* Accessed on June 2021 https://www.theiwsr.com/beverage-alcohol-in-2020-performs-better-than-expected/.

Lazcano, C., Decock, C., & Wilson, S. G. (2020). Defining and managing for healthy vineyard soils, intersections with the concept of terroir. *Frontiers in Environmental Science*. https://doi.org/10.3389/fenvs.2020.00068

Li, Y., & Bardají, I. (2017). Adapting the wine industry in China to climate change: Challenges and opportunities. *OENO One, 51*(2), 71—89.

Lorenzo, N., Diaz-Poso, A., & Roy, D. (2021). Heatwave intensity on the Iberian peninsula: Future climate projections. *Atmospheric Research, 258*, 105655, 15 August 2021.

Malvoni, M., Congedo, P. M., & Laforgia, D. (2017). Analysis of energy consumption: A case study of an Italian winery. In *Proceedings of the 72nd conference of the Italian thermal machines engineering association, ATI2017, 6—8 September 2017*. Lecce, Italy.

MAPA. (2019). *Encuesta sobre superficies y rendimientos de cultivos. Análisis de las plantacionesde viñedo en España. Gobierno de España.* https://www.mapa.gob.es/es/estadistica/temas/estadisticas-agrarias/agricultura/esyrce/.

Martínez-Gil, A., del Álamo-Sanza, M., Sánchez-Gómez, R., & Nevares, I. (2018). Different woods in cooperage for oenology: A review. *Beverages, 4*, 94—119.

Martucci, O., Arcese, G., Montauti, C., & Acampora, A. (2019). Social aspects in the wine sector: Comparison between social life cycle assessment and VIVA sustainable wine project indicators. *Resources, 8*(2), 69. https://doi.org/10.3390/resources8020069

Mas, A. (2018). Microbial challenges in sustainable winemaking. In *Opportunities and challenges for vine and wine production by preserving resources and environment. Proceedings of 7th international symposium* (pp. 38—43). OENOVITI International Network.

Matos, C., & Pirra, A. (2020). Water to wine in wineries in Portugal Douro region: Comparative study between wineries with different sizes. *Science of the Total Environment, 732*, 139332.

Meissner, G., Athmann, M., Fritz, J., Kauer, R., Stoll, M., & Schultz, H. (2019). Conversion to organic and biodynamic viticultural practices: Impact on soil, grapevine development and grape quality. *OENO One, 53*(4). https://doi.org/10.20870/oeno-one.2019.53.4.2470

Mira, H., de Pinho, M., Guiomar, A., & Geraldes, V. (2017). Membrane processing of grape must for control of the alcohol content in fermented beverages. *Journal of Membrane Science and Research, 3*(4), 308—312.

Mirás-Avalos, J. M., & Intrigliolo, D. S. (2017). Grape composition under Abiotic Constrains: Water stress and salinity. *Frontiers Plant Sciences, 8*, 851. https://doi.org/10.3389/fpls.2017.00851

Moscovici, D., & Reed, A. (2018). Comparing wine sustainability certifications around the world: History, status and opportunity. *Journal of Wine Research*. https://doi.org/10.1080/09571264.2018.1433138

Navarro, A., Puig, R., Kılıç, E., Penavayre, S., & Fullana-i-Palmer, P. (2017). Eco-innovation and benchmarking of carbon footprint data for vineyards and wineries in Spain and France. *Journal of Cleaner Production, 142*(4), 1661—1671.

Nielsen. (2020). *Rebalancing the "COVID-19 effect" on alcohol sales*. Chicago, IL: Nielsen Accessed on June 2021 https://nielseniq.com/global/en/insights/2020/rebalancing-the-covid-19-effect-on-alcohol-sales/.

Nordestgaard, S. (2013). *Improving winery refrigeration efficiency winery. Case study report*. The Australian Wine Research Institute. https://www.awri.com.au/wp-content/uploads/WineryB-CaseStudyReport2.pdf.

Notarnicola, B., Sala, S., Anton, A., Mclaren, S., Saouter, E., & Sonesson, U. (2017). The role of life cycle assessment in supporting sustainable agri-food systems: A review of the challenges. *Journal of Cleaner Production, 140*, 399–409.

Novara, A., Cerdà, A., & Gristina, L. (2018). Sustainable vineyard floor management: An equilibrium between water consumption and soil conservation. *Current Opinion in Environmental Science & Health, 5*, 33–37. https://doi.org/10.1016/j.coesh.2018.04.005

NZ Foreign Affairs & Trade. (2020). *United States: Covid-19 – opportunities in the wine sector – market report* Accessed on April 2021 https://www.mfat.govt.nz/assets/Trade-General/Trade-Market-reports/United-States-Covid-19-Opportunities-in-the-US-Wine-Sector-25-November-2020-PDF.pdf.

OIV. (2004). *Development of sustainable vitiviniculture*. Resolution CST 1/2004. Paris: International Organisation of Vine and Wine.

OIV. (2016). *OIV general principles of sustainable vitiviniculture-environmental-social-economic and cultural aspects OIVCST518-2016*. Retrieved from Accessed on 16 April 2019 http://www.oiv.int/en/technical-standards-and-documents/resolutions-of-the-oiv/resolution-cst.

OIV. (2019). *Statistical report on world vitiviniculture – 2019* Accessed on July 2021 https://www.oiv.int/public/medias/6782/oiv-2019-statistical-report-on-world-vitiviniculture.pdf.

OIV. (2020a). *Current situation of the vitivinicultural sector at a global level*. Press Release Accessed on July 2021 http://www.oiv.int/js/lib/pdfjs/web/viewer.html?file=/public/medias/7260/en-oiv-press-conference-april-2020-press-release.pdf.

OIV. (2020b). *OIV focus the global sparkling wine market* Accessed on July 2021 http://www.oiv.int/public/medias/7291/oiv-sparkling-focus-2020.pdf.

OIV. (2021a). *State of the world vitivinicultural sector in 2020*. https://www.oiv.int/public/medias/7909/oiv-state-of-the-world-vitivinicultural-sector-in-2020.pdf.

OIV. (2021b). *SO2 and wine: A review. OIV collective expertise document*. Paris: International Organization of Vine and Wine Accessed on July 2021 https://www.oiv.int/public/medias/7840/oiv-collective-expertise-document-so2-and-wine-a-review.pdf.

OIV. (2021c). *Sustainable use of water in winegrape vineyards* Accessed on July 2021.

Oliveira, M., Costa, J. M., Fragoso, R., & Duarte, E. (2019). Challenges for modern wine production in dry areas: Dedicated indicators to preview wastewater flows. *Water Supply, 19*(2), 653–661.

Oliveira, M., & Duarte, E. (2016). Integrated approach to winery waste: Waste generation and data consolidation. *Frontiers of Environmental Science & Engineering, 10*, 168–176.

Ollat, N., Peccoux, A., Papura, D., Esmenjaud, D., Marguerit, E., Tandonnet, J.-P., Bordenave, L., Cookson, S., Barrieu, F., Rossdeutsch, L., Lecourt, J., Lauvergeat, V., Vivin, P., Bert, P.-F., & Delrot, S. (2015). Rootstock as a component of adaptation to environment. In H. Gerós, M. Chaves, H. Medrano, & S. Delrot (Eds.), *Grapevine in a changing environment: A molecular and ecophysiological perspective* (pp. 68–108). Hoboken, NJ: Wiley-Blackwell.

Ollat, N., Touzard, J. M., & Van Leeuwen, C. (2016). Climate change impacts and adaptations for the wine industry. *Journal of Wine Economics, 11*(1), 139–149.

Pérez-Expósito, J. P., Fernández-Caramés, T. M., Fraga-Lamas, P., & Castedo, L. (2017). VineSens: An eco-smart decision-support viticulture system. *Sensors, 17*, 465.

Phogat, V., Cox, J. W., Simunek, J., & Hayman, P. (2020). Impact of long-term recycled water irrigation on crop yield and soil chemical properties. *Journal of Water and Climate Change, 11*(3), 901–915.

Pomarici, E. (2016). Recent trends in the international wine market and arising research questions. *Wine Economics and Policy, 5*(1), 1–3. https://doi.org/10.1016/j.wep.2016.06.001

Pomarici, E., Corsi, A., Mazzarino, S., & Sardone, R. (2021). The Italian wine sector: Evolution, structure, competitiveness and future challenges of an enduring leader. *Italian Economic Journal.* https://doi.org/10.1007/s40797-021-00144-5

Recarte, I. S. (2020). *Wine package: A first step in the long route towards recovery" secretary general — comité Européen des Entreprises Vins (CEEV).* https://www.europarl.europa.eu/cmsdata/214558/CEEV_Sanchez%20Recarte_Covid%20ISR%20summary%2020201026_v2.pdf.

Renaud-Gentié, C., Perrin, A., Rouault, A., Garrigues-Quéré, Renouf, M., Julien, S., Czyrnek-Delêtre, M., & Jourjon, F. (2018). Eco-quali-conception©: Agroecological transition in viticulture through life cycle assessment. In *Opportunities and challenges for vine and wine production by preserving resources and environment. Proceedings of 7th international symposium.* Villenave d'Ornon, France: OENOVITI International Network.

Ribéreau-Gayon, P., Dubourdieu, D., Doneche, B., & Lonvaud, A. (2006). *Handbook of Enology. The microbiology of wine and vinifications* (Vol. 1). New York, NY: Wiley.

Rinaldi, S., Bonamente, E., Scrucca, F., Merico, M. C., Asdrubali, F., & Cotana, F. (2016). Water and carbon footprint of wine: Methodology review and application to a case study. *Sustainability, 8,* 621.

Romero, P., & Garcia-Garcia, J. (2020). The productive, economic, and social efficiency of vineyards using combined drought-tolerant rootstocks and efficient low water volume deficit irrigation techniques under Mediterranean semiarid conditions. *Sustainability, 12,* 1930. https://doi.org/10.3390/su12051930

Russell, I. (2003). *Whisky: Technology, production and marketing.* Boston, MA: Academic Press.

Santillán, D., Garrote, L., Iglesias, A., & Sotes, V. (2020). Climate change risks and adaptation: New indicators for Mediterranean viticulture. *Mitigation and Adaptation Strategies for Global Change, 25,* 881899. https://doi.org/10.1007/s11027-019-09899-w

Santos, J. A., Fraga, H., Malheiro, A. C., Moutinho-Pereira, J., Dinis, L.-T., Correia, C., Moriondo, M., Leolini, L., Dibari, C., Costafreda-Aumedes, S., Kartschall, T., Menz, C., Molitor, D., Junk, J., Beyer, M., & Schultz, H. R. (2020). A review of the potential climate change impacts and adaptation options for European viticulture. *Applied Sciences, 10*(9), 3092. https://doi.org/10.3390/app10093092

Schmitt, M., & Christmann, M. (2019). Alcohol reduction by physical methods. In A. Morata, & I. Loira (Eds.), *Advances in grape and wine biotechnology* (pp. 251–267). London: IntechOpen.

Schultz, H. R., & Jones, G. V. (2010). Climate induced historic and future changes in viticulture. *Journal of Wine Research, 21*(2), 137–145.

Sheridan, C. M., Glasser, D., Hildebrandt, D., Peterson, J., & Rohwer, J. (2011). An annual and seasonal characterization of winery effluent in South Africa. *South African Journal of Enology and Viticulture, 32,* 1–8.

Silva, M. A., Julien, M., Jourdes, M., & Teissedre, P.-L. (2011). Impact of closures on wine post-bottling development: A review. *European Food Research and Technology, 233,* 905–914.

Silva, V., Mol, H. G. J., Zomer, P., Tienstra, M., Ritsema, C. J., & Geissen, V. (2018). Pesticide residues in European agricultural soils — A hidden reality unfolded. *Science of the Total Environment, 25*(653), 1532–1545.

Silvia, F., Ferrero, A., & Vidotto, F. (2020). *Current and future scenarios of glyphosate use in Europe: Are there alternatives? Advances in agronomy.* Amsterdam: Elsevier Inc. https://doi.org/10.1016/bs.agron.2020.05.005. ISSN 0065-2113.

Simonneau, T., Lebon, E., Coupel-Ledru, A., Marguerit, E., Rossdeutsch, L., & Ollat, N. (2017). Adapting plant material to face water stress in vineyards: Which physiological targets for an optimal control of plant water status? *OENO One, 51*(2), 167–179. https://doi.org/10.20870/oeno-one.2017.51.2.1870

Spadoni, R., Nanetti, M., Bondanese, A., & Rivaroli, S. (2019). Innovative solutions for the wine sector: The role of startups. *Wine Economics and Policy, 8*(2), 165–170. https://doi.org/10.14601/web-8207

Spigno, G., Marinoni, L., & Garrido, G. D. (2017). State of the art in grape processing by-products. In *Handbook of grape processing by-products sustainable solutions* (pp. 1–27). London: Academic Press. https://doi.org/10.1016/B978-0-12-809870-7.00001-6

Strub, L., Kurth, A., & Loose, S. (2021). The effects of viticultural mechanization on working time requirements and production costs. *American Journal of Enology and Viticulture, 72*, 46–55. https://doi.org/10.5344/ajev.2020.20027

Suciu, N., Gallo, A., Capri, E., Zambito Marsala, R., Russo, E., De Crema, M., Peroncini, E., Tomei, F., Antolini, G., Marcaccio, M., Marletto, V., & Colla, R. (2020). Evaluation of groundwater contamination sources by plant protection products in hilly vineyards of Northern Italy. *Science of the Total Environment, 749*. https://doi.org/10.1016/j.scitotenv.2020.141495

SUDVINBIO. (2019). MillésimeBIO press pack, MontPellier, France Accessed on July 2021 https://www.millesime-bio.com/app/millesime/files-module/local/documents/Dossier-de-Presse-Mill%C3%A9simeBio_2019-GB.pdf.

Taylor, S. (2017). *The business of sustainable wine. How to build a brand equity in a 21st Century wine industry* (p. 264). San Francisco, CA: Board and Bench.

Teissèdre, P. L. (2018). Wine quality production and sustainability. In *Opportunities and challenges for vine and wine production by preserving resources and environment. Proceedings of 7th international symposium* (pp. 31–37). OENOVITI International Network.

Tortosa, I., Douthe, C., Pou, A., Balda, P., Hernandez-Montes, E., Toro, G., Escalona, J. M., & Medrano, H. (2019). Variability in water Use efficiency of grapevine tempranillo clones and stability over Years at field conditions. *Agronomy, 9*(11), 701. https://doi.org/10.3390/agronomy9110701

UNEP. (2014). https://www.unepfi.org/fileadmin/documents/UNEPFI_SustainabilityMetrics_Web.pdf.

UNESCO. (2020). *Montado cultural landscape*. https://whc.unesco.org/en/tentativelists/6210/.

USDA. (2018). *Lessons learned from China's wine producing regions: Implications for U.S. Exporters*. GAIN report number GAIN0063. USDA Foreign Agricultural Service.

USDA. (2019). *India wine production and trade update*. GAIN report. Number:IN9073. USDA Foreign Agricultural Service https://apps.fas.usda.gov/newgainapi/api/report/downloadreportbyfilename?filename=Wine%20Production%20and%20Trade%20Update_New%20Delhi_India_8-7-2019.pdf.

Van Leeuwen, C., & Darriet, F. (2016). The impact of climate change on viticulture and wine quality. *Journal of Wine Economics, 11*, 150–167.

Vela, R., Mazarrón, F. R., Fuentes-Pila, J., Baptista, F., Silva, L. L., & García, J. L. (2017). Improved energy efficiency in wineries using data from audits. *Ciência e Técnica Vitivinícola, 32*, 62–71. https://doi.org/10.1051/ctv/20173201062

Verones, F., Bare, J., Bulle, C., Frischknecht, R., Hauschild, M. Z., Hellweg, S., Henderson, A., Jolliet, O., Laurent, A., Liao, X., Lindner, J. P., Maia de Souza, D., Michelsen, O., Patouillard, L., Pfister, S., Posthuma, L., Prado-Lopez, V., Ridoutt, B., Rosenbaum, R. K., & Fantke, P. (2017). LCIA framework and cross-cutting issues guidance within the UNEP-SETAC Life Cycle Initiative. *Journal of Cleaner Production, 161*, 957–967. https://doi.org/10.1016/j.jclepro.2017.05.206

Villanueva-Rey, P., Quinteiro, P., Vázquez-Rowe, I., Rafael, S., Arroja, L., Moreira, M. T., Feijoo, G., & Dias, A. C. (2018). Assessing water footprint in a wine appellation: A case study for Ribeiro in Galicia, Spain. *Journal of Cleaner Production, 172*, 2097–2107.

Waite, H., Whitelaw-Weckert, M., & Torley, P. (2015). Grapevine propagation: Principles and methods for the production of high-quality grapevine planting material. *New Zealand Journal of Crop and Horticultural Science, 43*(2), 144–161.

WineGB. (2021). *UK vineyard figures since 1975* Accessed on April 2021 https://www.winegb.co.uk/trade/industry-data-and-stats-2/.

Wineserver, UCD. (2020) Accessed on May 2021 https://wineserver.ucdavis.edu/about/facilities/jess-s-jackson-sustainable-winery-building.
WoS. (2021). *WoS abstract & citation database. 2021* Accessed on June 2021 https://apps.webofknowledge.com/WOS_GeneralSearch_input.do?product=WOS&search_mode=GeneralSearch&SID=E1CuPGLUVGqBNwL4Z6x&preferencesSaved.
WSET Alumni. (2015). *Wine production: A global overview*. Wine & Spirit Education Trust. https://www.wsetglobal.com/media/3009/wine-production-a-global-overview.pdf/acessedx.10.20.
Zheng, W., del Galdo, V., García, J., Balda, P., & Martínez de Toda, F. (2017). Use of minimal pruning to delay fruit maturity and improve berry composition under climate change. *American Journal of Enology and Viticulture, 68*(1), 136–140. https://doi.org/10.5344/ajev.2016.16038
Zhu, X., Li, X., Jiu, S., Zhang, K., Wang, C., & Fang, J. (2018). Analysis of the regulation networks in grapevine reveals response to waterlogging stress and candidate gene-marker selection for damage severity. *Royal Society Open Sciences 27, 5*(6), 172253. https://doi.org/10.1098/rsos.172253

Further reading

Baiano, A. (2021). An overview on sustainability in the wine production chain. *Beverages, 7*(1), 15. https://doi.org/10.3390/beverages7010015
Gutierrez-Gamboa, G., Zheng, W., & Martínez de Toda, F. (2021). Current viticultural techniques to mitigate the effects of global warming on grape and wine quality: A comprehensive review. *Food Research International*. https://doi.org/10.1016/j.foodres.2020.109946
Romero, P., Navarro, J. M., & Botía, P. (2022). Towards a sustainable viticulture: The combination of deficit irrigation strategies and agroecological practices in Mediterranean vineyards. A review and update. *Agricultural Water Management, 259*(C). In press.
White, R., & Krstic, M. (2019). *Healthy soils for healthy vines - Soil Management for Productive Vineyards*. Australia: CSIRO Publishing.
Zava, A., Sebastião, P. J., & Catarino, S. (2020). Wine traceability and authenticity: approaches for geographical origin, variety and vintage assessment. *Ciência e Técnica Vitivinícola, 35*(2), 133–147. https://doi.org/10.1051/ctv/20203502133

CHAPTER 2

Exploiting genetic diversity to improve environmental sustainability of Mediterranean vineyards

Josefina Bota[1], Rosa Arroyo-Garcia[2], Ignacio Tortosa[1] and Hipólito Medrano[1]

[1]*Research Group on Plant Biology Under Mediterranean Conditions, Departament de Biologia, Universitat de les Illes Balears (UIB) - Agro-Environmental and Water Economics Institute (INAGEA), Palma, Illes Balears, Spain;* [2]*Centro de Biotecnología y Genómica de Plantas (CBGP-INIA), Madrid, Spain*

2.1 Introduction

The availability of water in the Mediterranean is a recurring problem that has increased in recent years due to climate change, representing a limitation and a threat to the future of agriculture in this area (Hannah et al., 2013; Jones, 2007). It is clear that the sustainability in Mediterranean crops is a top priority in the current and future climatic scenario as is reflected in current research programs. In this context, we will need new approaches in order to ensure the correct development of the Mediterranean viticulture and its adaptation to future changes.

Some of the consequences of climate change are already visible in terms of vine development, grape ripening, and final wine quality (van Leeuwen & Destrac-Irvine, 2017). Late ripening, drought tolerant and/or more water use efficient cultivars, clones and rootstocks can contribute to solve these problems.

In this sense, different experts have proposed several measures to mitigate the effects of climate change or adapt to new conditions to maintain wine production and quality (Fraga, 2019; van Leeuwen et al., 2019). At long term, grapevine cultivation can be moved to cooler areas, whereas the actual existing regions will have to apply other "in farm" adaption measures. These include, in medium term, the substitution of plant material, rootstocks, cultivars, and clones, selected for their better adaptation to expected climatic conditions (Fraga et al., 2013; Santos et al., 2020). The success of selection programs depends on the existing genetic variability in the selection traits. Recently, Morales-Castilla et al. (2020) pointed out that cultivar diversity can decrease the loss of agricultural areas by over 50% in a climate change scenario. Thus, exploring the existing diversity in cultivars, clones, and rootstocks is presented as an interesting option to solve the threats of climate change. Moreover, the adaption through the use of more adapted genotypes is cost effective as it does not increase production cost.

The diversity of genetic resources of grapevine covers approximately 10,000 cultivars (OIV, 2009) and includes clones of different commercial cultivars that have different degrees of tolerance to water stress (Chaves et al., 2010). On the other hand, the use of rootstocks is required in most regions since the introduction of phylloxera to Europe in the late 19th century. Rootstocks are part of the root system responsible for water and nutrient uptake and, at time, hormone responses (Keller, 2015). Hence, the search or development of more drought tolerant or more water use efficient genotypes also concerns the rootstocks.

The knowledge of agronomical, physiological, and molecular response of the different genotypes under unfavorable conditions is critical for understanding the possible adaptation to future scenarios. For instance, for water use efficiency (WUE) extensive knowledge exists about the genetic variability, mainly at the leaf level (Bota et al., 2001, 2016; Chaves et al., 2010; Costa et al., 2012; Prieto et al., 2010; Soar et al., 2006; Tomás et al., 2012, Tomás, Medrano, Escalona, et al., 2014), but the physiological and molecular mechanisms associated with this variability are still largely unknown.

On the other hand, the advances in molecular and genomic approaches and the next generation sequencing tools, provide a valuable information about these responses or characteristics that confers better adaptation to environmental stress. Understanding the molecular basis of complex plant processes and the identification of key genes involved, could provide a rich source of information for breeding programs.

The present chapter summarizes recent contributions, which could contribute to exploit the existing grapevine genetic diversity to overcome climatic change challenges to this important, extensive, and typically Mediterranean crop. The interest of the wild grapevines, the intercultivar and also, the intracultivar variability, including the role of rootstocks, as the new developments in genomics which could accelerate the selection programs are presented.

2.2 Origin of cultivated grapevine and actual grapevine diversity

The Eurasian grapevine (*Vitis vinifera* L) is the most widely cultivated and economically important fruit crop in the world (Mattia et al., 2008). *Vitis vinifera* L includes the cultivated form *V. vinifera* ssp vinifera and the wild form *V. vinifera* ssp sylvestris, considered as two subspecies based on morphological differences. The most conspicuous differential trait is plant sex: wild grapevines are dioecious (male and female plants), while cultivated forms are mostly hermaphrodite plants, with self-fertile hermaphrodite flowers (This et al., 2006). However, it can be argued that those differences are the result of the domestication process (This et al., 2006). The wild form, considered as being the putative ancestor of the cultivated form, represents the only endemic taxon of the Vitaceae in Europe and the Maghreb (Heywood & Zohary, 1991). Although wild grapevines were spread over Southern Europe and Western and Central Asia during the Neolithic period, archeological and historical evidence suggest that primo domestication events would have occurred in the Near-East (McGovern et al., 1996). In addition, several studies have shown evidence supporting the existence of secondary domestication events along the Mediterranean basin (Arroyo-García et al., 2006; De Andres et al., 2012; De Lorenzis et al., 2015; Grassi et al., 2003; Lopes et al., 2009; Riaz et al., 2018). Recent genetic analyses using a large SNP platform provide genetic evidence supporting the Eastern origin of most cultivated germplasm as well as the existence of introgression from wild germplasm in Western regions, likely as the consequence of those predicted secondary domestication events (Myles et al., 2011).

Wild grapevines can still be found in Eastern and Western Europe (Arnold et al., 1998). The South Caucasus (Azerbaijan, Armenia, and Georgia), together with Eastern Anatolia, has been considered for a longtime as the birth place for viticulture with the earliest examples of wine-making (Levadoux, 1956; McGovern, 2003; Olmo, 1995; This et al., 2006). A 1998 census (Arnold et al., 1998) showed that wild grapevine was present in Spain, Italy, Switzerland, Romania, Bulgaria, Hungary, Austria, and in the countries of former Yugoslavia. Apparently, Spain and Italy harbor the highest number of recorded populations and they were proposed to work as shelters for *V. vinifera* during the last glaciation (about 12,000 years ago) as well as putative sources of postglacial colonization and diversification (Levadoux, 1956). Wild vines were abundant in their indigenous range in Europe until the middle of the 19th century, when the arrival of North American pests (*Phylloxera*) and pathogens (downy and powdery mildews) and the destruction of their habitats drove European wild vines close to extinction (IUCN, 1997). The solution to generate resistance to Phylloxera was the use of American species and hybrids as rootstocks and many varieties of rootstocks were developed by breeders at the end of the 19th and the beginning of the 20th century (Arrigo & Arnold, 2007). Currently, vines found in natural habitats are considered being mixture of wild forms, naturalized cultivated forms, and rootstocks escaped from vineyards as well as hybrids derived from spontaneous hybridizations among those species and forms (Lacombe et al., 2003; Laguna, 2003; This et al., 2006). Arrigo and Arnold (2007) compared ecological features and genetic diversity among populations of naturalized rootstocks and native wild grapevines and did not detect the existence of genetic flux between them. The genetic analysis of wild grapevine populations from France and Spain (De Andres et al., 2012; Di Vecchi et al., 2009) evidenced gene flow between cultivated and wild grapevine, estimating up to 3% of pollen migration between the cultivated fields and closely located wild grape. These pollen fluxes may have a significant effect on the evolution of those populations. Currently, wild grapevine is endangered throughout all its distribution range and conservation efforts are required to maintain the genetic integrity and survival of the remnant populations.

Genetic diversity has been studied for a long time in grapevine with different sets of molecular markers, mainly directed to neutral evolving genome sequences, what has provided partial views of the genetic relationships among cultivated and wild germplasm as well as on the expansion and evolution of domesticates (This et al., 2006). More recently, with the completion of the reference genome sequence, the possibilities of resequencing genomes and the development of SNP and SNP sets permitting the rapid genotyping of hundreds or thousands genotypes at thousands of loci, the genetic view of the grapevine domestication process is evolving into a more general integrative theory (Magris et al., 2021).

The picture arising today is one of a low but clear genetic differentiation of cultivars and wild grape, based either on chloroplast markers (Arroyo-Garcia et al., 2006; Grassi et al., 2006), nuclear microsatellites (De Andres et al., 2012; Ergul et al., 2011; Grassi et al., 2003; Lopes et al., 2009; Riaz et al., 2018; Snoussi et al., 2004) or a combination of both (Grassi et al., 2003; Sefc et al., 2003). The wild individuals also cluster according to their populations (Grassi et al., 2008). The positive Fis values observed in the wild grapevine accessions suggest a high level of genetic relationship among the individuals of the same wild populations. In fact, the detection of potential parent-progeny relationships within wild populations supports that possibility (De Andres et al., 2012). At the same time, the detection of gene flow between both compartments (De Andres et al., 2012; Di Vecchi et al., 2009) could have strong consequences in the future.

Several studies have analyzed the genetic diversity of wild and cultivated forms of grapevine in the Mediterranean basin, frequently considered as the wild and cultivated compartments of the species (Arroyo-García & Revilla, 2013; Riaz et al., 2018). The genetic relationship between cultivated varieties and wild grapevine populations from Spain suggests a genetic contribution of Southern wild populations in the autochthonous grapevine cultivars varieties (De Andres et al., 2012). Therefore, it seems that in opposition to the established dominant theory on the origin of the domestication of grapevine, many of the varieties of the Iberian Peninsula and from other European countries could have local origins.

Scientific interest in the highly endangered ancestor of cultivated grapevine, *V. vinifera* subsp. sylvestris, has centered on questions of conservation genetics, and deepening our understanding of the domestication history of the cultivated crop (Arroyo-Garcia et al., 2006). However, since domestication traits such as higher yield, larger berries, and higher sugar content are often accompanied by a loss of resistance to abiotic and biotic stress, it is beneficial to search for such factors in the wild forms of the crop's ancestors. In fact, the presence of salt-tolerant grape accessions in the North African sylvestris population (Askri et al., 2012), and the recent identification of wild and cultivated accessions from Germany, Iran, and Georgia with tolerance to mildew diseases, supports the potential of this wild ancestor as a genetic resource for disease-resistance breeding (Bitsadze et al., 2015; Duan et al., 2015; Riaz et al., 2013; Toffolatti et al., 2016). Given that Eurasian and North Africa wild *V. vinifera* germplasm and Asian *Vitis* germplasm are largely unexplored, their identification, preservation, and characterization for biotic and abiotic resistance and berry quality (Revilla et al., 2010) traits are very important for the future of the wine and grape industry.

2.3 Intercultivar variability in the physiological response to water stress

Even while grapevines are considered being well-adapted to semiarid environments, the negative effects of climate change maintain the study of drought tolerance in vines as a major topic for research. The study of morphoanatomical, physiological, and metabolomic characteristics and mechanisms that confers tolerance to drought and high temperature is crucial to understand how grapevine may cope with expected environmental changes (Carvalho et al., 2015; Chaves et al., 2010; Costa et al., 2016). Nevertheless, an integrative definition of grapevine drought tolerance and the plasticity of the key drought-tolerance traits within and among cultivars are still under discussion (Gambetta et al., 2020).

Thousands of grapevine cultivars have been described around the world (OIV, 2009; This et al., 2006), which can be explored for tolerant traits to finally select "elite" genotypes for direct use or in breeding programs. Large genotype variability has been described in responses to drought, regarding leaf photosynthesis, stomatal conductance, and WUE (Bota et al., 2001; Costa et al., 2012; Prieto et al., 2010; Soar et al., 2006; Tomás et al., 2012, Tomás, Medrano, Escalona, et al., 2014a). Stomatal control is the key mechanism that regulates the compromise between water loss and CO_2 uptake. Hence, it is the first mechanism in response to water deficit stress that needs to be studied (Chaves et al., 2016). Classically, two different behaviors have been described. Those grapevine cultivars with better stomatal control under drought than others were classified as isohydric ("pessimistic"), whereas others with a less marked stomatal regulation under drought were classified as anisohydric ("optimistic") (Schultz, 2003; Soar et al., 2006). Isohydry is commonly associated with strict stomatal regulation of transpiration under drought, which in turn is believed to minimize hydraulic risk at the expense of reduced carbon assimilation. However, this classification is now under debate as many authors believe that this differentiation may not be completely clear and accurate (Hochberg et al., 2018 and references there in). In fact, after evaluating a large number of cultivars under different conditions (controlled and field), the dichotomy classification of cultivars as being iso- or anisohydric is not always clear (Bota et al., 2016; Villalobos-González et al., 2019). Recently, in a study with 17 different cultivars, Levin et al. (2020) reported that in general, cultivars responded similarly under high and low water availability, but their stomatal behavior may differ at moderate water deficits. Nevertheless, some cultivars seem to have greater robustness in their response type, since they behave in the same way across years and experiments (Bota et al., 2001, 2016; Tomàs et al., 2012). This point suggests that the genetic basis of this trait could be stronger in some cultivars than others, which is a hypothesis worth being explored.

Even though several questions about the mechanism of many drought responses mechanisms still remain unanswered (Gambetta et al., 2020), it is clear that control of water loss by stomatal regulation is one of the essential points in drought-tolerance strategies (Hochberg et al., 2018). The key hormone involved in stomatal control and probably responsible for the different levels of anisohydry in response to environmental changes is abscisic acid (ABA). A recent review summarizes current knowledge on the role of ABA in mediating mechanisms responding to abiotic stresses, suggesting to focus future investigation in the basal level of ABA and on the modulations of ABA content in the different grapevine cultivars to characterize abiotic stress tolerance (Marusig & Tombesi, 2020). In fact, several studies pointed out the differences in ABA basal concentration, ABA sensitivity, and hydraulic-related processes among cultivars with contrasted response to water deficit (Coupel-Ledru et al., 2017; Dayer et al., 2020; Martorell, Medrano, et al., 2015; Soar et al., 2006). Part of these different responses have a genetic background (Rosedeutsch et al., 2016). This issue should be addressed in future experimentations to find the genotypes with better performance under drought.

However, drought response variability among cultivars is the result of numerous traits and the links among them and not stomatal regulation alone. Variability in xylem architecture has been described between cultivars and related to differences in hydraulic conductivity and responses to water deficit (Hochberg, Degu, Gendler, et al., 2015). Nevertheless, Alsina et al. (2007), in a study where eight different cultivars were compared, concluded that there was no relationship between embolism vulnerability and leaf drought tolerance traits. More recently, Albuquerque et al. (2020) found similar vulnerabilities to drought-induced xylem embolism in two cultivars (Cabernet Sauvignon and Chardonnay) with different stomatal behavior.

Another player in stomatal regulation as a response to water availability is osmotic adjustment. Several studies have shown differences in this trait in grapevine in response to water stress (During, 1984; Hochberg et al., 2017; Martorell, Diaz-Espejo, et al., 2015; Patakas et al., 2002; Patakas & Noitsakis, 1999), and evidence for differences in osmotic adjustment among cultivars has been reported (Martorell, Diaz-Espejo, et al., 2015). Levin et al. (2020) also suggested that differences in osmotic adjustment among cultivars may play primary major role in determining the observed differences in stomatal behavior of 17 grapevine cultivars. New studies including more cultivars are needed to corroborate this point.

Diffusive limitation to CO_2 under drought is not only imposed by the stomata but also by mesophyll conductance (g_m) (Flexas et al., 2002). Tomas, Medrano, Brugnoli, et al. (2014) found a significant variability of g_m among several grapevine cultivars and associated them with WUE variability. The use of WUE as a selection trait was explored recently (Tortosa et al., 2016), but it should be noted that

WUE is a complex multitrait phenotype related with not only stomatal control but also with leaf structure, leaf biochemistry, and leaf diffusive properties (Tomás, Medrano, Brugnoli, et al., 2014).

At the biochemical level, plant hormones, secondary metabolites, and other key molecules such as carbohydrates, amino acids, and polyamines play crucial roles in stress tolerance mechanism. Also, cultivar differences at metabolic level in response to water stress have been described (Florez-Sarasa et al., 2020; Hochberg, Degu, Cramer, et al., 2015). Hochberg, Degu, Cramer, et al. (2015) studied berry skin metabolism in response to water-deficit stress in two cultivars with different hydraulic behaviors, Syrah and Cabernet Sauvignon and concluded that the alteration in primary metabolism is linked with hydraulic behavior. At leaf level, Florez-Sarasa et al. (2020) found differences in primary metabolism between three local cultivars from Balearic Islands and three international varieties (Merlot, Syrah, and Muscat) with differences in physiological performances.

Undoubtedly, all these traits were shown to have high environmental plasticity (Hochberg, Degu, Gendler, et al., 2015, 2017; Lovisolo et al., 2010; Martorell, Medrano, et al., 2015). Hence, it must be clarified to what extent differences in water stress tolerance among cultivars result from innate genotypic differences or environmental factors.

2.4 Intracultivar variability in the physiological response to changing environments

Although a high number of cultivars are around the world, only a few elite cultivars occupy the majority of the planted area (Anderson & Aryal, 2013). For example, in Spain, only four cultivars, Airen, Tempranillo, Bobal, and Grenache, account more than 60% of total cultivated area (Ibañez et al., 2015). This situation is due to multiple factors. One of them is related to the specificity of wine market. Consumers use to prefer the same cultivars because they relate the name of a given variety with their expectations on wine quality (Eibach & Töpfer, 2015). Secondly, planting a vineyard is a capital consuming operation because vines will start to produce wine only after three or four years following plantation, but continue to do so for 25–50 years more. For this reason, winegrowers tend to be conservative about the choice of cultivars. Moreover, in most of the highly reputed wine production regions, regulation rules include a restrictive list of authorized cultivars.

Cultivated grapevines are vegetatively propagated. As a result, the genome of each cultivar accumulate somatic mutations that resulted distinguishable clones in a range of clones from a given cultivar (Ramu et al., 2017; Vondras et al., 2019). Clonal selection programs aim at exploiting this intracultivar diversity to improve some agronomic traits.

The origins of clonal selection programs go back to the end of 19th century in Germany, and it spread rapidly throughout Europe. The introduction of certified clones to the market was considered being a revolution; vineyards with mixed cultivars and usually with high levels of viruses were replaced by a single clone fields, with virus-free plants, higher uniformity, and supposed improved performance (Rülh et al., 2003).

Sometimes, clonal selection was a tool for the adaptation of world-famous varieties to the local production environments. In this sense, clonal selection allows to increase the climatic distribution range of one single cultivar, but always under certain limits. In the recent past, most of the production regions developed public clonal selections programs to standardize different genotype performances in yield or quality parameters, and in parallel, commercial nurseries started to certificate their own genotypes. For example, only in Spain more than 70 certified clones of Tempranillo cultivar are currently available (Ibañez et al., 2015).

In the actual Global Change context, clonal selection is a possible way to adapt vineyards to new climatic conditions, with a relative easy market acceptability.

The actual success of certified clones, however, implies an increasing genetic erosion because a limited number of clones are replacing old vineyards with great genetic diversity. In an attempt to conserve part of this variability, many public and private institutions are creating accession collections, introducing plants from old or particular vineyards. These collections also allow the characterization of the more interesting accessions for the wine industry. For example, Muñoz et al. (2014) found variability in anthocyanin profiles among Malbec clones, and found some genes related with these differences, and Tello et al. (2018) found differences in pollen viability among Tempranillo clones that could possibly explain differences in yield and wine composition. Other groups focus on the interaction between climate change and agronomic responses of clones. For example, Arrizabalaga-Arriazu et al. (2020) studied genotypic performance differences between Tempranillo clones grown at high temperatures and high CO_2 concentration, and found differences in biomass accumulation and photosynthesis among others characters.

The genetic variability among clones could be exploited to identify water stress tolerant or high WUE genotypes. Fig. 2.1 illustrates the wide intracultivar variability in comparison to intercultivar one. The total WUE variability among clones and accessions was up to 80% of the total WUE variability measured across 23 cultivars (Tortosa et al., 2016). The important influence of environmental conditions (location and year) was also evaluated (Tortosa, Douthe, et al., 2019, Tortosa, Escalona, et al., 2019). Later studies analyzing data from different growing conditions, years and locations showed that some genotypes performed consistently better than others regarding their WUE (Tortosa, Douthe, et al., 2019). Moreover, these differences identified under field conditions were confirmed when the same clones were grown in pots under semicontrolled conditions (Tortosa et al., 2020).

2.5 Rootstocks selection for better performance under semiarid conditions

Since grapevine cultivar is usually grafted onto a rootstock, the rootstock selection has been proposed as a powerful adaptation to climate change adaptation (Berdeja et al., 2015; Fraga et al., 2013; Ollat et al., 2016). A deeper and denser root system would provide the vine access to increased water resources, thus enhancing drought tolerance. Grapevine rootstocks have also been proposed as a solution to face heat stress and adverse edaphic conditions, such as salinity or other mineral toxicities (Berdeja et al., 2015; Delrot et al., 2020; Fraga et al., 2013). Moreover, the changing climatic conditions, together with globalization, are spreading new pests and diseases, and rootstocks could ensure an ecofriendly defense tool for sustainable wine production. Nevertheless, the relations between rootstock, scion, and environment (climate, soil conditions, and irrigation management) are complex (Morton, 1979). Even so, although more than 80% of vineyards are grafted, only a few rootstock genotypes are being used worldwide (Ollat et al., 2016). It is recognized that rootstocks can influence scion phenotypes in different ways: affecting the rhizosphere interaction, changing the water and the nutrient uptake capacity (Fig. 2.2). Due to this range of general mechanisms, there is a wide large of references showing the effect of rootstocks on plant water status, biomass accumulation, phenology, plant health, and grape composition (Cuneo et al., 2020; Gauthier et al., 2020; Zhang et al., 2016 and the references there in).

Chapter 2 Exploiting genetic diversity

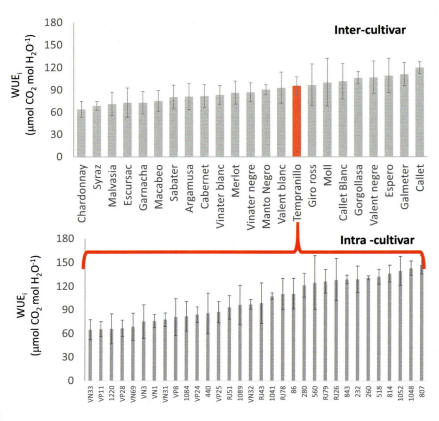

FIGURE 2.1

A comparison between phenotypic WUE$_i$ variability observed between a cultivar collection (up) and Tempranillo clonal collection (down).

Data from Tortosa, I., Escalona, J. M., Bota, J., Tomas, M., Hernandez, E., Escudero, E. G., & Medrano, H. (2016). Exploring the genetic variability in water use efficiency: evaluation of inter and intra cultivar genetic diversity in grapevines. Plant Science, 251, *35–43.*

The rhizosphere interactions are affected by root architecture and roots exudates, and this equilibrium between plant and soil is a primary defensive barrier against different pests, such as bacteria, fungi, or insects (Berendsen et al., 2012). The phylloxera pest (*Daktulosphaira vitifoliae*) introduced to Europe in the 1850s forced winegrowers to use native American rootstocks resistant to phylloxera. Protection against phylloxera is mandatory in any new rootstock developed, but other parameters are also being considered for in selection programs. An example is the resistance against different nematodes, such as *Xiphinema index*, *Meloidogyne javanica*, *M. hapla*, *M. javanica*, and *M. chitwoodi*, among others (Aballay & Vilches, 2016; Nguyen et al., 2020; Rubio et al., 2020; Smith et al., 2017; Zasada et al., 2019).

The influence of the grafted rootstock on plant drought tolerance has also been investigated. This response is linked to water uptake capacity (Carbonneau, 1985). These differences could be due to a

2.5 Rootstocks selection for better performance under semiarid conditions

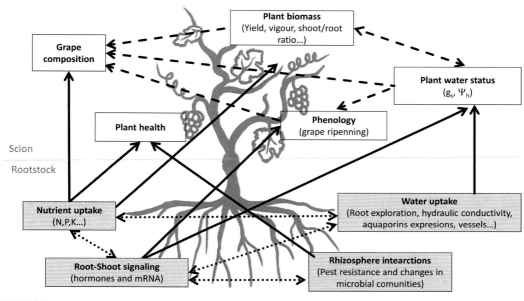

FIGURE 2.2

Summary of the relationships between rootstock and scion in grafted grapevines. In gray squares: influence and mechanism affecting the root system; in white squares: the scion physiological traits and mechanisms involved.

greater root hydraulic conductivity or a direct consequence of the volume of soil explored (De Herralde et al., 2010; Prinsi et al., 2018; Smart et al., 2006; Tandonnet et al., 2018). Variability in hydraulic conductivity may be related to root architecture or morphology. For instance, the presence of a suberized barrier can increase the accumulation of solutes, increasing the root osmotic potential (Barrios-Masias et al., 2015; Cuneo et al., 2020). Other authors showed the central role of aquaporins and their influence on drought tolerance in grapevine rootstocks (Gambetta et al., 2012; Lovisolo et al., 2007; Prinsi et al., 2018), among other factors (e.g., the number or size xylem vessels, see Santarosa et al., 2016 or De Micco et al., 2018). In addition, changes observed in the architecture of the root system (root length, number and length of lateral roots, number of fine roots) may be also influencing the root hydraulic conductivity and the overall access to soil water (Peiró et al., 2020; Yildirim et al., 2018). Nutrient (or mineral) uptake also showed variability among rootstocks, probably related to the previously mentioned differences in water uptake and root exudates (Cochetel et al., 2017; Covarrubias et al., 2016; Gauthier et al., 2020; Romero et al., 2018).

Another source of variability among rootstocks was related to the root-shoot signaling, both by hormones and transcriptomes. First studies focused on ABA because of its well-known influence on stomatal closure and rooting (Kracke et al., 1981; Soar et al., 2006). Moreover, ABA is implicated in berry ripening and induce changes in flavonoids biosynthesis and phenylpropanoid pathway (Keller et al., 2012; Stoll et al., 2000; Villalobos-Gonzales et al., 2016). Important differences in grape ripening induced by the rootstock also have been described (Corso et al., 2016; Keller et al., 2012;

Romero et al., 2018). Today, in the omics era, an important variability in the regulation of genes related with carbohydrate metabolism, protein metabolism, sugar transporters, and hormone synthesis and functionality were also highlighted (Cookson et al., 2013; Lucini et al., 2020; Yildirim et al., 2018).

Rootstocks provide a huge source of genetic variability to adapt wine production to a changing environment. However, the effect of the scion and "Terroir" (and the interactions between them) makes a local evaluation necessary to determine the suitable rootstocks for a specific cultivar and production area. At the same time, fundamental research has been focusing on key traits for the selection of new rootstock genotypes and finding molecular markers that could facilitate high throughput screening.

2.6 Progress in genomics tools and new breeding technologies

Innovation in plant genetics provides the opportunity to examine the genic basis of the traits that will support the breeding of genotypes better adapted to new environments (Gomès et al., 2021; Henry, 2020). The genetic origins of the differences in grapevine water stress response is still under debate (Marguerit et al., 2012). Generally, drought is associated with many morphological and physiological changes in plants over a range of spatial and temporal scales (Chaves et al., 2002). Accordingly, response to abiotic stress is highly complex and involves the interplay of different responses at crop, plant, and cell levels. A succession of molecular, cellular, and physiological events can occur simultaneously and at very small time steps (Cramer, 2010; Rocheta et al., 2014). Although large amounts of data on the expression of genes related to abiotic stress are available, the challenge now is to connect those genetic profiles to changes in plant metabolism and physiology. The analysis of global transcriptome is a key resource for the discovery of genes involved in responses to abiotic stress. Some studies have used large-scale gene expression microarrays analysis to study the water stress response of grapevine to improve the knowledge of transcriptional regulation during drought (Corso et al., 2015; Cramer et al., 2007; Deluc et al., 2009; Rocheta et al., 2014, 2016; Savoi et al., 2017). Haider et al. (2017) using a RNAseq analysis provided evidence that grapevine adaptation to drought stress is a multistep component system consisting of multiple genes that regulate several pathways. They have found a significant increase in the activity of ROS enzymes and hormone levels, revealing the defensive function of these enzymes and hormones during drought stress in grapevine leaves. However, precise information on regulatory mechanisms of these defense-related genes in grapevine under drought stress is still work in progress.

Comparative analysis of gene expression profiles among cultivars with different water stress responses evidenced the intraspecies diversity in stomatal control, WUE, and other drought tolerance traits (Dal Santo et al., 2016; Rocheta et al., 2016). More recently, Catacchio et al. (2019) combined physiological, transcriptomic, and genomic data to study the water deficit effects in two table grape cultivars with contrasted adaptation to drought and were able to identify candidate genes related to genotype-specific response to water deficit in grapevine. Nevertheless, more studies combining physiology and "omic" analyses for the identification of the genetic traits responsible for the genotypic-specific response to water deficit stress are still needed in grapevine.

Another hot research topic in plant functional genomics is epigenetics. Several studies have provided new insights into the epigenetic control of stress adaptation (Chang et al., 2020; Kim et al., 2015; Sahu et al., 2013). Growing evidence suggests that methylation of DNA in response to stress leads to the variation in phenotype. Transposon mobility, siRNA-mediated methylation, and host methyl transferase activation are implicated in this process (reviewed in Ashapkin et al. (2020)). However, most of these studies are carried out in model plants, such as Arabidopsis and there is still little information on how the environment affects grapevine's or other crop's epigenome.

Breeding programs can be based on classical genetic improvement or on so-called New Breeding Technologies (NBTs). Dalla-Costa et al. (2019) reviewed the state-of-the-art of these NBTs and other biotechnologies in grapevine. These techniques appear as an alternative to genetically modified organism approaches. They can be more easily accepted by the wine industry and consumers as they produce minimal and precise modifications in selected genotypes of interest maintaining the identity and agronomic characteristics of the cultivars. CRISPR/Cas9 technology has been successfully applied to generate edited grapevine plants (Osakabe et al., 2018). Nevertheless, the use of these NBTs in current breeding programs remains limited, especially in abiotic stress response and other climate change adaptation traits (e.g., late version) as the different adaptation traits are determined by many genes and its regulation. However, these techniques remain of interest for breeding programs, as they offer an approach to maintain the identity of cultivars and their appreciated agronomic characteristics, which increases acceptability by the wine production sector.

2.7 Concluding remarks

Climatic change endangers vineyard sustainability in different ways. Water availability is a major limiting factor in wine production, as most of the current vineyards are located in semiarid areas. Together with agronomic approaches to face climatic change, the impressive genetic variability of the grapevine is a major source for adaptation. Among these genetic resources, the wild grapevine populations can offer interesting adaptations. There is also an impressive variability among the large list of existing cultivars and clones. The emergence of genomics and omics era offer new tools to identify adaptive characters. As the present review shows, the search for genotypes better adapted to climatic change conditions seems a promising field to face the climatic change challenges in a sustainable and cost effective way.

Acknowledgments

This work was performed with financial support from the *Ministerio de Ciencia, Innovación y Universidades* (Spain) (projects RTI2018-094470-R-C21, RTI2018-094470-R-C22, AGL2017-83738-C3-1-R). I. Tortosa was supported by a predoctoral fellowship BES-2015-073331. The authors would like to acknowledge Belén Escutia for the illustration of Fig. 2.2.

References

Aballay, E., & Vilches, O. (2016). Resistance assessment of grapevine rootstocks used in Chile to the root-knot nematodes *Meloidogyne ethiopica*, *M. hapla*, and *M. javanica*. *International Journal of Agriculture and Natural Resources, 42*(3), 407–413.

Albuquerque, C., Scoffoni, C., Brodersen, C. R., Buckley, T. N., Sack, L., & McElrone, A. J. (2020). Coordinated decline of leaf hydraulic and stomatal conductances under drought is not linked to leaf xylem embolism for different grapevine cultivars. *Journal of Experimental Botany, 71*(22), 7286–7300.

Alsina, M. M., de Herralde, F., Aranda, X., Savé, R., & Biel, C. (2007). Water relations and vulnerability to embolism are not related: Experiments with eight grapevine cultivars. *Vitis, 46*, 1–6.

Anderson, K., & Aryal, N. R. (2013). *Which winegrape varieties are grown where? A global empirical picture*. University of Adelaide Press.

Arnold, C., Gillet, F., & Gobat, J. M. (1998). Situation de la vigne sauvage Vitis vinifera ssp silvestris en Europe. *Vitis, 41*, 159–170.

Arrigo, N., & Arnold, C. (2007). Naturalised vitis rootstocks in Europe and consequences to native wild grapevine. *PLoS One, 2*(6), 521e.

Arrizabalaga-Arriazu, M., Morales, F., Irigoyen, J. J., Hilbert, G., & Pascual, I. (2020). Growth performance and carbon partitioning of grapevine Tempranillo clones under simulated climate change scenarios: Elevated CO_2 and temperature. *Journal of Plant Physiology, 252*, 153–226.

Arroyo-García, R., & Revilla, E. (2013). The current status of wild grapevine populations (*Vitis vinifera* ssp sylvestris) in the Mediterranean basin. Agricultural and biological sciences. In D. Poljuha, & B. Sladonja (Eds.), *The Mediterranean genetic-code-grapevine and olive* (pp. 51–72). Rijeka: InTech.

Arroyo-Garcia, R., Ruiz-Garcia, L., Bolling, L., Ocete, R., Lopez, M. A., Arnold, C., Ergul, A., Soylemezoglu, G., Uzun, H. I., Cabello, F., Ibanez, J., Aradhya, M. K., Atanassov, A., Atanassov, I., Balint, S., Cenis, J. L., Costantini, L., Goris-Lavets, S., Grando, M. S., … Martinez-Zapater, J. M. (2006). Multiple origins of cultivated grapevine (*Vitis vinifera* L. ssp. sativa) based on chloroplast DNA polymorphisms. *Molecular Ecology, 15*(12), 3707–3714.

Ashapkin, V. V., Kutueva, L. I., Aleksandrushkina, N. I., & Vanyushin, B. F. (2020). Epigenetic mechanisms of plant adaptation to biotic and abiotic stresses. *International Journal of Molecular Sciences, 21*(20), 7457.

Askri, H., Daldoul, S., Ben Ammar, A., Rejeb, S., Jardak, R., Rejeb, M. N., Mliki, A., & Ghorbel, A. (2012). Short-term response of wild grapevines (*Vitis vinifera* L. ssp. sylvestris) to NaCl salinity exposure: Changes of some physiological and molecular characteristics. *Acta Physiologiae Plantarum, 34*, 957–968.

Barrios-Masias, F. H., Knipfer, T., & McElrone, A. J. (2015). Differential responses of grapevine rootstocks to water stress are associated with adjustments in fine root hydraulic physiology and suberization. *Journal of Experimental Botany, 66*(19), 6069–6078.

Berdeja, M., Nicolas, P., Kappel, C., Dai, Z. W., Hilbert, G., Peccoux, A., Lafontaine, M., Ollat, N., Gomes, E., & Delrot, S. (2015). Water limitation and rootstock genotype interact to alter grape berry metabolism through transcriptome reprogramming. *Horticulture Research, 2*(1), 1–13.

Berendsen, R. L., Pieterse, C. M., & Bakker, P. A. (2012). The rhizosphere microbiome and plant health. *Trends in Plant Science, 17*(8), 478–486.

Bitsadze, N., Aznarashvili, M., Vercesi, A., Chipashvili, R., Failla, O., & Maghradze, D. (2015). Screening of Georgian grapevine germplasm for susceptibility to downy mildew (*Plasmopara viticola*). *Vitis, 54*, 193–196.

Bota, J., Flexas, J., & Medrano, H. (2001). Genetic variability of photosynthesis and water use in Balearic grapevine cultivars. *Annals of Applied Biology, 138*(3), 353–361.

Bota, J., Tomás, M., Flexas, J., Medrano, H., & Escalona, J. M. (2016). Differences among grapevine cultivars in their stomatal behavior and water use efficiency under progressive water stress. *Agricultural Water Management, 164*, 91–99.

Carbonneau, A. (1985). The early selection of grapevine rootstocks for resistance to drought conditions. *American Journal of Enology and Viticulture, 36*(3), 195–198.

Carvalho, L. C., Coito, J. L., Colaço, S., Sangiogo, M., & Amancio, S. (2015). Heat stress in grapevine: The pros and cons of acclimation. *Plant, Cell and Environment, 38*(4), 777–789.

Catacchio, C. R., Alagna, F., Perniola, R., Bergamini, C., Rotunno, S., Calabrese, F. M., Crupi, P., Antonacci, D., Ventura, M., & Cardone, M. F. (2019). Transcriptomic and genomic structural variation analyses on grape cultivars reveal new insights into the genotype-dependent responses to water stress. *Scientific Reports, 9*(1), 2809.

Chang, Y. N., Zhu, C., Jiang, J., Zhang, H., Zhu, J. K., & Duan, C. G. (2020). Epigenetic regulation in plant abiotic stress responses. *Journal of Integrative Plant Biology, 62*(5), 563–580.

Chaves, M. M., Costa, J. M., Zarrouk, O., Pinheiro, C., Lopes, C. M., & Pereira, J. S. (2016). Controlling stomatal aperture in semi-arid regionsâ€"the dilemma of saving water or being cool? *Plant Science, 251*, 54–64.

Chaves, M. M., Maroco, J. P., & Pereira, J. S. (2002). Understanding plant responses to drought â€" from genes to the whole plant. *Functional Plant Biology, 30*, 239–264.

Chaves, M. M., Zarrouk, O., Francisco, R., Costa, J. M., Santos, T., Regalado, A. P., Rodrigues, M. L., & Lopes, C. M. (2010). Grapevine under deficit irrigation: Hints from physiological and molecular data. *Annals of Botany, 105*(5), 661–676.

Cochetel, N., Escudié, F., Cookson, S. J., Dai, Z., Vivin, P., Bert, P. F., Muñoz, M. S., Delrot, S., Klopp, C., Ollat, N., & Lauvergeat, V. (2017). Root transcriptomic responses of grafted grapevines to heterogeneous nitrogen availability depend on rootstock genotype. *Journal of Experimental Botany, 68*(15), 4339–4355.

Cookson, S. J., Jane, S., & Ollat, N. (2013). Grafting with rootstocks induces extensive transcriptional reprogramming in the shoot apical meristem of grapevine. *BMC Plant Biology, 13*(1), 147.

Corso, M., Vannozzi, A., Maza, E., Vitulo, N., Meggio, F., Pitacco, A., Telatin, A., Angelo, M., Feltrin, E., Negri, A. S., Prinsi, B., Valle, G., Ramina, A., Bouzayen, M., Bonghi, C., & Lucchin, M. (2015). Comprehensive transcript profiling of two grapevine rootstock genotypes contrasting in drought susceptibility links the phenylpropanoid pathway to enhanced tolerance. *Journal of Experimental Botany, 66*(19), 5739–5752.

Corso, M., Vannozzi, A., Ziliotto, F., Zouine, M., Maza, E., Nicolato, T., Vitulo, N., Meggio, F., Valle, G., Bouzayen, M., Müller, M., Munné-Bosch, S., Lucchin, M., & Bonghi, C. (2016). Grapevine rootstocks differentially affect the rate of ripening and modulate auxin-related genes in Cabernet Sauvignon berries. *Frontiers of Plant Science, 7*, 69.

Costa, J. M., OrtuÃ±o, M. F., Lopes, C. M., & Chaves, M. M. (2012). Grapevine varieties exhibiting differences in stomatal response to water deficit. *Functional Plant Biology, 39*(3), 179–189.

Costa, J. M., Vaz, M., Escalona, J., Egipto, R., Lopes, C., Medrano, H., & Chaves, M. M. (2016). Modern viticulture in southern Europe: Vulnerabilities and strategies for adaptation to water scarcity. *Agricultural Water Management, 164*, 5–18.

Coupel-Ledru, A., Tyerman, S. D., Masclef, D., Lebon, E., Christophe, A., Edwards, E. J., & Simonneau, T. (2017). Abscisic acid down-regulates hydraulic conductance of grapevine leaves in isohydric genotypes only. *Plant Physiology, 175*(3), 1121–1134.

Covarrubias, J. I., Retamales, C., Donnini, S., Rombolà, A. D., & Pastenes, C. (2016). Contrasting physiological responses to iron deficiency in Cabernet Sauvignon grapevines grafted on two rootstocks. *Scientia Horticulturae, 199*, 1–8.

Cramer, G. R. (2010). Abiotic stress and plant responses from the whole vine to the genes. *Australian Journal of Grape and Wine Research, 16*, 86–93. https://doi.org/10.1111/j.1755-0238.2009.00058.x

Cramer, G. R., Ergül, A., Grimplet, J., Tillett, R. L., Tattersall, E. A., Bohlman, M. C., Vincent, D., Sonderegger, J., Evans, J., Osborne, C., Quilici, D., Schlaud, K. A., Schooley, D. A., & Cushman, J. C. (2007). Water and salinity stress in grapevines: Early and late changes in transcript and metabolite profiles. *Functional and Integrative Genomics, 7*(2), 111–134.

Cuneo, I. F., Barrios-Masias, F., Knipfer, T., Uretsky, J., Reyes, C., Lenain, P., Brodersen, C. R., Walker, M. A., & McElrone, A. J. (2020). Differences in grapevine rootstock sensitivity and recovery from drought are linked to fine root cortical lacunae and root tip function. *New Phytologist, 229*(1), 272–283.

Dal Santo, S., Palliotti, A., Zenoni, S., Tornielli, G. B., Fasoli, M., Paci, P., Tombesi, S., Frioni, T., Silvestroni, O., Bellicontro, A., Onofrio, C., Matarese, F., Gatti, M., Poni, S., & Pezzoti, M. (2016). Distinct transcriptome responses to water limitation in isohydric and anisohydric grapevine cultivars. *BMC Genomics, 17*, 815.

Dalla-Costa, L., Malnoy, M., Lecourieux, D., Deluc, L., Ouaked-Lecourieux, F., Thomas, M. R., & Torregrosa, L. (2019). The state-of-the art of grapevine biotechnology and new breeding technologies (NBTs). *OENO One, 53*, 189–212.

Dayer, S., Herrera, J. C., Dai, Z., Burlett, R., Lamarque, L. J., Delzon, S., Bortolami, G., Cochard, H., & Gambetta, G. A. (2020). The sequence and thresholds of leaf hydraulic traits underlying grapevine varietal differences in drought tolerance. *Journal of Experimental Botany, 71*(14), 4333–4344.

De Andres, M. T., Benito, A., PÃ©rez-Rivera, G., Ocete, R., Lopez, M. A., Gaforio, L., MuÃ±oz, G., Cabello, F., MartÃnez Zapater, J. M., & Arroyo-GarcÃa, R. (2012). Genetic diversity of wild grapevine populations in Spain and their genetic relationships with cultivated grapevines. *Molecular Ecology, 21*(4), 800–816.

De Herralde, F., SavÃ©, R., Aranda, X., & Biel, C. (2010). Grapevine roots and soil environment: Growth, distribution and function. In *Methodologies and results in grapevine research* (pp. 1–20). Dordrecht: Springer.

De Lorenzis, G., Chipashvili, R., Failla, O., & Maghradze, D. (2015). Study of genetic variability in *Vitis vinifera* L. Germplasm by high-throughput Vitis18kSNP array: The case of Georgian genetic resources. *BMC Plant Biology, 15*, 154.

De Micco, V., Zalloni, E., Battipaglia, G., Erbaggio, A., Scognamiglio, P., Caputo, R., & Cirillo, C. (2018). Rootstock effect on tree-ring traits in grapevine under a climate change scenario. *IAWA Journal, 39*(2), 145–155.

Delrot, S., Grimplet, J., Carbonell-Bejerano, P., Schwandner, A., Bert, P. F., Bavaresco, L., Costa, L. D., Gaspero, G., Duchêne, E., Hausmann, L., Malnoy, M., Morgante, M., Ollat, N., Pecile, M., & Vezzulli, S. (2020). Genetic and genomic approaches for adaptation of grapevine to climate change. In *Genomic designing of climate-smart fruit crops* (pp. 157–270). Cham: Springer.

Deluc, L. G., Quilici, D. R., Decendit, A., Grimplet, J., Wheatley, M. D., Schlauch, K. A., Mérillon, J. M., Cushman, J. C., & Cramer, G. R. (2009). Water deficit alters differentially metabolic pathways affecting important flavor and quality traits in grape berries of Cabernet Sauvignon and Chardonnay. *BMC Genomics, 10*(1), 212.

Di Vecchi, M., Lucou, V., Bruno, G., Lamcombe, T., Gerber, S., Bourse, T., Boselli, M., & This, P. (2009). Low level of pollen-mediated gene flow from cultivated to wild grapevine: Consequences for the evolution of the endangered subspecies *Vitis vinifera* L. subsp. silvestris. *Journal of Heredity, 100*(1), 66–75.

Duan, D., Halter, D., Baltenweck, R., Tisch, C., Tröster, V., Kortekamp, A., Hugueney, P., & Nick, P. (2015). Genetic diversity of stilbene metabolism in *Vitis sylvestris*. *Journal of Experimental Botany, 66*(11), 3243–3257.

During, H. (1984). Evidence for osmotic adjustment to drought in grapevines. *Vitis, 23*, 1–10.

Eibach, R., & Töpfer, R. (2015). Traditional grapevine breeding techniques. In *Grapevine breeding programs for the wine industry* (pp. 3–22). Woodhead Publishing.

Ergul, A., Perez-Rivera, G., Soybelezoglu, G., Kazan, K., & Arroyo-Garcia, R. (2011). Genetic diversity in Anatolian wild grapes (*Vitis vinifera* subsp. sylvestris) estimated by SSR markers. *Plant Genetic Resources, 9*(3), 375–383.

Flexas, J., Bota, J., Escalona, J. M., Sampol, B., & Medrano, H. (2002). Effects of drought on photosynthesis in grapevines under field conditions: An evaluation of stomatal and mesophyll limitations. *Functional Plant Biology, 29*, 461–471.

Florez-Sarasa, I., Clemente-Moreno, M. J., Cifre, J., Capó, M., Llompart, M., Fernie, A. R., & Bota, J. (2020). Differences in metabolic and physiological responses between local and widespread grapevine cultivars under water deficit stress. *Agronomy, 10*(7), 1052.

Fraga, H. (2019). Viticulture and winemaking under climate change. *Agronomy, 9*(12), 783.

Fraga, H., Malheiro, A. C., Moutinho-Pereira, J., & Santos, J. A. (2013). Future scenarios for viticultural zoning in Europe: Ensemble projections and uncertainties. *International Journal of Biometeorology, 57*(6), 909–925.

Gambetta, G. A., Herrera, J. C., Dayer, S., Feng, Q., Hochberg, U., & Castellarin, S. D. (2020). The physiology of drought stress in grapevine: Towards an integrative definition of drought tolerance. *Journal of Experimental Botany, 71*, 4658−4676.

Gambetta, G. A., Manuck, C. M., Drucker, S. T., Shaghasi, T., Fort, K., Matthews, M. A., & McElrone, A. J. (2012). The relationship between root hydraulics and scion vigour across vitis rootstocks: What role do root aquaporins play? *Journal of Experimental Botany, 63*(18), 6445−6455.

Gauthier, A., Cookson, S. J., Lagalle, L., Ollat, N., & Marguerit, E. (2020). Influence of the three main genetic backgrounds of grapevine rootstocks on petiolar nutrient concentrations of the scion, with a focus on phosphorus. *OENO One, 54*(1), 1−13.

Gomès, E., Maillot, P., & Duchêne, E. (2021). Molecular tools for adapting viticulture to climate change. *Frontiers of Plant Science, 12*, 633846.

Grassi, F., De Mattia, F., Zecca, G., Sala, F., & Labra, M. (2008). Historical isolation and quaternary range expansion of divergent lineages in wild grapevine. *Biological Journal of the Linnean Society, 95*, 611−619.

Grassi, F., Labra, M., Imazio, S., Ocete, R. R., Failla, O., Scienza, A., & Sala, F. (2006). Phylogeographical structure and conservation genetics of wild grapevine. *Conservation Genetics, 7*, 837−845.

Grassi, F., Labra, M., Imazio, S., Spada, A., Sgorbati, S., Scienza, A., & Sala, F. (2003). Evidence of a secondary grapevine domestication centre detected by SSR analysis. *Theoretical and Applied Genetics, 107*, 1315−1320.

Haider, M. S., Zhang, C., Kurjogi, M. M., Pervaiz, T., Zheng, T., Zhang, C., Lide, C., Shangguan, L., & Fang, J. (2017). Insights into grapevine defense response against drought as revealed by biochemical, physiological and RNA-seq analysis. *Scientific Reports, 7*(1), 13134.

Hannah, L., Roehrdanz, P. R., Ikegami, M., Shepard, A. V., Shaw, M. R., Tabor, G., Zhi, L., Marquet, P. A., & Hijmans, R. J. (2013). Climate change, wine, and conservation. *Proceedings of the National Academy of Sciences, 110*(17), 6907−6912.

Henry, R. J. (2020). Innovations in plant genetics adapting agriculture to climate change. *Current Opinion in Plant Biology, 56*, 168−173.

Heywood, V., & Zohary, D. (1991). A catalogue of wild relatives of cultivated plants native to Europe. *Flora Mediterranea, 5*, 375−415.

Hochberg, U., Degu, A., Cramer, G. R., Rachmilevitch, S., & Fait, A. (2015). Cultivar specific metabolic changes in grapevines berry skins in relation to deficit irrigation and hydraulic behavior. *Plant Physiology and Biochemistry, 88*, 42−52.

Hochberg, U., Degu, A., Gendler, T., Fait, A., & Rachmilevitch, S. (2015). The variability in the xylem architecture of grapevine petiole and its contribution to hydraulic differences. *Functional Plant Biology, 42*(4), 357−365.

Hochberg, U., Herrera, J. C., Degu, A., Castellarin, S. D., Peterlunger, E., Alberti, G., & Lazarovitch, N. (2017). Evaporative demand determines the relative transpirational sensitivity of deficit-irrigated grapevines. *Irrigation Science, 35*(1), 1−9.

Hochberg, U., Rockwell, F. E., Holbrook, N. M., & Cochard, H. (2018). Iso/anisohydry: A plant€"environment interaction rather than a simple hydraulic trait. *Trends in Plant Science, 23*(2), 112−120.

Ibañez, J., Carreño, J., Yuste, J., & Martínez-Zapater, J. M. (2015). Grapevine breeding and clonal selection programs in Spain. In *Grapevine breeding programs for the wine industry* (pp. 183−209). Woodhead Publishing.

IUCN. (1997). In J. Thorsell, & T. Sigaty (Eds.), *A global overview of forest protected areas on the world heritage list*. IUCN.

Jones, G. V. (2007). Climate change and the global wine industry. *Proceedings of the Thirteenth Australian Wine Industry Technical Conference*, 1−8.

Keller, M. (2015). *The science of grapevines: Anatomy and physiology* (2nd ed.). Academic Press.

Keller, M., Mills, L. J., & Harbertson, J. F. (2012). Rootstock effects on deficit-irrigated winegrapes in a dry climate: Vigor, yield formation, and fruit ripening. *American Journal of Enology and Viticulture, 63*(1), 29–39.

Kim, J. M., Sasaki, T., Ueda, M., Sako, K., & Seki, M. (2015). Chromatin changes in response to drought, salinity, heat, and cold stresses in plants. *Frontiers in Plant Science, 6*, 114.

Kracke, H., Cristoferi, G., & Marangoni, B. (1981). Hormonal changes during the rooting of hardwood cuttings of grapevine rootstocks. *American Journal of Enology and Viticulture, 32*(2), 135–137.

Lacombe, T., Laucou, V., Di Vecchi, M., Bordenave, L., Bourse, T., Siret, R., David, J., Boursiquot, J. M., Bronner, A., Merdinoglu, D., & This, P. (2003). Inventory and characterization of *Vitis vinifera* ssp. silvestris in France. *Acta Horticulturae, 603*, 553–557.

Laguna, A. (2003). Sobre las formas naturalizadas de Vitis en la Comunidad Valenciana I. Las especies. *Flora Montiberica, 23*, 46–82.

van Leeuwen, C., & Destrac-Irvine, A. (2017). Modified grape composition under climate change conditions requires adaptations in the vineyard. *OENO One, 51*(2), 147–154.

van Leeuwen, C., Destrac-Irvine, A., Dubernet, M., Duchêne, E., Gowdy, M., Marguerit, E., Pieri, P., Parker, A., Rességuier, L., & Ollat, N. (2019). An update on the impact of climate change in viticulture and potential adaptations. *Agronomy, 9*(9), 514.

Levadoux, L. (1956). *Les populations sauvages et cultivÃ©es des* Vitis vinifera *L* (Vol. 1). Institut national de la recherche agronomique.

Levin, A. D., Williams, L. E., & Matthews, M. A. (2020). A continuum of stomatal responses to water deficits among 17 wine grape cultivars (*Vitis vinifera*). *Functional Plant Biology, 47*(1), 11–25.

Lopes, M. S., MendonÃ§a, D., Rodrigues, M., Eiras-Dias, J. E., & Machado, A. (2009). New insights on the genetic basis of Portuguese grapevine and on grapevine domestication. *Genome, 52*(9), 790–800.

Lovisolo, C., Perrone, I., Carra, A., Ferrandino, A., Flexas, J., Medrano, H., & Schubert, A. (2010). Drought-induced changes in development and function of grapevine (*Vitis* spp.) organs and in their hydraulic and non-hydraulic interactions at the whole-plant level: A physiological and molecular update. *Functional Plant Biology, 37*, 98–116.

Lovisolo, C., Secchi, F., Nardini, A., Salleo, S., Buffa, R., & Schubert, A. (2007). Expression of PIP1 and PIP2 aquaporins is enhanced in olive dwarf genotypes and is related to root and leaf hydraulic conductance. *Physiologia Plantarum, 130*(4), 543–551.

Lucini, L., Miras-Moreno, B., Busconi, M., Marocco, A., Gatti, M., & Poni, S. (2020). Molecular basis of rootstock-related tolerance to water deficit in *Vitis vinifera* L. cv. Sangiovese: A physiological and metabolomic combined approach. *Plant Science, 299*, 110600.

Magris, G., Jurman, I., Formasiero, A., Paparelli, E., Schwope, R., Marroni, F., di Gaspero, G., & Morgante, M. (2021). The genomes of 204 *Vitis vinifera* accessions reveal the origin of European wine grapes. *Nature Communications, 12*, 7240.

Marguerit, E., Brendel, O., Lebon, E., Van Leeuwen, C., & Ollat, N. (2012). Rootstock control of scion transpiration and its acclimation to water deficit are controlled by different genes. *New Phytologist, 194*(2), 416–429.

Martorell, S., Diaz-Espejo, A., TomÃ s, M., Pou, A., El Aou-ouad, H., Escalona, J. M., & Medrano, H. (2015). Differences in water-use-efficiency between two *Vitis vinifera* cultivars (Grenache and Tempranillo) explained by the combined response of stomata to hydraulic and chemical signals during water stress. *Agricultural Water Management, 156*, 1–9.

Martorell, S., Medrano, H., TomÃ s, M., Escalona, J. M., Flexas, J., & Diaz-Espejo, A. (2015). Plasticity of vulnerability to leaf hydraulic dysfunction during acclimation to drought in grapevines: An osmotic-mediated process. *Physiologia Plantarum, 153*(3), 381–391.

Marusig, D., & Tombesi, S. (2020). Abscisic acid mediates drought and salt stress responses in *Vitis vinifera*-a review. *International Journal of Molecular Sciences, 21*(22), 8648.

Mattia, F., Imazio, S., Grassi, F., Doulati, H., Scienza, A., & Labra, M. (2008). Study of genetic relationships between wild and domesticated grapevine distributed from middle east regions to European countries. *Rendiconti Lincei. Scienze Fisiche e Naturali, 19*, 223–240.

McGovern, P. E. (2003). *Ancient wine. The search for the origins of viniculture.* Princeton, NJ: Princeton University Press.

McGovern, P. E., Glusker, D. L., Exener, L. J., & Voigt, N. M. (1996). Neolithic resinated wine. *Nature, 381*(6528), 480–481.

Morales-Castilla, I., de Cortázar-Atauri, I. G., Cook, B. I., Lacombe, T., Parker, A., van Leeuwen, C., Nicholas, K. A., & Wolkovich, E. M. (2020). Diversity buffers winegrowing regions from climate change losses. *Proceedings of the National Academy of Sciences, 117*(6), 2864–2869.

Morton, L. T. (1979). The myth of the universal rootstock [Grapes, varieties]. *Wines and Vines, 60*, 24–26.

Muñoz, C., Gomez-Talquenca, S., Chialva, C., Ibáñez, J., Martínez-Zapater, J. M., Peña-Neira, A., & Lijavetzky, D. (2014). Relationships among gene expression and anthocyanin composition of Malbec grapevine clones. *Journal of Agricultural and Food Chemistry, 62*(28), 6716–6725.

Myles, S., Boyko, A. R., Owens, C. L., Brown, P. J., Grassi, F., Aradhya, M. K., Prins, B., Reynolds, A., Chia, J. M., Ware, D., Bustamante, C. D., & Buckler, E. S. (2011). Genetic structure and domestication history of the grape. *Proceedings of the National Academy of Sciences, 108*(9), 3530–3535.

Nguyen, V. C., Tandonnet, J. P., Khallouk, S., Van Ghelder, C., Portier, U., Lafargue, M., Banora, M. Y., Ollat, N., & Esmenjaud, D. (2020). Grapevine resistance to the nematode *Xiphinema index* is durable in muscadine-derived material obtained from hardwood cuttings but not from in vitro. *Phytopathology, 110*(9), 1565–1571.

OIV. (2009). *Descriptor list for grape varieties and Vitis species.* Paris: OIV Publications.

Ollat, N., Peccoux, A., Papura, D., Esmenjaud, D., Marguerit, E., Tandonnet, J. P., Bordenave, L., Cookson, F. B., Rossdeutch, J. L., Lecourt, J. L., Lauvergeat, P., Vivin, P., Bert, P. F., & Delrot, S. (2016). Rootstocks as a component of adaptation to environment. Grapevine in. *A Changing Environment: A molecular and Ecophysiological Perspective, 1*, 68–108.

Olmo, H. P. (1995). The origin and domestication of the *vinifera* grape. In *The origins and ancient history of wine* (pp. 31–43). Amsterdam: Gordon and Breach.

Osakabe, Y., Liang, Z., Ren, C., Nishitani, C., Osakabe, K., Wada, M., Komori, S., Malnoy, M., Velasco, R., & Poli, M. (2018). CRISPR–Cas9-mediated genome editing in apple and grapevine. *Nature Protocols, 13*(12), 2844–2928.

Patakas, A., Nikolaou, N., Zioziou, E., Radoglou, K., & Noitsakis, B. (2002). The role of organic solute and ion accumulation in osmotic adjustment in drought-stressed grapevines. *Plant Science, 163*, 361–367.

Patakas, A., & Noitsakis, B. (1999). Osmotic adjustment and partitioning of turgor responses to drought in grapevines leaves. *American Journal of Enology and Viticulture, 50*, 76–80.

Peiró, R., Jiménez, C., Perpiñà, G., Soler, J. X., & Gisbert, C. (2020). Evaluation of the genetic diversity and root architecture under osmotic stress of common grapevine rootstocks and clones. *Scientia Horticulturae, 266*, 109283.

Prieto, J. A., Lebon, Ã‰., & Ojeda, H. (2010). Stomatal behavior of different grapevine cultivars in response to soil water status and air water vapor pressure deficit. *Journal International de Sciences de la Vigne et du Vin, 44*(1), 9–20.

Prinsi, B., Negri, A. S., Failla, O., Scienza, A., & Espen, L. (2018). Root proteomic and metabolic analyses reveal specific responses to drought stress in differently tolerant grapevine rootstocks. *BMC Plant Biology, 18*(1), 126.

Ramu, P., Esuma, W., Kawuki, R., Rabbi, I. Y., Egesi, C., Bredeson, J. V., Bart, R. S., Verma, S., Buckler, E. S., & Lu, F. (2017). Cassava haplotype map highlights fixation of deleterious mutations during clonal propagation. *Nature Genetics, 49*(6), 959–963.

Revilla, E., Carrasco, D., Benito, A., & Arroyo-García, R. (2010). Anthocyanin composition of several wild grape accessions. *American Journal of Enology and Viticulture, 61*(4), 536–543.

Riaz, S., Boursiquot, J. M., Dangl, G. S., Lacombe, T., Laucou, V., Tenscher, A. C., & Walker, M. A. (2013). Identification of mildew resistance in wild and cultivated Central Asian grape germplasm. *BMC Plant Biology, 13*(1), 149.

Riaz, S., De Lorenzis, G., Velasco, D., Koehmstedt, A., Maghradze, D., Bobokashvili, Z., Musayev, M., Zdunic, G., Laucou, V., Walker, M. A., Failla, O., Preece, J. E., Aradhya, M., & Arroyo-Garcia, R. (2018). Genetic diversity analysis of cultivated and wild grapevine (*Vitis vinifera* L.) accessions around the Mediterranean basin and Central Asia. *BMC Plant Biology, 18*(1), 137.

Rocheta, M., Becker, J. D., Coito, J. L., Carvalho, L., & Amâncio, S. (2014). Heat and water stress induce unique transcriptional signatures of heat-shock proteins and transcription factors in grapevine. *Functional & Integrative Genomics, 14*(1), 135–148.

Rocheta, M., Coito, J. L., Ramos, M. J., Carvalho, L., Becker, J. D., Carbonell-Bejerano, P., & Amâncio, S. (2016). Transcriptomic comparison between two *Vitis vinifera* L. varieties (Trincadeira and Touriga Nacional) in abiotic stress conditions. *BMC Plant Biology, 16*(1), 224.

Romero, P., Botía, P., & Navarro, J. M. (2018). Selecting rootstocks to improve vine performance and vineyard sustainability in deficit irrigated Monastrell grapevines under semiarid conditions. *Agricultural Water Management, 209*, 73–93.

Rossdeutsch, L., Edwards, E., Cookson, S. J., Barrieu, F., Gambetta, G. A., Delrot, S., & Ollat, N. (2016). ABA-mediated responses to water deficit separate grapevine genotypes by their genetic background. *BMC Plant Biology, 16*, 91.

Rubio, B., Lalanne-Tisné, G., Voisin, R., Tandonnet, J. P., Portier, U., Van Ghelder, C., Lafargue, M., Petit, J. P., Donnart, M., Joubard, B., Bert, P. F., Papura, D., Cunff, L. L., Ollat, N., & Esmenjaud, D. (2020). Characterization of genetic determinants of the resistance to phylloxera, *Daktulosphaira vitifoliae*, and the dagger nematode *Xiphinema index* from muscadine background. *BMC Plant Biology, 20*, 1–15.

Rühl, E., Konrad, H., Lindner, B., & Bleser, E. (2003). Quality criteria and targets for clonal selection in grapevine. In *I international symposium on grapevine growing, commerce and research 652* (pp. 29–33).

Sahu, P. P., Pandey, G., Sharma, N., Puranik, S., Muthamilarasan, M., & Prasad, M. (2013). Epigenetic mechanisms of plant stress responses and adaptation. *Plant Cell Reports, 32*(8), 1151–1159.

Santarosa, E., de Souza, P. V. D., de Araujo Mariath, J. E., & Lourosa, G. V. (2016). Physiological interaction between rootstock-scion: Effects on xylem vessels in Cabernet Sauvignon and merlot grapevines. *American Journal of Enology and Viticulture, 67*(1), 65–76.

Santos, J. A., Fraga, H., Malheiro, A. C., Moutinho-Pereira, J., Dinis, L. T., Correia, C., Moriondo, L., Leolini, C., Dibari, S., Costafreda-Aumedes, T., Kartschall, C., Menz, D., Molitor, J., Junk, M., Beyer, M., & Schultz, H. R. (2020). A review of the potential climate change impacts and adaptation options for European viticulture. *Applied Sciences, 10*(9), 3092.

Savoi, S., Wong, D. C., Degu, A., Herrera, J. C., Bucchetti, B., Peterlunger, E., Fait, A., Mativi, F., & Castellarin, S. D. (2017). Multi-omics and integrated network analyses reveal new insights into the systems relationships between metabolites, structural genes, and transcriptional regulators in developing grape berries (*Vitis vinifera* L.) exposed to water deficit. *Frontiers of Plant Science, 8*, 1124.

Schultz, H. R. (2003). Differences in hydraulic architecture account for near-isohydric and anisohydric behaviour of two field-grown *Vitis vinifera* L. cultivars during drought. *Plant, Cell and Environment, 26*(8), 1393–1405.

Sefc, K. M., Steinkellner, H., Lefort, F., Botta, R., Machado, A. D., Borrego, J., Maletic, E., & Glossl, J. (2003). Evaluation of the genetic contribution of local wild vines to European grapevine cultivars. *American Journal of Enology and Viticulture, 54*, 15−21.

Smart, D. R., Schwass, E., Lakso, A., & Morano, L. (2006). Grapevine rooting patterns: A comprehensive analysis and a review. *American Journal of Enology and Viticulture, 57*(1), 89−104.

Smith, B. P., Morales, N. B., Thomas, M. R., Smith, H. M., & Clingeleffer, P. R. (2017). Grapevine rootstocks resistant to the root-knot nematode *Meloidogyne javanica*. *Australian Journal of Grape and Wine Research, 23*(1), 125−131.

Snoussi, H., Slimane, M. H., Ruiz-Garcia, L., Martinez-Zapater, J. M., & Arroyo-Garcia, R. (2004). Genetic relationship among cultivated and wild grapevine accessions from Tunisia. *Genome, 47*(6), 1211−1219.

Soar, C. J., Dry, P. R., & Loveys, B. R. (2006). Scion photosynthesis and leaf gas exchange in *Vitis vinifera* L. Cv. Shiraz: Mediation of rootstock effects via xylem sap ABA. *Australian Journal of Grape and Wine Research, 12*(2), 82−96.

Stoll, M., Loveys, B., & Dry, P. (2000). Hormonal changes induced by partial rootzone drying of irrigated grapevine. *Journal of Experimental Botany, 51*(350), 1627−1634.

Tandonnet, J. P., Saubignac, L., Gautier, A., Marguerit, E., Cookson, S. J., Laurent, V., Molllier, A., Pagès, L., Ollat, N., & Vivin, P. (2018). Screening and modelling the diversity of root system architecture in vitis genotypes: New opportunity for rootstock selection?. In *International conference on grapevine breeding and genetics. (Geneva, New York, USA)*.

Tello, J., Montemayor, M. I., Forneck, A., & Ibáñez, J. (2018). A new image-based tool for the high throughput phenotyping of pollen viability: Evaluation of inter-and intra-cultivar diversity in grapevine. *Plant Methods, 14*(1), 3.

This, P., Lacombe, T., & Thomas, M. R. (2006). Historical origins and genetic diversity of wine grapes. *Trends in Genetics, 22*, 511−519.

Toffolatti, S. L., Maddalena, G., Salomoni, D., Maghradze, D., Bianco, P. A., & Failla, O. (2016). Evidence of resistance to the downy mildew agent *Plasmopara viticola* in the Georgian *Vitis vinifera* germplasm. *Vitis, 55*, 121−128.

Tomás, M., Medrano, H., Brugnoli, E., Escalona, J. M., Martorell, S., Pou, A., Ribas-Carbó, M., & Flexas, J. (2014). Variability of mesophyll conductance in grapevine cultivars under water stress conditions in relation to leaf anatomy and water use efficiency. *Australian Journal of Grape and Wine Research, 20*(2), 272−280.

Tomás, M., Medrano, H., Escalona, J. M., Martorell, S., Pou, A., Ribas-CarbÃ[3], M., & Flexas, J. (2014). Variability of water use efficiency in grapevines. *Environmental and Experimental Botany, 103*, 148−157.

Tomás, M., Medrano, H., Pou, A., Escalona, J. M., Martorell, S., Ribas-CarbÃ[3], M., & Flexas, J. (2012). Water-use efficiency in grapevine cultivars grown under controlled conditions: Effects of water stress at the leaf and whole-plant level. *Australian Journal of Grape and Wine Research, 18*(2), 164−172.

Tortosa, I., Douthe, C., Pou, A., Balda, P., Hernandez-Montes, E., Toro, G., Escalona, J. M., & Medrano, H. (2019). Variability in water use efficiency of grapevine tempranillo clones and stability over years at field conditions. *Agronomy, 9*(11), 701.

Tortosa, I., Escalona, J. M., Bota, J., Tomas, M., Hernandez, E., Escudero, E. G., & Medrano, H. (2016). Exploring the genetic variability in water use efficiency: Evaluation of inter and intra cultivar genetic diversity in grapevines. *Plant Science, 251*, 35−43.

Tortosa, I., Escalona, J. M., Douthe, C., Pou, A., Garcia-Escudero, E., Toro, G., & Medrano, H. (2019b). The intracultivar variability on water use efficiency at different water status as a target selection in grapevine: Influence of ambient and genotype. *Agricultural Water Management, 223*, 105648.

Tortosa, I., Escalona, J. M., Toro, G., Douthe, C., & Medrano, H. (2020). Clonal behavior in response to soil water availability in Tempranillo grapevine cv: From plant growth to water use efficiency. *Agronomy, 10*(6), 862.

Villalobos-González, L., Muñoz-Araya, M., Franck, N., & Pastenes, C. (2019). Controversies in the midday water potential regulation and stomatal behauvior might result by the environment, genotype and/or roostock: Evidence from Carménère and Syrah grapevine varieties. *Frontiers in Plant Science, 10*, 1522.

Villalobos-González, L., Peña-Neira, A., Ibáñez, F., & Pastenes, C. (2016). Long-term effects of abscisic acid (ABA) on the grape berry phenylpropanoid pathway: Gene expression and metabolite content. *Plant Physiology and Biochemistry, 105*, 213–223.

Vondras, A. M., Minio, A., Blanco-Ulate, B., Figueroa-Balderas, R., Penn, M. A., Zhou, Y., Seymour, D., Ye, Z., Liang, D., Espinoza, L. K., Anderson, M. M., Walker, M. A., Gaut, B., & Cantu, D. (2019). The genomic diversification of grapevine clones. *BMC Genomics, 20*(1), 972.

Yildirim, K., Yağcı, A., Sucu, S., & Tunç, S. (2018). Responses of grapevine rootstocks to drought through altered root system architecture and root transcriptomic regulations. *Plant Physiology and Biochemistry, 127*, 256–268.

Zasada, I. A., Howland, A. D., Peetz, A. B., East, K., & Moyer, M. (2019). Vitis spp. rootstocks are poor hosts for *Meloidogyne hapla*, a nematode commonly found in Washington winegrape vineyards. *American Journal of Enology and Viticulture, 70*(1), 125–131.

Zhang, L., Marguerit, E., Rossdeutsch, L., Ollat, N., & Gambetta, G. A. (2016). The influence of grapevine rootstocks on scion growth and drought resistance. *Theoretical and Experimental Plant Physiology, 28*(2), 143–157.

CHAPTER 3

Optimizing conservation and evaluation of intravarietal grapevine diversity

Elsa Gonçalves and Antero Martins
LEAF- Linking Landscape, Environment, Agriculture and Food, Instituto Superior de Agronomia, Universidade de Lisboa, Lisboa, Portugal

3.1 Introduction

The grapevine is one of the most economically important horticultural crops in the world. The antiquity and diversity of traditional grapevine varieties guarantee a strong historical and natural character and add high economic value to viticulture and wine. Concern with these varieties has usually been more focused on intervarietal diversity, which already benefits from well-established programmes for preservation and characterization in ampelographic collections. A reference example is the largest collection of grapevine varieties maintained in Vassal (France), but there are also national ampelographic collections maintained all over the world. It is now known that the most popular varieties are all close relatives connected to one another by first- or second-degree relationships (Myles et al., 2011). The genetic diversity among varieties in quantitative traits in germoplasm collections has also been explored, and these collections are a highly valuable resource for genetic association studies in grapevines (Nicolas et al., 2016). Additionally, genetic, epigenetic, and genomic effects on variations in gene expression among grapevine varieties (Magris et al., 2019) have been studied.

However, a high level of other type of diversity, less known but not less important, exists within each variety, known as the intravarietal diversity of quantitative traits.

The grapevine is a native plant from the Eurasian Mediterranean (territory from the Iberian Peninsula to Afghanistan). That territory must have been the target of the first domestications. At a given time after domestication, the plants started to be vegetatively multiplied, with each plant serving as an origin to a population that was initially homogenous but that later created and accumulated natural intravarietal diversity of quantitative traits until it became a highly heterogeneous population: the population that today we call variety, "cépage" (France), "vitigno" (Italy), or "casta" (Portugal). These differences have arisen from the accumulation of somatic mutations during centuries of asexual propagation.

As stated above, an ancient variety is a genetically heterogeneous population in relation to the most important agronomic and oenological traits. Notable levels of variation within the variety are observed, which can reach, more than tenfold for yield or twofold for soluble solids (Martins & Gonçalves, 2015). This variation is highly advantageous because it is this diversity that has ensured stable behavior of the variety over time and through different environments (low genotype×environment interaction) and that constitutes today the raw material for carrying out the selection within the variety of genotypes with better performance for several useful traits, that is, with high genetic and economic gains.

However, all this potential of the ancient heterogeneous varieties has been seriously undermined since the middle of the last century due to three main factors: (1) the generalization of the selection with a narrow genetic basis (focused on the homogeneous clone), (2) the abandonment of the technique of grafting in the field (with buds of diverse origins) in favor of planting bench-grafted plants (made with buds from a few homogeneous plots dedicated to this purpose), and (3) plant certification laws that tend to favor the use of homogeneous materials, harming diversity.

Stopping this process of genetic erosion is strategic and urgent since diversity is a natural feature of Europe and is still concentrated in its older vineyards. In other words, preserving intravarietal diversity means consolidating a competitive advantage provided by the evolutionary history of grapevine varieties and strengthening the sustainability of the vine and wine sector. The greater the intravarietal diversity, the greater the gains obtained from selection and the greater the capacity of the variety to respond to present and future challenges to viticulture.

The molecular understanding of intravarietal diversity has been the subject of many studies (Pelsy et al., 2010), including in recent years. For example, Meneghetti et al. (2012) described a method to obtain DNA polymorphisms of *Vitis vinifera* genotypes from the same variety using amplified fragment length polymorphism (AFLP), selective amplification of microsatellite polymorphic loci (SAMPL), microsatellites AFLP (M-AFLP), and ISSR molecular markers. This work identified differences within varieties, such as Primitivo, Garnacha Tinta, Callet, Manto Negro, and Moll, and correlated the genetic differences to their geographical origins (i.e., Garnacha Tinta; Malvasia Nera di Brindisi/Lecce) or morphological traits (i.e., Malvasia of Candia). Roach et al. (2018) produced a high-quality genome assembly for the grapevine variety Chardonnay; additionally, using resequencing data for 15 popular clones, they were able to identify a large selection of markers that are unique to at least one clone. These authors also identified mutations that may confer phenotypic effects and were able to identify clones from material independently sourced from nurseries and vineyards. In a study conducted by Vondras et al. (2019), 15 Zinfandel clone genomes were sequenced and compared to one another using a highly contiguous genome reference produced from one of the clones. This work concluded that repetitive intergenic space is a major driver of clone genome diversification, clones accumulate putatively deleterious mutations, and selection occurs against deleterious variants in coding regions, or there is some mechanism by which mutations are less frequent in coding than noncoding regions of the genome.

However, the purpose of exploiting intravarietal variability for selection and fulfilling the immediate needs of the grape and wine sector has been insufficiently considered in approaches to grapevine diversity worldwide. Notwithstanding, concern with intravarietal variability has been largely handled in Portugal during the past 43 years, and methodological approaches have been developed for efficient conservation, evaluation, and selection with the prediction of genetic gains (Martins & Gonçalves, 2015).

The main objective of this chapter is precisely to outline the conservation and evaluation of intravarietal genetic variability as a crucial strategy for preserving traditional viticulture and facing future challenges (climate change, biotic and abiotic stresses, consumer demands, etc.). These circumstances are of particular benefit to European vine-growing countries, whose grapevine varieties have accumulated more diversity since ancient times.

The chapter is organized in three sections. Section 3.1 addresses the grapevine methodology for intravarietal diversity conservation, evaluation, and selection within a variety. Section 3.2 conducts a methodological study to provide new insights concerning the accuracy and precision of genetic intravarietal grapevine diversity evaluation and the efficiency of selection. Section 3.3 describes the practical application of conservation of intravarietal variability and selection in Portugal.

3.2 Grapevine methodology for conservation, evaluation, and selection within a variety

3.2.1 Representative sampling of intravarietal diversity

To efficiently conserve intravarietal genetic variability, representative samples of the diversity within all autochthonous varieties must be considered. The sample size depends on the total growing area and the number of discontinuous ancient growing regions of the variety. Simulation studies found a minimum sample size per region with approximately 70 genotypes. As varieties are usually grown in several regions with different evolutionary histories and diversity patterns, samples with 400 or more genotypes are frequently used. Therefore, the first stage of a well-conducted conservation of intravarietal diversity and selection within an ancient variety is the prospection of plants in the old vineyards to obtain a representative sample (randomly marked) of the variety in its main growing regions. The process to obtain that representative sample is described in detail in Martins and Gonçalves (2015). The marked plants are multiplied and planted or grafted to install a germplasm collection.

3.2.2 Conservation of intravarietal diversity

Depending on the type of conservation, different objectives will be achieved. To maximize the information that can be extracted from a germplasm collection, in its installation, the use of efficient experimental designs is crucial. In fact, for the evaluation of traits controlled by multiple genes, all acting collectively and affected by the environment (i.e., quantitative traits, such as yield, must quality traits, and tolerance to abiotic stresses), data sets must have been collected from efficient and well-designed experiments, which rely on the well-known basic principles of randomization, replication, and blocking. Replication enables estimation of experimental error variance, and the precision of estimates increases with the number of replications. Randomization ensures that all experimental units are equally likely to receive any genotype, minimizing systematic errors or bias induced by the experimenter. Finally, blocking controls the different sources of natural variation among experimental units and, when applied appropriately, controls field variation and helps to reduce background noise. These principles are strictly followed in agronomic experiments (Giesbrecht & Gumpertz, 2004; Mead et al., 2012; Welham et al., 2015) and are the well-known basis for the study of quantitative genetics and for plant breeding (Brown et al., 2014).

Conservation can be performed through two processes: (1) conservation without other immediate objectives; and (2) conservation that simultaneously ensures the evaluation of important quantitative traits (such as yield, must quality traits, and tolerance to abiotic stresses), including quantification of intravarietal genetic variability for those traits, and selection of a superior group of genotypes. The first method of conservation is performed, for example, in pots or/and in the field with a single plot per genotype, while the second method includes the plantation of field trials (which involves experimental design). Regarding the latter, the ultimate goal is to ensure that these collections of genotypes are not only repositories of genetic variability but can also be simultaneously evaluated and utilized for selection immediately.

When conservation is performed in pots or in the field with a single experimental unit (plot) per genotype (each genotype is usually represented by three to six plants), the only evaluation that can be obtained is the mean phenotypic value of the experimental unit for a target trait. Therefore, without replication, there is no way to assess the error and genotypic variance estimates or the predicted genotypic values. Hence, there is no basis for any useful quantification of intravarietal genetic variability and the selection of quantitative traits. This method of conservation is completely inefficient for later purposes because the genotypic value is overwhelmed by environmental deviations. Therefore, under such conditions, the plants are not selected for their genetic value, as intended. In grapevines, these inefficiencies have already been indicated through several types of results (Martins & Gonçalves, 2015). Namely, with simulation studies, estimates of broad-sense heritability of yield between 0 and 0.25 were obtained for phenotypic values from one plant. Therefore, as previously mentioned, the usefulness of this type of collection is only in conservation.

Field conservation with experimental design comprises partially replicated designs (augmented designs and alpha-alpha designs) or fully replicated designs (typically randomized complete block designs or designs of the family of incomplete block designs, such as alpha and row-column designs).

Numerous variants of partially replicated designs (also known as unreplicated trials) are frequently used in plant breeding, mainly for preliminary assessment of the available germplasm (Kempton & Gleeson, 1997). These field trials use two types of treatments—check and test—which correspond to replicated and nonreplicated treatments, respectively. Therefore, there is sufficient replication, based on check treatments, to allow a valid estimation of the error variance. The main objective is to select a subgroup of genotypes (normally approximately 1/3) for later more detailed studies. In grapevines, this type of experimental design was studied for the conservation of rarely grown varieties (Gonçalves, St.Aubyn, & Martins, 2013). The objective was to develop a method of preserving and quantifying intravarietal genetic diversity that provides some guidelines for future experimental procedures. Through simulation, several experimental situations were generated, varying the numbers of test and check treatments and the type of experimental design associated with the check treatments (augmented randomized complete block design and alpha-alpha design). A greater precision in the quantification of the intravarietal genetic diversity was achieved with collections with an alpha-alpha design of over 250 genotypes and a minimum of 33% of plots containing check genotypes. These authors also concluded that partially replicated designs are useful in the context of ancient grapevine varieties to evaluate intravarietal genetic variability but not to make successful genetic selection. The main advantage of this type of experimental design is the smaller experimental area required.

Regarding fully replicated field trials, their efficiency depends on the randomization process used to control environmental variation (particularly spatial variation) and the number of replicates. Traditionally, they were planted according to a randomized complete block design. However, when the number of genotypes used becomes large, field trials are laid out on a rectangular grid of plots. Under these conditions, experimental designs with incomplete but smaller blocks are preferred, such as alpha designs (Patterson & Williams, 1976), which constitute a particular class of generalized lattice designs. Alternatively, to control for environmental gradients in both row and column directions by blocking, row-column designs with blocks in both rows and columns are used (Wiliams & John, 1989), which correspond to groups of more complex Latin square designs. If incomplete blocks can be grouped into complete blocks, the design is called resolvable. Efficient row-column designs are readily available (Piepho et al., 2018; Whitaker et al., 2007; Williams et al., 2006).

Generating efficient experimental designs is a current research subject for plant breeding trials (Mramba et al., 2018; Piepho, 2015). In grapevines, the use of fully replicated designs means that the field trial would cover a large area (from 0.75 to 1.5 ha), which, by itself, causes large environmental variation. Therefore, the importance of the experimental design to control spatial variability is crucial to reach the objectives of quantification of genetic variability and selection. With the aim of identifying the most suitable experimental designs for these objectives, a study via simulation was carried out to assess the comparative efficiency of various experimental designs (Gonçalves et al., 2010): randomized complete block design, alpha design, and row-column design. A higher efficiency of resolvable row-column designs was found when compared with randomized complete blocks, or even with alpha designs, enabling more precise estimates of genotypic variance, greater precision in the prediction of genetic gain and, consequently, greater efficiency in polyclonal selection. Additionally, regarding the genetic selection goal, the number of replicates of the trial (no less than four replicates for yield data) proved to be a very important issue in the accuracy and precision of the performed selection, particularly in the composition of the group of genotypes selected (Gonçalves & Martins, 2019a). Most quality traits are laborious and costly to measure, so obtaining plot data for all replicates may be prohibitive. Therefore, it has become common practice to take berry samples from three-four replicates of the field trial.

3.2.3 Evaluation of intravarietal diversity and polyclonal selection

The field trial established with a representative sample of the intravarietal diversity of a variety according to a fully replicated design constitutes the second stage for selection and aims to quantify the intravarietal diversity and to perform polyclonal selection.

Data collection for yield, soluble solids, acidity, anthocyanins, and other traits (biotic and abiotic tolerances) generally takes place from the third year after plantation. This is followed by data analysis based on mixed models, considering data from several years.

For quantitative genetics analyses focused on selection, linear mixed models are fitted to large data sets using REML estimation (Lynch & Walsh, 1998; Searle et al., 1992). Only a combined approach comprising field design and mixed models for data analysis permits the separation of genotypic and nongenotypic variability and, consequently, the prediction of genetic gains of selection. This is the common procedure in plant breeding in general (Isik et al., 2017), and for selection within ancient grapevine varieties, several variants of linear mixed models have been implemented to respond to specific issues related to grapevines (Carvalho et al., 2020; Gonçalves et al., 2007, 2016, 2020; Gonçalves, Carrasquinho, StAubyn, & Martins, 2013; Martins & Gonçalves, 2015). In these studies, mixed models were fitted to data from yield, quality traits of the must, and surface leaf temperature.

Intravarietal variability for those traits was observed, and higher variability for yield than for quality traits was found. Similarly, predicted genetic gains of selection were observed for all the abovementioned traits. Specifically, Martins and Gonçalves (2015) presented the quantification of intravarietal genetic variability of the yield, expressed by the coefficient of genotypic variation, for 42 Portuguese grapevine varieties and the results of polyclonal selection in 38 Portuguese grapevine varieties, in which the predicted genetic gains of the yield varied from 5.9% to 46.0%. In another study conducted for 11 grapevine varieties, Gonçalves and Martins (2019b) reported predicted genetic gains of polyclonal selection for yield ranging from 16.7% to 58.4%; for quality traits, the genetic gains were lower, ranging between 2.8% and 13.2% for soluble solids, 2.0% and 28.3% for acidity, and 2.0% and 16.4% for anthocyanins.

Another important result is provided by a field trial established with representative samples of the intravarietal diversity of a variety in its main growing regions and the subsequent fitting of mixed models allowing unequal genotypic variances among the regions. As an example, Martins and Gonçalves (2015) examined yield data from a large field trial of Garnacha (or Grenache or Cannonau) with representative samples of genotypes from four regions in three different countries (Toledo and Zaragoza in Spain, Vaucluse in France, and Sardinia in Italy). According to the results of the genotypic coefficients of variation, the greatest genetic variability was found in Sardinia (Italy). From Sardinia, the variety was probably subsequently exported to the other regions of Spain and France, likely through selected material. These results regarding intravarietal diversity are of great interest because they highlight the cultural context of grapevine varieties and provide new rational guidance for selection and conservation.

In short, at this stage, the objectives of data analysis are to (1) separate the total variance in the nongenotypic and genotypic components; (2) calculate broad-sense heritability; and (3) obtain the empirical best linear unbiased predictors (EBLUPs) of the genotypic effects, which are used to select outstanding genotypes. The process ends with the ranking of the genotypes by their EBLUPs of the genotypic effects for each trait and the selection from 7 to 20 superior genotypes, with the respective prediction of the genetic gains for the target traits.

The focus on intravarietal diversity conservation, selection based on the predicted genotypic values of the genotypes rather than on phenotypic values, and the achievement and prediction of high genetic gains are strong hallmarks of this procedure. This methodology of polyclonal selection was expressly recognized in 2019 by the International Organization of Vine and Wine through the "Resolution OIV-VITI 564B-2019: OIV Process for the recovery and conservation of the intravarietal diversity and the polyclonal selection in grape varieties with wide genetic variability."

In short, the greatest innovation in this methodology is to afford polyclonal selected material that shows lower sensitivity to genotype×environment (G×E) interaction, ensures the maintenance of a certain level of diversity in vineyards (mitigates erosion), and provides high and accurately predicted genetic gains of selection. In addition, the selection criteria are adaptable to new contexts. The fact that the grapevine is perennial and field trials with a representative sample of intravarietal diversity are maintained for many years allows us to make various different polyclonal selections over time. In this way, the selection can be performed and modified according to the current objectives of viticulture: yield and quality traits (Gonçalves & Martins, 2019b), tolerance to abiotic stress (Carvalho et al., 2020) and, no less importantly, tolerance to biotic stress. These are crucial differences compared to traditional clonal selection methodologies still used worldwide.

3.2.4 Establishment of multienvironmental trials for clonal selection

Traditionally, the widespread methodology to select within a variety is clonal selection. In this context, one must be well aware that the phenotypic value of an individual for a given trait is controlled by its genotypic effect, the effect of the environment, and the genotype-by-environment (G×E) interaction effect. Therefore, the development of an efficient clonal selection process requires the study of G×E interaction, and another stage of selection must be executed. This stage consists of several field trials planted in the variety's ancient growing regions, with 20–40 clones selected from the previous stage (second stage for polyclonal selection).

The G×E interaction is a complex phenomenon and becomes perceptible when the comparative performance of genotypes varies according to the environment. It is thus understandable that a rigorous study of this phenomenon requires a large number of different environments, which with grapevine clones has some hurdles. One is related to the difficulty of field experimentation with this perennial crop, which is time consuming and has a high cost. Therefore, few locations are usually available (commonly 2 to 5), but the genotypes are evaluated for several years in each of the locations. The experimental design in each location is preferentially of the family of incomplete block designs (alpha design or row-column design), with 6 to 10 repetitions and 5 to 8 plants per plot. The number of repetitions of each field trial depends on the number of different locations: the lower the number of locations, the higher the number of replications in each trial.

Another key concept in clonal selection is the fitting of linear mixed models and obtaining the EBLUP of the genotypic effects for the several traits assessed to perform selection (Gonçalves et al., 2016, 2020). In the context of grapevines, Gonçalves et al. (2020) proposed a measure for comparative evaluation of the G×E interaction among genotypes within a variety (interaction sensitivity, IS) based on the variance of the values of the EBLUPs of G×E interaction effects across environments. The proposed measure permitted the evaluation of the variability of the G×E interaction among genotypes and the identification of the less sensitive genotypes.

However, the unpredictable behavior of a clone under unknown environmental conditions always presents possible problems and risks. Yet, clonal material is homogeneous; thus, with its cultivation, intravarietal genetic variability erosion is strongly accelerated, and the genetic gains obtained with selection are less accurate and precise. Despite these weaknesses, the clonal selection methodology continues to be widely applied worldwide, and clones are the most multiplied material by nurseries and the most grown in commercial vineyards. Hence, the complementary strategy to overcome the G×E interaction and the other mentioned problems is to select several different clones and enhance the cultivation of several selected clones. In Portugal, this strategy is used, and the methodology is completed with the selection of a plural number of different clones (usually 7). According to the results presented by Martins and Gonçalves (2015), it is reasonable to assume that the utilization of mixtures of 7 clones could ensure stability close to that observed for the traditional variety itself.

The scheme of the methodology of intravarietal variability conservation, evaluation, and selection within an ancient grapevine variety is summarized in Fig. 3.1.

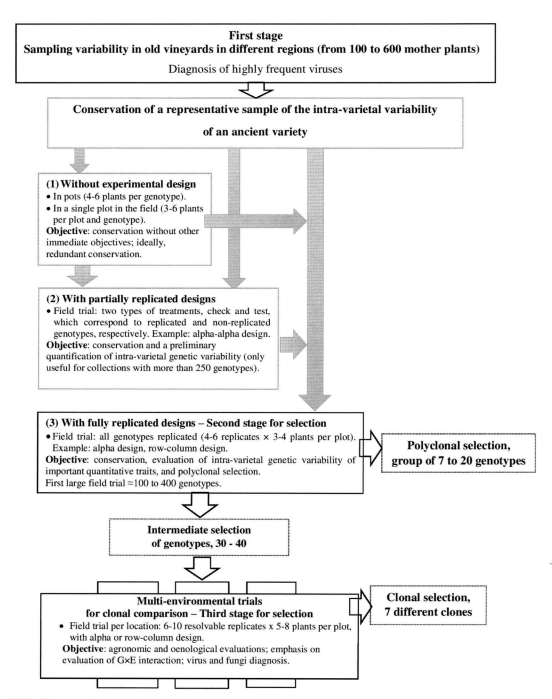

FIGURE 3.1

Methodology of intravarietal variability conservation, evaluation, and selection within an ancient grapevine variety: different steps and different objectives.

3.3 Advances in the methods for evaluation of genetic intravarietal grapevine diversity

Equally important for the evaluation of intravarietal variability are well-designed experiments and the fitting of appropriate models for data analysis. The best efforts to study intravarietal diversity from the perspective of its immediate use by vine growers almost always lead to the development of complementary studies to improve the accuracy and precision of the results of the quantification of diversity and selection. In the previous section, some references were made concerning some of those developments.

When linear mixed models are fitted to a data set, the corresponding genotypic variance component is estimated. In a field trial, the estimate of the genotypic variance component obtained for a given trait will differ among years. This occurs for several reasons: (1) data scales differ among years; (2) several sources of error variation are present (such as sampling, evaluation errors, and other environmental deviations among years); and (3) the range of genetic differences among genotypes differs with the year due to G×E interaction (different genes may be operative under the conditions of one year and not in another). Therefore, an overall view of the genetic diversity within a variety requires more than one year of evaluation, and usually the average of years is used to quantify genetic diversity and to perform selection. However, one question remains: how many years of evaluations are necessary for better quantification of diversity and the efficiency of selection?

The following methodological study intends to answer this question.

Let us consider the yield trait since it reveals the greatest amount of intravarietal diversity among the traits usually evaluated. Yield data from several years obtained in field trials of the second stage of selection of four ancient Portuguese varieties (Encruzado, Tinta Miúda, Touriga Nacional, and Viosinho) were used. A preliminary univariate analysis for yield in each year was conducted, and only data sets that showed significant genotypic variability (P - value < 0.05) were selected for the average of the years. The details of the field trials are shown in Table 3.1.

To establish the minimum number of years to quantify the genetic variability within a variety, linear mixed models were fitted to yield data for individual years, to yield mean data of all evaluated years, and to yield mean data for all possible combinations of 2, 3, 4, …, $a-1$ years (where a is the number of evaluated years in each field trial).

In all studied cases, the fitted models assumed block effects (resolvable replicate effects) as fixed effects. The other effects associated with the experimental design in the case of resolvable row-column design (the effects of rows and columns nested within complete replicates), and the genotypic effects were assumed to be random. Random errors and all random effects were assumed to be independent and identically distributed normal random variables. All random effects were assumed to be mutually independent. The REML estimation method was used for the estimation of the variance components (Patterson & Thompson, 1971). The genotypic variance component was tested using a REML ratio test. Because the null hypothesis was on the boundary of the parameter space, the P-value of the test was assumed to be half of the reported P-value from the chi-squared distribution with one degree of freedom (Self & Liang, 1987; Stram & Lee, 1994). Linear mixed models were fitted using ASReml-R package (Butler et al., 2018) within R (R Core Team).

Table 3.1 Description of the field trials for the four grapevine varieties studied (RCBD, complete block design; RCD, resolvable row-column design; v, number of genotypes; r, number of complete blocks; k, number of rows nested within complete block; s, number of columns nested within a complete block) and quantification of intravarietal genetic variability for yield: genotypic variance component estimate ($\hat{\sigma}_G^2$) and respective standard error (SE), broad-sense heritability (H^2), and genotypic coefficient of variation (CV_G) for individual years and for the average of all evaluated years.

Variety/field trial	Evaluations	$\hat{\sigma}_G^2$ [a](SE)	H^2	$CV_G(\%)$
Encruzado RCD: $v = 180$, $r = 4$, $k = 10$, $s = 18$, 3 plants per experimental unit (plot)	Year 1	0.196 (0.074)	0.30	13.3
	Year 2	0.075 (0.027)	0.31	14.9
	Year 3	0.290 (0.046)	0.68	29.5
	Year 4	0.383 (0.087)	0.49	17.1
	Year 5	0.488 (0.107)	0.50	18.7
	Year 6	0.178 (0.049)	0.40	15.7
	Year 7	0.348 (0.078)	0.50	17.0
	Year 8	0.453 (0.090)	0.55	20.3
	Year 9	0.084 (0.024)	0.39	16.5
	Average of 9 years	0.249 (0.041)	0.65	17.6
Tinta Miúda RCBD: $v = 100$, $r = 4$, 4 plants per plot	Year 1	0.259 (0.043)	0.85	33.7
	Year 2	0.739 (0.138)	0.77	22.9
	Year 3	0.440 (0.088)	0.72	37.3
	Year 4	0.096 (0.019)	0.74	32.3
	Year 5	0.989 (0.160)	0.88	37.3
	Year 6	0.320 (0.058)	0.80	29.3
	Year 7	1.083 (0.177)	0.87	32.8
	Year 8	1.463 (0.232)	0.90	32.6
	Average of 8 years	0.537 (0.080)	0.96	30.1
Touriga Nacional RCBD: $v = 196$, $r = 5$, 4 plants per plot	Year 1	0.058 (0.008)	0.76	34.9
	Year 2	0.058 (0.009)	0.66	25.1
	Year 3	0.078 (0.011)	0.71	27.3
	Year 4	0.030 (0.004)	0.78	35.0
	Year 5	0.201 (0.024)	0.86	37.5
	Year 6	0.084 (0.011)	0.75	28.1
	Year 7	0.118 (0.016)	0.73	24.9
	Year 8	0.029 (0.004)	0.68	36.6
	Year 9	0.066 (0.010)	0.69	20.4
	Year 10	0.016 (0.002)	0.71	37.3
	Year 11	0.016 (0.003)	0.53	23.8
	Year 12	0.061 (0.008)	0.75	27.2
	Average of 12 years	0.052 (0.006)	0.88	26.6

3.3 Evaluation of intravarietal grapevine diversity

Table 3.1 Description of the field trials for the four grapevine varieties studied (RCBD, complete block design; RCD, resolvable row-column design; v, number of genotypes; r, number of complete blocks; k, number of rows nested within complete block; s, number of columns nested within a complete block) and quantification of intravarietal genetic variability for yield: genotypic variance component estimate ($\hat{\sigma}_G^2$) and respective standard error (SE), broad-sense heritability (H^2), and genotypic coefficient of variation (CV_G) for individual years and for the average of all evaluated years.—cont'd

Variety/field trial	Evaluations	$\hat{\sigma}_G^2$ [a](SE)	H^2	$CV_G(\%)$
Viosinho RCBD: $v = 203$, $r = 5$, 3 plants per plot	Year 1	0.029 (0.005)	0.59	35.8
	Year 2	0.016 (0.002)	0.67	50.4
	Year 3	0.504 (0.061)	0.84	35.4
	Year 4	0.605 (0.100)	0.62	21.9
	Year 5	0.914 (0.139)	0.67	26.4
	Year 6	0.472 (0.077)	0.68	25.1
	Year 7	0.672 (0.093)	0.73	28.3
	Year 8	0.738 (0.101)	0.74	28.9
	Average of 8 years	0.339 (0.042)	0.82	25.4

[a] The residual likelihood ratio test for the genotypic variance component was performed ($H_0 : \sigma_G^2 = 0$ vs $H_1 : \sigma_G^2 > 0$). For all studied cases, P-value<0.05.

With the results of the fitted model for each yield data set, the following genetic indicators were computed: genotypic coefficient of variation (CV_G) to compare the intravarietal genetic diversity among different years and different averages of years; and a generalized measure of broad-sense heritability (H^2) (Gonçalves, Carrasquinho, et al., 2013) to evaluate how much phenotypic variability is due to genetic causes and to judge the efficiency of selection based on individual years and on different averages of years.

To evaluate the effect of the number of years on the accuracy and precision of quantification of intravarietal variability and, consequently, in genetic selection, the interpretation of the results joined the information provided by the individual years, the average of all evaluated years, and the average of all possible combinations of evaluated years.

To assess the variability among the values obtained for H^2 and CV_G among individual years and for each of the possible combinations of averages of years, two indicators were calculated: the range, i.e., the size of the interval over which the values of H^2 and the values of CV_G are distributed; and the coefficient of variation (CV) of the values of H^2 and the values of CV_G.

To evaluate the accuracy and precision of the quantification of the intravarietal variability and efficiency of selection compared to the average of all evaluated years, two indicators were also computed: the relative bias (RB) and the relative mean square error (RMSE) associated with the estimates of H^2 and CV_G for the individual years and for the average yield data for all possible combinations of years compared to the estimates obtained for the average of all evaluated years (9 years in Encruzado, 8 years in the case of Tinta Miúda, 12 years in Touriga Nacional, and 8 years in Viosinho). That is, for each studied case (individual years, or averages of 2, 3,...$a-1$ years), the bias and the mean square error were computed as:

$$Bias(\hat{\theta}) = \overline{\hat{\theta}} - \theta$$

$$MSE(\hat{\theta}) = \left[Bias(\hat{\theta})\right]^2 + Var(\hat{\theta})$$

where:

θ is the value for broad-sense heritability, or for the genotypic coefficient of variation, obtained for the average of all evaluated years;

$\overline{\hat{\theta}}$ is the mean of the values obtained for broad-sense heritability, or for the genotypic coefficient of variation, for the studied case;

$Var(\hat{\theta})$ is the variance of the values obtained for the broad-sense heritability, or for genotypic coefficient of variation, for the studied case.

The relative bias (RB) and relative MSE (RMSE) were defined as:

$$RB(\%) = \frac{Bias(\hat{\theta})}{\theta} \times 100$$

$$RMSE(\%) = \frac{MSE(\hat{\theta})}{\theta^2} \times 100.$$

The results obtained for the quantification of intravarietal genetic variability for each individual year and for the average of all evaluated years are shown in Table 3.1. For all studied varieties, the applied methodology permitted the identification of significant intravarietal genetic variability for yield (P-value<0.05), although the amount for each variety depended on the year, as expected. For the average of all evaluated years, the values obtained for H^2 were higher, which was accomplished by an increase in the precision of the estimate of genotypic variance (higher value for the ratio $\hat{\sigma}_G^2/SE$). This indicates that the quantification of intravarietal genetic variability should be based on the analysis of the average of all evaluated years.

The results to evaluate the effect of the number of years on the accuracy and precision of quantification of intravarietal variability and genetic selection are shown in Table 3.2.

Intravarietal genetic variability for yield depended on the year and on the combinations of the averages of years. When individual years were considered, the values of and CV_G showed higher variability compared to the averages of years, which is supported by the values obtained for the range and the coefficient of variation. With the averages of years, the values obtained for the range and the coefficient of variation of H^2 and CV_G decreased, with a substantial decrease up to an average of four

Table 3.2 Ranges and coefficients of variation (CV) of the values of H^2 and CV_G among individual years and for each of all the possible combinations of averages of years, and relative bias (RB) and relative mean square error (RMSE) values associated with the estimates of H^2 and CV_G for the individual years and the average yield data for all possible combinations of years compared to the estimates obtained for the average of all evaluated years.

Variety	Data set[a]	H^2 Range	H^2 CV (%)	H^2 RB	H^2 RMSE	CV_G(%) Range	CV_G(%) CV (%)	CV_G(%) RB	CV_G(%) RMSE
Encruzado	Individual years	0.37	26.0	−29.8	12.2	16.20	26.1	3.1	7.3
	Average of 2 years	0.29	11.8	−16.2	3.6	9.32	13.6	1.2	1.9
	Average of 3 years	0.22	7.4	−10.0	1.4	6.89	9.6	0.6	0.9
	Average of 4 years	0.16	5.2	−6.5	0.7	5.51	7.4	0.3	0.5
	Average of 5 years	0.12	3.9	−4.3	0.3	4.35	5.8	0.2	0.3
	Average of 6 years	0.08	2.9	−2.7	0.2	3.37	4.5	0.1	0.2
	Average of 7 years	0.05	2.1	−1.6	0.1	2.39	3.4	0.1	0.1
	Average of 8 years	0.03	1.3	−0.7	0.0	1.39	2.2	0.0	0.0
Tinta Miúda	Individual years	0.18	7.7	−14.6	2.6	14.4	13.4	7.3	2.6
	Average of 2 years	0.12	3.3	−6.9	0.6	12.6	11.6	3.8	1.6
	Average of 3 years	0.06	1.7	−3.9	0.2	10.8	9.5	2.2	1.0
	Average of 4 years	0.04	1.0	−2.4	0.1	8.8	7.7	1.4	0.6
	Average of 5 years	0.03	0.7	−1.4	0.0	6.7	6.1	0.8	0.4
	Average of 6 years	0.02	0.5	−0.8	0.0	4.7	4.6	0.5	0.2
	Average of 7 years	0.01	0.3	−0.3	0.0	3.3	3.1	0.2	0.1
Touriga Nacional	Individual years	0.32	10.4	−18.8	4.3	17.1	19.4	12.2	6.2
	Average of 2 years	0.21	5.7	−10.2	1.3	15.8	14.0	5.2	2.4
	Average of 3 years	0.16	4.1	−6.0	0.5	14.0	11.2	4.4	1.6
	Average of 4 years	0.12	2.9	−4.7	0.3	11.2	8.4	1.5	0.7
	Average of 5 years	0.10	2.3	−3.2	0.2	10.3	7.3	1.3	0.6
	Average of 6 years	0.08	1.9	−2.5	0.1	7.7	6.1	0.6	0.4
	Average of 7 years	0.07	1.5	−1.6	0.0	7.0	5.1	0.8	0.3
	Average of 8 years	0.06	1.3	−1.2	0.0	5.3	4.4	0.2	0.2
	Average of 9 years	0.05	1.1	−1.1	0.0	4.5	3.7	0.2	0.1
	Average of 10 years	0.03	0.9	−0.5	0.0	3.4	2.9	0.1	0.1
	Average of 11 years	0.02	0.6	−0.2	0.0	2.1	2.0	0.1	0.0

Continued

Table 3.2 Ranges and coefficients of variation (CV) of the values of H^2 and CV_G among individual years and for each of all the possible combinations of averages of years, and relative bias (RB) and relative mean square error (RMSE) values associated with the estimates of H^2 and CV_G for the individual years and the average yield data for all possible combinations of years compared to the estimates obtained for the average of all evaluated years.—cont'd

Variety	Data set[a]	H^2 Range	H^2 CV (%)	H^2 RB	H^2 RMSE	CV_G(%) Range	CV_G(%) CV (%)	CV_G(%) RB	CV_G(%) RMSE
Viosinho	Individual years	0.25	10.5	−15.5	3.2	28.4	26.7	24.3	16.9
	Average of 2 years	0.20	7.7	−8.7	1.3	16.0	13.6	8.2	2.8
	Average of 3 years	0.16	5.4	−5.0	0.5	11.6	8.7	4.2	1.0
	Average of 4 years	0.13	4.1	−3.0	0.2	7.2	6.4	2.4	0.5
	Average of 5 years	0.11	3.2	−1.8	0.1	5.4	4.8	1.4	0.3
	Average of 6 years	0.08	2.4	−1.0	0.1	3.8	3.5	0.8	0.1
	Average of 7 years	0.04	1.6	−0.4	0.0	1.8	2.3	0.3	0.1

[a]For Encruzado, 9 individual years and 36, 84, 126, 126, 84, 36, and 9 possible combinations for averages of 2, 3, 4, 5, 6, 7, and 8 years, respectively. For Tinta Miúda, 8 individual years and 28, 56, 70, 56, 28, and 8 possible combinations for averages of 2, 3, 4, 5, 6, and 7 years, respectively. For Touriga Nacional, 12 individual years and 66, 220, 495, 792, 924, 792, 495, 220, 66, and 12 possible combinations for averages of 2, 3, 4, 5, 6, 7, 8, 9, 10, and 11 years, respectively. For Viosinho, 8 individual years and 28, 56, 70, 56, 28, and 8 possible combinations for averages of 2, 3, 4, 5, 6, and 7 years, respectively.

years. From then on, differences among averages of years were smaller, although they depended on the quality of the field trial (smaller differences when higher values of heritability were observed).

The RB and RMSE associated with the estimates of H^2 and CV_G in relation to the averages of all evaluated years decreased with the increase in the number of years. For individual years, the values of H^2 and CV_G showed higher variability, leading to higher levels of RB and RMSE. However, greater decreases in RB and RMSE values up to the average of three years were observed. From then on, the decreases were less pronounced, and with averages of 5–6 years, the results for H^2 and CV_G were close to the values obtained with averages of all evaluated years (the RB and RMSE values were all close to zero). This finding was observed for all studied cases, including the one with 12 years of evaluation. That is, with averages of 5–6 years, the results for efficiency of selection and quantification of genetic variability were close to the values obtained with the average of the 12 evaluated years.

In sum, based on this study, for the quantification of intravarietal diversity, the average of several years should be used. The results obtained for the quantification of intravarietal diversity and for the efficiency of selection start to stabilize using the average of at least 5–6 years.

3.4 Practical applications in Portugal

The strategies of conservation, evaluation of intravarietal diversity, and selection within varieties described in the previous sections are currently implemented in Portugal by the Portuguese Association for Grapevine Diversity (PORVID). PORVID is a private nonprofit association that brings together the skills of universities, companies in the wine industry, and other stakeholders in the areas of wine and biodiversity. The overall goal of PORVID is to halt intravarietal diversity erosion of ancient Portuguese grapevine varieties to secure their sustainable use by future generations and to deploy an efficient system for polyclonal and clonal selection.

This work is supported by a network of field trials planted all over the country and by PORVID's Experimental Center for the Conservation of Grapevine Diversity, which is a farm with an area of 140 ha dedicated to the conservation of the intravarietal diversity of all autochthonous Portuguese varieties. At present, 30,000 genotypes of over 204 ancient varieties are conserved, 64 varieties are undergoing selection, and more than 185 field trials for polyclonal and clonal selection have been planted. This strategy provides added value to wine companies that increasingly want to make original and different wines to increase their competitiveness in the wine market. At present, PORVID's collection of intravarietal diversity already has an unusual dimension, and the goal is to reach 50,000 preserved genotypes of the 250 Portuguese autochthonous varieties.

These grapevine genetic resources are preserved in two ways: in pots and in field trials. In the conservation in pots, each genotype is represented by four plants. It is constituted essentially by rarely grown varieties at present, and the establishment of a representative sample of intravarietal diversity of each one is being carried out. Field trials are meant for the varieties that are in selection. These trials can be partially replicated trials (Fig. 3.2) or fully replicated trials (Fig. 3.3), which are used for several agronomic and oenological studies directly focused on grape and wine sector needs. This reflects not only new knowledge about intravarietal diversity but also the selection of new polyclonal materials that are immediately available to nurseries and winegrowers.

0076	5101	1362	1458	1446	2261	1557	1086	5094	0026	2063	1388	2258	2261	2023	1061	2373	1383	1533	1084	1313	5112
1101	0011	2358	1407	0111	2268	1459	1313	2333	1448	2352	2303	2312	2363	0112	2045	0011	0051	2064	1389	0076	2296
1533	2021	0112	1105	0062	2323	1332	2277	5125	2074	1387	5054	0095	5021	5024	1086	5045	1311	1305	1105	1455	1491
5042	5053	2043	5131	1453	5054	0081	1337	5121	2312	5081	5094	1499	1484	1411	1527	1228	1576	1337	2314	2251	5002
5066	5073	1314	1241	2335	0106	2251	2376	1481	2272	5004	0096	5004	0091	0013	0103	2268	1421	1344	5044	1139	5085
5082	1391	1406	0014	5045	0012	1371	2291	5025	1084	2303	1446	1391	1319	5061	1407	1303	5155	2061	2376	1387	5083
5141	1401	1061	2351	1335	5093	1278	1073	1344	1033	1527	0084	1335	1171	1317	1425	1422	0014	5134	1314	5053	2323
1484	1055	2364	5152	1043	0093	0125	0096	0052	1303	5024	1241	1497	5121	1055	5135	5075	1033	0062	1459	1125	1309
1111	1285	1311	5065	1454	5074	1319	1478	2023	0103	2327	2285	0081	1401	2291	5066	2371	0036	2358	1352	5065	1043
5085	1383	0041	1499	2336	1334	1491	2354	2045	0084	2366	1409	1512	1515	1453	1066	2366	1332	5101	5086	5074	1481
1422	2298	1309	2064	5044	1576	1001	1515	5092	5145	1245	1141	1101	5145	1278	0121	1362	5025	1438	0093	5105	2327
1453	1532	5004	1154	2064	1333	5085	1541	1313	5025	1319	2312	1043	2251	1362	2072	0062	2311	1311	2362	1516	5004
2303	2251	5034	1335	0066	1086	0093	5145	1339	1232	1533	2355	5024	5044	0076	1498	1056	2261	2366	5115	1335	5065
5113	2323	1459	2315	0096	2312	2316	2023	2366	1061	1209	2291	1422	1491	1261	5094	1313	1309	1489	1344	5064	0102
1362	1108	0025	5074	1332	1391	1527	1241	1197	0032	5044	1319	2041	1315	0096	1314	1383	0081	6001	0093	1087	5101
1576	2262	2326	2327	5121	0034	2291	1562	1314	0076	0011	1033	1445	1101	1515	1387	2317	2303	5085	1364	1401	2357
1055	1344	2261	5066	1192	0014	0082	1387	0094	0071	0084	5045	5121	5132	1273	1495	0112	2268	1391	1061	0115	1481
1102	5065	1303	1522	1101	1309	2045	0033	0062	5054	0116	5074	1459	2253	1278	1082	0084	5102	2356	1446	2045	1576
0081	1176	5001	1084	1446	2259	1311	1422	1401	1204	1481	5091	2367	2023	1442	1086	2327	2064	1055	1499	2376	1382
1407	2376	0112	1491	0016	1484	1502	5101	0015	1033	1506	2271	1407	5022	1084	2358	1527	5103	5054	1161	2323	5145
5003	5045	1515	1186	5024	1529	5053	5111	0103	1499	2358	5066	1331	0011	5023	1332	2334	5053	1177	1484	5025	1105
1277	1383	5094	1337	0092	1278	5005	2268	1105	1043	2276	1185	1533	0103	0014	1303	2304	2328	1337	1453	1241	2359
1021	1387	0081	1422	1337	0055	0112	1554	5104	5044	5025	0035	5053	1362	0096	5066	0045	1491	0062	1407	0044	2279
5053	1312	1533	2376	1033	5004	1494	0105	1452	5054	1278	5044	5094	0511	1533	1582	2281	2366	0093	2291	0085	1033
1335	0011	2365	1344	1361	1184	5043	1311	1515	2327	1407	1387	2312	2023	0122	5055	0123	2358	1576	5085	1278	1341
1523	5154	1313	2312	5024	2358	0093	1241	5114	1461	1481	5045	2327	0081	2321	1383	1303	1237	1515	1241	1572	1221
2023	1393	1484	2303	1314	5074	1391	2256	1043	0114	1091	5145	5063	1344	5062	0076	0124	5071	1391	1314	5054	1313
1105	1319	1163	0103	2323	1055	5094	0076	2305	1281	1011	2045	5101	1342	1335	2268	5024	1473	1249	0073	1319	1337
0086	1499	2045	1332	1552	1084	5133	0101	2291	2268	0062	1055	0126	1105	1259	1101	0084	2303	1355	0516	1061	1481
5085	1157	0056	5142	1362	5121	5065	1453	0084	1086	0104	5052	1459	0074	5074	0075	5025	5004	2269	1499	1422	1086
1446	1309	2295	1381	0053	1383	5066	5145	2251	5035	2366	1403	0519	2282	1309	1453	1084	1484	2323	1392	5065	0011
1303	0096	1576	1012	0065	2073	2064	1491	1061	1401	0031	2286	1401	5121	0112	2294	2251	0014	0103	2284	2292	1446
2375	1527	5045	2022	1101	5143	2274	1459	5101	0014	2261	1043	2062	1527	1311	2376	2064	2267	1294	2261	1332	0072

FIGURE 3.2

Example of the structure of a partially replicated design of the Bastardo (Trousseau) variety. The field trial contained 77 check genotypes established according to a resolvable row-column design with six resolvable replicates with three plants per plot (red numbers and gray-shaded plots) and 264 test genotypes each with one plot with three plants (nonreplicated treatments; white plots).

For selection, an integrated approach comprising two types of selected material is undertaken—polyclonal and clonal. The first polyclonal selected materials began to be distributed to vine growers in 1984, and since 2005, clones have also been selected. Both types of selected material carry high genetic and economic gains for important agronomic and oenological traits, in the order of the values referred to in Section 3.1.

2621	2574	3810	2581	2019	3905	4107	0508	10504	2423	10207	1550	1410	4104	3601
3001	2573	2521	12501	3809	11309	1116	11602	2651	3702	0223	41303	11301	40506	6118
12401	1170	2531	8202	12609	10503	2662	4101	3803	8204	11308	2561	41206	2603	40502
11501	3910	12605	3103	10505	3020	2424	40709	12601	6701	3903	2652	41304	3806	11303
8807	2572	40503	12604	7503	11205	1501	1510	3112	0459	2642	11502	3502	12608	2403
41201	2701	10702	41204	2601	11902	3016	6116	2410	11905	11314	10501	12607	1515	3404
2672	12610	6407	4110	3203	4103	0503	2691	41202	3901	3504	2671	3902	11310	3701
10804	2562	2602	11002	3808	0498	11603	2613	0527	41207	1117	2571	3402	2692	4105
40802	40510	2027	40105	41205	3305	3501	2512	40507	11304	41203	12304	11201	3205	3812
2693	41305	12502	2641	10502	10312	2682	2425	2612	11601	11313	11202	2611	2404	11203
3605	3804	2664	2631	11305	2661	40509	3204	1632	10305	2681	3807	11306	2511	3105
3901	2691	11603	2701	12608	3810	2511	2602	11203	1410	10505	2410	2693	40709	0223
3105	1117	2641	41304	2027	1515	2572	3404	2652	3112	0508	11309	41202	12609	40509
11602	1632	2681	10702	11301	2425	41205	2423	8807	3205	3502	12607	3808	8202	10312
40506	2581	3203	3803	2642	3902	2531	11313	2521	3020	41201	40502	0459	3804	12304
3016	11502	12601	2613	41207	3807	0498	2603	2019	12501	2403	12502	40105	11501	40510
7503	11205	41206	4104	6116	3601	10501	2621	12610	4105	2631	2424	2561	41203	11601
3305	2404	1510	3001	2574	0527	8204	11202	10207	0503	10503	11305	3910	11902	4110
12604	11201	1550	2682	3605	40802	3504	3905	11304	4101	2651	11310	11314	2671	12605
2571	1501	3702	41303	3701	1170	2611	2573	3402	2664	3501	2512	2661	6701	2601
11308	11905	11303	3812	2662	1116	2672	11306	3806	10502	2562	11002	6407	10504	3204
40507	3809	4103	10305	2612	3903	6118	3103	41204	10804	2692	41305	4107	40503	12401
8204	3812	0459	2691	12601	12610	12604	10312	3810	40506	3605	40509	2701	2424	2651
0223	2671	3905	3806	2611	11502	3701	10305	41201	40105	2602	8807	4101	2641	11305
2661	11313	3501	3502	40502	3205	3001	2561	8202	40709	1510	40507	0498	2019	2027
2512	4107	0527	2531	11304	3016	12605	2403	10502	3601	41202	2511	1550	1117	3103
4105	3402	12609	3702	40503	3504	41305	1515	3910	10702	11201	6407	2631	41207	2425
10504	4110	2652	10804	2621	12502	10503	2681	11310	2572	2601	11301	3903	10501	3901
10505	11309	12501	11202	11308	3404	3808	2574	11601	1501	3804	2613	1632	40802	2672
2603	3112	12607	11314	12304	0508	41203	11002	4103	3203	11203	1170	2662	11602	41303
41204	6116	6118	40510	2404	2693	1410	2664	2571	11306	1116	11303	2423	3803	11205
2682	2692	12608	2410	2562	6701	3807	3809	41304	12401	3305	2581	7503	3020	10207
3902	41205	3204	3105	0503	41206	11905	11501	2642	2612	4104	11902	2521	2573	11603
41205	3905	41204	40709	41202	40509	3903	41207	41303	41201	2664	11308	2404	3103	41206
2681	2572	41304	2573	2613	2423	40502	2693	2410	12605	40507	4105	2671	4103	3806
3810	3001	3504	3112	11305	2612	2561	2601	11303	12610	2692	2531	12604	1632	11602
3016	2641	2574	2631	10501	2512	12608	3812	3902	12607	2621	2651	12401	6407	3804
2562	2662	2425	11310	1410	2403	11902	41203	40105	11313	8204	2652	2701	3105	3809
11205	12502	11309	3808	2603	8202	40510	0503	3605	3702	3203	2581	0527	0223	3601
3803	1515	2602	10207	6701	11314	10502	40506	40503	1501	3204	4107	11301	40802	0498
8807	10505	11905	12304	3501	3402	3901	1550	10503	4101	3910	11306	3404	4104	2027
11202	2691	7503	6118	2642	3807	11201	1170	3205	10504	2424	11501	11603	2611	4110
12501	2571	12601	10305	2521	2672	0508	3502	1117	3305	11203	41305	2661	6116	0459
3701	11304	10804	1116	10312	3020	10702	11002	11601	2511	11502	1510	2019	12609	2682
2424	40503	2651	11905	2601	2571	41205	40802	11203	0527	3806	1410	2572	2603	2561
11301	2692	4104	1170	0459	2019	12605	11308	2512	10312	2641	2410	41201	3504	6118
10503	11002	40709	3105	11309	2531	2701	2423	10501	41207	11201	2682	3204	8807	3203
4105	40509	2693	3803	3103	1501	3501	12502	11501	11310	6116	2574	3402	3305	11304
41202	4101	3807	2621	10702	0223	4103	3809	2027	11303	6701	40506	11202	2631	3502
10305	2573	4110	0498	2404	3605	11502	11603	3702	0508	3810	6407	41203	40502	3020
40510	3901	1117	1510	2662	12501	1515	41204	41304	3804	3808	12304	11305	2612	10504
2672	2652	10502	40105	3902	3812	11602	2511	3905	7503	0503	3404	12601	2664	41305
1550	2642	8204	10505	12610	2425	11306	3701	1632	11314	12401	11205	3205	4107	2613
12609	2661	12607	2671	40507	41206	10207	3112	2562	3001	2403	3903	2581	11601	2602
11313	3601	2521	12608	2681	11902	10804	3910	12604	2611	41303	8202	1116	2691	3016
3402	6118	2027	12401	3804	11601	2651	4101	2681	2691	11002	12601	3001	11902	11502
2631	1501	2603	3910	3205	0527	2662	41305	2581	10305	10501	2671	11308	3810	3901
3404	3902	2611	41203	2561	2572	3305	3803	2425	2423	3605	12501	10804	2692	40709
2531	2621	3020	41205	40506	3701	2613	1116	2661	40503	40510	2693	12607	7503	2641
0503	0508	2410	1515	8807	11313	12609	3103	0498	0223	3807	2512	11203	11306	2574
40507	4107	3204	2602	41304	11602	11310	2403	2424	1170	2404	10702	11205	3702	2521
10504	41207	12610	40502	12605	6407	11202	2571	4104	3501	3112	3809	40105	2672	11314
41303	3806	1510	1632	11603	41204	3601	3504	1410	10502	0459	10503	40509	12502	2562
12608	11305	3808	2682	8204	11501	2664	11301	11304	41206	4103	3105	1550	1117	2601
40802	2701	3903	11303	3905	3502	2019	4110	2612	11309	3016	2642	6701	12304	4105
10312	8202	3812	2652	6116	11201	11905	3203	41201	12604	10505	10207	2573	41202	2511

FIGURE 3.3

Example of the structure of a fully replicated design of the Arinto variety: resolvable row-column design: $v = 165$ genotypes, $r = 6$ resolvable replicates (complete blocks; different colors for numbers), $k = 11$ rows nested within a complete block; $s = 15$ columns nested within a complete block. Plots in each resolvable replicate are arranged in rows and columns. For four genotypes (shaded), their positions within each complete block are exemplified.

3.5 Concluding remarks

Quality wine is a product whose value is strongly based on traditions and historical and natural factors. Therefore, in viticulture, the use of varieties obtained by modern plant breeding techniques, without history, is relatively low. However, there are numerous traditional varieties domesticated from wild relatives by the first viticulturists thousands of years ago that are different from each other and have high intravarietal diversity. It is this diversity both between and within varieties, both of natural origin and with evolutionary histories intertwined with history and human civilization, that can fully support the production of high-quality wines, different from each other and with a historical and natural character.

In general, a very wide range of intravarietal diversity is observed, which allows selection within a variety to adapt to the most diverse environmental, agricultural, and market contexts. However, this approach faces a serious drawback that is represented by the genetic erosion of the intravarietal diversity, which has been extraordinarily accelerated since the middle of the last century and threatens to destroy the diversity created over centuries and millennia within a few decades.

However, efficient methodologies for the conservation of diversity and for its use through selection (clonal and polyclonal) have been developed and applied on a large scale by our group in Portugal. It is this combined strategy between conservation and selection that was described in this chapter and that can contribute to the resolution of critical problems in viticulture today.

It is hoped that these techniques will be widely applied in European vine-growing countries, in which grapevine varieties have accumulated more diversity since ancient times.

Acknowledgments

The authors would like to acknowledge the colleagues of National Network for Grapevine Selection and Portuguese Association for Grapevine Diversity (PORVID) for their collaboration in management of field experiments and data collection.

References

Brown, J., Caligari, P., & Campos, H. (2014). *Plant breeding*. Wiley Blackwell.

Butler, D. G., Cullis, B. R., Gilmour, A. R., Gogel, B. J., & Thompson, R. (2018). *ASReml-R reference manual, version 4*. Wollongong, NSW, Australia: University of Wollongong.

Carvalho, L., Gonçalves, E., Amâncio, A., & Martins, A. (2020). Selecting Aragonez genotypes able to outplay climate change driven abiotic stress. *Frontiers in Plant Science, 11*.

Giesbrecht, F. G., & Gumpertz, M. L. (2004). *Planning, construction and analysis of comparative experiments*. New York, NY: Wiley.

Gonçalves, E., Carrasquinho, I., Almeida, R., Pedroso, V., & Martins, A. (2016). Genetic correlations in grapevine and their effects on selection. *Australian Journal of Grape and Wine Research, 22*, 52–63.

Gonçalves, E., Carrasquinho, I., & Martins, A. (2020). A measure to evaluate the sensitivity to genotype-by-environment interaction in grapevine clones. *Australian Journal of Grape and Wine Research, 26*, 259–270.

Gonçalves, E., Carrasquinho, I., StAubyn, A., & Martins, A. (2013). Broad-sense heritability in the context of mixed models for grapevine initial selection trials. *Euphytica, 189*, 379–391.

References

Gonçalves, E., & Martins, A. (2019). Methods for conservation of intra-varietal genetic variability in ancient grapevine varieties. *BIO Web of Conferences, 15*, 01029.

Gonçalves, E., & Martins, A. (2019b). Genetic gains of selection in ancient grapevine cultivars. *Acta Horticulturae, 1248*, 47−54.

Gonçalves, E., StAubyn, A., & Martins, A. (2007). Mixed spatial models for data analysis of yield on large grapevine selection field trials. *Theoretical and Applied Genetics, 115*(5), 653−663.

Gonçalves, E., StAubyn, A., & Martins, A. (2010). Experimental designs for evaluation of genetic variability and selection of ancient grapevine varieties: A simulation study. *Heredity, 104*, 552−562.

Gonçalves, E., StAubyn, A., & Martins, A. (2013). The utilization of unreplicated trials for conservation and quantification of intravarietal genetic variability of rarely grown ancient grapevine varieties. *Tree Genetics and Genomes, 9*, 65−73.

Isik, F., Holland, J., & Maltecca, C. (2017). *Genetic data analysis for plant and animal breeding*. Springer.

Kempton, R. A., & Gleeson, A. C. (1997). Unreplicated trials. In R. A. Kempton, & P. N. Fox (Eds.), *Statistical methods for plant variety evaluation* (pp. 86−100). Chapman & Hall.

Lynch, M., & Walsh, B. (1998). *Genetics and analysis of quantitative traits*. Sunderland, England: Sinauer Associates.

Magris, G., Di Gaspero, G., Marroni, F., Zenoni, S., Tornielli, G. B., Celii, M., De Paoli, E., Pezzotti, M., Conte, F., Paci, P., & Morgante, M. (2019). Genetic, epigenetic and genomic effects on variation of gene expression among grape varieties. *The Plant Journal, 99*(5), 895−909.

Martins, A., & Gonçalves, E. (2015). Grapevine breeding programmes in Portugal. In A. G. Reynolds (Ed.), *Grapevine breeding programs for the wine industry: Traditional and molecular techniques* (pp. 159−182). UK: Woodhead Publishing Elsevier.

Mead, R., Gilmour, S. G., & Mead, A. (2012). *Statistical principles for the design of experiments*. Cambridge, UK: Cambridge University Press.

Meneghetti, S., Caló, A., & Bavaresco, L. (2012). A strategy to investigate the intravarietal genetic variability in *Vitis vinifera* L. for clones and biotypes identification and to correlate molecular profiles with morphological traits or geographic origins. *Molecular Biotechnology, 52*, 68−81.

Mramba, L. K., Peter, G. F., Whitaker, V. M., & Gezan, S. A. (2018). Generating improved experimental designs with spatially and genetically correlated observations using mixed models. *Agronomy, 8*, 40.

Myles, S., Boyko, A., Owen, C., Brown, P., Grassi, F., Aradhya, B., Reynolds, A., Chia, J., Ware, D., Bustamante, C., & Buckler, E. (2011). Genetic structure and domestication history of the grape. *Proceedings of the National Academy of Sciences of the United States of America, 108*(9), 3530−3535.

Nicolas, S. D., Péros, J. P., Lacombe, T., Launay, A., Le Paslier, M. C., Bérard, A., Mangin, B., Valière, S., Martins, F., Le Cunff, L., Laucou, V., Bacilieri, R., Dereeper, A., Chatelet, P., This, P., & Doligez, A. (2016). Genetic diversity, linkage disequilibrium and power of a large grapevine (*Vitis vinifera* L) diversity panel newly designed for association studies. *BMC Plant Biology, 16*, 74.

Patterson, H. D., & Thompson, R. (1971). Recovery of inter-block information when block sizes are inequal. *Biometrika, 58*, 545−554.

Patterson, H. D., & Williams, E. R. (1976). A new class of resolvable incomplete block designs. *Biometrika, 63*, 83−92.

Pelsy, F., Hocquigny, S., Moncada, X., Barbeau, G., Forget, D., Hinrichsen, P., & Merdinoglu, D. (2010). An extensive study of the genetic diversity within seven French wine grape variety collections. *Theoretical and Applied Genetics, 120*(6), 1219−1231.

Piepho, H. P. (2015). Generating efficient designs for comparative experiments using the SAS procedure OPTEX. *Communications in Biometry and Crop Science, 10*, 96−114.

Piepho, H. P., Michel, V., & Williams, E. (2018). Neighbor balance and evenness of distribution of treatment replications in row-column designs. *Biometrical Journal, 60*, 1172−1189.

Roach, M. J., Johnson, D. L., Bohlmann, J., van Vuuren, H. J. J., Jones, S. J. M., Pretorius, I. S., Schmidt, S. A., & Borneman, A. R. (2018). Population sequencing reveals clonal diversity and ancestral inbreeding in the grapevine cultivar Chardonnay. *PLoS Genetics, 14*(11).

Searle, S., Casella, G., & McCulloch, C. (1992). *Variance components*. Hoboken, NJ, USA: John Wiley.

Self, S. G., & Liang, K. Y. (1987). Asymptotic properties of maximum likelihood estimators and likelihood ratio tests under nonstandard conditions. *Journal of the American Statistical Association, 82*, 605−610.

Stram, D. O., & Lee, J. W. (1994). Variance components testing in the longitudinal mixed effects model. *Biometrics, 50*, 1171−1177.

Vondras, A. M., Minio, A., Blanco-Ulate, B., Balderas, R. F., Penn, M. A., Zhou, Y., Seymour, D., Ye, Z., Liang, D., Espinoza, L. K., Michael, M., Anderson, M. M., Walker, M. A., Gaut, B., & Cantu, D. (2019). The genomic diversification of grapevine clones. *BMC Genomics, 20*, 972.

Welham, S. J., Gezan, S. A., Clark, S. J., & Mead, A. (2015). *Statistical methods in biology*. Boca Raton, FL: CRC Press.

Whitaker, D. D., Williams, E. R., & John, J. A. (2007). *CycDesigN: A package for computer generation of experimental designs*. Canberra, Australia: CSIRO.

Wiliams, E. R., & John, J. A. (1989). Construction of row and column designs with contiguous replicates. *Applied Statistics, 38*, 149−154.

Williams, E., John, J., & Whitaker, D. (2006). Construction of resolvable spatial row-column designs. *Biometrics, 62*(1), 103−108.

CHAPTER 4

Phenotyping for drought tolerance in grapevine populations: the challenge of heterogeneous field conditions

Aude Coupel-Ledru[1], Eric Lebon[1,a], Jean-Pascal Goutouly[2], Angélique Christophe[1], Pilar Gago[3], Charlotte Brault[4,5,6], Patrice This[4,6], Agnès Doligez[4,6] and Thierry Simonneau[1]

[1]LEPSE, University of Montpellier, INRAE, Institut Agro, Montpellier, France; [2]EGFV, Bordeaux Sciences Agro, INRAE, University of Bordeaux, ISVV, Villenave d'Ornon, France; [3]Misión Biológica de Galicia, Consejo Superior de Investigaciones Científicas, Pontevedra, Spain; [4]AGAP Institut, University of Montpellier, CIRAD, INRAE, Institut Agro, Montpellier, France; [5]Institut Français de la Vigne et du Vin, Montpellier, France; [6]UMT Geno-Vigne®, IFV-INRAE-Institut Agro, Montpellier, France

4.1 Introduction

Compared to annual crops, breeding fruit perennials for abiotic stress tolerance is a long-lasting and space-consuming process. In addition to the several year period needed to reach production maturity for each selection step, phenotyping requires extended fields or networks of experimental sites (McClure et al., 2014). Grapevine is no exception to the rule, and there is an urgent need to identify existing cultivars or create new varieties better adapted to climate change.

Few studies have analyzed the diversity of grapevine responses to drought and its genetic bases on large populations (Coupel-Ledru et al., 2014, 2016; Marguerit et al., 2012; Trenti et al., 2021). In these reports as in most studies on other species controlled environments and high-throughput phenotyping platforms have been preferred to stabilize and reproduce water deficit scenarios with fine control of other environmental variables. These conditions allow for detailed studies of functional traits and their interactions. However, the volume of soil explored by root systems in pots is considerably smaller than under field conditions resulting in very rapid water depletion in pots and difficulties to mimic actual drying dynamics observed in field soils (Passioura, 2006; Turner, 2019). Furthermore, the canopy structure of young, easy to manipulate plants grown in individual pots strongly differs from that of plant canopies in orchard/vineyard planting systems (Araus & Cairns, 2014). Finally, differences in vegetative and fruit growth potential of field-grown compared to young potted plants should be another reason for possible errors and difficulties when extrapolating from controlled to field conditions (Geny et al., 1998; Poorter et al., 2012). Yet, plant responses in the field

---

[a] Deceased.

are of upmost importance. We propose here ways to take into account spatial heterogeneity in soil water conditions for comparison of genotypes grown under field conditions, allowing identification of QTLs in field condition for $\delta^{13}C$.

4.2 Phenotyping large populations in the field: the challenge of soil heterogeneity

Understanding the genetic control of drought-related traits is a prerequisite for speeding up the breeding of grapevine varieties in the face of climate change. In this view, quantitative trait locus (QTL) studies are precious approaches that allow to dissect the phenotypic variation and determine the contribution of each QTL to this variation (Vezzulli et al., 2019). Biparental progenies issuing from the cross between two cultivars have been used with this aim, either for the scion (e.g., *Vitis vinifera* cv. Syrah × cv. Grenache pseudo-F1 population in Coupel-Ledru et al., 2014, 2016) or the rootstock (e.g., *V. vinifera* cv. Carbernet Sauvignon × *V. riparia* cv. Gloire de Montpellier in Marguerit et al., 2012). More recently, Genome Wide Association Studies (GWAS) have been devised using diversity panels and appear as an alternative method of choice, since it allows identifying the main QTLs with high resolution by exploiting past recombination events between cultivars (Rafalski, 2010; Rincent et al., 2014). As an example, a diversity panel of 279 cultivars with limited relatedness and including the major founders of modern cultivars has recently been designed (Nicolas et al., 2016). In either case, large panels are essential to detect reliable QTLs and to reduce their positional interval of confidence on the genome (respectively 188, 135, and 279 individuals in the three above-mentioned populations). However, the larger the experimental setting, the higher the risk of undesirable within-field variability.

4.2.1 Variations in soil characteristics hinder drought tolerance studies

Considerable variations in soil characteristics may be encountered in field conditions. Specifically, spatial heterogeneity in soil water properties may derive from variations in texture (Romano & Santini, 1997) or topography. Mechanical characteristics also interact with rooting depth (Gilad et al., 2007). Finally, spatial variation in the plant cover itself may also locally impact the availability of soil water and the period when soil water deficit develops all across the field (Masuka et al., 2012). Soil heterogeneity includes both qualitative heterogeneity and configurational heterogeneity, where the former refers to the differences in these properties between locations in the soil, and the latter reflects the patch size of these locations (Dufour et al., 2006; Kelly & Canham, 1992; Liu et al., 2017; Maestre & Cortina, 2002). Heterogeneous texture specifically causes spatial variation in soil water reserve and drainage while sloped fields induce run-off during heavy rains or intensive watering regimes. Perennials like grapevine may largely contribute to the amplification of soil heterogeneity due to their root activity over years. This may contribute to explain why the magnitude of intrafield variation for yield may reach up to 8 to 10-fold (Bramley & Hamilton, 2004).

Development of suited methodologies to map soil characteristics is required, together with appropriate statistical procedures to consider soil heterogeneity for further analyses of large populations under field conditions. Implementation of proximal sensing technology for soil mapping recently renewed our capacity to monitor soil characteristics with high spatial resolution. Specifically, electrical resistivity measured in soils (Brillante et al., 2015; Yu & Kurtural, 2020) provides information which can be related to instantaneous soil water content or, on the longer term, to soil water storage capacity. Soil resistivity is tightly related to soil physical and chemical properties. In cultivated soils, this trait has been shown to be affected by stone content (Tetegan et al., 2012) and soil texture and especially clay content (Andrenelli et al., 2013; Mertens et al., 2008) so that electrical resistivity measurements (ERa) could be used as a proxy for soil water storage capacity (Brillante et al., 2014, 2015; Yu & Kurtural, 2020). However, the relationship between electrical resistivity and water content in soils remains sensitive to soil structure, texture, and mineral composition. Alternatively, predawn leaf water potential has long been used as a proxy for soil water availability although with some restrictions (Améglio et al., 1999). This last method may be preferred to more direct measurements in the soil as it characterizes how the plant senses soil water availability.

4.2.2 Statistical methods to handle soil heterogeneity and spatial variations

Correcting for spatial soil heterogeneity is a key preliminary step for estimating genotype and/or water deficit scenario effects. In statistical methods related to analysis of variance, errors arising from spatial variations can be modeled as independent effects or as spatially correlated effects with an appropriate variance—covariance structure (Hartung et al., 2019; Rodríguez-Álvarez et al., 2018). Using a suitable experimental design (e.g., blocks or lattice designs), a part of the variance can be captured through blocking of the experimental units, including variance due to rows and columns (Collins et al., 2015; Hartung et al., 2019; Klassen et al., 2014). However, these methods usually ignore the actual pattern of spatial variation in soil water, resulting in a loss of statistical power in further analyses of plant responses.

As an alternative before analyzing plant response to drought, geostatistics can help to unravel patterns of soil heterogeneity (Jackson & Caldwell, 1993; Klassen et al., 2014). Contrary to methods related to analysis of variance, geostatistics takes into account possible spatial autocorrelations in soil properties in one or several directions (Rossi et al., 1992). It has been applied to soil sciences (Webster & Burgess, 1983), agronomy (Brillante et al., 2014, 2015; Yu & Kurtural, 2020), and ecology (Jackson & Caldwell, 1993) but has rarely been combined with genetics except in large scale, spatial analysis of genetic structures of wild populations (Monestiez et al., 1994).

Among geostatistical methods, kriging can generate maps of any soil characteristics via semivariograms to account for spatial autocorrelation (Webster & Oliver, 1990). Provided that a relationship can be drawn between soil water content and these characteristics, it can be used to correct each measurement of plant response for local deviation in soil water availability. Effects of undesired spatial variation in soil water availability can thereby be removed prior to statistical analysis and QTL detection.

4.2.3 Phenotyping for plant performance under water deficit: which traits for high-throughput measurements in the field?

Breeding grapevine for better performance under water deficit requires identifying appropriate phenotypic targets. A lot of such relevant traits have been identified including reduced water use, better water extraction and, more importantly, maintained production (Simonneau et al., 2017). Water use efficiency (WUE), that is the amount of biomass produced per g of water used, therefore arose as a target of particular interest combining traits related to production and to parsimonious use of water.

Moreover, a proxy for WUE, carbon isotope composition ($\delta^{13}C$) in plants can be determined at high-throughput allowing for the analysis of the genetic architecture of WUE in large genetic populations. Genetic variability of $\delta^{13}C$ among different grapevine varieties has already been observed on berry juice or must (Gaudillère et al., 2002) and leaf tissue (Bota et al., 2016; Gibberd et al., 2001; Tomás et al., 2012). However, the link between $\delta^{13}C$ in plant organs and the whole plant WUE is still debated (Bchir et al., 2016). While Gibberd et al., (2001) showed a negative correlation between both traits under well-irrigated conditions, other studies recently suggested no consistent correlation with few exceptions (Medrano et al., 2015; Pou et al., 2012). Given the sensitivity of $\delta^{13}C$ to environmental conditions and particularly to soil water deficit (Gaudillère et al., 2002; Santesteban et al., 2012), a possible explanation for such discrepancy when comparing plants in one same field experiment may be due to incomplete control of soil water status across the field. This may introduce uncontrolled noise when measuring drought-related traits, which ultimately hinders physiological and genetic analyses. In the following, we developed a case study where we explored the potential benefit of geostatistics to solve this issue.

4.3 Detection of genetic variability for water-use efficiency in field conditions: a case study

In the following study, we modeled the spatial variability of soil water status using geostatistics in a field experiment on a grapevine progeny (population obtained from a reciprocal cross between *V. vinifera* cvs. Syrah and Grenache). We used the carbon isotope composition ($\delta^{13}C$) of must, determined at fruit maturity, as a proxy for long-term plant WUE. The overall experimental and analytical workflow is summarized in Fig. 4.1.

4.3.1 Experimental setup

Experiments were carried out in 2010 and 2011 at Montpellier SupAgro Domaine du Chapitre experimental vineyard, near Montpellier, in south France (43°31N, 3°50E). The experimental trial (0.7 ha) was planted in 2003 (7 and 8-year-old plants when the experiment was carried out) at 3300 plants ha^{-1} (2.2 × 1.0 m, row × vine spacing) and trained as a vertical shoot positioning system. The vineyard was managed according to standard viticultural practices except the watering regime. Additionally, the inter-rows were not tilled but seeded with cover crops at leaf fall and chemically weeded at flowering to avoid excess soil water storage during winter prior to experiments.

The mapping population was composed of the pseudo-F1 progeny of 188 genotypes obtained as the first generation from a reciprocal cross between two grapevine cultivars, Syrah and Grenache (Fournier-Level et al., 2009), grafted on 110 Richter rootstock (*Vitis berlandieri* × *Vitis rupestris*).

The experimental design consisted of two complete blocks (Fig. 4.1), one irrigated (I) and one non-irrigated (D), where water deficit developed. Each block was subdivided in 220 elementary plots comprising five contiguous plants each on a same row, from one same genotype. Thirty-two elementary plots of Grenache ("control" plots) were planted on a regular grid pattern in each block of the experimental field and were used to quantify spatial heterogeneity of predawn leaf water potential. Each offspring genotype was randomly assigned to the remaining plots within each block. Drip irrigation was applied in the irrigated block twice a week, at approximately 50% of maximal evapotranspiration in July and August.

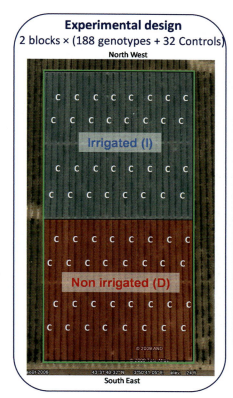 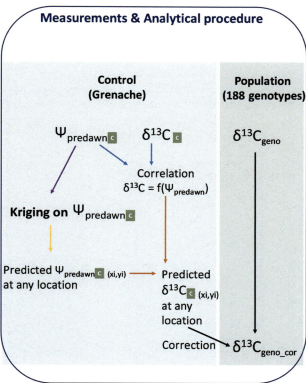

FIGURE 4.1

Overview of the experimental design and analytical procedure carried out in a case study on a grapevine progeny of Syrah and Grenache made of 188 genotypes. The left panel shows how 32 Grenache control plots ("c") were regularly distributed within the two blocks of the field (irrigated and non-irrigated). The right panel illustrates the workflow used to correct for spatial heterogeneity in water availability.

Climate variables were recorded continuously with a weather station (Campbell Scientific, Inc., Logan, Utah, USA) installed at the center of the trial, between the two blocks. Daily evolution of rainfall and evapotranspiration ET_0 (Allen et al., 1998), calculated from these data in 2010 and 2011 remained remarkably stable over the two growing seasons, reaching, respectively, 154 and 155 mm for cumulated rainfall and 756 and 745 mm for cumulated ET_0. However, the distribution of rainfall throughout the two seasons was markedly different. The 2010 growing season was characterized by a typical Mediterranean rainfall regime (rainy spring followed by a dry summer) whereas the 2011 growing season was marked by a dry spring followed by a wet summer.

4.3.2 Contrasted soil water scenarios established over the two years

Predawn leaf water potential ($\Psi_{predawn}$) was measured on control plots (cv. Grenache) approximately every two weeks, beginning just before the irrigation period on the irrigated block, until the end of

70 Chapter 4 Phenotyping for drought tolerance in grapevine populations

August, using four cross-calibrated pressure chambers (Soil moisture Equipment Corp. Santa Barbara, CA, USA). All measurements were completed in 2 h just before dawn on three outermost, fully expanded primary leaves taken from the three central plants of the control plots (one leaf per plant). A total of 192 (i.e., 96 per watering regime) leaves were sampled on each measurement date.

$\Psi_{predawn}$ showed different temporal evolutions depending on irrigation treatment and year (Fig. 4.2A and B). Average $\Psi_{predawn}$ on the irrigated block was maintained above −0.2 MPa over the two years, except for the last measurement carried out on September 01, 2010. This range corresponds

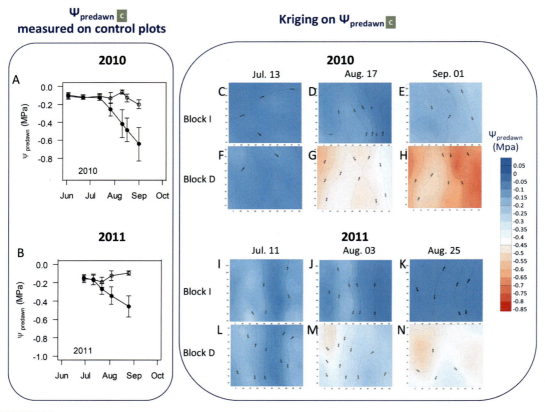

FIGURE 4.2

Evolution of predawn leaf water potential ($\Psi_{predawn}$) on control plots planted with cv. Grenache, and $\Psi_{predawn}$ maps built by kriging, during growing seasons 2010 and 2011. In (A) and (B), white and black circles represent values recorded, respectively, in the irrigated and nonirrigated blocks (means ± SDs, $n = 96$ leaves for each block). Tick marks on the x axis represent the beginning of each month. $\Psi_{predawn}$ was mapped for three dates in 2010 (C–H) and in 2011 (I–N) for both irrigated (C–E; I–K) and non-irrigated (F–H; L–N) blocks.

to well-watered to slight water deficit conditions. However, detailed examination of these results revealed that water deficit experienced by grapevines in the irrigated block from July until early August 2011 was slightly stronger compared to 2010. In the non-irrigated block, onset of water deficit was detected by comparison with the irrigated block in the beginning of August and increased until early September in both years. The maximal drought intensity achieved was higher in 2010 (−0.64 MPa) than in 2011 (−0.45 MPa) and reflected differences in rainfall regime (see above). Overall, these results indicated that the treatments were effective in imposing contrasting soil water scenarios.

4.3.3 Spatial distribution of predawn leaf water potential within the field

To evaluate the spatial heterogeneity of the experimental field, we first performed determinations of soil apparent resistivity using an automatic on-the-go DC recording resistivity meter (ARP©, Automatic Resistivity Profiling. Geocarta, Paris, France) towed by a ground vehicle through the trial along parallel transects at a distance of 4.4 m apart the row. The resulting map exhibits large resistivity contrasts ranging from 25 to 55 Ωm with a marked gradient perpendicular to the length of the plot (Fig. 4.3).

To account for this variation, $\Psi_{predawn}$ was used as a proxy for soil water availability. $\Psi_{predawn}$ values were estimated for each plot with a kriging interpolation method using data measured on control plots (Fig. 4.1). This procedure generates predicted $\Psi_{predawn}$ values via semivariograms, which characterize the spatial autocorrelation for the measured variable (Webster & Oliver, 1990). This approach makes it possible to estimate soil water deficit experienced by grapevines on any plot. Provided that a relationship can be drawn between $\Psi_{predawn}$ and $\delta^{13}C$, this relationship can be used to correct each measurement of $\delta^{13}C$ for local deviation in soil water availability compared to the mean soil water availability of the block. Effects of undesired spatial variation in soil water availability can thereby be removed prior to statistical analysis and QTL detection.

Prediction of the spatial distribution of $\Psi_{predawn}$ data was performed for each block on the basis of the 96 georeferenced measurements in control plots of each block by Ordinary Kriging method using the software package Surfer 8.03 (Golden Software, Inc, Col, USA). Experimental semivariograms were fitted to data and cross validated for each soil water scenario (watering regime and year) with a linear model taking into account geometric anisotropy. The semivariogram parameters are given in Table 4.1.

Fig. 4.2C−N shows the time-course and spatial variability of $\Psi_{predawn}$ measured on *control* plants over years 2010 and 2011. Due to water supply, the irrigated block did not show a marked spatial structure in spite of spatial variation detected by resistivity measurement except for August 03, 2011 where a slight spatial pattern was clearly apparent (Fig. 4.2J). By contrast, the non-irrigated block exhibited an increasingly marked spatial structure during drying out. Predawn leaf water potential measured during the driest period in the non-irrigated block varied from −1.20 to −0.36 MPa in 2010 and from −0.73 to −0.37 MPa in 2011, respectively, which corresponded to severe and moderate soil water deficit stress (Leeuwen et al., 2009) (Fig. 4.2H and N). Spatial patterns remained roughly stable between dates and years even if some apparent discrepancies were noted. A comparison of the maps

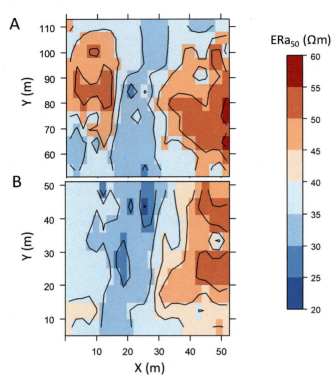

FIGURE 4.3

Apparent electrical resistivity (ERa$_{50}$) maps measured on March 2011 in irrigated (A) and non-irrigated (B) blocks. Data correspond to the 0–0.5 m depth soil layer.

drawn for the non-irrigated block on August 17 and September 01, 2010 shows that the zone with the highest $\Psi_{predawn}$ on the first date corresponded to the lowest $\Psi_{predawn}$ in the block on the second date. This indicates dynamic changes in spatial pattern of soil water balance likely resulting from spatial variability of water storage capacities.

4.3.4 Relationship between carbon isotope composition (δ^{13}C) and predawn leaf water potential

In grapevine, the pulp is known to be the most suitable organ for δ^{13}C determination in relation with intrinsic WUE and soil water regime (de Souza, 2005; Santesteban et al., 2015). Carbon isotope analyses were performed on must obtained from berries harvested at fruit maturity (35 days after

Table 4.1 Results of kriging analysis of predawn leaf water potential ($\Psi_{predawn}$) measured in control plots regularly spaced across the field. Analysis was performed on each year in irrigated (I) and nonirrigated (D) blocks. Column "model" indicates that a linear geostatistical model was used in all cases. Characteristics of the semivariograms are given in the following column (Nugget effect, anisotropy parameters and slope which are used for prediction of predawn leaf water potential at any location in the field).

	Block	Model	N	Nugget effect (MPa²)	Anisotropy angle (°)	Anisotropy ratio (°)	Slope (MPa² m⁻¹)	RMSE (MPa)
2010	I	Linear	96	0.0015	121.9	1.795	3.3E-05	0.04
	D	Linear	96	0.026	95.7	1.798	4.2E-04	0.17
2011	I	Linear	96	0.002	95.93	2.000	9.3E-06	0.04
	D	Linear	96	0.010	119.7	1.181	2.0E-04	0.105

RMSE, root mean square error of prediction for each semivariogram. *See details in Webster, R., & Burgess, T. M. (1983). Spatial variation in soil and the role of kriging. Agricultural Water Management, 6(2–3), 111–122.*

veraison stage, corresponding to the onset of ripening) using a continuous flow isotope ratio mass spectrometer (Model IsoPrime, GV Instruments, Manchester, UK). Must was obtained from bunches of the three central vines of each plot. The measurements were performed on both years (2010 and 2011) on each plot (including control plots). Results were expressed as $\delta^{13}C$ values using a secondary standard calibrated against the Vienna Pee Dee Belemnite as follows:

$$\delta^{13}C = [(Rs/Rb) - 1] \times 1000,$$

where Rs is the ratio $^{13}C/^{12}C$ of the sample and Rb is the $^{13}C/^{12}C$ of the standard (Farquhar et al., 1989).

In order to remove the impact of undesired spatial variations in soil water deficit on $\delta^{13}C$ determined for each genotype, a relationship first had to be drawn between $\delta^{13}C$ and $\Psi_{predawn}$ for control plants. Carbon isotope composition measured on must was plotted against predawn leaf water potential measured at different periods of the cycle in 2010 and 2011 (Fig. 4.4A and B). The best fit with a linear regression model was obtained with $\Psi_{predawn}$ measurements performed at the date, specific to each year, when minimum seasonal values of $\Psi_{predawn}$ were noted (Fig. 4.4C).

4.3.5 $\delta^{13}C$ correction procedure and effect of irrigation regimes on $\delta^{13}C$ measured on the whole progeny

This relationship was then used for all other plot locations where $\Psi_{predawn}$ was interpolated, yielding $\delta^{13}C_{krig\_(xi,yi)}$, an estimate of $\delta^{13}C$ that would be observed for Grenache with $\Psi_{predawn}$ as predicted at location (xi,yi):

$$\delta^{13}C_{krig\_(xi,yi)} = a \times \psi_{predawn(xi,yi)} + b$$

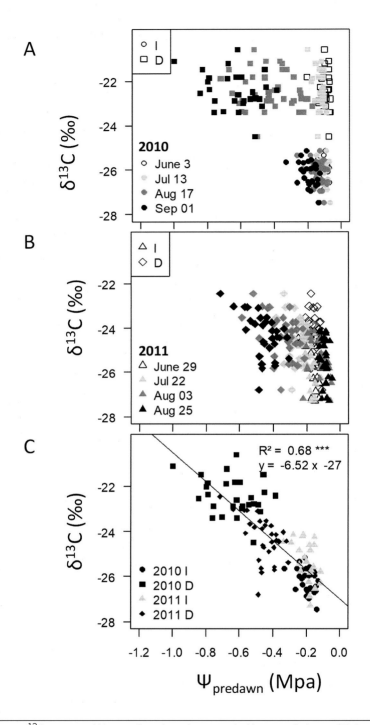

FIGURE 4.4

Relationship between $\delta^{13}C$ measured in must from bunches sampled in control plots at 35 days after *veraison* and predawn leaf water potential ($\Psi_{predawn}$) measured at four dates in 2010 (A) and 2011 (B) on the same control plots. (C) Regression of $\delta^{13}C$ on minimal predawn leaf water potential, for each year × watering scenario, which was used to calculate the correction to apply on $\delta^{13}C$ from local soil water status ($\delta^{13}C = -6.52*\Psi_{predawn} -27$, $n = 118$). 2010I and 2010D correspond to 2010-09-01, 2011I corresponds to 2011-07-22, and 2011D corresponds to 2011-08-25. *I*, irrigated; *D*, non-irrigated.

where $\Psi_{predawn\ (xi,yi)}$ is $\Psi_{predawn}$ predicted by kriging at location (xi, yi) and a and b are the regression coefficients drawn from the relationship between $\Psi_{predawn}$ and $\delta^{13}C$ in control plots ($a = -6,52$ and $b = -27$, Fig. 4.4C). All $\delta^{13}C_{krig\_(xi,yi)}$ values predicted in one year in one block were averaged to give $\delta^{13}C_{block}$, a prediction at the block level of the mean effect of all soil water scenarios in individual plots.

Final correction on raw $\delta^{13}C$ values for each genotype was applied by removing the predicted, local deviation in $\delta^{13}C$ ($\delta^{13}C_{krig\_(xi,yi)}$) relative to $\delta^{13}C_{krig\_block}$, due to local deviation in $\Psi_{predawn}$ compared to mean $\Psi_{predawn}$ value in the block, as follows:

$$\delta^{13}C_{geno\_cor\_(xi,yi)} = \delta^{13}C_{geno\_raw\_(xi,yi)} - \left(\delta^{13}C_{krig\_(xi,yi)} - \delta^{13}C_{block}\right)$$

where $\delta^{13}C_{geno\_cor\_(xi,yi)}$ and $\delta^{13}C_{geno\_raw\_(xi,yi)}$ are the carbon isotope ratio corrected and measured for a genotype at location (xi,yi).

The two irrigation regimes resulted in large differences in raw values of $\delta^{13}C$ measured on the progeny ranging in 2010 from -27.7 to -24.5‰ in the irrigated block I and from -26 to -21‰ in the non-irrigated block D (from -28 to -23‰ in I and from -27.2 to -22.8‰ in D in 2011). Differences between irrigated and non-irrigated blocks were significant in 2010 and 2011 (Fig. 4.5A). Results were qualitatively similar after correction for the estimated effect of spatial variations in soil water deficit and the scattering of data was maintained (Fig. 4.5B). Correction for spatial variation in $\Psi_{predawn}$ mostly affected the ranking of genotypic values within the non-irrigated block as indicated by a higher scattering of points when plotting raw versus corrected values of $\delta^{13}C$ (Fig. 4.5C).

4.3.6 Genetic variability of $\delta^{13}C$ and QTL detection

An ANOVA was conducted for the whole progeny on each year to test the effect of genotype on carbon isotope composition in must considering either raw ($\delta^{13}C\_raw$) or corrected ($\delta^{13}C\_cor$) values (with one replicate per block for each of the 188 genotypes). Broad-sense heritability was calculated from genotypic and residual variances obtained by fitting mixed linear models including a fixed effect of the block, respectively, for each year (R/lme4). Best linear unbiased predictors (BLUPs) of genotypic values were then extracted on each year for $\delta^{13}C\_raw$ and $\delta^{13}C\_cor$ using mixed-models. Models tested included a random effect of the genotype (G), completed or not by the fixed effects of the block. The best model was chosen based on the lowest Akaike Information Criterion. The variance components were then used to estimate broad-sense heritability (H^2) as: $H^2 = varG/(varG + varRes/n)$ with varG the genotypic variance, varRes the residual variance, n the number of replicates per genotype. In 2010, a significant genotypic effect could be detected on corrected $\delta^{13}C\_cor$ values ($P = .04$) (Fig. 4.6) which could not be revealed on raw values. This was consistent with an increase in broad-sense heritability (H^2) on $\delta^{13}C\_cor$ values as compared to raw values (Fig. 4.6). However, in 2011, no significant effect of the genotype on $\delta^{13}C$ was detected whether $\delta^{13}C$ was corrected or not. This suggests that the higher contrast in soil water scenarios between I and D blocks experienced in 2010 compared to 2011 favored a better characterization of genotypic effects.

QTL analyses were further carried out by Simple Interval Mapping (Lander & Botstein, 1989) for $\delta^{13}C\_raw$ and $\delta^{13}C\_cor$ using 3961 fully informative markers derived from GBS (consensus genetic map previously built in Brault et al. (2021)). Analyses were performed (i) on each year, using the genotypic means (BLUPs) calculated from data collected on the two blocks jointly with a mixed-model

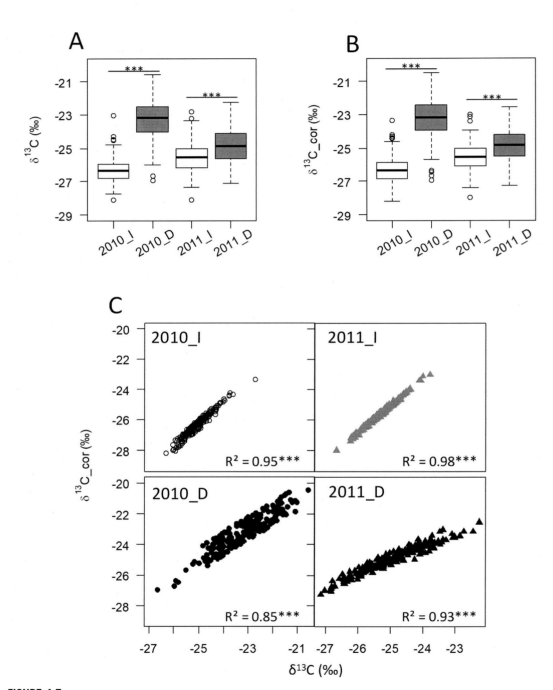

FIGURE 4.5

Distribution of $\delta^{13}C$ observed in two years (2010 and 2011) for irrigated (I) and non-irrigated (D) blocks in the Syrah × Grenache progeny. (A) Boxplots of raw values of carbon isotope composition ($\delta^{13}C$). (B) Boxplots of carbon isotope composition values corrected for local deviation from mean soil availability in the block ($\delta^{13}C\_cor$). (C) Relationship between $\delta^{13}C\_cor$ and $\delta^{13}C$ for each year and block. Significance is indicated as follows: ***$P < 10^{-3}$.

4.3 Detection of genetic variability

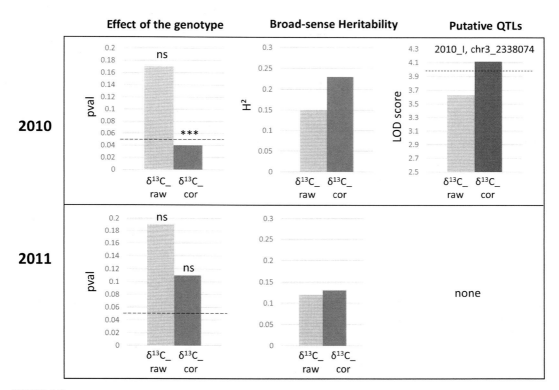

FIGURE 4.6

Effect of the genotype on raw $\delta^{13}C$ values measured in musts for the whole progeny as compared to corrected data ($\delta^{13}C\_cor$) for each year (2010 and 2011), broad-sense heritability (H^2), and QTL detection. P-values are indicated with their significance (*, Pval<.05). QTL analyses were performed using R/qtl version 1.46–2 (Broman et al., 2003). QTL positions were determined as those with LOD peaks above the threshold, with LOD-1 confidence intervals. On linkage groups 3, LOD score was substantially increased for $\delta^{13}C\_cor$ as compared to $\delta^{13}C$, yielding one putative QTL in the analysis carried out on genotypic values in the irrigated block (2010_I). Chr3_2338074 is the name of the nearest markers for the corresponding locus.

accounting for the block effect, (ii) on each year, using the genotypic values from each block separately. A LOD (logarithm of the odds) score was computed every 0.1 cM, then 1000 permutations were performed to determine the LOD threshold so that the family-wise error rate was controlled at 5%. QTLs were declared significant if their LOD score were above the LOD threshold, and putative if their LOD score were above 4.

In agreement with the analysis of genotypic effects on $\delta^{13}C$, no QTL was detected for raw $\delta^{13}C$ values. By contrast, after the correction accounting for spatial variation in soil water deficit, LOD score was substantially increased yielding one putative QTL (detected in the driest year 2010 and corresponding to the irrigated block) on linkage group 3 (Fig. 4.6).

4.4 Main outcomes
4.4.1 Taking into account spatial heterogeneity of soil water deficit within blocks improves statistical power for QTL detection

As a main outcome of this study, it is shown that modeling spatial heterogeneity within blocks may help to detect genotypic effect. This is illustrated for WUE, a trait of huge interest when breeding for drought tolerance in many crops including grapevine (Morison et al., 2008). Kriging offers several ways for building statistical models of spatial heterogeneity based on nested observations spread over an area of interest. Thereby, soil heterogeneity can be handled in extended, cultivated surface as required to develop quantitative genetics on large populations of perennials. Here, we used predawn water potential as a proxy for soil water availability as sensed by the plant itself, but other methods like networks of soil water probes can be implemented (Vereecken et al., 2014). To our knowledge, this study is the first attempt to incorporate kriging methods in quantitative genetics analysis in grapevine. It can be regarded as successful considering the gain in variance explained in ANOVA (Fig. 4.6) once raw $\delta^{13}C$ data were corrected for variation in local soil water conditions. Detection of QTLs should therefore be facilitated as evidenced with the gain in LOD score (Fig. 4.6). However, QTL detection was limited to certain irrigation regimes and years, and only reached putative but not absolute statistical thresholds. It can therefore be questioned whether the spatial distribution of control plots was dense enough to fully capture variations in soil water availability. Moreover, application of other kriging methods than the one used in this study could have further improved the genetic analysis. Co-kriging methods, using co-variable measured on the same field, could also help to better predict $\Psi_{predawn}$ for plots where it was not characterized (Knotters et al., 1995). The soil resistivity map that was established in the present study could be of valuable interest in that sense.

4.4.2 QTL detection under field conditions reveals new genomic regions as compared to those obtained on potted plants in phenotyping platforms with controlled conditions

Putative QTLs detected for a proxy of WUE in field conditions in this study were localized on regions that were not identified in a previous study in controlled conditions for WUE on potted plants (Coupel-Ledru et al., 2016). Pot conditions, especially the rooting volume, can influence plant response to drought (Passioura, 2006). Specific underlying physiological processes may dominate in pots with early soil water deficit, possibly resulting in detection of QTLs different from those detected in field conditions on aged plants. Moreover, in the phenotyping experiment on potted plants, WUE was calculated as the ratio of vegetative biomass increase to the amount of water used by plants during a period of several days (Coupel-Ledru et al., 2016) whereas WUE was estimated using $\delta^{13}C$ in must in the present study. Physiological mechanisms grounding the use of $\delta^{13}C$ mostly rest on stomatal control of gas exchange and may therefore reveal genetic differences in WUE occurring in the daytime. By contrast, direct determination of WUE as g of dry matter produced per g of water transpired over several days may also depend on water-use during the night (Coupel-Ledru et al., 2016). This may also explain why different QTLs were detected in field and phenotyping platform conditions.

In the present study, a QTL for WUE was detected in 2010 in irrigated conditions, but not in the non-irrigated block. This result was not expected considering the low correction provided by the weakly uneven distribution of soil water availability in irrigated compared to non-irrigated conditions. However, other studies on grapevine evidenced genetic contrasts in WUE under well-watered conditions (Bota et al., 2016; Coupel-Ledru et al., 2016; Gibberd et al., 2001; Tomás et al., 2014). A possible reason why genetic determinism of WUE could be more easily detected in well-watered than in soil water deficit conditions is that stomatal conductance is decreased by soil drying and relative differences between genotypes are reduced while the relative error in measurement increases. It may also be due to differences in maximal stomatal conductance between genotypes which were not accompanied by parallel differences in photosynthesis rate. Experiments are worth being performed to test this hypothesis. Correction of $\delta^{13}C$ for spatial variation in soil water availability within the blocks in 2011, which was the year with milder summer drought, did not improve the genetic analyses. This suggests that precipitations may have blurred the relationship between $\delta^{13}C$ and WUE.

4.4.3 Minimal predawn leaf water potential in control plots was the best predictor of $\delta^{13}C$ measured in must

In this study, several correlations were tested between $\delta^{13}C$ measured in must sampled in control plots and $\Psi_{predawn}$ measured on leaves in the same plots (planted with cv Grenache) at different times following *veraison*. Rough correlations were logically observed in all cases suggesting that the $\delta^{13}C$ signature in must integrated the behavior all over the period following *veraison*. This period corresponds to sharp accumulation of sugars in berries and is strongly dependent on photosynthesis rate. Any change in soil water availability during this period is therefore expected to impact photosynthesis and hence $\delta^{13}C$ in must. However, the best linear correlation between $\delta^{13}C$ in must and $\Psi_{predawn}$ for control plots was observed at the time when minimal $\Psi_{predawn}$ values were observed in each scenario (irrigation treatment × year) (Fig. 4.4). This suggests that minimal $\Psi_{predawn}$ was representative of mean soil water availability over the whole period following *veraison* in each condition. By contrast, at any other given date, rainfall may have restored soil water reserves resulting in transiently high $\Psi_{predawn}$ non-representative of soil drying over the plant cycle. This result may be specific to the present study and remains to be confronted to other climate and soil conditions, although four contrasting soil water scenarios were explored in this study.

4.5 Conclusions

This study is a preliminary attempt to model and take into account spatial heterogeneity in soil water conditions for comparison of genotypes grown under field conditions. Although further developments are required, this procedure opens promising ways to detect QTLs for traits related to drought tolerance like $\delta^{13}C$ using corrected values while no QTL was detected on raw values. This confirmed the relevance of geostatistical methods to correct for spatial heterogeneity of soil characteristics intrinsic of field experiments.

Acknowledgments

This work was supported by the French program ANR-09-GENM-024-002. AC-L received a PhD grant from the French Government.

In memory of our dear colleague and friend Eric Lebon.

References

Allen, R. G., Pereira, L. S., Raes, D., & Smith, M. (1998). Crop evapotranspiration-Guidelines for computing crop water requirements-FAO Irrigation and drainage paper 56. *FAO, 300*(9), 1–15.

Améglio, T., Archer, P., Cohen, M., Valancogne, C., Daudet, F., Dayau, S., & Cruiziat, P. (1999). Significance and limits in the use of predawn leaf water potential for tree irrigation. *Plant and Soil, 207*(2), 155–167.

Andrenelli, M. C., Magini, S., Pellegrini, S., Perria, R., Vignozzi, N., & Costantini, E. A. C. (2013). The use of the ARP© system to reduce the costs of soil survey for precision viticulture. *Journal of Applied Geophysics, 99*, 24–34. https://doi.org/10.1016/j.jappgeo.2013.09.012

Araus, J. L., & Cairns, J. E. (2014). Field high-throughput phenotyping: The new crop breeding frontier. *Trends in Plant Science, 19*(1), 52–61. https://doi.org/10.1016/j.tplants.2013.09.008

Bchir, A., Escalona, J. M., Gallé, A., Hernández-Montes, E., Tortosa, I., Braham, M., & Medrano, H. (2016). Carbon isotope discrimination ($\delta^{13}C$) as an indicator of vine water status and water use efficiency (WUE): Looking for the most representative sample and sampling time. *Agricultural Water Management, 167*, 11–20. https://doi.org/10.1016/j.agwat.2015.12.018

Bota, J., Tomás, M., Flexas, J., Medrano, H., & Escalona, J. M. (2016). Differences among grapevine cultivars in their stomatal behavior and water use efficiency under progressive water stress. *Agricultural Water Management, 164*, 91–99. https://doi.org/10.1016/j.agwat.2015.07.016

Bramley, R. G. V., & Hamilton, R. P. (2004). Understanding variability in winegrape production systems. *Australian Journal of Grape and Wine Research, 10*(1), 32–45. https://doi.org/10.1111/j.1755-0238.2004.tb00006.x

Brault, C., Doligez, A., Cunff, L. le, Coupel-Ledru, A., Simonneau, T., Chiquet, J., This, P., & Flutre, T. (2021). Harnessing multivariate, penalized regression methods for genomic prediction and QTL detection to cope with climate change affecting grapevine. *G3 Genes|Genomes|Genetics, 11*(9). https://doi.org/10.1101/2020.10.26.355420

Brillante, L., Bois, B., Mathieu, O., Bichet, V., Michot, D., & Lévêque, J. (2014). Monitoring soil volume wetness in heterogeneous soils by electrical resistivity. A field-based pedotransfer function. *Journal of Hydrology, 516*, 56–66. https://doi.org/10.1016/j.jhydrol.2014.01.052

Brillante, L., Mathieu, O., Bois, B., van Leeuwen, C., & Lévêque, J. (2015). The use of soil electrical resistivity to monitor plant and soil water relationships in vineyards. *Soil, 1*(1), 273–286. https://doi.org/10.5194/soil-1-273-2015

Broman, K. W., Wu, H., Sen, Ś., & Churchill, G. A. (2003). R/qtl: QTL mapping in experimental crosses. *Bioinformatics, 19*(7), 889–890. https://doi.org/10.1093/bioinformatics/btg112

Collins, D., Benedict, C., Bary, A., Cogger, C., Collins, D., Benedict, C., Bary, A., & Cogger, C. (2015). Soil parameter mapping and ad hoc power analysis to increase blocking efficiency prior to establishing a long-term field experiment. *The Scientific World Journal*, 8. https://doi.org/10.1155/2015/205392

Coupel-Ledru, A., Lebon, E., Christophe, A., Doligez, A., Cabrera-Bosquet, L., Péchier, P., Hamard, P., This, P., & Simonneau, T. (2014). Genetic variation in a grapevine progeny (*Vitis vinifera* L. cvs Grenache×Syrah) reveals inconsistencies between maintenance of daytime leaf water potential and response of transpiration rate under drought. *Journal of Experimental Botany, 65*(21), 6205–6218. https://doi.org/10.1093/jxb/eru228

References

Coupel-Ledru, A., Lebon, E., Christophe, A., Gallo, A., Gago, P., Pantin, F., Doligez, A., & Simonneau, T. (2016). Reduced nighttime transpiration is a relevant breeding target for high water-use efficiency in grapevine. *Proceedings of the National Academy of Sciences, 113*(32), 8963−8968. https://doi.org/10.1073/pnas.1600826113

Dufour, A., Gadallah, F., Wagner, H. H., Guisan, A., & Buttler, A. (2006). Plant species richness and environmental heterogeneity in a mountain landscape: Effects of variability and spatial configuration. *Ecography, 29*(4), 573−584. https://doi.org/10.1111/j.0906-7590.2006.04605.x

Farquhar, G. D., Ehleringer, J. R., & Hubick, K. T. (1989). Carbon isotope discrimination and photosynthesis. *Annual Review of Plant Physiology and Plant Molecular Biology, 40*, 503−537.

Fournier-Level, A., Le Cunff, L., Gomez, C., Doligez, A., Ageorges, A., Roux, C., Bertrand, Y., Souquet, J.-M., Cheynier, V., & This, P. (2009). Quantitative genetic bases of anthocyanin variation in grape (*Vitis vinifera* L. ssp. *sativa*) berry: A quantitative trait locus to quantitative trait nucleotide integrated study. *Genetics, 183*(3), 1127−1139. https://doi.org/10.1534/genetics.109.103929

Gaudillère, J. P., Van Leeuwen, C., & Ollat, N. (2002). Carbon isotope composition of sugars in grapevine, an integrated indicator of vineyard water status. *Journal of Experimental Botany, 53*(369), 757−763.

Geny, L., Ollat, N., & Soyer, J.-P. (1998). Grapevine fruiting cuttings: Validation of an experimental system to study grapevine physiology. II. Study of grape development. *OENO One, 32*(2), 83−90. https://doi.org/10.20870/oeno-one.1998.32.2.1052

Gibberd, M. R., Walker, R. R., Blackmore, D. H., & Condon, A. G. (2001). Transpiration efficiency and carbon-isotope discrimination of grapevines grown under well-watered conditions in either glasshouse or vineyard. *Australian Journal of Grape and Wine Research, 7*(3), 110−117.

Gilad, E., Shachak, M., & Meron, E. (2007). Dynamics and spatial organization of plant communities in water-limited systems. *Theoretical Population Biology, 72*(2), 214−230. https://doi.org/10.1016/j.tpb.2007.05.002

Hartung, J., Wagener, J., Ruser, R., & Piepho, H.-P. (2019). Blocking and re-arrangement of pots in greenhouse experiments : Which approach is more effective? *Plant Methods, 15*(1), 143. https://doi.org/10.1186/s13007-019-0527-4

Jackson, R. B., & Caldwell, M. M. (1993). Geostatistical patterns of soil heterogeneity around individual perennial plants. *Journal of Ecology, 81*(4), 683−692. https://doi.org/10.2307/2261666

Kelly, V. R., & Canham, C. D. (1992). Resource heterogeneity in old fields. *Journal of Vegetation Science, 3*(4), 545−552. https://doi.org/10.2307/3235811

Klassen, S. P., Villa, J., Adamchuk, V., & Serraj, R. (2014). Soil mapping for improved phenotyping of drought resistance in lowland rice fields. *Field Crops Research, 167*, 112−118. https://doi.org/10.1016/j.fcr.2014.07.007

Knotters, M., Brus, D. J., & Oude Voshaar, J. H. (1995). A comparison of kriging, co-kriging and kriging combined with regression for spatial interpolation of horizon depth with censored observations. *Geoderma, 67*(3−4), 227−246. https://doi.org/10.1016/0016-7061(95)00011-C

Lander, E. S., & Botstein, D. (1989). Mapping Mendelian Factors underlying quantitative traits using RFLP linkage maps. *Genetics, 121*(1), 185−199.

van Leeuwen, C., Trégoat, O., Choné, X., Bois, B., Pernet, D., & Gaudillère, J.-P. (2009). Vine water status is a key factor in grape ripening and vintage quality for red Bordeaux wine. How can it be assessed for vineyard management purposes? *OENO One, 43*(3), 121−134. https://doi.org/10.20870/oeno-one.2009.43.3.798

Liu, Y., Boeck, H. J. D., Wellens, M. J., & Nijs, I. (2017). A simple method to vary soil heterogeneity in three dimensions in experimental mesocosms. *Ecological Research, 32*(2), 287−295. https://doi.org/10.1007/s11284-017-1435-6

Maestre, F. T., & Cortina, J. (2002). Spatial patterns of surface soil properties and vegetation in a Mediterranean semi-arid steppe. *Plant and Soil, 241*(2), 279−291. https://doi.org/10.1023/A:1016172308462

Marguerit, E., Brendel, O., Lebon, E., Leeuwen, C. V., & Ollat, N. (2012). Rootstock control of scion transpiration and its acclimation to water deficit are controlled by different genes. *New Phytologist, 194*(2), 416–429. https://doi.org/10.1111/j.1469-8137.2012.04059.x

Masuka, B., Araus, J. L., Das, B., Sonder, K., & Cairns, J. E. (2012). Phenotyping for abiotic stress tolerance in maize. *Journal of Integrative Plant Biology, 54*(4), 238–249. https://doi.org/10.1111/j.1744-7909.2012.01118.x

McClure, K. A., Sawler, J., Gardner, K. M., Money, D., & Myles, S. (2014). Genomics : A potential panacea for the perennial problem. *American Journal of Botany, 101*(10), 1780–1790. https://doi.org/10.3732/ajb.1400143

Medrano, H., Tomás, M., Martorell, S., Flexas, J., Hernández, E., Rosselló, J., Pou, A., Escalona, J.-M., & Bota, J. (2015). From leaf to whole-plant water use efficiency (WUE) in complex canopies : Limitations of leaf WUE as a selection target. *The Crop Journal, 3*(3), 220–228. https://doi.org/10.1016/j.cj.2015.04.002

Mertens, M. F., Pätzold, S., & Welp, G. (2008). Spatial heterogeneity of soil properties and its mapping with apparent electrical conductivity. *Journal of Plant Nutrition and Soil Science, 171*(2), 146–154. https://doi.org/10.1002/jpln.200625130

Monestiez, P., Goulard, M., & Charmet, G. (1994). Geostatistics for spatial genetic structures: Study of wild populations of perennial ryegrass. *Theoretical and Applied Genetics, 88*(1), 33–41.

Morison, J. I. L., Baker, N. R., Mullineaux, P. M., & Davies, W. J. (2008). Improving water use in crop production. *Philosophical Transactions of the Royal Society Biological Sciences, 363*(1491), 639–658. https://doi.org/10.1098/rstb.2007.2175

Nicolas, S. D., Péros, J.-P., Lacombe, T., Launay, A., Le Paslier, M.-C., Bérard, A., Mangin, B., Valière, S., Martins, F., Le Cunff, L., Laucou, V., Bacilieri, R., Dereeper, A., Chatelet, P., This, P., & Doligez, A. (2016). Genetic diversity, linkage disequilibrium and power of a large grapevine (*Vitis vinifera* L) diversity panel newly designed for association studies. *BMC Plant Biology, 16*(1), 74. https://doi.org/10.1186/s12870-016-0754-z

Passioura, J. B. (2006). The perils of pot experiments. *Functional Plant Biology, 33*(12), 1075–1079.

Poorter, H., Bühler, J., van Dusschoten, D., Climent, J., & Postma, J. A. (2012). Pot size matters : A meta-analysis of the effects of rooting volume on plant growth. *Functional Plant Biology, 39*(11), 839. https://doi.org/10.1071/FP12049

Pou, A., Medrano, H., Tomàs, M., Martorell, S., Ribas-Carbó, M., & Flexas, J. (2012). Anisohydric behaviour in grapevines results in better performance under moderate water stress and recovery than isohydric behaviour. *Plant and Soil, 359*(1–2), 335–349. https://doi.org/10.1007/s11104-012-1206-7

Rafalski, J. A. (2010). Association genetics in crop improvement. *Current Opinion in Plant Biology, 13*(2), 174–180. https://doi.org/10.1016/j.pbi.2009.12.004

Rincent, R., Moreau, L., Monod, H., Kuhn, E., Melchinger, A. E., Malvar, R. A., Moreno-Gonzalez, J., Nicolas, S., Madur, D., Combes, V., Dumas, F., Altmann, T., Brunel, D., Ouzunova, M., Flament, P., Dubreuil, P., Charcosset, A., & Mary-Huard, T. (2014). Recovering power in association mapping panels with variable levels of linkage disequilibrium. *Genetics, 197*(1), 375–387. https://doi.org/10.1534/genetics.113.159731

Rodríguez-Álvarez, M. X., Boer, M. P., van Eeuwijk, F. A., & Eilers, P. H. C. (2018). Correcting for spatial heterogeneity in plant breeding experiments with P-splines. *Spatial Statistics, 23*, 52–71. https://doi.org/10.1016/j.spasta.2017.10.003

Romano, N., & Santini, A. (1997). Effectiveness of using pedo-transfer functions to quantify the spatial variability of soil water retention characteristics. *Journal of Hydrology, 202*(1), 137–157. https://doi.org/10.1016/S0022-1694(97)00056-5

Rossi, R. E., Mulla, D. J., Journel, A. G., & Franz, E. H. (1992). Geostatistical Tools for modeling and Interpreting Ecological spatial dependence. *Ecological Monographs, 62*(2), 277–314. https://doi.org/10.2307/2937096

Santesteban, L. G., Miranda, C., Barbarin, I., & Royo, J. B. (2015). Application of the measurement of the natural abundance of stable isotopes in viticulture: A review. *Australian Journal of Grape and Wine Research, 21*(2), 157–167. https://doi.org/10.1111/ajgw.12124

Santesteban, L. G., Miranda, C., Urretavizcaya, I., & Royo, J. B. (2012). Carbon isotope ratio of whole berries as an estimator of plant water status in grapevine (*Vitis vinifera* L.) cv. 'Tempranillo. *Scientia Horticulturae, 146*, 7–13. https://doi.org/10.1016/j.scienta.2012.08.006

Simonneau, T., Lebon, E., Coupel-Ledru, A., Marguerit, E., Rossdeutsch, L., & Ollat, N. (2017). Adapting plant material to face water stress in vineyards : Which physiological targets for an optimal control of plant water status? *OENO One, 51*(2), 167–179. https://doi.org/10.20870/oeno-one.2016.0.0.1870

de Souza, C. R. (2005). Impact of deficit irrigation on water use efficiency and carbon isotope composition (13C) of field-grown grapevines under Mediterranean climate. *Journal of Experimental Botany, 56*(418), 2163–2172. https://doi.org/10.1093/jxb/eri216

Tetegan, M., Pasquier, C., Besson, A., Nicoullaud, B., Bouthier, A., Bourennane, H., Desbourdes, C., King, D., & Cousin, I. (2012). Field-scale estimation of the volume percentage of rock fragments in stony soils by electrical resistivity. *Catena, 92*, 67–74. https://doi.org/10.1016/j.catena.2011.09.005

Tomás, M., Medrano, H., Escalona, J. M., Martorell, S., Pou, A., Ribas-Carbó, M., & Flexas, J. (2014). Variability of water use efficiency in grapevines. *Environmental and Experimental Botany, 103*, 148–157. https://doi.org/10.1016/j.envexpbot.2013.09.003

Tomás, M., Medrano, H., Pou, A., Escalona, J. M., Martorell, S., Ribas-Carbó, M., & Flexas, J. (2012). Water-use efficiency in grapevine cultivars grown under controlled conditions: Effects of water stress at the leaf and whole-plant level. *Australian Journal of Grape and Wine Research, 18*(2), 164–172. https://doi.org/10.1111/j.1755-0238.2012.00184.x

Trenti, M., Lorenzi, S., Bianchedi, P. L., Grossi, D., Failla, O., Grando, M. S., & Emanuelli, F. (2021). Candidate genes and SNPs associated with stomatal conductance under drought stress in *Vitis*. *BMC Plant Biology, 21*(1), 7. https://doi.org/10.1186/s12870-020-02739-z

Turner, N. C. (2019). Imposing and maintaining soil water deficits in drought studies in pots. *Plant and Soil, 439*, 45–55. https://doi.org/10.1007/s11104-018-3893-1

Vereecken, H., Huisman, J. A., Pachepsky, Y., Montzka, C., van der Kruk, J., Bogena, H., Weihermüller, L., Herbst, M., Martinez, G., & Vanderborght, J. (2014). On the spatio-temporal dynamics of soil moisture at the field scale. *Journal of Hydrology, 516*, 76–96. https://doi.org/10.1016/j.jhydrol.2013.11.061

Vezzulli, S., Doligez, A., & Bellin, D. (2019). Molecular mapping of grapevine genes. In D. Cantu, & M. A. Walker (Eds.), *The grape genome* (pp. 103–136). Springer International Publishing. https://doi.org/10.1007/978-3-030-18601-2_7

Webster, R., & Burgess, T. M. (1983). Spatial variation in soil and the role of kriging. *Agricultural Water Management, 6*(2–3), 111–122.

Webster, R., & Oliver, M. A. (1990). *Statistical methods in soil and land resource survey* (p. 316). Oxford: Oxford University Press. CABDirect2.

Yu, R., & Kurtural, S. K. (2020). Proximal sensing of soil electrical conductivity provides a link to soil-plant water relationships and supports the identification of plant water status zones in vineyards. *Frontiers in Plant Science, 11*. https://doi.org/10.3389/fpls.2020.00244

CHAPTER 5

Soil management in sustainable viticultural systems: an agroecological evaluation

Johanna Döring[1], Matthias Friedel[1], Maximilian Hendgen[2], Manfred Stoll[1] and Randolf Kauer[1]

[1]*Hochschule Geisenheim University, Department of General and Organic Viticulture, Geisenheim, Germany;*
[2]*Hochschule Geisenheim University, Department of Soil Science and Plant Nutrition, Geisenheim, Germany*

5.1 Introduction

Ecosystems provide a wide range of goods and services to the benefits of mankind (Costanza et al., 1997). "These include provisioning services such as food, water, timber, and fiber; regulating services that affect climate, floods, disease, wastes, and water quality; cultural services that provide recreational, aesthetic, and spiritual benefits; and supporting services such as soil formation, photosynthesis, and nutrient cycling." (Reid et al., 2005) The concept of ecosystem services (ESs) aims to provide an informational framework to ascribe value to these benefits. The attributed values of the ESs should facilitate decision support for the implementation of sustainable practices. The preservation and enhancement of soil-based ES is the goal of sustainable soil management. Under agricultural use, these services comprise the ability to sustainably provide plant yield, conserve biodiversity, ensure the supply of high-quality water, control erosion, pests, and diseases, and regulate the climate (Bünemann et al., 2018).

The ability of soils to deliver ES depends on a complex interplay of soil processes or functions (carbon and nutrient cycling, water cycling, habitat provision, and soil structure maintenance) and local conditions (geology, soil type, topography, and climate). The interaction of these factors is strongly influenced by land use and its associated management decisions (Fig. 5.1).

Agricultural and viticultural use of soils shift the provision of ES mainly toward the provision of plant yield and health, and rearrange soil functions to achieve these aims. In the context of viticulture, the result of the interaction of topography, geology, and human intervention that shapes ES provision additionally forms an important part of the terroir concept. The ability to deliver one individual ES often stands in conflict to provide others (Costanza et al., 1997), and it remains "unclear how and to what extent agriculture can meet all expectations relating to environmental sustainability simultaneously" (Schulte et al., 2014). Overall sustainability assessment of management practices is extremely complicated in a practical context, as there are strong interactions between the static (geology, topography, climate, and environment) and the dynamic (management) factors that influence ES provision. In addition, the contribution of individual ES to overall sustainability is hard to define.

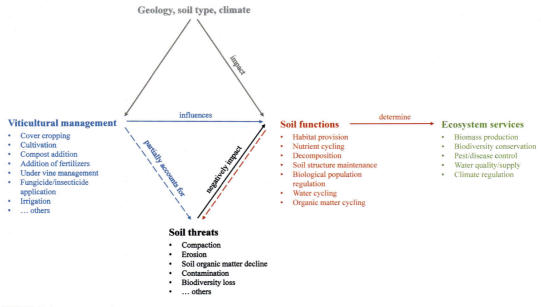

FIGURE 5.1

Linkages between viticultural management, soil threats, soil functions, and soil-based ecosystem services (ESs).

Further developed from the scheme presented by Kibblewhite, M. G., Ritz, K., Swift, M. J. (2008). Soil health in agricultural systems. Philosophical Transactions of the Royal Society of London - B, *363 (1492), 685–701. Doi: 10.1098/rstb.2007.2178 and modified by Brussaard, L. (2012). Ecosystem services provided by the soil biota. In: D. H. Wall, R. D. Bardgett, V. Behan-Pelletier, J. E. Herric, H. Jones, K. Ritz, J. Six, D. R. Strong, W. H. van der Putten.* Soil Ecology and Ecosystem Services *(pp. 45–58). Oxford, UK: Oxford University Press, and Bünemann, E. K., Bongiorno, G., Bai, Z., Creamer, R. E., de Deyn, G., de Goede, R., et al. (2018. Soil quality - a critical review.* Soil Biology and Biochemistry, *120, 105–125. Doi: 10.1016/j.soilbio.2018.01.030.*

In a holistic approach, the performance and sustainability of an economic sector can be judged by the extent to which the needs of people are met without transgressing planetary boundaries (Raworth, 2018).

To ensure the sustainability of agricultural production, the integration of good agricultural practices (GAPs) into national legislation of member states has been demanded by the European Union. A "GAP approach aims at applying available knowledge addressing environmental, economic, and social sustainability dimensions for on-farm production and postproduction processes, resulting in safe and healthy food and nonfood agricultural products." (FAO Agriculture Department, 2003). GAPs form the foundation of present integrated agriculture (Flint and van den Bosch, 1981). Meanwhile concepts of integrated pest management encompass other elements of agricultural production including integrated soil management, integrated crop nutrition, integrated crop protection, as well as the integration of practices toward water use and protection, energy efficiency, and environmental conservation (European Initative for Sustainable Development in Agriculture EISA, 2015). The guidelines of the *code of good practice* are compulsive for all winegrowers in Europe, thus all the principles concerning soil management in viticulture are also essential for organic as well as biodynamic growers.

Beyond integrated (INT) crop production, the management systems of organic (ORG) and biodynamic (BD) farming address sustainability in historically grown ways that have already exerted a large influence on the practices of integrated crop production, but remain different in many key aspects of crop protection and soil management. Worldwide, a growing number of countries have adopted mutually accepted legal standards on organic production and certification (European Parliament and Council of the European Union, 2018; Schlatter et al. 2020). The emphasis on the delivery of individual ES in organic farming is shifted away from the rather productivity-oriented approach of integrated farming and takes into account social and planetary boundaries (European Parliament and Council of the European Union, 2018). The definition of organic farming by the FAO underlines the importance of agroecosystem health, including biodiversity, biological cycles, and soil biological activity (FAO Committee on Agriculture, 1999). As a result, the use of herbicides, synthetic nitrogen (N) fertilizers, genetically modified organisms (GMOs), and products derived from GMOs is forbidden in organic farming. The use of synthetic materials and off-farm inputs is restricted and reduced to a minimum, while local nutrient cycling plays an important role. Cover crops increase biodiversity and play an important role within soil management strategies in ORG/BD viticulture (Kauer & Fader, 2015). The cover cropping strategy is highly dependent on climate, soil type, soil water-holding capacity, and the distribution of precipitation throughout the growing season. Strategies can vary from a complete green cover within the rows to the use of winter cover crops combined with open soil throughout the summer in case water availability is restricted during the growing season. Cover crop rotation, in which perennial cover crops and winter cover crops are planted in alternating rows, is one characteristic strategy for cool climate ORG/BD viticulture. The rows of the perennial cover crops serve as driving rows, while the rows planted with winter cover crops are tilled in spring to ensure soil organic matter (SOM) mineralization and limit water consumption. Rows of winter and perennial cover crop are swapped every three to five years. This system results in a comparatively high SOM content in all rows while providing a sufficient nutrient supply to the vines. By the selection of adapted species mixtures or by the cover crop management (e.g., mulching or rolling), water consumption of cover crops can be regulated throughout the season (Delpuech & Metay, 2018; Kauer & Fader, 2015).

All the described principles and practices of organic viticulture are also valid for biodynamic viticulture (European Commission, 2021; European Parliament and Council of the European Union, 2018). Biodynamic farming was first mentioned in 1924 by Rudolf Steiner in his agricultural course and was one of the first movements toward organic agriculture (Steiner, 2005). One major difference between organic and biodynamic management is the application of the characteristic biodynamic preparations. They consist of field spray (horn manure and horn silica) and compost preparations (yarrow blossoms, chamomile, stinging nettle, oak bark, dandelion, and valerian). Respecting and including natural rhythms into the everyday work is another essential characteristic of biodynamic farming (Masson & Masson, 2013).

The simultaneous variation of a large set of management parameters in ORG, INT, and BD systems, their economic importance, and the worldwide comparable legal frameworks regulating their implementation have sparked significant research efforts on system comparisons around the globe.

Data from these experiments provide a wealth of comparable data sets allowing for the evaluation of the sustainability of management regimes and individual practices. In this chapter, we will attempt to evaluate the effects associated with INT, ORG, and BD management regimes on the provision of individual soil ES and overall sustainability of soil management in viticulture.

5.2 Sustainable management systems and their properties toward the avoidance of soil threats and the provision of soil ES

Vineyard sites are often established on less fertile and shallow soils, often on more or less steep slopes, characterized by low water-holding capacity, high sun exposure, and low N availability (Hendgen et al., 2018; Lazcano et al., 2020). Particularities such as frequent tillage, mechanical undervine weeding, a high frequency of tractor overpasses, high pesticide inputs, and irrigation add to the threats to soil ES generally prevailing in agriculture, forming a complex pattern of risks associated with viticultural management. The threats for soil health and the provision of ES associated with viticultural management comprise the decline of SOM, soil compaction and erosion, the contamination of soils with plant protection agents, salinization and overfertilization, and a loss of biodiversity (Bindraban et al., 2012). These threats can be linked to individual management practices and their interactions and inhibit a range of soil functions that are vital to the provision of ES. Taken together, viticultural soils are particularly at risk for degradation. The threats exhibited by certain viticultural practices interact with other practices and among each other, and the site- and climate-specific conditions. Thus, an assessment of the sustainability of individual management practices needs to be case specific.

In addition to the particular threats associated with viticultural management, there are some opportunities for ES provision, which should be considered in the sustainability assessment of viticultural practices. The first point is that the grapevine is a relatively stress-tolerant, deep rooting plant that does not have a very high nutrient demand. In fact, a certain degree of stress might benefit wine sensory quality (Gambetta et al., 2020). This facilitates practices like cover cropping which are crucial to the provision of ES. The second point might be labeled the yield opportunity: Unlike in most other crops, upper limits for yields have historically been set in many wine regions mostly in Europe, but also worldwide to combat oversupply and improve wine quality on a regional scale (Robinson, 2015). Yield is indeed associated to correlate negatively with wine quality. The diminished demand to provide yield opens new possibilities for the provision of alternative ES in viticulture.

Sustainable management of soil as a resource and thus avoidance of soil threats and maintenance of soil ES are major goals of GAPs on soil management and soil protection (FAO Agriculture Department, 2003). This is achieved by reducing the threats to soil functions and favoring soil ES by the adaptation of viticultural management.

5.2.1 Avoidance of soil compaction

Avoidance of soil compaction is crucial to improve infiltration, percolation, and root penetration (Li et al., 2009; Zimmermann & Kardos, 1961) and reduce erosion (Capello et al., 2020). Compaction can be reduced by decreasing the number of overpasses implementing machine combinations, by the use of optimized equipment (i.e., above row machines or multiline sprayers), by the reduction of tire pressure, and by the use of cover crops, which stabilize soils against compaction and enhance soil life, and break up compacted soil layers (Alakukku et al., 2003; Chen & Weil, 2010; Lagacherie et al., 2006; Way et al., 2009). Due to the increased frequency of plant protection applications in ORG compared to INT viticulture, a higher risk for soil compaction might be expected in these systems (Beni & Rossi, 2009; Polge de Combret-Champart et al., 2013). In line with this hypothesis, Coll et al. (2011) reported an increase in soil bulk density and Beni and Rossi (2009) described a higher penetration resistance in ORG compared to INT vineyards in southern France and Italy, respectively. However, soil compaction depends not only on the frequency of tractor passes but also on the ratio of absolute load to the resistance of the soil structure as well as on soil loosening countermeasures.

ORG and BD farming did not lead to higher soil compaction in a long-term systems comparison trial in Germany comparing INT, ORG, and BD viticulture (INBIODYN trial) (Döring et al., 2015; Hendgen et al., 2020) although ORG and BD treatments both had on average one more tractor overpass per year and row compared to the INT, comparable to other systems comparisons (Beni & Rossi, 2009). Instead, bulk density was highest in the soil of INT plots (1.53 g cm^3 in average) and differed from BD management (1.47 g cm^3 in average), which showed the lowest bulk density. This was attributed to the type of cover crops used in ORG and BD (multispecies, legume-rich mixture) and INT (sward) (Döring et al., 2015) (Fig. 5.2).

Species-rich cover crops of the ORG/BD treatment not only reduced soil compaction but were also able to restore soil structure more effectively compared to sward, probably due to the more complex and deeper root system. A positive effect of cover crops on soil structure, which is based on a combination of aboveground cushioning effect of the plant biomass, a soil aggregate stabilization effect by the root system, and a regular input of organic carbon is widely recognized (Capello et al., 2019; Hartwig & Ammon, 2002). The differences between these studies may be explained by the limited cover cropping options in warmer and drier climates, as well as a more frequent and deeper tillage in the ORG treatments of Coll et al. (2011).

FIGURE 5.2

Integrated management on the left versus BD management on the right in the long-term systems comparison trial in Geisenheim, Germany, in July 2017.

Credit: © J. Döring.

5.2.2 Erosion control

The risk of water erosion is highly increasing with extreme weather events considered as a future climate risk factor and is of particular importance in steep slope viticulture where usually the fraction of plant-available water will be reduced (Novara et al., 2018). Erosion can cause serious soil and nutrient losses on slopes, resulting in lasting damage to soil functions, such as habitat provision, element cycling, maintenance of soil structure, and water infiltration and percolation (Bünemann et al., 2018). It is known that cultivated soils are more susceptible than undisturbed, vegetated soils. Since mechanical undervine management is widely used in ORG/BD viticulture in contrast to INT viticulture, where herbicides may also be applied, it can be hypothesized that ORG/BD are more susceptible to erosion, especially in steep slope sites. Therefore, undervine management in ORG and BD viticulture must find alternatives to mechanical treatment such as usage of different covers, i.e., straw, mulch, or undervine cover cropping. Several mixtures including different clover species or hawkweed are often used as competitive undervine cover crops in ORG viticulture (Moneyron et al., 2017), and their selection can be adapted to local conditions (Chou & Vanden Heuvel, 2019).

5.2.3 Water quality and supply

Water quality and supply are threatened by water use through irrigation and agricultural inputs of nitrogen (N) and phosphate.

The environmental safety margins for planetary N and phosphorus (P) flows have already been exceeded and, in particular in the case of N, soils under agricultural use are the major contributor to this excess (Kopittke et al., 2021). This is aggravated by a seemingly decreasing N use efficiency of worldwide agriculture (Lassaletta et al., 2014). The N demand and use efficiency of grapevines is low compared to other crops. To prevent leaching of nitrates, N should be provided to the vines when demand is high, i.e., between flowering and veraison (Löhnertz, 1988) or after harvest, if the timespan between harvest and leaf fall is long enough. This can be ensured by timely tillage to provide enough time for SOM mineralization, or by applying mineral fertilizers. The timing of tillage and fertilization depends on local conditions, especially expected rainfall and soil temperatures before flowering. Repeated and fine tillage accelerates SOM mineralization. Mineral fertilizers have the advantage to be more readily plant available and can thus be deployed in a more targeted way. On the other hand, fertilization via (soil) organic matter ensures a continuous, slow supply of N, which might be taken up better by the vines. When plant demand is low and water balance is positive (i.e., winter in most viticultural regions), nitrate may leach into the groundwater. Climate change contributes to the process, as rising winter soil temperatures increase SOM mineralization (Jabloun et al., 2015) by stimulation of microbial activity. N_{min} monitoring data from the long-term systems comparison trial in Geisenheim, Germany, suggest that the risk of fall N_{min} values surpassing the critical level of 50 kg/ha may indeed be elevated in ORG and BD systems, especially when tillage or seeding of winter cover crops was conducted late in the season. Hence, vineyard managers working in ORG and BD systems need to take special care to avoid tillage at the end of the season and allow for the timely establishment of winter cover crops with a high ability to take up N, preferably sown by direct drill.

Phosphorus has a poor mobility in soils, but presents a major threat to freshwater ecosystems and oceans if topsoil is lost by erosion. Consequently, all measures reducing erosion also lower the risk of P contamination of water bodies. Historic vineyard sites are often oversupplied with P, so P fertilization must be avoided. This must be respected when organic fertilization is conducted. Manures, especially swine manure, contain high concentrations of P, and their use should be minimized in soils containing high amounts of P. When P_2O_5 content was assessed in the German trial four and 10 years after conversion, it did not differ among treatments (Di Giacinto et al., 2020; Döring et al., 2015). Similar results were found by Probst et al. (2008), who compared conventional and BD vineyard pairs in Alsace, France.

Irrigation of grapevines can be necessary to ensure yield in some areas even under moderate climatic conditions. However, the available water is also needed for other agricultural crops, human use, and other ecosystems. If irrigation is applied, best practices (monitoring or modeling of plant water status, drip or subsurface irrigation, adapted irrigation frequencies and quantities) must be followed to ensure maximum use efficiency of irrigation water. It remains to be elucidated whether the different management systems differ in their irrigation demand. It can be hypothesized that this highly depends on management, especially on the type of cover crop used, as well as on climatic factors.

5.2.4 Avoidance of contamination for habitat provision

In viticulture, copper (Cu) is used to protect the vines against downy mildew (*Plasmopara viticola*). Copper also enters the soil via drift, drip losses, or leaf fall, where as a heavy metal it is not subject to any degradation. Instead, Cu is immobilized relatively quickly by sorption to organic matter and clay minerals, resulting in an accumulation of Cu in the topsoil over time (Parat et al., 2002). As a result of immobilization, the bioavailability and thus the toxicity of Cu in the soil is reduced (Rooney et al., 2006). The toxic effect of free Cu ions on microorganisms in soil is due to its interference with enzymes, nucleic acids of proteins, and membranes, where it damages the cell metabolism (Trevors & Cotter, 1990). The bioavailability of Cu in soil strongly depends on soil type, pH, amount and structure of organic matter, and redox potential (Chaignon et al., 2003).

To reduce the toxic effect of high Cu concentrations on soil organisms and plants, there is a legal upper limit of an average of 4 kg per ha and year (28 kg of Cu over a period of 7 years) for the use of Cu in plant protection in the European Union (Brun et al., 2001; Council of the European Union, 1986; European Commission, 2021; Giller et al., 1998). Beyond this, Cu application in viticulture in Germany is limited to 3 kg/ha a due to the application instructions in the registration reports of Cu-containing fungicides (Federal Office of Consumer Protection and Food Safety 2011, 2020).

In INT viticulture, Cu is largely replaced by synthetic fungicides. In ORG/BD viticulture, however, where such plant protection products are not permitted, the use of Cu is still without any alternative—at least in the cultivation of traditional grape varieties sensitive to diseases such as powdery and downy mildew. Due to the higher use of Cu in crop protection, ORG/BD viticulture are criticized with regard to Cu contamination of the soil. However, several studies comparing Cu contents in soils of ORG and conventionally managed vineyards did not detect differences among the systems (Probst et al., 2008; Radić et al., 2014; Strumpf et al., 2011). In general, many studies on vineyard soils show high levels of Cu contamination, regardless of the management system (Brun et al., 2001; Fernández-Calviño et al., 2010; Parat et al., 2002). The reason might be that Cu contents in vineyard soils have accumulated over the past century, thereby masking recent management effects (Probst et al., 2008; Strumpf et al., 2011). As Cu accumulation in the soil is a gradual process, it might only be detectable in the long-term. A study comparing ORG and conventional management over nine years in Italy detected higher Cu contents in ORG managed soils, which underlines the necessity of long-term field trials to assess changes in soil parameters (Beni & Rossi, 2009).

The concentrations of total, as well as of bioavailable Cu were higher in the soils of the ORG and BD managed plots compared to the INT management in the Geisenheim trial 12 years after conversion (Hendgen et al., 2020). With mean values of 74, 87, and 87 mg/kg soil of total Cu and 19, 25, and 22 mg/kg soil of bioavailable Cu for INT, ORG and BD management, respectively, all three systems ranged below the European Union's critical limit of 140 mg/kg soil (Council of the European Union, 1986) but above the precautionary value of 40 mg/kg in Germany (Federal Ministry of Justice and Consumer Protection, 2017). The average difference of 13 mg/kg total Cu in topsoil between INT and ORG/BD plots after 12 years corresponds to an annual increase in total Cu concentration of about 1 mg/kg due to ORG and BD management. This result is in line with the theoretical Cu input into the topsoil by plant protection measures, calculated on the basis of the maximum annual application rate of 3 kg Cu/ha a as permitted in Germany (Federal Office of Consumer Protection and Food Safety, 2011, 2020). Actual Cu input in the INBIODYN trial in Geisenheim averaged 2.5 kg/ha a over the same period (Hendgen et al., 2020). Although the annual increase is comparatively small [Beni and Rossi (2009), administering the high amount of 6.8 kg/ha a over a period of 10 years, recorded an increase of 85 mg Cu/kg of topsoil], limited to the soil surface, and a harmful effect is difficult to establish at a specific concentration threshold (Flemming & Trevors, 1989), Cu input must nevertheless be viewed critically, since remediation of Cu-contaminated soils is costly with today's means (Apori et al., 2018; Mackie et al., 2012). It is therefore necessary to reduce the amount of Cu required in ORG farming, either by developing other bioactive substances to replace Cu in plant protection, or by promoting cultivation of disease-tolerant grape varieties.

5.2.5 Biodiversity conservation

Loss of biodiversity in soil is a major threat, since soils represent one of the richest habitats on earth. Soil microbial, floral and faunal biodiversity contribute to a large extent to biomass production, pest and disease control, and erosion control (Bünemann et al., 2018). Biodiversity loss due to viticultural activity can be reduced by providing habitat and sustenance for soil and aboveground life. In terms of soil management, a reduction of tillage frequency may improve habitat quality for soil biota. A larger diversity of cover crops provides a more diverse habitat and feed sources for soil and aboveground life. This effect is especially pronounced when cover crops are allowed to blossom, and if mulching frequency is decreased or mulching is replaced by mowing or rolling of cover crops.

Native cover crop mixtures do possibly not provide as much biomass as industrially produced seed mixtures, but should be preferred to preserve genetic diversity. To allow for the establishment of slow germinating and perennial cover crop species and stabilize soil biological communities, cover crop rows should ideally be left untilled for several years. The use of contaminating substances such as Cu should ideally be avoided to prevent shifts in the microbial community.

Since management practices of INT, ORG, and BD viticultural systems have a strong impact on soil physical and chemical composition, they also have a strong impact on microbial communities in the soil. The richness, composition, and abundance of microbial communities are shaped by the abundance and diversity of C and N sources, such as root exudates and detritus, and edaphic properties. In a viticultural context, a strong relation of the microbiome and fertilization and cover cropping strategies can be expected. A sequence of enzymatic activity analyses, PLFA, and a metabarcoding approach in the Geisenheim long-term trial suggested that while soil enzymatic activities are little affected by management during the conversion phase (Meissner et al., 2019), stable microbial communities are established after conversion. Communities in ORG and BD after conversion seemed to be shaped by a higher SOM turnover, driven by a higher and more diverse biomass production (Hendgen et al., 2020). This manifested in higher bacterial and fungal PLFA marker concentration, higher urease, glucosidase, dehydrogenase, catalase activities, higher N_{min} concentration (Di Giacinto et al., 2020), a different fungal community composition, and a higher bacterial species richness (Hendgen et al., 2018). INT showed a higher biomass of mycorrhizae (Di Giacinto et al., 2020; Hendgen et al., 2018). The latter finding is plausible as mycorrhizae are negatively affected by soil disturbance, higher N_{min} levels, and high biomass turnover (Nouri et al., 2014; Oehl et al., 2003). Nevertheless, it stands in contrast to other studies comparing INT and ORG viticulture (Freitas et al., 2011; Radić et al., 2014).

Other studies showed a higher microbial biomass in ORG or BD compared to INT viticulture (Coll et al., 2011; Freitas et al., 2011; Okur et al., 2009). Elevated enzymatic activities were also found in ORG and BD as compared to INT crop rotations (Mäder et al., 2002) and vineyards (Okur et al., 2009) and BD vineyards show a different fungal community compared to INT vineyards (Morrison-Whittle et al., 2017). ORG or BD management in most of the studies, however, was characterized by elevated external C inputs compared to INT, thus increasing microbial activity (García-Ruiz et al., 2008). It is remarkable that although similar C and N inputs were administered to all management systems in the German systems comparison, these effects persisted, and differences in microbial community composition were most pronounced in the cover cropped row (Hendgen et al., 2018), indicating that the communities were strongly shaped by the cover crop management regime (Peregrina et al., 2014).

Fungal and bacterial community composition in the undervine area differed strongly from the one in the inter-row space, probably because of differences in bulk density, plant-available water capacity, herbicide use, and soil organic carbon (SOC) (Hendgen et al., 2020). When the microbial community associated with soil-living arthropods was analyzed in 2017, the composition of the fungal community associated with soil-living arthropods was different between INT and ORG/BD treatments, and both ORG and BD showed a higher richness. Differential abundance analysis revealed an elevated abundance of wine-relevant fungal genera in both ORG and BD aboveground trap samples (Agerbo Rasmussen et al., 2021).

Apart from soil microbiological data, positive effects of ORG and BD farming on nematofauna have been reported (Coll et al., 2011), while data on earthworms and arthropods are not always in agreement. Agerbo Rasmussen et al. (2021) reported a higher diversity of soil arthropods in ORG/BD, while Collins et al. (2015) reported higher invertebrate numbers in INT. Coll et al. (2011) described higher earthworm numbers and biomass for INT, while Collins et al. (2015) found higher numbers for ORG and BD. Puig-Montserrat et al. (2017) reported a higher diversity of butterflies and moths for ORG, associated with a higher diversity of vascular plants. It is likely that the discrepancy among the described effects is associated with individual management practices (e.g., tillage strategy in the case of the study of Coll et al., 2011) rather than the management systems themselves. The effects are also harder to estimate with increasing mobility of the investigated organisms.

Taken together, there is strong evidence that ORG and BD management in viticulture has a positive effect on the biological performance of the soil. Soil biological communities are strongly shaped by the cover cropping strategy and tillage. On the other hand, N_{min} values as a result of the differences in cover cropping, seem to influence biomass of arbuscular mycorrhizal fungi. The extent to which the altered microbial community also affects aboveground plant tissues has yet to be unraveled.

5.2.6 Biomass production

Grape yield is ensured by providing sufficient nutrients and water to the vines. Nutrient supply can be achieved by mineral (INT viticulture) and organic fertilization, and by the use of cover crops via the SOM cycle. A good practice of fertilization should ensure sufficient plant growth and yield for high quality produce while protecting soil and water resources by minimizing nutrient loss and leaching (FAO Agriculture Department, 2003). The concept of fertilization in ORG agriculture differs substantially from the one of conventional agriculture. While in INT agriculture nutrients may be supplied to the plants using highly water soluble mineral fertilizers, ORG farming is heavily reliant on fertilizers which are more complex and must often be made plant-available by the activity of soil organisms (Scheller, 2013). Examples of such fertilizers are composts, farmyard manure, or rock phosphate (Kauer & Fader, 2015). Biodynamic composts generally include cattle manure and the BD compost preparations (yarrow blossoms, chamomile, stinging nettle, oak bark, dandelion, and valerian, mostly buried underground for six months in animal organs).

Numerous studies have already addressed the question of how farming practices affect the amount of SOC and the SOM cycle (Brock et al., 2011; Gehlen et al., 1988; Hepperly et al., 2006; Reganold et al., 1993). In general, it is assumed that ORG farming promotes the accumulation of SOC in the soil and thus fertility compared to conventional systems (Leifeld & Fuhrer, 2010). Studies comparing commercial vineyards of ORG and conventional management in the Moselle valley in Germany (Gehlen et al., 1988) and in southern France (Coll et al., 2011) showed higher SOM content in ORG viticulture. A sequence of SOC analyses in a long-term trial in Germany showed inconclusive results (Di Giacinto et al., 2020; Hendgen et al. 2018, 2020) with INT showing the highest SOC 9 years and BD showing the highest SOC 12 years after conversion. In this trial, compost was added in similar amounts to all treatments and the cover cropping strategy differed only in its species composition and diversity. SOC in the undervine space of the BD treatment was also enhanced, indicating a positive effect of mechanical undervine treatment in ORG/BD compared to herbicide application in the INT treatment (Hendgen et al., 2020).

Although SOM content did not seem to be strongly influenced by management systems in the latter trial, long-term data showed strong effects in N mineralization: Mineralized nitrogen content (N_{min}) in soil in the first three years of conversion was higher in the INT treatment in Germany (Meissner et al., 2019). After the conversion phase, regular assessments of N_{min} showed higher levels in ORG/BD in topsoil for cultivated (33 and 34 kg/ha on average) and greened rows (19 kg/ha on average) compared to INT management (cultivated rows 28 kg/ha and greened rows 8 kg/ha on average) in the period of 2010–21 (Di Giacinto et al., 2020) (Döring et al. unpublished). Similarly, Coll et al. (2011) reported higher total N levels for ORG compared to INT viticulture, with differences increasing over time. As the differences among INT and ORG/BD management systems were most evident in the cover cropped rows, it is likely that N fixation by the diverse and legume-rich cover crops in ORG and BD is the factor explaining these differences. The fact that N_{min} was consistently elevated in ORG and BD, while differences in SOM were not consistent was attributed to the microbial community in ORG and BD being more active in the turnover of organic matter (see 5.3.3). Mineral fertilization was required to keep the N supply of the INT treatment at a comparable level.

ESs provided by extensive cover cropping, however, are achieved at the expense of a higher water deficit (Döring et al., 2015) and decreased soil moisture (Collins et al., 2015) in ORG or BD viticulture, but also in INT viticulture if cover crops are used (Celette et al., 2008). The consequences are a decrease of physiological performance and a reduced growth under ORG and BD management (Döring et al., 2015).

In a meta-analysis, ORG and BD treatments showed 21% lower pruning weight compared to conventional/INT viticulture (Döring et al., 2019). Long-term data from the INBIODYN trial showed an average reduction of pruning weights of 18% (ORG) and 20% (BD), respectively, compared to the in the INT treatment (35.8 dt/ha) between 2006 and 2020, despite a better N supply. Leaf area, measured as leaf area index, secondary shoot leaf area, and shoot length were reduced under ORG and BD viticulture (Döring et al., 2015; Meissner et al., 2019). Yield under ORG and BD viticulture decreased by 18% compared to conventional/INT viticulture in a meta-analysis (Döring et al., 2019) and in long-term data from the INBIODYN trial (ORG -17.1 and BD -16.5 % compared to INT 8.84 t/ha). In all studies in which data were available, these yield differences were linked to reduced berry and cluster weights (Döring et al. 2015, 2019; Meissner et al., 2019), which are likely related to early season water deficits due to cover crop competition (Hardie & Considine, 1976; McCarthy, 1997). In addition, there is evidence that bunch numbers might be slightly decreased under ORG (−4.5%) and BD (−8%) compared to INT management (Döring, unpublished), possibly linked to vine water status (Matthews & Anderson, 1989).

Competition by cover crops may influence water availability in the soil profile spatially and temporally (Collins et al., 2015; Monteiro & Lopes, 2007), and root growth of grapevines seems to be redirected to zones with lower competition, i.e., deeper soil layers and the undervine zone (Celette et al., 2008). This might benefit the vines during drought periods. During a prolonged drought period in Germany (2018−20), ORG and BD showed yields and pruning weights which were comparable to INT. It can be speculated that through the sustained mild water deficit in the years with higher rainfall, vines in ORG and BD adapted better to drought conditions. Another possibility is that the process of rolling the cover crop may have produced a dense layer of mulch, which protected the soils from heat and increased evaporation.

Despite the reduced yields, TSS, total titratable acidity, and pH in berries, juice, and wine do not seem to be affected by management systems (Collins et al., 2015; Danner, 1985; Döring et al. 2015, 2019; Hofmann, 1991; Kauer, 1994; Linder et al., 2006; Malusà et al., 2004; Meissner et al., 2019; Tassoni et al. 2013, 2014). There is, however, evidence that other juice and wine compounds are influenced by management systems. As N compounds are continuously released from SOM, and N supply is often increased in ORG or BD systems, YAN concentration seems to be elevated in ORG and BD (Döring et al., 2015). Anthocyanin and flavonoid content in berry skin, as well as polyphenol content, antioxidant potential, and phenolic acid content in juice and wine from ORG/BD management were elevated compared to conventional treatments (Döring et al., 2019), possibly related to lower canopy volume, reduced berry size, and lower water availability in ORG and BD.

5.2.7 Climate regulation

Viticultural soils are an important source for the emission of GHG like CO_2 and N_2O but also act as a strong sink for CO_2 in the form of SOM. Tillage and SOM degradation lead to substantial CO_2 emissions into the atmosphere (Reicosky, 1997), and SOM is further depleted when plant residues are not returned to the field. An increase of SOM levels, and thus C sequestration is possible by organic fertilization or cover cropping (Abad et al., 2021). Green manuring with legumes offers a high potential of C sequestration, but also increases N_2O emissions to a level that might outweigh the C sequestration effect once a certain level of SOM is reached (Lugato et al., 2018; Muhammad et al., 2019). N_2O emissions also seem to be correlated to the use of mineral N fertilizers (Kopittke et al., 2021;

Skinner et al., 2019). Herbicide application in the undervine area seems to enhance N_2O emissions as well (Steenwerth & Belina, 2010). A higher microbial activity enhances CO_2 emissions (Muhammad et al., 2019). N use efficiency of grapevines should thus be increased and mineral N fertilization reduced to a minimum. The dimensions of the single emissions, their proportion, and the impact of the management systems is still widely unknown.

5.3 Implications for future soil management of vineyards

Choice and management of cover crops revealed to be crucial for several soil functions and their related ES in INT, ORG, and BD viticultural systems. Therefore, legume-based cover crop mixtures are one major trait of ORG/BD systems fulfilling multiple functions at the same time, such as soil structure maintenance, habitat provision and regulation of the microbial community, decomposition and nutrient cycling. Especially the enhanced soil fungal and bacterial activity and the sufficient N supply of the vines in opposition to input of synthetic N fertilizer in INT seem to be the two most advantageous traits associated to legume-based cover crop use. This is why the use of legume-rich cover crop mixtures in viticulture should be propagated and adapted to a wide range of different climates, cultivars, and soils. Of course, other management practices associated to ORG/BD viticulture such as addition of composted farmyard manure might as well partially account for the management effects observed. Still the use of legume-based cover crops in viticulture might pose two different risks: On one hand, the risk of nitrate leaching is enhanced. This can be prevented by adapting tillage and establishment of winter cover crops to climatic conditions and grapevine phenological development. On the other hand, legume-based cover crops with a high biomass production might compete with the vines for water and nutrients potentially leading to reduced grapevine growth and yield. Therefore, adaptation of cover crop implementation to local conditions and adapted management of cover crops (e.g., rolling) is necessary.

Concerning productivity, one major ES of agroecosystems, ORG and BD treatments showed 18% lower yield ha^{-1} on an area-scale, although it can be hypothesized that established ORG/BD systems have a better drought tolerance and showed a similar productivity in hot and dry years, presumably due to an adapted root system and a highly active soil microbial community.

This highly active soil microbiome revealed to be a characteristic trait of ORG and BD systems substantially differing from INT viticulture, which showed less microbial activity, but enhanced mycorrhizal biomass. It should be further investigated how the diverse and active microbial community in ORD/BD influences grapevine performance in order to use it in a targeted manner.

The BD treatment showed positive effects on soil structure maintenance (lower bulk density compared to INT), whereas INT viticulture showed an increased risk of soil compaction. The cover crop mixture rich in legumes can definitely account for part of these differences. On the other hand, the application of the BD preparations on plants, soil, and compost seems to have positive effects concerning soil functions, although their mode of action remains unclear.

The erosion risk under ORG/BD relying on mechanical undervine management is potentially increased and alternatives such as undervine cover cropping should thus be promoted.

There is the absolute necessity to develop alternatives to Cu in the ORG plant protection strategy. Furthermore, remediation of Cu should be investigated in a viticultural context, since background Cu levels in numerous traditional vineyard sites are high due to excessive input in past decades.

More effort should be made to propagate disease-tolerant grapevine varieties, not only for ORG and BD viticulture but also for INT viticulture, since these varieties allow a significant reduction of spraying events and can therefore make INT and especially ORG/BD viticulture much more sustainable.

Soil-based climate regulation linked to a soil's capacity of nutrient and organic matter cycling is one soil ES gaining more and more importance in a climate change scenario. Providing guidance about how to minimize soil-borne GHG emissions in viticulture while sustainably producing high-quality grapes will be one major future task.

Acknowledgments

Authors would like to thank Dr. M. Reiss, Hochschule Geisenheim University, Department of Landscape Planning and Nature Conservation, for reading and improving the manuscript.

References

Abad, J., Hermoso de Mendoza, I., Marín, D., Orcaray, L., & Santesteban, L. G. (2021). Cover crops in viticulture. A systematic review (1): Implications on soil characteristics and biodiversity in vineyard. *OENO One, 55*(1), 295–312. https://doi.org/10.20870/oeno-one.2021.55.1.3599

Agerbo Rasmussen, J., Nielsen, M., Mak, S. S. T., Döring, J., Klincke, F., Gopalakrishnan, S., Dunn, R. R., Kauer, R., & Gilbert, M. T. P. (2021). eDNA-based biomonitoring at an experimental German vineyard to characterize how management regimes shape ecosystem diversity. *Environmental DNA, 3*(1), 70–82. https://doi.org/10.1002/edn.3.131

Alakukku, L., Weisskopf, P., Chamen, W. C. T., Tijink, F. G. J., van der Linden, J. P., Pires, S., Sommer, C., & Spoor, G. (2003). Prevention strategies for field traffic-induced subsoil compaction: A review. *Soil and Tillage Research, 73*(1–2), 145–160. https://doi.org/10.1016/S0167-1987(03)00107-7

Apori, O. S., Hanyabui, E., & Asiamah, Y. J. (2018). Remediation technology for copper contaminated soil: A review. *Asian Soil Research Journal, 1*(3), 1–7.

Beni, C., & Rossi, G. (2009). Conventional and organic farming: Estimation of some effects on soil, copper accumulation and wine in a Central Italy vineyard. *Agrochimica, 53*, 145–159.

Bindraban, P. S., van der Velde, M., Ye, L., van den Berg, M., Materechera, S., Kiba, D. I., Tamene, L., Ragnarsdottir, K. V., Jongschaap, R., Hoogmoed, M., Hoogmoed, W., van Beek, Ch., & van Lynden, G. (2012). Assessing the impact of soil degradation on food production. *Current Opinion in Environmental Sustainability, 4*(5), 478–488. https://doi.org/10.1016/j.cosust.2012.09.015

Brock, C., Fließbach, A., Oberholzer, H.-R., Schulz, F., Wiesinger, K., Reinicke, F., Koch, W., Pallutt, B., Dittman, B., Zimmer, J., Hülsbergen, K.-J., & Leithold, G. (2011). Relation between soil organic matter and yield levels of nonlegume crops in organic and conventional farming systems. *Journal of Plant Nutrition and Soil Science, 174*(4), 568–575.

Brun, L. A., Maillet, J., Hinsinger, P., & Pépin, M. (2001). Evaluation of copper availability to plants in copper-contaminated vineyard soils. *Environmental Pollution, 111*(2), 293–302. https://doi.org/10.1016/S0269-7491(00)00067-1

Brussaard, L. (2012). Ecosystem services provided by the soil biota. In D. H. Wall, R. D. Bardgett, V. Behan-Pelletier, J. E. Herrick, H. Jones, K. Ritz, J. Six, D. R. Strong, & W. H. van der Putten (Eds.), *Soil ecology and ecosystem services* (pp. 45–58). Oxford, UK: Oxford University Press.

Bünemann, E. K., Bongiorno, G., Bai, Z., Creamer, R. E., de Deyn, G., de Goede, R., Fleskens, L., Geissen, V., Kuijper, T. W. M., Mäder, P., Pulleman, M. M., Sukkel, W., van Groenigen, J. W., & Brussaard, L. (2018). Soil quality - a critical review. *Soil Biology and Biochemistry, 120*, 105−125. https://doi.org/10.1016/j.soilbio.2018.01.030

Capello, G., Biddoccu, M., & Cavallo, E. (2020). Permanent cover for soil and water conservation in mechanized vineyards: A study case in Piedmont, NW Italy. *Italian Journal of Agronomy, 15*(4), 323−331. https://doi.org/10.4081/ija.2020.1763

Capello, G., Biddoccu, M., Ferraris, S., & Cavallo, E. (2019). Effects of tractor passes on hydrological and soil erosion processes in tilled and grassed vineyards. *Water, 11*(10), 2118. https://doi.org/10.3390/w11102118

Celette, F., Gaudin, R., & Gary, C. (2008). Spatial and temporal changes to the water regime of Mediterranean vineyard due to the adoption of cover cropping. *European Journal of Agronomy, 29*, 153−162.

Chaignon, V., Sanchez-Neira, I., Herrmann, P., Jaillard, B., & Hinsinger, P. (2003). Copper bioavailability and extractability as related to chemical properties of contaminated soils from a vine-growing area. *Environmental Pollution, 123*(2), 229−238. https://doi.org/10.1016/S0269-7491(02)00374-3

Chen, G., & Weil, R. R. (2010). Penetration of cover crop roots through compacted soils. *Plant and Soil, 331*(1−2), 31−43. https://doi.org/10.1007/s11104-009-0223-7

Chou, M.-Y., Vanden, H., & Justine, E. (2019). Annual under-vine cover crops mitigate vine vigor in a mature and vigorous Cabernet franc vineyard. *American Journal of Enology and Viticulture, 70*(1), 98−108. https://doi.org/10.5344/ajev.2018.18037

Collins, C., Penfold, C. M., Johnston, Luke, Marschner, P., & Bastian, S. (2015). *The relative sustainability of organic, biodynamic and conventional viticulture*. The University of Adelaide (Project No. UA 1102).

Coll, P., Le Cadre, E., Blanchart, E., Hinsinger, P., & Villenave, C. (2011). Organic viticulture and soil quality: A long-term study in Southern France. *Applied Soil Ecology, 50*, 37−44. https://doi.org/10.1016/j.apsoil.2011.07.013

Costanza, R., d'Arge, R., de Groot, R., Farber, S., Grasso, M., Hannon, B., Limburg, K., Naeem, S., O'Neill, R. V., Paruelo, J., Raskin, R. G., Sutton, P., & van den Belt, M. (1997). The value of the world's ecosystem services and natural capital. *Nature, 387*(6630), 253−260. https://doi.org/10.1038/387253a0

Council of the European Union. (June 12, 1986). Council Directive 86/278/EEC of 12 June 1986 on the protection of the environment, and in particular of the soil, when sewage sludge is used in agriculture. 86/278/EEC. *Official Journal of the European Union L 181*.

Danner, R. (1985). *Vergleichende Untersuchungen zum konventionellen, organisch-biologischen und biologisch-dynamischen Weinbau* (Doctoral dissertation).

Delpuech, X., & Metay, A. (2018). Adapting cover crop soil coverage to soil depth to limit competition for water in a Mediterranean vineyard. *European Journal of Agronomy, 97*, 60−69. https://doi.org/10.1016/j.eja.2018.04.013

Di Giacinto, S., Friedel, M., Poll, C., Döring, J., Kunz, R., & Kauer, R. (2020). Vineyard management system affects soil microbiological properties. *OENO One, 54*(1). https://doi.org/10.20870/oeno-one.2020.54.1.2578

Döring, J., Collins, C., Frisch, M., & Kauer, R. (2019). Organic and biodynamic viticulture affect biodiversity, vine and wine properties: A systematic quantitative review. *American Journal of Enology and Viticulture, 70*(3), 221−242. https://doi.org/10.5344/ajev.2019.18074

Döring, J., Frisch, M., Tittmann, S., Stoll, M., & Kauer, R. (2015a). Growth, yield and fruit quality of grapevines under organic and biodynamic management. *PLoS One, 10*(10), e0138445. https://doi.org/10.1371/journal.pone.0138445

European Commission. (2021). Commission Implementing Regulation (EU) 2021/1165of 15 July 2021 authorising certain products and substances for use in organic production and establishing their lists. *Official Journal of the European Union*, (L 253/13). https://eur-lex.europa.eu/legal-content/EN/TXT/PDF/?uri=CELEX:32021R1165&from=DE. (Accessed 5 January 2022).

European Initative for Sustainable Development in Agriculture EISA. (2015). *Annual report 2014. Hg. v.* Berlin: EISA e.V.

European Parliament and Council of the European Union. (2018). Regulation (EU) 2018/848 of the European Parliament and of the Council of 30 May 2018 on organic production and labelling of organic products and repealing Council Regulation (EC) No 834/2007. *Official Journal of the European Union*, (L 150/1). https://eur-lex.europa.eu/legal-content/EN/TXT/PDF/?uri=CELEX:32018R0848&from=DE [Accessed 05 January 2022].

FAO Agriculture Department. (2003). *Report of the expert consultation on a good agricultural practices (GAP) approach*. Rome, Italy: FAO Agriculture Department. Retrieved from http://www.fao.org/tempref/GI/Reserved/FTP_FaoRlc/old/foro/bpa/pdf/good.pdf. (Accessed 24 February 2021).

FAO, Committee on Agriculture. (1999–2021). *FAO position paper on organic agriculture*. Rome, Italy: Definition of Organic Agriculture. Online verfügbar unter. Retrieved from http://www.fao.org/3/X0075e/X0075e.htm, zuletzt aktualisiert am 09.02. (Accessed 17 February 2021).

Federal Ministry of Justice and Consumer Protection. (September 27, 2017). *Federal soil protection and contaminated sites ordinance*. BBodSchV.

Federal Office of Consumer Protection and Food Safety (Hg). (2011). *Registration report - copper hydroxide*. SPU-02720-F. Online verfügbar unter. Retrieved from https://www.bvl.bund.de/SharedDocs/Downloads/04_Pflanzenschutzmittel/01_zulassungsberichte/006896-00-00.pdf?__blob=publicationFile&v=4. (Accessed 2 September 2020).

Federal Office of Consumer Protection and Food Safety (Hg). (2020). *Registration report - cuprozin progress. Art. 51 Extension of authorisation for minor uses*. Online verfügbar unter. Retrieved from https://www.bvl.bund.de/SharedDocs/Downloads/04_Pflanzenschutzmittel/01_zulassungsberichte/006895-00-23.pdf?__blob=publicationFile&v=2. (Accessed 2 September 2020).

Fernández-Calviño, D., Soler-Rovira, P., Polo, A., Díaz-Raviña, M., Arias-Estévez, M., & Plaza, C. (2010). Enzyme activities in vineyard soils long-term treated with copper-based fungicides. *Soil Biology and Biochemistry, 42*(12), 2119–2127. https://doi.org/10.1016/j.soilbio.2010.08.007

Flemming, C. A., & Trevors, J. T. (1989). Copper toxicity and chemistry in the environment: A review. *Water, Air and Soil Pollution, 44*, 143–158.

Flint, M. L., & van den Bosch, R. (1981). *Introduction to integrated pest management*. Boston, MA: Springer US.

Freitas de, N. O., Yano-Melo, A. M., da Silva, F. S. B., Melo de, N. F., & Maia, L. C. (2011). Soil biochemistry and microbial activity in vineyards under conventional and organic management at Northeast Brazil. *Scientia Agricola, 68*(2), 223–229. https://doi.org/10.1590/S0103-90162011000200013

Gambetta, G. A., Herrera, J. C., Dayer, S., Feng, Q., Hochberg, U., & Castellarin, S. D. (2020). The physiology of drought stress in grapevine: Towards an integrative definition of drought tolerance. *Journal of Experimental Botany, 71*(16), 4658–4676. https://doi.org/10.1093/jxb/eraa245

García-Ruiz, R., Ochoa, V., Hinojosa, M. B., & Carreira, J. A. (2008). Suitability of enzyme activities for the monitoring of soil quality improvement in organic agricultural systems. *Soil Biology and Biochemistry, 40*(9), 2137–2145. https://doi.org/10.1016/j.soilbio.2008.03.023

Gehlen, P., Neu, J., & Schröder, D. (1988). Bodenchemische und bodenbiologische Vergleichsuntersuchungenkonventionell und biologisch bewirtschafteter Weinstandorte an der Mosel. *Weinwissenschaft, 43*, 161–173.

Giller, Ken E., Witter, Ernst, & McGrath, S. P. (1998). Toxicity of heavy metals to microorganisms and microbial processes in agricultural soils: A review. *Soil Biology and Biochemistry, 30*(10–11), 1389–1414. https://doi.org/10.1016/S0038-0717(97)00270-8

Hardie, W. J., & Considine, J. A. (1976). Response of grapes to water-deficit stress in particular stages of development. *American Journal of Enology and Viticulture, 27*(2), 55–61.

Hartwig, N. L., & Ammon, H. U. (2002). Cover crops and living mulches. *Weed Science, 50*(6), 688–699. https://doi.org/10.1614/0043-1745(2002)050[0688:AIACCA]2.0.CO;2

Hendgen, M., Döring, J., Stöhrer, V., Schulze, F., Lehnart, R., & Kauer, R. (2020). Spatial differentiation of physical and chemical soil parameters under integrated, organic, and biodynamic viticulture. *Plants (Basel, Switzerland), 9*(10). https://doi.org/10.3390/plants9101361

Hendgen, M., Hoppe, B., Döring, J., Friedel, M., Kauer, R., Frisch, M., Dahl, A., & Kellner, H. (2018). Effects of different management regimes on microbial biodiversity in vineyard soils. *Scientific Reports, 8*(1), 9393. https://doi.org/10.1038/s41598-018-27743-0

Hepperly, P. R., Douds, D., & Seidel, R. (2006). The Rodale Institute Farming System Trial 1981 to 2005: Long-term analysis of organic and conventional maize and soybean cropping systems. In J. Raupp, C. Pekrun, M. Oltmanns, & U. Köpke (Eds.), *Long-term field experiments in organic farming* (pp. 15–31). Berlin: Verlag Dr. Köster (Scientific Series).

Hofmann, U. (1991). *Untersuchungen über die Umstellungsphase auf ökologische Bewirtschaftungssysteme im Weinbau im Vergleich zur konventionellen Wirtschaftsweise am Beispiel Mariannenaue – Erbach* (Doctoral dissertation). Gießen: Justus-Liebig-Universität.

Jabloun, M., Schelde, K., Tao, F., & Olesen, J. E. (2015). Effect of temperature and precipitation on nitrate leaching from organic cereal cropping systems in Denmark. *European Journal of Agronomy, 62*, 55–64. https://doi.org/10.1016/j.eja.2014.09.007

Kauer, R. (1994). *Vergleichende Untersuchungen zum integrierten und ökologischen Weinbau in den ersten drei Jahren der Umstellung: Ergebnisse von 12 Standorten im Anbaugebiet Rheinhessen bei den Rebsorten Müller-Thurgau und Riesling* (Doctoral dissertation). Gießen: Justus-Liebig-Universität.

Kauer, R., & Fader, B. (2015). *Praxis des ökologischen Weinbaus*. Darmstadt ((Keine Angabe), 506).

Kibblewhite, M. G., Ritz, K., & Swift, M. J. (2008). Soil health in agricultural systems. *Philosophical Transactions of the Royal Society of London - B, 363*(1492), 685–701. https://doi.org/10.1098/rstb.2007.2178

Kopittke, P. M., Menzies, N. W., Dalal, R. C., McKenna, B. A., Husted, S., Wang, P., & Lombi, E. (2021). The role of soil in defining planetary boundaries and the safe operating space for humanity. *Environment International, 146*, 106245. https://doi.org/10.1016/j.envint.2020.106245

Lagacherie, P., Coulouma, G., Ariagno, P., Virat, P., Boizard, H., & Richard, G. (2006). Spatial variability of soil compaction over a vineyard region in relation with soils and cultivation operations. *Geoderma, 134*(1–2), 207–216. https://doi.org/10.1016/j.geoderma.2005.10.006

Lassaletta, L., Billen, G., Grizzetti, B., Anglade, J., & Garnier, J. (2014). 50 year trends in nitrogen use efficiency of world cropping systems: The relationship between yield and nitrogen input to cropland. *Environmental Research Letters, 9*(10), 105011. https://doi.org/10.1088/1748-9326/9/10/105011

Lazcano, C., Decock, C., & Wilson, S. G. (2020). Defining and managing for healthy vineyard soils, intersections with the concept of terroir. *Frontiers in Environmental Science, 8*, 68. https://doi.org/10.3389/fenvs.2020.00068

Leifeld, J., & Fuhrer, J. (2010). Organic farming and soil carbon sequestration: What do we really know about the benefits? *Ambio, 39*(8), 585–599. https://doi.org/10.1007/s13280-010-0082-8

Linder, Ch, Viret, O., Spring, J.-L., Droz, P., & Dupuis, D. (2006). Viticulture intégrée et bio-organique: Synthèse de sept ans d'observations. *Revue Suisse Viticulture, Arboriculture, Horticulture, 38*(4), 235–243.

Li, Z., Wu, P., Feng, H., Zhao, X., Huang, J., & Zhuang, W. (2009). Simulated experiment on effect of soil bulk density on soil infiltration capacity. *Transactions of the Chinese Society of Agricultural Engineering, 25*(6), 40–45.

Löhnertz, O. (1988). *Untersuchungen zum zeitlichen Verlauf der Nährstoffaufnahme bei* Vitis vinifera L. cv. *Riesling. 1st*. Geisenheim: Veröffentlichungen der Forschungsanstalt Geisenheim.

Lugato, E., Leip, A., & Jones, A. (2018). Mitigation potential of soil carbon management overestimated by neglecting N$_2$O emissions. *Nature Climate Change, 8*(3), 219–223. https://doi.org/10.1038/s41558-018-0087-z

Mackie, K. A., Müller, T., & Kandeler, Ellen (2012). Remediation of copper in vineyards—a mini review. *Environmental Pollution, 167*, 16–26. https://doi.org/10.1016/j.envpol.2012.03.023

Mäder, P., Fließbach, A., Dubois, D., Gunst, L., Fried, P., & Niggli, U. (2002). Soil fertility and biodiversity in organic farming. *Science, 296*(5573), 1694–1697. https://doi.org/10.1126/science.1071148

Malusà, E., Laurenti, E., Ghibaudi, E., & Rolle, L. (2004). Influence of organic and conventional management on yield and composition of grape cv. 'Grignolino'. *Acta Horticulturae, 240*, 135–141.

Masson, P., & Masson, V. (2013). *Landwirtschaft, Garten- und Weinbau biodynamisch*. Aarau, München: AT-Verl.

Matthews, M. A., & Anderson, M. M. (1989). Reproductive development in grape (*Vitis vinifera* L.): Responses to seasonal water deficits. *American Journal of Enology and Viticulture, 40*(1), 52–60.

McCarthy, M. G. (1997). The effect of transient water deficit on berry development of cv. Shiraz (*Vitis vinifera* L.). *Australian Journal of Grape and Wine Research, 3*(3), 2–8. https://doi.org/10.1111/j.1755-0238.1997.tb00128.x

Meissner, G., Athmann, M., Fritz, J., Kauer, R., Stoll, M., & Schultz, H. R. (2019). Conversion to organic and biodynamic viticultural practices: Impact on soil, grapevine development and grape quality. *OENO One, 53*(4), 639–659.

Moneyron, A., Lallemand, J. F., Schmitt, C., Perrin, M., Soustre-Gacougnolle, I., & Masson, J. E. (2017). Linking the knowledge and reasoning of dissenting actors fosters a bottom-up design of agroecological viticulture. *Agronomy for Sustainable Development, 37*(5), 1–14. https://doi.org/10.1007/s13593-017-0449-3

Monteiro, A., & Lopes, C. M. (2007). Influence of cover crop on water use and performance of vineyard in Mediterranean Portugal. *Agriculture, Ecosystems and Environment, 121*, 336–342.

Morrison-Whittle, P., Lee, S. A., & Goddard, M. R. (2017). Fungal communities are differentially affected by conventional and biodynamic agricultural management approaches in vineyard ecosystems. *Agriculture, Ecosystems & Environment, 246*, 306–313. https://doi.org/10.1016/j.agee.2017.05.022

Muhammad, I., Sainju, U. M., Zhao, F., Khan, A., Ghimire, R., Fu, X., & Wang, J. (2019). Regulation of soil CO$_2$ and N$_2$O emissions by cover crops: A meta-analysis. *Soil and Tillage Research, 192*, 103–112. https://doi.org/10.1016/j.still.2019.04.020

Nouri, E., Breuillin-Sessoms, F., Feller, U., & Reinhardt, D. (2014). Phosphorus and nitrogen regulate arbuscular mycorrhizal symbiosis in *Petunia hybrida*. *PLoS One, 9*(6), e90841. https://doi.org/10.1371/journal.pone.0090841

Novara, A., Pisciotta, A., Minacapilli, M., Maltese, A., Capodici, F., Cerdà, A., & Gristina, L. (2018). The impact of soil erosion on soil fertility and vine vigor. A multidisciplinary approach based on field, laboratory and remote sensing approaches. *The Science of the Total Environment, 622–623*, 474–480. https://doi.org/10.1016/j.scitotenv.2017.11.272

Oehl, F., Sieverding, E., Ineichen, K., Mäder, P., Boller, T., & Wiemken, A. (2003). Impact of land use intensity on the species diversity of arbuscular mycorrhizal fungi in agroecosystems of Central Europe. *Applied and Environmental Microbiology, 69*(5), 2816–2824. https://doi.org/10.1128/aem.69.5.2816-2824.2003

Okur, N., Altindişli, A., Çengel, M., Göçmez, S., & Kayikçioğlu, H. H. (2009). Microbial biomass and enzyme activity in vineyard soils under organic and conventional farming systems. *Turkish Journal of Agriculture and Forestry, 33*, 413–423.

Parat, C., Chaussod, R., Lévêque, J., Dousset, S., & Andreux, F. (2002). The relationship between copper accumulated in vineyard calcareous soils and soil organic matter and iron. *European Journal of Soil Science, 53*(4), 663–670. https://doi.org/10.1046/j.1365-2389.2002.00478.x

Peregrina, F., Pilar Pérez-Álvarez, E., & García-Escudero, E. (2014). Soil microbiological properties and its stratification ratios for soil quality assessment under different cover crop management systems in a semiarid vineyard. *Zeitschrift für Pflanzenernährung und Bodenkunde, 177*(4), 548–559. https://doi.org/10.1002/jpln.201300371

Polge de Combret-Champart, L., Guilpart, N., Mérot, A., Capillon, A., & Gary, C. (2013). Determinants of the degradation of soil structure in vineyards with a view to conversion to organic farming. *Soil Use & Management, 29*(4), 557–566. https://doi.org/10.1111/sum.12071

Probst, B., Schüler, C., & Joergensen, R. G. (2008). Vineyard soils under organic and conventional management - microbial biomass and activity indices and their relation to soil chemical properties. *Biology and Fertility of Soils, 44*(3), 443–450. https://doi.org/10.1007/s00374-007-0225-7

Puig-Montserrat, X., Stefanescu, C., Torre, I., Palet, J., Fàbregas, E., Dantart, J., Arrizabalaga, A., & Flaquer, C. (2017). Effects of organic and conventional crop management on vineyard biodiversity. *Agriculture, Ecosystems & Environment, 243*, 19–26. https://doi.org/10.1016/j.agee.2017.04.005

Radić, T., Likar, M., Hančević, K., Bogdanović, I., & Pasković, I. (2014). Occurrence of root endophytic fungi in organic versus conventional vineyards on the Croatian coast. *Agriculture, Ecosystems & Environment, 192*, 115–121. https://doi.org/10.1016/j.agee.2014.04.008

Raworth, K. (2018). *Dougnout economics. Seven ways to think like a 21st-century economist.* White River Junction: Chelsea Green Publishing.

Reganold, J. P., Palmer, A. S., Lockhart, J. C., & Macgregor, A. N. (1993). Soil quality and financial performance of biodynamic and conventional farms in New Zealand. *Science, 260*, 344–349.

Reicosky, D. C. (1997). Tillage-induced CO 2 emission from soil. *Nutrient Cycling in Agroecosystems, 49*(1/3), 273–285. https://doi.org/10.1023/A:1009766510274

Reid, W. V., Mooney, H. A., Cropper, A., Capistrano, D., Carpenter, S. R., Chopra, K., Dasgupta, P., Dietz, T., Duraiappah, A. K., Hassan, R., Kasperson, R., Leemans, R., May, R. M., McMichael, T.(A. J.), Pingali, P., Samper, C., Scholes, R., Watson, R. T., Zakri, A. H., … Lee, M. J. (2005). *Ecosystems and human well-being: Synthesis. Millennium ecosystem Assessment.* World Resources Institute. Retrieved from https://www.millenniumassessment.org/documents/document.356.aspx.pdf.

Robinson, Jancis (2015). In *The Oxford companion to wine* (4th rev. ed.). Corby, Oxford Companions: Oxford University Press.

Rooney, C. P., Zhao, F.-J., & McGrath, S. P. (2006). Soil factors controlling the expression of copper toxicity to plants in a wide range of European soils. *Environmental Toxicology and Chemistry, 25*(3), 726–732. https://doi.org/10.1897/04-602r.1

Scheller, E. (2013). *Grundzüge einer Pflanzenernährung des ökologischen Landbaus - ein Fragment.* Darmstadt: Verlag Lebendige Erde.

Schlatter, B., Trávníček, J., Lernoud, J., & Willer, H. (2020). *Organic agriculture worldwide: Current statistics.* Frick, Switzerland (The world of organic agriculture - Statistics and emerging trends 2020).

Schulte, R. P. O., Creamer, R. E., Donnellan, T., Farrelly, N., Fealy, R., O'Donoghue, C., & O'hUallachain, D. (2014). Functional land management: A framework for managing soil-based ecosystem services for the sustainable intensification of agriculture. *Environmental Science & Policy, 38*, 45–58. https://doi.org/10.1016/j.envsci.2013.10.002

Skinner, C., Gattinger, A., Krauss, M., Krause, H.-M., Mayer, J., van der Heijden, M. G. A., & Mäder, P. (2019). The impact of long-term organic farming on soil-derived greenhouse gas emissions. *Scientific Reports, 9*(1), 1702. https://doi.org/10.1038/s41598-018-38207-w

Steenwerth, K. L., & Belina, K. M. (2010). Vineyard weed management practices influence nitrate leaching and nitrous oxide emissions. *Agriculture, Ecosystems & Environment, 138*(1–2), 127–131. https://doi.org/10.1016/j.agee.2010.03.016

References

Steiner, R. (2005). *Geisteswissenschaftliche Grundlagen zum Gedeihen der Landwirtschaft*. Dornach: Rudolf Steiner Verlag.

Strumpf, T., Steindl, A., Strassemeyer, J., & Riepert, F. (2011). Erhebung von Kupfergesamtgehalten in ökologisch und konventionell bewirtschafteten Böden.Teil 1: Gesamtgehalte in Weinbergsböden deutscher Qualitätsanbaugebiete. *Journal für Kulturpflanzen, 63*(5), 131–143.

Tassoni, A., Tango, N., & Ferri, M. (2013). Comparison of biogenic amine and polyphenol profiles of grape berries and wines obtained following conventional, organic and biodynamic agricultural and oenological practices. *Food Chemistry, 139*(1), 405–413.

Tassoni, A., Tango, N., & Ferri, M. (2014). Polyphenol and biogenic amine profiles of Albana and Lambrusco grape berries and wines obtained following different agricultural and oenological practices. *Food and Nutrition Sciences, 5*, 8–16.

Trevors, J. T., & Cotter, C. M. (1990). Copper toxicity and uptake in microorganisms. *Journal of Industrial Microbiology, 6*, 77–84.

Way, T. R., Kishimoto, T., Allen Torbert, H., Burt, E. C., & Bailey, A. C. (2009). Tractor tire aspect ratio effects on soil bulk density and cone index. *Journal of Terramechanics, 46*(1), 27–34. https://doi.org/10.1016/j.jterra.2008.12.003

Zimmermann, R. P., & Kardos, L. T. (1961). Effect of bulk density on root growth. *Soil Science, 91*(4), 280–288.

CHAPTER 6

Vineyard water balance and use

Ignacio Buesa[1], Pascual Romero-Azorín[2], José M. Escalona[3] and Diego S. Intrigliolo[4]

[1]*University of the Balearic Islands (UIB), Research Group of Plant Biology under Mediterranean Conditions, Palma, Spain;* [2]*Instituto Murciano de Investigación y Desarrollo Agrario y Alimentario (IMIDA), Unidad "Fertirriego y Calidad Hortofrutícola", Unidad Asociada de I+D+I al CSIC, La Alberca, Spain;* [3]*Agro-Environmental and Water Economics Institute, University of the Balearic Islands, Palma, Spain;* [4]*Spanish National Research Council (CSIC), Desertification Research Center (CIDE), CSIC-UV-GV, Valencia, Spain*

6.1 The water balance concept: from the single leaf to the whole vineyard

Sustainable viticultural activity implies the rational and efficient use of resources. Among them, water is of particular importance, due to its scarcity and/or its cost of access in most wine-growing regions (Schultz & Stoll, 2010). Given the projections of climate change (IPCC, 2016), introducing irrigation could partially compensate for the detrimental effects of severe water stress on grapevine performance (van Leeuwen et al., 2019). Therefore, both under rainfed and irrigated conditions, water-saving practices are decisive in ensuring the sustainability of water use and the profitability of the grapevine industry. The improvement of water balances is therefore an objective to be pursued.

First of all, the different components that determine water balances must be defined. These components are different depending on the temporal and spatial scale of the water balance concept (Medrano et al., 2015). Thus, at the leaf level, the water balance is closely related to transpiratory water consumption, which largely depends on plant water status which in turn depends on soil-water content in the root zone, and to the transpiration rate determined by atmospheric demand and the plant genotype. Transpiration is also linked to radiation interception by the plant, and therefore to the canopy size, structure, and spatial disposition (Escalona et al., 2003). In the specific case of vineyards under vertically shoot positioning, vineyard rows' orientation is also an aspect to consider (Buesa et al., 2020). The ability to regulate water consumption by transpiration has very relevant physiological connotations associated with stoma regulation mechanisms (hormonal and chemical), foliar anatomical characteristics associated with the energy dissipation capacity, hydraulic aspects of the vascular system, etc (Dayer et al., 2020). This regulatory capacity is strongly determined by the genotype, with great differences being observed between rootstocks, varieties, and clones within the same variety (Bota et al., 2016; Romero et al., 2018).

On a larger scale of definition, we must consider other components that allow us to rigorously establish water balances. Thus, at the plot level, the inputs of the water balance are precipitation,

106 Chapter 6 Vineyard water balance and use

irrigation, runoff (influx), capillary rise, and dew deposition; and the outputs of the balance are water not retained by the soil that is lost by leaching processes (deep percolation), surface runoff (outflux), water directly evaporated from the soil, and the water used by the plant by transpiration processes, as well as the water accumulated in the soil layer and crop tissues. Nonetheless, at the vineyard scale, the main components of water balance to be considered are shown in Fig. 6.1.

The sum of the two main processes whereby water is constantly lost, on the one hand plant transpiration and on the other soil evaporation, define the concept of evapotranspiration (ET). The FAO-56 recommended the Penman-Monteith method as the standardized reference crop evapotranspiration (ET_o) calculation (Allen et al., 1998). This equation requires data of radiation, air temperature,

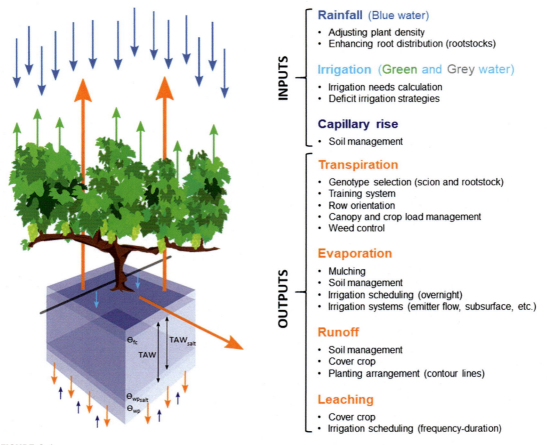

FIGURE 6.1

Components of the vineyard water balance and its optimization options. Θ_{fc}, volumetric water content at field capacity; Θ_{wp}, volumetric water content at the permanent wilting point; *TAW*, total available water content. The subscript "salt" is used when the variables are under saline conditions.

Based on FAO-56 Allen, R. G., Pereira, L. S., Raes, D., & Smith, M. (1998). Crop evapotranspiration-guidelines for computing crop water requirements-FAO Irrigation and drainage paper 56. Rome, Italy: FAO.

air humidity, and wind speed. The evapotranspiration rate is normally expressed in millimeters (L/m^2) per unit time, but can also be expressed in terms of energy (latent heat) required to vaporize free water (MJ/m^2). In order to consider the crop-related factors on ET, the FAO recommends the use of an empirical crop coefficient (K_c). Briefly, the procedure estimates water requirements or evapotranspiration of the crop (ET_c) based on the climate variables (ETo) and a factor linked to the crop (K_c). Nevertheless, this simple model ($ET_c = ET_o \times K_c$) is recommended to be adjusted taking into account management factors and environmental conditions such as soil salinity, nutritional deficiencies, pests, etc (Katerji & Rana, 2014). Other factors to be considered when assessing ET_c, particularly in woody crops such as grapevine, are ground cover, canopy architecture, plant density, crop management practices, and phenology. For this purpose, the seasonal variation of K_c is split into two separate coefficients, one for crop transpiration, the basal crop coefficient (K_{cb}), and another one for soil evaporation (K_e).

A direct way to improve water balances is therefore to reduce water losses by direct evaporation (K_e). This can be achieved through the use of drip irrigation systems, where the percentage of wet soil is relatively low. Or further improved by subsurface drip irrigation (Pisciotta et al., 2018) and direct root-zone irrigation (DRZ) (Ma et al., 2020). Regarding mulching and other soil management practices, recent studies that compare different materials used for mulching, from pruning waste, straw, or plastic located in the line of vines, can reduce the total water consumption by 16%—30% and consequently improving vine water status (Buesa et al., 2021). Moreover, the potential transpiration cost of the vineyard can be significantly reduced by setting the training system and orientation of the vine's rows (Buesa et al., 2020), or by green pruning, for instance, with shoot trimming and leaf removal (Buesa et al., 2019). Aiming to reduce other outputs of the vineyard water balance such as runoff losses, the proper soil and cover crop management can improve it through increases in water infiltration and reductions in soil penetration resistance. The implementation of cover crops as an alternative to soil tillage improves the physical properties of the soil associated with the infiltration rate of rainwater, and largely avoids losses due to runoff (Ruiz-Colmenero et al., 2011). However, cover crops, as weeds, also compete with grapevine for water, which can reduce yield without any benefits to grape quality (Lopes et al., 2011). Moreover, these field practices can have a direct effect not only on the soil water balance (evaporation, infiltration rate, water-holding capacity, etc.) but also on plant transpiration via vineyard energy balance, vine microclimate, and also vine root growth and activity. This affects the physiology of the plant and thus the yield and quality of the grapes.

On the other hand, in order to improve water use efficiency (WUE) of the vineyard water balance (i.e., output/input ratio) of its individual component at the respective levels (Hsiao et al., 2007), different ways to increase the efficiency of using inputs such as rainfall and irrigation are proposed. The optimization of rainwater use, the so-called blue water, can be achieved indirectly by enhancing the exploration of the soil by the roots. It should be noted that plants responses cannot be fully explained only in terms of soil water and nutrients availability, but it is also necessary to consider the plant's ability to access available soil resources. This can be enhanced, for instance, by selecting more vigorous rootstocks (Zhang et al., 2016), but also through proper preparation of the soil before planting. It is in the planting when the crop spacing is also decided, which will determine the potential rainwater stored in the soil per plant (van Leeuwen et al., 2019). If the planting density is too low, there will be water stored in the soil that will not be used by the crop and will flow out of the wine-growing system by deep percolation (aquifer recharge on a river basin scale). On the contrary, if the planting density is too high, the vineyard will rely on irrigation dosages. In this sense, irrigation is the other main water input to the vineyard balance which can also be optimized through adequate management. These aspects will be addressed in detail in the subsequent sections.

6.2 Grapevine water status assessment: from soil to atmosphere

It is widely accepted that the movement of water through the plant is based on the concept of the soil-plant-atmosphere continuum based on the stress-cohesion theory proposed by Dixon and Jolly in 1894. These authors stated that the absorption and transport of water is due to a difference in water tension between the soil and the atmosphere and the cohesion of the water molecules associated with the characteristics of the capillaries that make up the vascular system, from the absorbent hairs of the root to the leaf blade. This process is the driving force for water flow through the plant, known as transpiration. The intensity of the leaf transpiration activity depends therefore on the following main factors: the availability of soil water for the plant (Ψ_{soil}), the water vapor pressure deficit (VPD) in the atmosphere and the plant.

In relation to the first factor, the physical properties of the soil such as its texture and aggregation determine its hydraulic characteristics associated with the retention energy of the water by the soil particles that affect its conductivity and availability for plants. Texture is a very steady soil property, while state of aggregation is more variable. These properties, together with the soil depth, define soil water storage capacity, a key characteristic to understand the viability of a vineyard in semiarid conditions and with little precipitation. The water storage capacity of farming soils can vary from 100 to 300 mm m-1. The soil water potential depends mainly on the adsorption forces that appear between the soil particles and the water ($-0.8 > \Psi_{soil} > 0.3$ MPa), depending, among other factors, on the water volumetric content of the soil, which ranges from field capacity to the wilting point (Fig. 6.1). Furthermore, the ability of the roots to explore the soil and the association or nonassociation with arbuscular mycorrhizal fungi will determine the soil water that can be used (Torres et al., 2018). Under irrigation, usually in a localized manner, the inherent heterogeneity of soil water content (SWC) and thus, the root system distribution, also affect plant responses to atmosphere demand.

Regarding the second factor, the atmospheric demand is mainly determined by temperature, solar radiation, relative humidity, and wind speed (Allen et al., 1998). As stated above, the water movement in the soil-plant-atmosphere continuum is due to differences in gradient of water potential (Ψ_w). Water flow will occur spontaneously in favor of a gradient of water potential, from the places with the highest Ψ_w to the places with the lowest Ψ_w. The flow of water will be directly proportional to the difference in water potential but will also depend on the resistance it encounters along its pathway. At a larger spatial scale, for example, at the individual vine or vineyard scale, the total amount of transpired water also depends on the leaf surface exposed to the sun and its distribution, which depends on the vigor of the crop, the training system and pruning, thus as well as the orientation of the plant canopy (Escalona et al., 2003).

The third factor, the plant, was already addressed when describing the characteristics of the soil and the atmosphere, since the soil-plant-atmosphere continuum is intrinsically linked. The flow of water from the soil into the atmosphere takes place through the whole plant, and can also be explained on the basis of the gradient of water potential along the pathway. The plant's water potential (Ψ_p) results from the combination of four components, the osmotic potential (p), the turgor pressure (*P*), the matric potential (t), and the gravitational potential (Turner, 1981). In grapevines the gravitational component can be neglected as it is only 0.01 MPa/m. Therefore, the plant water uptake is the result of the combined effect of osmotic and hydrostatic forces. In practice, the matric potential breaks down into either a pressure term or osmotic term, resulting in a root pressure generated by the active transport of solutes and biosynthesis of new osmolytes, and the cohesive tension generated by the leaf transpiration on the whole water column. In this regard, the resistance along water pathway depends on the hydraulic properties of the soil but also on the hydraulic capacity of the plant vascular system, which is determined by the genotype (Lovisolo et al., 2016). In the case of the vineyard, the appropriate choice of the rootstock and, to a lesser extent, the variety, can allow to increase this hydraulic capacity (Romero et al., 2018; Zhang et al., 2016). Nevertheless, the plant also has mechanism of stomatal regulation in order to avoid situations of excessive foliar dehydration (Bota et al., 2016). Grapevine responds to soil water shortage through stomata closure to minimize transpiration and consequently, decreasing its photosynthetic capacity (Flexas et al., 2002). Under conditions of high evaporative demand, a nontight stomatal regulation may provoke decrease in cell turgor and xylem embolism formation, which are necessary for water transport capacity and gas exchange (Knipfer et al., 2015). However, the stomata opening is necessary for the plant not only to capture carbon dioxide for photosynthesis but also for cooling the canopy through evaporation (due to the high latent heat of water vaporization). Even at nigh-time, the stomata remain active, taking part in the refilling process of depleted stem and organ (Dayer et al., 2021). Consequently, there is transpiration at night-time, although uncoupled from photosynthesis. Night-time transpiration is not usually taken into account in the vineyard water balance, but may have some relevance (Escalona et al., 2013; Fuentes et at., 2014). Recently Montoro et al. (2020) using a weighting lysimeter quantified nocturnal transpiration ranging from 3% up to 35% of the total daily water use, depending on soil water availability.

6.2.1 Main indicators of soil-plant-atmosphere water status

There are several methodologies for the measurement of vineyard water status. Most of these methodologies are shown in Fig. 6.2, including the ones based on the determination of the SWC, plant water status, and hydraulics and the atmospheric demand conditions. However, all these measurement methodologies are based on interrelated parameters, and often overlapped.

SWC can be measured directly by gravimetry and Richard plates or indirectly, by measuring soil matric potential (tensiometers), soil electrical resistance (watermark probes), soil dielectric properties (capacitive sensors such as time domain reflectometry [TDR] and frequency domain reflectometry [FDR]), soil electromagnetism (induction sensors), or by neutron probes. Direct methods are very accurate, but have the disadvantage that they are destructive, labor intensive, and unsuitable for continuous soil moisture monitoring. Therefore, they are used to calibrate indirect methods. Among the latter, TDR and FDR are widely used because of its ease of continuous monitoring, relatively good accuracy, and low cost. Nevertheless, these probes are sensitive to salinity, temperature, air gaps, and percentage of stone. Furthermore, it may not account for water consumption in vines with extensive root systems grown in deep soils.

The plant water potential (Ψ_p) has gained prominence as the most widely adopted measurement of plant water status. It is the most traditional indicator used to monitor the water status of a plant, which involves the use of a pressure chamber. The Ψ_p is a physiological parameter that indicates the balance between the flows of water absorption and transpiration by means of the potential required to extract the water contained in the leaf tissues. It is a consequence of the interaction among the three factors of the soil-water-atmosphere continuum. It is a suitable indicator of plant water status because as the water stress suffered by the plant intensifies, Ψ_p decreases. If measured before dawn, it is called predawn water potential (Ψ_{pd}), which can be used as surrogate or Ψ_{soil}, at midday on sun-exposed

FIGURE 6.2

Measurement methodologies of the soil water content, plant hydraulics, plant water status, energy balance, and the atmospheric demand conditions. (1) near infrared reflectance; (2) weighting lysimeter; (3) TDR probe; (4) dendrometer; (5) TDR probes; (6) pressure chamber; (7) infrared gas analyzer (IRGA) at whole vine; (8) IRGA at whole vine and leaf level; (9) Eddy covariance method; (10) meteorological station; (11) high pressure flow meter (HPFM); (12) sap-flow gauges; (13) electromagnetic induction sensors; (14) drone with hyperspectral-camera; and (15) chlorophyll fluorescence.

leaves (Ψ_{leaf}) and in bag-covered leaves (Ψ_{stem}), when leaf water potential equilibrates with the stem (Intrigliolo & Castel, 2006). Furthermore, in order to standardize these determinations among different atmospheric conditions, the relationship between Ψ_{stem} or Ψ_{leaf} and VPD can be used to establish a nonwater-stressed baseline under semiarid grape-growing regions.

Among the other methodologies based on the measurement of plant parameters, it should be highlighted first, the destructive ones, the relative water content, the isotopic carbon ratio determination, pressure-volume curves, and hydraulic conductivity. The most interesting among the nondestructive ones are dendrometry, psycrhometry, sap-flow measurement, leaf turgor pressure, thermography, infrared gas analysis, reflectance indices, and chlorophyll fluorescence.

In order to accurately capture the spatial variability of the water status of the vineyard and its evolution throughout the season, ground-based determinations are required at such high number and frequency of measurements that challenges its feasibility (Ortega-Farias & Intrigliolo, 2021). In this sense, there are other more cost-effective techniques that can provide spatially distributed information of vine water status and soil hydraulic dynamics, such as remote sensing and electrical resistivity tomography (Araya Vargas et al., 2021).

Nowadays, many growers dispose of several useful utilities for monitoring vineyard water needs based in classical methodologies but also it is more and more common to use new technologies based on mobile apps and remote sensing data acquisition and processing. For instance, in the last years, some apps as Viticanopy© and Flir one thermal cameras connected to mobile smart phones has been developed to be use in the field for the estimation of plant growth and plant water status (Petrie et al., 2019). Based on canopy temperatures, a crop water stress index has been developed to quantify the level of water stress of crop canopies using thermal difference between the canopy and surrounding air which is normalized for differences in climatic conditions using the VPD of the air (Zarco-Tejada et al., 2013). Also, some "on the go" equipment's based on the capture of infrared and hyper-spectral images by using high-definition cameras have been used for mapping the degree of variability of different plant water status parameters (Diago et al., 2018). Finally, technologies based on remote sensing have been developed in the last years providing high-quality and fine-resolution weather characterization for extensive areas (Carrasco-Benavides et al., 2014). From unmanned aerial vehicle (UAV) that are able to estimate with a very high precision plant water status, which allow covering the spatial variability within the vineyard in a time and cost-effective way (Gago et al., 2020).

For the measurement of atmospheric conditions, the most widely used methodology is the weather station. For agrometeorological purposes, these must have thermometer, hygrometer, pyranometer, anemometer, and pluviometer. These sensors allow the estimation of such important parameters for vineyard water needs determination as the VPD or the ET_o.

6.3 Vineyard water needs: crop coefficients in relation to vegetative development (LAI) and reproductive cycle. Crop stress coefficients

Determining the potential grapevine water needs is required for establishing efficient irrigation scheduling programs and for determining the vineyard impacts in the agroecosystem water balance. This is particularly relevant in those areas where grapevine cultivation covers a large area and it is the predominant farming system heavily impacting the water cycle at the entire water basin level. Even if grapevine is considered a drought-tolerant species, the potential water needs under absence of soil water limitation, or any other pedoclimatic or vine endogenous factor that can impair the potential transpiration, can be relatively high. The total water consumption of vineyards ranges from 300 to 700 mm per season depending on water availability and grapevine varieties-environment interactions (Medrano et al., 2015).

Grapevine water use has been quantified with different techniques such as lysimeters, micrometeorological methods, sap flow sensors, thermal remote sensing, radiometric temperature measurements and two-source energy balance, surface renewal, and remote sensing energy balance for application over large areas. Specific robust research employing weighting lysimeters have reported crop coefficient values for vertical training systems ranging from 1.0 to 1.2 for a ground cover of 60% of the soil allotted per vine (Picón-Toro et al., 2012). López-Urrea et al. (2012) carried out a similar study with less vigorous vines that did not reach ground cover values higher than 45%, and reported K_c up to 0.68 for this ground cover value. The most valuable results of the mentioned researches are the linear equations there reported for estimating K_c depending on several vegetation indexes. It has been obtained similar slopes for the increase in K_c with vary ground cover percentages ranging from 0.017 to 0.02 (López-Urrea et al., 2012). But there are also researches that report greater slopes than those already mentioned (Munitz et al., 2019), suggesting the importance that can have the disposition of the leaves within the canopy.

Since vine vegetative vigor can be quantified using remote sensing techniques with the possibility of covering large vineyard areas, nowadays estimating grapevine water needs using, for instance, the Normalized Different Vegetation Index (NDVI) is a common approach and a linear relationship between NDVI and K_c has been obtained (Campos et al., 2010). This approach has the important advantage to be able to determine the expected evapotranspiration for large areas of the vineyard and even to establish different management zones based on the remote sensing data obtained. But on the other hand, it also has drawbacks such as the fact that the relationship between NDVI and K_c is site-specific and does not adequately perform for water-stressed plants. Moreover, as vineyards are characterized by discontinuous canopies leaving important portion of the soil areas uncovered by the vine vegetation, when satellites (e.g., Sentinel-II) are used to obtain NDVI, it is not possible to distinguish between the grapevine canopy and the inter-row vineyard soil. Therefore, the soil management used in the vineyard might indeed affect the relationship between the obtained NDVI and the K_c. However, the indexes based in visible (380–780 nm) and near infrared electromagnetic spectrum (780–2500 nm) such as NDVI are not suitable for direct ET estimation (Ballesteros et al., 2015). Moreover, the feasibility of one- and two-source energy balance models to analyze the spatio-temporal variability of ET using thermal infrared and multispectral sensors placed on satellite and UAV platforms should be highlighted (Sánchez et al., 2019).

While the grapevine potential water needs have been reasonably well-established, grapevine ET values data available under certain stress conditions (soil water deficit or salinity) are much more scarce (Hochberg et al., 2017). This is an important limitation, since grapevine is a crop that, particularly in the so-called old-world viticulture, is often rainfed or irrigated under deficit irrigation (DI) and therefore subject to certain degrees of soil water deficit (Sadras & Schultz, 2012). Certainly, it is more difficult to quantify grapevine water use under drought conditions because vine water stress not only affects the vine canopy growth but also the stomatal conductance and therefore the transpiration capacity for unit of leaf area. As a consequence, it is more difficult to model grapevine transpiration under certain degrees of soil water deficit and, so far, the most common approach is to derive empiric stress coefficients to reduce the potential plant transpiration according to the degree of water stress suffered by the vines. Several authors have reported relationships between the expected reduction in vine evapotranspiration depending on the midday leaf water potential (Williams et al., 2012), predawn water potential (Ferreira et al., 2012), and midday stem water potential (Picón-Toro et al., 2012).

It is therefore possible nowadays to fairly well estimate grapevine water needs by considering three main drivers: (1) the evaporative demand by computing the reference evapotranspiration, (2) vegetative vigor by employing for instance ground cover measurements, and (3) the plant water status in case soil water is a limiting factor. Finally, local adjustments will have to be in any case employed depending on the soil wetted area and the irrigation frequency, both factors affecting soil evaporation, but also on the different genotype sensitivities of stomatal conductance, berry growth, and grape ripening to changes in Ψ_{soil}. In fact, irrigation programs, including deficit irrigated strategies, have been stablished taking into account the variety-rootstock combination as explained in the following section.

6.4 Water-saving strategies and irrigation scheduling

DI consists on applying water rates to replace only a percentage of the potential vine evapotranspiration (Intrigliolo & Castel, 2010). When the water shortage is applied steadily throughout the season, the DI strategy is known as sustained deficit irrigation (SDI), whereas if applied during specific phenological stages is referred as regulated deficit irrigation (RDI). The latter is a particularly promising management irrigation technique in arid and semiarid areas as it offers greater potential to increase WUE and improve berry and wine quality (Fereres & Soriano, 2007). Numerous DI studies carried out under different edaphoclimatic and experimental conditions, with several winegrapes varieties-rootstocks combinations and with different RDI strategies tested and water volumes applied have revealed significant improvements in WUE and berry quality, although also with yield losses compared to conventional or standard irrigation, SDI and full irrigation practices. Moreover, it must be noted that the application of irrigation in the vineyard is usually performed by efficient technologies such as pressurized irrigation systems (Keller, 2005) and therefore dependent on significant energy inputs for its pumping. In this regard, it should be borne in mind that any reduction in irrigation water to the vineyard will lead to energy savings.

The physiological base of DI approaches consists in applying irrigation based upon optimum water status in order to maintain vine water status within the range of optimum thresholds proposed (Romero et al., 2013). On the other hand, the principle of RDI is that plant sensitivity to water stress, in terms of yield and berry quality, is different depending on vine phenological stage and the severity of the stress imposed (McCarthy et al., 2002). If water deficit is applied from fruit set to veraison (preveraison water deficit), it will mainly control berry size and reduce vine vigor (McCarthy et al., 2002). If applied after veraison, during berry ripening (Phase III) (postveraison water deficit), it will mainly increase the biosynthesis of anthocyanins and other phenolic compounds (Castellarin et al., 2007). In general, preveraison water deficit is more effective than postveraison water deficit in reducing berry growth (Levin et al., 2020). Preveraison water deficit can also accelerate berry color change (Herrera & Castellarin, 2016). Both pre- and postveraison water deficits have the potential to benefit berry and wine quality in different ways, including: (1) a reduced berry weight and size and higher skin/pulp ratio, (2) an improvement in the cluster microclimate (due to reduced canopy leaf area and more open canopies), (3) altered endogenous hormonal response (i.e., increased abscisic acid (ABA) in roots, leaves, and/or berries), (4) changes in root to shoot ratio, and/or (5) greater gene expression regulating flavonoid biosynthesis.

The effect of RDI depends on the phenological stage and the severity of the stress imposed in each stage, which in turn is variety specific (Mirás-Avalos & Intrigliolo, 2017). An efficient scheduling of RDI requires maintaining the soil and plant water status within a narrow tolerance range, and defining several optimum thresholds values for soil–plant stress indicators to avoid severe root and leaf function damage and drastic yield and berry quality losses (Gambetta et al., 2020). Nonetheless, these indicators depend on the iso/anisohydric behavior of the variety resulting from its interaction with the environment.

Negative carryover effects of some RDI strategies can be also important to maximize grapevine yields or to maintain a sustained long-term high productivity is the main goal (Buesa et al., 2017). A recent study in 15 field-grown grapevine varieties reported that preveraison water deficits not only reduced yield in the current season by reducing berry size but can also reduce yield the following season by reducing bud fruitfulness through a reduction in clusters vine^{-1} (Levin et al., 2020). Thus, to maximize grapevine yields, water should be applied early to avoid preraraison water deficits that can inhibit berry growth in the current season and bud fruitfulness for the following season (Levin et al., 2020). Probably, preveraison water deficits of Ψ_{leaf} averaged ≥ -1.5 MPa (as reported in Levin's study, -1.5 to -1.6 MPa), indicator of severe water stress, was enough to disrupt development of cluster primordia in the bud.

As a general guide, most of RDI strategies that have shown benefits apply moderate or low seasonal irrigation water volumes, between 50 and 150 mm season^{-1}. Besides, the physiologically based irrigation scheduling studies carried out in different grapevine varieties and edaphoclimatic conditions have found significant correlations between midday Ψ_{stem} and berry quality components and have showed Ψ_{stem} as a suitable physiological indicator to apply accurate water supply and irrigation scheduling in RDI winegrapes. These studies have revealed that by maintaining moderate levels of mid-day stem water potential, within an optimum Ψ_{stem} within -1.2 and -1.4 MPa (never $\Psi_s \geq -1.4$ MPa) during the pre- and postveraison periods (mainly from fruit set to harvest) could improve WUE, berry quality, and irrigation management in RDI grapevines (Levin et al., 2020; Romero et al., 2013). Moreover, the decoupling between grape primary metabolism (e.g., synthesis of sugars) and secondary metabolism (e.g., synthesis of phenolic substances) observed in field-grown grapevines under heat stress (Sadras & Moran, 2012) may also suggest that RDI strategies will have to be adapted to new climatic conditions as consequence of global warming to minimize these undesirable effects on berry quality.

RDI strategies has been developed and applied more frequently by using surface drip irrigation, but other physiologically based irrigation methods have also been successfully used in RDI winegrapes such as partial root zone drying irrigation (PRI). Intrinsic changes in vine physiology and berry metabolism caused specifically by PRI have been reported in different grapevine varieties (Romero et al., 2016). Previous research in PRI field-grown vines showed an increased xylem sap ABA concentration following reirrigation (Romero et al., 2014) and altered endogenous hormonal status of the berries (e.g., ABA) at harvest, compared to RDI (Romero et al., 2015). As ABA is implicated in berry ripening process and grapevine response to stress (Ferrandino & Lovisolo, 2014), berry ABA accumulation may explain the increase in the concentration of flavonoids and stilbenes in PRI (Romero et al., 2018). PRI shows a great potential to improve berry and wine quality when is applied in combination with optimized DI strategies in semiarid areas, but the successful of PRI is dependent on several factors such as variety-rootstock interaction, soil-climate characteristics, cultural practices, and irrigation management. All these factors, plus the extra costs of the irrigation installation and management, point to the importance of carefully evaluating the circumstances in which the PRI may be economically justified.

Other additional factors that may be important in increasing irrigation efficiency and vineyard performance under DI strategies are the frequency of irrigation, emitter spacing and water distribution patterns, and water volume applied in each irrigation (discharge flow rate). The effect of irrigation frequency seems to be more relevant than that of emitter spacing and water distribution pattern (Sebastian et al., 2016). The results obtained in Syrah vineyards indicate that applying a small irrigation dose with a high irrigation frequency (every two days) in a heavy clay soil may lead to an efficiency loss (compared to every four days). This is because as the wetted soil volume created is small and close to the soil surface, it favors soil water evaporation, overall under low water irrigation volume (137 mm season^{-1}) as usually applied in DI strategies (Sebastian et al., 2015). Besides, under low water availability conditions, plant irrigated every four days had higher average net assimilation than plants irrigated every two days (Sebastian et al., 2016). In contrast, in table grapes in sandy soils, more frequently irrigation resulted more beneficial than low irrigation frequency (Myburgh, 2012). These contrasts evidence the need to adapt the agronomic design of the irrigation system to the particular conditions of each vineyard (soil texture and depth, meteorology, irrigation water characteristics, rootstock, etc.) as well as, of course, to the winemaking objective.

Moreover, the economic evaluation of the irrigation system must be carried out before adopting any irrigation strategy. In this sense, the energy efficiency of the pressurized system plays a crucial role. Both proper hydraulic design and maintenance are essential for effective operation (Moreno et al., 2016). Simple actions such as cleaning filters and emitters periodically are key to achieving energy savings and ensuring optimal operating pressures. In this regard, it is worth noting the usefulness that soil moisture sensors can have in monitoring the application efficiency of irrigation (Intrigliolo & Castel, 2006). But also improve the scheduling time and frequency of irrigation by adjusting the soil profile recharge where the roots are active and avoiding excessive water percolation. Finally, modeling of irrigation systems can also play an important role in precision irrigation, as it allows the determination of irrigation water application uniformity (González-Perea et al., 2018).

6.5 Use of nonconventional water for irrigation: wastewater and saline water. Effects on vine performance and grape composition

As fresh water is scarce resource in most grape-growing regions, there is a need to use alternative water sources to mitigate drought stress such as using brackish water or wastewater often having also high salt concentration (1.5—3 g/L). Wastewater may contain high concentration of dissolved ions but also constituents of potential concern such as heavy metals, pathogens, and a high biological oxygen demand (Mosse et al., 2011). Nonetheless, heavy metals are also nutritionally necessary for plants and in the concentrations normally found after proper depuration, they should not result in phytotoxicities as they are mostly immobilized in the soil (Petousi et al., 2019). Currently there are several technologies to reclaim wastewater and anticipate these potential problems (Liu et al., 2020). Therefore, nonconventional waters might also be a source of nutrients, particularly of macronutrients such as nitrogen, phosphorous, and potassium potentially affecting grape composition and the WUE (Hirzel et al., 2017).

Aiming at a sustainable viticulture, the use of nonconventional water for irrigation must not compromise soil and plant health. Therefore, the short and midterm detrimental effects of saline waters on vine performance have to be evaluated (Costa et al., 2016), but also the long-term ones. Regardless of the water quality, drip irrigation practices might salinize vineyard soils due to the accumulation of solutes dissolved in the irrigation water (Aragüés et al., 2014), unless there is sufficient leaching beyond the root zone (Hsiao et al., 2007). As irrigation water is evapotranspired from the soil, the ions it contains accumulate (i.e., precipitation of salts and ion pair formation as a result of chemical interactions). Under drip irrigated systems, this may result in the accumulation of salts on the boundaries of the wetting zone, which can restrict root development. Therefore, when the rainfall is not sufficient to leach this accumulation from the root zone, mitigation practices of soil salinity must be implemented aiming sustainability of viticulture.

The initial effects of soil salinity on plant responses are similar to the effects of drought stress, including reduction in stomatal conductance and thus in transpiration and photosynthesis. However, over the long-term, the photosynthetic capacity declines leading to vegetative growth, yield and berry ripening constraints (Liu et al., 2020). Under high salt exposure, a simultaneous decrease in water availability occurs due to an osmotic effect (osmotic stress), whereas other effects are salt-specific (ionic stress), also called phytotoxicity (Chaves et al., 2009).

The *Vitis vinifera* L. is a crop classified as moderately sensitive to salinity with a threshold value (the value from where crop yield starts to be affected by salinity) of electrical conductivity in the saturated extract of the soil (ECe) that is between 1.8 and 4.0 dS/m (Cramer et al., 2007; Maas & Hoffman, 1977). This means that water production function (i.e., relative yield response to water salinity) decreases at a rate of 2.3%—15.0% for each unit increase in the CEe expressed in dS/m. For instance, Minhas et al. (2020) estimated that with an ECe of 6.7 dS/m, grapevine yield is reduced by 50%.

Coping with saline water in irrigated agriculture requires indeed a good understanding of the impacts of temporal and spatial dynamics of salinity on soil water balances and the development of optimal irrigation schedules and efficient methods (Allen et al., 1998). These include an excess of irrigation to leach salts down the soil profile, increasing the crop water requirements. On the other hand, salinity can also alter the physico-chemical properties of the soil through ion exchange and the consequent loss of soil structure (Shani & Ben-Gal, 2005), which could threaten the sustainability of irrigation. The latter depends on soil components determining its cation exchange capacity and the availability of nutrients, with the pH governing the chemical form in which the different elements are present. In this regard, the ionic chemistry of soil solutions/irrigation waters, soil texture and clay mineralogy, agronomical practices and climate require due consideration (Minhas et al., 2020). For instance, sodification processes leads to dispersion of soil colloids that indirectly impacts crops by deterioration of soil structure (infiltration, aeration, and mechanical strength). The dispersing effect of sodium can be offset by high levels of other free ions in soil solution, mainly Ca^{2+} and Mg^{2+}, which have an aggregating function. For assessing saline waters effects on soil properties, it is recommended to use the cation ratio of soil-structural stability (CROSS, Rengasamy & Marchuk, 2011) rather than the traditional sodium adsorption ratio (SAR) criteria (Richards, 1954).

When the vineyard water balance is such that there is no deep percolation and the salts accumulate in the root zone, a leaching fraction (LF) should be adopted. The estimation of leaching requires an accurate soil water balance to be able to define the required excess of irrigation water (Walker et al., 2005). However, this is much more complicated when it comes to DI because no leaching would occur during the irrigation season (Aragüés et al., 2014). Under drip irrigation, whether superficial or subsurface, it is critical to design irrigation systems taking into account the need to ensure overlap between wet soil zones so as not to force the roots through the high concentrations of salts generated at their boundaries. DI management should aim at keeping high soil water potential the root zone, so that salts do not concentrate where the most active roots are. This can be achieved with proper management of the frequency and duration of irrigation. In addition, occasional water applications (rainfall (R) plus irrigation (I)) that exceed vine water needs and wet bulb water storage capacity is required periodically in order to displace the accumulation of salts outside the boundaries of the wet zone. This is known as LF, and is defined as $LF = DP/(I + R)$, where DP is deep percolation. Guidelines for estimating appropriate LF have been established and are available from FAO-29 (Ayers & Westcot, 1985).

In regions where annual rainfall exceeds 300 mm season^{-1} and with soils with good infiltration rates, drainage should be enough to eliminate salt accumulation when irrigating with medium quality water and there is no ion exchange. However, particular attention should be paid to shallow soils or those with a very superficial water table, because then precipitation could cause an ascent of accumulated salinity beyond the root zone. Therefore, monitoring salinity is highly recommended aiming to sustain vineyards in the long term, mainly in arid areas and with poor quality irrigation water. Other possible mitigation practices include the application of mulching to limit evaporation, and less soil tilling to conserve soil structure but also grapevine rootstocks-variety combinations affecting osmotic adjustment and salt exclusion (Keller, 2010). New information about salinity tolerance conferred by rootstocks is needed, especially addressing the possible interactive effects of water quality and water deficit.

6.5.1 Effects on vine performance and grape composition

The use of saline and wastewater for vine irrigation can significantly affect vegetative development, yield components, and the composition of berries and wine (Mosse et al., 2013). These effects are the more pronounced, the greater the accumulation of salt in the soil and the longer the duration of exposure to the plant. In general terms, in a situation of moderate saline stress, shoot growth and yield can be reduced (Walker et al., 2002). Usually, the most affected parameters are bud fruitfulness and the berry weight. Although exceptions have also been reported, such as the reported by Netzer et al. (2014) in a six-year study on Superior Seedless table grapes, where grape yield was unaffected by salinity water despite Na^+ accumulation in vine tissues.

The effects of salt stress on grape composition are not very conclusive. For instance, in field-grown vines during six years, Stevens et al. (2011) found no consistent effects in the TSS accumulation in response to saline water application compared to control water. However, saline irrigation raised juice pH and this was associated with a rise in juice Na^+. In another study investigating the effects of wastewater irrigation on grape and wine chemical composition, Hirzel et al. (2017) found minor effects on grape and wine composition despite soil accumulation of Na^+ and K^+ cations. Other authors observed slights advances in veraison and thus, berries reaching a higher sugars concentration at harvest (Mosse et al., 2013). In view of these findings, it can be inferred that the effects of irrigation with saline water on grape composition cannot be generalized, since they depend on many variables such as the soil, the quality of the irrigation water and its management, the rootstock-variety combination, etc. In a meta-analysis performed assessing water and salinity stress in grapevine, Mirás-Avalos and Intrigliolo (2017) reported that up to 90% of the variability in the berry composition of both red and white cultivars, as well as on the final wine, was explained by the effects of cultivar, rootstock, and vine water status. In general, there is no consistent effect of irrigation with saline or wastewater on vine performance and grape and wine composition. More specific research is required to better assess the effects of nonconventional water use on grapevine performance and particularly on wine quality and safety.

6.6 Concluding remarks

Sustainable viticulture implies a rational and efficient use of water resources. For a sustainable vineyard management, all the components of the water balance must be considered. The appropriate use of continuous water status indicators, the accurate estimation of the vineyard's water requirements, and the scheduling of irrigation strategies as well as other field practices have proven to be useful. In this regard, remote sensing techniques have proven to be particularly effective for the spatial variability estimation of ET and Ψ_p. Nevertheless, the particularity of viticulture for winemaking requires paying special attention not only to vine performance but also to grape and wine composition. Therefore, the latest findings on DI strategies are presented as well as its optimal management. In this regard, RDI and PRI strategies are very promising for grapevine irrigation in arid and semiarid regions. However, the results of using DI strategies with nonconventional waters in vineyards are not fully consistent given the complexity of the osmotic and/or phytotoxic effects caused by its use, which vary according to soil type, management practices and, ultimately, the plant material (rootstock-variety combination).

Acknowledgments

This work was performed with financial support from the *Ministerio de Ciencia, Innovación y Universidades* (Spain) (project AGL2017-83738-C3-1-R).

References

Allen, R. G., Pereira, L. S., Raes, D., & Smith, M. (1998). *Crop evapotranspiration-guidelines for computing crop water requirements-FAO Irrigation and drainage paper 56*. Rome, Italy: FAO.

Aragüés, R., Medina, E. T., Clavería, I., Martínez-Cob, A., & Faci, J. (2014). Regulated deficit irrigation, soil salinization and soil sodification in a table grape vineyard drip-irrigated with moderately saline waters. *Agricultural Water Management, 134*, 84–93.

Araya Vargas, J., Gil, P. M., Meza, F. J., Yáñez, G., Menanno, G., García-Gutiérrez, V., Luque, A. J., Poblete, F., Figueroa, R., Maringue, J., Pérez-Estay, N., & Sanhueza, J. (2021). Soil electrical resistivity monitoring as a practical tool for evaluating irrigation systems efficiency at the orchard scale: A case study in a vineyard in Central Chile. *Irrigation Science, 39*(1), 123–143.

Ayers, S., & Westcot, D. (1985). *Water quality for agriculture (FAO 29)*. Rome.

Ballesteros, R., Ortega, J. F., Hernández, D., & Moreno, M.Á. (2015). Characterization of *Vitis vinifera* L. canopy using unmanned aerial vehicle-based remote sensing and photogrammetry techniques. *American Journal of Enology and Viticulture, 66*, 120–129.

Bota, J., Tomás, M., Flexas, J., Medrano, H., & Escalona, J. M. (2016). Differences among grapevine cultivars in their stomatal behavior and water use efficiency under progressive water stress. *Agricultural Water Management, 164*(Part 1), 91–99.

Buesa, I., Caccavello, G., Basile, B., Merli, M. C., Poni, S., Chirivella, C., & Intrigliolo, D. S. (2019). Delaying berry ripening of Bobal and Tempranillo grapevines by late leaf removal in a semi-arid and temperate-warm climate under different water regimes. *Australian Journal of Grape and Wine Research, 25*(1), 70–82. https://doi.org/10.1111/ajgw.12368

Buesa, I., Mirás-Avalos, J. M., De Paz, J. M., Visconti, F., Sanz, F., Yeves, A., Guerra, D., & Intrigliolo, D. S. (2021). Soil management in semi-arid vineyards: Combined effects of organic mulching and no-tillage under different water regimes. *European Journal of Agronomy, 123*, 126198. https://doi.org/10.1016/j.eja.2020.126198

Buesa, I., Mirás-Avalos, J. M., & Intrigliolo, D. S. (2020). Row orientation effects on potted-vines performance and water-use efficiency. *Agricultural and Forest Meteorology, 294*, 108148. https://doi.org/10.1016/j.agrformet.2020.108148

Buesa, I., Pérez, D., Castel, J., Intrigliolo, D. S., & Castel, J. R. (2017). Effect of deficit irrigation on vine performance and grape composition of *Vitis vinifera* L. cv. Muscat of Alexandria. *Australian Journal of Grape and Wine Research, 23*, 251–259. https://doi.org/10.1111/ajgw.12280

Campos, I., Neale, C. M. U., Calera, A., Balbontín, C., & González-Piqueras, J. (2010). Assessing satellite-based basal crop coefficients for irrigated grapes (*Vitis vinifera* L.). *Agricultural Water Management, 98*(1), 45–54.

Carrasco-Benavides, M., Ortega-Farias, S., Lagos, L. O., Kleissl, J., Morales-Salinas, L., & Kilic, A. (2014). Parameterization of the satellite-based model (METRIC) for the estimation of instantaneous surface energy balance components over a drip-irrigated vineyard. *Remote Sensing, 6*(11), 11342–11371. https://doi.org/10.3390/rs61111342

Castellarin, S. D., Matthews, M. A., Di Gaspero, G. D., & Gambetta, G. A. (2007). Water deficits accelerate ripening and induce changes in gene exprésion regulating flavonoid biosynthesis in grape berries. *Planta, 227*, 101–112.

Chaves, M. M., Flexas, J., & Pinheiro, C. (2009). Photosynthesis under drought and salt stress: Regulation mechanisms from whole plant to cell. *Annals of Botany, 103*(4), 551–560.

Costa, J. M., Vaz, M., Escalona, J., Egipto, R., Lopes, C., Medrano, H., & Chaves, M. M. (2016). Modern viticulture in Southern Europe: Vulnerabilities and strategies for adaptation to water scarcity. *Agricultural Water Management, 164*, 5–18.

Cramer, G. R., Ergül, A., Grimplet, J., Tillett, R. L., Tattersall, E. A. R., Bohlman, M. C., Vincent, D., Sonderegger, J., Evans, J., Osborne, C., Quilici, D., Schlauch, K. A., Schooley, D. A., & Cushman, J. C. (2007). Water and salinity stress in grapevines: early and late changes in transcript and metabolite profiles. *Functional & Integrative Genomics, 7*(2), 111–134. https://doi.org/10.1007/s10142-006-0039-y

Dayer, S., Herrera, J. C., Dai, Z., Burlett, R., Lamarque, L. J., Delzon, S., Bortolami, G., Cochard, H., & Gambetta, G. A. (2020). The sequence and thresholds of leaf hydraulic traits underlying grapevine varietal differences in drought tolerance. *Journal of Experimental Botany, 71*, 4333–4344.

Dayer, S., Herrera, J. C., Dai, Z., Burlett, R., Lamarque, L. J., Delzon, S., Bortolami, G., Cochard, H., & Gambetta, G. A. (2021). Nighttime transpiration represents a negligible part of water loss and does not increase the risk of water stress in grapevine. *Plant, Cell & Environment, 44*(2), 387–398. https://doi.org/10.1111/pce.13923

Diago, M. P., Fernández-Novales, J., Gutiérrez, S., Marañón, M., & Tardaguila, J. (2018). Development and validation of a new methodology to assess the vineyard water status by on-the-go near infrared spectroscopy. *Frontiers of Plant Science, 9*(59).

Escalona, J., Bota, J., & Medrano, H. (2003). Distribution of leaf photosynthesis and transpiration within grapevine canopies under different drought conditions. *VITIS-Journal of Grapevine Research, 42*(2), 57–64.

Escalona, J. M., Fuentes, S., Tomás, M., Martorell, S., Flexas, J., & Medrano, H. (2013). Responses of leaf night transpiration to drought stress in *Vitis vinifera* L. *Agricultural Water Management, 118*, 50–58.

Fereres, E., & Soriano, M. A. (2007). Deficit irrigation for reducing agricultural water use. *Journal of Experimental Botany, 58*, 147–159.

Ferrandino, A., & Lovisolo, C. (2014). Abiotic stress effects on grapevine (*Vitis vinifera* L.): Focus on abscisic acid-mediated consequences on secondary metabolism and berry quality. *Environmental and Experimental Botany, 103*, 138–147.

Ferreira, M. I., Silvestre, J., Conceinçao, N., & Malheiro, A. (2012). Crop and stress coefficients in rainfed and deficit irrigation vineyards using sap flow techniques. *Irrigation Science, 30*, 433−447.

Flexas, J., Bota, J., Escalona, J. M., Sampol, B., & Medrano, H. (2002). Effects of drought on photosynthesis in grapevines under field conditions: an evaluation of stomatal and mesophyll limitations. *Functional Plant Biology, 29*(4), 461−471. https://doi.org/10.1071/PP01119

Fuentes, S., De Bei, R., Collins, M. J., Escalona, J. M., Medrano, H., & Tyerman, S. (2014). Night-time responses to water supply in grapevines (*Vitis vinifera L.*) under deficit irrigation and partial root-zone drying. *Agricultural Water Management, 138*, 1−9.

Gago, J., Estrany, J., Estes, L., Fernie, A. R., Alorda, B., Brotman, Y., Flexas, J., Escalona, J. M., & Medrano, H. (2020). Nano and micro unmanned aerial vehicles (UAVs): A new grand challenge for precision agriculture? *Current Protocols in Plant Biology, 5*(1), e20103.

Gambetta, G. A., Herrera, J. C., Dayer, S., Feng, Q., Hochberg, U., & Castellarin, S. D. (2020). Grapevine drought stress physiology: Towards an integrative definition of drought tolerance. *Journal of Experimental Botany, 71*, 4658−4676.

González-Perea, R., Daccache, A., Rodríguez-Díaz, J. A., Camacho-Poyato, E., & Knox, J. W. (2018). Modelling impacts of precision irrigation on crop yield and in-field water management. *Precision Agriculture, 19*(3), 497−512.

Herrera, J. C., & Castellarin, S. D. (2016). Preveraison water deficit accelerates berry color change in merlot grapevines. *American Journal of Enology and Viticulture, 67*, 356−360.

Hirzel, D. R., Steenwerth, K., Parikh, S. J., & Oberholster, A. (2017). Impact of winery wastewater irrigation on soil, grape and wine composition. *Agricultural Water Management, 180*, 178−189.

Hochberg, U., Herrera, J. C., Asfaw, D., Castellarin, S. D., Peterlunger, E., Alberti, G., & Lazarovitch, N. (2017). Evaporative demand determines the relative transpirational sensitivity of deficit-irrigated grapevines. *Irrigation Science, 35*, 1−9.

Hsiao, T. C., Steduto, P., & Fereres, E. (2007). A systematic and quantitative approach to improve water use efficiency in agriculture. *Irrigation Science, 25*(3), 209−231.

Intrigliolo, D., & Castel, J. (2006). Vine and soil-based measures of water status in a Tempranillo vineyard. *VITIS-Journal of Grapevine Research, 45*(4), 157.

Intrigliolo, D., & Castel, J. (2010). Response of grapevine cv. 'Tempranillo' to timing and amount of irrigation: Water relations, vine growth, yield and berry and wine composition. *Irrigation Science, 28*(2), 113−125.

IPCC. (2016). *Intergovernmental panel on climate change (IPCC)*. http://www.ipcc.ch/. (Accessed November 2021).

Katerji, N., & Rana, G. (2014). FAO-56 methodology for determining water requirement of irrigated crops: Critical examination of the concepts, alternative proposals and validation in Mediterranean region. *Theoretical and Applied Climatology, 116*(3), 515−536.

Keller, M. (2005). Deficit irrigation and vine mineral nutrition. *American Journal of Enology and Viticulture, 56*(3), 267−283.

Keller, M. (2010). Managing grapevines to optimize fruit development in a challenging environment: A climate change primer for viticulturists. *Australian Journal of Grape and Wine Research, 16*, 56−69.

Knipfer, T., Eustis, A., Brodersen, C., Walker, A. M., & McElrone, A. J. (2015). Grapevine species from varied native habitats exhibit differences in embolism formation/repair associated with leaf gas exchange and root pressure. *Plant, Cell and Environment, 38*(8), 1503−1513.

van Leeuwen, C., Pieri, P., Gowdy, M., Ollat, N., & Roby, J.-P. (2019). Reduced density is an environmental friendly and cost effective solution to increase resilience to drought in vineyards in a context of climate change. *Oeno One, 53*(2), 129.

Levin, A., Mathews, M. A., & Williams, L. (2020). Impact of pre-veraison water deficits on the yield components of 15 winegrape cultivars. *American Journal of Enology and Viticulture*. https://doi.org/10.5344/ajev.2020.19073 (in press).

Liu, X., Wang, L., Wei, Y., Zhang, Z., Zhu, H., Kong, L., Meng, S., Song, C., Wang, H., & Ma, F. (2020). Irrigation with magnetically treated saline water influences the growth and photosynthetic capability of *Vitis vinifera* L. seedlings. *Scientia Horticulturae, 262*, 109056.

Lopes, C. M., Santos, T. P., Monteiro, A., Lucilia Rodrigues, M., Costa, J. M., & Manuela Chaves, M. (2011). Combining cover cropping with deficit irrigation in a Mediterranean low vigor vineyard. *Scientia Horticulturae, 129*(4), 603–612.

López-Urrea, R., Montoro, A., Mañas, F., López-Fuster, P., & Fereres, E. (2012). Evapotranspiration and crop coefficients from lysimeter measurements of mature "Tempranillo" wine grapes. *Agriculture Water Management, 112*, 13–20.

Lovisolo, C., Lavoie-Lamoureux, A., Tramontini, S., & Ferrandino, A. (2016). Grapevine adaptations to water stress: New perspectives about soil/plant interactions. *Theoretical and Experimental Plant Physiology, 28*(1), 53–66.

Maas, E. V., & Hoffman, G. J. (1977). Crop salt tolerance—current assessment. *Journal of the Irrigation and Drainage Division, 103*(2), 115–134.

Ma, X., Sanguinet, K. A., & Jacoby, P. W. (2020). Direct root-zone irrigation outperforms surface drip irrigation for grape yield and crop water use efficiency while restricting root growth. *Agricultural Water Management, 231*, 105993.

McCarthy, M. G., Lveys, B. R., Dry, P. R., & Stoll, M. (2002). *Regulated deficit irrigation and partial root zone drying as irrigation management techniques for grapevines. Deficit irrigation practices*. FAO Water Reports No. 22. Rome, Italy.

Medrano, H., Tomás, M., Martorell, S., Escalona, J. M., Pou, A., Fuentes, S., Flexas, J., & Bota, J. (2015). Improving water use efficiency of vineyards in semi-arid regions. A Review. *Agronomy for Sustainable Development, 35*(2), 499–517.

Minhas, P. S., Ramos, T. B., Ben-Gal, A., & Pereira, L. S. (2020). Coping with salinity in irrigated agriculture: Crop evapotranspiration and water management issues. *Agricultural Water Management, 227*, 105832.

Mirás-Avalos, J. M., & Intrigliolo, D. S. (2017). Grape composition under abiotic constrains: Water stress and salinity. *Frontiers of Plant Science, 8*, 851–872.

Montoro, A., Torija, I., Mañas, F., & López-Urrea, R. (2020). Lysimeter measurements of nocturnal and diurnal grapevine transpiration: Effect of soil water content, and phenology. *Agricultural Water Management, 229*, 105882.

Moreno, M. A., del Castillo, A., Montero, J., Tarjuelo, J. M., & Ballesteros, R. (2016). Optimisation of the design of pressurised irrigation systems for irregular shaped plots. *Biosystems Engineering, 151*, 361–373.

Mosse, K. P. M., Lee, J., Leachman, B. T., Parikh, S. J., Cavagnaro, T. R., & Patti, A. F. (2013). Irrigation of an established vineyard with winery cleaning agent solution (simulated winery wastewater): Vine growth, berry quality, and soil chemistry. *Agricultural Water Management, 123*, 93–102.

Mosse, K. P. M., Patti, A. F., Christen, E. W., & Cavagnaro, T. R. (2011). Review: Winery wastewater quality and treatment options in Australia. *Australian Journal of Grape and Wine Research, 17*(2), 111–122.

Munitz, S., Schwartz, A., & Netzer, Y. (2019). Water consumption, crop coefficient and leaf area relations of a *Vitis vinifera* cv. 'Cabernet Sauvignon' vineyard. *Agricultural Water Management, 219*, 86–94.

Myburgh, P. A. (2012). Comparing irrigation systems and strategies for table grapes in the weathered granite-gneiss soils of the lower Orange River region. *South African Journal of Enology and Viticulture, 33*, 184–197.

Netzer, Y., Shenker, M., & Schwartz, A. (2014). Effects of irrigation using treated wastewater on table grape vineyards: Dynamics of sodium accumulation in soil and plant. *Irrigation Science, 32*, 283–294.

Ortega-Farias, S., & Intrigliolo, D. S. (2021). Special issue: Multiscale technologies for irrigation management. *Irrigation Science, 39*(1), 1–3.

Petousi, I., Daskalakis, G., Fountoulakis, M. S., Lydakis, D., Fletcher, L., Stentiford, E. I., & Manios, T. (2019). Effects of treated wastewater irrigation on the establishment of young grapevines. *Science of the Total Environment, 658*, 485–492.

Petrie, P. R., Wang, Y., Liu, S., Lam, S., Whitty, M. A., & Skewes, M. A. (2019). The accuracy and utility of a low cost thermal camera and smartphone-based system to assess grapevine water status. *Biosystems Engineering, 179*, 126–139.

Picón-Toro, J., González-Dugo, V., Uriarte, D., Mancha, L. A., & Testi, L. (2012). Effects of canopy size and water stress over the crop coefficient of a "Tempranillo" vineyard in South-Western Spain. *Irrigation Science, 30*, 419–432.

Pisciotta, A., Di Lorenzo, R., Santalucia, G., & Barbagallo, M. G. (2018). Response of grapevine (Cabernet Sauvignon cv) to above ground and subsurface drip irrigation under arid conditions. *Agricultural Water Management, 197*, 122–131.

Rengasamy, P., & Marchuk, A. (2011). Cation ratio of soil structural stability (CROSS). *Soil Research, 49*(3), 280–285.

Richards, L. A. (1954). *Diagnosis and improvement of saline and alkali soils*. Agric Handbook No. 60. Washington, DC: USDA, US Govt Printing Office.

Romero, P., Botía, P., & Navarro, J. M. (2018). Selecting rootstocks to improve vine performance and vineyard sustainability in deficit irrigated Monastrell grapevines under semiarid conditions. *Agriculture Water Management, 209*, 73–93.

Romero, P., García-García, J., Fernández-Fernández, J. I., Gil-Muñoz, R., del Amor, F., & Martínez-Cutillas, A. (2016). Improving berry and wine quality attributes and vineyard economic efficiency by long-term deficit irrigation practices under semiarid conditions. *Science Horticulture, 203*, 69–85.

Romero, P., Gil Muñoz, R., Fernández-Fernández, J. I., Del Amor, F. M., Martínez-Cutillas, A., & García-García, J. (2015). Improvement of yield and grape and wine composition in field-grown Monastrell grapevines by partial root zone irrigation, in comparison with regulated deficit irrigation. *Agriculture Water Management, 149*, 55–73.

Romero, P., Gil-Muñoz, R., Del Amor, F., Valdés, E., Fernández-Fernández, J. I., & Martínez-Cutillas, A. (2013). Regulated deficit irrigation based upon optimum water status improves phenolic composition in Monastrell grapes and wines. *Agriculture Water Management, 121*, 85–101.

Romero, P., Pérez-Pérez, J. G., del Amor, F., Martínez-Cutillas, A., Dodd, I. C., & Botía, P. (2014). Partial root zone drying exerts different physiological responses on field-grown grapevine (*Vitis vinifera* cv. Monastrell) in comparison to regulated deficit irrigation. *Functional Plant Biology, 41*, 1087–1106.

Ruiz-Colmenero, M., Bienes, R., & Marques, M. (2011). Soil and water conservation dilemmas associated with the use of green cover in steep vineyards. *Soil and Tillage Research, 117*, 211–223.

Sadras, V. O., & Moran, M. A. (2012). Elevated temperature decouples anthocyanins and sugars in berries of Shiraz and Cabernet Franc. *Australian Journal of Grape and Wine Research, 18*(2), 115–122.

Sadras, V., & Schultz, H. (2012). *Crop yield response to water*. FAO Water Reports No. 66. Rome, Italy.

Sánchez, J. M., López-Urrea, R., Valentín, F., Caselles, V., & Galve, J. M. (2019). Lysimeter assessment of the Simplified Two-Source Energy Balance model and eddy covariance system to estimate vineyard evapotranspiration. *Agricultural and Forest Meteorology, 274*, 172–183.

Schultz, H. R., & Stoll, M. (2010). Some critical issues in environmental physiology of grapevines: Future challenges and current limitations. *Australian Journal of Grape and Wine Research, 16*, 4–24.

Sebastian, B., Baeza, P., Santesteban, L. G., Sánchez de Miguel, P., De la Fuente, M., & Lissarrague, J. R. (2015). Response of grapevine cv. Syrah to irrigation frequency and water distribution pattern in a clay soil. *Agricultural Water Management, 148*, 269–279.

Sebastian, B., Lissarrague, J. R., Santesteban, L. G., Linares, R., Junquera, P., & Baeza, P. (2016). Effect of irrigation frequency and water distribution pattern on leaf gas exchange of cv. Syrah grown on a clay soil at two levels of water availability. *Agricultural Water Management, 177*, 410–418.

Shani, U., & Ben-Gal, A. (2005). Long-term response of grapevines to salinity: Osmotic effects and ion toxicity. *American Journal of Enology and Viticulture, 56*(2), 148–154.

Stevens, R. M., Harvey, G., & Partington, D. L. (2011). Irrigation of grapevines with saline water at different growth stages: Effects on leaf, wood and juice composition. *Australian Journal of Grape and Wine Research, 17*(2), 239–248.

Torres, N., Goicoechea, N., & Carmen Antolín, M. (2018). Influence of irrigation strategy and mycorrhizal inoculation on fruit quality in different clones of Tempranillo grown under elevated temperatures. *Agricultural Water Management, 202*, 285–298.

Turner, N. C. (1981). Techniques and experimental approaches for the measurement of plant water status. *Plant and Soil, 58*(1), 339–366.

Walker, R., Blackmore, D. H., Clingeleffer, P. R., & Correll, R. L. (2002). Rootstock effects on salt tolerance of irrigated field-grown grapevines (*Vitis vinifera* L. Cv. Sultana): 1. Yield and vigour inter-relationships. *Australian Journal of Grape and Wine Research, 8*, 3–14.

Walker, R. R., Blackmore, D. H., Clingeleffer, P. R., Kerridge, G. H., Rühl, E. H., & Nicholas, P. R. (2005). Shiraz berry size in relation to seed number and implications for juice and wine composition. *Australian Journal of Grape and Wine Research, 11*(1), 2–8.

Williams, L. E., Baeza, P., & Vaughn, P. (2012). Midday measurements of leaf water potential and stomatal conductance are highly correlated with daily water use of Thompson Seedless grapevines. *Irrigation Science, 30*(3), 201–212.

Zarco-Tejada, P. J., González-Dugo, V., Williams, L. E., Suárez, L., Berni, J. A. J., Goldhamer, D., & Fereres, E. (2013). A PRI-based water stress index combining structural and chlorophyll effects: Assessment using diurnal narrow-band airborne imagery and the CWSI thermal index. *Remote Sensing of Environment, 138*, 38–50.

Zhang, L., Marguerit, E., Rossdeutsch, L., Ollat, N., & Gambetta, G. A. (2016). The influence of grapevine rootstocks on scion growth and drought resistance. *Theoretical and Experimental Plant Physiology, 28*(2), 143–157.

CHAPTER 7

Modern approaches to precision and digital viticulture

Sigfredo Fuentes[1] and Jorge Gago[2]

[1]*Digital Agriculture, Food and Wine Group, School of Agriculture and Food, Faculty of Veterinary and Agricultural Sciences, University of Melbourne, Parkville, VIC, Australia;* [2]*Plant Biology Research Under Mediterranean Conditions Group, University of the Balearic Islands, Palma, Spain*

7.1 Introduction

The field of artificial intelligence (AI) research had its formal origins from the mid-1950s (Kline, 2010). However, it has gained major attraction, interest, and applications using machine learning (ML) in the last decade to advance new algorithms, especially for artificial neural networks (ANNs), and computing power. Many times, AI and ML are referred to and addressed indistinctively, which is not accurate (Fig. 7.1). Specifically, the field of AI comprises subdisciplines and areas of applications, such as sensor networks, integrated sensor technology, ML, computer vision (CV), robotics, and biometrics (Gonzalez Viejo et al., 2019), which are of interest for this chapter since they have been recently applied to viticulture and winemaking.

It is also important to distinguish the difference between precision viticulture (PV) and digital viticulture (DV). In the case of PV, it is a specific application to viticulture derived from precision agriculture (PA), which has been around since the early 1980s originating from the USA and sprouting international attention from the early 1990s (Srinivasan, 2006). The more accepted PA definition is: "Precision agriculture is a management strategy that gathers, processes, and analyzes temporal, spatial, and individual data and combines it with other information to support management decisions according to estimated variability for improved resource use efficiency, productivity, quality, profitability, and sustainability of agricultural production" (International Society of Agriculture Precision, 2019). In general, PA deals with implementing technological advances to assess spatial and temporal differences within a field of interest (in our context, understanding field as a vineyard block or group of blocks). These technologies can be implemented using different approaches, including wireless sensor networks (more recently through the Internet of Things [IoT] connectivity), to assess the soil-plant-atmosphere continuum. These proximal sensors can be in direct contact with the soil (soil moisture/salinity sensors), based on the plant trunk (e.g., sap flow sensors, dendrometers, acoustic sensors), at the canopy level (i.e., leaf temperature, psychrometry, and leaf wetness) or within the environment (atmosphere) of interest (mesoclimate or microclimate), such as automatic meteorological stations (AMSs) (Gautam & Pagay, 2020). Most of these proximal and contact sensor technologies offer high temporal resolution data that can be used to assess grapevine growth, water status, irrigation

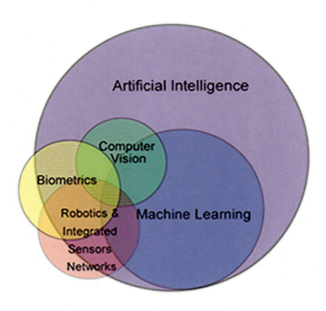

FIGURE 7.1

Venn's diagram showing artificial intelligence (AI) as the field of study, including other disciplines and fields of application, such as computer vision (CV), machine learning (ML), biometrics, robotics, and integrated sensor networks.

Modified from Gonzalez Viejo, C., Torrico, D. D., Dunshea, F. R., & Fuentes, S. (2019). Emerging technologies based on artificial intelligence to assess the quality and consumer preference of beverages. Beverages, 5(4), 62.

requirements, and detection of pests and diseases, among others. However, they have the same limitations in selecting sentinel grapevines to install contact sensors in their vicinity or directly in grapevines, which may not represent the vineyard's spatial variability. In the case of AMS, specifically for irrigation scheduling, they need to be installed in a reference area for irrigation scheduling purposes (Allen et al., 1998) or within the vineyard, close or within canopies for microclimate-based assessments. Remote and proximal sensing appears to solve this spatial resolution problem of sensed data, and its representation is mainly dependant on the resolution of instrumentation and platforms used, such as satellites, manned aircraft, unmanned aerial or terrestrial vehicles (UAVs/UTVs) (Weiss et al., 2020). Temporal resolution using the different monitoring platforms will depend on the sampling frequency-time available for each, currently from 7 to 8 days for satellite to a user-defined timeframe for UAV/UTV.

In the case of DV, it relates to creating information value from data for accurate and intelligent decision-making, obtained from any of the technologies discussed before (Robertson et al., 2020; Zhang et al., 2017). More recently, the processing of this information has been done by applying AI to high spatial and temporal resolution data to create sensible and accurate ML models. Furthermore, some of the sensor technology with the higher temporal resolution are meteorological data, combined with manual vineyard observations, such as phenological monitoring, yield, and berry quality records have become a powerful tool to use with DV in the context of big data as we will explore later in this chapter.

A novel combination using biological sensors and digital technologies has been recently applied with trained dogs to detect pests and diseases such as phylloxera. The dogs have small backpacks containing a smartphone with the Inspector Paw application (Fuentes et al. unpublished), which detect the dogs' gestures using ML algorithms and the GPS location of positive detection at the per-plant scale. The data gathered from the application can be later mapped to obtain the spatial variability of the insect presence.

7.2 Remote sensing for vineyard management

For thousands of years, viticulture has been mostly developed in the Mediterranean basin to produce wine and table grapes. Still, after a great expansion of vineyard plantations in the last years to all the continents of the planet (except Antarctica), its production is associated with arid and semiarid regions (Bouquet, 2011). These agricultural regions are the main ones that are affected by climate change in terms of increased ambient temperatures, heatwaves, variable weather conditions, including general reductions in the precipitation, and also characterized by torrential rain events and long and intense drought seasons that threaten grape productivity and quality (Masson-Delmotte et al., 2018).

Traditionally, viticultural practices have been driven by uniformed management strategies applied to the whole vineyard without taking into account the natural spatial and temporal heterogeneity. Furthermore, grape yields and quality traits often present significant variability between different areas of a vineyard due to physical and environmental factors, such as soil physicochemical characteristics, topography, orientation, and microclimate, among others (Arnó Satorra et al., 2009; Bramley, 2010). PV relies on avoiding such discordance between homogeneous management and heterogeneous environmental conditions that determine yield and quality within vineyards. In this sense, and considering climate change, it becomes almost mandatory for the industry and the scientific community to develop new tools and PV strategies. The employment of high-throughput sensing technologies for effective monitoring of the performance of different expressed phenotypes, productivity, and physiological stress status under field conditions (field phenotyping) has opened up new opportunities to manage the whole data-driven decision-making and variety selection. Altogether, the main objective is clear: to improve resource efficiency and sustainability while ensuring optimal crop yields and quality.

These PV technologies were highly linked to vineyards in earlier studies in which airborne and satellite-based remote sensing were used (Hall et al., 2002). But only after the mid-2000s, PV use in viticulture became more extensive due to the difficulties intrinsically associated with vineyards, such as heterogeneity produced by vineyard canopies and interrows without a uniform total canopy coverage compared to broadacre crops. This heterogeneity requires remote sensing technologies with a higher spatial resolution to avoid the "mixed pixel" noise problem with a not available solution in the 80s and 90s (Santesteban, 2019). These studies focused on critical viticulture management, such as water and nutrient requirements managed by irrigation and fertilization, plant physiological and biotic and abiotic stresses, soil and canopy management, among others (Santesteban, 2019). Manned aircraft and satellites were employed as remote sensing platforms to monitor productive grapevine traits, such as vigor, leaf area index (LAI), and canopy density but also disease detection at the block level (or management unit, i.e., irrigation block) and without the possibility to extract pure canopy pixels from individual vines (Hall et al., 2002; Tisseyre et al., 2007). A wide range of remote sensors was employed to evaluate the vineyard conditions and provide useful recommendations based on the plant status for precision management, such as RGB, multi-, and hyperspectral cameras. The main objective was to develop wide and narrow-band reflectance indexes to relate with plant vigor and stress, thermal imagery to assess plant water status, define irrigation strategies, and even chlorophyll fluorescence to associate with plant status photosynthetic capacity but also productivity and berry quality (Bramley, 2010; Santesteban, 2019; Zarco-Tejada, Catalina, et al., 2013).

Remote sensing technologies are becoming more affordable for users than earlier applications, thanks to miniaturized new sensors and cameras, making them cost-affordable. Furthermore, technological improvements in satellite geographical positioning combined with miniaturized electronic sensors led to the development of new, unmanned, automated platforms, which can equip remote sensors as payload (capacity to carry the weight of different equipment) (Alessandro Matese & Di Gennaro, 2015; Gago et al., 2015). Currently, there is no doubt that remote sensing from UAVs has revolutionized agricultural applications by their user-friendliness, low-cost, and increased spatial and temporal resolution compared to satellites, manned aircraft, or fixed towers (Gago et al., 2020). This increased capacity becomes more relevant considering the highly dynamic relationship between vegetation and its environment in the case of vineyards, in which the use of UAVs could facilitate precise management of vineyards throughout the whole season and even allowing intensive monitoring when most required for accurate irrigation management during warmer and dry periods, pest and disease detection and management, and crop production estimation before harvest, among others (Bramley, 2010).

However, UAVs are still limited by their compromise between reduced payload weight allowance in fixed-wing UAVs with longer operating times and the opposite for multirotor UAVs. Furthermore, the use of UAVs can be severely limited since flight campaigns for data acquisition need to be performed under strict national flight regulations in some countries to ensure public safety and privacy (Gago et al., 2015). Data obtained from UAVs can reach a subcentimeter spatial resolution, which is the main aspect that has revolutionized spatial plant ecophysiology monitoring and applications (Gago et al., 2015), and are currently widely used for vineyard research and management applied from pest and disease detection to plant water status monitoring throughout the whole season. Furthermore, it has even been proposed to use micro-UAVs, which can fly in swarms within vineyard rows to directly count and measure grape clusters (Albetis et al., 2019; Gago et al., 2017; Zarco-Tejada, González-Dugo et al., 2013).

It is worth mentioning another remote sensing platform that can provide high-resolution and useful information for vineyard management as terrestrial ground vehicles. Mobile terrestrial platforms have several advantages compared to all the aerial platforms mentioned before. The most evident is their higher payload capacity, and thus they can include into the same platform a higher number of different remote sensors (Andrenelli et al., 2013; Andújar et al., 2019). Their use does not depend on strict flight regulations, and, of course, they are much less affected by environmental weather conditions (windy periods). However, compared to UAVs, these platforms are more time-consuming and highly dependent on vibrations that can reduce the data's quality (Milella et al., 2019). These mobile terrestrial platforms can also retrieve highly essential information that is currently impossible to obtain this level of detail with aerial platforms, such as soil characteristics and crop status and quantity (Santesteban, 2019).

Altogether, this unprecedented resolution in spatial and temporal terms will help drive PV management to improve resource use efficiency, yield, and grape quality (Bramley, 2010; Santesteban, 2019).

Comparatively, even an open-access satellite with high-resolution imagery (such as 10 m pixel resolution for Sentinel-2) is not enough to detect plant physiology changes due to abiotic or biotic stress [19,27]. Satellite decametric resolution was found to be efficient for broadacre crops in important cultivated extensions (He et al., 2018). However, for vineyards, having discontinuous layouts, inter-row paths, and possible weed vegetation will finally affect vegetation indices (VIs) calculations if they cannot be avoided (Khaliq et al., 2019).

High-resolution remote sensing data can be used to monitor the vineyard's natural variability to provide specific recommendations for each vineyard block to improve resource use efficiency, quality, yield, and sustainability. Interestingly, much more work has been done in PV regarding canopy features (VI, leaf area, and shoot growth) associated with vigor, stress status, and yield parameters than grape composition and aroma traits, as it was observed recently after a review on the topic (Santesteban, 2019). Many technologies and methodologies have been developed for precise vineyard management (Bramley, 2010; Gago et al., 2015; Santesteban, 2019). The most widespread remote sensing technologies applied to vineyards are canopy reflectance in the visible range with RGB cameras (RGB, red, green, and blue) and nonvisible as near-infrared (IR), and the emission of far-IR (thermal imagery) (Santesteban, 2019; Zarco-Tejada, Catalina, et al., 2013). Canopy reflectance can be related to plant physiological status since the reflected light from the canopy is related to its interaction with leaves and their pigment composition (mostly chlorophylls, xanthophylls, carotenes, and anthocyanins) and the canopy architecture (usually measured by the LAI as the basic indicator of the vigor and growth of the vines). Changes in the leaves pigment profile are widely employed to understand plants' stress levels (Esteban et al., 2015). Most of the plant stress proxies developed for vineyards have used wide spectral bands to calculate VIs as the Normalized Differential Vegetation Index (NDVI= (IR − R)/(IR + R)), based on the near-IR band (ca. 800 nm) and the red band (R) (ca. 680 nm), originally coming from the satellite LandSat sensor bands in these wavelengths (Rouse et al., 1974). This index is related to plant "greenness" because of the high reflectance of the chlorophylls in the IR range and their elevated absorbance behavior in the visible spectra's red part (energy mostly employed for photosynthesis). Another wide-band indices were used to monitor more specifically vine vigor and status as the optimized index transformed chlorophyll absorption in reflectance index/optimized soil-adjusted vegetation index (TCARI/OSAVI), index more insensitive to the possible soil background (due to mixed information in lower resolution imagery), solar angle and vigor of the vines (using the LAI as proxy) (Haboudane et al., 2002).

Most of the high-resolution imagery applications from UAVs employing wide-band indices were related to monitoring vine vigor and growth, such as the LAI (Alessandro Matese et al., 2019; Campos et al., 2019). Additionally to VI based on high-resolution reflectance data, digital elevation models allowing the appraisal of vines' physical characteristics, such as canopy size, volume, and height, can be related grapevine vigor and vegetative growth (A Matese et al., 2017; Pádua et al., 2019, 2018). Image resolution used in these methodologies ranges from 1 to 5 cm/px, which constitutes a significant improvement in airborne and satellite imagery to avoid mixed information from the inter-row background such as weeds or herbaceous cover crops.

Wide-band indices were also employed to detect different diseases in vines, mostly symptomatologies that imply changes in the leaf pigment content and structure as grapevine leaf stripe disease (GLSD), *Flavescence dorée* (FD), and grapevine trunk diseases (GTDs) (Albetis et al., 2019; Di Gennaro et al., 2016). GLSD was successfully detected employing a UAV equipped with a multispectral camera, obtaining high-resolution multispectral images (0.05 m/px) to calculate the NDVI.

Another study was developed by Albetis et al. (2019) employing multispectral reflectance from a UAV to detect vines with the disease FD and GTDs (0.10 m/px resolution); in this case, different VI as Red Green Index (RGI)/Green Red Vegetation Index (GRVI) (green and red spectral bands) and Car (linked to carotenoid content) were the most effective spectral indicator to determine vines with infection levels higher than 50%.

Also, narrow-band indices can focus on more detailed plant biochemical responses to the incoming light, as the Photochemical Reflectance Index (PRI) ($R550 - R531/R550 + R531$), these specific wavelengths used to compute this index relate to the changes in the epoxidation state of the xanthophyll cycle to dissipate excessive energy (Gamon et al., 1992). Narrow-band indices were also used in vineyards through UAVs to assess water stress (Zarco-Tejada, Catalina, et al., 2013; Zarco-Tejada, Guillén-Climent et al., 2013). For example, to determine leaf carotenoid content through the index $R515/R570$, and chlorophyll content through the index TCARI/OSAVI to characterize vine plant status employing the leaf pigments functionality (Zarco-Tejada, Catalina, et al., 2013). Beyond pigment status, another narrow-band index proposed: the normalized PRI (an improvement of the PRI) the ($PRI_{norm} = PRI/[RDVI \cdot R700/R670]$), where the RDVI (renormalized differential vegetation index = $R800 - R600/R800 - R600$) is sensitive to the canopy structure, and the red edge index is sensitive to the chlorophyll content ($R700/R670$). This new index that considers canopy structure and greenness was demonstrated as more effective than the conventional PRI compared with plant-truth measurements of the vine water status as stomatal conductance and leaf water potential (Zarco-Tejada, Guillén-Climent et al., 2013).

Another technology widely employed for PV is thermal imagery. Canopy temperature relates with the vines' hydric status; when water availability is reduced, drought imposes significant physiological limitations, promoting stomata closure and thus reducing transpiration and its evaporative cooling effect in the leaves (Jones, 1999). The reduction of the thermal cameras (through uncooled microbolometers employing bands in the range of 7–14 μm) allowed using this remote sensing technology in fixed-wing UAVs. Multicopters can fly closer to the ground, enabling thermal imagery acquisition with an unprecedented high-resolution for the typical low image resolution of the thermal sensors (Gago et al., 2017; Santesteban et al., 2017). Increased thermal imagery resolution has been achieved thanks to UAV platforms, which has helped avoiding the usual problem of the "mixed pixel," to ensure that temperature retrieval came exclusively from the plant and not from the background (Jones & Sirault, 2014). As canopy temperature depends not only on the plant physiological status but also on the environmental conditions, thermal indexes were developed to consider the climate's effect and focus on plant stress status assessed by leaf water potential, stomatal conductance, and stem sap flow (Leinonen et al., 2006).

Additionally, when weather data are available, the leaf energy balance model can be applied to determine the stomatal conductance based on canopy temperature (Jones, 1999; Monteith & Unsworth, 1990) that can give more accurate information to estimate the plant's productivity stress status. This methodology was employed previously with a UAV obtaining significant correlations with gs and stem sap flow, facilitating precision vineyard water management (Gago et al., 2017). Typical coarser resolutions obtained from thermal sensors can be improved by airborne and most UAVs that reduce this problem by increasing spatial resolutions in orders of magnitude (30 cm/px airborne, 18–2.5 cm/px for UAVs). Higher resolutions let to extract pure canopy temperature data without background influence (Alessandro Matese & Di Gennaro, 2018; Santesteban et al., 2017; Zarco-Tejada, Guillén-Climent et al., 2013). In this sense, thermal imagery was employed further to assess additional parameters beyond plant water status and further comparisons between water stress vineyard mapping and other traits related to productivity and grape quality, such as yield, berry weight, and total soluble solids (Santesteban et al., 2017).

Furthermore, all the recent developments in terrestrial autonomous robotics have become available for viticulture applications [46], with vehicles adapted to move between the rows (Andújar et al., 2019). These platforms can equip a wide variety of sensors, not just focused on the individual vines but also retrieve soil characteristics (Andújar et al., 2019; Lanyon DM et al., 2007; Santesteban et al., 2017). For instance, Andújar et al. (2019) compared vine volumes calculated from UAV and on-ground imagery to determine precise fertilizer application. Both methodologies performed similar results, however on-ground technologies obtained by a manned vehicle equipped with light detection and ranging (LiDAR)-based measurements taken with an RTK-GNSS and a Kinect v2 sensor was also used as a low-cost depth camera reach a higher level of detail and precision but also at a higher cost than UAV RGB imagery models. Terrestrial vehicles were also employed to "on-the-go" detecting plant water status to improve irrigation scheduling, different sensors were used as NIR spectroscopy (Diago et al., 2017), and thermal imagery (S Gutiérrez et al., 2017) checked their performance by plant-truth measurements as leaf stomatal conductance and leaf water potential. This type of vehicle can also be equipped with specific remote sensors to determine grape quality as hyperspectral imagery evaluating TSS and anthocyanin content (Salvador Gutiérrez et al., 2018). Combination of hyperspectral and NIR-sensors on-the-go through manned and autonomous terrestrial vehicles can provide high-throughput grape composition at field conditions offering detailed information about chemistry, nutrition, and quality of the grapes (Power et al., 2019), an essential aspect for the future of PV (Ozdemir et al., 2017; Santesteban, 2019).

Terrestrial "on-the-go" data acquisition is key to address a relevant topic of PV, soil heterogeneity. Soil properties and agricultural practices will drive the final plant's response to the environment, and subsequently, it will be reflected in grape production and quality (Bramley, 2010). Several soil-based proximal sensors were proposed to assess different soil properties. For instance, soil electrical conductivity can be monitored by electromagnetic induction (EMI) and electrical resistivity. Soil water content can be monitored by ground-penetrating radar, soil organic carbon, and iron levels by optical sensors. Soil organic carbon, clay levels, mineral composition, and other soil properties can be monitored by visible, near- and mid-infrared reflectance spectroscopy; meanwhile, other parameters such as soil pH, lime requirement, and soil nutrients by ion-selective electrodes and ion-sensitive field-effect transistors (Adamchuk & Rossel, 2010). Andrenelli et al. (2013) showed an optimized procedure to minimize costs for soil survey-based into ARP(c) (Automatic Resistivity Profile); clay content was the most correlated parameter with apparent electrical resistivity (ERa), an important soil parameter to be considered in PV. Further relationships were explored between an automated method to capture soil electrical resistivity and ancillary topographic attributes (slope and elevation) compared to growth and yield in a commercial vineyard. Interestingly, the parameter survey relates to the trunk circumference spatial pattern and yield of the studied vineyard (Rossi et al., 2013). In this sense, automated terrestrial vehicles equipped with different sensors focused on the vines, grape clusters, and soil properties offer an exciting opportunity to develop technologies that could be finally available in the market.

7.3 Artificial intelligence and remote sensing

Recent reviews have been published on the use and implementation of AI in viticulture [54] and related to applications using remote sensing and ML modeling, specifically on CV on harvest yield estimations and management related to disease monitoring, berry quality evaluation, and grape phenology (Seng et al., 2018). These advances have been possible through the recent development of CV and ML. The volume of data acquired using remote sensing in viticulture and winemaking makes possible ML modeling and deep learning techniques to assess different aspects from the vineyard to the bottle and consumer acceptability.

In terms of computation requirements, new computers with integrated chips for several different components, including the CPU, GPU, unified memory architecture (RAM), and neural engines. This has allowed ML and AI computation and modeling easier for research purposes. AI applications with more demanding computation power requires virtual PCs and cloud computing with many commercial options in the market for different technical requirements. Real or near-real time computation for AI applications in agriculture using cloud computing requires a fast internet connection, which may not be available in many rural areas, even in developed countries, such as those with extensive land (i.e., USA and Australia). An alternative to this problem could be in-situ machine and deep learning computation using intensive AI local computation processors, which are more affordable day by day for specific AI applications.

7.3.1 Computer vision

CV refers to the multidisciplinary effort using computer science to process and analyze digital images or videos (Sonka et al., 2014). Images of importance for agricultural applications are in the range of light spectra from visible (320–700 nm), near-infrared (780–2500 nm), and infrared thermal ranges (9000–14,000 nm or 9–14 μm) as mentioned before. Cameras used to acquire these images or videos range from conventional RGB cameras for visible spectra capture, multispectral cameras comprised of RGB cameras and multiple cameras with sensitivity in different light spectral ranges of interest, and hyperspectral cameras with each pixel sensitive to a range of light spectra. The variation of cameras, sensitivity ranges, and configurations define the type of datasets that needs to be processed and analyzed using CV algorithms. In the case of visible cameras, the main layers, or channels of information are three: red, blue, and green, which results in a 2D by three-layer matrix. Each of the 2D layers is of the specific dimension of the camera resolution or number of pixels. However, this is only one of the more than 100 VIs that can be extracted from multispectral images of vineyards (Xue & Su, 2017). Once these indices are calculated at the per-pixel resolution, they can form a multilayer matrix increasing the information level. Hyperspectral cameras produce images with each pixel containing light spectra, which will include what is called a cube of information.

One of the main objectives when using CV analysis of images and videos is to automate image processing and analysis. As shown in the aerial image examples presented, they can contain a lot of information within the images that are not of interest for agricultural research, such as bare soil, grass areas between rows, trees, buildings, and water bodies, among others. When analyzing these images, it is good to filter out this nonplant material before processing images further, which reduces computer power and processing required of very high-resolution images and especially hyperspectral images.

There are a few approaches that can be applied to recognize plant material within rows using automatic segmentation of vine rows from UAV visible images (Nolan et al., 2015) and combined with digital survey model to differentiate grapevine canopies from grass cover by height (Baofeng et al., 2016; Xue et al., 2019).

7.3.2 Machine and deep learning in viticulture and winemaking

For ML modeling, there are two main processes of data and information: supervised and unsupervised. Supervised ML modeling deals with variables or parameters, which could be VIs, such as inputs to model specific targets of interest, from grapevine water status, yield estimations, grape composition, and aroma profiles the detection of pests and diseases. These models can be classification ML models using categorical targets, such as low, medium, and high water stress levels, or regression ML models using numerical targets, such as stem water potential (Ψs, MPa). Unsupervised ML leaves the process of feature identification to the modeling strategy, which requires a higher volume of data for the training, validation, and testing of models. Hence, unsupervised ML usually is referred to as deep learning.

ANNs can be considered one of the most implemented and used algorithms for ML composed of a multilayer neural network representing inputs, hidden layers, and output layers (Fig. 7.2). In supervised ML, inputs should be obtained through a parameter engineering process, from which the parameters selected are linked to the specific target responses. The latter is based on previous studies on plant physiology and interactions between the soil and atmosphere factors. Hence, it could be considered a trial and error process to obtain more sensible and accurate predictive models.

As mentioned before, there are many supervised ML algorithms available (Fig. 7.3), from which ANNs are inspired by the framework of the network of neurons in the brain (Mitchell, 2019), and they have produced the most accurate models in agricultural applications so far.

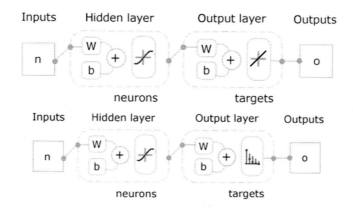

FIGURE 7.2

Diagrams representing machine learning modeling using (A) classification algorithms and (B) regression algorithms. Both are composed of inputs, hidden layers or inner layers, and output layer. The number of neurons corresponds to the number of hidden layers used with weights (*w*) and biases (*b*) changing during training until obtaining the best fitting options. N = number of inputs; o = number of outputs categorical for classification and numerical for regression.

134 Chapter 7 Modern approaches to viticulture

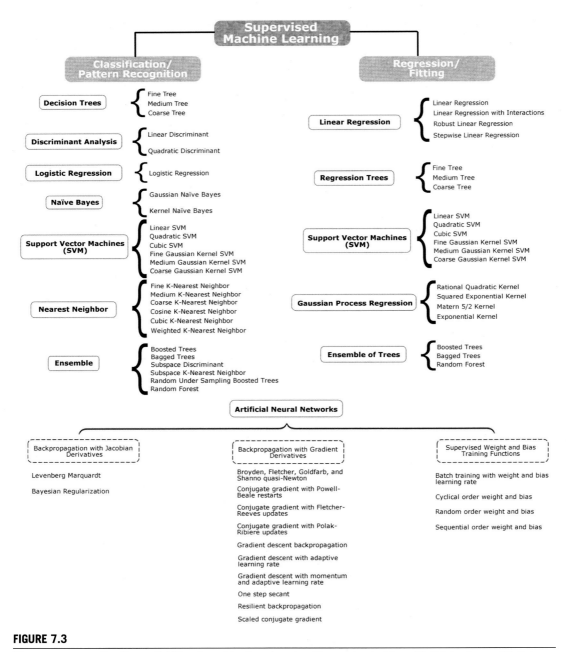

FIGURE 7.3

Diagram representing the availability of machine learning algorithms extracted from Matlab toolboxes (Matlab, Mathworks Inc. Matick. MA.USA).

Modified from Gonzalez Viejo, C., Torrico, D. D., Dunshea, F. R., & Fuentes, S. (2019). Emerging technologies based on artificial intelligence to assess the quality and consumer preference of beverages. Beverages, *5(4), 62.*

For unsupervised ML algorithms, it is frequently employed ANNs with multihidden layers, mainly applied to the image or video-based information. In this case, there is no parameter engineered inputs, and the actual network extract features of interest automatically by image transformation algorithms. Often, unsupervised ML modeling is called deep learning, which requires a high amount of data (big data) to be available as inputs compared to supervised ML modeling. As a rule of thumb, supervised ML requires inputs on the scale of hundreds; on the contrary, unsupervised ML could require thousands to million inputs (images). For unsupervised ML, the main examples can be applied to object recognition for unmanned automotive driving, face recognition, animal recognition, among others. Initial inputs can be accessed from public libraries available to do initial training of algorithms and include specific ones depending on the targeted output for particular studies.

Deep learning (Fig. 7.4) can also have supervised inputs. In this case, the user has the time-consuming task of selecting attributes or regions of interest assigned as targets, formally called labeling. The latter is considered a tedious task since the user may need to do the manual labeling, for example, to recognize bunches from latera imaging of canopies, which can amount to thousands to train the ML algorithm.

7.3.2.1 Soil-based applications using digital tools and machine learning

Soil mapping for physical and chemical properties is one of the most valuable information for DV, since its variability will determine critical grapevine factors such as root depth, soil water holding capacity, cation exchange capacity, and nutrient availability, among others, that ultimately may

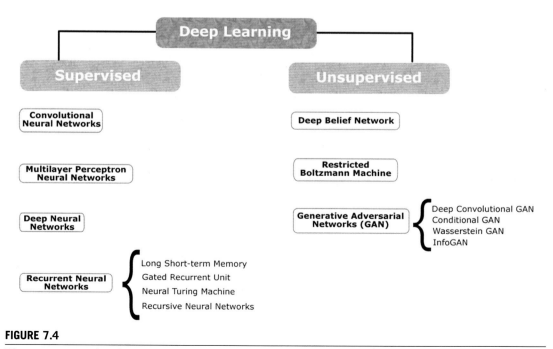

FIGURE 7.4

Diagram representing deep learning algorithms using either supervised (left) and unsupervised (right).

Diagram made by authors.

determine grapevine growth, productivity, and quality of grapes. Digital soil mapping can be performed: (i) directly through instrumentation, such as proximal geoelectrical sensing to assess soil variability (Tardaguila et al., 2018); (ii) digital predictive tools to estimate soil properties (Mashalaba et al., 2020), 3D mapping to determine soil compaction on steep slopes (Santos et al., 2018), to map soil carbon in vineyards (Bonfatti et al., 2016), and to assess soil erosion processes (Rodrigo-Comino & Cerdà, 2018); (iii) digital processing of collected information from different studies, such as the Soil and Landscape Grid project in Australia (Grundy et al., 2015). The latter offering freely available information about the whole of Australia's soil physical and chemical properties in a 90 × 90 m grid and up to 2 m depth, which is a great resource as a first approach for site selection, for example, and other similar studies. These techniques offer a great opportunity to implement machine learning techniques to obtain accurate and efficient models to obtain different levels of automation in the data gathering and processing.

The main problem with using remote sensing for soil carbon and nutrient availability is that most passive sensors (working on reflectance) can penetrate a maximum of a few centimeters of the soil, which can only represent the topsoil's spatial variability. The latter makes more practical use of proximal active sensors, such as those emitting EMI energy (i.e., EM-38, 36, 31, Geonics. Ltd., Mississauga, ON, Canada) that can reach deeper layers of the soil (meters) or electrical resistivity tomography (Brillante, Bois, Lévêque, & Mathieu, 2016). These systems can provide critical ancillary data as inputs for modeling related to soil texture, soil water content, cation exchange capacity, and organic matter, among others [70] that can be used for yield estimations, production classification quality zones within vineyards (Brillante, Bois, Mathieu, & Lévêque, 2016). Hence, an AI approach for soil-based applications should be focused on integrating these technologies (i.e., remote sensing and proximal sensor technology) to provide a continuous source of validation and adjustment of ML models when required.

7.3.2.2 Plant-based applications using digital tools and machine learning

For any plant-based AI application using aerial and satellite remote sensing, it is important to filter out nonleaf material. Some methodologies used to filter nonleaf material from CV have been previously discussed. However, other studies have successfully implemented CV and ML algorithms to remove shades (Poblete et al., 2018) and to obtain only grapevine canopy information from remotely sensed images, specifically using random forest supervised ML (Fig. 7.3) (Poblete-Echeverría et al., 2017) with an overall accuracy of 98%.

One of the main applications of CV on grapevine canopies is the assessment of canopy growth. One of the earliest canopy growth assessment using CV from upward-looking cover photography images and videos of grapevines (Fuentes et al., 2014), resulted in the development of a computer application (App) named VitiCanopy (De Bei et al., 2016). The Viticanopy App works based on CV algorithms for gap analysis of images to assess LAI, canopy porosity, and other canopy architecture parameters using the sky as a background. The advantages of using this application are that it is for free, results are in real-time, it required only smartphones capabilities, such as the cameras and GPS, results can be easily sent via email in comma-separated values files (.csv) for postprocessing or mapping purposes. A similar analysis can be performed from aerial images (downward-looking images) combined with automated recognition of grapevine rows (Nolan et al., 2015) to obtain LAI, canopy cover on a plant-by-plant scale. This approach has been applied in vineyards to detect plants affected by frosts and missing plants (Baofeng et al., 2016). Canopy architecture and vigor are related to grapevine's reproductive performance (Wang et al., 2019) and berry quality traits (Martínez-Lüscher et al., 2019; Scafidi et al., 2018). A study done in cocoa plants predicted aroma profiles of cocoa beans based on canopy architecture using CV algorithms from aerial images and supervised ML regression modeling (ANN) with correlations of $R = 0.82$ with no statistical signs of overfitting (Fuentes, Chacon, et al., 2019). Similar studies have been conducted for several grapevine cultivars to model grape quality traits as targets using canopy architecture parameters as inputs with similar preliminary accuracies (Universities of Adelaide and Melbourne, Australia, unpublished). Other canopy architecture monitoring techniques are based on terrestrial laser scanners or LiDAR (Arnó et al., 2013; del-Moral-Martinez et al., 2016). High-resolution LiDARs that can be used as UAV payload is still price-limiting and used primarily for research. These technologies offer a high volume of data at the plant-by-plant scale, which can be used for ML or deep learning modeling.

Plant water status assessment is critical for optimal irrigation scheduling in vineyards and to control water stress at critical phenological stages to increase berry quality. The main techniques available to manage water stress are regulated deficit irrigation and partial rootzone drying, which aim to increase water use efficiency with minimal detrimental effects on yield by plant hormonal manipulation, such as abscisic acid. However, to successfully apply these techniques, they require a narrow water stress threshold to be applied to grapevines. Therefore, precise monitoring methods based on AI need to develop accurate models with high spatial and temporal resolution. Several studies have focused on using remote sensing platforms with multispectral, hyperspectral, and infrared thermal cameras, as mentioned in detail. Automated analysis of grapevine water stress using proximal infrared thermal imagery has been proposed using CV algorithms (Fuentes et al., 2012) and ML modeling based on multispectral information (Romero et al., 2018).

Furthermore, researchers have proposed using UAV and remote sensing to estimate evapotranspiration (ET) at the submeter scale (Aboutalebi et al., 2020; Park, 2018) and yield (Aboutalebi et al., 2019; Maimaitiyiming et al., 2019). Further management strategies related to the detection of pests and diseases have been researched in grapevines using UAV and ML from multispectral images (Albetis et al., 2017), deep learning approaches (Kerkech et al., 2020), and also combined with multispectral images (Kerkech et al., 2019). Automated yield components recognition and yield estimations have recently implemented CV algorithms and deep learning techniques in an on-the-go fashion using proximal remote sensing, which has reported a high level of accuracies in recognition of buds (Díaz et al., 2018; Pérez et al., 2017), flowers (Liu et al., 2018; Palacios et al., 2020), bunches (Pérez-Zavala et al., 2018), cluster compactness (Palacios et al., 2019), and individual grapes for yield estimation purposes.

Other applications using supervised ML algorithms for grapevine cultivar classification purposes from leaf imagery (digital ampelography), fractal analysis, and chemical fingerprinting using NIR have shown highly accurate models (>90%) for 16 cultivars. These classification models can also be implemented to evaluate water stress, nutritional deficits, and biotic stresses. The integration of proximal and remote sensing techniques with AI could be a transformational tool for the future's viticulture and winemaking (Fig. 7.5).

7.3.2.3 *Winemaking and consumer acceptability using machine learning*
The implementation of AI in winemaking research has been recent and in many cases, a reactive strategy due to climate change and ambient temperatures, and climatic anomalies, such as droughts and bushfires (Sutton & Hawkins, 2020). Recent DV applications have focused on developing novel

138 **Chapter 7** Modern approaches to viticulture

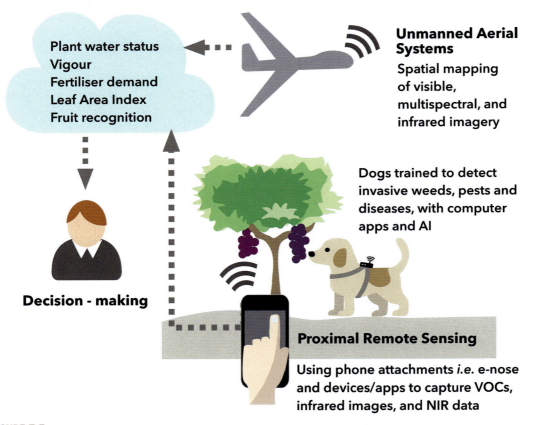

FIGURE 7.5

New and emerging technologies integration system (NETIS).

grape maturity assessments based on berry cell death (Fuentes et al., 2010) using NIR chemical fingerprinting (Fuentes, Tongson, et al., 2020). These approaches could offer maturity assessments based on the development of flavor and aroma from berries and final wines. Another ML approach has been proposed using historical weather and wine quality data from vineyards (Fuentes, Gonzalez Viejo, et al., 2019).

One of the main environmental threats affecting winemaking countries such as USA, Australia, South Africa, Portugal, Spain, and Italy are bushfires (Summerson et al., 2021), increasing in intensity and number within the growing seasons. Smoke-contaminated grapevines can produce glyco-conjugates with smoke-derived compounds passed to the wine in the fermentation process producing smoke taint (Wilkinson & Ristic, 2019). Recent advances in AI to assess in real-time smoke contamination in berries and taint levels in wines from the field to the winery (Fuentes, Tongson, et al., 2019) have also been based on sensor technology through low-cost electronic noses (e-noses), NIR spectroscopy (Fuentes, Summerson, et al., 2020).

7.3.2.4 Data access, ownership, and security using AI
Data access, ownership, and security concerns have been around since the start of sensor technology. The data accumulation in agriculture has been one of the most recurring questions, who owns the data, and who can use it? AI technologies and potential implementation in viticulture and oenology have not been exempt from these concerns from winegrowers. Possible solutions to this problem have been proposed using digital ledgers to handle AI systems' information as currency through blockchain technology (Kamilaris et al., 2019; Torky & Hassanein, 2020).

7.4 Conclusion
There is enough knowledge based on new and emerging technologies for DV and winemaking that need to be integrated through the production, winemaking, and commercialization chain for sustainability and efficiency. This integration will offer stakeholders from the whole chain to share information to make adjustments to maintain wine styles, increase quality, and respond better to environmental hazards. With the level of data that these technologies can produce and the implementation of big data from other sources, AI implementation can play a critical role in influencing the whole production chain transforming the viticultural and winemaking industry into a more predictive rather than reactive enterprise. Precision and DV have been recognized as the best approach to maintain and increase the wine industry's competitiveness in a changing climate and in a market where consumers are demanding higher quality and production standards.

The success of AI-based tools for viticulture and winemaking relies on the appropriate availability of target data or ground-truth and its availability to adjust ML models, making use of the "learning" capabilities to deal with new data. One of the main pitfalls of AI applications in agriculture in general to obtain information in real-time has been the velocity of data transmission and processing speed, whether these processes are in-situ or using cloud computing. However, recent advances in data communication speed such as 5G technology and processing speed and power using system-on-a-chip integrated technology could be a game changer to be implemented in IoT, AI, deep learning, and neural networks to expedite high-performance processing of big data. These processes could receive a further quantum leap with quantum computer power development shortly.

References

Aboutalebi, M., Torres-Rua, A. F., McKee, M., Kustas, W. P., Nieto, H., Alsina, M. M., White, A., Prueger, J. H., McKee, L., Alfieri, J., & Hipps, L. (2020). Incorporation of unmanned aerial vehicle (UAV) point cloud products into remote sensing evapotranspiration models. *Remote sensing, 12*(1), 50.

Aboutalebi, M., Torres-Rua, A. F., McKee, M., Nieto, H., Kustas, W. P., Sanchez, L., Alsina, M. M., White, W. A., Hipps, L., Prueger, J. H., McKee, L., Alfieri, J. G., Dokoozlian, N., & Coopmans, C. (2019). Incorporation of unmanned aerial vehicle (UAV) point cloud product into remote sensing evapotranspiration models and yield estimation in grapevine vineyards. *AGU Fall Meeting, 2019*. B42E-03.

Adamchuk, V., & Rossel, R. V. (2010). Development of on-the-go proximal soil sensor systems. In *Proximal soil sensing* (pp. 15–28). Springer.

Albetis, J., Duthoit, S., Guttler, F., Jacquin, A., Goulard, M., Poilvé, H., Féret, J. B., & Dedieu, G. (2017). Detection of Flavescence dorée grapevine disease using unmanned aerial vehicle (UAV) multispectral imagery. *Remote Sensing, 9*(4), 308.

Albetis, J., Jacquin, A., Goulard, M., Poilvé, H., Rousseau, J., Clenet, H., Dedieu, G., & Duthoit, S. (2019). On the potentiality of UAV multispectral imagery to detect Flavescence dorée and grapevine trunk diseases. *Remote Sensing, 11*(1), 23.

Allen, R. G., Pereira, L. S., Raes, D., & Smith, M. (1998). FAO irrigation and drainage paper no. 56. Rome. *Food and Agriculture Organization of the United Nations, 56*(97), e156.

Andrenelli, M., Magini, S., Pellegrini, S., Perria, R., Vignozzi, N., & Costantini, E. (2013). The use of the ARP© system to reduce the costs of soil survey for precision viticulture. *Journal of Applied Geophysics, 99*, 24–34.

Andújar, D., Moreno, H., Bengochea-Guevara, J. M., de Castro, A., & Ribeiro, A. (2019). Aerial imagery or on-ground detection? An economic analysis for vineyard crops. *Computers and Electronics in Agriculture, 157*, 351–358.

Arnó Satorra, J., Martínez Casasnovas, J. A., Ribes Dasi, M., & Rosell Polo, J. R. (2009). Precision viticulture. Research topics, challenges and opportunities in site-specific vineyard management. *Spanish Journal of Agricultural Research, 7*(4), 779–790.

Arnó, J., Vallès, J. M., Llorens, J., Sanz, R., Masip, J., Palacín, J., & Rosell-Polo, J. R. (2013). Leaf area index estimation in vineyards using a ground-based LiDAR scanner. *Precision Agriculture, 14*(3), 290–306.

Baofeng, S., Jinru, X., Chunyu, X., Yuyang, S., & Fuentes, S. (2016). Digital surface model applied to unmanned aerial vehicle based photogrammetry to assess potential biotic or abiotic effects on grapevine canopies. *International Journal of Agricultural and Biological Engineering, 9*(6), 119–130.

Bonfatti, B. R., Hartemink, A. E., Giasson, E., Tornquist, C. G., & Adhikari, K. (2016). Digital mapping of soil carbon in a viticultural region of Southern Brazil. *Geoderma, 261*, 204–221.

Bouquet, A. (2011). Grapevines and viticulture. *Genetics, Genomics, and Breeding of Grapes, 2011*, 1–29.

Bramley, R. (2010). Precision Viticulture: Managing vineyard variability for improved quality outcomes. In *Managing wine quality* (pp. 445–480). Elsevier.

Brillante, L., Bois, B., Lévêque, J., & Mathieu, O. (2016). Variations in soil-water use by grapevine according to plant water status and soil physical-chemical characteristics—a 3D spatio-temporal analysis. *European Journal of Agronomy, 77*, 122–135.

Brillante, L., Bois, B., Mathieu, O., & Lévêque, J. (2016). Electrical imaging of soil water availability to grapevine: A benchmark experiment of several machine-learning techniques. *Precision Agriculture, 17*(6), 637–658.

Campos, J., Llop, J., Gallart, M., García-Ruiz, F., Gras, A., Salcedo, R., & Gil, E. (2019). Development of canopy vigour maps using UAV for site-specific management during vineyard spraying process. *Precision Agriculture, 20*(6), 1136–1156.

De Bei, R., Fuentes, S., Gilliham, M., Tyerman, S., Edwards, E., Bianchini, N., Smith, J., & Collins, C. (2016). VitiCanopy: A free computer app to estimate canopy vigor and porosity for grapevine. *Sensors, 16*(4), 585.

Di Gennaro, S. F., Battiston, E., Di Marco, S., Facini, O., Matese, A., Nocentini, M., Palliotti, A., & Mugnai, L. (2016). Unmanned Aerial Vehicle (UAV)-based remote sensing to monitor grapevine leaf stripe disease within a vineyard affected by esca complex. *Phytopathologia Mediterranea*, 262–275.

Diago, M. P., Bellincontro, A., Scheidweiler, M., Tardáguila, J., Tittmann, S., & Stoll, M. (2017). Future opportunities of proximal near infrared spectroscopy approaches to determine the variability of vineyard water status. *Australian Journal of Grape and Wine Research, 23*(3), 409–414.

Díaz, C. A., Pérez, D. S., Miatello, H., & Bromberg, F. (2018). Grapevine buds detection and localization in 3D space based on structure from motion and 2D image classification. *Computers in Industry, 99*, 303–312.

Esteban, R., Barrutia, O., Artetxe, U., Fernández-Marín, B., Hernández, A., & García-Plazaola, J. I. (2015). Internal and external factors affecting photosynthetic pigment composition in plants: A meta-analytical approach. *New Phytologist, 206*(1), 268–280.

References

Fuentes, S., Chacon, G., Torrico, D. D., Zarate, A., & Gonzalez Viejo, C. (2019). Spatial variability of aroma profiles of cocoa trees obtained through computer vision and machine learning modelling: A cover photography and high spatial remote sensing application. *Sensors, 19*(14), 3054.

Fuentes, S., De Bei, R., Pech, J., & Tyerman, S. (2012). Computational water stress indices obtained from thermal image analysis of grapevine canopies. *Irrigation Science, 30*(6), 523–536.

Fuentes, S., Gonzalez Viejo, C., Wang, X., & Torrico, D. D. (2019). Aroma and quality assessment for vertical vintages using machine learning modelling based on weather and management information. In *Paper presented at the proceedings of the 21st GiESCO international meeting, Thessaloniki, Greece.*

Fuentes, S., Poblete-Echeverría, C., Ortega-Farias, S., Tyerman, S., & De Bei, R. (2014). Automated estimation of leaf area index from grapevine canopies using cover photography, video and computational analysis methods. *Australian Journal of Grape and Wine Research, 20*(3), 465–473. https://doi.org/10.1111/ajgw.12098

Fuentes, S., Sullivan, W., Tilbrook, J., & Tyerman, S. (2010). A novel analysis of grapevine berry tissue demonstrates a variety-dependent correlation between tissue vitality and berry shrivel. *Australian Journal of Grape and Wine Research, 16*(2), 327–336. https://doi.org/10.1111/j.1755-0238.2010.00095.x

Fuentes, S., Summerson, V., Gonzalez Viejo, C., Tongson, E., Lipovetzky, N., Wilkinson, K. L., Szeto, C., & Unnithan, R. R. (2020). Assessment of smoke contamination in grapevine berries and taint in wines due to bushfires using a low-cost E-nose and an artificial intelligence approach. *Sensors, 20*(18), 5108.

Fuentes, S., Tongson, E., Chen, J., & Gonzalez Viejo, C. (2020). A digital approach to evaluate the effect of berry cell death on pinot noir wines' quality traits and sensory profiles using non-destructive near-infrared spectroscopy. *Beverages, 6*(2), 39.

Fuentes, S., Tongson, E. J., De Bei, R., Gonzalez Viejo, C., Ristic, R., Tyerman, S., & Wilkinson, K. (2019). Non-invasive tools to detect smoke contamination in grapevine canopies, berries and wine: A remote sensing and machine learning modeling approach. *Sensors, 19*(15), 3335.

Gago, J., Douthe, C., Coopman, R. E., Gallego, P. P., Ribas-Carbo, M., Flexas, J., Escalona, J., & Medrano, H. (2015). UAVs challenge to assess water stress for sustainable agriculture. *Agricultural Water Management, 153*, 9–19.

Gago, J., Estrany, J., Estes, L., Fernie, A. R., Alorda, B., Brotman, Y., Flexas, J., Escalona, J. M., & Medrano, H. (2020). Nano and micro unmanned aerial vehicles (UAVs): A new grand challenge for precision agriculture? *Current Protocols in Plant Biology, 5*(1), e20103.

Gago, J., Fernie, A. R., Nikoloski, Z., Tohge, T., Martorell, S., Escalona, J. M., Ribas-Carbó, M., Flexas, J., & Medrano, H. (2017). Integrative field scale phenotyping for investigating metabolic components of water stress within a vineyard. *Plant Methods, 13*(1), 1–14.

Gamon, J., Penuelas, J., & Field, C. (1992). A narrow-waveband spectral index that tracks diurnal changes in photosynthetic efficiency. *Remote Sensing of Environment, 41*(1), 35–44.

Gautam, D., & Pagay, V. (2020). A review of current and potential applications of remote sensing to study the water status of horticultural crops. *Agronomy, 10*(1), 140.

Gonzalez Viejo, C., Torrico, D. D., Dunshea, F. R., & Fuentes, S. (2019). Emerging technologies based on artificial intelligence to assess the quality and consumer preference of beverages. *Beverages, 5*(4), 62.

Grundy, M., Rossel, R. V., Searle, R., Wilson, P., Chen, C., & Gregory, L. (2015). Soil and landscape grid of Australia. *Soil Research, 53*(8), 835–844.

Gutiérrez, S., Diago, M., Fernández-Novales, J., & Tardaguila, J. (2017). On-the-go thermal imaging for water status assessment in commercial vineyards. *Advances in Animal Biosciences, 8*(2), 520.

Gutiérrez, S., Diago, M. P., Fernández-Novales, J., & Tardaguila, J. (2018). Vineyard water status assessment using on-the-go thermal imaging and machine learning. *PLoS One, 13*(2), e0192037.

Haboudane, D., Miller, J. R., Tremblay, N., Zarco-Tejada, P. J., & Dextraze, L. (2002). Integrated narrow-band vegetation indices for prediction of crop chlorophyll content for application to precision agriculture. *Remote Sensing of Environment, 81*(2–3), 416–426.

Hall, A., Lamb, D., Holzapfel, B., & Louis, J. (2002). Optical remote sensing applications in viticulture-a review. *Australian Journal of Grape and Wine Research, 8*(1), 36–47.

He, M., Kimball, J. S., Maneta, M. P., Maxwell, B. D., Moreno, A., Beguería, S., & Wu, X. (2018). Regional crop gross primary productivity and yield estimation using fused landsat-MODIS data. *Remote Sensing, 10*(3), 372.

International Society of Agriculture Precision. (2019). *Precision agriculture latest definition.* Retrieved from https://www.ispag.org/about/definition#:~:text=%E2%80%9CPrecision%20Agriculture%20is%20a%20management,%2C%20productivity%2C%20quality%2C%20profitability%20and.

Jones, H. G. (1999). Use of infrared thermometry for estimation of stomatal conductance as a possible aid to irrigation scheduling. *Agricultural and Forest Meteorology, 95*(3), 139–149.

Jones, H. G., & Sirault, X. R. (2014). Scaling of thermal images at different spatial resolution: The mixed pixel problem. *Agronomy, 4*(3), 380–396.

Kamilaris, A., Fonts, A., & Prenafeta-Boldύ, F. X. (2019). The rise of blockchain technology in agriculture and food supply chains. *Trends in Food Science & Technology, 91*, 640–652.

Kerkech, M., Hafiane, A., & Canals, R. (2019). Vine disease detection in UAV multispectral images with deep learning segmentation approach. Preprint arXiv:1912.05281 *arXiv.*

Kerkech, M., Hafiane, A., & Canals, R. (2020). Vine disease detection in UAV multispectral images using optimized image registration and deep learning segmentation approach. *Computers and Electronics in Agriculture, 174*, 105446.

Khaliq, A., Comba, L., Biglia, A., Ricauda Aimonino, D., Chiaberge, M., & Gay, P. (2019). Comparison of satellite and UAV-based multispectral imagery for vineyard variability assessment. *Remote Sensing, 11*(4), 436.

Kline, R. (2010). Cybernetics, automata studies, and the Dartmouth conference on artificial intelligence. *IEEE Annals of the History of Computing, 33*(4), 5–16.

Lanyon DM, H. J., Goodwin, I., Whitfield, D., Gobbett, D. L., McClymont, L., Bramley, R. G. V., Mowat, D., & Christen, E. W. (2007). Capturing the variation in vine and edaphic properties using a mobile multi-functional platform. In *Paper presented at the proceedings of the thirteenth Australian wine industry technical conference, Australian wine industry technical conference, Inc., Adelaide, SA, Australia.*

Leinonen, I., Grant, O., Tagliavia, C., Chaves, M., & Jones, H. (2006). Estimating stomatal conductance with thermal imagery. *Plant, Cell and Environment, 29*(8), 1508–1518.

Liu, S., Li, X., Wu, H., Xin, B., Tang, J., Petrie, P. R., & Whitty, M. (2018). A robust automated flower estimation system for grape vines. *Biosystems Engineering, 172*, 110–123.

Maimaitiyiming, M., Sagan, V., Sidike, P., & Kwasniewski, M. T. (2019). Dual activation function-based Extreme Learning Machine (ELM) for estimating grapevine berry yield and quality. *Remote Sensing, 11*(7), 740.

Martínez-Lüscher, J., Brillante, L., & Kurtural, S. K. (2019). Flavonol profile is a reliable indicator to assess canopy architecture and the exposure of red wine grapes to solar radiation. *Frontiers in Plant Science, 10*, 10.

Mashalaba, L., Galleguillos, M., Seguel, O., & Poblete-Olivares, J. (2020). Predicting spatial variability of selected soil properties using digital soil mapping in a rainfed vineyard of central Chile. *Geoderma Regional, 22*, e00289.

Masson-Delmotte, V., Zhai, P., Pörtner, H. O., Roberts, D., Skea, J., Shukla, P. R., Pirani, A., Moufouma-Okia, W., Péan, C., Pidcock, R., & Connors, S. (2018). *Global warming of 1.5 C. An IPCC Special Report on the impacts of global warming of 1.5 C* (Vol. 1).

Matese, A., & Di Gennaro, S. F. (2015). Technology in precision viticulture: A state of the art review. *International Journal of Wine Research, 7*, 69–81.

Matese, A., & Di Gennaro, S. (2018). Practical applications of a multisensor uav platform based on multispectral, thermal and rgb high resolution images in precision viticulture. *Agriculture, 8*(7), 116.

Matese, A., Di Gennaro, S., Miranda, C., Berton, A., & Santesteban, L. (2017). Evaluation of spectral-based and canopy-based vegetation indices from UAV and Sentinel 2 images to assess spatial variability and ground vine parameters. *Advances in Animal Biosciences, 8*(2), 817–822.

Matese, A., Di Gennaro, S. F., & Santesteban, L. (2019). Methods to compare the spatial variability of UAV-based spectral and geometric information with ground autocorrelated data. A case of study for precision viticulture. *Computers and Electronics in Agriculture, 162*, 931−940.

Milella, A., Marani, R., Petitti, A., & Reina, G. (2019). In-field high throughput grapevine phenotyping with a consumer-grade depth camera. *Computers and Electronics in Agriculture, 156*, 293−306.

Mitchell, M. (2019). *Artificial intelligence: A guide for thinking humans*. UK: Penguin.

Monteith, J., & Unsworth, M. (1990). *Principles of environmental physics* (2nd ed.). London: Edward Arnold, 291 pp.

del-Moral-Martinez, I., Rosell-Polo, J. R., Sanz, R., Masip, J., Martínez-Casasnovas, J. A., & Arnó, J. (2016). Mapping vineyard leaf area using mobile terrestrial laser scanners: Should rows be scanned on-the-go or discontinuously sampled? *Sensors, 16*(1), 119.

Nolan, A., Park, S., Fuentes, S., Ryu, D., & Chung, H. (2015). Automated detection and segmentation of vine rows using high resolution UAS imagery in a commercial vineyard. In *Paper presented at the proceedings of the 21st international congress on modelling and simulation, Gold Coast, Australia*.

Ozdemir, G., Sessiz, A., & Pekitkan, F. G. (2017). Precision viticulture tools to production of high quality grapes. *Scientific Papers Series B Horticulture, 61*, 209−218.

Pádua, L., Marques, P., Adão, T., Guimarães, N., Sousa, A., Peres, E., & Sousa, J. J. (2019). Vineyard variability analysis through UAV-based vigour maps to assess climate change impacts. *Agronomy, 9*(10), 581.

Pádua, L., Marques, P., Hruška, J., Adão, T., Peres, E., Morais, R., & Sousa, J. J. (2018). Multi-temporal vineyard monitoring through UAV-based RGB imagery. *Remote Sensing, 10*(12), 1907.

Palacios, F., Bueno, G., Salido, J., Diago, M. P., Hernández, I., & Tardaguila, J. (2020). Automated grapevine flower detection and quantification method based on computer vision and deep learning from on-the-go imaging using a mobile sensing platform under field conditions. *Computers and Electronics in Agriculture, 178*, 105796.

Palacios, F., Diago, M. P., & Tardaguila, J. (2019). A non-invasive method based on computer vision for grapevine cluster compactness assessment using a mobile sensing platform under field conditions. *Sensors, 19*(17), 3799.

Park, S. (2018). *Estimating plant water stress and evapotranspiration using very-high-resolution (VHR) UAV imagery*.

Pérez-Zavala, R., Torres-Torriti, M., Cheein, F. A., & Troni, G. (2018). A pattern recognition strategy for visual grape bunch detection in vineyards. *Computers and Electronics in Agriculture, 151*, 136−149.

Pérez, D. S., Bromberg, F., & Diaz, C. A. (2017). Image classification for detection of winter grapevine buds in natural conditions using scale-invariant features transform, bag of features and support vector machines. *Computers and Electronics in Agriculture, 135*, 81−95.

Poblete-Echeverría, C., Olmedo, G. F., Ingram, B., & Bardeen, M. (2017). Detection and segmentation of vine canopy in ultra-high spatial resolution RGB imagery obtained from unmanned aerial vehicle (UAV): A case study in a commercial vineyard. *Remote Sensing, 9*(3), 268.

Poblete, T., Ortega-Farías, S., & Ryu, D. (2018). Automatic coregistration algorithm to remove canopy shaded pixels in UAV-borne thermal images to improve the estimation of crop water stress index of a drip-irrigated Cabernet Sauvignon vineyard. *Sensors, 18*(2), 397.

Power, A., Truong, V. K., Chapman, J., & Cozzolino, D. (2019). From the laboratory to the vineyard—evolution of the measurement of grape composition using NIR spectroscopy towards high-throughput analysis. *High-Throughput, 8*(4), 21.

Robertson, M., Moore, A., Barry, S., Lamb, D., Henry, D., Brown, J., Darnell, R., Gaire, R., Grundy, M., George, A., & Donohue, R. (2020). Digital agriculture. In *Australian Agriculture in 2020: From Conservation to Automat* (pp. 389−403).

Rodrigo-Comino, J., & Cerdà, A. (2018). Improving stock unearthing method to measure soil erosion rates in vineyards. *Ecological Indicators, 85*, 509−517.

Romero, M., Luo, Y., Su, B., & Fuentes, S. (2018). Vineyard water status estimation using multispectral imagery from an UAV platform and machine learning algorithms for irrigation scheduling management. *Computers and Electronics in Agriculture, 147*, 109–117.

Rossi, R., Pollice, A., Diago, M. P., Oliveira, M., Millan, B., Bitella, G., Amato, M., & Tardaguila, J. (2013). Using an automatic resistivity profiler soil sensor on-the-go in precision viticulture. *Sensors, 13*(1), 1121–1136.

Rouse, J. W., Haas, R. H., Schell, J. A., & Deering, D. W. (1974). NASA special publication. *Monitoring vegetation systems in the great plains with ERTS* (Vol. 351(1974), p. 309).

Santesteban, L. G. (2019). Precision viticulture and advanced analytics. A short review. *Food Chemistry, 279*, 58–62.

Santesteban, L., Di Gennaro, S., Herrero-Langreo, A., Miranda, C., Royo, J., & Matese, A. (2017). High-resolution UAV-based thermal imaging to estimate the instantaneous and seasonal variability of plant water status within a vineyard. *Agricultural Water Management, 183*, 49–59.

Santos, L., Ferraz, N., dos Santos, F. N., Mendes, J., Morais, R., Costa, P., & Reis, R. (2018). Path planning aware of soil compaction for steep slope vineyards. In *Paper presented at the 2018 IEEE international conference on autonomous robot systems and competitions (ICARSC)*.

Scafidi, P., Barbagallo, M., Pisciotta, A., Mazza, M., & Downey, M. (2018). Defoliation of two-wire vertical trellis: Effect on grape quality. *New Zealand Journal of Crop and Horticultural Science, 46*(1), 18–38.

Seng, K. P., Ang, L.-M., Schmidtke, L. M., & Rogiers, S. Y. (2018). Computer vision and machine learning for viticulture technology. *IEEE Access, 6*, 67494–67510.

Sonka, M., Hlavac, V., & Boyle, R. (2014). *Image processing, analysis, and machine vision*. Cengage Learning.

Srinivasan, A. (2006). *Handbook of precision agriculture: Principles and applications*. CRC Press.

Summerson, V., Gonzalez Viejo, C., Pang, A., Torrico, D. D., & Fuentes, S. (2021). Review of the effects of grapevine smoke exposure and technologies to assess smoke contamination and taint in grapes and wine. *Beverages, 7*(1), 7.

Sutton, R. T., & Hawkins, E. (2020). ESD Ideas: Global climate response scenarios for IPCC AR6. *Earth System Dynamics Discussions, 2020*, 1–4.

Tardaguila, J., Diago, M. P., Priori, S., & Oliveira, M. (2018). Mapping and managing vineyard homogeneous zones through proximal geoelectrical sensing. *Archives of Agronomy and Soil Science, 64*(3), 409–418.

Tisseyre, B., Ojeda, H., & Taylor, J. (2007). New technologies and methodologies for site-specific viticulture. *Journal International des Sciences de la Vigne et du Vin, 41*(2), 63–76.

Torky, M., & Hassanein, A. E. (2020). Integrating blockchain and the internet of things in precision agriculture: Analysis, opportunities, and challenges. *Computers and Electronics in Agriculture*, 105476.

Wang, X., De Bei, R., Fuentes, S., & Collins, C. (2019). Influence of canopy management practices on canopy architecture and reproductive performance of semillon and shiraz grapevines in a hot climate. *American Journal of Enology and Viticulture, 70*(4), 360–372.

Weiss, M., Jacob, F., & Duveiller, G. (2020). Remote sensing for agricultural applications: A meta-review. *Remote Sensing of Environment, 236*, 111402.

Wilkinson, K., & Ristic, R. (2019). Winemaking: Managing smoke taint: Understanding the effects of smoke taint on fruit and wine. *Australian and New Zealand Grapegrower and Winemaker*, (660), 42.

Xue, J., Fan, Y., Su, B., & Fuentes, S. (2019). Assessment of canopy vigor information from kiwifruit plants based on a digital surface model from unmanned aerial vehicle imagery. *International Journal of Agricultural and Biological Engineering, 12*(1), 165–171.

Xue, J., & Su, B. (2017). Significant remote sensing vegetation indices: A review of developments and applications. *Journal of Sensors, 2017*.

Zarco-Tejada, P. J., Catalina, A., González, M., & Martín, P. (2013). Relationships between net photosynthesis and steady-state chlorophyll fluorescence retrieved from airborne hyperspectral imagery. *Remote Sensing of Environment, 136*, 247–258.

Zarco-Tejada, P. J., González-Dugo, V., Williams, L., Suárez, L., Berni, J. A., Goldhamer, D., & Fereres, E. (2013). A PRI-based water stress index combining structural and chlorophyll effects: Assessment using diurnal narrow-band airborne imagery and the CWSI thermal index. *Remote Sensing of Environment, 138*, 38–50.

Zarco-Tejada, P. J., Guillén-Climent, M. L., Hernández-Clemente, R., Catalina, A., González, M., & Martín, P. (2013). Estimating leaf carotenoid content in vineyards using high resolution hyperspectral imagery acquired from an unmanned aerial vehicle (UAV). *Agricultural and Forest Meteorology, 171*, 281–294.

Zhang, A., Baker, I., Jakku, E., & Llewellyn, R. (2017). *Accelerating precision agriculture to decision agriculture: The needs and drivers for the present and future of digital agriculture in Australia.*

CHAPTER 8

Novel technologies and Decision Support Systems to optimize pesticide use in vineyards

Cristina C.R. Carlos[1,2] and Maria do Carmo M. Val[1]

[1]*ADVID — Associação para o Desenvolvimento da Viticultura Duriense, Edifício Centro de Excelência daVinha e do Vinho, Vila Real, Portugal;* [2]*Centre for the Research and Technology of Agro-Environmental and Biological Sciences, CITAB, Universidade de Trás-os-Montes e Alto Douro, Vila Real, Portugal*

8.1 Introduction

Grapevine is one of the most important crops worldwide, in relation to the production of both wine and table grapes. Several pests and diseases may affect grapevine, so an intensive pesticide schedule is often required to meet production standards. In general, fungicides account for the largest share of pesticide treatments in most vineyards (with an average of 12–15, up to 25–30 applications in the most problematic conditions) (Pertot et al., 2016).

The use of pesticides plays an important role in agricultural production by ensuring less damages to crops and a consistent yield. However, their use can have negative environmental impacts on water quality, terrestrial and aquatic biodiversity (persistence and toxic effects on nontarget species, etc.) (EUROSTAT, 2020a,b).

At the moment, harmonized statistical data on pesticides use are not available at European scale. The most recent data (EUROSTAT, 2020a,b) analyses the 2011–2018 period, giving general information about pesticide sales in EU. Data show that France, Spain, Italy, and Germany are the countries with the highest values of chemical use in agriculture. They also show that, among the 14 EU Member States which provided nonconfidential data for all the major groups, eight countries (Portugal, Ireland, Czechia, Italy, the Netherlands, Belgium, Romania, and Hungary) have decreased their total sales of pesticides up to 40% in some cases (EUROSTAT, 2020a,b).

Regarding pesticide categories, the same statistical source indicates that in 2018, the largest share of "fungicides and bactericides" sold in the EU were the inorganic fungicides. These are copper compounds, inorganic sulfur and other inorganic fungicides. More than half (53.1%) of all fungicide sales were inorganic fungicides. It should be noted that copper compounds and sulfur are also permitted for use in organic farming (EUROSTAT, 2020a,b).

Unfortunately, there is a lack of quantitative data about pesticide use in each crop, including in vineyards. The European Network for Durable Exploitation of crop protection strategies have published a report (Endure, 2010) showing that, during the period 1992–2003, fungicides represented more than 90% of the total mass of pesticides, due to an intensive use of inorganic sulfur (76% of fungicides). This report also states that viticulture used 80% more synthetic fungicides than fruit production, and 13 times more than arable crops, highlighting that viticulture is the agricultural activity with the most intensive use of pesticides in mass of active substances per unit area (Endure, 2010).

The Sustainable Use of Pesticide (SUP) Directive (Directive 2009/128/EC) establishes a framework to achieve a sustainable use of pesticides to reduce the risks and impacts of its use on human health and the environment. On its 14th article, it clearly states that "Member States shall take all necessary measures to promote low pesticide-input pest management, giving wherever possible priority to non-chemical methods, so that professional users of pesticides switch to practices and products with the lowest risk to human health and the environment among those available for the same pest problem. Low pesticide-input pest management includes integrated pest management as well as organic farming according to Council Regulation (EC) No 834/2007 of June 28, 2007 on organic production and labeling of organic products" (EU, 2009).

Thus, the SUP Directive promotes the use of integrated pest management (IPM) and of alternative approaches and techniques such as nonchemical alternatives to pesticides. Farmers must implement IPM principles (Table 8.1) and give preference to nonchemical methods if they provide satisfactory pest control. The main purpose is to reduce the dependency on pesticides in agriculture (EUROSTAT, 2020a,b).

Table 8.1 General principles of integrated pest management (Annex III, Directive 2009/128/EC).

1.	*The prevention and/or suppression of harmful organisms* should be achieved or supported among other options especially by: crop rotation, use of adequate cultivation techniques (e.g., stale seedbed technique, sowing dates and densities, undersowing, conservation tillage, pruning, and direct sowing), use, where appropriate, of resistant/tolerant cultivars and standard/certified seed and planting material, use of balanced fertilization, liming and irrigation/drainage practices, preventing the spreading of harmful organisms by hygiene measures (e.g., by regular cleansing of machinery and equipment), protection and enhancement of important beneficial organisms, e.g., by adequate plant protection measures or the utilization of ecological infrastructures inside and outside production sites.
2.	*Harmful organisms must be monitored by adequate methods and tools*, where available. Such adequate tools should include observations in the field as well as scientifically sound warning, forecasting and early diagnosis systems, where feasible, as well as the use of advice from professionally qualified advisors.
3.	Based on the results of the monitoring, *the professional user has to decide whether and when to apply plant protection measures. Robust and scientifically sound threshold values are essential components for decision making*. For harmful organism's threshold levels defined for the region, specific areas, crops, and particular climatic conditions must be taken into account before treatments, where feasible.
4.	*Sustainable biological, physical and other nonchemical methods must be preferred to chemical methods* if they provide satisfactory pest control.
5.	*The pesticides applied shall be as specific as possible for the target* and shall have the least side effects on human health, nontarget organisms and the environment.
6.	The professional user should *keep the use of pesticides and other forms of intervention to levels that are necessary*, e.g., by reduced doses, reduced application frequency or partial applications, considering that the level of risk in vegetation is acceptable and they do not increase the risk for development of resistance in populations of harmful organisms.
7.	Where the risk of resistance against a plant protection measure is known and where the level of harmful organisms requires repeated application of pesticides to the crops, *available antiresistance strategies should be applied to maintain the effectiveness of the products*. This may include the use of multiple pesticides with different modes of action.
8.	Based on the records on the use of pesticides and on the monitoring of harmful organisms, the professional user *should check the success of the applied plant protection measures*.

8.1 Introduction

The report on the implementation of the Directive 2009/128/EC on the Sustainable Use of Pesticides (SUD) dated October 21, 2009, among other things concludes: *Integrated Pest Management is a cornerstone of the Directive, and it is therefore of particular concern that Member States have not yet set clear targets and ensured their implementation, including for the more widespread use of land management techniques such as crop rotation …. the Commission will support the Member States in the development of methodologies to assess compliance with the eight IPM principles, taking into account the diversity of EU agriculture and the principle of subsidiarity* (PAN Europe et al., 2019).

As shown in IPM pyramid (Fig. 8.1), the general principle of IPM is that the prevention and/or suppression of harmful organisms should be achieved or supported especially by alternatives to synthetic chemical pesticides.

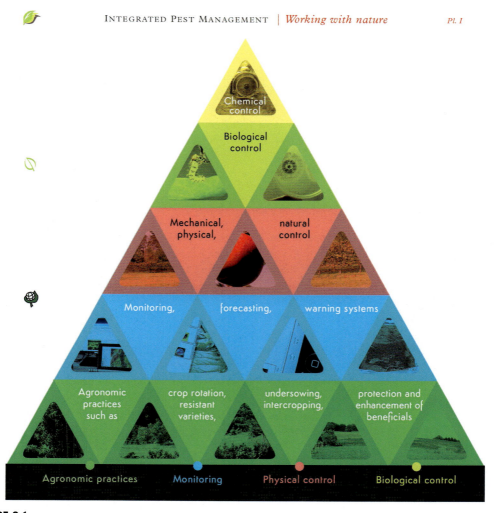

FIGURE 8.1

IPM Pyramid which represents the tools that can be used in IPM.

Adapted from IPM working with nature (PAN Europe et al., 2019).

According to the IPM pyramid, the agronomic practices—crop rotation, adequate cultivation techniques, use of resistant/tolerant cultivars and standard/certified seed and planting material, balanced fertilization, liming and irrigation/drainage practices, preventing the spread of harmful organisms by hygiene measures, protection and enhancement of important beneficial organisms, use of ecological infrastructures inside and outside production sites—represent the fundamentals of a healthy crop.

Warning, monitoring and forecasting systems and early diagnosis represent the second step to estimate the risk of crop damages or losses in order to optimize the use of the control measures. When an intervention is justified sustainable biological, physical, and other nonchemical methods must be preferred to chemical methods. The synthetic chemical pesticides represent the last choice to be used by farmers. The choice of the best active ingredient to be used should be made trying to minimize the risk for the environment (PAN Europe et al., 2019).

Reducing the dependence of agricultural production on synthetic pesticides and promoting the principles of IPM have long been central goals for the International Organization for Biological and Integrated Control (IOBC-wprs) (Wijnands et al., 2018). Although challenges remain, research efforts of scientists combined with innovative approaches of farmers and farm advisors have already produced inspiring success stories. New biocontrol solutions are becoming increasingly available and being strategically combined with other tools of plant health management (PAN Europe et al., 2019).

Failure to implement effective pest and disease management practices can result in complete loss of a hight value crop. To avoid this, it is essential that crop enemies are monitored and managed using the latest technologies. Some innovative techniques used either in disease or in pest management are summarized in the chapters toward.

8.2 Disease management

Since their introduction from America in the 19th century, downy mildew, caused by *Plasmopara viticola*, and powdery mildew, caused by *Erysiphe necator*, have been the most common and important global fungal diseases and are also the major contributors to the high consumption of fungicides (González-Dominguez, Caffi, & Rossi, 2019). There is also a growing concern on Flavescence dorée (FD), caused by the bacteria *Candidatus Phytoplasma vitis* and transmitted by the vector *Scaphoideus titanus*, and with grape trunk diseases, including Eutypa, Esca, and Black Dead Arm dieback (Micheloni, 2017).

The use of preventive control measures is a first step to reduce the use of pesticides. Some examples are soil management that facilitate drainage, balanced nitrogen fertilization to avoid excess vigor of the plants that leads to high susceptibility to downy and powdery mildew, training and pruning systems that facilitate air circulation in the canopy or leaves removal to facilitate bunch ventilation and reduce powdery mildew. The use of prophylactic measures is of outmost importance to prevent the spread of diseases. For example, major attention should be paid to the quality of the planting material (Waite et al., 2015). Also, the disinfection of pruning tools to avoid spreading of trunk diseases by workers, the removal of diseased plants in the vineyard to reduce the inoculum of diseases and sources of infected materials for vectors should be considered (Micheloni, 2017).

Grape varieties have low to high susceptibility to the different fungal diseases that result in significant production costs and economic losses (Pertot et al., 2016). Douro Demarcated Region (DDR), a winegrowing area (43,670 ha) located in the Northeast of Portugal, has more than 150 grapevine varieties, as well as a high diversity of climate conditions and vineyards systems installed. In this context, the choice of the most adapted varieties and rootstocks is crucial to minimize pesticide use.

New resistant/tolerant varieties have been created in the last few years (Saumon et al., 2018; Eiras-Dias et al., 2018; Herralde et al., 2018), but the advantages of their use are not confirmed in many traditional grape growing areas. In Portugal research conducted by the Portuguese Association for Grapevine Diversity (PORVID) is ongoing with the goal of better understanding how intra- and intervarietal diversity of local varieties can be explored to reduce the use of chemical inputs, particularly fungicides. It is expected that, in a few years, useful knowledge about the most disease-tolerant autochthones varieties will be available.

Having accurate and reliable information about pest and disease risk is essential when planning IPM strategies, especially in steep slope vineyards, where the different microclimates conditions induce high heterogeneity in the development of pests and diseases in vineyards agroecosystem (unpublished own data).

In the last decades, the availability of information technology (IT) tools, including wireless sensors to constantly monitor climatic data and vegetation development, as well as more developed and precise algorithms to forecast diseases development cycles allowed to implement in many regions/farms IPM and precision plant protection techniques (Micheloni, 2017).

Weather stations (Fig. 8.2A), to measure atmospheric weather data and soil probes (Fig. 8.2B) to measure soil humidity, are some of the most common examples of IT tools used by pioneer's winefarmers, in order to maximize or reduce the use of several inputs (e.g., water, pesticides).

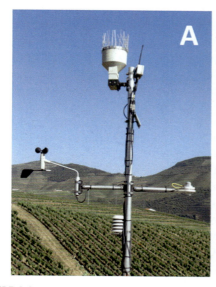 

FIGURE 8.2

(A) Weather station (C. Carlos/ADVID); (B) Soil sensor probe (I. Gonçalves/ADVID).

Forecasting systems and decision support systems (DSSs) to identify the risk level and decide when to act for plant protection have been developed for different diseases, but especially for downy and powdery mildew (Micheloni, 2017). DSS goal is to guide practitioners in the efficient implementation of plant protection schemes (if to spray, when and what to spray). IT and Internet of Things (IoT) technologies implementation made available several tools (Apps, web-based services, etc.) that facilitate their direct use by farmers and advisers (with no need of intermediate steps/actors). They rely on forecasting systems and constant monitoring, allowing high efficiency and savings of fungicides (Micheloni, 2017).

As a result, in last years, there has been an increased interest in implementing and calibrating prediction models for diseases, to support and provide the wine sector with such innovative tools, which give indications for diseases infection risks and consequently the best opportunity for treatments. Thus, these models contribute to a more precise application of pesticides, and consequently to reduce the environmental impacts of such inputs (González-Dominguez, Legler et al., 2019).

At the DDR, the Association for the Development of Viticulture in Douro Region (ADVID) has been testing vite.net®, a DSS developed and provided by the company Hort@ srl (https://www.horta-srl.it/en/portfolio-item/vite-net/).

DSSs are IT platforms that gather crop data in real-time via sensors and scouting tools. Data collected are then organized in cloud systems, analyzed and interpreted using advanced modeling techniques and big data, and automatically integrated to provide information and alarms to support decisions. Farmers use this information for precision agronomic crop management. Data relative to farming operations are also entered into the system to generate a continuous flow of up-to-date information between the crop, the DSS, and the user (Rossi et al., 2014). This DSS includes a model for plant development that simulates phenological stages (bud burst, leaves development on the main shoots, and inflorescences and bunches development), and different epidemiological models for main diseases and pest in the vineyard (mildews, black rot, bunch rot, European grapevine moth (EGVM), mealybugs, and *S. titanus* leafhopper). Through these predictive models, the system provides information about possible diseases infection risks and development of pests and thus it can help to assess the need of intervention and help for a better disease and pest control (Caffi et al., 2010, 2012; Lessio & Alma, 2021). ADVID is, at the moment, collaborating with HORT@ on the calibration of vite.net® to Douro vineyards conditions and for improved adaptation to the specific needs of Portuguese end-users.

In DDR, most common control strategy still relies in the use of fungicides, and an average number of five-six treatment are performed, according to the phenology of the crop. This implies preventive treatments to avoid disease outset and consequently spread in vineyards. In grapevine more than 95% of the phytosanitary products used are fungicides, and 70% of these are sulfur-based compounds, usually used for controlling powdery mildew (González-Dominguez, Caffi, & Rossi, 2019).

To maximize efficacy and reduce the use of sulfur-based compounds, trials have been conducted by ADVID in DDR to test the efficacy of different treatment programs against powdery mildew. Treatments were performed according the releasing rate of ascospores and at different phenological stages, from the bud burst to bunch closure (Val et al., 2014). It was shown that early treatments, conducted from the first leaf unfolded to the preflowering period performed with sulfur do not have significant increase in the effectiveness of controlling powdery mildew. Such results contributed to the reduction of early applications of the fungicide in vineyards, traditionally carried out in the region, with environmental and public health benefits leading also to a reduction in costs, labor, and soil compaction (Val et al., 2014).

In the case of copper-based products (the most widely used fungicide to control downy mildew), it is expected to be increasingly restricted due to its toxicity and related accumulation in the soil as a heavy metal (Lamichhane et al., 2018). Copper is included in the list of "substitution candidates," which means that it should be replaced whenever possible, since it is of special concern for public health or the environment. Currently, in view of the increasing restrictions on the permitted doses of copper by the European Commission's implementing Regulation 2018/1981, the maximum annual amount of copper allowed per hectare, turn to be 4 kg/year, in a total of 28 kg/ha distributed over a period of seven years. The threat of a total ban as lead to the search for effective alternatives (La Torre et al., 2018). Strategies to reduce the use of this active ingredient include the application of preventive phytosanitary measures, innovative formulations with reduced copper content, optimization of copper dosages, use of prediction models, use of resistant varieties, optimization of agricultural management, and natural alternatives to copper-based products (La Torre et al., 2018).

ADVID is participating in a SUDOE project called COPPEREPLACE—"Development and implementation of novel technologies, products and strategies to reduce the application of copper in vineyards and remediate contaminated soils in SUDOE regions" (www.coppereplace.com), in which several alternatives to copper for controlling downy mildew will be tested, including emerging products such as biocontrol agents, plant extracts, and/or elicitors. This project will also develop new viticulture management strategies optimizing phytosanitary products and pesticides application.

8.3 Pest management

Challenges posed by arthropods differ depending on localities. The European Focus Group expert's on Pests and diseases (Micheloni, 2017) has identified a list of key pests (Table 8.2), as representing the main challenges that winegrowers have usually to manage, especially in warmer climates.

Grape producers have always had to respond to changing pest complexes over the years, and they have learned new techniques for vineyard management that address these novel situations (Oliveira et al., 2019). It is apparent that the pace of change in crop pest management has increased, with the converging forces of increased global trade, climate change, consumer preference, new technology and regulations coming together to make the pest management approaches of only one generation ago seem obsolete. This pace of change argues strongly for continued vigilance against new pests, development of regionally appropriate management tactics, and investment in educational programs that focus on technology transfer to grape growers and other vineyard managers (Isaacs, Vincent, & Bostanian, 2012).

Without sampling plans and economic thresholds, implementation of complete IPM programs is not possible (Wilson & Daane, 2017). To ensure their greatest chance of adoption, development of these tools requires careful research and then subsequent validation to test them under commercial vineyard conditions. Despite their importance for IPM programs, there are relatively few widely used thresholds for grape pest management (Isaacs, Vincent, & Bostanian, 2012; Wilson & Daane, 2017).

The EGVM, *Lobesia botrana* (Den & Schiff.) is among the most economically important insect pests in Europe. At the DDR, where "Port" D.O.C. wine and other remarkably high-quality table wines are produced, this pest is feared by growers mostly for its impact on quality, while damages are highly variable among years, ranging from 0% to 90% of infested clusters at harvest (Carlos et al., 2014). As a result, it has received considerable attention by researchers attempting to develop effective control strategies against it.

Table 8.2 List of key pests in vineyards and major impact on crop performance, according to the PEI-AGRI Focus Group expert's on Pests and diseases (Micheloni, 2017).

Key — pests	Impacts on crop
• Grape moths, European grape moth (*Lobesia botrana*), and Cochylis grape moth (*Eupocilia ambiguella*)	Two lepidoptera of Tortricidae family that cause direct damage on bunches, as they feed on the grape content, and indirect damage as they open wounds that consequently offer opportunity for the development of diseases such as Botrytis
• Mites (different species such as *Calepitrimerus vitis*, *Eriophyes vitis*, *Eotetranychus pruni*, *Panonychus ulmi*)	More common in mild climates they damage leaves and shoots, decreasing the photosynthetic activity of the plant
• Green leafhopper (*Empoasca vitis*)	Feeds on the leaves of the vine decreasing the photosynthetic activity of the plant
• FD leafhopper (*Scaphoideus titanus*)	Vector of *Flavescence Dorée* phytoplasma
• Grape mealybug, an unarmored scale insect of the *Pseudococcidae* family	Damage grapes by contaminating clusters with cottony egg sacs, larvae, adults, and honeydew and can transmit grape viruses

To optimize IPM implement against pests, a correct monitoring, using traps and performing visual inspections, is a basic requirement to plan and implement an efficient monitoring system. Prediction of flight activity of EGVM during the growing season is also critical, for better timing of sampling or control operations. In DDR, flight activity of this pest was studied by analyzing data on male catches in sex pheromone traps recorded over a 20-year period, and degree-day models for predicting flights occurrence were developed (Fig. 8.3).

Average dates (Julian days-JD) and degree-days (°DD) corresponding to main flight events (beginning and peak of catches in each flight) of EGVM were obtained for DDR, as reported in Table 8.3.

After introduced in a DSS, such information can be useful for DDR growers, because it gives a first approach of EGVM flight behavior in this region and the average dates of the beginning and peak of the three-main flight are strategic for scheduling risk assessments, as well as to better timing the treatments against the pest, mainly if account is taken of the growing interest for biorational insecticides, for which this timing is very important (Carlos et al., 2018).

In the Douro region, monitoring of grapevine pests, their natural enemies, and diseases during the vine's phenological development is locally done by IPM teams, such as by ADVID technical department, which cooperates with local winegrowers and the agricultural administrations to determine the best moments to carry phytosanitary treatments. Therefore, results obtained would not replace the common monitoring programs or using economic thresholds, before deciding to apply insecticides. In practice, they specify the time to start careful vineyard monitoring, supporting growers to make better control decisions in a timely manner (Carlos et al., 2018).

The control of EGVM traditionally relies on the use of insect growth regulators or pyrethroid insecticides, applied once or twice a year, against the second and/or the third generation. Efforts to

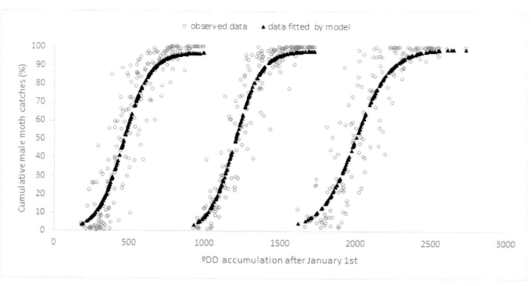

FIGURE 8.3

Observed and predicted cumulative percentage of male catches of *Lobesia botrana* for the first, second, and third flights in relation to accumulated DD, using January 1st as starting point for Degree-Day accumulation. DDR (1989–1991, 2000–2013), obtained for Douro region (Carlos et al., 2018).

Table 8.3 Observed values (means ± SE) on the occurrence of first catches of *Lobesia botrana* males (in Julian-days (JD), average date in brackets and degree-days (DD)) and 50% of catches (in DD) for each flight, by using January 1st (January 1st) as starting points for DD accumulation. DDR (1989–1991, 2000–2013) (Carlos et al., 2018).

Flight	N	First catches JD (Average date)	°DD	50% of catches °DD
First	23	81.3 ± 0.3 (21/03)	282.8 ± 2.2	470.0 ± 3.6
Second	37	162.1 ± 0.2 (10/06)	1034.9 ± 2.0	1257.5 ± 3.2
Third	36	210.5 ± 0.2 (28/07)	1798.7 ± 2.8	2090.7 ± 4.1
Fourth	17	253.1 ± 0.4 (9/09)	2539.8 ± 6.8	—

N, *Number of samples (years/sites).*

reduce pesticide use in the last years in the Douro region were mostly concentrated on implementing a conservation biological control strategy together with the use of mating disruption (MD) technique (Carlos, 2017).

Research about ecological-based pest management strategies has attracted increasing interest in recent decades about those aimed at promoting conservation biological control of pests (Thomson &

Hoffmann, 2009; Wilson & Daane, 2017). The loss of both agrobiodiversity and natural habitats that surround agroecosystems can lead to the loss of multiple ecosystem services, including biological control (Miles et al., 2012), that was estimated worldwide, in 1997, in approximately US$410 billion per year (Costanza et al., 1997; IPBES, 2019; Vallecillo et al., 2019; Winkler et al., 2017). While most vineyards in the world are nowadays typically extreme and intensive monocultures, with little remaining native vegetation and a suite of introduced weeds, whose provision of ecosystem services is, as a result, at a low level, in the "Alto Douro Vinhateiro" UNESCO classified region offer strong potential from this standpoint, due to the occurrence of a significant area of noncrop habitats (NCHs) (e.g., woodland remnants, grassy slopes, or terraces with natural vegetation and dry-stone walls) (Andresen et al., 2004).

These NCH has several advantages from the biodiversity point of view, in particular, in conservation biological control. Thus, they provide important resources for natural enemies, such as refugia, overwintering habitat, nectar, pollen and alternate hosts or prey, supporting natural enemy populations in nearby crop fields, which can lead to increased levels of biological control of pests. NCH have been reported to act as biodiversity reservoirs for plants, insects, birds, and mammals (Bianchi et al., 2006; Lourenço et al., 2021). Moreover, according to Böller et al. (2004), a high potential for a species-rich and natural green cover has been found in sloping vineyards, with small-scale terraces. Thus, as the plant community found in terraced vineyards contains several perennial plant species of value in fostering beneficial predators and parasitoids, it functions as an internal ecological infrastructure of the vineyard, which act at very short distances. The maintenance of such habitats into the vineyard, beyond biocontrol, can provide other benefits such as conserving wildlife, maintenance of water quality, and reducing erosion and runoff (Paiola et al., 2020; Viers et al., 2013; Winkler et al., 2017). In the case of the Douro region, the positive impact of NCH, as well as of ground cover vegetation of terraced vineyards on overall biodiversity of arthropods and on natural enemies (predators and parasitoids), was previously demonstrated (Carlos, 2017). The results obtained indicate that maintaining both NCH on the neighborhood of vineyards and vegetation of slopes and horizontal alleys enhanced functional biodiversity of this agroecosystem. These slopes could act as potential ecological infrastructures in the increase of populations of natural enemies of vineyard's pests, namely predators (spiders and coccinellids) and parasitoids (Fig. 8.4).

In a study held over a nine-year period (2002–2015) in 30 commercial vineyards of DDR a complex of 16 taxa of parasitoids of *L. botrana* was identified, the majority belonging to Hymenoptera (Carlos, 2017). The most abundant parasitoids found were *Elachertus* sp. (Eulophidae), *Campoplex capitator* Aubert (Ichneumonidae), and *Brachymeria tibialis* (Walker) (Chalcididae), which represented 62.5%, 12.6%, and 12.0% of the total assemblage of parasitoids emerged, respectively. It was found that the parasitism rate was positively related with ground cover management and negatively related with chemical treatments. When comparing the impact of chemical pesticides on parasitism of *L. botrana*, it was shown that those with higher chemical impact resulted in a substantial decrease in rates of parasitism of *L. botrana* in DDR (Carlos, 2017). Thus, it was concluded that to enhance conservation biological control of *L. botrana* in DDR, it is important to keep in mind the following two aspects: (1) to promote a high abundance and diversity of natural vegetation inside or at the edge of vineyards plots and (2) to select pesticides with minimal risks to natural enemies (Carlos, 2017).

In several steep sloped vineyards, soil cover of embankments and in the alleys is left to grow naturally being this cover managed mechanically. Traditionally, these vineyards were managed chemically, as growers perceived "weeds" as having a negative impact in vineyard ecosystem, so

FIGURE 8.4

(A) Slope in Douro vineyard covered by natural vegetation. Examples of conservation biological control agents present in Douro vineyards. (B) *Chrysoperla carnea* (Neuroptera: Chrysopidae), a predator of two vineyards key-pests (*Empoasca vitis* and *L. botrana*); (C) *Dictyna* sp. (Araneae: Dictynidae) preying *Lobesia botrana*; (D) Larva and (E) adult, respectively, of *Elachertus affinis* (Hymenoptera: Eulophidae), one of the most important parasitoids of *L. botrana* in Douro region (Credits photos: D. Carlos/ADVID).

herbicides were applied to the all surface of the productive area. In the last 15–20 years, with the development of research on ecosystem services, conservation biological control and agroecology, the pressure to reduce the use of pesticides and climate change impacts (e.g., erosion) the wine sector has changed the way they perceive the services/disservices provided by ground cover vegetation in vineyards (Bakıs et al., 2021). Gonçalves et al. (2019) review main results of research and academic works carried out over several years in the DDR in the field of functional agrobiodiversity, understood as the part of ecosystem biodiversity that provides ecosystem services, which support sustainable agricultural production and can also bring benefits to the regional and global environment and to society as a whole. Such studies specifically aimed to contribute knowledge about the diversity of arthropods in the vineyard ecosystem and about practices that can increase their abundance, diversity, and services provided.

The changing from chemical to mechanical management of ground cover was in fact one of the biggest changes in Douro vineyard management implemented in the last 15 years. This resulted in a significant reduction on herbicide amount used in vineyards of this region. Depending on the vineyard system (terraces, vertical rows), this reduction may have reached 70%, according to a consultation performed near several technicians of Douro wine companies (C. Carlos, unpublisded data).

The use of MD with hand-applied dispensers is one of the most effective and widely applied semiochemical control technique used against grapevine moths worldwide (Ioriatti & Lucchi, 2016). Pheromone-mediated MD is based on the release of synthetic sex attractants into a crop, interfering with mate finding of a given pest species (Lucchi et al., 2018; Mansour et al., 2017).

After being used since many years in DDR vineyards, Carlos (2017) found that MD can be an effective method to reduce EGVM presence in vineyards that contributes to reduce the use of conventional insecticides. However, the MD is more effective in years with lower pest population density, when applied in large areas, with more points of release per hectare, and after consecutive seasons (Carlos et al., 2010). Some constrains were identified and relate with: (a) the biology of the pest namely with its long life cycle and its high biotic potential; (b) the natural conditions of DDR region, such as the orography (high steepness), and the fragmentation, size and shape of the plots; (c) the climate conditions, such as the wind pattern and the high summer temperatures that, in some years, leads to the exhaustion of the pheromone in the dispensers by the end of July, just when *L. botrana* begin the third generation, which is its most damaging generation (Carlos et al., 2010). The adaptation of this innovative technique to steep slope conditions is the main subject of an ongoing project (PDR2020, operational group) named "CSinDouro"—Mating disruption of the European grapevine moth, *Lobesia botrana* (Denn. & Schiff.) in steep slope viticulture: the particular case of the DDR (ADVID, 2020).

The higher cost of MD technique, when compared to chemical pesticides, is one of the majors constrains to its wide application. According to a survey conducted near growers and providers of MD dispensers available in the market, only a maximum of 2%–3% of the total surface of vineyards in DDR are under MD (C. Carlos, unpublished data). As this technique leads to an effective reduction on the use of chemical insecticides, to increase its use in a widely manner, it is important that governments consider supporting its application in the implementation of future national CAP reforms.

Scaphoideus titanus Ball (Hemiptera: Cicadellidae) is a leafhopper widespread in European vineyards, responsible for transmitting the phytoplasma FD (EPPO Code: PHYP64), one of the most threatening among European grapevine yellow diseases (Tramontini et al., 2020). In regions where FD and its vector, *S. titanus*, was identified, one to three insecticides are mandatory to avoid the dissemination of the FD disease. Until a few years ago, the concept of mating disruption had been exclusively associated with the use of pheromones to reduce population density of insect pests (Cardé & Minks, 1995). Since the early 2000s, a novel approach has been proposed to the scientific community: vibrational mating disruption (VMD) (Mazonni et al., 2019). The novelty is the use of disturbance vibrations to disrupt the mating behavior of insect pests that communicate by means of substrate-borne vibrations (Mazonni et al., 2019). This technique was developed to disturb the behavior of *S. titanus*. In this case, VMD exploits vibrations used by *S. titanus*, both for mating and rivalry. Usually, a duet made of male's and female's signals is established and maintained until copula occurs, as reviewed by Mazzoni et al. (2019). In particular, the emission of a disturbance noise by rival males to interrupt the ongoing mating duet of a pair was crucial in developing the concept of VMD. Current knowledge about the sexual behavior of *S. titanus* and VMD experiments (reviewed by Mazzoni et al., 2019) suggest that playback of vibrational disturbance noise based on vibrational signals used by males during rivalry disrupts the courtship behavior and prevent mating. In 2012, the proof of concept was demonstrated also in field conditions: disturbance vibrations (i.e., disturbance noise playback) transmitted into plant tissues of rooted grapevines in a commercial vineyard can disrupt the mating behavior of a leafhopper species (Eriksson et al., 2012). The system was finally tested in 2017 in a large-scale field trial in S. Michele all'Adige (Italy) in which a total of 110 transducers (shakers) were installed on an area of 1.5 ha of Cabernet Franc, with a naturally occurring insect population. The field MVD experiments conducted starting from 2017 will require a long observation period, during which work on technological and methodological improvements and perform continuous monitoring of insects and plants/crops.

A considerable amount of research is still needed, not only to improve the efficacy of VMD technology but also to demonstrate its applicability into the field. The replication of the system in different environments and vineyards with different management practices will also help to standardize the method. However, authors emphasize that when the energetic issue is solved, an unbelievably high number of new utilities could be added to the mini-shaker prototype. Thus, this will make a vineyard a smart tech project, improving its general management (Mazzoni et al., 2019). Thus, VMD represents a promising technique for developing a nonchemical approach for controlling this invasive vector in Europe. Future developments of the technique will include development of a mechanism of signal transmission across large areas (Gordon & Krugner, 2019).

8.4 Concluding remarks

Wine systems are among the most demanding in terms of plant protection and soil and vegetation management operations. IPM is a well-tested approach that relies on knowledge of pest abundance and its relationship to crop damage to determine the point at which pest control is needed to prevent economic loss. The control method may be based in one or in a combination of strategies and the range of options available to grape producers is expected to continue to expand as new technologies are developed. However, according to Isaacs et al. (2012), there is a significant need for researchers to develop action thresholds for key pests that can guide decision-making using the foundational concepts that initiated IPM in the 1950s.

It is also important to adopt appropriate measures and promote the adoption of development models that allow to predict the occurrence of main pests and diseases and increase the resilience of the agroecosystem. The phytosanitary status of vineyards must be continuously monitored to precisely modulate management treatments based on real risks. The implementation of DSS is becoming more and more urgent as a goal to guide and support professionals in a more efficient plant protection schemes (if you spray, when you spray, what you spray) and to increase sustainability of the vineyard systems.

These innovative tools indicate the periods of risk and greater sensitivity of vines to the incidence of pests and diseases, the opportunity of treatments, contributing to a more accurate use of pesticides or other innovative solutions presented here, contributing for the reduction of environmental impacts of viticulture.

The viticulture sector is expected to continue moving toward increasing adoption of other production systems (e.g., organic). However, it is expected that these approaches will be implemented on a minority of vineyards around the globe for the foreseeable future, because of production goals. Nevertheless, the knowledge gained in these more innovative systems will be expected to have an influence on conventional viticulture (IPM), but it is likely that chemical methods of pest control will continue to dominate grape production. Farmers and agrochemical companies are under increasing societal and regulatory pressure to use and develop new pest control options with low nontarget impact. New classes of insecticides (e.g., microbial, botanical extracts) are providing unparalleled opportunities for insect and mite control without disrupting natural enemies. After the chemical-based vineyard pest management programs of the first half of the 20th century, and the integrated systems of management promoted in the late 20th century, there is now an increasing interest in biologically based components for use in vineyard management (Isaacs, Vincent, & Bostanian, 2012).

At the end of 2019, the European Commission adopted The European Green Deal, an ambitious vision for a sustainable, green transition that is just and socially fair. Biodiversity is one of its key policy areas. Most recently, on May 20, 2020 the European Commission adopted the EU Biodiversity Strategy for 2030 to halt the decline in biodiversity and bring nature back into our lives (EUROSTAT, 2020a,b). The "Farm to Fork Strategy"—for a fair, healthy, and environmentally friendly food system was also adopted by the European Commission on May 20, 2020 aiming to accelerate the transition to a more sustainable food system. In the strategy, the Commission presents actions to be taken forward in line with the better regulation principles, one of which is to propose a revision of the pesticides statistics regulation to overcome data gaps and reinforce evidence-based policy making, by 2023 (EUROSTAT, 2020a,b).

Finally, it is of outmost importance that knowledge generated by researchers reaches the main end-users (e.g., wine growers, technicians) and stakeholders' (policy makers and other researchers). This is one of the main roles of winegrowers' organization such as ADVID in the Douro region. Through the support given by technicians and advisers trained in IPM and sustainable use of pesticides, this organization provides growers with technical support in using DSS, implementing mating disruption technique, and promoting functional biodiversity in vineyards to enhance conservation biological control of pests. Finally, DSSs have a fundamental role in the implementation of an effective IPM strategy, reinforcing the strategies recently adopted by the European Commission.

References

ADVID. (2020). *CSinDouro — Confusão sexual contra a traça-da-uva em viticultura de montanha: caso particular da Região Demarcada do Douro*. https://www.advid.pt/pt./csindouro-confusao-sexual-contra-a-traca-da-uva-em-viticultura-de-montanha-caso-particular-da-regiao-demarcada-do-douro.

Andresen, T., Bianchi de Aguiar, F., & Curado, M. J. (2004). The Alto Douro Wine Region greenway. *Landscape and Urban Planning, 68*, 289—303.

Bakıs, A. L. P., Macovei, I., Barros, P., Gomes, C., Carvalho, D., Cabral, J. A., Travassos, P., Torres, L., Aranha, J., Galațchi, L. D., & Santos, M. (2021). Is biodiversity linked with farm management options in vineyard landscapes? A case study combining ecological indicators within a hybrid modelling framework. *Ecological Indicators, 121*(107012), 1—12. https://doi.org/10.1016/j.ecolind.2020.107012

Bianchi, F. J. J. A., Booij, C. J. H., & Tscharntke, T. (2006). Sustainable pest regulation in agricultural landscapes: A review on landscape composition, biodiversity and natural pest control. *Proceedings of the Royal Society of London B Biological Sciences, 273*, 1715—1727.

Caffi, T., Legler, S. E., & Rossi, V. (2012). Evaluation of a warning system for early-season control of grapevine powdery mildew. *Plant Disease, 96*, 104—110.

Caffi, T., Rossi, V., & Bugiani, R. (2010). *Evaluation of a Warning system for controlling Primary infections of grapevine downy mildew. 96* pp. 104—110). The American Phytopathological Society. https://doi.org/10.1094/PDIS-06-11-0484. Plant Disease.

Cardé, R., & Minks, A. (1995). Control of moth pests by mating disruption: Successes and constraints. *Annual Review of Entomology, 40*, 559—585. https://doi.org/10.1146/annurev.en.40.010195.003015

Carlos, C. C. R. (2017). *Towards a sustainable control of arthropod pests in Douro Demarcated Region vineyards with emphasis on the grape berry moth,* Lobesia botrana *(Denis & Schifermüller)*. PhD thesis, Universidade de Trás-os-Montes e Alto Douro, Vila Real (p. 164).

Carlos, C., Alves, F., & Torres, L. (2010). *Constrains to the application of mating disruption against* Lobesia botrana *in Douro Wine Region* (pp. 11—14). Sicilia: CERVIM. May 2010 (Cd-Rom).

Carlos, C., Gonçalves, F., Oliveira, I., & Torres, L. (2018). Is a biofix necessary for predicting the flight phenology of *Lobesia botrana* in Douro Demarcated Region vineyards? *Crop Protection, 110*, 57—64. https://doi.org/10.1016/j.cropro.2017.12.006

Carlos, C., Gonçalves, F., Sousa, S., Nóbrega, M., Manso, J., Salvação, J., Costa, J., Gaspar, C., Domingos, J., Silva, L., Fernandes, D., Val, M., Franco, J. C., Aranha, J., Thistlewood, H., & Torres, L. (2014). Success of mating disruption against the European grapevine moth, *Lobesia botrana* (den. & Schiff): A whole farm case-study in the Douro Wine Region. *IOBC-WPRS Bulletin, 105*, 93—102.

Costanza, R., d'Arge, R., de Groot, R., Farber, S., Grasso, M., Hannon, B., Limburg, K., Naeem, S., O'Neill, R. V., Paruelo, J., Raskin, R. G., Suttonkk, P., & van den Belt, M. (1997). The value of the world's ecosystem services and natural capital. *Nature, 387*, 253—260.

Eiras-Dias, J. E., Brazão, J., Cunha, J., Oliveira, H., Rodrigues, I., Carlos, C., & Almeida, F. (2018). *Livro Branco sobre as variedades resistentes. Estado da arte em Portugal*. Projecto VINOVERT (pp. 69—108).

Endure. (2010). Deliverable DR1.23. Pesticide use in viticulture, available data on current practices and innovations, bottlenecks and need for research in this field and specific leaflet analysing the conditions of adoption of some innovations. In *ENDURE — European network for durable exploitation of crop protection strategies, grapevine case study — guide number 5*. Published in July, 2010 http://www.endure-network.eu/.

Eriksson, A., Anfora, G., Lucchi, A., Lanzo, F., Virant-Doberlet, M., & Mazzoni, V. (2012). Exploitation of insect vibrational signals reveal a new method of pest management. *PLoS One, 7*, e32954. https://doi.org/10.1371/journal.pone.0032954

EU. (2009). Directive 2009/128/EC (sustainable use of pesticide). *Official Journal of the European Union*. Directives https://eur-lex.europa.eu/LexUriServ/LexUriServ.do?uri=OJ:L:2009:309:0071:0086:en:PDF.

EUROSTAT. (2020a). *Sales of pesticides in the EU.* https://ec.europa.eu/eurostat/web/products-eurostat-news/-/ddn-20200603-1.

EUROSTAT. (2020b). *Agri-environmental indicator − consumption of pesticides (2011−2018).* https://ec.europa.eu/eurostat/statistics-explained/index.php/Agri-environmental_indicator_-_consumption_of_pesticides#Key_messages. (Accessed 7 December 2020).

González-Dominguez, E., Caffi, T., & Rossi, V. (2019). Estamos controlando correctamente mildiu y oídio en viña? Descubrimientos recientes proponen câmbios en las estratégias de manejo. *Enoviticultura, 56*, 28−41. enero/febrero 2019.

González-Dominguez, E., Legler, S. E., Caballero, J. M., & Rossi, V. (2019). Desarrollo e implementación de un sistema de ayuda a la toma de decisiones en el manejo sostenible del viñedo: vite.net. *Enoviticultura, 56*, 42−49. enero/febrero 2019.

Gonçalves, F., Carlos, C., Crespí, A., Villemant, C., Trivellone, V., Goula, M., Canovai, R., Zina, V., Crespo, L., Pinheiro, L., Lucchi, A., Bagnoli, B., Oliveira, I., Pinto, R., & Torres, L. (2019). The functional agro-biodiversity in Douro demarcated region viticulture: Utopia or reality? Arthropods as a case-study − a review. *Ciência e Técnica Vitivinícola, 34*(2), 102−114.

Gordon, S. D., & Krugner, R. (2019). Mating disruption by vibrational signals: Applications for management of the glassy-winged sharpshooter. In P. S. M. Hill, R. Lakes-Harlan, V. Mazzoni, P. M. Narins, M. Virant-Doberlet, & A. Wessel (Eds.), *Biotremology: Studying vibrational behavior* (pp. 355−373). Cham: Springer International Publishing.

Herralde, F., Ruiz-Garcia, L., Savé, R., Elorduy, X., Bartra, E., Gil, J. M., & Escobar, C. (2018). *Libro blanco sobre variedades resistentes. Estado de situacion en España* (pp. 54−67). Projecto VINOVERT.

Ioriatti, C., & Lucchi, A. (2016). Semiochemical strategies for tortricid moth control in apple orchards and vineyards in Italy. *Journal of Chemical Ecology, 42*(7), 571−583.

IPBES. (2019). *Global assessment report on biodiversity and ecosystem services.* https://ipbes.net/sites/default/files/2020-02/ipbes_global_assessment_report_summary_for_policymakers_en.pdf.

Isaacs, R., Vincent, C., & Bostanian, N. J. (2012). Vineyard IPM in a changing world: Adapting to new pests, tactics and challenges. In N. J. Bostanian, C. Vincent, & R. Isaacs (Eds.), *Arthropod management in vineyards: Pests approaches, and future directions* (pp. 475−484). Dordrecht: Springer.

Isaacs, R., Saunders, M. C., & Bostanian, N. J. (2012b). Pest thresholds: Their development and use in vineyards for arthropod management. In N. J. Bostanian, C. Vincent, & R. Isaacs (Eds.), *Arthropod management in vineyards: Pests approaches, and future directions* (pp. 17−36). Dordrecht: Springer.

La Torre, A., Iovino, V., & Caradonia, F. (2018). Copper in plant protection: Current situation and prospects. *Phytopathologia Mediterranea, 57*(2), 201−236.

Lamichhane, J. R., Osdaghi, E., Behlau, F., Jürgen, K., Jones, J. B., & Aubertot, J.-N. (2018). Thirteen decades of antimicrobial copper compounds applied in agriculture. A review. *Agronomy for Sustainable Development, 38*, 28. https://doi.org/10.1007/s13593-018-0503-9

Lessio, F., & Alma, A. (2021). Models applied to grapevine pests: A review. *Insects, 12*(2), 169. https://doi.org/10.3390/insects12020169

Lourenço, R., Pereira, P. F., Oliveira, A., Ribeiro-Silva, J., Figueiredo, D., Rabaça, J. E., Mira, A., & Marques, J. T. (2021). Effect of vineyard characteristics on the functional diversity of insectivorous birds as indicator of potential biocontrol services. *Ecological Indicators, 122*, 107251. https://doi.org/10.1016/j.ecolind.2020.107251. ISSN1470-160X.

Lucchi, A., Ladurner, E., Iodice, A., Savino, F., Ricciardi, R., Cosci, F., Conte, G., & Benelli, G. (2018). Eco-friendly pheromone dispensers—a green route to manage the European grapevine moth? *Environmental Science and Pollution Research, 25*(10), 9426−9442.

Mansour, R., Grissa-Lebdi, K., Khemakhem, M., Chaari, I., Trabelsi, I., Sabri, A., & Marti, S. (2017). Pheromone-mediated mating disruption of *Planococcus ficus* (Hemiptera: Pseudococcidae) in Tunisian vineyards: Effect on insect population dynamics. *Biologia, 72*, 333—341. https://doi.org/10.1515/biolog-2017-0034

Mazzoni, V., Nieri, R., Eriksson, A., Virant-Doberlet, M., Polajnar, J., Anfora, G., & Lucchi, A. (2019). Mating disruption by vibrational signals: State of the field and perspectives. In P. Hill, R. Lakes-Harlan, V. Mazzoni, P. Narins, M. Virant-Doberlet, & A. Wessel (Eds.), *Biotremology: Studying vibrational behavior. Animal signals and communication, 6*. Cham: Springer. https://doi.org/10.1007/978-3-030-22293-2_17

Micheloni, C. (2017). *EIP-AGRI focus group diseases and pests in viticulture*. Starting paper. Version 19 January 2017.

Miles, A., Wilson, H., Altieri, M., & Nicholls, C. (2012). Habitat diversity at the field and landscape level: Conservation biological control research in California viticulture. In N. J. Bostanian, C. Vincent, & R. Isaacs (Eds.), *Arthropod management in vineyards: Pests, approaches, and future directions* (pp. 159—189). New York, NY: Springer.

Oliveira, M. J. R. A., Roriz, M., Vasconcelos, M. W., Bertaccini, A., & Carvalho, S. M. P. (2019). Conventional and novel approaches for managing "flavescence dorée" in grapevine: Knowledge gaps and future prospects. *Plant Pathology?, 68*, 3—17.

Paiola, A., Assandri, G., Brambilla, M., Zottini, M., Pedrini, P., & Nascimbene, J. (2020). Exploring the potential of vineyards for biodiversity conservation and delivery of biodiversity-mediated ecosystem services: A global-scale systematic review. *The Science of the Total Environment, 706*, 135839. https://doi.org/10.1016/j.scitotenv.2019.135839. ISSN 0048-9697.

PAN Europe, IOBC & IBMA. (2019). *Pesticide action Network (PAN) Europe, IOBC and IBMA*. IPM working with Nature.

Pertot, I., Caffi, T., Rossi, V., Mugnai, L., Hoffmann, C., Grando, M. S., Gary, C., Lafond, D., Duso, C., Thiery, D., Mazzoni, V., & Anfora, G. (2016). A critical review of plant protection tools for reducing pesticide use on grapevine and new perspectives for the implementation of IPM in viticulture. *Crop Protection, 97*, 70—84.

Rossi, V., Salinari, F., Poni, S., Caffi, T., & BettatI, T. (2014). Addressing the implementation problem in agricultural decision support systems: The example of vite.net. *Computers and Electronics in Agriculture, 100*, 88—99. https://doi.org/10.1016/j.compag.2013.10.011

Saumon, J. M., Hojeda, H., Hubert, A., Giraud Héraud, E., & Fuentes Spinoza, A. (2018). *Livre Blanc sur les variétés résistantes- États des lieux en France* (pp. 9—46). Projet VINOVERT.

Temperate zones of Europe. In Böller, E. F., Häni, F., & Poehling, H. M. (Eds.), *Ecological infrastructures: Ideabook on functional biodiversity at the farm level* (p. 212). Lindau: Swiss Centre for Agricultural Extension and Rural Development.

Thomson, L. J., & Hoffmann, A. A. (2009). Vegetation increases the abundance of natural enemies in vineyards. *Biological Control, 49*, 259—269.

Tramontini, S., Delbianco, A., & Vos, S. (2020). Pest survey card on flavescence dorée phytoplasma and its vector *Scaphoideus titanus*. *EFSA Supporting Publications, 17*(8). https://doi.org/10.2903/sp.efsa.2020.EN-1909

Vallecillo, S., La Notte, A., Ferrini, F., & Maes, J. (2019). How ecosystem services are changing: An accounting application at the EU level. *Ecosystem services, 40*, 101044. https://doi.org/10.1016/j.ecoser.2019.101044

Val, M. C., Silva, V., Amador, R., Manso, J., & Cortez, I. (2014). Eficacia del uso de diferentes fungicidas hasta pre-floración en control de Erysiphe necator, en la Region del Duero. *Enoviticultura, 29*, 14—23. July/August 2014.

Viers, J. H., Williams, J. N., Nicholas, K. A., Barbosa, O., Kotzé, I., Spence, L., Webb, L. B., Merenlender, A., & Reynolds, M. (2013). Vinecology: Pairing wine with nature. *Conserv. Lett., 6*, 287—299. https://doi.org/10.1111/conl.12011

Waite, H., Whitelaw-Weckert, M., & Torley, P. (2015). Grapevine propagation: Principles and methods for the production of high-quality grapevine planting material. *New Zealand Journal of Crop and Horticultural Science, 43*(2), 1–17. https://doi.org/10.1080/01140671.2014.978340

Wijnands, F., Malavolta, C., Alaphilippe, A., Gerowitt, B., & Baur, R. (Eds.). (2018). *Integrated production: IOBCWPRS objectives and principles*. Zurich: International Organisation for Biological and Integrated Control.

Wilson, H., & Daane, K. M. (2017). Review of ecologically based pest management in California vineyards. *Insects, 8*, 108. https://doi.org/10.3390/insects8040108

Winkler, K. J., Viers, J. H., & Nicholas, K. A. (2017). Assessing ecosystem services and multifunctionality for vineyard systems. *Front. Environ. Sci., 5*, 15. https://doi.org/10.3389/fenvs.2017.00015

CHAPTER 9

Processed kaolin particles film, an environment friendly and climate change mitigation strategy tool for Mediterranean vineyards

Lia-Tânia Dinis[1], Tommaso Frioni[2], Sara Bernardo[1], Carlos Correia[1] and José Moutinho-Pereira[1]

[1]*Centre for the Research and Technology of Agro-Environmental and Biological Sciences (CITAB), University of Trás-os-Montes e Alto Douro, Vila Real, Portugal;* [2]*Department of Sustainable Crop Production, Università Cattolica del Sacro Cuore, Piacenza, Italy*

9.1 Introduction

The geographical distribution of grapevine cultivation, as well as fruit ripening dynamics, has been challenged by global warming, resulting in changes in wine typicity (Santillan et al., 2019). Shifts in agricultural production patterns worldwide are ongoing due to climate change, which also impacts socioeconomic and cultural contexts of the wine industry (Fraga et al., 2013; Santos et al., 2020). The main observable effect of climate change is the enhancement in the growing season mean temperatures (Jones et al., 2005). Some climate models expect an increase up to 3.7 °C till the end of the century (Fraga et al., 2013; IPCC, 2013). Regarding precipitation, it is commonly accepted that the patterns will change in terms of periodicity and intensity depending on the region (IPCC, 2013). Displaying a temperate climate with mild and wet winters and warm and dry summers, the Mediterranean region is one of the most affected by these changes (Fraga et al., 2013). Furthermore, the combined effects of stresses are predictable to impair natural grapevine metabolisms, decreasing yield and berry quality, and resulting wine styles (van Leeuwen & Darriet, 2016). Among these stresses, drought, high temperature and irradiance, as well as salinity, are the most limiting factors for Mediterranean viticulture, exhibiting synergetic effects with irreversible consequences on growth, productivity, and consequently in berry composition. Scarce water availability has negative outcomes on nutrient uptake, water relationships and thermal regulation, photorespiration, carbon assimilation, hormone levels, and oxidative pathways (Brito, Dinis, Meijón et al., 2018; Chaves et al., 2010). Also, high temperatures (over 40°C) impact membrane stability and protein denaturation and aggregation (Wahid et al., 2007), and under high photosynthetically active radiation (PAR), photoinhibition and chlorosis are observed in grapevine plants as well as limitations in transpiration and photosynthesis (Moutinho-Pereira et al., 2009). High levels of ultraviolet-B radiation, often associated with these environmental stresses, also stimulate deleterious morphological, physiological, and biochemical changes (Correia et al., 2005).

In this sense, adopting the good agricultural practices (GAPs) throughout the world must be imperative and highly recommended for stakeholders (Martínez-Lüscher et al., 2016). Several potential GAPs could be adopted to vineyards under adverse conditions, such as the use of more resistant rootstocks and cultivars, adequate training systems, efficient irrigation strategies, implementation of breeding programmes, improvement of soil management, and application of sunlight protector compounds. In addition, given the scarce water resources and rugged topography of some arid and semiarid areas of the world, large-scale systems of water abstraction and distribution are costly and environmentally unsustainable (Martínez-Lüscher et al., 2016).

One of the numerous innovations for GAPs was the development of processed particle film technology. This review highlights the benefits/effects of processed kaolin particle film (PKPF) application in vines, covering its functions on abiotic stress mitigation, such as physiological performance and yield improvement (Bernardo et al., 2018; Dinis et al., 2020; Frioni, Saracino et al., 2019; Frioni, Tombesi et al., 2019).

9.2 Climate change effects
9.2.1 Phenology

The phenology of grapevines includes some stages or phenophases (bud break, flowering, and *veraison*) of vegetative and reproductive cycle that are mostly controlled by environmental conditions (Fraga et al., 2013). Phenological changes are one of the primary biological indicators of stress and are used to quantify the level of climate change impact in vines during the different stages till harvest (de Cortázar-Atauri et al., 2017). Phenology evolution models showed that all grapevine phenological stages would advance in the upcoming years, more noticeable in northern vineyards, leading to an earlier onset of flowering, *veraison*, and harvest (de Cortázar-Atauri et al., 2017). Therefore, earlier *veraison* onset suggests that the ripening stage may shift to the warmest period of the season, affecting yield and fruit composition, mostly sugars, organic acids, and phenolics (Ferrandino & Lovisolo, 2014), compromising the production of high quality wines (van Leeuwen & Darriet, 2016). Drought provokes yield losses by limiting photosynthesis, meaning that only a reduced quantity of berries can achieve the full ripeness. Overall, water deficit appeared to delay grape maturity, displaying significant interactions with both temperature and air CO_2 concentration (Martinez- Lüscher et al., 2016).

9.2.2 Physiology

Severe weather conditions impair several physiological processes, such as photosynthesis and water status (Fernández, 2014), disrupting cellular membrane stability, and increasing the permeability and leakage of ions (Elbasyoni et al., 2017). The most prominent summer stress effects on grapevine physiology include stomatal closure, consequently the reduction of photosynthetic rates (stomatal limitation), and photoinhibition of photosystem II (PSII) (nonstomatal limitation) (Dinis, Ferreira et al., 2016), which greatly reduce net carbon assimilation and impair berry ripening (Martínez-Lüscher et al., 2015). Changes in plant physiological processes due to water stress have also been linked to changes in plant reflectance including in the so-called photochemical reflectance index, which can be used to monitor dynamic changes in photochemical efficiency of photosynthesis in stressed plants (Evain et al., 2004). Light energy absorbed by the leaf cannot be used to drive photosynthetic electron transport under low CO_2 assimilation rates, and part of this energy is dissipated as heat, increasing the nonphotochemical fluorescence quenching (Baker & Rosenqvist, 2004). NPQ is the primary protective mechanism against photoinhibition, including xanthophylls to dissipate the excessive nonradiative energy (Hendrickson et al., 2004). High NPQ value suggests a higher dissipation of the excess energy by heat to avoid photosynthetic damages due to oxidative stress (Dinis, Malheiro et al., 2018). On the other hand, lower maximum (F_v/F_m), effective quantum efficiency of photosystem II (Φ_{PSII}), and apparent electron transport rate values, along with increasing basal fluorescence (F_0), indicate that the total electron flow through PSII was inhibited (photoinhibitory damage in response to high temperature and drought) (Dinis et al., 2014). Summer stress can also promote chlorophyll degradation, and seems to be related to the production of reactive oxygen species (ROS) (Camejo et al., 2006). Besides pigment degradation, high temperatures, light, and drought can also decrease soluble protein contents and alter the rate of rubisco regeneration (Todorov et al., 2003).

9.2.3 Leaf metabolites

Under severe environmental conditions, an uncontrolled increase of ROS was reported in different cellular components (Ahmad et al., 2010), boosting oxidative stress by ROS overloading. The imbalance of ROS also changes membrane fluidity and permeability, causing the denaturation and inactivation of some enzymes, proteins degradation, bleaching of pigments, and disruption of DNA strands, which ends in programmed cell death (Apel & Hirt, 2004). A decrease in soluble proteins by hydrolysis arises to overcome the boosted request of amino acids stimulated by high temperature and water stress (Dinis, Bernardo et al., 2018). At molecular level, DNA methylation is a good indicator of epigenetic answers of biotic and abiotic factors (Mirbahai & Chipman, 2014). The role of epigenetic mechanisms, like DNA methylation in many genes and histone acetylation is vital in managing acclimation, and therefore in plants adaptation to high temperatures (Correia et al., 2013) associated to increased solar radiation and low water availability, which leads to a DNA hypermethylation (Bernardo et al., 2017). The prompt response of hormones is crucial for plant physiology and biochemistry acclimation, being a serious condition for their survival (Jesus et al., 2015). One of the most important phytohormones due to their multiple functions is the abscisic acid (ABA) which activates a wide range of cellular and adaptive physiological and hydraulic responses and is a key regulator in the activation of plant cellular adaptation strategies to abiotic stresses (Takahashi et al., 2020), having a crucial function as a growth inhibitor (Golldack et al., 2014). High cellular levels of this hormone enable the reduction of water loss most via stomatal closure (Kim et al., 2010). In turn, the indoleacetic acid (IAA), which is the most considerable auxin in plants, is known to influence plant morphogenesis, as growth, development, and root formation (Zhao, 2010). Also, this hormone is involved in the regulation of some signaling pathways, such as cell viability, cell cycle progression, and programmed cell death (Zhao, 2010). Under water stress conditions, ABA synthesis occurs in leaves, with a restriction of IAA accumulation in the guard cells, which leads to stomatal closure (Dinis, Bernardo et al., 2018), revealing a possible crosstalk modulated by ABA and IAA levels that could affect the response of plants to drought (Sharma, Sharma et al., 2015).

9.2.4 Yield and berry quality attributes

Several biochemical pathways in the grapevine metabolism are both light and temperature sensitive (Spayd et al., 2002), having a strong effect on the secondary metabolism of grape berry (Conde et al., 2016, 2018; Dinis, Bernardo et al., 2016). Shaded grapevine berries were often 2.4°C above ambient air temperature, whereas sun-exposed clusters were up to 12.4°C above reference air temperature (Smart & Sinclair, 1976), which impacts berry composition. Secondary metabolites are indeed extremely essential for fruit and wine quality, namely phenolics content due to their contribution to color, flavor, aroma, texture, astringency, wine stabilization, and antioxidant properties (Dinis, Bernardo et al., 2016). In fact, grapevine fruits under severe weather conditions can have lower synthesis and accumulation of phenolics, mainly anthocyanins, as often this stress is associated with superimposed very high temperatures in the vineyards (Dinis, Bernardo et al., 2016; Spayd et al., 2002). Regarding organic acids, malic and tartaric acids are the most common in grapevine fruits, having variable regulation over the ripening stage (Conde et al., 2007). It is pointed out that severe climate conditions during summer induce alterations in aroma and reduced malic acid concentrations in berries (van Leeuwen & Destrac-Irvine, 2017). Additionally, reductions in total acidity are usually linked with higher pH, in spite of being affected by increased potassium accumulation, which is also temperature dependent (Coombe, 1987). Extreme temperatures may also negatively influence the synthesis of volatile compounds, which strongly contribute to the sensory character of wines (Robinson et al., 2014). Earlier maturation makes that grape ripening occurs under excessively warm conditions (Webb et al., 2008), resulting in increased alcohol content and decreased acidity (Neethling et al., 2012). Water deficit before *veraison* was also responsible to stimulate anthocyanins and phenolic concentrations biosynthesis (Deluc et al., 2007), but the timing and intensity of drought influences changes in berry metabolism and in wine organoleptic properties (color, aroma, and flavor) by altering berry size and/or the synthesis of berry compounds. These conditions may result in unbalanced wines, with undesirable high alcoholic content and low acidity, and consequently with low commercial value. Some authors (Martinez-Luscher et al., 2016) showed that grapes grown under both high air CO_2 (over 500 ppm) and high temperatures (over 40°C), attained the sugar criteria for harvest much earlier, decreasing anthocyanin accumulation, which can potentially threaten wine's typicity of traditional winemaking regions (Molitor & Junk, 2019).

9.3 Kaolin case: short-term adaptation strategy
9.3.1 Kaolin characterization

Kaolin is a white clay mostly formed from kaolinite, which is in turn formed by aluminosilicates ($Al_2Si_2O_5(OH)_4$) without isomorphic replacements, initially explored in ancient China (Garcia-Valles et al., 2020). This mineral is chemically inert, nonabrasive, nontoxic, and easily dissolves in water. In current times, kaolin is used for mitigating the damaging effect of extreme temperatures on leaves and fruits due to its reflective properties (Brito et al., 2019; Dinis, Bernardo et al., 2016; Frioni, Saracino et al., 2019; Frioni, Tombesi et al., 2019; Glenn et al., 2010). Kaolin is extremely important for the industry, mostly due to its specific properties that benefit a large number of functions. Kaolin mining began in the Neolithic, when their appropriateness for painting was discovered. Kaolin has several other specific requests including pharmaceutical and medical applications, due to its fine grain size, adsorptive properties, and whiteness (Hernández et al., 2019). Once sprayed as a suspension on leaf surface, water evaporates leaving a reflective particle film with an impact on plant microclimate as well as on pest and disease control (Glenn & Puterka, 2005). Nevertheless, this effectiveness as plant protector, including brightness and dispersibility, depends in large part on the average size (ideally diameter <2 µm) and uniformity of the kaolin particles, in order to form a uniform film layer, that transmits PAR but that excludes UV and IR radiation to some extent (Sharma, Sharma et al., 2015).

9.3.2 Reflection of radiation excess and reduction of organ temperature

One of the macroscopic effects once kaolin formulations are applied on the canopies is the high reflection of the incident radiation on plant structures. The first works investigating this aspect were conducted on *Citrus* trees (Abou-Khaled et al., 1970). Leaves of orange trees coated with kaolin had a higher fraction of reflected radiation and a lower fraction of transmitted radiation, as compared to untreated leaves (Abou-Khaled et al., 1970). The reflected radiation was mostly belonging to wavelengths of the visible, whereas the difference in transmitted radiation was almost totally ascribed to the infrared wavelengths. As expected, higher doses of the formulation increased the amount of reflected radiation (Abou-Khaled et al., 1970). More recently, in almond and walnut trees, it has been demonstrated that kaolin could change the PAR received by a leaf according to its position within the canopy and to the actual amount of PAR received prekaolin. In detail, if kaolin could reduce by 20%—40% the PAR absorbed by sunlit leaves, the inner canopy shaded leaves exhibited an increase of absorbed PAR by up to 80% (Rosati et al., 2007). Glenn et al. (2001) reported similar effects for apple trees and Campostrini et al. (2004) for papaya plants. In grapevine, despite the large variability of training systems in use, canopy shapes and vigor-driven leaf layers density, only a few amount of papers studied radiation distribution and reflection after kaolin applications. In Chile, Lobos et al. (2015) found that kaolin could increase PAR and UV reflection by 26%—155% in vines trained to vertical shoot positioned (VSP) systems, as compared to untreated canopies. In northern Italy, after spraying a suspension of 3% kaolin on VSP Pinot Noir canopies, fully exposed leaves showed 50% higher PAR reflectance and 17% lower PAR transmission than untreated leaves (Frioni et al., 2020). Analyzing spectral data, Tosin et al. (2019) reported that kaolin coating doubles up reflectance in the visible domain in both Touriga Nacional and Touriga Franca, whereas differences in the infrared can be lower. Also, in Portugal, kaolin-treated leaves showed high reflectance percentage than untreated ones (Fig. 9.1). The literature still lacks a proper evaluation of kaolin effects on radiation diffusion according to different canopy shapes and/or to leaf position and shading, though evidence on other species suggests that kaolin could increase the fraction of PAR reaching inner leaves, which is particularly relevant in dense canopies composed by multiple layers (Rosati et al., 2007). However, literature is consistently reporting that kaolin crushes energy excess in terms of radiation and temperature for fully exposed leaves and fruits (Frioni et al., 2020; Frioni, Saracino et al., 2019; Frioni, Tombesi et al., 2019; Lobos et al., 2015).

Coming to the effects of kaolin on organ temperature, a preliminary distinction is needed: independently of kaolin presence, leaves are very effective in dissipating thermal excess by evaporative cooling related to transpiration, if water is available in soil. Berries have instead lower transpiration rates and are slightly less efficient in dissipating heat, especially after veraison (Zhang & Keller 2015). Contrariwise, if water supply is reduced or not available, then leaf transpiration is suppressed together with heat dissipation via evaporative cooling. Consequently, leaf temperature rises to high values, as influenced by the surrounding air temperature. Glenn et al. (2010) and Shellie and King (2013a)

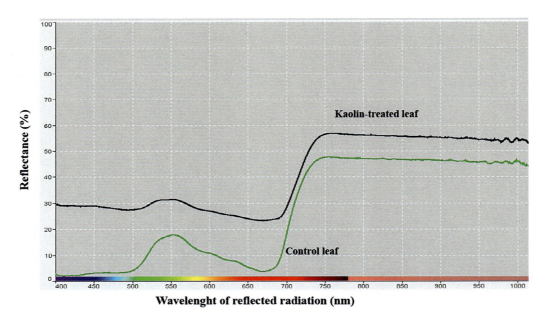

FIGURE 9.1

Kaolin effect on leaf reflectance percentage.

showed that kaolin reduces grapevines leaf temperature by 2°C as compared to untreated vines only in the hottest days, whereas differences were lower or absent in cooler days. Similarly, Brillante et al. (2016) reported that in kaolin vines, leaf temperatures were lower than in controls only in the seasons with lower mean air temperature. In Portugal, in field-grown Touriga Nacional vines, kaolin reduced midday leaf temperatures by up to 13.3%, as compared to nontreated control vines but differences were smaller in the cooler summer of 2012 than in the warmer 2013 (Dinis, Bernardo et al., 2018). Apart from air temperature, effects of kaolin on leaf temperatures are clearly linked to transpiration rates. Frioni et al. (2020) and Frioni, Saracino et al. (2019) demonstrated that on potted vines cv. Sangiovese, leaf temperature was reduced by kaolin coating only when stem water potential was lower than −1.2 MPa and transpiration rates reduced by 75%. The difference between kaolin-coated and untreated vines ranged from 1 to 3°C, with control leaves reaching 40°C with an air temperature of 32°C. In central Italy, and under air temperatures of 40°C, Pinot Noir kaolin-coated vines exhibited a leaf temperature of 6°C lower than control vines, which reached 49.1°C (Frioni, Tombesi et al., 2019). In synthesis, the literature provides the scenario that the efficacy of kaolin in leaf cooling directly relates to drought severity and air temperature, and it is maximized in hot days when leaves lose their capability of evaporative cooling via transpiration. Similarly, if kaolin is correctly applied on bunches, it reduces berry temperature. Lobos et al. (2015) found that kaolin could reduce Cabernet Sauvignon fruit temperature by up to 2.4°C. Shellie and King (2013b) showed that under standard irrigation, both sides of Malbec VSP canopies showed lower berry surface temperature, if kaolin coated. In detail, reduction in berry surface cumulated minutes at $T > 30°C$ due to kaolin was amplified under reduced irrigation (up to −2000 h at $T > 30°C$). In Uruguay, kaolin applied on Sauvignon Blanc vines subjected to basal leaf removal significantly reduced berry temperature postveraison by up to 6°C, when compared to leaf pulled but untreated vines (Coniberti et al., 2013). In central Italy, kaolin at 3% (w/v) reduced bunch temperature of cv. Sangiovese by 6°C (Palliotti et al., 2019). Similarly, in northern Italy, kaolin reduced the temperature of ripening fruits cv. Barbera by about 4°C (Fig. 9.2).

9.3.3 Kaolin effects on vine water status and photosynthetic activity

Canopy microclimatic conditions, together with water availability, play a key-role in driving grapevine photosynthetic capacity. In detail, and besides soil water availability, the main environmental parameters affecting leaf gas exchange rates are light intensity and air temperature. Mature leaves typically exhibit rapid photosynthetic rates increment when PAR increases from 10 to 400 μmol m^{-2} s^{-1}. At higher PAR, any marginal increase in light intensity leads to lower gains in terms of assimilation rates, up to the light saturation point, above which no increments occur anymore (Kriedemann, 1968). Similarly, optimal temperatures for grapevine leaf photosynthesis range between

FIGURE 9.2

Pictures (above) and thermal images (below) of untreated (left) and kaolin-coated (right) grapes cv. Barbera captured on a hot day, after a prolonged period of drought. Kaolin coating was very effective in preventing berry overheating and spread of sunburn symptoms.

25°C and 30°C (Greer & Weedon, 2012; Kriedemann, 1968). Considering the effects of kaolin coverage on leaf temperatures and radiation absorption mentioned in the previous section, one should expect the quite simple scheme where: (i) under nonlimiting conditions, treated canopies receive lower yet nonlimitation PAR and set at optimal temperatures; (ii) when multiple summer stress occur (i.e., air temperatures >30°C, reduced water supply, and radiation excess), the better physiological performances should be displayed as a function of better microclimatic conditions. Instead, the available literature depicts a more complex and composite scenario, as a function of several interacting factors.

The study of kaolin effects on plant physiological performance dates back to the 1970s when kaolin was first tested on herbaceous crops. Moreshet et al. (1977) found that kaolin coverage reduced the net photosynthesis of *Sorghum bicolor* by 23% as a result of reflected light. Many other papers reported lower transpiration rates and water use for annual crops covered by kaolinite formulations, so that kaolin was quickly listed as an antitranspirant. Interestingly, Moreshet et al. (1979) found a reduction in CO_2 uptake in cotton plants when kaolin was sprayed over the canopies, but, on the contrary, whole-canopy photosynthesis and transpiration rates were increased if kaolin was applied only to the soil. They were the first to understand the core concept that the canopy architecture and soil layout could interfere with kaolin effects on whole-canopy assimilation rates.

The first studies of kaolin effects on tree physiology were related to the renovated interest in the early 2000s in kaolin as a tool to prevent negative effects of warming trends. Glenn et al. (2001) found that kaolin sprays could improve whole-canopy CO_2 assimilation in apple plants. Similarly, Jifon and Syvertsen (2003) showed that kaolin increased grapefruit single-leaf photosynthesis and improved water use efficiency (WUE) by 25%. Then, Rosati et al. (2006) reported that kaolin reduced maximum single-leaf assimilation by 1–4 $\mu mol\ m^{-2}\ s^{-1}$ in walnut and almond trees, because of the 37% reduction of incident radiation on external leaves. By the way, the reduced individual leaf photosynthetic rates of external leaves were compensated by higher radiation reflection to the inner canopy leaves, so that, on a whole-canopy basis, the loss of photosynthetic radiation use efficiency was minimal (Rosati et al., 2006, 2007).

Effects of kaolin on grapevine photosynthetic capacity were first tested by Shellie & Glenn (2008) and Glenn et al. (2010) on cv. Cabernet Sauvignon, showing no changes in single-leaf gas exchange on well-watered vines, yet higher net photosynthetic rates and WUE under prolonged water deficit. The first attempt hinted that repercussions of kaolin sprays on grapevine physiology could vary based on plant water status, in agreement with previous findings on other species (Rosati et al., 2006, 2007). Subsequent works confirmed that under nonlimiting conditions, minimal or no effects on single-leaf assimilation rates can be expected in kaolin-coated vines (Brillante et al., 2016, Dinis, Bernardo et al., 2018; Frioni et al., 2020; Frioni, Tombesi et al., 2019). In certain situations, a slight reduction of stomatal conductance, transpiration rates, and WUE can be found because kaolin reflects a significant fraction of incident radiation on external leaves (Frioni et al., 2020; Shellie & Glenn 2008; Shellie & King 2013a). About that, the work of Palliotti et al. (2019), testing different dosages of kaolin on field-grown Sangiovese vines, seems of particular interest. They found that kaolin applied at 6% and 9% w/v was effective in preventing leaf chlorosis, but it was also reducing single-leaf photosynthesis and WUE, as compared to kaolin applied at 3% w/v. Therefore, correct doses when applying kaolin formulations could be pivotal to avoid undesired reduction in external single-leaf net CO_2 assimilation rates. Contrariwise, kaolin at high doses (6% or higher) acts as an antitranspirant, reducing single-leaf net CO_2 assimilation and transpiration rates, when this can be desirable, namely to avoid excessive sugars concentration in berries.

Undoubtedly, kaolin positive effects on grapevine physiological performances come out when multiple summer stresses occur, and environmental conditions become limiting (Fig. 9.3). Positive effects on single-leaf gas exchange parameters under water deficit conditions are consistently reported (Brillante et al., 2016; Dinis et al., 2018; Frioni et al., 2020; Frioni, Saracino et al., 2019; Frioni, Tombesi et al., 2019; Glenn et al., 2010, Palliotti et al., 2019). The magnitude of net CO_2 assimilation and transpiration rates maintenance varied among sites, cultivars, and experimental conditions. Brillante et al. (2016) reported an increase of single-leaf WUE in potted vines by 26% under water deficit conditions. In Portugal, field-grown kaolin-coated vines cv. Touriga Nacional had consistently higher stomatal conductance rates at midday leaf water potential between -0.7 and -1.5 MPa, as compared to control vines, as well as higher net CO_2 assimilation rates (+58.7%) and WUE (Dinis, Bernardo et al., 2018; Dinis, Malheiro et al., 2018). In central Italy, when control vines

FIGURE 9.3

Kaolin-coated vine cv. Barbera **(A)** showing no symptom of summer stresses and untreated vine **(B)** with basal leaves showing loss of cell turgor and nonreversible photoinhibitions.

Table 9.1 Leaf water potential (Ψ), gas exchange parameters, and water use efficiency for field-grown grapevines cv. Barbera subjected to preveraison kaolin application.

Treatment 6/08/2020	Ψ (MPa)	Photosynthesis (µmol/m² s)	Transpiration (mmol/m² s)	Stomatal conductance (mmol/m² s)	Water use efficiency (µmol/m²s/ mmol/m²s)
Kaolin	−1.37	9.04	6.24	225.1	1.47
Control	−1.37	6.92	5.16	166.7	1.36
t[a]	ns	***	*	**	***

[a],*,**, and *** indicate significant differences per P <.05, P <.005, and P <.001, respectively. ns, not significant.

cv. Pinot Noir exhibited full stomatal closure, kaolin-coated vines maintained significantly higher single-leaf photosynthesis (5.1 µmol CO_2 m^{-2} s^{-1}) and transpiration (2.2 mmol H_2O m^{-2} s^{-1}) (Frioni, Tombesi et al., 2019). In northern Italy, kaolin-protected field-grown grapevines cv. Barbera showed significantly higher single-leaf gas exchange and WUE at leaf water potentials of −1.37 MPa (Table 9.1). The above-mentioned works examined kaolin effects on vine's physiological performance by measuring single-leaf gas exchange parameters. Despite the quite clear scenario accounting for substantially neutral effects under nonlimiting conditions and positive effects if water or other factors become limiting, it is still unclear if the canopy architecture and density (number of leaf layers) could interfere with the effects on leaves other than the mature fully exposed ones. For instance, the kaolin-induced reflection of light could theoretically increase diffused light within the canopy and increase net assimilation rates of leaves remaining under nonsaturating light conditions (namely PAR<1000) during most of the day, as already reported for other species (Moreshet et al., 1979; Rosati et al., 2006, 2007). Moreover, considering that leaves of different ages have different response to limiting light conditions and that the percentage of shaded leaves within a canopy ranges between 10% and 70% of the total leaf area (Smart, 1974, 1985), the dilemma is if single-leaf readings are a good indicator of whole-canopy gas exchange. The recent work of Frioni, Saracino et al. (2019) provides some interesting insights. They measured kaolin-coated vines gas exchanges by including vines in chambers providing real-time readings of whole-canopy CO_2 assimilation and transpiration rates. Kaolin reduced whole-canopy assimilation rates by about 1 µmol CO_2 m^{-2} s^{-1} and transpiration rates by 0.3 mmol H_2O m^{-2} s^{-1} compared to untreated vines. Under water-deficit, despite no effects were found at moderate stress intensity, in agreement with single-leaf readings, kaolin helped to maintain active whole-canopy gas exchanges later when control vines exhibited full stomatal closure. The effects observed by several authors at single-leaf level were confirmed when up-scaling the study to the entire canopy (Frioni, Saracino et al., 2019).

The most prominent and relevant positive effect of kaolin application in vineyards exposed to drought and summer stresses is the protection of canopy photochemical efficiency and the full and prompt restoration of leaf gas exchange parameters once water supply is restored and temperature returns to optimal ranges. In central Italy regions, when air temperatures returned to the optimal threshold after an extremely hot summer, kaolin-coated vines resumed full single-leaf photosynthesis and stomatal conductance, whereas untreated vines showed significantly lower photosynthetic rates

(-72% than kaolin vines) (Frioni, Tombesi et al., 2019). Similarly, in the Douro Portuguese region, kaolin promoted higher single-leaf photosynthetic rates of vines cv. Touriga Nacional in September (Dinis, Malheiro et al., 2018). The improved physiological performance after rewatering of vines subjected to kaolin sprays, as compared to untreated vines, was also confirmed on a whole-canopy basis (Frioni, Saracino et al., 2019). Rewatering was always associated to the prevention of a decrease in F_v/F_m values, which represents the onset of irreversible leaf photoinhibition and chlorophylls degradation, chlorosis and leaf abscission in grapevine (Fig. 9.4) (Dinis, Bernardo et al., 2018; Dinis, Malheiro et al., 2018; Dinis, Bernardo et al., 2016; Frioni, Saracino et al., 2019; Frioni, Tombesi et al., 2019). The mechanisms behind the observed physiological changes and adaptations consist of a complex interaction of different direct and indirect effects. At the basis of the pyramid, there is the simple physical barrier effect to light, which results in lower leaf temperature and in a moderate

FIGURE 9.4

In northern Italy, after the very hot July 2020, a rain of 43 mm fell on August 5, 2020 in a vineyard where vines cv. Ortrugo were sprayed preveraison with kaolin at 6% w/v and compared to untreated vines. The rain completely washed off the kaolin clay from the canopies. On August 6, 2020, kaolin-protected vines had green intact canopies **(A, C)**, whereas control vines had large leaf portion with yellowing and degradation of chlorophylls **(B)**.

antitranspirant effect under full irrigation. When environmental conditions become limiting for single-leaf gas exchange, kaolin helps to maintain positive net CO_2 assimilation rates, which improve energy excess dissipation by transpiration, in addition to the physical radiation screen effect.

9.3.4 Impact on leaf metabolism

PKPF is effective in reducing sunburn damage in grapevine leaf (Bernardo et al., 2017; Dinis, Ferreira et al., 2016). The PKPF prevents grapevine leaf chlorophyll degradation and improves carotenoid concentrations under stressful conditions (Dinis, Bernardo et al., 2016; Dinis, Malheiro et al., 2018). The higher content of total chlorophyll found in treated leaves under severe conditions (Dinis, Malheiro et al., 2018) is a sign that plants are more tolerant to oxidative stress (Smirnoff, 1993). And the higher chlorophyll/carotenoids ratio denotes a reduced requirement for photoprotection of chlorophylls, as carotenoids play an important role in scavenging ROS and releasing the excess energy by thermal dissipation via the xanthophyll cycle (Sharma et al., 2012). In other species, was also observed a reduction in the chl a/b ratio, and an increase in the total chlorophyll/carotenoids ratio (Brito, Dinis, Ferreira et al., 2018; Shellie & King, 2013a), being typical features of low-light-adapted leaves (Lichtenthaler et al., 2007). Taken altogether, these results also indicate that the photosystems may have larger antenna sizes at the expense of reaction center pigment proteins to enhance their ability to capture and utilize photon energy (Lichtenthaler et al., 2007). Regarding carotenoids, the xanthophyll (VAZ) cycle involves the amount of violaxanthin, neoxanthin, and zeoxanthin (Vx + Nx + Zx), which increased in kaolin-treated vines, and could be linked to reduced ABA accumulation in leaves (Frioni et al., 2020). Supposedly, the kaolin treatment reduced conversion of zeoxanthin (Zx) into neoxanthin (Nx) by stimulating the activity of zeaxanthin epoxidase (ZEP) or by reducing the activity of neoxanthin synthase (NxS). ZEP activity is inhibited by high light intensities, and the epoxidation of Zx to Vx mainly occurs at night (Jahns & Miehe, 1996). Thus, reducing the synthesis of Nx from Zx, kaolin reduced ABA accumulation in leaves, resulting in faster recovery of single-leaf gas exchange under water-deficit conditions (Frioni et al., 2020). Similarly, under high light and temperature, Dinis, Bernardo et al. (2018) also found a decrease of ABA levels in kaolin-sprayed leaves. Additionally, there is a possible crosstalk controlled by ABA and auxin (IAA) concentrations that might affect the plants response to drought (Sharma, Sharma et al., 2015). IAA is one of the paramount hormones regulating cell growth and development. In drought periods, ABA synthesis is mandatory and IAA accumulation is limited in the guard cells, decreasing stomatal conductance by triggering stomatal closure (Dinis, Bernardo et al., 2018; Dodd, 2005). Nevertheless, under kaolin application (lower leaf temperature and better water status), this tendency was counteracted with decreasing ABA levels and higher IAA accumulation, leading to a better plant growth (Dinis, Bernardo et al., 2018). Therefore, overproduction of IAA, as it occurs in kaolin-treated leaves, suggests improved acclimation of grapevine to environmental stresses (Ke et al., 2015).

PKPF foliar application reduced the amount of ROS and stimulated changes in nonenzymatic antioxidants components, including phenols, flavonoids, anthocyanins, and carotene, part of them with well-known contribution to berry and wine quality (Dinis, Bernardo et al., 2016). Also, it leads to a reduction of the oxidative damage (Bernardo et al., 2017), as well as an increase of soluble proteins (Dinis, Bernardo et al., 2018) in sprayed organs. Furthermore, kaolin stimulates the grape leaf metabolome providing leaves a better capacity to tolerate the summer stress conditions. Likewise, it is essential to emphasize that a remarkable proportion of the main sugars, sugars phosphate, polyols, organic acids, and amino acids were more abundant in leaves subject to kaolin applications (Conde et al., 2018). Also, transcriptional analyses and enzymatic activity assays carried out under high temperature and irradiance reported that kaolin was able to increase sucrose concentration in leaves, sucrose transport, and phloem loading capacity (Conde et al., 2016, 2018).

9.3.5 Impact on berries and wine

As expected, the fruit cooling effect induced by the kaolin sprays contributed to higher fruit quality (Dinis et al., 2016, 2020). Particularly in the mature grape berry, the particle film boosts the amounts of phenolic compounds, including total phenolics and tannins, increasing the antioxidant activity. This happened due to a global stimulation of phenylpropanoid and flavonoid-flavonol pathways at the gene expression and/or protein activity levels in relation of increased phenols compounds and anthocyanins (Conde et al., 2016; Dinis, Bernardo et al., 2016). This effect was also observed in grapevine berries grown under arid conditions and high solar radiation (Shellie & King, 2013a,b). One of the most valued effects of kaolin is the increase of the anthocyanins obtained in mature fruits (Brillante et al., 2016; Dinis, Bernardo et al., 2016; Frioni, Saracino et al., 2019). Previous studies performed on a wide range of grapevine varieties, such as Cabernet Sauvignon, Pinot Noir, Merlot, Muscat Hamburg, and Touriga Nacional, showed that PKPF application enhanced total amount of berry anthocyanins (Brillante et al., 2016; Conde et al., 2016; Dinis, Bernardo et al., 2016; Shellie & King, 2013a; Song et al., 2012). Frioni, Saracino et al. (2019) support the statement that color accumulation in berries is sensitive to PKPF application since kaolin-water stress plants had similar levels of total anthocyanins and phenolics concentration when compared to control well-watered vines.

Concerning soluble sugars, inconsistent findings were obtained depending on the variety. In fact, Shellie and King (2013b) showed that PKPF increased soluble solids concentration of Cabernet Sauvignon, while Ferrari et al. (2017) showed that in both Malbec and Sauvignon Blanc cvs, no effect was noticeable. Equally, an increase in berry soluble solids contents in Viognier (Shellie & Glenn, 2008) was observed whereas Merlot was unaffected by kaolin. Furthermore, other studies did not report any changes after kaolin application on the concentration of soluble solids in Cabernet Sauvignon (Brillante et al., 2016; Glenn et al., 2010), Merlot (Shellie, 2015), and Viognier (Glenn et al., 2010).

There are few studies about PKPF effects on must or wine. Must organic acids concentration, and must and wine acidity and pH derived from kaolin-treated plants seem not to change in a study with the cv Sauvignon Blanc (Coniberti et al., 2013). However, higher malic acid concentration and total acidity were observed, which could be associated with a delay in maturation (Coniberti et al., 2013; Dinis et al., 2020; Ferrari et al., 2017). In a study from Ferrari et al. (2017), it is interesting that kaolin treatment induced higher scores by the experts associated with greater wine typicity and reasonable changes in the wine aroma and wine body. However, despite the lower absorbance (abs/420) obtained in the wine prevenient from kaolin-treated vines, no differences in wine color were detected during wine tasting. Also, Dinis et al. (2020) showed that kaolin provokes an increase of tartaric and malic acid, and consequently higher total acidity in the wine, while the sugar concentration decreased provoking a low wine alcohol level. Moreover, kaolin-treated vines presented higher content of esters associated with fruity notes, as well as with higher potassium, magnesium, and iron levels, and low copper and aluminum concentrations.

9.4 Kaolin impacts: pros and cons
9.4.1 For the environment
Among the enormous effort that has been made by the scientific community to test exogenous substances with a protective effect in alleviating the damage of summer stress in plants, e.g., phytohormones, signaling molecules, osmoprotectants, and minerals (Bernardo et al., 2018), the foliar application of processed kaolin particles has proved to be very promising. Indeed, due to its mineral, inert, and whitish nature, the use of kaolin technology, aimed to replace conventional pesticides for pest control, reducing pesticide load over the environment, which will improve safety of viticulturists, and providing a more sustainable food supply for the consumer. Later, it was found that kaolin could also increase the capacity to reduce dependency of agriculture on irrigation by increasing the WUE and by replacing evaporative cooling irrigation technique used to reduce sunburn and heat stress in horticultural crops (Frioni, Saracino et al., 2019).

Kaolin properties are important for understanding the performance of microorganisms in natural environments. It is obvious that research investment will improve our understanding of such complex interactions at many spatial and timescales, which would benefit our global society. Several significant geochemical processes in the environment are controlled by the interactions of ions, organics, and microorganisms at the aqueous interface with kaolin-group minerals. Kaolinite has both hydrophilic and relatively hydrophobic external surfaces that exhibit diverse adsorption phenomena. The contaminants adsorption on the surfaces of clay minerals is an important geochemical method for keeping the quality of groundwater. The fine-grained nature of clay minerals and their respective high surface area indorse quite large adsorption capacities allowing clays to be used as efficient migration barriers at contaminant sites and nuclear waste repositories (Cygan & Tazaki, 2014).

9.4.2 Costs
Currently, several particle film formulations are accessible in the global market for their commercial use in horticultural crops for several desirable effects (Sharma, Reddy et al., 2015). Future request for kaolin as an input into emerging technologies and products continues to increase. Prospects are emerging in the market as end users begin to recognize the importance of this "environmentally friendly" practice. Expanding population and rising consumer standards of welfare will potentially grow kaolin demand in several crops. The use of this reflective particle film has significant economic benefits in productivity performance of crops growing under arid and semiarid conditions (Glenn, 2012). Recent works in agriculture refer kaolin as a low cost-strategy, as treatment costs are around 20–30 € per hectare (Bernardo et al., 2017; Conde et al., 2018; Dinis, Ferreira et al., 2016; Dinis, Bernardo et al., 2018; Frioni, Saracino et al., 2019).

9.5 Concluding remarks and prospects
Although positive processed kaolin particles effects ideally occur under reasonably stressful conditions, its effectiveness in several aspects and numerous crops is clear and it is quantified (Fig. 9.5). This technology showed to be effective against insects and mites and in abiotic stresses alleviation, improving radiation reflection, leaf cooling, and increasing WUE. Therefore, processed kaolin

9.5 Concluding remarks and prospects   179

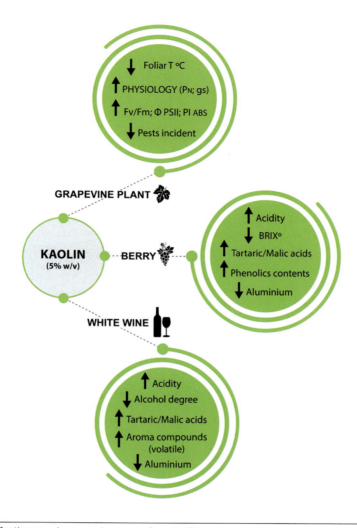

FIGURE 9.5

Kaolin effect on effectiveness in several aspects from leaf to wine.

particles application leads to several morphophysiological and biochemical changes, providing additional resilience to multiple stress factors. All together, these modifications stimulated an upsurge in growth, yield, and even fruit quality. Nonetheless, it is evident that this effect depends on the species and/or genotype. The use of this technology seems to be a suitable strategy for arid and semiarid areas, where incident PAR typically outdoes plant's capability. Beyond its physiological effects, processed kaolin particles application under severe water scarcity condition may also add significant economic advantages, by potentially reducing irrigation needs to mitigate stress and protecting berries from sunburn.

Acknowledgments

The Portuguese team would like to thank "*Herdade do Esporão*" (Alentejo Region), "*Quinta do Vallado*" (Douro Valley), and "*Real Companhia Velha*" (Douro Valley) companies for the collaboration and efforts in making the vineyard's facilities available for the research studies. We thank the *Fundação Maria Rosa* (2016, 2018) and *ADVID* (2018) for the honourable awards received under this topic.

References

Abou-Khaled, A., Hagan, R. M., & Davenport, D. C. (1970). Effects of kaolinite as a reflective antitranspirant on leaf temperature, transpiration, photosynthesis, and water-use efficiency. *Water Resources Research, 6*(1), 280–289.

Ahmad, P., Jaleel, C. A., Salem, M. A., Nabi, G., & Sharma, S. (2010). Roles of enzymatic and nonenzymatic antioxidants in plants during abiotic stress. *Critical Reviews in Biotechnology, 30*, 161–175.

Apel, K., & Hirt, H. (2004). Reactive oxygen species: Metabolism, oxidative stress, and signal transduction. *Annual Review of Plant Biology, 55*, 373–399.

Baker, N. R., & Rosenqvist, E. (2004). Applications of chlorophyll fluorescence can improve crop production strategies: An examination of future possibilities. *Journal of Experimental Botany, 55*, 1607–1621.

Bernardo, S., Dinis, L.-T., Luzio, A., Pinto, G., Meijón, M., Valledor, L., Conde, A., Gerós, H., Correia, C. M., & Moutinho-Pereira, J. (2017). Kaolin particle film application lowers oxidative damage and DNA methylation on grapevine (*Vitis vinifera* L.). *Environmental and Experimental Botany, 139*, 39–47.

Bernardo, S., Dinis, L.-T., Machado, N., & Moutinho-Pereira, J. (2018). Grapevine abiotic stress assessment and search for sustainable adaptation strategies in Mediterranean-like climates. A review. *Agronomy for Sustainable Development, 38*, 66.

Brillante, L., Belfiore, N., Gaiotti, F., Lovat, L., Sansone, L., Poni, S., & Tomasi, D. (2016). Comparing kaolin and pinolene to improve sustainable grapevine production during drought. *PLoS One, 11*, e0156631.

Brito, C., Dinis, L.-T., Ferreira, H., Rocha, L., Pavia, I., Moutinho-Pereira, J., & Correia, C. M. (2018). Kaolin particle film modulates morphological, physiological and biochemical olive tree responses to drought and rewatering. *Plant Physiology and Biochemistry, 133*, 29–39.

Brito, C., Dinis, L.-T., Luzio, A., Silva, E., Goncalves, A., Meijon, M., Escandon, M., Arrobas, M., Rodrigues, M. A., Moutinho-Pereira, J., & Correia, C. M. (2019). Kaolin and salicylic acid alleviate summer stress in rainfed olive orchards by modulation of distinct physiological and biochemical responses. *Scientia Horticulturae, 246*, 201–211.

Brito, C., Dinis, L.-T., Meijón, M., Ferreira, H., Pinto, G., Moutinho-Pereira, J., & Correia, C. M. (2018). Salicylic acid modulates olive tree physiological and growth responses to drought and rewatering events in a dose dependent manner. *Journal of Plant Physiology, 230*, 21–32.

Camejo, D., Jiménez, A., Alarcón, J. J., Torres, W., Gómez, J. M., & Sevilla, F. (2006). Changes in photosynthetic parameters and antioxidant activities following heat-shock treatment in tomato plants. *Functional Plant Biology, 33*, 177–187.

Campostrini, E., Reis, F. O., Souza, M. A., & Yamanishi, O. K. (2004). Processed-kaolin particle film on papaya leaves: A study related to gas exchange, leaf temperature and light distribution in canopy. In *Acta horticulturae 864 of the III international symposium on tropical and subtropical fruits* (pp. 195–200).

Chaves, M. M., Zarrouk, O., Francisco, R., Costa, J. M., Santos, T., Regalado, A. P., Rodrigues, M. L., & Lopes, C. M. (2010). Grapevine under deficit irrigation: Hints from physiological and molecular data. *Annals of Botany, 105*, 661–676.

Conde, A., Neves, A., Breia, R., Pimentel, D., Dinis, L. T., Bernardo, S., Correia, C. M., Cunha, A., Gerós, H., & Moutinho-Pereira, J. (2018). Kaolin particle film application stimulates photoassimilate synthesis and modifies the primary metabolome of grape leaves. *Journal of Plant Physiology, 223*, 47−56.

Conde, A., Pimentel, D., Neves, A., Dinis, L.-T., Bernardo, S., Correia, C. M., Gerós, H., & Moutinho-Pereira, J. (2016). Kaolin foliar application has a stimulatory effect on phenylpropanoid and flavonoid pathways in grape berries. *Frontiers of Plant Science, 7*, 1150.

Conde, C., Silva, P., Fontes, N., Dias, A. C., Tavares, R., Sousa, M., Agasse, A., Delrot, S., & Gerós, H. (2007). Biochemical changes throughout grape berry development and fruit and wine quality. *Food, 1*, 1−22.

Coniberti, A., Ferrari, V., Dellacassa, E., Boido, E., Carrau, F., Gepp, V., & Disegn, E. (2013). Kaolin over sun-exposed fruit affects berry temperature, must composition and wine sensory attributes of Sauvignon blanc. *European Journal of Agronomy, 50*, 75−81.

Coombe, B. (1987). Influence of temperature on composition and quality of grapes. *Acta Horticulturae, 206*, 23−35.

Correia, C. M., Pereira, J. M. M., Coutinho, J. F., Björn, L. O., & Torres-Pereira, J. M. G. (2005). Ultraviolet-B radiation and nitrogen affect the photosynthesis of maize: A Mediterranean field study. *European Journal of Agronomy, 22*, 337−347.

Correia, B., Valledor, L., Meijon, M., Rodriguez, J. L., Dias, M. C., Santos, C., Canal, M. J., Rodriguez, R., & Pinto, G. (2013). Is the interplay between epigenetic markers related to the acclimation of cork oak plants to high temperatures? *PLoS One, 8*(1), e53543.

de Cortázar-Atauri, G., Duchêne, E., Destrac-Irvine, A., Barbeau, G., de Rességuier, L., Lacombe, T., Parker, A. K., Saurin, N., & van Leeuwen, C. (2017). Grapevine phenology in France: From past observations to future evolutions in the context of climate change. *OENO One, 51*, 115−126.

Cygan, R. T., & Tazaki, K. (2014). Interactions of kaolin minerals in the environment. *Elements, 10*, 195−200.

Deluc, L. G., Grimplet, J., Wheatley, M. D., Tillett, R. L., Quilici, D. R., Osborne, C., Schooley, D. A., Schlauch, K. A., Cushman, J. C., & Cramer, G. R. (2007). Transcriptomic and metabolite analyses of Cabernet Sauvignon grape berry development. *BMC Genomics, 8*, 429.

Dinis, L.-T., Bernardo, S., Conde, A., Pimentel, D., Ferreira, H., Felix, L., Gerós, H., Correia, C. M., & Moutinho-Pereira, J. (2016). Kaolin exogenous application boosts antioxidant capacity and phenolic content in berries and leaves of grapevine under summer stress. *Journal of Plant Physiology, 191*, 45−53.

Dinis, L.-T., Bernardo, S., Luzio, A., Pinto, G., Meijon, M., Pinto-Marijuan, M., Cotado, A., Correia, C. M., & Moutinho-Pereira, J. (2018). Kaolin modulates ABA and IAA dynamics and physiology of grapevine under Mediterranean summer stress. *Journal of Plant Physiology, 220*, 181−192.

Dinis, L.-T., Bernardo, S., Matos, C., Malheiro, A., Flores, R., Alves, S., Costa, C., Rocha, S., Correia, C., Luzio, A., & Moutinho-Pereira, J. (2020). Overview of kaolin outcomes from vine to wine: Cerceal white variety case study. *Agronomy, 10*, 1422−1447.

Dinis, L.-T., Correia, C. M., Ferreira, H. F., Gonçalves, B., Gonçalves, I., Coutinho, J. F., Ferreira, M. I., Malheiro, A. C., & Moutinho-Pereira, J. (2014). Physiological and biochemical responses of Semillon and Muscat blanc a Petits grains winegrapes grown under Mediterranean climate. *Scientia Horticulturae, 175*, 128−138.

Dinis, L.-T., Ferreira, H., Pinto, G., Bernardo, S., Correia, C. M., & Moutinho-Pereira, J. (2016). Kaolin-based, foliar reflective film protects photosystem II structure and function in grapevine leaves exposed to heat and high solar radiation. *Photosynthetica, 54*, 47−55.

Dinis, L.-T., Malheiro, A., Luzio, A., Fraga, H., Ferreira, H., Gonçalves, I., Pinto, G., Correia, C. M., & Moutinho-Pereira, J. (2018). Improvement of grapevine physiology and yield under summer stress by kaolin-foliar application: Water relations, photosynthesis and oxidative damage. *Photosynthetica, 56*, 641−651.

Dodd, I. (2005). Root-to-shoot signalling: Assessing the roles of 'up' in the up and down world of long-distance signalling in planta. *Plant and Soil, 274*, 251−270.

Elbasyoni, I., Saadalla, M., Baenziger, S., Bockelman, H., & Morsy, S. (2017). Cell membrane stability and association mapping for drought and heat tolerance in a worldwide wheat collection. *Sustainability, 9*, 1606−1622.

Evain, S., Flexas, J., & Moya, I. (2004). A new instrument for passive remote sensing: 2. Measurement of leaf and canopy reflectance changes at 531 nm and their relationship with photosynthesis and chlorophyll fluorescence. *Remote Sensing of Environment, 91*, 175−185.

Fernández, J.-E. (2014). Understanding olive adaptation to abiotic stresses as a tool to increase crop performance. *Environmental and Experimental Botany, 103*, 158−179.

Ferrandino, A., & Lovisolo, C. (2014). Abiotic stress effects on grapevine (*Vitis vinifera* L.): Focus on abscisic acid-mediated consequences on secondary metabolism and berry quality. *Environmental and Experimental Botany, 103*, 138−147.

Ferrari, V., Disegna, E., Dellacassa, E., & Coniberti, A. (2017). Influence of timing and intensity of fruit zone leaf removal and kaolin applications on bunch rot control and quality improvement of Sauvignon blanc grapes, and wines, in a temperate humid climate. *Scientia Horticulturae, 223*, 62−71.

Fraga, H., Moutinho-Pereira, J., Malheiro, A. C., & Santos, J. A. (2013). An overview of climate change impacts on European viticulture. *Food and Energy Security, 1*(2), 94−110.

Frioni, T., Saracino, S., Squeri, C., Tombesi, S., Palliotti, A., Sabbatini, P., Magnanini, E., & Poni, S. (2019). Understanding kaolin effects on grapevine leaf and whole-canopy physiology during water stress and re-watering. *Journal of Plant Physiology, 242*, 153020−153032.

Frioni, T., Tombesi, S., Luciani, E., Sabbatini, P., Berrios, J. G., & Palliotti, A. (2019). Kaolin treatments on Pinot noir grapevines for the control of heat stress damages. In *BIO web of conferences* (Vol. 13, p. 04004). EDP Sciences.

Frioni, T., Tombesi, S., Sabbatini, P., Squeri, C., Lavado Rodas, N., Palliotti, A., & Poni, S. (2020). Kaolin reduces ABA biosynthesis through the inhibition of neoxanthin synthesis in grapevines under water deficit. *International Journal of Molecular Sciences, 21*(14), 4950−4965.

Garcia-Valles, M., Alfonso, P., Martinez, S., & Roca, N. (2020). Mineralogical and thermal characterization of kaolinitic clays from Terra Alta (Catalonia, Spain). *Minerals-Basel, 10*, 142−157.

Glenn, D. M. (2012). The Mechanisms of plant stress mitigation by kaolin-based particle films and applications in horticultural and agricultural crops. *HortScience, 47*, 710−711.

Glenn, D. M., Cooley, N. M., Walker, R. R., Clingeleffer, P. R., & Shellie, K. C. (2010). Impact of kaolin particle film and water deficit on wine grape water use efficiency and plant water relations. *HortScience, 45*(8), 1178−1187.

Glenn, D. M., & Puterka, G. J. (2005). Particle films: A new technology for agriculture. *Horticultural Reviews, 31*, 1−44.

Glenn, D. M., Puterka, G. J., Drake, S. R., Unruh, T. R., Knight, A. L., Baherle, P., Prado, E., & Baugher, T. A. (2001). Particle film application influences apple leaf physiology, fruit yield, and fruit quality. *Journal of the American Society for Horticultural Science, 126*(2), 175−181.

Glenn, D. M., van der Zwet, T., Puterka, G., Gundrum, P., & Brown, E. (2001). Efficacy of kaolin-based particle films to control apple diseases. *Plant Health Progress, 2*(1), 4−13.

Golldack, D., Li, C., Mohan, H., & Probst, N. (2014). Tolerance to drought and salt stress in plants: Unraveling the signaling networks. *Frontiers in Plant Scince, 5*, 151.

Greer, D. H., & Weedon, M. M. (2012). Modelling photosynthetic responses to temperature of grapevine (*Vitis vinifera* cv. Semillon) leaves on vines grown in a hot climate. *Plant, Cell and Environment, 35*(6), 1050−1064.

Hendrickson, L., Forster, B., Furbank, R. T., & Chow, W. S. (2004). Processes contributing to photoprotection of grapevine leaves illuminated at low temperature. *Physiologia Plantarum, 121*(2), 272−281.

Hernández, A. C., Sánchez-Espejo, R., Meléndez, W., González, G., López-Galindo, A., & Viseras, C. (2019). Characterization of Venezuelan kaolins as health care ingredients. *Applied Clay Science, 175*, 30−39.

IPCC. (2013). Climate change 2013: The physical science basis. In T. F. Stocker, D. Qin, G.-K. Plattner, M. Tignor, S. K. Allen, J. Boschung, A. Nauels, Y. Xia, V. Bex, & P. M. Midgley (Eds.), *Contribution of working group I to the fifth assessment report of the intergovernmental panel on climate change* (p. 1535). Cambridge and New York, NY: Cambridge University Press. ISBN 978-92-9169-138-8.

Jahns, P., & Miehe, B. (1996). Kinetic correlation of recovery from photoinhibition and zeaxanthin epoxidation. *Planta, 198*, 202–210.

Jesus, C., Meijon, M., Monteiro, P., Correia, B., Amaral, J., Escandon, M., Canal, M. J., & Pinto, G. (2015). Salicylic acid application modulates physiological and hormonal changes in Eucalyptus globulus under water deficit. *Environmental and Experimental Botany, 118*, 56–66.

Jifon, J. L., & Syvertsen, J. P. (2003). Kaolin Particle Film applications can increase photosynthesis and water use efficiency of Ruby Red Grapefruit leaves. *Journal of the American Society for Horticultural Science, 128*(1), 107–112.

Jones, G. V., White, M. A., Owen, R., & Storchmann, C. (2005). Climate change and global wine quality. *Climate Change, 73*, 319–343.

Ke, Q., Wang, Z., Ji, C. Y., Jeong, J. C., Lee, H. S., Li, H., Xu, B., Deng, X., & Kwak, S. S. (2015). Transgenic poplar expressing Arabidopsis YUCCA6 exhibits auxin-overproduction phenotypes and increased tolerance to abiotic stress. *Plant Physiology and Biochemistry, 94*, 19–27.

Kim, T. H., Bohmer, M., Hu, H., Nishimura, N., & Schroeder, J. I. (2010). Guard cell signal transduction network: Advances in understanding abscisic acid, CO_2, and Ca^{2+} signaling. *Annual Review of Plant Biology, 61*, 561–591.

Kriedemann, P. E. (1968). Photosynthesis in vine leaves as a function of light intensity, temperature, and leaf age. *Vitis, 7*, 213–220.

van Leeuwen, C., & Darriet, P. (2016). The impact of climate change on viticulture and wine quality. *Journal of Wine Economics, 11*(1), 150–167.

van Leeuwen, C., & Destrac-Irvine, A. (2017). Modified grape composition under climate change conditions requires adaptations in the vineyard. *OENO One, 51*, 147–154.

Lichtenthaler, H. K., Ac, A., Marek, M. V., Kalina, J., & Urban, O. (2007). Differences in pigment composition, photosynthetic rates and chlorophyll fluorescence images of sun and shade leaves of four tree species. *Plant Physiology and Biochemistry, 45*, 577–588.

Lobos, G. A., Acevedo-Opazo, C., Guajardo-Moreno, A., Valdés-Gómez, H., Taylor, J. A., & Laurie, V. F. (2015). Effects of kaolin-based particle film and fruit zone netting on Cabernet Sauvignon grapevine physiology and fruit quality. *OENO One, 49*(2), 137–144.

Martínez-Lüscher, J., Morales, F., Sánchez-Díaz, M., Delrot, S., Aguirreolea, J., Gomès, E., & Pascual, I. (2015). Climate change conditions (elevated CO_2 and temperature) and UV-B radiation affect grapevine (*Vitis vinifera* cv. Tempranillo) leaf carbon assimilation, altering fruit ripening rates. *Plant Science, 236*, 168–176.

Martínez-Lüscher, J., Kizildeniz, T., Vucetic, V., Dai, Z., Luedeling, E., van Leeuwen, Gomes, E., Pascual, I., Irigoyen, J. J., Morales, F., & Delrot, S. (2016). Sensitivity of grapevine phenology to water availability, temperature and CO_2 concentration. *Frontiers in Environmental Science, 4*, 48–60.

Mirbahai, L., & Chipman, J. K. (2014). Epigenetic memory of environmental organisms: A reflection of lifetime stressor exposures. *Mutation Research. Genetic Toxicology and Environmental Mutagenesis, 764–765*, 10–17.

Molitor, D., & Junk, J. (2019). Climate change is implicating a two-fold impact on air temperature increase in the ripening period under the conditions of the Luxembourgish grapegrowing region. *Oeno One, 53*, 409–422.

Moreshet, S., Cohen, Y., & Fuchs, M. (1979). Effect of increasing foliage reflectance on yield, growth, and physiological behavior of a dryland cotton crop. *Crop Science, 19*(6), 863–868.

Moreshet, S., Stanhill, G., & Fuchs, M. (1977). Effect of increasing foliage reflectance on the CO_2 uptake and transpiration resistance of a grain Sorghum crop. *Agronomy Journal, 69*(2), 246–250.

Moutinho-Pereira, J., Gonçalves, B., Bacelar, E., Cunha, J. B., Coutinho, J., & Correia, C. M. (2009). Effects of elevated CO_2 on grapevine (*Vitis vinifera* L.). Physiological and yield attributes. *Vitis, 48*, 159–165.

Neethling, E., Barbeau, G., Bonnefoy, C., & Quenol, H. (2012). Change in climate and berry composition for grapevine varieties cultivated in the Loire Valley. *Climate Research, 53*, 89–101.

Palliotti, A., Luciani, E., Sforna, A., Boco, M., Squeri, C., & Frioni, T. (2019). *Ondate di calore e protezione del vigneto con il caolino*. VVQ, 5/2019, 32–35.

Robinson, A. L., Boss, P. K., Solomon, P. S., Trengove, R. D., Heymann, H., & Ebeler, S. E. (2014). Origins of grape and wine aroma. Part 1. Chemical components and viticultural impacts. *American Journal of Enology and Viticulture, 65*, 1–24.

Rosati, A., Metcalf, S. G., Buchner, R. P., Fulton, A. E., & Lampinen, B. D. (2006). Physiological effects of kaolin applications in well-irrigated and water-stressed walnut and almond trees. *Annals of Botany, 98*(1), 267–275.

Rosati, A., Metcalf, S. G., Buchner, R. P., Fulton, A. E., & Lampinen, B. D. (2007). Effects of kaolin application on light absorption and distribution, radiation use efficiency and photosynthesis of almond and walnut canopies. *Annals of Botany, 99*(2), 255–263.

Santillan, D., Iglesias, A., La Jeunesse, I., Garrote, L., & Sotes, V. (2019). Vineyards in transition: A global assessment of the adaptation needs of grape producing regions under climate change. *The Science of the Total Environment, 657*, 839–852.

Santos, J. A., Fraga, H., Malheiro, A. C., Moutinho-Pereira, J., Dinis, L.-T., Correia, C. M., Moriondo, M., Leolini, L., Dibari, C., Costafreda-Aumedes, S., Kartschall, T., Menz, C., Molitor, D., Junk, J., Beyer, M., & Schultz, H. R. (2020). A review of the potential climate change impacts and adaptation options for European viticulture. *Applied sciences, 10*, 3092–3120.

Sharma, P., Jha, A. B., Dubey, R. S., & Pessarakli, M. (2012). Reactive oxygen species, oxidative damage, and antioxidative defense mechanism in plants under stressful conditions. *Journal of Botany, 2012*. Article ID 217037.

Sharma, R. R., Reddy, S. V., & Datta, S. C. (2015). Particle films and their applications in horticultural crops. *Applied Clay Science, 116–117*, 54–68.

Sharma, E., Sharma, R., Borah, P., Jain, M., & Khurana, J. (2015). Emerging roles of auxin in abiotic stress responses. In G. K. Pandey (Ed.), *Elucidation of abiotic stress signaling in plants* (pp. 1–29). New York, NY: Springer Science+Business Media. ISBN: 978-1-4939-2210-9.

Shellie, K. (2015). Foliar Reflective Film and water deficit increase anthocyanin to soluble solids ratio during berry ripening in Merlot. *American Journal of Enology and Viticulture, 66*, 348–356.

Shellie, K., & Glenn, D. M. (2008). Wine grape response to kaolin particle film under deficit and well-watered conditions. *Acta Horticulturae, 792*, 587–591.

Shellie, K., & King, B. A. (2013a). Kaolin particle film and water deficit influence Malbec leaf and berry temperature, pigments, and photosynthesis. *American Journal of Enology and Viticulture, 64*(2), 223–230.

Shellie, K., & King, B. A. (2013b). Kaolin particle film and water deficit influence red winegrape color under high solar radiation in an arid climate. *American Journal of Enology and Viticulture, 64*(2), 214–222.

Smart, R. E. (1974). Photosynthesis by grapevine canopies. *Journal of Applied Ecology, 11*, 997–1006.

Smart, R. E. (1985). Principles of grapevine canopy microclimate manipulation with implications for yield and quality. A review. *American Journal of Enology and Viticulture, 36*(3), 230–239.

Smart, R. E., & Sinclair, T. R. (1976). Solar heating of grape berries and other spherical fruits. *Agricultural Meteorology, 17*(4), 241–259.

Smirnoff, N. (1993). The role of active oxygen in the response of plants to water deficit and desiccation. *New Phytologist, 125*, 27–58.

Song, J., Shellie, K. C., Wang, H., & Qian, M. C. (2012). Influence of deficit irrigation and kaolin particle film on grape composition and volatile compounds in Merlot grape (*Vitis vinifera* L.). *Food Chemistry, 134*, 841–850.

Spayd, S. E., Tarara, J. M., Mee, D. L., & Ferguson, J. (2002). Separation of sunlight and temperature effects on the composition of *Vitis vinifera* cv. Merlot berries. *American Journal of Enology and Viticulture, 53*, 171–182.

Takahashi, F., Kuromori, T., Urano, K., Yamaguchi-Shinozaki, K., & Shinozaki, K. (2020). Drought stress responses and resistance in plants: From cellular Responses to long-distance intercellular communication. *Frontiers of Plant Science, 11*, 556972–556986.

Todorov, D. T., Karanov, E. N., Smith, A. R., & Hall, M. A. (2003). Chlorophyllase activity and chlorophyll content in wild type and eti 5 mutant of *Arabidopsis thaliana* subjected to low and high temperatures. *Biologia Plantarum, 46*, 633–636.

Tosin, R., Pôças, I., & Cunha, M. (2019). Spectral and thermal data as a proxy for leaf protective energy dissipation under kaolin application in grapevine cultivars. *Open Agriculture, 4*(1), 294–304.

Wahid, A., Gelani, S., Ashraf, M., & Foolad, M. R. (2007). Heat tolerance in plants: An overview. *Environmental and Experimental Botany, 61*(3), 199–223.

Webb, L. B., Whetton, P. H., & Barlow, E. W. (2008). Modelling the relationship between climate, winegrape price and winegrape quality in Australia. *Climate Research, 36*, 89–98.

Zhang, Y., & Keller, M. (2015). Grape berry transpiration is determined by vapor pressure deficit, cuticular conductance, and berry size. *American Journal of Enology and Viticulture, 66*(4), 454–462.

Zhao, Y. (2010). Auxin biosynthesis and its role in plant development. *Annual Review of Plant Biology, 61*, 49–64.

CHAPTER 10

Wine quality production and sustainability*

Pierre-Louis Teissedre[1], Sofia Catarino[2] and Piergiorgio Comuzzo[3]

[1]Université de Bordeaux, ISVV, Unité de Recherche Œnologie EA 4577, USC 1366 INRA, Bordeaux INP, Villenave d'Ornon, France; [2]LEAF — Centre Linking Landscape, Environment, Agriculture and Food Research Center, Instituto Superior de Agronomia, University of Lisbon, Lisbon, Portugal; [3]Department of Agricultural Food, Environmental and Animal Science, University of Udine, Udine, Italy

10.1 Introduction

As stated in the first chapter of the present book, in 1987, in the Brundtland Report, Our Common Future, the World Commission on Environment and Development defined sustainability as "development that meets the needs of the present without compromising the ability of future generations to meet their own needs" (Brundtland, 1987).

Wine is a product strictly connected to tradition and today the consumers are increasingly expecting wine to be produced in a sustainable way (Bisson et al., 2002; Capitello & Sirieix, 2019; European Commission, 2017; Forbes et al., 2009). The complexity in defining sustainability results in a multidisciplinary literature base. Certification, water use and quality, soil, air and climatic impacts, energy, chemicals, wildlife, materials, waste, and globalization are all important topics within the discussion about sustainable wine (OIV, 2008, 2016). The wine sector is facing increasing pressure in order to fulfill environmental requirements and all these factors are relevant and need to be considered.

There are many ways to be more sustainable; major challenges consist in reducing energy consumption, global greenhouse gas (GHG) emissions, raw material use, waste output, and water consumption. Wineries can be built with reclaimed materials, employing skylights for natural light, planting more trees for shade, and collecting and recycling water for its reuse (Boulton, 2017). One popular sustainable winery architecture strategy is earth sheltering, in which wineries or cellars are built partially or completely undergrounded, where it is naturally cooler and easier to moderate temperatures (Tinti et al., 2014). Gravity-flow wineries, which allow grapes and wine to be moved more energy-efficiently and gently during the winemaking process is also an example of sustainable architecture. Some of the green strategies that can be employed by the wineries include solar power,

---

*Revised and extended, this chapter is based on a conference proceeding published by the first author in 2018 (Teissedre, 2018).

recycled building materials, living soil roofs, natural ventilation, geothermal heating and cooling systems, drought-tolerant landscaping, and even insulation made of blue-jean scraps (Boulton, 2017; Carroquino et al., 2018; Gómez-Villarino et al., 2021).

The following paragraphs aim to give a first overview of these aspects, introducing what will be further discussed more in detail in the subsequent chapters.

10.2 Existing systems and initiatives at winery level

Toward a more sustainable wine production, several systems and initiatives at winery level exist, even if the winemaking processes produces lower environmental impacts compared to other production phases, namely viticulture production and packaging (Ferrara & De Feo, 2018).

In modern Oenology, sustainable winemaking starts from the moment in which the winery is designed. Indeed, sustainable winery architecture is a global trend that needs to be developed. Nigro (2010) reported that in North America, many producers are building their new facilities according to green approaches, basing on the standards set by the LEED (Leadership in Energy and Environmental Design) Green Building Rating System. LEED voluntary certification was developed by the U.S. Green Building Council (1998) and is an international point of reference for the construction of environmentally friendly buildings, designed to become healthy places to work or to live (Nigro, 2010). Several North American wineries have already earned LEED certification for one or more of their buildings (Nigro, 2010).

In addition to LEED, other specialized certifications are available, oriented toward specific aspects of sustainability; an overview is given by Clarke (2018). The paper reports different examples of certification, including that released by Demeter (for biodynamic standards) or the Sustainability in Practice designation, which takes in high regard the well-being of the employees. In addition, the author describes several virtuous examples of wineries which applied different sustainable solutions to solve practical problems (Clarke, 2018). An interesting model, from the point of view of sustainable building, is the Living Building Challenge (LBC) Sustainability Certification. The standards of the LBC are codified in seven "petals" related to design and function: (i) water, (ii) energy, (iii) materials, (iv) place, (v) health and happiness, (vi) equity, and (vii) beauty, meaning that buildings which earned the LBC certification recycle water, produce more energy than their own consumption, are built without using harmful materials for the environment and, furthermore, are nice and comfortable places for workers (O'Donnell, 2020).

Talking about buildings, examples of applied sustainability are represented by the University of California at Davis, with its new teaching and research winery (Boulton, 2017, 2019), and by the Napa Valley Vintners association, with its new headquarter; both these cases are models for the industry.

Winery architectures often reach incredible results, becoming landmarks, thanks to their innovative design; some examples are described at https://it.lazenne.com (Lazenne, 2015). Bodegas Ysios in Spain, Château Cheval Blanc in France, Marchese Antinori and Petra Winery in Italy are examples of structures perfectly integrated in the surrounding landscape.

In Europe, sustainability issues are becoming increasingly important. In France, starting from vintage 2019, the appellations St.-Emilion, St.-Emilion Grand Cru, Lussac St.-Emilion, and Puisseguin St.-Emilion produce wines made from grapes grown with sustainable farming methods, such as organic or biodynamic; this decision includes nearly 3.85 million cases of wine, yearly made within the St.-Emilion denomination and any bottle of wine not produced according to such standard may only be labeled as generic "Bordeaux" (Mustacich, 2017). The St.-Emilion case is the result of a continuity of projects aimed at reducing pesticides, or at preserving biodiversity and landscape (St.-Emilion territory is a UNESCO World Heritage site); this was achieved by different state-approved certifications, such as biodynamic, organic, or HVE 3 (*Haute Valeur Environmentale*) standards (Mustacich, 2017).

Another interesting example is represented by Castillon denomination (Mustacich, 2017). Castillon is the Bordeaux appellation with the greatest percentage of vineyards in conversion to organic; here, approx. 120 châteaux obtained the HVE 3 certification, the highest level of certified sustainable farming in France.

In Italy, Franciacorta (known for the production of sparkling wines) is moving in the same direction, aiming at creating a strong link between wine system and territory, developing synergies between wine production—strongly devoted to organic winemaking (today the 70% of the whole production and 2/3 of the vineyards are certified as organic)—and tourism (Chierchia, 2019). Similarly to St.-Emilion, also in this case the priorities are protecting biodiversity and environment, reducing the use of inputs in wineries and vineyards; this was achieved though different projects, self-regulations, and certifications (Consorzio per la tutela del Franciacorta, 2021). An interesting example is the "Ita.Ca® Model," a control tool used by Franciacorta companies for monitoring the impact of GHG emissions originated from winegrowing activities (Consorzio per la tutela del Franciacorta, 2021). The calculator and the method used for data management complies with the GHGAP protocol established by the International Organization of Vine and Wine (OIV, 2011) recognized worldwide as a specific reference method in the sector for this evaluation (Consorzio per la tutela del Franciacorta, 2021).

10.3 Principal aspects to consider for a sustainable wine production
10.3.1 Carbon dioxide reuse

In 2015, OIV adopted a resolution stating recommendations for GHG emissions in the vitivinicultural sector, providing specific information on methodologies for their evaluation (OIV, 2015).

The vine and wine sector involves a series of activities which sequestrate and emit carbon dioxide (CO_2), among other gases. The GHG balance and the net emissions can be estimated by summing the CO_2 uptake by grapevines with the gross emissions (Marchi et al., 2018). During the grape growing season, CO_2 is sequestered by the grapevines growth and production of sugar in the grapes, through the photosynthesis process (Nistor et al., 2018). This is more than the CO_2 emitted by the biomass and during the alcoholic fermentation process, meaning that production and alcoholic fermentation of 1 kg of grapes reduces the CO_2 by approximately 0.3 kg (Table 10.1). Interestingly, the production of wines with a higher alcoholic strength sequesters greater amounts of CO_2 (CIVB, 2018).

It is possible to further increase these amounts by implementing systems useful to collect CO_2 from alcoholic and malolactic fermentations in the wineries. This GHG is mainly produced during alcoholic fermentation of the grape must, in amounts depending on sugars content, among other factors (Ribéreau-Gayon et al., 2006). According to the literature, 48.9 g of CO_2 are dispersed in the atmosphere per 100 g of fermented glucose (Marchi et al., 2018).

Table 10.1 Sequestration of CO_2 in function of sugars content of grapes.

Sugars content of grapes (°Brix)	CO_2 generated/sequestered per kg of grapes
25	−0.94 kg total CO_2 sequestered via photosynthesis +0.53 kg total biomass CO_2 emission +0.1 kg CO_2 produced in fermentation −0.31 kg net sequestration of CO_2
21	−0.79 kg total CO_2 sequestered via photosynthesis +0.45 kg total biomass CO_2 emission +0.1 kg CO_2 produced in fermentation −0.24 kg net sequestration of CO_2

Adapted from CIVB. (2018). Plan climat 2020 des vins de Bordeaux. Les cahiers techniques du CIVB, 64, 12, Avril 2018.

Fermentation processes begin as soon as the skin of the grapes is split and the temperature exceeds 12°C. The sugars (glucose and fructose) then come into contact with the yeasts present on the skin of the berry or in the air, being gradually transformed into ethanol. During alcoholic fermentation, many secondary metabolites are released such as CO_2, ethanol, glycerol, succinic acid, acetic acid, and volatile compounds (Ribéreau-Gayon et al., 2006). Malolactic fermentation accounts only for approximately 2 g of CO_2 per L of wine (Lonvaud-Funel & Ribéreau-Gayon, 1977).

In the field of Oenology, the uses of CO_2 are numerous, first and foremost, the production of the frigories required to keep the must at suitable temperatures. Carbon dioxide has also a blanketing/protective activity, contrasting oxidations and allowing to control the development of microorganisms (e.g., yeasts), so that it is used in the refrigeration of harvested grapes (especially for mechanical harvesting), as well as in the continuous cooling of the mash after crushing (CO_2 in dry ice form). These operations allow reducing the temperature of the crushed grapes to around 5°C, obtaining an immediate and homogeneous state of coldness.

Another important point is the use of CO_2 as a technical and antioxidant gas during all the operations ranging from fermentation to bottling, and in particular to replace oxygen in the headspace of wine containers. Indeed, CO_2 can be used as inert gas for wine aging, potentially allowing the reduction of sulfur dioxide (SO_2) use.

Carbon dioxide can be used also in cryoextraction, a process through which the partial freezing of grapes before pressing results in higher quality white wines. Enzymatic processes that can lead to degeneration of the aromatic components are inhibited, preserving the aromas and thereby producing a better *bouquet* that can be maintained in the finished product. With CO_2 it is also possible to freeze only the less-ripened grapes (with lower sugar content) in order that the must that is released during pressing comes only from the ripest grapes. Carbonic maceration exploits CO_2 to induce intercellular fermentation of the whole grape, which is the basis for the production of new wines (*Primeurs* wine) (Marchi et al., 2018). Finally, it can be used as food grade CO_2, e.g., in carbonated drinks. In fact, while CO_2 from alcoholic fermentation is normally released in the atmosphere, significant amounts of CO_2 produced in certain chemical industries are purchased in tanks or in form of dry ice. It is estimated that 70% of the demand for food grade CO_2 comes from the beverage sector (Vicenzi, 2013). Bearing this in mind, the use of CO_2 of biogenic origin and high purity, recovered from wine production, could be suitable for this sector.

FIGURE 10.1

Example of a CO_2 captation system: process VALECARB, by Alcion (2021).

Credits photos Alcion/SEDE Environnement.

In recent years, innovations in CO_2 capture technologies, namely produced using alcoholic fermentation as CO_2 emission source, have been developed (Alonso-Moreno & García-Yuste, 2016). An interesting example for capturing the CO_2 emitted by fermentation above the vinification tanks and transforming it into baking soda is the principle proposed by Alcion Environnement, a company operating in Bordeaux and the Paris region (Fig. 10.1). This capture system is an innovative project, labeled by Inno'vin. According to this company from 1000 hL of wine produced, 8—10 tons of CO_2 are captured, allowing the production of 12—18 tons sodium bicarbonate (Alcion, 2021). At the scale of a Château of 28 ha, 20 tons of sodium bicarbonate can be produced from the vineyard. This sequestration of CO_2, resulting from winemaking, will allow the manufacture of compounds for local uses. Thus, sodium bicarbonate can be used as cleaning agent in water softeners, as well as for agri-food and pharmaceutical applications. The capture of this CO_2 and its recovery will have an impact on the carbon footprint of the vineyard. For 100 ha of vineyards, the estimated reduction of CO_2 emitted is of 37 tons/year in the Bordeaux appellation (CIVB, 2018). The carbon footprint of the Bordeaux wine sector has been evaluated to 744,000 ton/year of CO_2; the reduction would be 147,000 tons/year of CO_2 (CIVB, 2018).

10.3.2 Water management and saving

The ISO-14001 Standard is considered the first necessary step for a winery toward a more environmental friendly management of its operations (ISO, 2015). It measures and allows wineries to work toward an optimal efficiency in use of water (in some parts of the world, more than 10 L of water is used for each L of wine produced), energy, raw materials, maximizing recycling and waste treatment (Matos & Pirra, 2020; Mosse et al., 2011; Oliveira et al., 2019). For example, the use of water within the winery itself accounts for about 2 L per shipped bottle of Champagne (4.1 L for combined vineyard/winery operations (CIVC, 2015). This is slightly higher than the global average use for wine production—close to the average for soft drink production—but significantly lower than the amounts consumed by brewing industry (CIVC, 2015).

Water preservation remains a priority, without ever compromising high standards of hygiene in wineries and other work premises. Winemakers use various methods to reduce water consumption. These include the ecodesign or ecorefurbishment of buildings, improved systems of water purification, recycling and/or collection, and a general reduction in water wastage wherever possible (Costa et al., 2020; GWRDC, 2011; Mosse et al., 2011; OIV, 1999; Rochard, 2017). Collection of rainwater with storage in tanks is a good way to do water economy, and to be able to use it during dry period (Boulton, 2017, 2019). Wastewater treatment and/or closed loop reuse of cooling water also contributes significantly to improving the environmental impact of winery operations. Regarding sustainable water consumption, three key points must be addressed: (i) to generalize the installation of differential submetering utilities for measuring/monitoring water consumption in the different departments of the winery; (ii) to promote the collection of rainwater; (iii) to optimize cleaning practices.

10.3.3 Renewable energy

The practice of viticulture and winemaking is highly dependent upon the weather and climate. Any future changes in the seasons, their duration, local maximum, minimum and mean temperatures, frost occurrence and heat accumulation could have a major impact on the winegrowing areas of the world. Given that the wine industry has substantial energy requirements and it is directly influenced by any changes in climate, the industry should be at the forefront in promoting the case of energy efficiency and the adoption of renewable technologies (Garcia-Casarejos et al., 2018; Gómez-Villarino et al., 2021). Renewable solar energy in the form of solar thermal and photovoltaic (PVs) offer a complimentary solution to many winegrowing processes. Distinct initiatives can be adopted to achieve a more sustainable energy use: (i) identify the energy consumption of each equipment, including fuel; (ii) promote the installation of PV energy; (iii) improve the efficiency of winery operations and the use of renewable energy and biofuels; (iv) generalize the installation of electric differential submeters.

10.3.4 Good practices in Oenology and winemaking process

The force of gravity has been exploited by wineries for racking and wine handling since the 1960s; this approach has regained importance in the design of new sustainable wine buildings. There are no formal definitions for "gravity-flow" or "gravity-fed" wineries, but these terms indicate a particular winery-design style, suggesting that the winery production will take place on at least two different floors or levels. If entire winery operation is on one floor, every time it is intended to move a wine from a crusher or press to a tank or barrel, there will be a need for pumps, conveyors, or other materials. Gravity-flow winery designs take advantage of gravity, allowing wine to be moved around in an energy-saving way and much more gently. Too much force, too much rough-and-tumble handling and a wine might become over extracted or too tannic, as well as too oxidized. It is also harder (e.g., during racking) to leave grape solids behind if there is a hose on at full blast. Ideally, a gravity-flow winery could have a different floor for every stage of wine production. The grapes are delivered from vineyard on the top floor and move from crushing to fermentation, to aging and bottling; every time the wine is moved along, gravity helps to move it. For oenologists the idea is that gravity flow provides a gentler, less interventionist approach to winemaking, taking less work moving around pumps and hoses and requiring less electricity. But gravity-flow wineries can be expensive to build, and require a winemaking team willing to walk up and down stairs and ladders. Their designs typically work best when built into an existing hillside and some of the best wine regions are in valleys.

Another fundamental concept connected to a sustainable wine production approach is related to the management of hygiene issues. Indeed, this is a relevant question because it directly impacts on the use of chemicals (cleaning agents and sanitizers) and consequently on the production of effluents and water pollution/purification issues. In addition, a high hygienic standard in a winery is a key-factor for allowing the reduction of the use of certain additives, in particular, sulfur dioxide.

The "hygienic question" is also related to a responsible production approach, managed through transparent production practices, which guarantee the protection and the health of the consumer. In France, a recent official validation of the Guide of Good Hygiene Practices for the wine sector (GBPH, April 1st 2017) responds to changes in hygiene and food safety practices and regulations while taking into account the specificities of the sector of wines and wine spirits (IFV, 2021). National and Community regulations lay down obligations to ensure consumer safety and traceability of foodstuffs. In particular, producers are responsible for the sanitary quality of the food they put on the market. The obligations which concern more particularly the "hygiene package" are defined in terms of results, the means to be implemented being left to the choice of the operators. The document was designed to help all operators of the French wine industry in the implementation of good hygiene practices and HACCP. The GBPH is a guide of voluntary application and a reference document that can be used as proof for official controls if the operator follows the principles described. The guide allows each company to assess its own risks and define the means of control to be implemented as part of its food safety policy, taking into account the main regulations, i.e., the general regulations on food safety and the specific regulations for wine products.

Concerning the technologies, the additives and the processing aids applied to wine production, the good practices in Oenology need to refer to the International Code of Oenological Practices and to the International Oenological Codex, where practices and products whose use is allowed in Oenology are described (OIV, 2018). It is necessary to insist on the fact that any product/technique with new functionality must comply with the two codes with tests validated by national/international bodies.

Despite the awareness of companies and institutions toward a more responsible production approach increased in recent years, there is still a long way to go to achieve more sustainable wine production. However, several strategies and tools are going to be implemented and the near future will certainly see several changes and innovations in winery practice. Current trends are moving toward the introduction of cross-cutting actions for harmonizing input reduction, good winemaking practices, the use of rapid in-site analytical methods and devices integrated with rapid decision support tools. Similarly to what happens in viticulture, the concept of Precision Oenology is rapidly growing. In addition, research and innovation are being developed toward the selection of yeasts for specific functionalities and the creation of tools and products tailored for specific applications, such as fermentation management and stabilization.

10.3.5 Functional biodiversity

Wine production in most countries is based on the use of commercial yeast and bacteria strains. There is a certain debate about the use of selected microorganisms in winemaking, especially among organic and biodynamic producers. On one hand, the advantages connected to the use of commercial strains are not in question and their introduction may favor the possibility of significantly reducing the use of sulfites during the early stages of vinification process. However, the widespread use of commercial microorganisms may lead to the colonization of wineries and vineyards by those strains, with the potential consequent reduction of autochthonous biodiversity. This also implies that wine styles could become standardized.

Vineyard could be an important source of native yeasts and bacteria of oenological interest (Loureiro et al., 2012; Belda et al., 2017). A better knowledge about the functional role of biodiversity in the vineyard and wine ecosystems is required, including the spatial and temporal interactions between native microorganisms (yeasts and bacteria), natural enemies and pests (fungal infestation) and climate and management factors (e.g., irrigation, temperature, perennial cover crops, use of agrochemicals, harvesting practices, or fermentation performance).

Biodiversity conservation and agricultural sustainability should also take into account "ecoagriculture" landscapes. It will be important to conserve wild biodiversity in agricultural landscapes; this will require increased research, policy coordination and strategic support to agricultural communities and conservationists. Recently, a collective expert report was published by OIV presenting an overview about functional biodiversity in vineyards and illustrating major aspects of functional biodiversity in the viticultural sector (OIV, 2018a).

10.3.6 Management and use of by-products in Oenology

There is an increasing interest in the management and valorization of secondary products generated at different stages of wine production chain, with a significant number of studies being developed on the reuse of grape by-products, namely grape pomace, seeds and wine lees (Bordiga et al., 2019; Gómez-Brandón et al., 2019; Molina-Alcaide et al., 2008; OIV, 2018b; Troilo et al., 2021).

In 2018, OIV published a collective expertise report on the management of by-products of vitivinicultural origin, including a definition of by-products and their classification as wastes, residues, subproducts and by-products (OIV, 2018b). Yet, the classification of wastewater and other residues as by-products is not consensual and more often grape marc or pomace, skins, stalks, seeds, and lees are considered the main by-products from wine production (Bordiga et al., 2019).

The principal residues from wine industry are organic wastes (pomace, seeds, stems, and leaves), wastewater, GHGs (CO_2, volatile organic compounds), and inorganic wastes (perlite, diatomaceous earth, and bentonite) (Oliveira et al., 2013; Teixeira et al., 2014). A reference to the production of 14.5 million tons of by-products per year just in Europe can be found in the literature (Chouchouli et al., 2013).

Traditionally, the valorization of these winemaking by-products is obtained through different options: (i) distillation of the marc (or pomace) and lees, which are generally delivered to authorized distilleries (Lempereur & Penavayre, 2014); (ii) production of fertilizers (e.g., compost) or livestock feeds, as well as (iii) production of fermentation substrates for obtaining biomass (Yilmaz & Toledo, 2004; Arvanitoyannis, 2006; Harsha et al., 2013).

However, these conventional "ways of use" of winery by-products have some side-effects. For instance, some of them may contain phytotoxic compounds which display antimicrobial effects during composting, affecting their utilization for this purpose (Teixeira et al., 2014). Regarding their use in livestock feed, certain components (e.g., proanthocyanidins) may generate intolerances in some animals, negatively affecting digestibility (González-Centeno et al., 2014; Teixeira et al., 2014).

The management, diversification, and valorization of by-products in Oenology need to be developed taking into account their content in value-added compounds, as well as their extractability, degradability, digestibility and nutrient content.

The valorization of residues as a source of bioactive phytochemicals might constitute an efficient, profitable and environment-friendly alternative (Makris et al., 2007). In fact, a wide range of compounds can be recovered for recycling, with potential application in processed foods, cosmetics, and human health products, or used as animal feed ingredients: ethanol for industrial use and motor fuel, grape-seed oil, polyphenols, antioxidants and natural color pigments, tartaric acid, co-products used as organic fertilizers and carbon-based additives. An interesting review about possible alternative uses was published by Teixeira et al. (2014), concerning the phytochemicals present in winery by-products, their extraction techniques, industrial uses and biological activities.

With this approach, the concept of biorefineries would have sense economically for the products obtained from the vineyard, to develop bioplastics, biocompatible polymers, fertilizers (organic and inorganic), phenolic extracts, biofuels and gases, color extracts, etc. Otherwise, the use of grape derivatives and bioactive products in the formulation of nutraceuticals/cosmetics, ensuring a feature of strong antioxidant activity must be encouraged (e.g., seed flours or grape skins). In particular, proanthocyanidins recovered from grape seeds have medical applications (Spranger et al., 2008). An excellent example of the biorefinery concept and valorization of by-products is the extraction and trade of "enocyanine," whose main use is addressed toward the food industry in addition to the cosmetics and pharmaceutical fields (Bordiga et al., 2019).

Another possible field to explore respects the nutritional value. As an example, the use of by-products as potential sources of nutrients for ruminants was investigated (Molina-Alcaide et al., 2008), and concentrations of 45–203 and 122–741 g/kg dry matter were reported respectively for crude protein and neutral detergent fibers.

Hence, several actions should be encouraged, including: (i) valorization of by-products in compounds of interest, via distilleries or others specialized enterprises; (ii) development of composting of shoots and effluents; (iii) valuing of biomass.

Fig. 10.2 shows an overview of the main by-products generated during wine production and the corresponding value-added compounds. The yields shown are indicative values and may considerable vary depending on grape variety, environmental conditions, grape maturation stage, and winemaking technology.

Finally, reducing and recovering waste along the whole oenological chain should be also an ongoing priority for the wine industry. Consumers discard tons of waste per year, mainly glass (90%), cardboard, paper, aluminum, steel, cork and plastic. Two special measures should be put in place to address the issue of postconsumer waste: (i) waste prevention (also known as source reduction), mainly by switching to ecofriendly packaging and lighter glass bottles when possible; (ii) setting-up of a membership scheme for all wine producers and stakeholders via subscriptions to firms providing community-based, domestic waste collection, and reuse services (ecopackaging).

10.4 Concluding remarks

Sustainable initiatives require planning, monitoring and assessment of knowledge. It is a constantly evolving process and, as such, it requires continuous evaluation and improvement. For an innovative sustainable Oenology, several points have to be considered: CO_2 reuse solutions, water management and saving, renewable energy, good practices in Oenology, functional biodiversity, management and

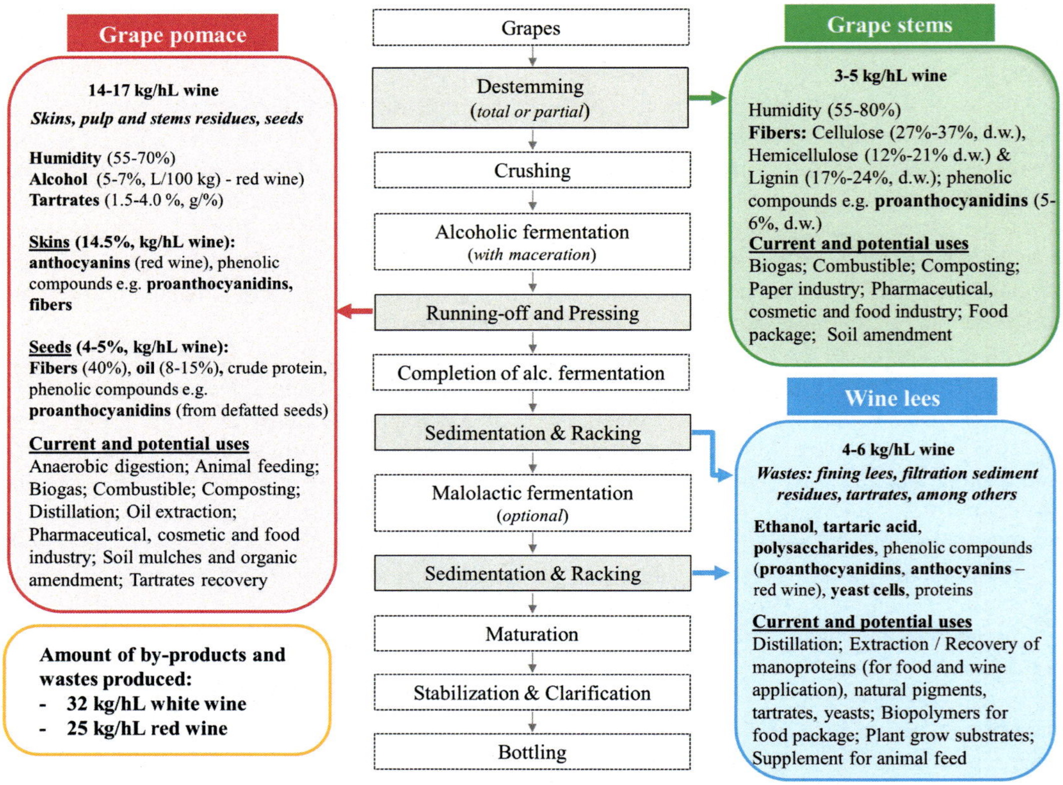

FIGURE 10.2

Overview of the main by-products generated throughout the winemaking process of red wine, value-added compounds liable to be recovered and current and potential uses. The yields and chemical composition shown are based on the literature and may considerable vary according to environmental, viticultural and winemaking aspects (Belchior & Costa, 1974; Bordiga et al., 2019; OIV, 2018; Troilo et al., 2021).

use of by-products in Oenology and climate change adaptation. Many questions and challenges need to be explored to suggest advances for a more sustainable Oenology production. Interactions of oenologists, winemakers, chemists, microbiologists, renewable energy and water specialists, process specialists, computer and electronic scientists need to be encouraged to develop wineries adapted to climate changes scenarios with a sustainable, qualitative and economically compatible Oenology production.

References

Alcion. (2021). http://www.alcion-env.com/fr/wine/co2.html. (Accessed 30 July 2021).

Alonso-Moreno, C., & García-Yuste, S. (2016). Environmental potential of the use of CO_2 from alcoholic fermentation processes. The CO_2-AFP strategy. *Science of the Total Environment, 568*, 319–326.

Arvanitoyannis, I. S., Ladas, D., & Mavromatis, A. (2006). Potential uses and applications of treated wine waste: A review. *International Journal of Food Science & Technology, 41*, 475–487. https://doi.org/10.1111/j.1365-2621.2005.01111.x

Belchior, A. P., & Costa, Y. E. (1974). Élaboration et utilisation des produits secondaires de la vinification. *Bulletin de l'OIV, 47*(517), 232–245.

Belda, I., Zarraonaindia, I., Perisin, M., Palacios, A., & Acedo, A. (2017). From vineyard soil to wine fermentation: Microbiome approximations to explain the "terroir" concept. *Frontiers in Microbiology, 8*, 821.

Bisson, L. F., Waterhouse, A. L., Ebeler, S. E., Walker, M. A., & Lapsley, J. T. (2002). The present and future of the international wine industry. *Nature, 418*, 696–699.

Bordiga, M., Travaglia, F., & Locatelli, M. (2019). Valorisation of grape pomace: An approach that is increasingly reaching its maturity – a review. *International Journal of Food Science and Technology, 54*, 933–942.

Boulton, R. (2017). A self-sustainable winery, an advanced passive building and remote monitoring of environments in wineries. *Journal of Agricultural Engineering, XLVIII*(s1), 735.

Boulton, R. (2019). *The design solutions for a self-sustainable zero-carbon winery*. https://climatechange-porto.com/wp-content/uploads/2019/03/Session-4-Wineries-of-the-Future-Roger-Boultton-CCL2019.pdf (Accessed 24 June 2021).

Brundtland, G. (1987). *Report of the world commission on environment and development: Our common future*. United Nations General Assembly document A/42/427 – Development and International Economic Co-operation: Environment.

Capitello, R., & Sirieix, L. (2019). Consumer's perceptions of sustainable wine: An exploratory study in France and Italy. *Economies, 7*, 1–20.

Carroquino, J., Roda, V., Mustata, R., Yago, J., Valino, L., Lozano, A., & Barreras, F. (2018). Combined production of electricity and hydrogen from solar energy and its use in the wine sector. *Renewable Energy, 122*, 251–263.

Chierchia, V. (2019). *Franciacorta Docg: Obiettivo total bio e sviluppo in Cina e Canada*. https://www.ilsole24ore.com/art/franciacorta-docg-obiettivo-total-bio-e-sviluppo-cina-e-canada-ABuCCMTB?refresh_ce=1. (Accessed 26 July 2021).

Chouchouli, V., Kalogeropoulos, N., Konteles, S. J., Karvela, E., Makris, D. P., & Karathanos, V. T. (2013). Fortification of yoghurts with grape (*Vitis vinifera*) seed extracts. *LWT-Food Science and Technology, 53*, 522–529.

CIVB. (2018). Plan climat 2020 des vins de Bordeaux. *Les cahiers techniques du CIVB, 64*, 12. April 2018.

CIVC. (2015). *Viticulture durable en Champagne*. Rapport 2015. Comité Interprofessionnel du Vin de Champagne.

Clarke, S. (2018). *The most sustainable wineries in America*. https://www.winecountry.com/blog/sustainable-wineries-usa/. (Accessed 24 July 2021).

Consorzio per la tutela del Franciacorta. (2021). https://www.franciacorta.net/it/viticultura/sostenibilita/. (Accessed 26 July 2021).

Costa, J. M., Oliveira, M., Egipto, R. J., Cid, J., Fragoso, R. A., Lopes, C. M., & Duarte, E. N. (2020). Water and wastewater management for sustainable viticulture. *Ciência e Técnica Vitivinícola, 35*(1), 1–15.

European Commission. (2017). *Attitudes of European citizens towards the environment*. Report Special Eurobarometer 468– Wave EB88.1 – TNS opinion & social https://op.europa.eu/pt./publication-detail/-/publication/018fcab9-e6d6-11e7-9749-01aa75ed71a1/language-en. (Accessed 24 June 2021).

Ferrara, C., & De Feo, G. (2018). Life cycle assessment application to the wine sector: A critical review. *Sustainability, 10*, 395.

Forbes, S. L., Cohen, D., Cullen, R., Wratten, S. D., & Fountain, J. (2009). Consumer attitudes regarding environmentally sustainable wine: An exploratory study of the New Zealand marketplace. *Journal of Cleaner Production, 17*, 1195–1199.

Garcia-Casarejos, N., Gargallo, P., & Carroquino, J. (2018). Introduction of renewable energy in the Spanish wine sector. *Sustainability, 10*, 3157. https://doi.org/10.3390/su10093157

Gómez-Brandón, M., Lores, M., Insam, H., & Domínguez, J. (2019). Strategies for recycling and valorization of grape marc. *Critical Reviews in Biotechnology, 39*(4), 437–450.

Gómez-Villarino, M. T., Barbero-Barrera, M. D. M., Mazarrón, F. R., & Cañas, I. (2021). Cost-effectiveness evaluation of nearly zero-energy buildings for the aging of red wine. *Agronomy, 11*, 627.

González-Centeno, M. R., Knoerzer, K., Sabarez, H., Simal, S., Rosselló, C., & Femenia, A. (2014). Effect of acoustic frequency and power density on the aqueous ultrasonic-assisted extraction of grape pomace (*Vitis vinifera* L.) – A response surface approach. *Ultrasonics Sonochemistry, 21*, 2176–2184.

GWRDC. (2011). *Winery wastewater management and recycling – Operational guidelines. Grape and wine research and development corporation* (p. 79). Australia Government. www.gwrdc.com.au/www.

Harsha, P. S. C. S., Gardana, C., Simonetti, P., Spigno, G., & Lavelli, V. (2013). Characterization of phenolics, *in vitro* reducing capacity and anti-glycation activity of red grape skins recovered from winemaking by-products. *Bioresource Technology, 140*, 263–268. https://doi.org/10.1016/j.biortech.2013.04.092

IFV. (2021). https://www.vignevin.com/process-vins/guides-des-bonnes-pratiques-dhygiene/. (Accessed 31 July 2021).

ISO. (2015). *ISO 14001:2015. Environmental management systems – requirements with guidance for use*. International Organization for Standardization.

Lazenne. (2015). *Grand designs! Europe's incredible wineries*. https://it.lazenne.com/blogs/lazenneblog/19103531-grand-designs-europe-s-incredible-wineries. (Accessed 26 July 2021).

Lempereur, V., & Penavayre, S. (2014). Grape marc, wine lees and deposit of must: How to manage oenological by-products? *Bio Web of Conferences, 3*, 01011.

Lonvaud-Funel, A., & Ribéreau-Gayon, P. (1977). Le gaz carbonique des vins II. – Aspect Technologique. *Connaissance de la Vigne et du Vin, 11*(2), 165.

Loureiro, V., Ferreira, M. M., Monteiro, S., & Ferreira, R. B. (2012). The microbial community of grape berry. In H. Gerós, M. M. Chaves, & S. Delrot (Eds.), *The biochemistry of the grape berry* (pp. 241–268). Bentham Science Publishers.

Makris, D. P., Boskou, G., & Andrikopoulos, N. K. (2007). Polyphenolic content and *in vitro* antioxidant characteristics of wine industry and other agri-food solid waste extracts. *Journal of Food Composition and Analysis, 20*, 125–132. https://doi.org/10.1016/j.jfca.2006.04.010

Marchi, M., Neri, E., Pulselli, F. M., & Bastianoni, S. (2018). CO_2 recovery from wine production: Possible implications on the carbon balance at territorial level. *Journal of CO_2 Utilization, 28*, 137–144.

Matos, C., & Pirra, A. (2020). Water to wine in wineries in Portugal Douro region: Comparative study between wineries with different sizes. *Science of the Total Environment, 732*, 139332.

Molina-Alcaide, E., Moumen, A., & Martín-García, A. I. (2008). By-products from viticulture and the wine industry: Potential as sources of nutrients for ruminants. *Journal of the Science of Food and Agriculture, 88*, 597–604. https://doi.org/10.1002/jsfa.3123

Mosse, K. P. M., Patti, A. F., Christen, E. W., & Cavagnaro, T. R. (2011). Review: Winery wastewater quality and treatment options in Australia. *Australian Journal of Grape and Wine Research, 17*, 111–122. https://doi.org/10.1111/j.1755-0238.2011.00132.x

Mustacich, S. (2017). *Bordeaux's St.-Emilion Mandates sustainable viticulture*. https://www.winespectator.com/articles/st-emilion-mandates-sustainable-viticulture. (Accessed 26 July 2021).

Nigro, D. (2010). *LEED winery projects in North America—Expanded profiles and photos*. https://www.winespectator.com/articles/leed-winery-projects-in-north-america-expanded-profiles-and-photos. (Accessed 24 July 2021).

Nistor, E., Dobrei, A. G., Dobrei, A., Camen, D., Sala, F., & Prundeanu, H. (2018). N_2O, CO_2, production, and C sequestration in vineyards: A review. *Water Air & Soil Pollution, 229*, 299.

O'Donnell, B. (2020). *Exclusive: Silver Oak becomes first winery to Earn living building challenge sustainability certification.* https://www.winespectator.com/articles/silver-oak-becomes-first-winery-to-earn-living-building-challenge-certification. (Accessed 26 July 2021).

OIV. (1999). *OIV "Scientific and technical notebook": Gestion des effluents de cave et de destillerie. Cahier scientifique et technique* (p. 79). Office International de la Vigne et du Vin.

OIV. (2008). *OIV guidelines for sustainable vitiviniculture: Production, processing and packaging of products.* Resolution CST 1/2008 https://www.oiv.int/public/medias/2089/cst-1-2008-en.pdf. (Accessed 24 June 2021).

OIV. (2011). *General principles of the OIV greenhouse gas accounting protocol for the vine and wine sector.* Resolution OIV-CST 431-2011 https://www.oiv.int/public/medias/2107/oiv-cst-431-2011-en.pdf. (Accessed 26 June 2021).

OIV. (2015). *Greenhouse gases accounting in the vine and wine sector — recognized gases and inventory of emissions and sequestrations.* Resolution OIV-CST 503AB-2015.

OIV. (2016). *OIV General principles of sustainable vitiviniculture-environmental-social-economic and cultural aspects.* Resolution OIV-CST 518-2016 http://www.oiv.int/en/technical-standards-and-documents/resolutions-of-the-oiv/resolution-cst. (Accessed 24 June 2021).

OIV. (2018a). *OIV collective expertise: Functional biodiversity in the vineyard.* International Organisation of Vine and Wine.

OIV. (2018b). *OIV collective expertise: Managing by-products of vitivinicultural origin.* International Organisation of Vine and Wine.

Oliveira, M., Costa, J. M., Fragoso, R., & Duarte, E. (2019). Challenges for modern wine production in dry areas: Dedicated indicators to preview wastewater flows. *Water Supply, 19*(2), 653–661.

Oliveira, D. A., Salvador, A. A. A. S., Smânia, E. F. A., Maraschin, M., & Ferreira, S. R. S. (2013). Antimicrobial activity and composition profile of grape (*Vitis vinifera*) pomace extracts obtained by supercritical fluids. *Journal of Biotechnology, 164*, 423–432. https://doi.org/10.1016/j.jbiotec.2012.09.014

Ribéreau-Gayon, P., Dubourdieu, D., Donèche, B., & Lonvaud, A. (2006). *Handbook of enology. Vol. 1. The microbiology of wine and vinifications* (2nd ed.). Wiley.

Rochard, J. (2017). *Processes of treatment of wineries effluents adapted to the organic wine sector: Current situation and prospects.* 40th World Congress of Vine and Wine.

Spranger, I., Sun, B., Mateus, A. M., de Freitas, V., & Ricardo-da-Silva, J. M. (2008). Chemical characterization and antioxidant activities of oligomeric and polymeric procyanidin fractions from grape seeds. *Food Chemistry, 108*, 519–532.

Teissedre, P.-L. (2018). Wine quality production and sustainability. In *OENOVITI 7th International Symposium "Opportunities and challenges for vine and wine production by preserving resources and environment* (pp. 31–37). France: OENOVITI International Network, Villenave d'Ornon.

Teixeira, A., Baenas, N., Dominguez-Perles, R., Barros, A., Rosa, E., Moreno, D. A., & Garcia-Viguera, C. (2014). Natural bioactive compounds from winery by-products as health promoters: A review. *International Journal of Molecular Sciences, 15*, 15638–15678. https://doi.org/10.3390/ijms150915638

Tinti, F., Barbaresi, A., Benni, S., Torreggiani, D., Bruno, R., & Tassinari, P. (2014). Experimental analysis of shallow underground temperature for the assessment of energy efficiency potential of underground wine cellars. *Energy and Buildings, 80*, 451–460.

Troilo, M., Difonzo, G., Paradiso, V. M., Summo, C., & Caponio, F. (2021). Bioactive compounds from vine shoots, grape stalks, and wine lees: Their potential use in agro-food chains. *Foods, 10*, 342.

U.S. Green Building Council. (1998). https://new.usgbc.org/about. (Accessed 31 July 2021).

Vicenzi, N. (2013). Consorzio di Tutala del Soave, F. Bussola, Gruppo Collis, Recupero della CO_2 di fermentazione. *Mille Vigne — Enologia*, 24–27.

Yilmaz, Y., & Toledo, R. T. (2004). Major flavonoids in grape seeds and skins: Antioxidant capacity of catechin, epicatechin, and gallic acid. *Journal of Agricultural and Food Chemistry, 52*, 255–260. https://doi.org/10.1021/jf030117h

CHAPTER 11

Water management toward regenerative wineries

Margarida Oliveira[1,2], Artur Saraiva[1,2], Milena Lambri[3], Joel Rochard[4], Rita Fragoso[1], Elia Romanini[3], Pedro Hipólito[5,6], Capri Ettore[3] and Elizabeth Duarte[1]

[1]*LEAF - Linking Landscape, Environment, Agriculture and Food, TERRA, Instituto Superior de Agronomia, Universidade de Lisboa, Lisboa, Portugal;* [2]*ESAS, UIIPS, CIEQV, Instituto Politécnico de Santarém, Quinta do Galinheiro, Santarém, Portugal;* [3]*DiSTAS, Department for Sustainable Food Process, Università Cattolica del Sacro Cuore, Piacenza, Italy;* [4]*VitisPlanet, Bouilly, France;* [5]*Instituto Superior de Agronomia, Universidade de Lisboa, Lisboa, Portugal;* [6]*HUVA, SA — Herdade da Mingorra, Beja, Portugal*

11.1 Introduction

For a long time considered as a craft world, the wine sector, while preserving its traditional know-how, is gradually becoming more professional, in particular to meet the new challenges linked to food safety, social/societal and environmental responsibility. It is a holistic approach, which obviously first of all integrates the adaptation of the different technical viticulture and oenological itineraries, but also management within the company. Thus, training, organization, traceability, adherence to recognized and certified (national or international) procedures, and communication are all imperatives that are now part of the daily life of wine professionals. At the same time, environmental aspects and more generally sustainable dimensions are the subject of growing expectations on the part of the public, which puts the development of ecoenotourism into perspective. Sustainable development also applies to the design, construction, and operation of the wineries through the ecodesign concept. Water is becoming an important issue, particularly in Mediterranean areas, which justifies optimal management within the winery, which also facilitates the wastewater treatments. Until now, these wastewaters were mainly treated by aerobic systems, which globally made it possible to meet local environmental regulatory requirements, but with constraints, particularly with regard to potential noise, odor, landscape nuisances, and high energy consuming. In connection with the increasing water scarcity in Mediterranean areas, the prospect of reusing treated wastewater justifies the development of additional finishing systems, in a more technological or environmentally friendly approach.

The Europe 2020 strategy aims to increase the share of energy from renewable sources to 20% to reduce the primary energy consumption up to 20% by improving energy efficiency and to reduce the greenhouse gases emissions. The small and medium-sized enterprises are generally driven by economic reasons to implement energy efficiency strategies to combine the energy costs decrease with an increase of the company competitiveness. Beyond the strictly economic aspects, the impact of fossil fuels on the greenhouse effect has led to the implementation of carbon balance assessments on the scale of wineries or wine-growing regions, resulting in climate plans in which energy is an important component. This is an important issue in a context of climate change associated with earlier harvests and hot conditions, resulting in higher energy consumption (Rochard, 2017). In this context, the energy audit represents a powerful tool to evaluate energy consumption in the wine production in order to provide energy indicators and identify the energy saving actions (Parmenter, 2019).

The conventional concepts of sustainability, already adopted and highlighted, find a new paradigm: regenerative sustainability (Gibbons, 2020). The term *regenerative* associated with an economic activity dates back to the 80s when Francis and Harwood (1985) defended a new vision for agriculture, adopting farming practices with a holistic systems approach, promoting innovation for environmental, social, and economic well-being. More recently, due to climate change, this vision has been gaining attention again. Moreover, the circular economy aims to promote closing biological cycles, regenerating living systems and providing renewable resources for the economy. Transposing the concept of regenerative agriculture to the wine production sector, a *regenerative winery* is one that has practices that not only guarantee the efficient use of resources, but whenever possible, also regenerates the resources it uses recovering value from wastewater and waste streams. This approach can be applied more effectively to farm-winery systems that have been widely used in this value chain (Carroquino et al., 2020; Oliveira et al., 2019) as a model for design criteria, water management, or key sustainable indicators definition. In this concept, vineyard and wine production are considered as a single-stream grape system, where all the production is carried out on the farm. This model is specifically applied to small and medium dimension wineries.

This chapter reports cutting-edge science, technology, and practices that will inspire wineries to implement a regenerative approach. The chapter provides detailed information about water uses and management in wine production. Furthermore, it addresses potential environmental impacts and best available techniques to avoid/mitigate those impacts. It highlights technologies to treat winery wastewater and strategies to travel the road to regeneration. Case studies located in Mediterranean region are presented to help readers and researchers understanding the relevance of the water-energy nexus in a region under climate change threats.

11.2 Environmental impacts

Regarding the environmental impacts, several studies have been focusing on the three main wine production phases (viticulture, vinification, and bottling/packaging). The application of life cycle assessment (LCA) methodology to the wine sector, considering the whole life cycle or a part of it, allows the evaluation of wine production from an environmental point of view throughout a quantitative way (D'Ammaro et al., 2021; Ferrara & De Feo, 2018). LCA methodology is a powerful tool for both the identification of hotspots and the evaluation of the environmental performance of alternative processes or procedures (Buccino et al., 2019). This methodology is also relevant as a communication tool regarding health and safety, protection of biodiversity and territorial values enforcement (Luzzani et al., 2021). Regarding winery activities, bottling/packaging phase presents a high environmental impact mainly due to the glass bottle production processes (Amienyo et al., 2014). The vinification phase is responsible for fewer environmental impacts, when comparing to viticulture and bottling/packaging phases, being energy consumption its highest contributor (Vázquez-Rowe et al., 2012). Regarding the wastewater treatment process, the energy demand is considered the dominant component of carbon footprint (Rosso & Bolzonella, 2009), besides the possible significant impact that the discharge of the wastewater may have (Vázquez-Rowe et al., 2012).

Other important methodology when evaluating the environmental performance of one product is *water footprint*, once it corresponds to the volumetric measure of water consumption and pollution impact involved in the production of one unit of the product (Borsato et al., 2019; Hoekstra et al., 2011). This indicator is particularly important in the Mediterranean region, once the impact of climate change on water resources are not favorable to the wine sector (Carroquino et al., 2020; Saraiva et al., 2020). The treatment of the winery wastewater was therefore found to be a key factor to achieve sustainable wine production, due to the impact it can have on the environment (Saraiva et al., 2019). The potential environmental impacts caused by winery wastewater are mainly due to its high organic load, salinity and pH (Iannou et al., 2015) but, if adequate treatment is carried out, winery wastewater can, and should, be considered as a potential new source of water (Oliveira & Duarte, 2016).

11.3 Regenerative wineries
11.3.1 The water cycle

A strategic approach to water use efficiency in wine production must involve a holistic analysis, including not only the winery system but also landscape: vineyard or other crops, and respective water needs. This is particularly relevant for dry climate regions such as the South of Europe, Australia, and USA, where the amount of water used can vary between 88 and 400 L of water per litter of wine produced (Engel et al., 2015; Saraiva et al., 2020). In a regenerative perspective, the 5 Rs principles can be applied: Reduce the amounts used; Reuse water in a fit for purpose perspective; Recycle materials as nutrients or organic matter; Recover thermal energy, organic energy or hydraulic energy; Replenish the environment, discharging only what can be absorbed by the natural environment.

11.3.1.1 Water use efficiency

In the winery, the water use is mainly related to cleaning operations. However, each winery has different practices and therefore distinct water consumption profiles. Water consumption depends on the size of the winery, on the type of wine (red, white, or special wines) and related winemaking technology (Andreottola et al., 2009; GWRDC, 2011), but the water consumption patterns along the year are independent of winery's size (Oliveira & Duarte, 2016), where the peak water consumption occurs during vintage and first racking period. In turn, the produced wastewater flow is proportional to the vintage duration (GWRDC, 2011; Pirra, 2005). The usual combination of the referred periods can partly explain the large variation found in literature for water use in the wineries, 0.4−14 L water/L wine produced (Andreottola et al., 2009; Mosse et al., 2011; Oliveira et al., 2019) in different wine regions and countries.

An appropriate assessment should be carried out in order to identify hotspots and respective solutions to save water and decrease environmental impacts. Summarizing and evaluating collected data will create a diagram of all major uses of water, the location of all on-site water meters, the points where domestic, industrial wastewater enter the drainage system, identify hotspots and report gaps to prepare further assessment. Regular training is also important. By doing so, it will be possible to hierarchize winery operations in terms of water needs, contribution for wastewater flows and organic loads.

Simple and easy quantitative indicators are essential to diagnose purposes and to optimize wastewater treatment infrastructures. For example, the Italian "VIVA" certification framework, the government standard of sustainability, developed and in use in Italy, apply a water balance method from a LCA perspective that reflects a volumetric measure of water consumption, providing more precise recommendations for the optimal management of water use, requiring less data compared to the classical LCA-based approach, which investigates both the freshwater consumption and depletion using different impact indicators. This approach easily allows also the comparison of sustainability performance of different management strategies in both winery and vineyard (Borsato et al., 2019, 2020).

11.3.1.2 Wastewater treatment technologies toward water reuse

The wastewater produced during winery activities varies both in quality as in quantity throughout the year, due to the different winemaking phases (Bolzonella et al., 2019; Ioannou et al., 2015). The seasonality of the wastewater produced has been reported by several authors, with higher pollution load in vintage period (Bolzonella et al., 2010; Saraiva et al., 2019). But to select an appropriate winery wastewater treatment system (WWTS), the final destination and the legal requirements are also imperative to be considered (Bolzonella et al., 2019).

There are several different technological solutions available according to the dimension of the winery, infrastructure, accessibility to sewer and quality standards (Table 11.1).

Table 11.1 Treatment technologies for winery wastewaters according to winery size.

WWTS	Small	Small/medium	Medium	Large	References
Activated sludge	X	X	X	X	Ioannou et al. (2015), Oliveira and Duarte (2016)
Membrane bioreactor (MBR)		X	X	X	Bolzonella et al. (2010), Valderrama et al. (2012)
Up-flow anaerobic blanket (UASB)		X	X	X	Moletta (2005), Petropoulos et al. (2016)
Sequencing batch reactor (SBR)	X	X			López-Palau et al. (2012), Pirra (2005)
Aerated lagoons	X	X			Kerner et al. (2004), Masi et al. (2002), Montalvo et al. (2010)
Constructed wetlands (CWs)	X	X			Johnson and Mehrvar (2019), Rochard (2017), Skornia et al. (2020)
Integrative systems	X	X	X	X	Bolzonella et al. (2019), Solís et al. (2018), Tanzi and Mazzei (2009)

WWTS, winery wastewater treatment system; winery dimension: Small <2000 hL/year; Small/medium 2000–5000 hL/year; Medium 5000–10,000 hL/year; Large >10,000 hL/year (IVV, 2016; Oliveira et al., 2019); Integrative systems-combination of different treatment systems.

WWTS usually involves at least one physical treatment step, used to remove solids such as grape seeds, stalks and leaves. Other physical treatment widely used in Mediterranean region is evaporation (natural or forced). The use of evaporation lagoons has reduced investment and maintenance costs, whenever land is available (Masi et al., 2002; Mosse et al., 2011). Nevertheless, high-tech systems as membrane technologies, such as microfiltration, ultrafiltration (UF), electrodialysis and reverse osmosis, can also be applied to these WWTS for water reuse, disinfection, irrigation and biological processes (Sahar et al., 2011). In many cases, these technologies are also coupled with the recovery of value-added compounds like polyphenolic and polysaccharides (Giacobbo et al., 2013). Reverse osmosis is one of the most effective method for water purification but requires pretreatments (Ioannou et al., 2015). Physicochemical treatments can be used for WWTS or as a pretreatment in the conversion of recalcitrant pollutants into biodegradable compounds. Advanced oxidation processes, such as radiation photolysis and photocatalysis, sonolysis, electrochemical oxidation technologies, Fenton-based reactions, and ozone-based processes, are among the options (Khan et al., 2019; Lucas et al., 2010).

Biological treatments are largely used in WWTS, due to its high content of soluble and easily biodegradable organic matter. In fact, biological processes proved to be the most suitable treatment for winery wastewater (Bolzonella et al., 2010; Pirra et al., 2004). Aerobic treatment systems are normally used by their versatility, high efficiency, and ease of use (Mosse et al., 2011). There are several types of aerobic treatment systems, being aerated lagoons, activated sludge, sequencing batch reactors, membrane bioreactors (MBR) and aerobic biofilm reactors some of the most common, with reported COD removal varying from 91% to 98% (Andreottola et al., 2009; Bolzonella et al., 2019; Ioannou et al., 2015). Anaerobic treatment systems have the advantage of minimizing energetic costs through the nonexistence of energy consumption with aeration and the existence of biogas recovery/valorization. The anaerobic systems are also better adapted to winery wastewater than aerobic systems due to the low nitrogen and phosphorus content (Bolzonella et al., 2010).

Constructed wetlands are another biological treatment based on the principle of infiltration-percolation and biodegradation. Constructed wetlands can provide a low cost, low maintenance and low energy WWTS adapted to handling seasonal flows, with COD and total suspended solids removal efficiencies varying between 85% and 98% (Johnson & Mehrvar, 2019; Kerner & Rochard, 2004). This system is also applied to remove nitrogen (94%) and phosphorous (99%) (Skornia et al., 2020).

In wine producing countries, there are several small and medium wineries (IVV, 2016; Litaor et al., 2015), which are usually in the vicinity of the vineyards. In this case, irrigation could be a strategy for recycling nutrients and water in a winey system focused on the circular economy. The implementation of a new WWTS must consider the energy expenditure, the cost of its construction, maintenance, and qualified personnel needs. A robust and simple WWTS should therefore be the choice. Due to its ease of use, aerated treatment systems are normally implemented, although they are characterized by high energy requirements (Bolzonella et al., 2019). Nowadays, several wineries have changed old aeration systems by micro or ultrafine bubble aerators, with reductions in energetic consumption of up to 78% (Bolzonella et al., 2019; SWBC, 2018), with both economic and environmental gains. The use of a less intensive aerobic treatment followed by constructed wetlands can be an environmental and economical way to treat winery wastewater with lower energetic costs, while still maintaining the quality needed for its reuse. This WWTS may also have an important role in the ecological environment of the winery, allowing for a harmonious integration of the treatment system in an ecoenotourism perspective (Rochard, 2017). In recent years, several advances have been reported in the constructed wetlands use

(Johnson & Mehrvar, 2019; Skornia et al., 2020), including the direct treatment of winery wastewater by constructed wetlands, with recirculation of the wastewater through the constructed wetlands (until adequate quality is achieved) or the use of a highly absorbent material (zeolite support) also without the need for a previous aeration tank (Rochard, 2017). Anyway, if the winery is integrated in the zero-waste concept, the use of membrane technology for the recovery of value-added products can be an option, providing high quality water for reuse.

11.3.2 Strategies toward regenerative wineries

The wine production sector has been pursuing adaptation to climate change and adoption of sustainable practices and therefore has been implementing several strategies that contribute to systems regeneration. These strategies can target (i) the efficient use of resources and/or (ii) recovering value from waste.

11.3.2.1 Efficient use of resources (water and energy)

Water use efficiency has been a concern of wineries for a long time. Therefore, it is one of the key aspects included in the Californian, New Zealand and Chilean sustainable wine certification systems that have been used as benchmarking for wineries all over the world.

The first action toward minimizing water use in the winery is to adopt self-assessment, monitoring water use along winemaking activities to be able to identify hotspots for implementing best practices. Therefore, installing water meters should be the first step of the plan. If possible, water meters should be placed in different sites. There are several strategies to cut down water consumption, among which the most relevant are: (a) having a maintenance program to quickly identify and repair leaks; (b) attaching low volume/high pressure nozzles to the end of the hoses; (c) avoiding changing preset valve positions; (d) using block valves instead of control valves; (e) removing solid waste before water cleaning the equipment; (f) adopting pigging of pipes during transfer operations, leading to 50%–80% reduction of water consumption associated with transfer operations; (g) using clean-in-place; (h) using steam for cleaning/sanitation; (i) promoting continuous loops, reusing clean water before discharging; (j) screening out solids larger than 0.5–1.0 mm; (k) reducing contact time between solids and wastewater; using ozone as a disinfectant (CSWA, 2014; EPA, 2011; Navarro et al., 2017). Applying these strategies will also have an impact on wastewater quality. This aspect should be considered to avoid failure of the WWTS. On another hand, these measures can contribute for reducing energy consumption.

Understanding energy consumption throughout operations and identifying opportunities for reduction is essential. The winery should perform a self-energy assessment or hire an energy audit service, to identify the energy saving actions. Similarly to water, energy consumption is higher during vintage period. Due to cooling needs, the most energy-intensive processes are fermentation, cold stabilization and wine storage (Malvoni et al., 2017). Several strategies can be applied to reduce energy consumption and to adopt renewable energy sources, displacing the use of fossil fuels and therefore mitigating carbon emissions. The Australian Wine Research Institute (Nordestgaard et al., 2012) provides an overview of winery refrigeration and addresses potential improvement opportunities. According to Malvoni et al. (2017), insulation of storage tank and integration of solar cooling system can reduce annual average primary energy consumption for cooling by, respectively, 11%–21% and up to 41%. Solar technologies are the most common renewable energy sources implemented in wineries and can be used to supply process heat, to preheat boiler-heated water or to drive a chiller. Photovoltaic (PV) technology is used to generate electricity for winery self-consumption or to inject into the electrical grid. The main strategies toward energy efficiency in the winery can be summarized as (Malvoni et al., 2017; Nordestgaard et al., 2012): equipments' maintenance plan to ensure their efficient operation; night-harvest, as at night it is cooler and therefore grapes have reduced heat energy, avoiding potential need for refrigeration at the winery; insulation of storage; insulation of refrigeration supply piping; adoption of electrodialysis or contact stabilization techniques to reduce tartaric acid crystallization time, recovering the crystals by cross flow filtration; use heat exchangers to recover the lower temperatures of freshly stabilized wine to chill the next batch of wine; optimizing the chilled water system, by reducing the storage size to what is needed and setting the water temperature in accordance to what is required for the process; keeping the chiller tank and refrigeration plant shaded from direct sunlight, reducing radiant heat gain and therefore increasing energy efficiency; recover heat wasted from the refrigeration systems to preheat hot water; insulate hot water lines.

11.3.2.2 Recovering water and nutrients from wastewater

As winery wastewater production is related to cleaning operations, the monitoring plan mentioned for water use self-assessment will also allow to identify wastewater sources, flows and physical-chemical standards. The selection of WWTS (discussed in Section 11.3.1, Section 11.3.2) should take into account fluctuations of volumes and loads, allowing an efficient removal of contaminants during the peak season. Furthermore, treatment solution should be designed according to the new trends of circular bioeconomy, recovering value from wastewater. In this particular case, treatment strategy can encompass recovering phenolic compounds (although other winery waste streams have better recovery potential), polysaccharides, nutrients and water.

Recycling of treated wastewater for irrigation purposes is one of the most commonly applied options, recovering water and nutrients. Taking this in mind, the following strategies should be implemented in the winery: separating domestic wastewater collection from industrial flows (to avoid the need for disinfection treatment step); minimizing COD by screening out solids larger than 0.5–1.0 mm (with basket screens) and reducing the contact time between solids and wastewater, minimizing mass transfer; using alternative cleaning agents based on potassium hydroxide and magnesium hydroxide to reduce sodium adsorption ratio (SAR) (GWRDC, 2011; Mosse et al., 2011; Oliveira et al., 2019).

11.3.2.3 Production of value-added products from solid waste

As mentioned above, the circular bioeconomy approach should be applied to waste management/treatment strategies. In the case of solid waste streams produced in the winery, this can be achieved by implementing a cascade treatment system, generating value-added products and additional economic benefits.

Winery solid waste streams include grape stalks, grape marc and wine lees that have different physical-chemical characteristics and consequently diverse potential for value-added products recovery.

Grape stalks contain large amounts of proteins, hemicellulose, cellulose and lignin; grape marc contains water-soluble carbohydrates, protein, cellulose and pectin; wine lees' solid fractions contains cellulose, hemicellulose and lignin, whereas the liquid fraction is a major source of polyphenol compounds (Bassani et al., 2020; Kontogiannopoulos et al., 2016; Sirohi et al., 2020).

Grape marc can be used to recover phytochemicals, including several phenolic compounds, pigments, antioxidants, fatty acids, sugars and lignocellulosic compounds (Muhlack et al., 2018). Papadaki and Mantzouridou (2019) reported that grape marc could be mixed with olive processing wastewater and used for the production of citric acid using *Aspergillus niger*. More recently, Ferri et al. (2020) proposed the valorization of grape marc to recover phenolic compounds by a solvent-based technique and using the obtained extraction residue to incorporate in a biocomposite material.

Wine lees have been widely used to recover tartaric acid, for example, as reported by Kontogiannopoulos et al. (2016) using cation exchange resin with the reduction of undesirable potassium content. Moreover, the use of a powder derived from roasted grape seeds recovered form grape marc and suitable to be used as clarifying and fining agent in white winemaking has been proposed by Romanini et al. (2021). Alternatively to its use for value-added compounds production, winery solid wastes can be used as feed and for agricultural or environmental applications, such as soil remediation (Rivera et al., 2021).

11.3.2.4 Recovering bioenergy from solid waste streams

Another valorization route for winery solid waste is the production of biofuels, namely bioethanol and biogas (Ahmad et al., 2020; Muhlack et al., 2018).

Corbin et al. (2015) assessed different pretreatments of wine marc to promote lignocellulose hydrolysis, followed by enzymatic saccharification and fermentation. The authors have concluded that wine mark could produce up to 400 L of bioethanol per tonne of white marc.

Anaerobic digestion (AD) technology is another alternative to obtain bioenergy from winery solid waste. Several authors have studied AD of these wastes to produce biomethane, concluding that they show an interesting potential (Muhlack et al., 2018; Parralejo et al., 2019).

Cáceres et al. (2012) through a dynamic process simulation model established the operating conditions of an AD digester and a microturbine system, estimating that it is possible to recover about 94 MWh per 1000 t of grapes crushed, contributing to supply the energy demand for refrigeration during fermentation.

The possibility of decentralized production of renewable energy could largely contribute to winery sustainability, namely by mitigating GHG emissions associated with fossil energy.

11.4 Case studies

The winemaking is a very complex process. All the steps involve energy and water consumption, as well as wastewater generation. The main objective of this section is to present an analysis of water and energy consumption of four wineries, located in Portugal, France and Italy to define some strategies to reduce consumption and improve efficiency. The main results included in the case studies are also useful to establish reference indicators for the wineries.

11.4.1 Case study—Portugal

11.4.1.1 Case study characterization

Case study one (CS1) is located in the Baixo Alentejo, 15 km South of Beja, has 160 ha of vineyards. The farm-winery, with an area of 2000 m^2, processes 80% of red wine, 16% of white wine, and 4% of rosé wine (Table 11.2). The grapes are from its own vineyard, designed to adopt traditional techniques,

Table 11.2 Management practices toward a regenerative winery.

Country	Case study	Production (hL/year)	Type of wine (%) Red	White	Other[a]	Source of process water WC	PWS	RW	Wastewater collection system All	S	Wastewater treatment system PT + MS	CT AE	M	Key performance indicators (KPI'ss) Water rate (L/Lwine)	Water reuse (%)	Energy rate (kWh/hLwine)	RE (%)
Portugal	CS1-Y1	6750	80	16	4	X	X			X		X		1.4	100	8.0	15[b]
Portugal	CS1-Y2	7875	80	16	4	X	X			X		X		1.6	100	6.3	15[b]
France	CS2-Y1	7000	100	0	0				X		X			4.4	0	13	20
France	CS3-Y1	6600	0	100	0	X		X	X		X			1.8	0	10	20
Italy	CS4-Y1	27,650	57	42	1	X	X		X			X	X	4.0	0	3.4	22

AE, aerobic; CS1-Y1, case study 1, year 1; CS1-Y2, case study 1, year 2; CS2-Y1, case study 2, year 1; CS3-Y1, case study 3, year 1; CS4-Y1, case study 4, year 1; CT, completed treatment; M, membrane; MS, municipal sewer; PT, pretreated; PWS, public water supply; RE, renewable energy; RW, rain water; S, segregated; WC, water catchment.
[a]Rosé.
[b]Installed capacity in 2020.

integrating them, whenever possible, with innovative technology. The monitoring plan applied to this case study took place during (CS1-Y1) 2018−2019 and 2019−2020 (CS1-Y2).

The climate in the region of Beja has an average temperature of 14.5°C with maximum values of 40°C (July−August) and minimum values of 5°C or less (in January). July is the month with the highest number of sunshine hours (360 and 390 h), reaching between 80% and 85% of the maximum value of hours of sunshine. In contrast, December is the month with lowest number of sunshine hours. During November to March, since the greater cloudiness occurs a reduction of the sunshine hours to values around 50%. Also, the lowest values of relative humidity are in July and August. The weather station used to characterize CS1 is located in the center of Beja, about 17 km north of the winery.

11.4.1.2 Water use

Concerning the water use, the specific water consumption has been 1.4 L/L$_{Wine}$ to 1.6 L/L$_{Wine}$, with 30% to 35% of the total water consumption, taking place during the vintage period. In CS1, the water used is from the company own catchment and from the public water supply. The wastewater treatment plant, implemented in situ, consists of an AMBB (air microbubble bioreactor) with a chemical oxygen demand (COD) removal efficiency of 98%, achieving a treated effluent that fulfills the requirements for incorporation into the irrigation water system. However, this flow represents only 0.5% of the water needs in the vineyard.

11.4.1.3 Electrical energy consumption rate

In the last two years (2017−2018 and 2018−2019), an energy audit has been carried out to assess the performance of the winery. By analyzing the historical consumption and the state of the system/equipment to identify opportunities to improve energy efficiency, thus reducing the energy costs as well as the carbon footprint of the processes and product. This framework was applied to the physical limits of the winery, considering only the electricity consumed in the wine production process. For the evaluation of the winery's performance, the authors highlighted the following considerations: (a) data disaggregation to fulfill the objective of determining the energy consumption by type of grape processed and by wine production volume (hL/year), this approach allows the identification of the respective energy intensity of each type of wine contributing to implement energy efficiency measures; (b) an energy performance indicator (EPI) was defined as the specific electricity consumption, which is the ratio of energy consumption to the amount of product obtained (kWh/L$_{Wine}$) or grapes processed (kWh/ton$_{Grape}$). It was found that the most energy-intensive processes are, respectively: pressing and fermentation (53%); bottling, storage and delivery (19%); auxiliary processes (17%); stabilization and finning (7%); grape reception, destemming and crushing (4%). The type of wine with higher electricity consumption is red wine, representing 60% of the consumption, followed by white wine with 29% and rosé with 11%. However, in the case study, only 16% of the produced wine is white, but in terms of energy, it represents 29%, meaning that white wine is more energy consuming. The EPI indicator ranges from 8.0 to 6.3 kWh/hL, for the considered years. Nevertheless, the increase of global wine production in CS1-Y2, the total energy consumption decreased, shown by a 10% reduction of total energy and 21% reduction of EPI.

With the aim of achieving the reduction of energy and carbon footprint of wine production at CS1, in 2020, a PV solar system was implemented and connected to the national electricity grid. The incorporation of renewable energy in the mix considered for production reduces the energy intensity of the wine production. Attending to National Legislation, the production of electrical energy from renewable sources for customers with connection capacities above 5750 kW is called mini-generation. The installation implemented has a 20,240 kW power representing an investment of around 40,000 euros with an approximate payback of 6.7 years.

11.4.2 Case study—France
11.4.2.1 Cases studies characterization
Located in the Burgundy region, one of the most productive regions, case study 2 (CS2-Y1) produces 7000 hL of red wine per year and case study 3 (CS3-Y1) produces around 6600 hL of white wine per year (Table 11.2).

These wineries, only process the grapes from the own vineyard, having been designed to adopt traditional techniques, integrating them, whenever possible, with innovative technology. These case studies were monitored for two years and the results are the average values of the performance.

The annual values of sunshine in the Burgundy region vary between 1750 and 2250 h. In this temperate climate region, sunshine helps both in summer as in winter, limiting frost damage risks in winter and contributing to ripen grapes in summer (with 1300 sunshine hours between April and September). Mild temperatures in the summer, around 20°C, and cold winters, with 1.5°C average, are also characteristic of this region. The precipitation occurring mainly during May and June, with an annual average around 700 mm, are also ideal for vine growth.

11.4.2.2 Water usage
In France, the obligation to treat on-site effluent from most medium and large wineries and the rising cost of water are leading to a high level of consumption awareness. Water management approaches are generally associated with environmental management. Many wineries are ISO 14 001 certified and from a technological point of view, there has been an abandonment of the open cooling system and also the use of stainless-steel equipment which allows to reduce the volume of water in cleaning operations. The optimization of floor design and wastewater treatment is also an important strategy.

Concerning the water usage in CS2 and CS3, the average specific water consumption has been 4.4 L/L$_{Wine}$ and 1.8 L/L$_{Wine}$, respectively. These differences are related to the winemaking process, as the CS2 is a red wine cellar and the CS3 produces only white wine. It is important to note that studies developed in this region show that the average water consumption in red wine production is 6.4 L/L$_{Wine}$. However, these values are very variable and can be between 1.8 and 12.5 L/L$_{Wine}$. As far as the production of white wine is concerned, water consumption is significantly lower, with an average value of 2.3 L/L$_{Wine}$. Other wineries may present ratios between 0.4 and 5.0 L/L$_{Wine}$.

Regarding the wastewater, during the harvest period an equalization storage is required in order to limit the impact of the peak flow and the destabilization of the municipal biological treatment plant that receives this wastewater.

In medium and large wineries, the WWTS are mainly carried out by aerobic processes due to the ease management. However, in the last 10 years, a new approach has been taken with phytoremediation finishing treatments, mostly on beds planted with reeds. This principle now applies to most of the new plants. The reuse of treated wastewater remains at an experimental stage but is part of the strategy for new treatment plants.

11.4.2.3 Electrical energy consumption rate

The energies taken into account are electricity, natural gas, fuel oil, and propane. Electrical energy is used for lighting, operation of electrical equipment such as the press, sorting table, destemmer, pumping systems, etc. and, in some cases, for temperature regulation. Thermal energy is used for the thermal regulation of the winery, for the bottle storage room, or as an additional purpose in the sales room and the tasting cellar.

The analysis of energy consumption in these case studies was done considering both, the electrical and thermal energies (Table 11.2). In the CS2, all the energy is electrical energy, but in the CS3, the electrical energy stands for 60% of the total energy consumed. Although these two case studies do not show great variations in total energy consumption, this is not the case in the Burgundy region. Analysis of energy consumption in wineries producing white wine also shows variable consumption (10—160 kWh/hL), but considerably lower than in wineries producing red wine: the average amount of electrical energy consumed per hectoliter is around 30 kWh/hL and the thermal energy used per hectoliter is about 15 kWh/hL. As with electricity, there are very large variations regarding thermal energy: from the absence of heating systems to values above 280 kWh/hL. Concerning farms that produce only red wine, the average amount of electrical energy consumed per hectoliter is around 75 kWh/hL. However, there are large variations, as there are wineries consuming around 5 kWh/hL and wineries consuming more than 325 kWh/hL. The average amount of thermal energy used per hectoliter is around 68 kWh/hL. These presented case studies are among the most energy-efficient wineries in this region, operating without high thermal requirements for the winemaking phase and in a "bioclimatic" way, i.e., using the freshness or the "bioclimatic." The average value of energy consumption for red winemaking was therefore 143 and 45 kWh/hL for white winemaking. Although the harvest period is considered to be a high energy consumption phase, the results obtained did not allow for a segregated analysis by production phases.

Energy audits are widespread enough to limit losses and increase energy efficiency. In an ecodesign vision, the development of green roofs and the use of underground thermal inertia is in the forefront. These guidelines are subsidized at European level, transmitted by the organization France Agrimer for the creation or renovation of cellars, which integrates environmental aspects in the assessment process. Also, a government-funded solar installation program, with subsidized purchase of electricity by the grid operator, has contributed to the development of solar panels in agriculture and viticulture. As part of the restructuring of cooperatives, solar systems have been installed in old buildings to ensure their sustainability. In parallel with solar energy, some projects are being considered with the use of low enthalpy geothermal energy and the participation of wineries in wind energy programmes.

11.4.3 Case study—Italy
11.4.3.1 Case study characterization
Located in Emilia Romagna region, case study 4 (CS4-Y1) produces around 27,650 hL of wine per year (Table 11.2).

The winery from CS4-Y1 only processes the grapes from its own vineyards and conceives to adopt traditional techniques integrated with innovative technology. Thanks to a partnership with a big company dealing with wine market and export trading, the winery has introduced research and innovation across all grape processing and winemaking with a big qualitative improvement while maintaining compliance with authentic processes. This case study was monitored through one year.

11.4.3.2 Water usage

The results of specific water consumption had been 4.0 L/L$_{Wine}$ in CS4. These differences are related to the winemaking processes as the CS4 is a winery producing 57% as red and 42% as white wines mainly of still type. In CS4, the water was caught from public water supply system and the average consumption rate considering both red and white wines was 4.0 L/L$_{Wine}$. This value is low if compared to other case studies in the region as the winery is large in terms of wine production and thus allows the variable costs to be minimized.

In order to get an optimal decrease in microbial population in terms of yeast, molds and bacteria, the winery of CS4 performed some tests on various types of tanks where the use of a UVT6K lamp was compared with the standard cleaning/sanitizing procedure using peracetic acid, caustic soda and water. Even in presence of some tartrate residues, a good effectiveness was shown especially on the prebottling tanks (70% decrease in Unit Forming Colonies with UV technology against the 99% obtained with chemicals). Surprisingly, the total volume of water saved with UV sanitization was about 67% more than required with the standard procedure; also, the process and management costs for cleaning and sanitization were lowered by 63%.

The wastewater collection system has been greatly improved in recent years in the winery belonging to the CS4. The treatment plant is equipped with an MBR which can produce low-polluting wastewater. Although this plant was already sufficient to return purified wastewater to the environment, its residual polluting load was avoided with the installation of a final UF step. Trials done in the aim of a regional project in collaboration with DiSTAS Department of Università Cattolica del Sacro Cuore have shown that UF reduced the residual concentrations of COD/BOD and suspended solids by more than 95% on post-MBR wastewater in full agreement with literature reports (Ioannou et al., 2015; Sahar et al., 2011). When the impact of production peaks (e.g., during vintage time) on the performances of the UF system was verified, data showed that the efficiency of the UF plant was well maintained (Fig. 11.1). As pending the approval for water reuse in the winery or for irrigation purposes, the winery currently drains the treated wastewater into the municipal sewer.

11.4.3.3 Electrical energy consumption rate and renewable resources

The production of electrical energy from renewable sources in the CS4 winery enabled it to get 204,000 kWh/year from PV solar panels in respect to a consumption from national electricity grid of 939.000 kWh/year (22%). Such strong incorporation of renewable energy leads to a big reduction in the fossil energy needed for winemaking reaching an average value of only 0.034 kWh per liter of wine produced. This represents a large cost saving also considering that the cost of electricity to wineries in Italy is about 0.15 €/kWh.

11.4.4 General overview and future challenges

Water scarcity is becoming very serious in recent years. In fact, the search for new water sources is a constant challenge and needs to be adopted urgently. Also, the energy transition should be properly adapted to the regions and the country in order to achieve carbon neutrality targets.

Management practices and the CS performance presented toward regenerative wineries showed that water efficiency measures are already implemented, which is corroborated by the low water consumption ratios (Table 11.2). These case studies could be considered for benchmarking.

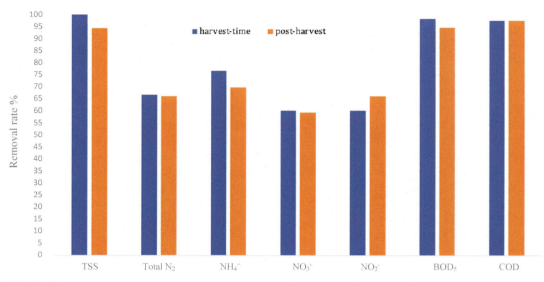

FIGURE 11.1

Reduction of pollutant load after ultrafiltration (UF) system downstream of the membrane bioreactor (MBR) purification plant. Analyses at harvest and postharvest time. *BOD₅*, biochemical oxygen demand; *COD*, Chemical oxygen demand; *TSSs*, total suspended solids.

The consumption indicators (water, energy) are related to the type of wine produced. The highest values of water consumption were detected in the red wine production, being within the lower values reported by the literature for these regions (Andreottola et al., 2009; Engel et al., 2015; Saraiva et al., 2020). Concerning WWTS the selection of treatment systems is closely related to the winery's sustainability strategy and wastewater's destination. While in Portugal the greatest concern is the production of treated wastewater for irrigation, in France, an ecological approach of phytoremediation is being made to consider water recycling. In Italy and derived from the size of the winery, the approach is more technological combining MBR with UF. Although only the Portuguese CS reuses 100% of the treated wastewater, the remaining CS intend to implement this measure soon. The climatic conditions of the different CS imply different energy needs in wine production. In Portugal (CS1), it is necessary to cool the white wine musts during the harvest, given the high temperatures in August, while in France, it is the fermentation of red wines that requires heating. The adopted renewable energy depends on the region, in countries with longer sunshine periods, there is a propensity to adopt PV solar panels, but in other regions, it may be more favorable to adopt wind energy.

Despite the fact that only four case studies were addressed, the methodology used in this study can be replicated in other regions, allowing a broader assessment. Similar studies, carried out in other regions, could be used as comparison with the Mediterranean countries performance. It would also be interesting to quantify the impacts of viticulture and winemaking guided by organic, natural, or biodynamic production criteria, to obtain objective indicators of sustainability. The scores obtained by the indicators could be used posteriorly to classify the wineries.

11.5 Conclusions

This chapter presented a systematic analysis regarding regenerative wineries feasibility, focusing on efficient water use, natural-based treatment solutions and energy intensity reduction. The key performance indicator water consumption varies according to the type of wine produced, 1.4–4.0 L/L$_{Wine}$, with the highest values for red wine production. Despite the variability between regions, it is important to highlight the commitment of wineries to reach the water/wastewater management goals, what can be explained by the water scarcity in the Mediterranean region, where WWTS tend to close the water cycle.

The study also focuses on how wineries use energy in regions with different edaphoclimatic conditions, in order to clearly identify the type of energy and associated costs. The EPI to evaluate the effectiveness of energy use in the winery ranged from 3 to 13 kWh/hL, also revealing a dependence on the type of wine.

The adoption of renewable energy in wineries is now a reality, but the adopted source depends on the region where it is located. The incorporation of renewable energy leads to a reduction in nonrenewable energy consumption during winemaking, as exemplified by the Italian case study.

References

Ahmad, B., Yadav, V., Yadav, A., Rahman, M. U., Yuan, W. Z., Li, Z., & Wang, X. (2020). Integrated biorefinery approach to valorize winery waste: A review from waste to energy perspectives. *Science of the Total Environment, 719*, 137315. https://doi.org/10.1016/j.scitotenv.2020.137315

Amienyo, D., Camilleri, C., & Azapagic, A. (2014). Environmental impacts of consumption of Australian red wine in the UK. *Journal of Clean Production, 72*, 110–119. https://doi.org/10.1016/j.jclepro.2014.02.044

Andreottola, G., Foladori, P., & Ziglio, G. (2009). Biological treatment of winery wastewater: An overview. *Water Science and Technology, 60*, 1117–1125.

Bassani, A., Alberici, N., Fiorentini, C., Giuberti, G., Dordoni, R., & Spigno, G. (2020). Hydrothermal treatment of grape skins for sugars, antioxidants and soluble fibers production. *Chemical Engineering Transactions, 79*, 127–132. https://doi.org/10.3303/CET2079022

Bolzonella, D., Fatone, F., Pavan, P., & Cecchi, F. (2010). Application of a membrane bioreactor for winery wastewater treatment. *Water Science and Technology, 62*, 2754–2759. https://doi.org/10.2166/wst.2010.645

Bolzonella, D., Papa, M., Da Ros, C., Muthukumar, L. A., & Rosso, D. (2019). Winery wastewater treatment: A critical overview of advanced biological processes. *Critical Reviews in Biotechnology*. https://doi.org/10.1080/07388551.2019.1573799

Borsato, E., Giubilato, E., Zabeo, A., Lamastra, L., Criscione, P., Tarolli, P., Marinello, F., & Pizzol, L. (2019). Comparison of water-focused life cycle assessment and water footprint assessment: The case of an Italian wine. *Science of the Total Environment, 666*, 1220–1231.

Borsato, E., Zucchinelli, M., D'Ammaro, D., Giubilato, E., Zabeo, A., Criscione, P., Pizzol, L., Cohen, Y., Tarolli, P., Lamastra, L., & Marinello, F. (2020). Use of multiple indicators to compare sustainability performance of organic vs conventional vineyard management. *Science of the Total Environment, 711*, 135081.

Buccino, C., Ferrara, C., Malvano, C., & De Feo, G. (2019). LCA of an ice cream cup of polyethylene coated paper: How does the choice of the end-of-life affect the results? *Environmental Technology, 40*(5), 584–593. https://doi.org/10.1080/09593330.2017.1397771

Cáceres, G. X., Cáceres, R. E., Hein, D., Molina, M. G., & Pia, J. M. (2012). Biogas production from grape pomace: Thermodynamic model of the process and dynamic model of the power. *International Journal of Hydrogen Energy, 37*(13), 1011–1017. https://doi.org/10.1016/j.ijhydene.2012.01.178

Carroquino, J., Garcia-Casarejos, N., & Gargallo, P. (2020). Classification of Spanish wineries according to their adoption of measures against climate change. *Journal of Cleaner Production*. https://doi.org/10.1016/j.jclepro.2019.118874

Corbin, K. R., Hsieh, Y. S. Y., Betts, N. S., Byrt, C. S., Henderson, M., Stork, J., DeBolt, S., Fincher, G. B., & Burton, R. A. (2015). Grape marc as a source of carbohydrates for bioethanol: Chemical composition, pretreatment and saccharification. *Bioresource Technology, 193*, 76–83. https://doi.org/10.1016/j.biortech.2015.06.030

CSWA – California Sustainable Winegrowing. (2014). *Sustainable water management handbook for small wineries*. San Francisco, CA: Alliance (CSWA).

D'Ammaro, D., Capri, E., Valentino, F., Grillo, S., Fiorini, E., & Lamastra, L. (2021). Benchmarking of carbon footprint data from the Italian wine sector: A comprehensive and extended analysis. *Science of the Total Environment*. Early accepted paper.

Engel, M., Hörnlein, T., Jacques, F., & Ohlsson, A. (2015). *Manual de produção mais limpa para adegas* (p. 43). Comissão Vitivinícola Regional Alentejana & International Institute for Industrial Environmental Economics.

EPA - Environment Protection Agency. (2011). *Lean & water toolkit*. Adobe Digital Editions. https://www.epa.gov/sites/production/files/2013-10/documents/lean-water-toolkit.pdf.

Ferrara, C., & De Feo, G. (2018). Life cycle assessment application to the wine sector: A critical review. *Sustainability, 10*, 395. https://doi.org/10.3390/su10020395

Ferri, M., Vannini, M., Ehrnell, M., Eliasson, L., Xanthakis, E., Monari, S., Sisti, L., Marchese, P., Celli, A., & Tassoni, A. (2020). From winery waste to bioactive compounds and new polymeric biocomposites: A contribution to the circular economy concept. *Journal of Advanced Research, 24*, 1–11. https://doi.org/10.1016/j.jare.2020.02.015

Francis, C. A., & Harwood, R. R. (1985). *Enough food. Special publ*. Emmaus, PA: Rodale Press and Regenerative Agriculture Association.

Giacobbo, A., Oliveira, M., Duarte, E., Mira, H., & Bernardes, A. M. (2013). Ultrafiltration based process for the recovery of polysaccharides and polyphenols from winery effluents. *Separation Science and Technology, 48*(3), 438–444.

Gibbons, L. V. (2020). Regenerative – The new sustainable? *Sustainability, 12*, 5483. https://doi.org/10.3390/su12135483

GWRDC. (2011). *Winery wastewater management and recycling – operational guidelines* (p. 79). Adelaide, SA: Grape and Wine Research and Development Corporation. Australia Government www.gwrdc.com.au/www.

Hoekstra, A. Y., Chapagain, A. K., Aldaya, M. M., & Mekonnen, M. M. (2011). *The water footprint assessment manual setting the global standard*. London: Earthscan.

Ioannou, L. A., Puma, G. Li, & Fatta-Kassinos, D. (2015). Treatment of winery wastewater by physicochemical, biological and advanced processes: A review. *Journal of Hazardous Materials, 286*, 343–368. https://doi.org/10.1016/j.jhazmat.2014.12.043

IVV. (2016). *Anuário de Vinhos e Aguardentes de Portugal 2016*. Lisboa: Instituto da Vinha e do Vinho.

Johnson, M. B., & Mehrvar, M. (2019). Winery wastewater management and treatment in the Niagara region of Ontario, Canada: A review and analysis of current regional practices and treatment performance. *The Canadian Journal of Chemical Engineering, 98*, 5–24. https://doi.org/10.1002/cjce.23657

Kerner, S., & Rochard, J. (2004). Winery wastewater treatment by constructed wetlands: Principles and prospects. In *Proceedings of the 3rd international specialised conference on sustainable viticulture and winery wastes management* (pp. 137–142). Barcelona: University of Barcelona.

Kerner, S., Sabatier, R., & Rochard, J. (2004). Impact environnemental de différentes techniques de filtration. In *Proceedings 3rd international specialized conference on sustainable viticulture and winery wastes management, Barcelona, Spain, May 24–26* (pp. 331–332).

Khan, S., Sayed, M., Sohail, M., Shah, L. A., & Raja, M. A. (2019). Advanced oxidation and reduction processes. *Advances in Water Purification Techniques*, 135–164.

Kontogiannopoulos, K., Patsios, S. I., & Karabelas, A. J. (2016). Tartaric acid recovery from winery lees using cation exchange resin: Optimization by response surface methodology. *Separation and Purification Technology, 165*. https://doi.org/10.1016/j.seppur.2016.03.040

Litaor, M. I., Meir-Dinar, N., Castro, B., Azaizeh, H., Rytwo, G., Levi, N., Levi, M., & Mar Chaim, U. (2015). Treatment of winery wastewater with aerated cells mobile system. *Environmental Nanotechnology, Monitoring and Management, 4*, 17–26.

López-Palau, S., Pinto, A., Basset, N., Dosta, J., & Mata-Álvarez, J. (2012). ORP slope and feast–famine strategy as the basis of the control of a granular sequencing batch reactor treating winery wastewater. *Biochemical Engineering Journal, 68*, 190–198. https://doi.org/10.1016/j.bej.2012.08.002

Lucas, M. S., Peres, J. A., & Puma, G. L. (2010). Treatment of winery wastewater by ozone-based advanced oxidation processes (O_3, O_3/UV and O_3/UV/H_2O_2) in a pilot-scale bubble column reactor and process economics. *Separation and Purification Technology, 72*(3), 235–241.

Luzzani, G., Lamastra, L., Valentino, F., & Capri, E. (2021). Development and implementation of a qualitative framework for the sustainable management of wine companies. *Science of the Total Environment, 759*, 143462.

Malvoni, M., Congedo, P. M., & Laforgia, D. (2017). Analysis of energy consumption: A case study of an Italian winery. *Energy Procedia, 126*, 227–233. https://doi.org/10.1016/j.egypro.2017.08.144

Masi, F., Conte, G., Martinussi, N., & Pucci, B. (2002). Winery high organic content wastewaters treated by constructed wetlands in Mediterranean climate. In *Proceedings of the 8th IWA international conference on wetland systems for water pollution control, Arusha, Tanzania* (Vol. 1, pp. 274–282).

Moletta, R. (2005). Winery and distillery wastewater treatment by anaerobic digestion. *Water Science and Technolog, 51*(1), 137–144.

Montalvo, S., Guerrero, L., Rivera, E., Borja, R., Chica, A., & Martín, A. (2010). Kinetic evaluation and performance of pilot-scale fed batch aerated lagoons treating winery wastewaters. *Bioresource Technology, 101*, 3452–3456.

Mosse, K. P. M., Patti, A. F., Christen, E. W., & Cavagnaro, T. R. (2011). Review: Winery wastewater quality and treatment options in Australia. *Australian Journal of Grape and Wine Research, 17*, 111–122. https://doi.org/10.1111/j.1755-0238.2011.00132.x

Muhlack, R. A., Potumarthi, R., & Jeffery, D. W. (2018). Sustainable wineries through waste valorisation: A review of grape marc utilisation for value-added products. *Waste Management, 72*, 99–118. https://doi.org/10.1016/j.wasman.2017.11.011

Navarro, A., Puig, R., Kılıç, E., Penavayre, S., & Fullana-i-Palmer, P. (2017). Eco-innovation and benchmarking of carbon footprint data for vineyards and wineries in Spain and France. *Journal of Cleaner Production, 142*(4), 1661–1671. https://doi.org/10.1016/j.jclepro.2016.11.124

Nordestgaard, S., Forsyth, K., Roget, W., & O'Brien, V. (2012). *Improving winery refrigeration efficiency* (p. 16). AWRI and GWRDC Edition.

Oliveira, M., Costa, J. M., Fragoso, R., & Duarte, E. (2019). Challenges for modern wine production in dry areas: Dedicated indicators to preview wastewater flows. *Water Science and Technology: Water Supply, 19*, 653–661. https://doi.org/10.2166/ws.2018.171

Oliveira, M., & Duarte, E. (2016). Integrated approach to winery waste: Waste generation and data consolidation. *Frontiers of Environmental Science and Engineering, 10*, 168–176. https://doi.org/10.1007/s11783-014-0693-6

Papadaki, E., & Mantzouridou, F. T. (2019). Citric acid production from the integration of Spanish style green olive processing wastewaters with white grape pomace by *Aspergillus niger*. *Bioresource Technology, 280*, 59–69.

Parmenter, D. (2019). *Key performance indicators: Developing, implementing, and using winning KPIs* (4th ed.). Hoboken, NJ: John Wiley & Sons.

Parralejo, A. I., Royano, L., González, J., & González, J. F. (2019). Small scale biogás production with animal excrement and agricultural residues. *Industrial Crops and Products, 131*, 307–314. https://doi.org/10.1016/j.indcrop.2019.01.059

Petropoulos, E., Cuff, G., Huete, E., Garcia, G., Wade, M., Spera, D., Aloisio, L., Rochard, J., Torres, A., & Weichgrebe, D. (2016). Investigating the feasibility and the limits of high rate anaerobic winery wastewater treatment using a hybrid-EGSB bio-reactor. *Process Safety and Environmental Protection, 102*, 107–118.

Pirra, A. (2005). *Caracterização e tratamento de efluentes vinícolas da Região Demarcada do Douro* (Ph.D. thesis). Vila Real, Portugal: UTAD.

Pirra, A., Arroja, L., & Capela, I. (2004). Winery effluents aerobic treatability in the Oporto wine region. In *Conference: New trends in farm buildings – international symposium of the CIGR, Évora, Portugal*.

Rivera, O. M. P., Leos, M. D. S., Solis, V. E., & Domínguez, J. M. (2021). Recent trends on the valorization of winemaking industry wastes. *Current Opinion in Green and Sustainable Chemistry, 27*, 100415. https://doi.org/10.1016/j.cogsc.2020.100415

Rochard, J. (2017). Processes of treatment of wineries effluents adapted to the organic wine sector: Current situation and prospects. In *40th world congress of vine and wine, Sofia, Bulgaria*.

Romanini, E., McRae, J. M., Bilogrevic, E., Colangelo, D., Gabrielli, M., & Lambri, M. (2021). Use of grape seeds to reduce haze formation in white wines. *Food Chemistry, 341*, 128250.

Rosso, D., & Bolzonella, D. (2009). Carbon footprint of aerobic biological treatment of winery wastewater. *Water Science and Technology, 60*, 1185–1189. https://doi.org/10.2166/wst.2009.556

Sahar, E., Ernst, M., Godehardt, M., Hein, A., Herr, J., Kazner, C., Melin, T., Cikurel, H., Aharoni, A., Messalem, R., Brenner, A., & Jekel, M. (2011). Comparison of two treatments for the removal of selected organic micropollutants and bulk organic matter: Conventional activated sludge followed by ultrafiltration versus membrane bioreactor. *Water Science and Technology, 63*(4), 733–740.

Saraiva, A., Presumido, P., Silvestre, J., Feliciano, M., Rodrigues, G., Silva, P. O.e., Damásio, M., Ribeiro, A., Ramôa, S., Ferreira, L., Gonçalves, A., Ferreira, A., Grifo, A., Paulo, A., Ribeiro, A. C., Oliveira, A., Dias, I., Mira, H., Amaral, A., Mamede, H., & Oliveira, M. (2020). Water footprint sustainability as a tool to address climate change in the wine sector: A methodological approach applied to a Portuguese case study. *Atmosphere, 11*, 934. https://doi.org/10.3390/atmos11090934

Saraiva, A., Rodrigues, G., Mamede, H., Silvestre, J., Dias, I., Feliciano, M., Oliveira e Silva, P., & Oliveira, M. (2019). The impact of the winery's wastewater treatment system on the winery water footprint. *Water Science and Technology, 80*, 1823–1831. https://doi.org/10.2166/wst.2019.432

Sirohi, R., Tarafdar, A., Singh, S., Negi, T., Gaur, V. K., Gnansounou, E., & Bharathiraja, B. (2020). Green processing and biotechnological potential of grape pomace. *Current Trends and Opportunities for Sustainable Biorefinery, 314*, 123771. https://doi.org/10.1016/j.biortech.2020.123771

Skornia, K., Safferman, S. I., Rodriguez-Gonzalez, L., & Ergas, S. J. (2020). Treatment of winery wastewater using bench-scale columns simulating vertical flow constructed wetlands with adsorption media. *Applied Sciences, 10*, 1063. https://doi.org/10.3390/app10031063

Solís, R. R., Rivas, F. J., Ferreira, C. L., Pirra, A., & Peres, J. A. (2018). Integrated aerobic biological-chemical treatment of winery wastewater diluted with urban wastewater. LED-based photocatalysis in the presence of monoperoxysulfate. *Journal of Environmental Science and Health: A – Toxic/Hazardous Substances and Environmental Engineering, 53*(2), 124–131. https://doi.org/10.1080/10934529.2017.1377584

SWBC – Sustainable Winegrowing British Columbia. (2018). *Winery process wastewater management handbook: Best practices and technologies*. https://www.bcwgc.org/sites/default/files/uploads/Wastewater%20Management%20-%20Final%20Digital.pdf.

Tanzi, G., & Mazzei, A. (2009). Combination of advanced aeration system and membrane bioreactor for winery wastewater treatment. In *Proceedings 5th international specialized conference on sustainable viticulture and winery wastes management, Trento-Verona, Italy*.

Valderrama, C., Ribera, G., Bahí, N., Rovira, M., Giménez, T., Nomen, R., Lluch, S., Yuste, M., & Martinez-Lladó, X. (2012). Winery wastewater treatment for water reuse purpose: Conventional activated sludge versus membrane bioreactor (MBR): A comparative case study. *Desalination, 306*, 1–7. https://doi.org/10.1016/j.desal.2012.08.016

Vázquez-Rowe, I. M., Villanueva-Rey, P., & Moreira, M. T. (2012). Environmental analysis of Ribeiro wine from a timeline perspective: Harvest year matters when reporting environmental impacts. *Journal of Environmental Management, 98*, 73–83. https://doi.org/10.1016/j.jenvman.2011.12.009

CHAPTER 12

Energy use and management in the winery

Matia Mainardis[1] and Rino Gubiani[2]

[1]*Department Polytechnic of Engineering and Architecture, University of Udine, Udine, Italy;* [2]*Department of Agricultural, Food, Environmental and Animal Sciences, University of Udine, Udine, Italy*

12.1 Introduction

Worldwide, the wine sector produces about 266 millions of hL of product per year; among the most important producers, Italy accounts for about 44.1 million of hL of wine, being the second most largest producer after France (Istat, Annual Italian Statistics on Wine Production, 2019). Electricity represents more than 90% of the overall energy request in wineries (Vela et al., 2017), while thermal energy accounts for the residual 10% (as a mean value). In the European Union (EU), electricity consumption for the whole sector was estimated as 1750 million of kWh/y (France: 500 million of kWh/y, Italy: 500 million of kWh/y, Spain: 400 million of kWh/y, Portugal: 75 million of kWh/y) (Vela et al., 2017). In California, the electricity consumption of 1100 wineries was reported to be as high as 400 GWh/y, with prevalent utilization of fossil fuels such as natural gas, liquefied petroleum gas (LPG), and propane (Jia et al., 2018). In a more general perspective, energy consumption in the food sector accounted for about 17% of EU gross energy consumption in 2013 (Catrini et al., 2020); thus, energy consumption optimization appears fundamental to improve wineries' efficiency.

The agro-food sector is aiming to improve its overall environmental sustainability, as strongly recommended by recent EU acts and directives (including the European Green Deal [EGD]), and following sustainable development goals (SDGs) promoted by United Nations Organization (UNO). The wine sector, in order to thoroughly address the energy and environmental aspects embedded in the production chain, should consider both the operations in the vineyard and those executed in the winery. As an example, different management practices in the vineyard are characterized by diverse efficiency, cost, and biological effects (Mainardis, Boscutti, et al., 2020; Mainardis, Buttazzoni, & Goi, 2020). The energy production profile, i.e., the provision of energy request throughout time, is as relevant as the energy embedded in the inputs required for wine production system (such as chemicals).

The European Directives stimulate the implementation of energy consumption assessment measures in residential and industrial sectors to identify the issues and the potentialities to improve energy efficiency (Malvoni et al., 2017). In order to ameliorate energy management, the wineries must be equipped with meters to evaluate electricity consumption profile in each process phase (Gubiani et al., 2019). A number of projects and methodologies have been recently developed throughout the world to analyze the most important aspects related to sustainability in wineries, including carbon footprint, air and water pollutant emissions, energy consumption, life cycle assessment, and water usage (Gubiani et al., 2019). Water management and sustainability aspects are specifically discussed in other chapters of the present book.

The significant energy demand in wineries (4–16 kWh/hL) (Malvoni et al., 2017) pushes for the adoption of low-carbon and more sustainable energy models, mostly based on renewable energy sources (RESs) (Campos et al., 2019). A sustainable viticulture strategy, in fact, should strongly encourage RES, aiming to a greener wine production process, able to lower carbon emissions and energy costs. The local production and self-consumption of energy, commonly known as "prosumerism," should be widely incentivized, considering the large economic, social, and ecologic RES benefits (Campos et al., 2019). Wineries can largely profit from RES implementation, because much of the waste generated in the wine production process, being of organic source, can be fully recycled (Gómez-Lorente et al., 2017). Among the wide RES category, anaerobic digestion (AD) process is one of the most widely applied technologies worldwide for production of valuable biogas from waste and wastewater streams (Mainardis, Boscutti, et al., 2020; Mainardis, Buttazzoni, & Goi, 2020). Biogas can either be used to generate electricity and heat in combined heat and power (CHP) units, or even upgraded to biomethane, that can fully substitute fossil-derived natural gas.

Considering this general framework, the present chapter addresses the broad topic of energy consumption and management in the winery, considering the most recent literature outcomes and presenting the results of meaningful research projects in the field. In Section 12.2, energy audits are briefly presented as a starting point to improve energy management in wineries. In Section 12.3, the energy consumption in a "standard" winery is analyzed and split into fundamental operations, while in Section 12.4 the methodologies currently available on the market to reduce energy consumption in the wine production chain are presented. In Section 12.5, renewable energy production and utilization in the sector is discussed, given the outstanding importance of increasing clean energy exploitation to substitute fossil fuels and reduce greenhouse gases (GHGs) emissions. In Section 12.6, some examples on energy assessment and optimization in wineries are proposed and discussed, with useful insights for stakeholders working in the field. The chapter is finally completed with a brief conclusive section, forecasting future perspectives for an efficient energy production and utilization in the wine sector, with improved sustainability and adoption of circular economy models.

12.2 Energy audit in wineries

An energy audit is a systematic and standardized procedure that allows to determine inefficienies of plants and buildings, suggesting at the same time technical solutions able to reduce (or even remove) the related issues. The EN 16247 is the European legislative standard that provides the requirements and describes the methodologies for conducting high-quality energy audits, with definition of obtainable outputs (Malvoni et al., 2017). The energy audits are applicable to all types of processes and sectors as well as to different energy sources. In addition, the energy audits generate as outputs useful energy indicators, that connect total energy consumption in the studied processes with company production (Malvoni et al., 2017). Conducting these audits allows as well comparing the performances of a selected plant with other relevant plants (benchmarking). Energy audits are fundamental to assess the starting point in terms of energy consumption in a selected industry (in this specific case study, a winery), and are structured in a series of successive phases, substantially including:

1. Acquisition of technical information and data organization;
2. Acquisition of energy consumption data related to the previous years and information gathering regarding the way the spaces are divided inside the plant;
3. Technical inspections aimed at collecting missing information and execution of specific campaigns to evaluate how devices and buildings are employed in the current operations mode;
4. Definition of a basic energy balance using a static or a dynamic calculation software;
5. Definition of retrofit or upgrading interventions, such as building insulation and plant improvements, RES utilization, including also management optimization;
6. Techno-economic evaluation of the proposed solutions, with analysis of different scenarios having a progressively increasing economic weight, simulating the outputs, and discussing the results with company owners, to select the best tailored solution;
7. Drafting the technical diagnosis report.

The energy audits identify customized measures that require mathematical analysis and calculations to determine the specific potential energy savings, unique to customer's situation (Casentini & Rushforth, 2008). If appropriate, additional resource saving potentials (e.g., water efficiency, renewable energy, and demand response) can be calculated as well throughout the energy audit. Financial considerations are coupled with technical diagnosis, including commonly analyzed parameters such as pay-back time (PBT), net present value (NPV), and internal rate of return (IRR). The final report format should highlight the key points for decision-makers, providing a list of prioritized recommendations (Casentini & Rushforth, 2008).

The solutions that are commonly applied in wineries to increase energy efficiency include building insulated walls and tanks, with reduction of heat losses, or specific improvements more dedicated to the process, such as substitution of inefficient electro-mechanical devices and installation of meters to improve the knowledge about the most energy-intensive operations.

12.3 Energy consumption in the winery

In this section, considering that Italy is one of the largest wine-producing countries worldwide, the Italian case study will be considered as a relevant example to analyze the energy consumption throughout the different processing phases. In a traditional perspective, the energy needed in wineries is mainly obtained in three ways: (i) electricity withdrawal from the national grid, (ii) utilization of on-site diesel generators, and (iii) diesel fuel consumption for agricultural mechanical machinery and mobility (Carroquino et al., 2017).

As previously introduced, electricity is the main energy source needed in the processes (>90%) and is necessary for pneumatic presses, refrigeration units, bottling machines, and conditioning plants. In Italy, the overall electric energy consumption for the wine production sector is about 500 million kWh/y; consequently, wineries are considered as energy-intensive processes (Gubiani et al., 2019).

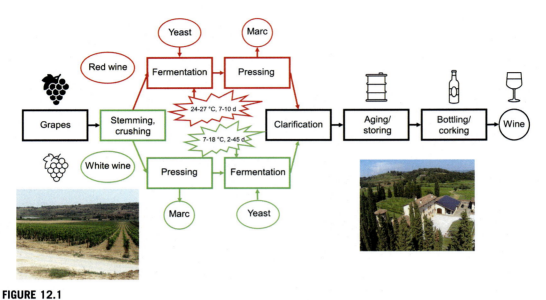

FIGURE 12.1

Simplified process scheme commonly applied in wineries.

Modified from Jia, T., Dai, Y., & Wang, R. (2018). Refining energy sources in winemaking industry by using solar energy as alternatives for fossil fuels: A review and perspective. Renewable and Sustainable Energy Reviews, 88, *278–296. https://doi.org/10.1016/j.rser.2018.02.008.*

A simplified scheme of traditional operations normally executed in wineries is reported in Fig. 12.1, highlighting the basic differences between red and white wine production process. In the following, the electricity consumption throughout the different process phases will be deepened.

Wine production in a generic winery starts with receiving grapes from the vineyard and successive juice extraction. Grape quality is typically assessed through sugar content and acidity analysis. In this phase, most of the energy consumption is due to electric engines connected to different mechanical devices such as augers, air compressors, and destemmers. In the successive fermentation step, a significant amount of electricity is needed by pumps (used for material handling), but also to keep the optimum temperature required by fermentative microorganisms. It should be reminded that fermentation is an exothermic process, so fermentation tanks must be cooled down to maintain the desired temperature. Cooling systems that employ food glycol are typically employed to this purpose.

The pressing section requires electricity as well: in medium to high wineries, marc pressing is typically conducted by means of pneumatic presses. The pressing phase is performed in different times, depending on the applied vinification method: in white wine vinification, pressing is applied just after receiving the raw material, while in red wine vinification, the pressing phase is carried out after alcoholic fermentation, to allow a good color and polyphenol extraction from grape skins.

The final operations that are executed to obtain the final product are clarification, stabilization, bottling, and conservation. Wine cooling under 0°C for two weeks allows physical sedimentation of colloids and tartrate crystals. Wine must be kept at stable humidity and temperature levels: often, air-conditioning systems or steam generators are used to maintain optimal environmental conditions, unless the winery is positioned in an extremely favorable natural environment.

The auxiliary processes, finally, include the utilization of a number of electrical equipment's that are not strictly included in wine production chain, but are necessary for smooth operations, such as computers and controllers, as well as accumulators for charging forklift trucks.

Considering a simplified energy balance in a "standard" winery, calculated as mean values of different wineries included in the TESLA (Transferring Energy Save Laid on Agroindustry, https://teslaproject.chil.me/) research project, total electric energy consumption can be split as follows (Fig. 12.2).

In literature, mean total electricity consumption in wineries is typically expressed in a standard manner as kWh/L wine: recent studies report a wide consumption range (0.21–2.14 kWh/L), mostly depending on local climatic conditions (Smyth & Nesbitt, 2014). As an example, wineries located in California and Australia are characterized by a significantly higher electricity consumption than wineries positioned in the United Kingdom. In Italy, mean electricity consumption in wineries is in the range of 0.2–0.35 kWh/L, lower than other reported literature values (Gubiani et al., 2019). Even lower specific electricity consumptions (0.11 kWh/L) were claimed in the TESLA project results, together with a moderate thermal energy request (0.01 kWh/L) (Catrini et al., 2020). Small and medium scale wineries (wine production below 50,000 hL/y), which produce high-quality wines, typically report a higher specific electricity consumption, given the less advanced technological devices that are installed in comparison with larger scale plants.

Thermal energy, on the other hand, accounts for less than 10% of total energy requirement, and is mostly needed for thermal conditioning of the premises and product inside the tanks, as well as for bottling phase and auxiliary processes (e.g., hot water and steam production for sterilization and cleaning). Consequently, in most of the wineries, installed thermal power is significantly lower than electric power. In Southern Europe, the combination of air heat pumps, covering peak cooling loads, and liquified petroleum gas (LPG)-fired boilers, used for winter heating, is the most common utilized solution for thermal energy production in wineries located in the countryside (i.e., off-grid) (Tinti et al., 2017).

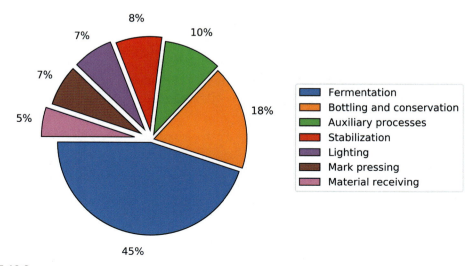

FIGURE 12.2

Simplified electric energy balance of the different operations in a "standard" winery.

It could be concluded that total energy consumption in a winery strongly depends on its geographic position, which affects local climate conditions. Thermal energy need, despite being lower than electricity request, shows higher seasonal fluctuations. Thermal regulation expenses, aimed at keeping a controlled and constant temperature throughout alcoholic fermentation, stabilization and maceration processes, are directly connected to local climate. Cold energy request to cool the premises is obviously higher in temperate or tropical areas, when compared to cold climates. When projecting a winery, the peculiar wine production chain (i.e., red or white wine production) must be considered together with actual environmental conditions, as each processing step is characterized by specific needs.

12.4 Methodologies for reduction of energy demand

This paragraph presents some basic considerations on the applicable methodologies for energy efficiency improvement in wineries, referring to recent literature studies and to authors' experience in the field. Electricity is mostly analyzed in this section, while the main investigated process is bottled wine production.

Generally, as previously introduced, to improve energy efficiency in daily operations, an energy audit (Section 12.3) should be applied, considering the following points: (i) reduction of energy usage in unnecessary operations; (ii) optimization of unavoidable energy usage through careful monitoring of energy demand and by selecting efficient electro-mechanical devices (pumps, tanks). Considering that electric energy is the most utilized energy form in wineries, it is fundamental to develop energy audits specifically tailored to electricity. The combination of detailed process measurements and a correct energy management enables to increase energy efficiency of the analyzed plant or facility.

It is important to verify long-term installation performances and devices' efficiency, constantly evaluating the improvements and questioning deviations from predicted behaviors. Benchmarking can be useful to compare the overall energy consumption of a selected winery with facilities having a similar scale (Celorrio et al., 2017). This is particularly true for small-scale plants, where limited energy meters are often installed: an effort to move towards a better comprehension of existing inefficiencies is the first step to improve the overall energy performances.

Cooling has been shown to be one of the most intensive energy operations in wineries; potential actions that can reduce the energy required for cooling include, among others, thermal insulation of storage tanks and installation of solar cooling systems (Malvoni et al., 2017). In wine processing chain, cooling is mostly needed in five phases:

1. At raw material receiving, to cool the grapes to the desired processing temperature;
2. During alcoholic fermentation, where in absence of refrigeration, exothermal reactions (converting sugars to alcohols) are compromised, with significant loss of quality in the final product;
3. During wine tartaric stabilization, that is a necessary operation to avoid cloudiness in the bottle or to prevent deposit formation on the bottom of commercial containers;
4. During the aging process in tanks (required temperature of 15–18°C) and in barriques (same required temperature, but need to control also humidity to 75%–85%);
5. During bottling phase, where the temperature of sparkling wines must be near 0°C.

Cold stabilization is one of the most widely applied processes in wineries, and thus its optimization is fundamental to improve overall energy efficiency (Celorrio et al., 2017). Cold pre-fermentation has grown in popularity in recent years, as a technique able to improve electric energy balance in wineries by extracting phenolic compounds at low temperature (3–10°C for 3–8 d). Furthermore, initial grape cooling is energy-intensive, considering the important temperature difference that must be covered in a short time. As previously stated, the optimization of this process phase is fundamental to avoid product deterioration, such as oxidation reactions, and delay the beginning of unwanted fermentation reactions (Celorrio et al., 2015).

A recent study compared different methods for wine stabilization in a winery located in Northern Spain, with the aim of increasing process efficiency, reducing at the same time CO_2 emissions, thanks to photovoltaic solar energy installation (Celorrio et al., 2017). A preliminary monitoring campaign was carried out, measuring product and winery temperature together with wine characteristics throughout the year, and analyzing different applicable energy solutions. The most efficient techniques to reduce electricity consumption were shown to be thermal insulation of tanks and installation of a heat exchanger to cool the input material. It was demonstrated that environmental temperature played a major role in lowering energy consumption: thermally insulated underground rooms allowed to reduce in a substantial way energy request. When a longer wine stabilization was required (>10 d), the installation of thermally insulated tanks was particularly favorable, with energy savings of about 16%. In addition, cooling the inlet raw material could lead to meaningful electricity savings (in the range of 30%–55%), depending on tank characteristics, treatment duration, and winery temperature (Celorrio et al., 2017).

The actual energy saving achievable using efficient cooling technologies is often estimated using simplified approaches that do not consider real winery operating conditions as well as daily and seasonal variations (Catrini et al., 2020). A novel bottom-up procedure was recently proposed in literature to develop reliable refrigeration and air-conditioning load profiles: by applying this approach to a medium-scale winery, it was shown that the use of standard indicators could lead to a significant overestimation of potential energy savings, sustaining the importance of applying customized seasonal indicators, based on actual dynamic conditions (Catrini et al., 2020).

Furthermore, by studying energy performances in new wineries planning, it was found that underground buildings allow to reduce both energy demand for heating (−14%) and, mostly, that required for cooling (−50%), if compared to aboveground solutions (Benni et al., 2013). A particular benefit in energy demand is observed in wine aging, when executed underground. It was seen that underground buildings and partially underground constructions with sun-shading devices, despite increasing capital costs (in the range of 12%–27%), allow to reduce energy consumption up to 63% when compared to traditional aboveground buildings (Benni et al., 2013). It has to be pointed out, anyway, that since the 90's, winery planning underwent deep changes, focusing on a new conception, more open to customers (and in general to stakeholders) and energy efficiency-devoted (Barbaresi et al., 2017). A significant improvement of energy efficiency in wineries can almost undo GHG emissions of winemaking process.

Beside adopting innovative solutions for energy saving, also age and efficiency of installed electromechanical devices should considered, when evaluating total electricity consumption in a winery. Thus, maintenance plans must be adopted in each winery, substituting old and inefficient devices. Moreover, electricity price has a defined price pattern on 24-h basis, with peak values typically observed in the morning (around 9 a.m.) and in the late afternoon (around 7 p.m.) (Cottes et al., 2020).

Considering that the peak factor (i.e., ratio between maximum and mean price) is high, if electricity consumption is consistent when energy price is low, a significant saving in total energy price can be achieved at the same energy consumption conditions. Consequently, shifting energy-intensive operations from peak hours to off-peak periods, following the so-called demand response principle, is another possibility to reduce energy costs.

Finally, energy provision is fundamental, considering that the energy sources are characterized by a different carbon footprint. However, it should be considered as well that small and medium wineries, where often there is a lack of design criteria specifically addressed to energy efficiency, account for a significant share of total enterprises in the Mediterranean area (Barbaresi et al., 2017), and thus a particular effort should be devoted to small-scale plants.

12.5 Renewable energy utilization

RESs (including wind, solar, and hydroelectric power, ocean and geothermal energy, biomass, and biofuels) are alternative sources to traditional fossil fuels, that contribute to reduce GHG emissions, diversifying energy supply and lowering the dependence on volatile fossil fuel markets (particularly oil and gas). EU legislation in the field of RES promotion significantly evolved in the last years: in 2018, the target of energy consumption share from RES (to be reached by 2030) was fixed at 32%. In 2019, EU proposed the EGD, a set of 50 actions across all sectors to be performed in the upcoming five years to help EU economy to reach climate neutrality by 2050. More specifically, regarding renewable energy generation, the urgent need for power system flexibility (i.e., RES interconnection, smart grids, demand response, and energy storage) is expected to further accelerate, also considering that at present, this task is still underdeveloped. SDG promoted by UNO specifically address the energy topic in SDG7, related to affordable and clean energy. In the near future, national incentives for RES should conform to the proposed European guidelines.

Local electricity production from RES has a number of benefits, including economic advantages due to avoided electricity withdrawal from the grid and to local incentives for renewable energy production. Moreover, the environmental balance is undoubtedly positive, because electricity grid losses are reduced, and CO_2 emissions are abated. A recent study carried out in Portugal highlighted that the high capital cost of photovoltaic systems and storage technologies (batteries) is one of the main constraints for a further renewable energy implementation in the region (Campos et al., 2019). Moreover, the presence or absence of additional support schemes, such as feed-in-tariffs, play an important role in evaluating the economic viability of the investment. The main advantages in installing photovoltaic systems consist in the increased productivity, with progressive capital cost reduction (Campos et al., 2019).

12.5.1 Anaerobic digestion

Both red and white wine production results in the generation of high amounts of solid organic waste, including grape marc (pomace) and stalks (Muhlack et al., 2018). Considering a simplified mass balance, 10%–30% of grapes total mass is converted to grape marc, containing unfermented sugars, polyphenols, alcohols, and pigments. Despite their high inner potential, at present advanced technologies are not widely implemented in wineries to valorize these organic subproducts. Traditional approaches commonly involve composting (i.e., aerobic stabilization of an organic residue, with production of a high-value fertilizer for agricultural land application (Pinter et al., 2019)) and utilization as animal feed.

Among the thermochemical processes, AD is a mature technology that converts organic waste in an oxygen-free environment into a gaseous product (biogas) with high calorific value (Mainardis, Flaibani, Trigatti, & Goi, 2019). AD has been widely applied to produce renewable energy in biogas form from a variety of solid and liquid organic substrates, both using traditional digesters and innovative high-rate anaerobic systems, such as up-flow anaerobic sludge blanket (UASB) reactor (Mainardis, Boscutti, et al., 2020; Mainardis, Buttazzoni, & Goi, 2020). The residues from food and beverage production are typically characterized by a substantial biodegradability, that translates into a consistent methane production potential, boosting for full-scale AD implementation (Mainardis, Flaibani, Mazzolini, et al., 2019).

Energy valorization of organic residues from wine production chain is attracting significant interest in the scientific literature: recently, it was shown that grapes have a higher methane yield in AD process, if compared to other organic wastes (Guerini Filho et al., 2018). A meaningful study (Da Ros et al., 2017) investigated the organic residues produced by a winery with a production of 300,000 hL wine per year: considering a mass balance, per each hL of wine, 196 L of wastewater, 0.1 kg of activated sludge, and 1.6 kg of wine lees were generated. Co-digestion (i.e., simultaneous treatment of complementary substrates, aimed at stabilizing operating conditions and enhancing energy yield (Mainardis, Boscutti, et al., 2020; Mainardis, Buttazzoni, & Goi, 2020)) of wine lees and waste activated sludge, originated from wastewater treatment, was proposed at pilot-scale, highlighting a stable biogas production at mesophilic conditions (37°C), with biogas yield of 0.386 m^3/kg fed chemical oxygen demand (COD$_{fed}$) (Da Ros et al., 2017).

In order to evaluate the anaerobic biodegradability of the substrates, typically biochemical methane potential (BMP) tests are conducted at laboratory scale: from these preliminary assays, it is possible to investigate not only total methane production from a given substrate, but also degradation kinetics and effectiveness of different pretreatments (including physical, chemical, and biological ones) to enhance the energy yield (Mainardis, Flaibani, Mazzolini, et al., 2019). It was recently demonstrated that, among the different produced organic wastes in wineries, grape must, the mixture of all produced biomass and bagasse are the substrates able to produce the highest methane yield, respectively up to 1,151, 289, and 199 m^3 biogas/ton volatile solids (VS) (Guerini Filho et al., 2018). Nonetheless, a lack of standardization in the reported biogas yields emerges from the analysis of existing literature, also due to the heterogeneity of the different organic substrates produced in wineries, so an effort toward standardization is needed to allow an easier benchmarking between literature results.

12.5.2 Thermochemical conversion processes

Considering a wider biorefinery approach, sustained by EU acts, innovative technologies allow nowadays both the extraction of useful chemicals and the production of bioenergy through thermochemical conversion processes (such as hydrothermal carbonization-HTC, combustion, gasification, and pyrolysis) and biological treatment (alcoholic fermentation, bioethanol, and biogas production) (Muhlack et al., 2018). Beside traditional approaches, the biorefinery perspective, specifically applied to wineries, potentially allows to extract from the produced organic waste a pool of high-value secondary products, such as phenolic compounds, flavonoids, tartaric acids, lignocellulosic substances, hydroxybenzoic acids, and hydroxycinnamic acids (Ahmad et al., 2020).

Alternatively to direct combustion, the wood biomass from pruning operations in vineyards can be used to replace pine pellets used in boilers, with thorough exploitation of circular economy principles: in the work by (Fernández-Puratich et al., 2015), the calorific value and the ash content of chip pruning residues from five different wineries were analyzed to this purpose. Nonetheless, winery subproducts (marcs, stalks, peels, and grapeseeds) are commonly characterized by an extreme heterogeneity that requires a complex physicochemical characterization (including moisture content, ash, lower heating value, and volatile fraction) to assess their gross energy content before energy recovery (Toscano et al., 2013). The potential for full-scale energy recovery from these streams was assessed in literature by analyzing the peculiar characteristics of each fraction, estimating a gross energy content of approximately 19 GJ from 1 hectare of grapevine (Toscano et al., 2013).

Another meaningful study investigated the energy valorization of pruning residues to produce thermal energy, supplying winery operation requirements (Bacenetti, 2019). A combination of geothermal energy and heat pumps was instead proposed in Tinti et al. (2017) to reduce GHG emissions in the winemaking sector by applying a tailored mathematical simulation program. From a more general perspective, on-site valorization of produced organic residues from wineries, through biological or thermochemical processes, can be beneficial to reduce global energy need, both in terms of electricity and heat, as well as to improve the overall process sustainability, with exploitation of circular economy principle and an indirect positive outcome also for customers and stakeholders.

12.5.3 Solar systems

One of the strategies most commonly applied in wineries to reduce the economic burdens due to electric energy consumption, as previously introduced, is the implementation of solar systems, that can be installed in large ground or roof areas (Catrini et al., 2020). A solar winery is defined as a winery where its building or processes directly or indirectly utilize solar energy, gathered from solar radiation (Jia et al., 2018). The energy from the sun can be exploited as solar thermal energy, which includes solar water heaters, solar cookers, or solar driers (ELkhadraoui et al., 2015), but also as photovoltaic energy, that utilizes dedicated cells to transform solar irradiation into electricity, thanks to the photovoltaic effect. The electrical efficiency of photovoltaic modules depends on module construction and climatic parameters, such as solar irradiance, packing factor, and temperature (Kumar & Rosen, 2011). The integration of photovoltaic modules into existing architectural elements (winery roofs and carports) brings additional benefits when compared to ground-mounted panels, including avoided need for land-use or land-cover change (Hernandez et al., 2014).

In wineries, as previously discussed, the prevalence of electricity request, rather than thermal energy, boosts for photovoltaic modules installation. However, the mismatch between energy request profile and solar source availability throughout the seasons should be considered (Jia et al., 2018). A noticeable research highlighted that photovoltaic systems implementation in Spanish wineries with grid connection could reduce electrical power costs by 4%–36%, with renewable energy generation up to 57% (Gómez-Lorente et al., 2017). More complex systems may be adopted to thoroughly eliminate both power generation from diesel oil and the need for electricity withdrawal from the grid: a recent study proposed electricity and hydrogen production from solar energy, supplying the produced energy also to winery wastewater treatment plant and to vineyard's irrigation system (Carroquino et al., 2018). These innovative solutions exploit the so-called water-energy nexus, that is gaining momentum in the scientific literature.

Climate change, by increasing the temperature on Earth due to a constant augment in anthropogenic CO_2 emissions, is known to severely impact winemaking process: thus, it is fundamental for the wine sector to move towards sustainability and emission reduction. Stakeholders' positive perception is mandatory to enhance RES utilization in wineries. As an example, a survey conducted on a number of Spanish wineries in (Garcia-Casarejos et al., 2018) highlighted that 20% of the facilities were willing to produce on-site renewable energy, while 40% of them, despite being generally favorable, did not believe that it was worth investing. Finally, the residual 40% was not interested at all in implementing renewable energy generation. In order to improve this partially negative outcome, it was suggested to increase the available know-how for decision-makers and to allow, in addition, an easier access to financial support schemes.

12.6 Energy consumption and optimization in wineries: some case studies

12.6.1 Energy audit of an Italian winery

As highlighted in the previous sections, a detailed energy audit is generally required to optimize energy consumption in a selected winery. In the following, the case study of a winery located in Friuli-Venezia Giulia region (North-East of Italy), having 25 ha of cultivated grape area, is presented to give a meaningful (despite simple) example. The winery is cultivating a mixture of grapes for white (Sauvignon, Friulano, Riesling, Pinot, and Ribolla) and red (Merlot, Cabernet Sauvignon, Refosco, and Franconia) wine production. Total wine production is about 2000 hL/y, coherent with other small wineries located in the same geographic area.

Winery building is composed of thick stone walls: consequently, it is possible to keep a constant temperature throughout the year without the need for conditioning systems. The studied winery installed a photovoltaic power system in 2011 for a total covered area of 240 m^2 (and a related electric power of 35 kW). Considering a mean operational time of 1,500 h per year of photovoltaic unit, the produced solar energy was calculated as 37,000 kWh/y (corresponding to an economic saving of about 7,000–8,000 €/y).

In grape receiving area, most of the electricity is used for electric motors that feed conveyor belts and augers, as well as for machines that produce compressed air. In fermentation area, a huge electricity consumption is needed both for refrigeration (to keep the desired temperature level) and for wine handling. The winery is equipped with 65 stainless steel fermenters, with a capacity of 10–100 hL each. Regarding the cooling group, the installed cooling power is 38.7 kW (with a coefficient of performance of 2.7) and R-404A is used as cooling gas; the system is able to control the temperature of up to 1,200 hL of wine. The wine aging room is particularly important, because in this premise temperature needs to be carefully monitored to allow a better aging in barrique. In the aging room, humidity level is around 80%–85%, while temperature is in the range of 11–15°C. In the analyzed winery, every year about 200 hL of fine wines are refined (of which 75% are red wines and 25% white wines).

Table 12.1 Correlation between grape harvest temperature and total electricity consumption in the studied winery (years 2015–18).

Year	Total electricity consumption (kWh)	Harvest temperature (°C)
2015	46,353	23
2016	60,259	27
2017	45,023	22
2018	61,696	28

The winery is not equipped of a bottling section, but rather entrusts the service to an external company, as commonly done by other Italian wineries; the available area allows to store about 250,000 bottles. In this area, the temperature is kept in the range of 12–16°C throughout the whole year, and energy consumption is mainly related to electric forklifts (for bottle handling) and to lighting.

The energy analysis was coupled with a climatic analysis that highlighted a significant temperature increase in Friuli-Venezia Giulia region in the last years, due to global climate change. Considering that white wine represents most of the production, an advance of the harvest time to the month of August (while previously it was typically conducted in September) unavoidably led to an increase in energy consumption needed for grape cooling. In fact, total winery consumption in the years 2015–2018 was positively related to the harvest temperature (Table 12.1): it was seen, as previously detailed, that grape cooling was a significant factor affecting total energy consumption.

The electricity price for the studied winery was about 0.18 €/kWh: total electricity cost for the year 2018 was 11,100 €, corresponding to a specific cost of about 0.04 € per bottle of wine (considering total wine production of 1900 hL). Analyzing more in detail energy consumption throughout the year, it was seen that two distinct peaks appeared, the first one (more evident) from August to October, related to grape harvest, wine production and handling, while the second one was registered in May, and was linked to red wine bottling. Also, in the months of November and December, a significant energy consumption arose due to tartaric stabilization process of wines.

The simplified energy audit in the different production phases was reported in Table 12.2, highlighting that most of the energy was needed in vinification stage, while grape receiving, storage, and management accounted for a similar consumption (in the range of 14%–16% of total value).

Table 12.2 Simplified energy audit results from the selected Italian winery.

Process	Total working hours (h/y)	Electricity consumption (kWh/y)	Energy share (%)
Grape receiving	400	8637	14
Vinification	750	27,763	45
Wine aging	150	6169	10
Storage	75	9254	15
Management	120	9871	16
Total	–	61,694	100

12.6.2 TESLA research project

The TESLA research project, funded by "Intelligent Energy Europe" program of EU, was specifically intended to monitor and optimize energy consumption of 39 wineries located in France, Italy, Portugal, and Spain (Vela et al., 2017). A comparative analysis of different energy efficiency measures was carried out through a techno-economic assessment, calculating the return of investment of each proposed solution. The results showed that the measures having the greatest impact on energy saving were: (i) old chillers' replacement by modern devices; (ii) installation of variable speed drives on pumps; (iii) replacement of electric fermentation motors with high-efficiency engines; (iv) replacement of conventional lighting by light emitting diode (LED) lighting; (v) introduction of management and control measures (e.g., implementation of building management systems); and (vi) installation of energy analyzers to identify users' consumption (Vela et al., 2017).

In Table 12.3, the results of an energy audit conducted on an Italian winery, producing 30,000 hL of wine per year, were reported as a meaningful example: both total installed electric and thermal powers were considered, together with specific energy consumption (expressed as kWh/hL wine). As evidenced in Section 12.6.1, the alcoholic fermentation consumed a relevant quote of total energy request, while thermal energy need globally accounted for less than 10%, if compared to electricity consumption.

12.6.3 Energy assessment related to wineries located in Veneto (Italy)

Another recent investigation, aimed at evaluating electricity consumption profile of nine small and medium wineries (including a social cooperative winery), located near the Piave river, was carried out in Veneto Region (North-East of Italy). Five-year electricity consumption was considered, together with the presence or absence of local photovoltaic generation and water consumption. Moreover, some general data regarding winery characteristics were included in the analysis, such as cultivated area (ha), total power of agricultural machines (kW), diesel consumption for grape cultivation, and size of company premises (area, m^2, and volume, m^3).

The results of the survey are summarized in Table 12.4: the significant difference encountered in agricultural machines' power could be related to the fact that some of the studied wineries entrusted some operations to external contracting firms. Apart from the cooperative winery, it could be seen that larger wineries had a lower specific electricity consumption (expressed as kWh/hL) than smaller wineries, due to more efficient processes and devices. Larger wineries are the ones that are typically more inclined to install renewable energy production systems, such as photovoltaic power, given the huge capital costs that need to be covered for implementing these technical solutions. Winery B had a similar wine production than that reported for the winery analyzed in Table 12.3; however, its specific electricity consumption was significantly lower, probably due to the installation of photovoltaic modules.

Analyzing more in detail the data regarding the cooperative winery, in the year 2018 electricity production from the installed photovoltaic system was 491,833 kWh, corresponding to an economic saving of about 80,000 €. Mean electricity price from the national grid, considering Italian market, was around 0.18 €/kWh (Cottes et al., 2020). A "standard" winery consuming 20 kWh/hL of electricity has an energy cost of about 0.036 €/L of wine: considering a mean value of wine production of 30,000 hL/y (Table 12.3), the yearly energy expense can be thus calculated as 108,000 €/y.

Table 12.3 Energy consumption in a "standard" winery producing 30,000 hL wine per year (results from TESLA project).

Process phase	Applied devices	Installed electric power (kW)	Electricity consumption (kWh/hL)	Installed thermal power (kW)	Heat consumption (kWh/hL)
Grape reception	Hoppers, cochleae, and electric motor devices	57	0.55	0	0
Destemming and crushing	Mechanical destemming, rollers, and electric motors	64			
Alcoholic fermentation	Cooling systems and electric motors	276	5	0	0
Pressing	Cooling systems for malolactic fermentation, pumps, and electric motors	76	0.75	0	0
Stabilization	Cooling systems for stabilization (pasteurization, pumps, and electric motors)	91	0.90	0	0
Bottling, conservation and distribution	Electric motors and lifts	102	1.95	50	0.5
Lighting	Fluorescent lamps	10	0.75	0	0
Auxiliary processes	Conditioned air and boiler for hot water (sanitary use)	124	1.10	20	0.5
Total		800	11	70	1

The presence of bottling phase in wineries significantly increases specific electricity consumption; most of the reported wineries in Table 12.4 externalize this phase, due to a general technical and economic convenience. Mobile bottling plants are commonly used, generally equipped of electric generators to ensure a constant electricity supply; the connection with the grid is typically avoided, as this operation is highly energy-intensive (particularly when using isobaric fillers). In addition, it should be considered that the type of produced wine influences energy consumption as well: as an example, the wineries F and G in Table 12.4 produce more red wine in comparison with the other studied wineries. Regarding photovoltaic power installation, four out of the nine wineries reported in Table 12.4 (particularly the larger ones) exploit solar energy for electricity production.

Table 12.4 Results of the energy survey conducted in nine wineries located in Veneto region (Italy).

Name	Winery area (ha)	Grape area (Ha)	Machine power (kW)	Premise area (m^2)	Premise volume (m^3)	Wine production (hL/y)	Bottling (yes/no)	Electric consumption (kWh/y)	Electric consumption (kWh/hL)	Photovoltaic power (kW)
A	80	65	270	3200	18,000	10,000	No	160,833	16	100
B	120	101	1000	2000	18,600	28,000	No	120,620	4.5	150
C	30	25.50	180	1300	9200	8550	No	50,000	9	20
Coop	–	1200	–	8400	33,100	300,000	Yes	5,651,786	19	1000
D	30	22	161	1800	11,000	10,100	Yes	220,850	21	0
E	70	43	100	1000	9950	2500	No	19,000	8	0
F	45	42	264	700	2400	5500	No	6400	2	0
G	25	20	235	800	5600	12,000	No	43,000	4	0
H	4	3.5	44	730	2230	930	No	17,000	18	0
Total	404	322.7	2254	–	–	377,580	–	6,289,489	–	1270
Mean	45	36	250	2214	12,230	41,953	–	698,832	16	141

Considering only the wineries with a complete production chain (included bottling phase), from Table 12.4 a mean wine production of 10,000 hL/y arises, coupled with a specific electricity consumption of 20 kWh/hL, comparable to typical literature values, and coherent with what reported in Section 12.3. Studying more in detail electricity consumption throughout the years, it could be seen that specific energy consumption is positively correlated with environmental temperature. As an example, by studying electricity consumption profile of the cooperative winery in the years 2013−2017, a sensible increase in specific electricity consumption was observed in the warmest years (2014−2016).

For five selected wineries among those presented in Table 12.4, which presented a similar wine production chain, electricity consumption was related to vinification type, i.e., the relative proportion between red and white wine production. It could be seen that an increase in the relative proportion of white wine generally augmented electricity consumption. A remarkable exception was winery E, where despite white wine production was predominant, electricity need was lower than the mean calculated value.

12.7 Concluding remarks

This chapter analyzed energy consumption in wineries, considering the outcomes of specific research projects in the field together with the most recent literature evidences on the topic. The main aim of the chapter was to illustrate energy need in the different winery process phases, together with the solutions that may be applied to standardize energy assessments and improve the overall energy efficiency. It was highlighted that a detailed energy audit is a fundamental step to assess the starting point in energy management in existing wineries, detecting in an easier way the process operations where a significant improvement can be achieved. Energy audits can also be a good approach for benchmarking purposes. All wineries, including both large scale and small-to-medium scale enterprises, are strongly encouraged to conduct internal audits to assess their actual energy efficiency level. Regarding the technical solutions to enhance energy efficiency level, it was seen that a good thermal insulation of tanks and walls is fundamental to improve energy efficiency, considering that cooling is one of the most energy-intensive operations. The adoption and integration of RESs, including on-site biogas generation from AD of organic residues, photovoltaic generation, and solar thermal generation, can create positive economic and environmental effects, reducing at the same time the dependence on fossil fuels. Finally, given the prevalence of electricity consumption over thermal request, it is recommended at first to focus the energy audits and the technical improvements on electric energy, leaving thermal energy request to a successive phase.

References

Ahmad, B., Yadav, V., Yadav, A., Rahman, M. U., Yuan, W. Z., Li, Z., & Wang, X. (2020). Integrated biorefinery approach to valorize winery waste: A review from waste to energy perspectives. *The Science of the Total Environment, 719*, 137315. https://doi.org/10.1016/j.scitotenv.2020.137315

Bacenetti, J. (2019). Heat and cold production for winemaking using pruning residues: Environmental impact assessment. *Applied Energy, 252*, 113464. https://doi.org/10.1016/j.apenergy.2019.113464

Barbaresi, A., Torreggiani, D., Tinti, F., & Tassinari, P. (2017). Analysis of the thermal loads required by a small-medium sized winery in the Mediterranean area. *Journal of Agricultural Engineering*, 9–20. https://doi.org/10.4081/jae.2017.670

Benni, S., Torreggiani, D., Barbaresi, A., & Tassinari, P. (2013). Thermal performance assessment for energy-efficient design of farm wineries. *Transactions of the ASABE, 56*(6), 1483–1491.

Campos, I., Marín-González, E., Luz, G., Barroso, J., & Oliveira, N. (2019). Renewable energy prosumers in Mediterranean viticulture social–ecological systems. *Sustainability, 11*(23), 6781. https://doi.org/10.3390/su11236781

Carroquino, J., García-Casarejos, N., & Gargallo, P. (2017). Introducing renewable energy in vineyards and agricultural machinery: A way to reduce emissions and provide sustainability. *Wine Studies, 6*(1). https://doi.org/10.4081/ws.2017.6975. Article 1.

Carroquino, J., Roda, V., Mustata, R., Yago, J., Valiño, L., Lozano, A., & Barreras, F. (2018). Combined production of electricity and hydrogen from solar energy and its use in the wine sector. *Renewable Energy, 122*, 251–263. https://doi.org/10.1016/j.renene.2018.01.106

Casentini, D., & Rushforth, L. (2008). Harvesting savings: Energy efficiency roots sustainability in the wine industry. *2008 ACEEE Summer Study on Energy Efficiency in Buildings*, 37–49.

Catrini, P., Panno, D., Cardona, F., & Piacentino, A. (2020). Characterization of cooling loads in the wine industry and novel seasonal indicator for reliable assessment of energy saving through retrofit of chillers. *Applied Energy, 266*, 114856. https://doi.org/10.1016/j.apenergy.2020.114856

Celorrio, R., García, J. L., Martínez, E., Jiménez, E., & Blanco, J. (2017). Methodology for the reduction of energy demand during cold stabilisation in the wine industry. *Energy and Buildings, 142*, 31–38. https://doi.org/10.1016/j.enbuild.2017.03.005

Celorrio, R., Martínez, E., Saenz-Díez, J. C., Jiménez, E., & Blanco, J. (2015). Methodology to decrease the energy demands in wine production using cold pre-fermentation. *Computers and Electronics in Agriculture, 117*, 177–185. https://doi.org/10.1016/j.compag.2015.08.009

Cottes, M., Mainardis, M., Goi, D., & Simeoni, P. (2020). Demand-response application in wastewater treatment plants using compressed air storage system: A modelling approach. *Energies, 13*(18), 4780. https://doi.org/10.3390/en13184780

Da Ros, C., Cavinato, C., Pavan, P., & Bolzonella, D. (2017). Mesophilic and thermophilic anaerobic co-digestion of winery wastewater sludge and wine lees: An integrated approach for sustainable wine production. *Journal of Environmental Management, 203*, 745–752. https://doi.org/10.1016/j.jenvman.2016.03.029

ELkhadraoui, A., Kooli, S., Hamdi, I., & Farhat, A. (2015). Experimental investigation and economic evaluation of a new mixed-mode solar greenhouse dryer for drying of red pepper and grape. *Renewable Energy, 77*, 1–8. https://doi.org/10.1016/j.renene.2014.11.090

Fernández-Puratich, H., Hernández, D., & Tenreiro, C. (2015). Analysis of energetic performance of vine biomass residues as an alternative fuel for Chilean wine industry. *Renewable Energy, 83*, 1260–1267. https://doi.org/10.1016/j.renene.2015.06.008

Garcia-Casarejos, N., Gargallo, P., & Carroquino, J. (2018). Introduction of renewable energy in the Spanish wine sector. *Sustainability, 10*(9), 3157. https://doi.org/10.3390/su10093157

Gómez-Lorente, D., Rabaza, O., Aznar-Dols, F., & Mercado-Vargas, M. J. (2017). Economic and environmental study of wineries powered by grid-connected photovoltaic systems in Spain. *Energies, 10*(2), 222. https://doi.org/10.3390/en10020222

Gubiani, R., Pergher, G., & Mainardis, M. (2019). The winery in a perspective of sustainability: The parameters to be measured and their reliability. In *2019 IEEE international workshop on metrology for agriculture and forestry (MetroAgriFor)* (pp. 328–332). https://doi.org/10.1109/MetroAgriFor.2019.8909221

Guerini Filho, M., Lumi, M., Hasan, C., Marder, M., Leite, L. C. S., & Konrad, O. (2018). Energy recovery from wine sector wastes: A study about the biogas generation potential in a vineyard from Rio Grande do Sul, Brazil. *Sustainable Energy Technologies and Assessments, 29*, 44–49. https://doi.org/10.1016/j.seta.2018.06.006

Hernandez, R. R., Easter, S. B., Murphy-Mariscal, M. L., Maestre, F. T., Tavassoli, M., Allen, E. B., Barrows, C. W., Belnap, J., Ochoa-Hueso, R., Ravi, S., & Allen, M. F. (2014). Environmental impacts of utility-scale solar energy. *Renewable and Sustainable Energy Reviews, 29*, 766−779. https://doi.org/10.1016/j.rser.2013.08.041

Istat. (2019). *Annual Italian statistics on wine production.* www.istat.it.

Jia, T., Dai, Y., & Wang, R. (2018). Refining energy sources in winemaking industry by using solar energy as alternatives for fossil fuels: A review and perspective. *Renewable and Sustainable Energy Reviews, 88*, 278−296. https://doi.org/10.1016/j.rser.2018.02.008

Kumar, R., & Rosen, M. A. (2011). A critical review of photovoltaic−thermal solar collectors for air heating. *Applied Energy, 88*(11), 3603−3614. https://doi.org/10.1016/j.apenergy.2011.04.044

Mainardis, M., Boscutti, F., Cebolla, M. del M. R., & Pergher, G. (2020). Comparison between flaming, mowing and tillage weed control in the vineyard: Effects on plant community, diversity and abundance. *PLoS One, 15*(8), e0238396. https://doi.org/10.1371/journal.pone.0238396

Mainardis, M., Buttazzoni, M., & Goi, D. (2020). Up-flow anaerobic sludge blanket (UASB) technology for energy recovery: A review on state-of-the-art and recent technological advances. *Bioengineering, 7*(2), 43. https://doi.org/10.3390/bioengineering7020043

Mainardis, M., Flaibani, S., Mazzolini, F., Peressotti, A., & Goi, D. (2019). Techno-economic analysis of anaerobic digestion implementation in small Italian breweries and evaluation of biochar and granular activated carbon addition effect on methane yield. *Journal of Environmental Chemical Engineering, 7*(3), 103184. https://doi.org/10.1016/j.jece.2019.103184

Mainardis, M., Flaibani, S., Trigatti, M., & Goi, D. (2019). Techno-economic feasibility of anaerobic digestion of cheese whey in small Italian dairies and effect of ultrasound pre-treatment on methane yield. *Journal of Environmental Management, 246*, 557−563. https://doi.org/10.1016/j.jenvman.2019.06.014

Malvoni, M., Congedo, P. M., & Laforgia, D. (2017). Analysis of energy consumption: A case study of an Italian winery. *Energy Procedia, 126*, 227−233. https://doi.org/10.1016/j.egypro.2017.08.144

Muhlack, R. A., Potumarthi, R., & Jeffery, D. W. (2018). Sustainable wineries through waste valorisation: A review of grape marc utilisation for value-added products. *Waste Management, 72*, 99−118. https://doi.org/10.1016/j.wasman.2017.11.011

Pinter, I. F., Fernández, A. S., Martínez, L. E., Riera, N., Fernández, M., Aguado, G. D., & Uliarte, E. M. (2019). Exhausted grape marc and organic residues composting with polyethylene cover: Process and quality evaluation as plant substrate. *Journal of Environmental Management, 246*, 695−705. https://doi.org/10.1016/j.jenvman.2019.06.027

Smyth, M., & Nesbitt, A. (2014). Energy and English wine production: A review of energy use and benchmarking. *Energy for Sustainable Development, 23*, 85−91. https://doi.org/10.1016/j.esd.2014.08.002

Tinti, F., Barbaresi, A., Torreggiani, D., Brunelli, D., Ferrari, M., Verdecchia, A., Bedeschi, E., Tassinari, P., & Bruno, R. (2017). Evaluation of efficiency of hybrid geothermal basket/air heat pump on a case study winery based on experimental data. *Energy and Buildings, 151*, 365−380. https://doi.org/10.1016/j.enbuild.2017.06.055

Toscano, G., Riva, G., Duca, D., Pedretti, E. F., Corinaldesi, F., & Rossini, G. (2013). Analysis of the characteristics of the residues of the wine production chain finalized to their industrial and energy recovery. *Biomass and Bioenergy, 55*, 260−267. https://doi.org/10.1016/j.biombioe.2013.02.015

Vela, R., Mazarrón, F. R., Fuentes-Pila, J., Baptista, F., Silva, L. L., & García, J. L. (2017). Improved energy efficiency in wineries using data from audits. *Ciência e Técnica Vitivinícola, 32*(1), 62−71. https://doi.org/10.1051/ctv/20173201062

CHAPTER 13

Microbiological control of wine production: new tools for new challenges

M. Carmen Portillo and Albert Mas
Department de Bioquímica i Biotecnologia, Facultat d'Enologia, Universitat Rovira i Virgili, Tarragona, Spain

13.1 Introduction

Alcoholic fermentation (AF) is an essential step to produce any kind of wine in which the sugars present in the grapes (mainly glucose and fructose) are biotransformed by microorganisms to ethanol and carbon dioxide (Ribéreau-Gayon et al., 2006). In addition to sugars, grapes contain other compounds, like amino acids, polyphenols, or acids, also susceptible of being metabolized and impact the flavor and aroma of the wine (Pretorius, 2016). The main microorganism of AF is the yeast *Saccharomyces cerevisiae* due to its adaptation to the harsh environmental conditions occurring during the winemaking process (low pH, high osmotic pressure, unbalanced concentrations of nutrients, high ethanol concentration, etc.) and its rapid transformation of sugars from the grape must. However, many other microbes including filamentous fungi, yeasts, and bacteria, are present during the winemaking process. The complex and highly diverse microbial communities associated with the fermentation of the grape must are known as wine microbiome.

Under certain conditions, some species of yeasts and bacteria can cause spoilage of the wine affecting its quality (Bartowsky, 2009). Wine susceptibility to spoilage depends on the species of yeast and bacteria present and their population size. Additionally, wine physical-chemical characteristics like ethanol content, residual sugar concentration, pH, amount and composition in main acids or oxygen, also condition wine spoilage (Bartowsky, 2009; Loureiro & Malfeito-Ferreira, 2003).

The microbiological stability of wine is fundamental to preserve its quality and produce sustainable wines avoiding economical losses. This stability may be achieved using chemical preservatives and/or physical treatments, aimed at killing microorganisms or at least at inhibiting their proliferation, or at physically removing them from wine by filtration. However, these treatments are not specific and may be detrimental for the desirable and beneficial microorganisms during fermentation. Thus, the very first step to control wine microbiome during wine production is to know its composition and its functional attributes.

In the last decade, a plethora of studies about wine microbiome have redefined our understanding of the microorganisms involved in the winemaking process. The combination of affordable high-throughput sequencing (HTS) technologies generating large datasets with insightful bioinformatic tools that enable analysis and interpretation of complex patterns has enhanced our

240 Chapter 13 Microbiological control of wine production

understanding of wine microbiome composition and function. In particular, genomics, transcriptomics, metabolomics, and proteomics have been broadly implemented to characterize microbial genes, transcripts, and proteins, respectively, during wine production (Sirén, Mak, Fischer, et al., 2019).

Recent rapid advances in HTS and DNA synthesis techniques are enabling the design and construction of new genes, gene networks and biosynthetic pathways, and the redesign of cells and organisms for useful purposes (Pretorius, 2017). Additionally, with the advent of the CRISPR/Cas9 (Clustered Regularly Interspaced Short Palindromic Repeats/CRISPR associated protein 9) genome editing methods, yeast strain engineering has become rapid, efficient, and multiplexed (Zhang et al., 2019).

This chapter will cover the role and future potential of such recent techniques in the microbial control of wine production and highlight the potential challenges that will be faced.

13.2 New tools
13.2.1 "Omics" technologies: genomics, metagenomics, transcriptomics, metatranscriptomics, proteomics, and metabolomics

"Omics" technologies are primarily aimed at the universal detection of genes and transcribed genes in a single organism (genomics and transcriptomics, respectively) or in a microbiome (metagenomics and metatranscriptomics, respectively). Additionally, the term "omics" includes the technologies for the study of protein function, structure, and differential expression level (proteomics) and the metabolites generated from cellular processes (metabolomics) in a specific biological sample.

The "omics" analyses offer potential with regards to microbial control of wine production, and they have been applied in a plethora of wine-related studies (Fig. 13.1 and Table 13.1). Within the *metagenomics* analysis, the PCR amplification and the later sequencing of gene-marker specific regions is an approach known as metabarcoding or amplicon-based metagenomics. Alternatively, the shotgun sequencing retrieves the information from the whole metagenome of a sample (all genes from all genomes in the community) without including any primer selection and, thus, is less biased by the PCR step. In fact, when comparing metabarcoding and shotgun metagenomics analysis of five spontaneous fermentations, metabarcoding analysis biased the overabundance of the genus *Metschnikowia* (Sternes et al., 2017). However, the combination of both metagenomic procedures has demonstrated to be useful to study the influence of vineyard community composition on the fermentation of Riesling and revealed the putative role of *Metschnikowia* as biocontrol agent against bacteria (Sirén, Mak, Melkonian, et al., 2019). Metagenomic analysis has created the notion that apart from lactic acid bacteria (LAB) and acetic acid bacteria (AAB), other bacteria, not previously described, may be present during the process (Godálová et al., 2016). Although the possible impact of these newly described bacterial genera is still to be demonstrated, Sirén, Mak, Melkonian, et al. (2019) detected an increase of functions assigned to class Actinobacteria at the end of fermentation, pointing to a putative role in winemaking. The metagenomic analysis has been mainly used to describe which microbes are present and relevant in wine-related samples, to reveal the relationship between the microbial communities and the wine terroir (reviewed in Belda et al. (2021)), to monitor wine fermentations under different conditions (reviewed in Kioroglou et al. (2018)), to relate the microbial communities with wine chemical composition (Bokulich et al., 2016), or to monitor the changes in the grape must and wine microbiota due to vineyard influence and different winemaking practices (reviewed in Stefanini and Cavalieri (2018)). Another important question that metagenomic analysis has been

13.2 New tools   241

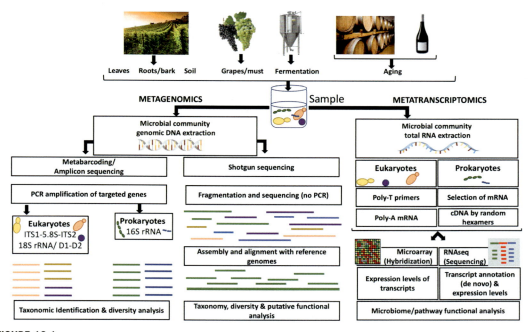

FIGURE 13.1

Schematic representation of the workflow followed during metagenomic and metatrascriptomic analysis from different wine-related samples. Within metabolomics, the comparison between metabarcoding and shotgun sequencing is presented, whereas the metatranscriptomics indicates the main differences between microarrays and RNAseq outputs.

called to answer is whether grapes are the source of spoilage microorganisms (Renouf et al., 2005), or the wine-making equipment (Couto et al., 2005). Even though there is no clear answer to this debate, studies from Pinto et al. (2015) and Suárez et al. (2007) seem to support the latter hypothesis.

Knowing the composition of the microbial community during wine production is crucial to control it. That is the main reason why the first applications of "omics" techniques to wine research aimed the characterization of the bacterial, yeast, and fungal communities. However, revealing the interactions of microbial communities in different stages of the winemaking process and the metabolic pathways involved is of paramount importance to determine the microbial influence in wine quality. In this sense, metatranscriptomics, proteomics, and metabolomics are the applied techniques to complement the metagenomics information.

One of the concerns that can be raised in metagenomics is that is a technology based on DNA. It is well-known the resilience of this molecule and that can be present for long time after the microorganisms are dead. This could lead to an overestimation of the population. Some approaches have been proposed to tackle this aspect: one of them could be the use of RNA (see metatranscriptomics), known to be less stable and thus, could reflect the real live population and another possibility is the use of DNA binding dyes as ethidium monoazide and propidium monoazide, which would prevent the amplification of DNA (Andorrà et al., 2010; Navarro et al., 2020; Rizzotti et al., 2015).

Table 13.1 Summary of omics technologies applied to oenology.

Omic technology	Target	Aim	References
Metagenomics	DNA	Microbial taxonomic identification, genes, and genetic pathways	Belda et al. (2021)* Kioroglou et al. (2018)* Stefanini and Cavalieri (2018)*
Metatranscriptomics	Total RNA and mRNA	Gene expression and functions	Alonso-del-Real et al. (2019) Barbosa et al. (2015) Curiel et al. (2017) Rossouw et al. (2015) Sadoudi et al. (2017) Shekhawat et al. (2019) Sunyer-Figueres et al. (2020) Tronchoni et al. (2017)
Metaproteomics	Protein	Protein function, structure, and differential expression level	González-Jiménez et al. (2020) Mencher et al. (2020) Peng et al. (2019)
Metabolomics	Metabolites	Produced metabolites	Alañón et al. (2015) Alves et al. (2015) Arapitsas et al. (2018) Bordet et al. (2020) Cozzolino (2016) Kioroglou et al. (2020) López-Malo et al. (2013) Mazzei et al. (2013) Peng et al. (2018) Petitgonnet et al. (2019) Richter et al. (2015) Roullier-Gall et al. (2020)* Sirén, Mak, Fischer, et al. (2019)

The asterisks indicate bibliographic review about the topic. Due to large number of articles in using metagenomics only reviews are indicated.

Metatranscriptomics refers to the measurement of total gene expression in a target sample by extracting messenger RNA (mRNA) and then converting it to cDNA using random hexamers or, in the case of Eukaryotes, poly-T primers that target the poly-A mRNA tail (Zepeda-Mendoza et al., 2018). This analysis gives information about the gene activity of the target organisms within the sample (Belda et al., 2017; De Filippis et al., 2018). Analyses can also be performed using stable isotope probing targeting a specific microbial group in the samples to enrich its transcriptome and then using the RNASeq in NGS platforms (Dumont et al., 2013). RNA-seq is the methodology that recently has become predominant in metatranscriptomics studies because it offers several advantages over microarrays. However, recent studies comparing both techniques pointed to the high consistence between both platforms, encouraging the use of microarray as a versatile tool for differential gene expression analysis (Nookaew et al., 2012). The metatranscriptomics analysis has been extensively used in wine research in recent years to elucidate, for example, interactions between microorganisms during wine AF (Alonso-del-Real et al., 2019; Barbosa et al., 2015; Curiel et al., 2017), the effect of different stresses over gene transcription in wine microorganisms (Shekhawat et al., 2019; Tronchoni et al., 2017), or even to reveal the protective role of some compounds during the oxidative stress of wine yeasts (Sunyer-Figueres et al., 2020). Complete metabolic pathways are affected by altered gene expression, as shown by Sadoudi et al. (2017) with a change in acetic acid and glycerol metabolism in *S. cerevisiae* in the presence of *Metschnikowia pulcherrima*. Furthermore, in the case of direct cell contact between two populations of distinct species, a change in the expression of FLO genes has been described, leading to a modification of population dynamics (Rossouw et al., 2015). The main challenge of the interpretation of the metatranscriptomics results during the study of yeast interactions is that the growth of yeasts in mixed fermentations may be affected by several factors other than the specific used strains, as, for example, the grape must composition, nutrient limitations, or fermentation temperature. All these factors should be considered to extrapolate the results from this analysis.

Metaproteomics is the identification and quantification of the expressed proteins in any matrix, which improves the functional gene annotations and provides better understanding of the microorganism interactions within that matrix. Generally, all mass spectrometry-based proteomic workflows comprise first the isolation of proteins from their source and can be further fractionated. After digestion, the peptides are analyzed by mass-spectrometry qualitatively and quantitatively. Then, the large amount of generated data is analyzed by appropriate software tools to deduce the amino acid sequence and, if applicable, to quantify the proteins in a sample. Recently, Peng et al. (2019) evaluated the proteomic response of *S. cerevisiae* during AF when it was co-inoculated with *Lachancea thermotolerans*. Additionally, metaproteomics could be used to investigate the transcription of taste-active peptides in wine (González-Jiménez et al., 2020) or the possible involvement of extracellular vesicles in the complexity of wine sensory features (Mencher et al., 2020). Similarly to metatranscriptomics analysis, the biggest limitation of the metaproteomics technology is to evaluate the effect of the external factors over the results during the experimentation making difficult the prediction of the transcriptome under the semi-industrial or industrial scale.

Metabolomics approaches aim to identify and quantify multiple metabolites or chemical compounds in a single matrix using nuclear magnetic resonance or mass spectrometry-based methods (Cozzolino, 2016; Sirén, Mak, Fischer, et al., 2019). Metabolomics data can provide general proof of gene function and complement the information gathered through metagenomics and transcriptomics studies. Both volatile and nonvolatile metabolites can be studied in either targeted or nontargeted fashion. It is known that environmental factors and winemaking decisions have a strong impact on the microbial metabolic profiles and metabolomics is useful in the investigation of dynamics between microbial communities and the matrix (Cozzolino, 2016). Metabolomics has already been applied to wine production to study questions ranging from the cultivar differences, monitoring of the fermentation process, and guiding of winemaking decision making, as well as the exploration of aroma and flavor variation by vintage (Alañón et al., 2015; Arapitsas et al., 2018) or aging conditions (Kioroglou et al., 2020). The literature includes various studies in which the specific composition of wine enables distinguishing between wines on the basis of fermentations with different yeast species and strains (Alves et al., 2015; López-Malo et al., 2013; Mazzei et al., 2013) and with single cell co-cultures (Peng et al., 2018; Petitgonnet et al., 2019; Richter et al., 2015). Significant metabolic changes have been identified at each stage of the fermentation studied (Peng et al., 2018; Richter et al., 2015) highlighting that sampling time is an essential point for understanding interaction phenomena (reviewed in Bordet et al. (2020)). Furthermore, some studies explore the differences between the endometabolome and the exometabolome associated with microorganisms involved in fermentation processes (Richter et al., 2015). It should also be noted that the identification of compounds detected during the metabolic profiling of wine remains difficult at present due to the incomplete databases that frequently do not allow identifying all the biomarkers (Roullier-Gall et al., 2020).

13.2.2 Genome editing: CRISPR/Cas9

The CRISPR/Cas9 genome editing tool has been successfully implemented both in *Saccharomyces* and non-*Saccharomyces* genome modification attempts and it is evident it will become more routine (Raschmanová et al., 2018). In short, CRISPR/Cas9 involves utilizing the natural mechanism that has been described in bacteria and archaea to develop a tool capable of conducting precise genome editing of any organism. Most CRISPR/Cas9 editing systems require two components, i.e., a guide RNA, which is a chimeric RNA molecule, and an RNA-guided DNA endonuclease like Cas9. Part of the guide RNA is bound by the Cas9 and directs it to the complementary genomic DNA region causing a double strand break upstream of a protospacer adjacent region (that in the case of the commonly used Cas9 from *Streptococcus pyogenes*, it is an NGG sequence). A double strand break would often be lethal for an organism if not repaired rapidly. The endogenous repair machineries allow for the introduction of a variety of genomic modifications. This tool has been fine-tuned and streamlined for yeast DNA editing (Jakočiunas et al., 2016; Weninger et al., 2018). Advantages of a CRISPR/Cas9 tool include that changing the target locus can be done simply by modifying a 20-bp sequence of the guide RNA and, once supplied with an appropriate repair template, large insertions and deletions can be done. Also, the selection marker can easily be removed from the resulting strain, a great concern for any genetic modification in food applications. The genetic modification of wine strains of *S. cerevisiae* has shown tremendous potential in improving many oenological aspects albeit mostly restricted to laboratory level (Van Wyk et al., 2019), as summarized in Table 13.2.

Table 13.2 Recent applications of the CRISPR-Cas9 technique in *Saccharomyces cerevisiae* to improve some aspects of wine making.

Winemaking goal	Gene edited	Result	Reference
Reduction of urea	*CAN1* (arginine permease)	25%–40% urea reduction	Vigentini et al. (2017)
Reduction of urea and ethyl carbamate	*DUR3* (urea transporter)	92% urea reduction 52% ethyl carbamate	Wu et al. (2020)
Fermentation of high sugar concentration and glycerol production	*STL1* (sugar transporter)	Low fermentation activity, increased glycerol	Muysson et al. (2019)
High glycerol production	*GPD1* (glycerol 3-phosphate dehydrogenase)	High production of glycerol	van Wyk et al. (2020)
Aroma production (esters and acetates)	*ATF1* (alcohol acetyltransferase)	High concentration of several acetates	van Wyk et al. (2020)

Wine yeasts are known to produce a broad array of compounds, not all of them with a positive character in wines. One of them is the generation of urea that can combine with ethanol and produce the carcinogenic compound ethyl carbamate. This has been the target of the first application of this technique in wine yeast. Vigentini et al. (2017) have eliminated the arginine permeases pathway (the *CAN1* gene) to reduce urea production in two different commercial strains of *S. cerevisiae*. They have obtained reductions between 20% and 35%, depending on the strain. Reduction of urea has also been obtained by a different strategy using also the CRISPR-Cas9 editing tool: overexpression of the *DUR3* gene (Urea active transporter, Wu et al., 2020). They observed that the modified *S. cerevisiae* also reduced the level of urea by 92% and those of ethyl carbamate by 52% in Chinese rice wine.

Another successful application has focused on the glycerol response to high sugar concentration that is required in yeast fermenting special wines with this high sugar content. In this case, Muysson et al. (2019) deleted the functional *STL1* gene to analyze their effect in ice-wine fermentations and the resulting mutant yeasts presented reduced fermentation performance and elevated concentrations of glycerol and acetic acid, compared to parental strains. It has to be emphasized that genes involved in ethanol and glycerol modulation will be the target for genetic modifications, in order to get wines with reduced alcohol content (Goold et al., 2017). In the same pathway (production of glycerol), van Wyk et al. (2020) overexpressed the gene GPD1 (Glycerol 3-phosphate dehydrogenase) by changing the promoter. The resulting strain had significantly higher production of glycerol but also acetic acid that the parental strain. In the same work, they also focused on the production of aromas (acetate esters), overexpressing alcohol acetyltransferase (*ACT1*). The double mutant had also increased levels of glycerol, and very high concentrations of the different acetates analyzed (ethyl acetate, isoamyl acetate, isobutyl acetate, phenylethyl acetate, and hexyl acetate).

However, this technique is open to be used to many other non-*Saccharomyces* yeasts (Raschmanová et al., 2018). So far, its application to other yeast has been mostly for other applications (production of products with pharmaceutical or nutrition interest, or production of biofuel, for instance). Only the wine-related yeast *Brettanomyces bruxellensis* has been successfully modified (Varela et al., 2020). However, the applicability of this modification is mostly in brewing, as *Brettanomyces* is used for the development of some beer aromas.

Another interest in genome modification is to expand on the aroma-producing capabilities of wine yeast. This includes overexpressing genes involved in the synthesis of esters like the alcohol acetyltransferases 1 and 2, which promote increased condensation between alcohols and acetyl-CoA resulting in more acetate esters being produced (see above, work of van Wyk et al., 2020).

Despite some drawbacks, the value of the CRISPR/Cas9 tool in generating wine yeast strains remains largely unexploited. Of the current genome editing tools available, CRISPR/Cas9-based editing have been shown to be the most adaptable, versatile, and cost-effective. This methodology has opened a new era for the improvement and genetic modification of the wine yeasts. The process should be seen in two different ways, on one side to improve the knowledge acquisition but in another to improve wine quality. It is evident, though, consumer acceptance to these methodologies requires still a communicative effort with educational purposes from researchers and innovators. Legislation will probably follow the consumer's concerns but, most interestingly, it should be shifted to food safety, clearly stating the benefits and risks of using this methodology.

13.3 New challenges
13.3.1 Grape microbiome and its control

Grape berries harbor a wide range of microbes including bacteria, fungi, and yeasts originated from the vineyard environment (Zarraonaindia et al., 2015), many of which are recognized for their role in the must fermentation process shaping wine quality. Experimental analyses suggest that microbes colonizing berries could significantly affect grapevine and fruit health and development (Barata et al., 2012). Furthermore, grape microbiome also contribute to shaping phenotypic characteristics, such as flavor, color, and sugar content (Belda et al., 2017), thus influencing the winemaking process as well (Capozzi et al., 2015).

HTS techniques have being used to characterize bacterial communities of the grapevine plant (Belda et al., 2021) and to assess the provenance of some microbial groups (Bokulich et al., 2013; Zarraonaindia et al., 2015). It has been revealed that soil serves as a primary source of microorganisms with edaphic factors influencing the native grapevine microbiome (Zarraonaindia et al., 2015) and that the grape microbiome biogeography is nonrandomly associated with regional, varietal, and climatic factors across multiscale viticultural zones (Bokulich et al., 2014). Moreover, Bokulich et al. (2016) suggested a strong association involving grapevine microbiota, fermentation characteristics, and wine chemical composition. The beneficial effect of certain microbial taxa on host plants as growth promoters and stress resistance inducers has been reported in several articles and some of them addressed their influence on grape and wine quality (Huang et al., 2018; Yang et al., 2016). Thus, the control of the grape microbiome through physical, chemical, or biological treatment of the grapevine to promote certain taxa could affect both the health of the plant and the quality of the wine. Since microbiome metabolism can contribute to that of the plant host and the biochemical composition of its fruits, the nature of grapevine microbiome taxa identities, ecological attitudes, potential toxicity, and clinical relevance are all aspects worthy of a thorough investigation and the new technologies and tools explained in Section 13.2 are the most promising right now.

13.3.2 Reduction of SO_2 use

Sulfites are considered the main additives in winemaking for their antimicrobial and antioxidant activities. The most important role of this compound lies in its antimicrobial action against Acetic Acid Bacteria (AAB) and Lactic Acid Bacteria (LAB), and molds to prevent spoilage and to determine the microbiological stabilization of wines to enhance aging potential. Furthermore, sulfur dioxide (SO_2) addition prior to the onset of AF also exerts a selective antimicrobial activity against spoilage yeasts, by inhibiting their growth and promoting the rapid development of *S. cerevisiae*. The current concern about the potential negative effects of SO_2 on consumer health has motivated the interest on replacing or reducing SO_2 use. Thus, research is focused on looking for other preservatives and innovative technologies, harmless to health, to reduce SO_2 content in wine. Recently, numerous alternatives have been proposed to replace the activity of SO_2 by the use of chemical additives and physical treatments, aimed at the microbiological stability of wine (reviewed by Lisanti et al. (2019)).

There are many different chemical solutions (antimicrobial compounds), some of them approved by the EU authorities and/or OIV legislation. The most used chemical alternative to SO_2 is the dimethyl dicarbonate (DMDC), which is active on the inhibition of some microbial critical enzymes and is hydrolyzed to CO_2 and methanol. It kills yeast cells almost immediately and later the residue is minimal, without any health concern (Ribéreau-Gayon et al., 2006). The effectiveness of DMDC could be jeopardized in musts with high microbial load, but it is considered very effective in final wines, especially sweet and semisweet wines, once the viable load of microorganisms is reduced (Bartowsky, 2009) The effect on bacteria is more limited, and when bacteria is the main microbial problem, the use of lysozyme could be another alternative. Lysozyme acts by hydrolysis of the cell wall in gram-positive bacteria (for instance, LAB), but it does not have any action against gram-negative bacteria (for instance, AAB) or yeast. Sorbic acid has been traditionally used in the food industry as antifungal compound and in wines has been considered effective to inhibit refermentation by *S. cerevisiae* in bottled sweet wines (Zoecklein et al., 1995) and toward the growth of film-forming yeasts (*Candida* spp.) on the wine surface (Ribéreau-Gayon et al., 2006). Nowadays it is hardly used for its limited effect and the possible negative effects on consumer's health. Some of these treatments are not really alternatives because of their limited microbial effects but are recommended to be used together to reduce the SO_2 dosage (Ribéreau-Gayon et al., 2006).

Some other additives, also common in winemaking for other reasons are also known to have some antimicrobial action against wine spoilage microorganisms. Among them, we can mention the phenolic compounds (Silva et al., 2018) or chitosan (Ferreira et al., 2013; Valera et al., 2017). Due to the interest to reduce or eliminate the use of SO_2, many other compounds are being tested, although they are not yet authorized in the EU. Among them, we can mention nisin, basically for the treatments against LAB (Rojo-Bezares et al., 2007), silver nanomaterials, active against yeasts, LAB and AAB (Garde-Cerdán et al., 2014), or hydroxytyrosol active also against the three kinds of wine microorganisms (Ruiz-Moreno et al., 2015). Finally, saturated short-chain fatty acids were also used to control the growth of some spoilage yeasts (Ribéreau-Gayon et al., 2006).

An option that is gaining interest is the use of some microorganisms able to inhibit the growth of other microorganisms through several mechanisms, among them, cell-to-cell contact (Nissen & Arneborg, 2003) or antimicrobial peptides (Albergaria et al., 2010). This option is named as biocontrol. In fact, it has been described that the interaction between yeasts induces the *viable but not culturable* states as a mechanism to overgrow the other yeasts and take over the AF (Branco et al., 2015; Wang et al., 2016). Even *S. cerevisiae* can enter this state in presence of other non-*Saccharomyces* species (Navarro et al., 2020). Thus, biocontrol, or the use of certain yeasts to limit the growth of others is a very attractive line of research.

Furthermore, some other alternatives for microbial stabilization have been considered, mostly physical treatments. Among them, microfiltration is probably the most useful at cellar level. However, several concerns have been raised regarding wine quality as microfiltration will also remove colorant matter, other macromolecules, and even volatile compounds, which will be very detrimental for wine quality due to its sensory impact (Lisanti et al., 2019). Thermal treatments are also a possibility, although their impact on sensory attributes limits its application to low quality wines exclusively (Ribéreau-Gayon et al., 2006). Other physical methods, such as high hydrostatic pressure, ultrasound, pulsed electric fields, ultraviolet irradiation, and microwave, successfully used in the last few years for the microbiological stabilization of wine as alternative to the use of SO_2 should be considered still far from a routine use in cellars.

Although exhibiting a certain microbial inhibition, no physical or chemical treatment has to date shown to be able to replace the efficiency and the broad spectrum of antimicrobial action of SO_2

(Santos et al., 2012). Thus, the main challenge when reducing SO_2 or substituting it by chemical compounds or physical treatments would be the microbial control during and after fermentation in addition to the control of the organoleptic properties of the produced wine.

The improvement of the tools for microbial monitoring described in the previous sections could be good help for the microbial control. However, those tools are still far from being useful at cellar level, as they are costly, time-consuming, and with complex interpretation. Adequation of those methodologies to cellar level is far from being practical, although it might be a research and transfer objective.

13.3.3 Spontaneous versus inoculated fermentations

Traditionally, AFs have proceeded spontaneously, with the microbiota that was already present on the grapes or resident in the winery. The spontaneous fermentations are normally slow and with unpredictable outcome, as it depends on the microbiota present and its capability to overcome the other yeasts. The wine is normally considered that reflects the "terroir" typicality, but it might have many risks of spoilage. The control of all fermentative processes is normally done by starter cultures that could be from a fermenting substrate or pure microbiological cultures. Thus, in wine making, we might have spontaneous fermentation (without any starter culture) or inoculated fermentations when a starter is used.

Either a fermenting substrate or from pure microbiological culture the inoculation has been traditionally done by pied-de-cuve. In those cases, the name comes from the "bottom of the deposit" that means that a 5%—10% of the total volume of the deposit is filled with an actively fermenting must and the rest of the deposit is filled up with fresh must. In this way, as the fermenting must have a very high concentration of yeast that are very active (typically could be between 10^7-10^8 cells/mL) could easily take over the fermentation of the whole deposit (the population reduction of one log unit is not relevant, as yeasts are already active and growing). With this mechanism, the winemakers ensure a quick fermentation start and a good rate. If the pied-de-cuve is derived from a single culture (normally a selected yeast strain), this yeast must be propagated in optimal culture medium until it achieves a volume that can be used as pied-de-cuve. Often the last passages are done with must either sterilized or with low indigenous population. In this way, the selected strain will take over the fermentation and provide the final wine with the characteristics that the strain can develop in the wine, although this is not the case with the pied-de-cuve from fermenting vats, as they are the result of a mixed inoculum. However, the propagation needs a laboratory where minimal sterile conditions could be kept as well as it is a slow process that may take several days.

During the last century, many different strains of *S. cerevisiae* have been selected to be used as starter cultures to repress the wild microorganisms and achieve more predictable and desired outcomes. A big step forward in the use of starter cultures was the development of the active dry wine yeasts, where yeasts are dehydrated maintaining their full activity that is restored quickly after rehydration (Fleet, 1993). This must be considered a cellar-friendly procedure, as yeasts could be rehydrated in less than 30 min in the same cellar, facilitating the seeding of high numbers of yeast cells that are fully active and can initiate the AF quickly and effectively. In this way, the fermentation proceeds very fast and with good fermentation rate (Fig. 13.2).

However, these inoculated fermentations present the risk of uniformity, as selected yeasts provide a limited diversity of the final wines (Fleet, 1993). Against this "uniformization," several strategies have been in use: selection of local yeasts or mixed inoculation with selected non-*Saccharomyces* yeast.

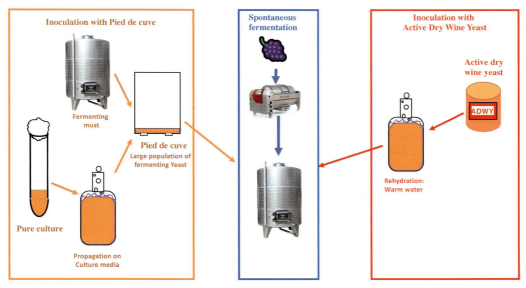

FIGURE 13.2

Inoculated and spontaneous alcoholic fermentations.

Recent movements of nonconventional wine making (organic, biodynamic, natural, etc.) have challenged the use of active dry yeast. A good alternative in these cases could be the use of pied-de-cuve that can be derived from small volumes of fermenting early musts that could be selected according to some variables (good fermentation activity and good sensory attributes), although the microbiological control will not be optimal, as there will be a mixed microbial population.

Grapes harbor a complex microbial community of fungi, bacteria, and between 10^4–10^6 yeasts cells per gram of grapes (Fleet, 2003), which are mainly non-*Saccharomyces* yeasts. The populations of *Saccharomyces* are indeed very low in grapes (Beltran et al., 2002). These populations change when they enter in contact with the cellar environment where they join the resident microbiota. In fact, the cellar is a good niche for *S. cerevisiae*, which becomes the main cellar-resident yeast (Beltran et al., 2002). Although the grape must is a very complex medium and can provide support for many different microorganisms, there are some characteristics that transform such universal medium into a very restrictive one. The high sugar concentration that derives in high osmotic pressure and low water activity; the high concentration of organic acids, with pH between 3 and 4; and the unbalance between nitrogen carbon sources makes the grape must a very selective medium. Thus, the initial grape juice only supports the growth of certain microbial species favoring the development of fermentative yeasts. Overall, species of *Hanseniaspora*, *Candida*, and *Metschnikowia* genera begin the fermentation process. Species of *Pichia*, *Issatchenkia*, and *Kluyveromyces* can also develop during this stage. These yeasts species may grow up to 10^6–10^7 cfu/mL of must until midfermentation when their population sharply decay. At this moment, *S. cerevisiae* becomes predominant, reaching populations of 10^7–10^8 cfu/mL, until the fermentation is completed. Nevertheless, the microbial succession occasionally can lead to stuck or sluggish fermentations as a result of an excessive proliferation of nonfermentative yeasts that consume nutrients needed for the development of the fermentative ones (Ciani et al., 2006; Medina et al., 2012; Padilla et al., 2016).

Thus, the inoculation of *Saccharomyces* starters is a tool for the wine maker to define wine production and quality. In fermentations without use of starters (spontaneous fermentations), the native microbiota, mostly non-*Saccharomyces*, proliferate for several days, producing different compounds that could improve the organoleptic quality of the wines, although it also includes a risk of spoilage and sluggish or stuck fermentations. The improvement has been correlated to the presence of interesting enzymatic activities, some of them of technological interest (pectinolytic activities that facilitate procedures in the cellar) or to improve the final wine (esterases, beta-glucosidase, etc.) (Jolly et al., 2014). Additionally, these Non-*Saccharomyces* yeast may be able to reduce ethanol (Gonzalez et al., 2013), which has been proposed as a key objective in current winemaking due to the increased concentration of sugars, among other effects, derived from climate change (Mira de Orduña, 2010). Nevertheless, the return to spontaneous wine fermentations may have considerable drawbacks especially in terms of economic losses, as these wines have much higher risks of presenting different levels of spoilage (presence of unwanted compounds that will be organoleptically detectable) that will not be acceptable for the consumer. Alternative microbial starters used in mixed or sequential fermentations, mainly non-*Saccharomyces*, have received increasing attention for their potential to produce wines with more distinctive and typical features (Jolly et al., 2014). This topic will be covered in the next section.

In order to take advantages of both inoculated and spontaneous fermentations and to improve certain wine characteristics, mixed and sequential fermentations using *S. cerevisiae* and different yeast strains or malolactic bacteria have attracted recent research interest (reviewed in Petruzzi et al., 2017). For example, it has been shown that *Torulaspora delbrueckii* enhances the complexity and fruity notes of wines (Renault et al., 2015). *Hanseniaspora vineae* enriches wines with fruity and flowery aromas (Lleixà, Martín, et al., 2016), *L. thermotolerans* increases the total acidity (Gobbi et al., 2013) and *M. pulcherrima* reduces the ethanol levels and enhances varietal aromas (Medina et al., 2012; Quirós et al., 2014). The increasing number of species used, often associated to new isolations from spontaneous fermentations (Garofalo et al., 2015; Padilla et al., 2016) introduces a relevant challenge in terms of interspecific interactions (Ciani & Comitini, 2015; Tronchoni et al., 2017; Wang et al., 2016). For example, to optimize the use of non-*Saccharomyces* yeasts or bacteria in mixed or sequential fermentations with *Saccharomyces* spp., it is necessary to better understand their metabolism and nutrient requirements. During a sequential inoculation, the initial consumption of nutrients by non-*Saccharomyces* yeasts could affect the growth and survival of *Saccharomyces* yeasts, inoculated later (Lleixà, Manzano, et al., 2016; Medina et al., 2012; Roca-Mesa et al., 2020). Furthermore, we have to consider that different grape varieties and batch volumes could influence the growth and final biomass of yeasts in mixed fermentations (Gobbi et al., 2013; Padilla et al., 2017). Thus, the field of interspecific interactions is of particular interest and necessary to scale from laboratory to industrial or semiindustrial scale.

13.3.4 The search for new strains

More than 200 commercial strains of wine yeast available on the market are used by winemakers to produce different types, varieties, and brands of wines. However, due to the highly competitive wine market with new demands for improved wine quality, it has become increasingly critical to develop new wine strains (Bisson, 2004).

Besides the isolations of new species of yeasts mainly from spontaneous fermentations (Garofalo et al., 2015; Padilla et al., 2016; Torija et al., 2001), the new genetic tools allow the metabolic engineering of known strains. Classical strain improvement methods based on the repeated alternation of successive stages of mutagenesis and selection have frequently been used to obtain starter cultures of wine strains (Steensels et al., 2014). These methods are quite lengthy and time-consuming because they require screening of a significant number of isolates. In fact, they have now been replaced by adaptive or directed laboratory evolution methods (ALE) which are more targeted and convenient (Sandberg et al., 2019). ALE technique is based on the selection of candidate strains through serial or continuous culturing of a particular yeast strain for many generations under selective pressure (i.e., high ethanol or high osmolarity) and has been applied successfully in previous studies (Betlej et al., 2020; Kutyna et al., 2012; McBryde et al., 2006).

Recently, novel methodologies for precise wine strain engineering based on better molecular knowledge have emerged due to the rapid progress in genomic studies with wine yeast strains, especially in *S. cerevisiae* strains (reviewed in Eldarov and Mardanov (2020)). An example of this new approaches would be the CRISPR/Cas9 tool.

Nevertheless, when taking in consideration the real commercial implementation of all these and other advances, a barrier arises: engineered yeasts are usually considered genetically modified organisms (GMOs) and legal issues impede its use. To the best of our knowledge, so far only two strains have been allowed for commercial implementation (Coulon et al., 2006; Husnik et al., 2006) although they are not extendedly used. In the European Union, Regulation (EC) 1829/2003 sets the legislation on genetically modified food and feed and postpone the use of GMO until better times.

13.4 Concluding remarks

Wine making is characterized for being a microbiological driven process where the biological control is a requirement for safety, reproducibility, and consumer acceptance. Although the process is open to incorporate the new tools that have been developed in recent years, the winemaker and consumer reluctance to some of those novelties as well as the reality of the cellars and the technological and expertise requirements of some of these methodologies makes their use in cellars still very limited. For instance, massive sequencing could be a very helpful methodology to monitor fermentations or postfermentative processes (aging, for instance) as well as safety control of the product. However, present development, costs, expertise, and timing makes their cellar application almost inviable. On the other hand, the use of CRISPR-Cas9 methodology that could generate mutant strains that incorporate winemaking improvements (for instance, reduction of urea and ethyl carbamate, increase of glycerol, reduced ethanol, improved aromatic expression) face the challenge of being considered GM and thus, the consumer rejection or the regulation limitation.

Acknowledgments

This manuscript has been prepared a within the framework of the SUMCULA and OENOBIO Erasmus + Program Projects (2017-1-SE01-KA203-034570 and 2018-1-FR01-KA203-047839, respectively).

References

Alañón, M. E., Pérez-Coello, M. S., & Marina, M. L. (2015). Wine science in the metabolomics era. *Trends in Analytical Chemistry, 74*, 1–20.

Albergaria, H., Francisco, D., Gori, K., Arneborg, N., & Gírio, F. (2010). *Saccharomyces cerevisiae* CCMI 885 secretes peptides that inhibit the growth of some non-*Saccharomyces* wine-related strains. *Applied Microbiology and Biotechnology, 86*, 965–972.

Alonso-del-Real, J., Pérez-Torrado, R., Querol, A., & Barrio, E. (2019). Dominance of wine *Saccharomyces cerevisiae* strains over *S. kudriavzevii* in industrial fermentation competitions is related to an acceleration of nutrient uptake and utilization. *Environmental Microbiology, 21*, 1627–1644.

Alves, Z., Melo, A., Figueiredo, A. R., Coimbra, M. A., Gomes, A. C., & Rocha, S. M. (2015). Exploring the *Saccharomyces cerevisiae* volatile metabolome: Indigenous versus commercial strains. *PLoS One, 10*, e0143641.

Andorrà, I., Esteve-Zarzoso, B., Guillamón, J. M., & Mas, A. (2010). Determination of viable wine yeast using DNA binding dyes and quantitative PCR. *International Journal of Food Microbiology, 144*, 257–262.

Arapitsas, P., Guella, G., & Mattivi, F. (2018). The impact of SO_2 on wine flavanols and indoles in relation to wine style and age. *Scientific Reports, 8*, 858.

Barata, A., Malfeito-Ferreira, M., & Loureiro, V. (2012). The microbial ecology of wine grape berries. *International Journal of Food Microbiology, 153*, 243–259.

Barbosa, C., García-Martínez, J., Pérez-Ortín, J. E., & Mendes-Ferreira, A. (2015). Comparative transcriptomic analysis reveals similarities and dissimilarities in *Saccharomyces cerevisiae* wine strains response to nitrogen availability. *PLoS One, 10*, e0122709.

Bartowsky, E. J. (2009). Bacterial spoilage of wine and approaches to minimize it. *Letters in Applied Microbiology, 48*, 149–156.

Belda, I., Gobbi, A., Ruiz, J., de Celis, M., Ortiz-Álvarez, R., Acedo, A., & Santos, A. (2021). Microbiomics to define wine terroir. In A. Cifuentes (Ed.), *Comprehensive foodomics* (pp. 438–451). Elsevier. https://doi.org/10.1016/B978-0-08-100596-5.22875-8

Belda, I., Ruiz, J., Esteban-Fernández, A., Navascués, E., Marquina, D., Santos, A., & Moreno-Arribas, M. (2017). Microbial contribution to wine aroma and its intended use for wine quality improvement. *Molecules, 22*, 189.

Beltran, G., Torija, M. J., Novo, M., Ferrer, N., Poblet, M., Guillamón, J. M., Rozès, N., & Mas, A. (2002). Analysis of yeast populations during alcoholic fermentation: A six year follow-up study. *Systematic & Applied Microbiology, 25*, 287–293.

Betlej, G., Bator, E., Oklejewicz, B., Potocki, L., Górka, A., Slowik-Borowiec, M., Czarny, W., Domka, W., & Kwiatkowska, A. (2020). Long-term adaption to high osmotic stress as a tool for improving enological characteristics in industrial wine yeast. *Genes (Basel), 11*, 576.

Bisson, L. F. (2004). The biotechnology of wine yeast. *Food Biotechnology, 18*, 63–96.

Bokulich, N. A., Collins, T. S., Masarweh, C., Allen, G., Heymann, H., Ebeler, S. E., & Millsa, D. A. (2016). Associations among wine grape microbiome, metabolome, and fermentation behavior suggest microbial contribution to regional wine characteristics. *mBio, 7*. e00631–16.

Bokulich, N. A., Ohta, M., Richardson, P. M., & Mills, D. A. (2013). Monitoring seasonal changes in winery-resident microbiota. *PLoS One, 8*, e66437.

Bokulich, N. A., Thorngate, J. H., Richardson, P. M., & Mills, D. A. (2014). Microbial biogeography of wine grapes is conditioned by cultivar, vintage, and climate. *Proceedings of the National Academy of Sciences, 111*, E139–E148.

Bordet, F., Joran, A., Klein, G., Roullier-Gall, C., & Alexandre, H. (2020). Yeast-yeast interactions: Mechanisms, methodologies and impact on composition. *Microorganisms, 8*, 600.

Branco, P., Viana, T., Albergaria, H., & Arneborg, N. (2015). Antimicrobial peptides (AMPs) produced by *Saccharomyces cerevisiae* induce alterations in the intracellular pH, membrane permeability and culturability of Hanseniaspora guilliermondii cells. *International Journal of Food Microbiology, 205*, 112−118.

Capozzi, V., Garofalo, C., Chiriatti, M. A., Grieco, F., & Spano, G. (2015). Microbial terroir and food innovation: The case of yeast biodiversity in wine. *Microbiological Research.*

Ciani, M., Beco, L., & Comitini, F. (2006). Fermentation behaviour and metabolic interactions of multistarter wine yeast fermentations. *International Journal of Food Microbiology, 108*, 239−245.

Ciani, M., & Comitini, F. (2015). Yeast interactions in multi-starter wine fermentation. *Current Opinion In Food Science, 1*, 1−6.

Coulon, J., Husnik, J. I., Inglis, D. L., van der Merwe, G. K., Lonvaud, A., Erasmus, D. J., & van Vuuren, H. J. J. (2006). Metabolic engineering of *Saccharomyces cerevisiae* to minimize the production of ethyl carbamate in wine. *American Journal of Enology and Viticulture, 57*, 113−124.

Couto, J. A., Neves, F., Campos, F., & Hogg, T. (2005). Thermal inactivation of the wine spoilage yeasts *Dekkera/Brettanomyces*. *International Journal of Food Microbiology, 104*, 337−344.

Cozzolino, D. (2016). Metabolomics in grape and wine: Definition, current status and future prospects. *Food Analytical Methods, 9*, 2986−2997.

Curiel, J. A., Morales, P., Gonzalez, R., & Tronchoni, J. (2017). Different non-*Saccharomyces* yeast species stimulate nutrient consumption in *S. cerevisiae* mixed cultures. *Frontiers in Microbiology, 8*, 2121.

De Filippis, F., Parente, E., & Ercolini, D. (2018). Recent past, present, and future of the food microbiome. *Annual Review of Food Science and Technology, 9*, 589−608.

Dumont, M. G., Pommerenke, B., & Casper, P. (2013). Using stable isotope probing to obtain a targeted metatranscriptome of aerobic methanotrophs in lake sediment. *Environmental Microbiology Reports, 5*, 757−764.

Eldarov, M. A., & Mardanov, A. V. (2020). Metabolic engineering of wine strains of *Saccharomyces cerevisiae*. *Genes (Basel), 11*, 964.

Ferreira, D., Moreira, D., Costa, E., Silva, S., Pintado, M., & Couto, J. (2013). The antimicrobial action of chitosan against the wine spoilage yeast *Brettanomyces/Dekkera*. *Journal of Chitin and Chitosan Science, 1*, 240−245.

Fleet, G. H. (1993). *Wine microbiology and biotechnology*. Chur, Switzerland: Harwood Academic Publishers.

Fleet, G. H. (2003). Yeast interactions and wine flavour. *International Journal of Food Microbiology, 86*, 11−22.

Garde-Cerdán, T., López, R., Garijo, P., González-Arenzana, L., Gutiérrez, A. R., López-Alfaro, I., & Santamaría, P. (2014). Application of colloidal silver versus sulfur dioxide during vinification and storage of Tempranillo red wines. *Australian Journal of Grape and Wine Research, 20*, 51−61.

Garofalo, C., El Khoury, M., Lucas, P., Bely, M., Russo, P., Spano, G., & Capozzi, V. (2015). Autochthonous starter cultures and indigenous grape variety for regional wine production. *Journal of Applied Microbiology, 118*, 1395−1408.

Gobbi, M., Comitini, F., Domizio, P., Romani, C., Lencioni, L., Mannazzu, I., & Ciani, M. (2013). *Lachancea thermotolerans* and *Saccharomyces cerevisiae* in simultaneous and sequential co-fermentation: A strategy to enhance acidity and improve the overall quality of wine. *Food Microbiology, 33*, 271−281.

Godálová, Z., Kraková, L., Puškárová, A., Bučková, M., Kuchta, T., Piknová, Ľ., & Pangallo, D. (2016). Bacterial consortia at different wine fermentation phases of two typical Central European grape varieties: Blaufränkisch (Frankovka modrá) and Grüner Veltliner (Veltlínske zelené). *International Journal of Food Microbiology, 217*, 110−116.

González-Jiménez del, M. C., Moreno-García, J., García-Martínez, T., Moreno, J. J., Puig-Pujol, A., Capdevilla, F., & Mauricio, J. C. (2020). Differential analysis of proteins involved in ester metabolism in two *Saccharomyces cerevisiae* strains during the second fermentation in sparkling wine elaboration. *Microorganisms, 8*, 403.

Gonzalez, R., Quirós, M., & Morales, P. (2013). Yeast respiration of sugars by non-Saccharomyces yeast species: A promising and barely explored approach to lowering alcohol content of wines. *Trends in Food Science & Technology, 29*, 55–61.

Goold, H. D., Kroukamp, H., Williams, T. C., Paulsen, I. T., Varela, C., & Pretorius, I. S. (2017). Yeast's balancing act between ethanol and glycerol production in low-alcohol wines. *Microbial Biotechnology, 10*, 264–278.

Huang, L. H., Yuan, M. Q., Ao, X. J., Ren, A. Y., Zhang, H. B., & Yang, M. Z. (2018). Endophytic fungi specifically introduce novel metabolites into grape flesh cells in vitro. *PLoS One, 13*, e0196996.

Husnik, J. I., Volschenk, H., Bauer, J., Colavizza, D., Luo, Z., & van Vuuren, H. J. J. (2006). Metabolic engineering of malolactic wine yeast. *Metabolic Engineering, 8*, 315–323.

Jakočiunas, T., Jensen, M. K., & Keasling, J. D. (2016). CRISPR/Cas9 advances engineering of microbial cell factories. *Metabolic Engineering, 34*, 44–59.

Jolly, N. P., Varela, C., & Pretorius, I. S. (2014). Not your ordinary yeast: Non-*Saccharomyces* yeasts in wine production uncovered. *FEMS Yeast Research, 14*, 215–237.

Kioroglou, D., Lleixá, J., Mas, A., & Portillo, M. C. (2018). Massive sequencing: A new tool for the control of alcoholic fermentation in wine? *Fermentation, 4*, 7.

Kioroglou, D., Mas, A., & Portillo, M. C. (2020). High-throughput sequencing approach to analyze the effect of aging time and barrel usage on the microbial community composition of red wines. *Frontiers in Microbiology, 11*, 2142.

Kutyna, D. R., Varela, C., Stanley, G. A., Borneman, A. R., Henschke, P. A., & Chambers, P. J. (2012). Adaptive evolution of *Saccharomyces cerevisiae* to generate strains with enhanced glycerol production. *Applied Microbiology and Biotechnology, 93*, 1175–1184.

Lisanti, M. T., Blaiotta, G., Nioi, C., & Moio, L. (2019). Alternative methods to SO_2 for microbiological stabilization of wine. *Comprehensive Reviews in Food Science and Food Safety, 18*, 455–479.

Lleixà, J., Manzano, M., Mas, A., & Portillo del, M. C. (2016). *Saccharomyces* and non-*Saccharomyces* competition during microvinification under different sugar and nitrogen conditions. *Frontiers in Microbiology, 7*, 1959. https://doi.org/10.3389/fmicb.2016.01959

Lleixà, J., Martín, V., Portillo del, M. C., Carrau, F., Beltran, G., & Mas, A. (2016). Comparison of fermentation and wines produced by inoculation of Hanseniaspora vineae and *Saccharomyces cerevisiae*. *Frontiers in Microbiology, 7*, 338.

López-Malo, M., Querol, A., & Guillamon, J. M. (2013). Metabolomic comparison of *Saccharomyces cerevisiae* and the cryotolerant species *S. bayanus* var. uvarum and *S. kudriavzevii* during wine fermentation at low temperature. *PLoS One, 8*, e60135.

Loureiro, V., & Malfeito-Ferreira, M. (2003). Spoilage yeasts in the wine industry. *International Journal of Food Microbiology, 86*, 23–50.

Mazzei, P., Spaccini, R., Francesca, N., Moschetti, G., & Piccolo, A. (2013). Metabolomic by 1H NMR spectroscopy differentiates "fiano di Avellino" white wines obtained with different yeast strains. *Journal of Agricultural and Food Chemistry, 61*, 10816–10822.

McBryde, C., Gardner, J. M., de Barros Lopes, M., & Jiranek, V. (2006). Generation of novel wine yeast strains by adaptive evolution. *American Journal of Enology and Viticulture, 57*, 423–430.

Medina, K., Boido, E., Dellacassa, E., & Carrau, F. (2012). Growth of non-*Saccharomyces* yeasts affects nutrient availability for *Saccharomyces cerevisiae* during wine fermentation. *International Journal of Food Microbiology, 157*, 245–250.

Mencher, A., Morales, P., Valero, E., Tronchoni, J., Patil, K. R., & Gonzalez, R. (2020). Proteomic characterization of extracellular vesicles produced by several wine yeast species. *Microbial Biotechnology, 13*, 1581–1596.

Mira de Orduña, R. (2010). Climate change associated effects on grape and wine quality and production. *Food Research International, 43*, 1844–1855.

Muysson, J., Miller, L., Allie, R., & Inglis, D. L. (2019). The use of CRISPR-Cas9 genome editing to determine the importance of glycerol uptake in wine yeast during icewine fermentation. *Fermentatio, 5*, 93.

Navarro, Y., Torija, M. J., Mas, A., & Beltran, G. (2020). Viability-PCR allows monitoring yeast population dynamics in mixed fermentations including viable but non-culturable yeasts. *Foods, 9*, 1373.

Nissen, P., & Arneborg, N. (2003). Characterization of early deaths of non-*Saccharomyces* yeasts in mixed cultures with *Saccharomyces cerevisiae*. *Archives of Microbiology, 180*, 257−263.

Nookaew, I., Papini, M., Pornputtapong, N., Scalcinati, G., Fagerberg, L., Uhlén, M., & Nielsen, J. (2012). A comprehensive comparison of RNA-seq-based transcriptome analysis from reads to differential gene expression and cross-comparison with microarrays: A case study in *Saccharomyces cerevisiae*. *Nucleic Acids Research, 40*, 10084−10097.

Padilla, B., García-Fernández, D., González, B., Izidoro, I., Esteve-Zarzoso, B., Beltran, G., & Mas, A. (2016). Yeast biodiversity from DOQ Priorat uninoculated fermentations. *Frontiers in Microbiology, 7*, 930.

Padilla, B., Zulian, L., Ferreres, À., Pastor, R., Esteve-Zarzoso, B., Beltran, G., & Mas, A. (2017). Sequential inoculation of native non-*Saccharomyces* and *Saccharomyces cerevisiae* strains for wine making. *Frontiers in Microbiology, 8*, 1293.

Peng, C., Andersen, B., Arshid, S., Larsen, M. R., Albergaria, H., Lametsch, R., & Arneborg, N. (2019). Proteomics insights into the responses of *Saccharomyces cerevisiae* during mixed-culture alcoholic fermentation with *Lachancea thermotolerans*. *FEMS Microbiology Ecology, 95*, fiz126.

Peng, C., Viana, T., Petersen, M. A., Larsen, F. H., & Arneborg, N. (2018). Metabolic footprint analysis of metabolites that discriminate single and mixed yeast cultures at two key time-points during mixed culture alcoholic fermentations. *Metabolomics, 14*, 93.

Petitgonnet, C., Klein, G. L., Roullier-Gall, C., Schmitt-Kopplin, P., Quintanilla-Casas, B., Vichi, S., Julien-David, D., & Alexandre, H. (2019). Influence of cell-cell contact between *L. thermotolerans* and *S. cerevisiae* on yeast interactions and the exo-metabolome. *Food Microbiology, 83*, 122−133.

Petruzzi, L., Capozzi, V., Berbegal, C., Corbo, M. R., Bevilacqua, A., Spano, G., & Sinigaglia, M. (2017). Microbial resources and enological significance: Opportunities and benefits. *Frontiers in Microbiology, 8*, 995.

Pinto, C., Pinho, D., Cardoso, R., Custódio, V., Fernandes, J., Sousa, S., Pinheiro, M., Egas, C., & Gomes, A. C. (2015). Wine fermentation microbiome: A landscape from different Portuguese wine appellations. *Frontiers in Microbiology, 6*, 905.

Pretorius, I. (2016). Conducting wine symphonics with the aid of yeast genomics. *Beverages, 2*, 36.

Pretorius, I. S. (2017). Synthetic genome engineering forging new frontiers for wine yeast. *Critical Reviews in Biotechnology, 37*, 112−136.

Quirós, M., Rojas, V., Gonzalez, R., & Morales, P. (2014). Selection of non-*Saccharomyces* yeast strains for reducing alcohol levels in wine by sugar respiration. *International Journal of Food Microbiology, 181*, 85−91.

Raschmanová, H., Weninger, A., Glieder, A., Kovar, K., & Vogl, T. (2018). Implementing CRISPR-Cas technologies in conventional and non-conventional yeasts: Current state and future prospects. *Biotechnology Advances, 36*, 641−665.

Renault, P., Coulon, J., de Revel, G., Barbe, J.-C., & Bely, M. (2015). Increase of fruity aroma during mixed *T. delbrueckii/S. cerevisiae* wine fermentation is linked to specific esters enhancement. *International Journal of Food Microbiology, 207*, 40−48.

Renouf, V., Claisse, O., & Lonvaud-Funel, A. (2005). Understanding the microbial ecosystem on the grape berry surface through numeration and identification of yeast and bacteria. *Australian Journal of Grape and Wine Research, 11*, 316−327.

Ribéreau-Gayon, P., Dubourdieu, D., Donèche, B., & Lonvaud, A. (2006). *Handbook of enology: the microbiology of wine and vinifications, handbook of enology* (Vol. 2). John Wiley & Sons.

Richter, C. L., Kennedy, A. D., Guo, L., & Dokoozlian, N. (2015). Metabolomic measurements at three time points of a chardonnay wine fermentation with *Saccharomyces cerevisiae*. *American Journal of Enology and Viticulture, 68*, 294—301.

Rizzotti, L., Levav, N., Fracchetti, F., Felis, G. E., & Torriani, S. (2015). Effect of UV-C treatment on the microbial population of white and red wines, as revealed by conventional plating and PMA-qPCR methods. *Food Control, 47*, 407—412.

Roca-Mesa, H., Sendra, S., Mas, A., Beltran, G., & Torija, M. J. (2020). Nitrogen preferences during alcoholic fermentation of different non-*Saccharomyces* yeasts of oenological interest. *Microorganisms, 8*, 157.

Rojo-Bezares, B., Sáenz, Y., Zarazaga, M., Torres, C., & Ruiz-Larrea, F. (2007). Antimicrobial activity of nisin against *Oenococcus oeni* and other wine bacteria. *International Journal of Food Microbiology, 116*, 32—36.

Rossouw, D., Bagheri, B., Setati, M. E., & Bauer, F. F. (2015). Co-flocculation of yeast species, a new mechanism to govern population dynamics in microbial ecosystems. *PLoS One, 10*, e0136249.

Roullier-Gall, C., David, V., Hemmler, D., Schmitt-Kopplin, P., & Alexandre, H. (2020). Exploring yeast interactions through metabolic profiling. *Scientific Reports, 10*, 6073.

Ruiz-Moreno, M. J., Raposo, R., Moreno-Rojas, J. M., Zafrilla, P., Cayuela, J. M., Mulero, J., Puertas, B., Guerrero, R. F., Piñeiro, Z., Giron, F., & Cantos-Villar, E. (2015). Efficacy of olive oil mill extract in replacing sulfur dioxide in wine model. *Lebensmittel-Wissenschaft und -Technologie- Food Science and Technology, 61*, 117—123.

Sadoudi, M., Rousseaux, S., David, V., Alexandre, H., & Tourdot-Maréchal, R. (2017). Metschnikowia pulcherrima influences the expression of genes involved in PDH bypass and glyceropyruvic fermentation in *Saccharomyces cerevisiae*. *Frontiers in Microbiology, 8*, 1137.

Sandberg, T. E., Salazar, M. J., Weng, L. L., Palsson, B. O., & Feist, A. M. (2019). The emergence of adaptive laboratory evolution as an efficient tool for biological discovery and industrial biotechnology. *Metabolic Engineering, 56*, 1—16.

Santos, M. C., Nunes, C., Saraiva, J. A., & Coimbra, M. A. (2012). Chemical and physical methodologies for the replacement/reduction of sulfur dioxide use during winemaking: Review of their potentialities and limitations. *European Food Research and Technology, 234*, 1—12.

Shekhawat, K., Patterton, H., Bauer, F. F., & Setati, M. E. (2019). RNA-seq based transcriptional analysis of *Saccharomyces cerevisiae* and *Lachancea thermotolerans* in mixed-culture fermentations under anaerobic conditions. *BMC Genomics, 20*, 145.

Silva, V., Igrejas, G., Falco, V., Santos, T. P., Torres, C., Oliveira, A. M. P., Pereira, J. E., Amaral, J. S., & Poeta, P. (2018). Chemical composition, antioxidant and antimicrobial activity of phenolic compounds extracted from wine industry by-products. *Food Control, 92*, 516—522.

Sirén, K., Mak, S. S. T., Fischer, U., Hansen, L. H., & Gilbert, M. T. P. (2019). Multi-omics and potential applications in wine production. *Current Opinion in Biotechnology, 56*, 172—178.

Sirén, K., Mak, S. S. T., Melkonian, C., Carøe, C., Swiegers, J. H., Molenaar, D., Fischer, U., & Thomas P Gilbert, M. (2019). Taxonomic and functional characterization of the microbial community during spontaneous in vitro fermentation of riesling must. *Frontiers in Microbiology, 10*, 697.

Steensels, J., Snoek, T., Meersman, E., Nicolino, M. P., Voordeckers, K., & Verstrepen, K. J. (2014). Improving industrial yeast strains: Exploiting natural and artificial diversity. *FEMS Microbiology Reviews, 38*, 947—995.

Stefanini, I., & Cavalieri, D. (2018). Metagenomic approaches to investigate the contribution of the vineyard environment to the quality of wine fermentation: Potentials and difficulties. *Frontiers in Microbiology, 9*, 991.

Sternes, P. R., Lee, D., Kutyna, D. R., & Borneman, A. R. (2017). A combined meta-barcoding and shotgun metagenomic analysis of spontaneous wine fermentation. *Gigascience, 6*, gix040.

Suárez, R., Suárez-Lepe, J. A., Morata, A., & Calderón, F. (2007). The production of ethylphenols in wine by yeasts of the genera *Brettanomyces* and *Dekkera*: A review. *Food Chemistry, 102*, 10—21.

Sunyer-Figueres, M., Vázquez, J., Mas, A., Torija, M. J., & Beltran, G. (2020). Transcriptomic insights into the effect of melatonin in *Saccharomyces cerevisiae* in the presence and absence of oxidative stress. *Antioxidants, 9*, 947.

Torija, M. J., Rozès, N., Poblet, M., Guillamón, J. M., & Mas, A. (2001). Yeast population dynamics in spontaneous fermentations: Comparison between two different wine-producing areas over a period of three years. *Antonie van Leeuwenhoek, 79*, 345–352.

Tronchoni, J., Curiel, J. A., Morales, P., Torres-Pérez, R., & Gonzalez, R. (2017). Early transcriptional response to biotic stress in mixed starter fermentations involving *Saccharomyces cerevisiae* and *Torulaspora delbrueckii*. *International Journal of Food Microbiology, 241*, 60–68.

Valera, M. J., Sainz, F., Mas, A., & Torija, M. J. (2017). Effect of chitosan and SO_2 on viability of *Acetobacter* strains in wine. *International Journal of Food Microbiology, 246*, 1–4.

Van Wyk, N., Grossmann, M., Wendland, J., Von Wallbrunn, C., & Pretorius, I. S. (2019). The whiff of wine yeast innovation: Strategies for enhancing aroma production by yeast during wine fermentation. *Journal of Agricultural and Food Chemistry, 67*, 13496–13505.

Varela, C., Bartel, C., Onetto, C., & Borneman, A. (2020). Targeted gene deletion in *Brettanomyces bruxellensis* with an expression-free CRISPR-Cas9 system. *Applied Microbiology and Biotechnology, 104*, 7105–7115.

Vigentini, I., Gebbia, M., Belotti, A., Foschino, R., & Roth, F. P. (2017). CRISPR/Cas9 system as a valuable genome editing tool for wine yeasts with application to decrease urea production. *Frontiers in Microbiology, 8*, 2194.

Wang, C., Mas, A., & Esteve-Zarzoso, B. (2016). The interaction between *Saccharomyces cerevisiae* and non-*Saccharomyces* yeast during alcoholic fermentation is species and strain specific. *Frontiers in Microbiology, 7*, 502.

Weninger, A., Fischer, J. E., Raschmanová, H., Kniely, C., Vogl, T., & Glieder, A. (2018). Expanding the CRISPR/Cas9 toolkit for *Pichia pastoris* with efficient donor integration and alternative resistance markers. *Journal of Cellular Biochemistry, 119*, 3183–3198.

Wu, D., Xie, W., Li, X., Cai, G., Lu, J., & Xie, G. (2020). Metabolic engineering of *Saccharomyces cerevisiae* using the CRISPR/Cas9 system to minimize ethyl carbamate accumulation during Chinese rice wine fermentation. *Applied Microbiology and Biotechnology, 104*, 4435–4444.

van Wyk, N., Kroukamp, H., Espinosa, M. I., von Wallbrunn, C., Wendland, J., & Pretorius, I. S. (2020). Blending wine yeast phenotypes with the aid of CRISPR DNA editing technologies. *International Journal of Food Microbiology, 324*, 108615.

Yang, M. Z., Di Ma, M., Yuan, M. Q., Huang, Z. Y., Yang, W. X., Zhang, H. B., Huang, L. H., Ren, A. Y., & Shan, H. (2016). Fungal endophytes as a metabolic fine-tuning regulator for wine grape. *PLoS One, 11*, e0163186.

Zarraonaindia, I., Owens, S. M., Weisenhorn, P., West, K., Hampton-Marcell, J., Lax, S., Bokulich, N. A., Mills, D. A., Martin, G., Taghavi, S., van der Lelie, D., & Gilbert, J. A. (2015). The soil microbiome influences grapevine-associated microbiota. *mBio, 6*. e02527–14.

Zepeda-Mendoza, M. L., Edwards, N. K., Madsen, M. G., Abel-Kistrup, M., Puetz, L., Sicheritz-Ponten, T., & Swiegers, J. H. (2018). Influence of *Oenococcus oeni* and *Brettanomyces bruxellensis* on wine microbial taxonomic and functional potential profiles. *American Journal of Enology and Viticulture, 69*, 321–333.

Zhang, Yueping, Wang, J., Wang, Z., Zhang, Yiming, Shi, S., Nielsen, J., & Liu, Z. (2019). A gRNA-tRNA array for CRISPR-Cas9 based rapid multiplexed genome editing in *Saccharomyces cerevisiae*. *Nature Communications, 10*, 1053.

Zoecklein, B. W., Fugelsang, K. C., Gump, B. H., & Nury, F. S. (1995). In B. W. Zoecklein, K. C. Fugelsang, B. H. Gump, & F. S. Nury (Eds.), *Sorbic acid, benzoic acid, and dimethyldicarbonate BT - wine analysis and production* (pp. 209–215). Boston, MA: Springer US.

Further reading

Comitini, F., Agarbati, A., Canonico, L., & Ciani, M. (2021). Yeast Interactions and molecular mechanisms in wine fermentation: a comprehensive review. *International Journal of Molecular Sciences, 22*, 7754. https://doi.org/10.3390/ijms22147754

Di Gianvitto, P., Englezos, V., Rantsiou, K., & Cocolin, L. (2022). Bioprotection strategies in winemaking. *International Journal of Food Microbiology.* https://doi.org/10.1016/j.ijfoodmicro.2022.109532

Gonzalez, R., & Morales, P. (2021). Truth in wine yeast. *Microbial Biotechnology.* https://doi.org/10.1111/1751-7915.13848

Mencher, A., Morales, P., Tronchoni, J., & Gonzalez, R. (2021). Mechanisms Involved in Interspecific Communication between Wine Yeasts. *Foods, 10*, 1734. https://doi.org/10.3390/foods10081734

Morata, A., Loira, I., González, C., & Escott, C. (2021). Non-*saccharomyces* as biotools to control the production of off-flavors in wines. *Molecules, 26*, 4571. https://doi.org/10.3390/molecules26154571

Zilelidou, E. A., & Nisiotou, A. (2021). Understanding wine through yeast interactions. *Microorganisms, 9*, 1620. https://doi.org/10.3390/microorganisms9081620

CHAPTER 14

Sustainable use of wood in wine spirit production

Sara Canas[1,2], Ilda Caldeira[1,2], Tiago A. Fernandes[3,4], Ofélia Anjos[5,6,7], António Pedro Belchior[1] and Sofia Catarino[8,9]

[1]*Instituto Nacional de Investigação Agrária e Veterinária, Dois Portos, Portugal;* [2]*MED — Mediterranean Institute for Agriculture, Environment and Development, Instituto de formação avançada, Universidade de Évora, Évora, Portugal;* [3]*CQE — Centro de Química Estrutural, Associação do Instituto Superior Técnico para a Investigação e Desenvolvimento (IST-ID), Universidade de Lisboa, Lisboa, Portugal;* [4]*DCeT — Departamento de Ciências e Tecnologia, Universidade Aberta, Lisboa, Portugal;* [5]*Instituto Politécnico de Castelo Branco, Quinta da Senhora de Mércules, Castelo Branco, Portugal;* [6]*CEF — Centro de Estudos Florestais, Instituto Superior de Agronomia, Universidade de Lisboa, Lisboa, Portugal;* [7]*Centro de Biotecnologia de Plantas da Beira Interior, Castelo Branco, Portugal;* [8]*LEAF — Centre Linking Landscape, Environment, Agriculture and Food Research Center, Instituto Superior de Agronomia, University of Lisbon, Lisbon, Portugal;* [9]*CEFEMA — Center of Physics and Engineering of Advanced Materials, Instituto Superior Técnico, Universidade de Lisboa, Lisboa, Portugal*

14.1 Introduction

The aged wine spirit (WS), a spirit drink derived from the wine, holds a prominent position worldwide owing to its production, trade (IBISWorld, 2010), and consumption (Grigg, 2004) together with a long history and a relevant socioeconomic role, particularly in traditionally wine-producing countries (Belchior et al., 2015; Garreau, 2008).

Considering the WS production process, the aging stage stands out for its remarkable contribution to drink refinement given the specific physicochemical and related sensory characteristics imparted by the wood. In many wine-producing countries, tradition and legislation impose the use of wooden barrels at this stage. However, this technology has some drawbacks. Actually, economic, social, and environmental issues are associated with low production efficiency, high cost, and high demand for wood, which is a natural resource with limited availability. Besides, in the face of a global and more competitive market with more informed and demanding consumers, product differentiation based on innovation (Mitry & Smith, 2009) and sustainable processes (Whelan & Kronthal-Sacco, 2019) are increasingly relevant. This context supports the search for alternatives toward a more sustainable wood-based aging technology.

Sustainability is a comprehensive approach that involves environmental, social, and economic dimensions and their linkages aiming to balance ecology and prosperity in a lasting way (Purvis et al., 2019). This concept emerged in the late 20th century (Purvis et al., 2019) and has been gradually adopted by several economic sectors, including the beverages one. For this sector, particularly in

Europe, this challenge has been addressed together with food security, quality, nutrition and health, and consumer satisfaction, implying commitments in three main areas: sustainable supply, resource efficiency, and sustainable production and consumption patterns (FoodDrinkEurope, 2012). Regarding the subsector of spirit beverages, the European strategy for achieving sustainable production by 2050 should be highlighted given that the main WS production regions are located in Europe. For this purpose, the main focuses are: farming practices, water, energy, reuse of by-products, packaging; and transport (spiritsEUROPE). Interestingly, the wood, which is a pivotal resource for the aging of spirits, was not included because "In Europe, only oaks of at least 150 years can be used to make barrels. Coopers optimize the use of the raw material so that the entire tree is used, and waste is reduced. For instance, in France, the leading producer of barrels in the EU, wine and spirits producers work hand-in-hand with barrel producers to ensure oak used for casks is sourced from certified, sustainably managed forests which meet the Forest Stewardship Council standard. According to Forest Europe, the forest area in the EU increased over the last 25 years, due to sustainable management practices, such as those seen with French oak. Sustainable management means, amongst others, that cut trees will be replaced by new ones, which store more CO_2 than the 150-year-old trees used to make barrels" (spiritsEUROPE, 2020). However, such an assumption does not reflect the European forest situation reported by Ceccherini et al. (2020), which show an abrupt rise in the harvested forest area (49%) and biomass loss (69%) over EU26 countries for the period of 2016–2018 *vs.* 2011–2015. The increase in the rate of forest harvest is mainly ascribed to the wood demand. In addition, according to the European Green Deal (European Commission, 2019), an expansion of EU's forested area and related quality (biodiversity and management) are imperative to reach the intended environmental improvement. Lastly, it should be stressed that many other countries exist in which the forest, the cooperage industry, and the WS industry are not so interconnected as in France, and the wood from many forest species other than oaks are increasingly used in WS' aging.

Taking these facts into account, the amount of wood used in WS' aging is a matter of concern, and, therefore, it is a mandatory topic in sustainability strategies.

In this scenario, sustainability of wood-based WS aging implies the production of high-quality and differentiated aged beverage through a process that increases the producer's profitability, based on the efficient use of wood and its reuse to obtain new products (circular economy)—Fig. 14.1.

This chapter provides an overview of these topics and the research conducted over the last 30 years on WS aging technologies. The studies have been focused on the phenomena underlying aging, the factors that govern them, and their impact on physicochemical and sensory characteristics of the aged WS. In this approach, they will be examined in the light of their contribution to the aging's sustainability.

14.2 The aged wine spirit and its production process
14.2.1 Wine spirit definition

According to the International Organisation of Vine and Wine (OIV), whose guidelines are applicable at the international level, the WS is defined as "a spirit beverage obtained exclusively by the distillation of wine, fortified wine or wine added with wine distillate, or by re-distillation of a wine distillate whose final product retains the aroma and the flavour of the raw materials. The alcoholic strength of the end product must not be less than 37.5% v/v" (OIV, 2020). Nevertheless, more restrictive conditions are imposed by some producing regions, such as the European Union (Reg. EU 2019/787).

FIGURE 14.1

Main components of a sustainable wood-based technology for the aging of WS.

14.2.2 Technological process of aged wine spirit production

The production process of aged WS includes four main stages: winemaking, distillation, aging, and finishing. At the end, the aged WS is bottled (Fig. 14.2). Each stage may comprise some variations according to the production region and the producer in order to afford specific characteristics to the final product.

Hence, the quality of the final product is related to the characteristics of the wine, imparted by the grapes and by the winemaking stage, and to the characteristics of the WS before and especially after aging.

The main objective of winemaking is to obtain a suitable wine for distillation without any particular differentiation (Belchior et al., 2015).

The distillation has a comprehensive role based on the concentration of ethanol and volatile compounds, the formation of new volatiles and the elimination of other volatiles (Léauté, 1990). Despite the research performed aiming to improve the distillation systems to save energy, time, and ameliorate the WS quality (Cantagrel, 1992), few technological changes have been done on the traditional distillation (in alembic or in column still).

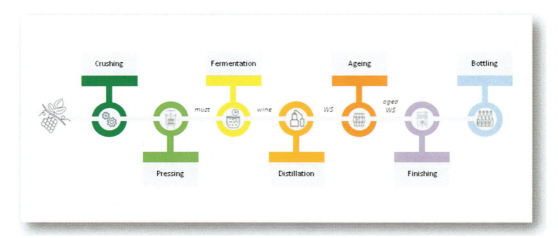

FIGURE 14.2

Technological diagram of the aged WS production.

After distillation, the WS has high concentration of ethanol and richness of volatile compounds, such as esters, terpenes, acids, aldehydes, alcohols, and acetals, proceeding from the raw material, fermentation, and distillation, but is devoid of phenolic compounds other than some volatile phenols (Caldeira et al., 2010). It also presents some metals such as copper and iron (Canas et al., 2020; Catarino et al., 2003). Consequently, the nonaged WS is colorless, has specific flavor attributes, and low sensory complexity.

The aging stage plays a crucial role in enhancing sensory properties and refining the beverage.

The finishing stage comprises blending, dilution, and filtration operations (Belchior et al., 2015).

14.2.3 Main production regions worldwide

WS production exists in many countries, however, the main designations of origin (DOs) are found in the traditionally wine-producing countries, located in Europe (Regulation EC 110/2008)—Fig. 14.3.

Among the European countries, it is worth mentioning the French regions of "Armagnac" and "Cognac," which are exclusive DOs for aged WS and that date back to the 15th and 16th centuries, respectively (Cantagrel, 2008; Garreau, 2008), producing the most prestigious and top-selling products. Similarly, the Portuguese DO "Lourinhã," whose historical references on WS production date back to the early 20th century, was delimited in 1992 as an exclusive DO for aged WS (Belchior et al., 2015). Indeed, these three regions assign DO only to aged WS while the others assign DO to several vitivinicultural products, such as WS and wines.

14.2.4 Regulations

The aging time stipulated for WS by the regulations of the main producing regions is shown in Table 14.1.

14.2 The aged wine spirit and its production process

FIGURE 14.3
WS producing regions.

Table 14.1 Aging time stipulated for WS in the main producing regions.

Producing region	Minimum aging time	Regulation
Europe	Wine spirit: General At least one year in wood containers with a capacity of at least 1000 L or for at least six months in wood containers with a capacity of less than 1000 L DOs "Armagnac": at least one year "Cognac" and "Lourinhã": at least two years	Reg UE 2019/787 Décret nº 2009-1285 Décret nº 2009-1146, Dec.-Lei nº 323/94
Canada	"Brandy": at least one year in wood containers or for at least six months in wood containers of small capacity	FDR (2020)
USA	"Grape brandy": at least two years in oak containers	CFR (2020)
South Africa	"Pot still brandy": at least three years and up to eight years in oak barrels	Liquor Products Act 60 (1989)
Australia	"Brandy": not less than two years in wooden containers	Wine Australia Compliance Guide (2016)
Chile	"Aged Pisco": at least one year in wooden containers	Decreto 521 (2000)
Peru	Not mentioned	NTP 211.001 (2006)

The term "brandy" used in many countries outside Europe means WS; it does not correspond to the definition of "brandy" prevailing in Europe (Reg UE 2019/787).

14.3 The aging stage

Aging "means the storage of a spirit drink in appropriate receptacles for a period of time for the purpose of allowing that spirit drink to undergo natural reactions that impart specific characteristics to that spirit drink" (Reg UE 2019/787).

As aforementioned, after distillation the WS has low sensory complexity. For this reason, the wood contact during the aging stage is pivotal to trigger several physicochemical phenomena that make the aged WS a complex mixture with hundreds of compounds (Canas, 2017) leading to sensory fullness and remarkable improvement of its quality.

14.3.1 Main physicochemical phenomena and determining factors

The physicochemical phenomena underlying the WS aging involve the wood, the wine distillate, and the environment surrounding the barrel or the tank:

(i) Direct extraction of wood constituents;
(ii) Decomposition of wood biopolymers (lignin, hemicelluloses, and cellulose) followed by the release of derived compounds into the distillate;
(iii) Chemical reactions involving only the wood extractable compounds, involving only the distillate compounds, involving both of them;
(iv) Evaporation of volatile compounds and concentration of volatile and nonvolatile compounds;
(v) Formation of a hydrogen-bonded network between ethanol and water.

Among these phenomena, the most studied are the direct extraction of compounds and the decomposition of wood biopolymers, which are responsible for the enrichment of WS in wood compounds, influencing its color, aroma, and flavor (Caldeira et al., 2006; Canas, 2017).

Regarding the reactions occurring in the liquid medium, evidence exists on the oxidation of wood compounds (Cernîsev, 2017), in which some mineral elements such as iron may be involved (Canas et al., 2020). In addition, several reactions involving the distillate compounds have been described such as the oxidation of alcohols to aldehydes and aldehydes to carboxylic acids followed by esterification with ethanol giving rise to esters and acetals (Puech et al., 1984). These reactions point out the remarkable role of oxygen in the aging of WS, which is supplied by the air surrounding the barrel and diffuses slowly and continuously through the wood and the space between staves (del Álamo-Sanza et al., 2017) or by micro-oxygenation (MOX) (*vide* 3.5.2).

Evaporation of volatile compounds through the barrel, including ethanol and water, causes a noticeable loss of WS (Canas et al., 2002), which is called the "Angel's share" and is closely related to the cellar conditions and the wood porosity (Roussey et al., 2021). Considering only "Cognac," it represents a waste estimated at about 22 million bottles per year (BNIC, 2009).

The aforementioned phenomena depend on the following factors (Belchior et al., 2015):

- The wood features—botanical species, toasting level, size, among others;
- The cellar conditions—temperature, relative humidity, and air circulation;
- The technological operations performed during the aging period such as the refilling, the addition of water, and the stirring.

14.3.2 The wood
14.3.2.1 Botanical species used in cooperage

Several oak species have been used in cooperage to make barrels for the aging of alcoholic beverages. Oak wood from *Quercus robur* L. species is traditionally used in the aging of WS (Puech et al., 1998), while other kinds of oak wood, from *Quercus sessiliflora* Salisb., *Quercus alba* L., and *Quercus pyrenaica* Willd. species, are mainly aimed at wine aging (Martínez-Gil et al., 2018). Chestnut wood (*Castanea sativa* Mill.) has also been exploited for these purposes (Canas et al., 2018).

Despite their differences, oak and chestnut wood present adequate anatomical characteristics (Table 14.2) and chemical composition (Fig. 14.4) that make them suitable for barrel making and for the production of wood fragments for WS' aging (Anjos et al., 2013; Canas et al., 2018; Keller, 1987).

Regarding the chemical composition, the wood is best characterized as a composite of three-dimensional biopolymers (high molecular weight compounds) consisting of an interconnected cellulose, hemicelluloses, and lignin network with limited amount of extractives and inorganics. Water is the main chemical component of a living tree, but all wood cell walls, on a dry weight basis, consist mostly of carbohydrates (sugar-based polymers, including holocellulose, cellulose, hemicelluloses, and minor polysaccharides) that are combined with lignin (Higuchi, 1990; Pettersen, 1984)—Fig. 14.4.

Low molecular weight compounds and hydrolysable tannins are the main extractives. In particular, this fraction of wood will decisively determine its oenological quality, as aforementioned (*vide* 3.1). In oak heartwood, hydrolysable tannins (ellagitannins) constitute ca. 10% of the dry weight and are responsible for astringency at relatively low threshold concentrations ranging from 0.2 to 6.3 μmol/L, whereas at threshold concentrations between 410 and 1650 μmol/L, bitter taste is perceived (Glabasnia & Hofmann, 2006). *Trans* and *cis* isomers of β-methyl-γ-octalactone, phenolic aldehydes, such as vanillin and syringaldehyde, and volatile phenols, such as eugenol, guaiacol, ethyl- and vinylphenols, furfural and its related compounds, are the major volatile compounds susceptible to migration from oak wood to WS (Caldeira et al., 2006; Canas, 2017). In general, due to the heterogeneity within the species (influenced by many other factors such as geographical origin, silviculture practices, and the tree), the ranges observed for ellagitannins and low molecular weight compounds in

Table 14.2 Anatomical characteristics of oak wood (*Quercus robur*) and chestnut wood.

Quercus robur	*Castanea sativa*
Ring-porous	Ring-porous
Earlywood ring with one-to-many rows of pores	Large growth rings
Latewood pores solitary or radially oriented in dendritic groups	Latewood pores radially oriented in flame-like groups
Pore density in latewood sparse to dense	Pore density in latewood sparse
Earlywood vessels with tyloses	Earlywood vessels with thin-walled tyloses
Apotracheal parenchyma either diffuse or in uniseriate diagonal and tangential bands	Apotracheal parenchyma both diffuse and in short, indistinct uniseriate tangential bands
Uni- and multiseriate rays	Uniseriate rays, rarely biseriate

Adapted from Schoch, W., Heller, I., Schweingruber, F. H., & Kienast, F. (2004). Wood anatomy of central European species. http://www.woodanatomy.ch. *(Accessed on 29 December 2020).*

266 Chapter 14 Sustainable use of wood in wine spirit production

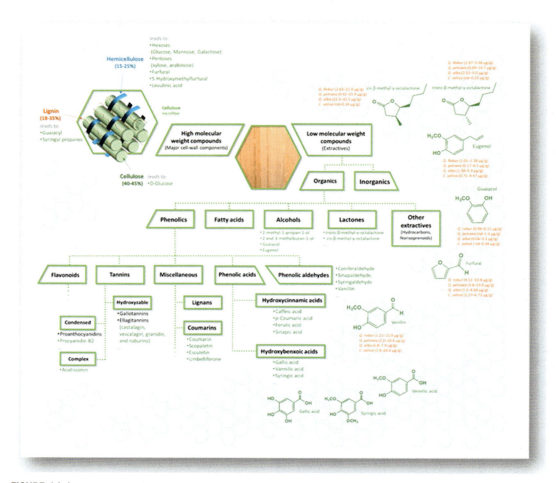

FIGURE 14.4

Wood chemical constituents' classification with structures of molecules addressed in the text. On the left side: schematic representation of lignocellulosic biomass cell walls highlighting the structure of cellulose (dark green), hemicelluloses (blue), and lignin (orange).

the same species are wide (Alañón et al., 2011; Flamini et al., 2007); as an illustration, the concentration ranges of some representative volatile compounds present in wood species are shown in Fig. 14.4.

The chemical composition of the aged WS has a strong dependence on these conditions as well as on those imparted by the cooperage operations (*vide* 3.2.3) and the aging process (*vide* 3.1).

14.3.2.2 Main forest ecosystems producing wood for cooperage and their sustainability

The main forests producing wood for cooperage are located in the northern hemisphere, especially in Europe and North America (Keller, 1987). Several oak species (white oaks) are grown in the USA and Canada, including *Q. alba, Quercus bicolour* Willd., *Quercus stellata* Wangenh., and *Q. lyrata* Walt

(Solomon, 1983). The greatest area in Europe is occupied by the native species *Quercus robur* and *Q. sessiliflora* (Leuschner & Ellenberg, 2017), and the most iconic forests are located in France (Chatonnet, 1995); Eastern Europe (Russia, Ukraine, Romania, among others) forests are also relevant suppliers of wood from these species. *Q. pyrenaica* is mainly found in the Iberian Peninsula (Martínez-Gil et al., 2018). The Mediterranean basin and southern Central European comprise the main chestnut (*C. sativa*) forest ecosystems (López-Sáez et al., 2017).

Some studies reflect the current concern with these forests management toward sustainability to overcome the pressure caused by the supply of wood and other threats. Molder et al. (2019) present an approach to sustain both the ecological value and the economic viability of silviculture in Central European forests of *Q. robur* and *Q. sessiliflora*. Integrative management is proposed including several measures such as the development of forest regeneration techniques, retention forestry, and adapted historical management techniques (coppicing and wood pasture). Tkach et al. (2019) address the importance of natural regeneration processes to assure a new generation of valuable *Q. robur* forests and to optimize the age structure of Ukrainian forest. Prada et al. (2016) show that carbon sequestration by chestnut coppices can increase by reducing the number of thinning and lengthening the rotation period.

Indeed, certification of sustainable forest management practices already exists through the "Program for the Endorsement of Forest Certification" or the "Forest Stewardship Council" (Paluš et al., 2018).

14.3.2.3 Cooperage process and sustainability

The wooden barrel dates back to antiquity, and has been used by Asian, European, and American civilizations. It is believed that in Europe, the Celts and the Romans played a key role in its manufacture (Twede, 2005). Nowadays, the barrel making process in cooperage comprises several steps from cutting the logs to testing the integrity and strength of the barrel (Puech et al., 1998). Among them, the wood seasoning and especially the heat treatment of the barrel are of remarkable importance for the aging result due to the changes made on the structure and chemical composition of the wood (Caldeira et al., 2006; Canas, 2017).

From the sustainability point of view, some particularities of this process should be highlighted: (i) energy consumption, especially for the manufacture of staves and heads, has a high environmental impact (Flor et al., 2017); (ii) higher waste of wood is associated with barrels made from European oaks than from American oaks and chestnut due to the need to split (instead of sawn) the logs because of their anatomical features (Carpena et al., 2020); (iii) a considerable waste of wood results from the entire process (Smailagić et al., 2019), including staves with defects and those broken during heat treatment.

Taking into account that millions of barrels are produced/exported annually worldwide (more than 90,000 t in 2019; Trade Map, 2020), several cooperages reflect a growing awareness of the need for sustainable production (e.g., https://www.iscbarrels.com; https://www.worldcooperage.com), developing strategies that involve the use of wood from sustainably managed forests, internal reuse of wood waste (e.g., to meet energy needs and to manufacture pieces of wood for alternative aging technologies), and sale of wood waste to external reuse.

14.3.2.4 Contribution of the wood reuse to a more sustainable process

The wood waste from cooperage can be externally reused by many ways, making circular economy a reality. Several works show the potential interest of wood compounds for other industrial applications. Tannins can be used in leather, inks, and dyestuffs production (Krisper et al., 1992). Lignin can be a source of several products, such as adhesives, dispersants, and animal feed additives (Abdelaziz et al., 2016). Obtaining vanillin, which is one of the most appreciated and sought natural food flavorings worldwide, is also an appealing option (Gallage & Moller, 2018). In addition, the wood and the bark are remarkable sources of bioactive compounds (especially phenols) of growing interest in the replacement of synthetic ones that are widely used in food, cosmetics, and pharmaceutical industries (Burlacu et al., 2020). Pellets, pallets, and composites for the furniture industry can also be obtained from wood residues.

14.3.3 The aging technology

The aging of WS is traditionally performed in oak barrels. Besides its mechanical strength, easy handling and versatility of shape (Twede, 2005), the barrel is not a simple vessel but an active interface between the liquid contained therein and the environment (*vide* 3.1).

In this aging technology, the WS is first lodged in new barrels for several months. Thereafter, the aged WS is transferred to used barrels in which the aging progresses, from one year to several decades according to the producing region regulation (*vide* 2.4) and the intended quality.

Thus, despite the high quality attained by the WS through this aging technology, the process is time-consuming and costly, resulting in low production efficiency and low profitability for the producer (the invested capital in WS and wood returns over the long term). In addition, barrels take up a lot of space in the cellar and their lifetime is limited, and there is a great loss of WS by evaporation (described above as "Angel's share"). For these reasons, new approaches to a more sustainable aging using barrels have been studied. Besides, in recent years, special attention has been devoted to alternative aging technologies to meet the sustainability criteria (Fig. 14.1).

14.3.4 How to assure a more sustainable aging using wooden barrels?

As aforementioned, the traditional aging of WSs using barrels presents some economic and environmental drawbacks. Furthermore, the regular replacement of barrels is required, being a common practice, as after a number of uses, the barrels are no longer capable of providing an adequate contribution of extractable compounds to the beverage because of the exhaustion of their transfer capacity. In this context, different strategies to extend the life of barrels for the aging of beverages or for other purposes can be followed. The most traditional one is the regeneration of the inner surfaces of the barrels, often involving retoasting. More recent options involve the insert of oak fragments through the bunghole of exhausted barrels to prolong their useful life (Flor-Montalvo et al., 2020; García-Alcaraz et al., 2020), taking advantage of air diffusion through the barrel and the consequent oxygen supply. In apparent contradiction to the detrimental effects of reuse, barrels can be reused and go from the aging of one beverage to another. Also, barrels can be reused for purposes other than beverage aging, such as crafting furniture and accessories as a best practice example of circular economy. Furthermore, specific technological options, such as barrel size, wood species, and toasting level, are relevant factors for the use of barrels to achieve more sustainable aging.

14.3.4.1 Alternative woods, toasting level, and barrel size

A more sustainable aging using barrels can be achieved by exploring alternative woods. Specifically, chestnut wood, of particular significance in the countries bordering the Mediterranean Sea, has been proved to be an interesting choice for the aging of WSs as summarized by Canas et al. (2018). More recently, chestnut wood was used for the aging of "Brandy de Jerez" (Garcia-Moreno et al., 2020), promoting a faster and cheaper aging and affording high quality and differentiated spirits. Using barrels with higher toasting intensity (heavy *vs.* medium and light levels) is also an available option to accelerate the aging process resulting from a faster extraction (higher pool of extractable compounds in the wood together with greater anatomical changes) (Caldeira et al., 2006; Canas, 2017). Another technological factor ruling the aging process is barrel size (*vide* 3.1). The use of barrels of higher size reduces the quantity of wood with the advantage of environmental sustainability to be developed. In addition, higher size barrels can be reused more often because of the slower exhaustion of their transfer capacity, as indicated by a comparative study on WS aging in barrels of 250 and 650 L (Canas et al., 2008). Indeed, the surface area to volume ratio decreases with barrel size (85 m^2/L in 250 L barrels, 57 m^2/L in 650 L), resulting in different extraction kinetics and explaining the comparative extended life of higher size barrels (Canas et al., 2016; Nocera et al., 2020).

14.3.4.2 Barrels regeneration

Decline in barrels has led to several methods of rejuvenation/regeneration. A frequently used method involves scraping the inner surface of barrels, typically removing several millimeters of the wood layer, possibly in conjunction with retoasting (Mosedale, 1995). Scrapping is an effective method for rejuvenation, as it exposes new unextracted wood to the distillate to be aged. However, this technology of mechanical working of inner surface causes thinning of the wall, and sometimes results in loss of its mechanical strength, thus cannot be repeated frequently. A number of oak barrels become rejected for reuse because they are not subject to regeneration with the help of such technology. Hence, alternative technologies for treatment and regeneration of inner surface of the oak barrels have been proposed. Most of them are aimed to barrels used for fermentation or wine aging, in which coloring matter, tartrate crystals and other deposits/contaminations, such as those of microbiological origin (e.g., *Brettanomyces*), on/inside the wood (García-Alcaraz et al., 2020). However, having in mind the WS aging, these technologies are expected appropriated for regeneration of barrels for this purpose.

A curious technology without any mechanical working, by using oak extracts, is claimed to enable multiple use of the oak barrels (Sula et al., 2011). According to the authors, the extract components are absorbed in the inner surface of the oak staves, causing full reproduction of basic chemical, physicochemical, and biochemical transformations.

The so-called "Barena method" consists of spraying an abrasive (of mineral origin) against at least a portion of the inner surface of the barrel, and preferably the entire surface, so as to remove the area of wood spoiled by deposits and microorganisms (Brunateu et al., 2012).

A number of treatments are used to regenerate barrels in terms of microbiological contamination, such as chemical agents (sulfur dioxide, ozone) or physical agents, with UV radiation, hot water, microwaves, and ultrasounds, showing varying levels of efficiency (Breniaux et al., 2019). The ability of these treatments to ensure both wood sterilization and wood structure preservation is of utmost importance.

14.3.4.3 Barrels reuse for the aging of other beverages

The barrels can be reused and go from the aging of one beverage to another. For instance, Scotch whisky production resorts to woods previously used in other beverages' aging such as Bourbon or Sherry, being also reported the reuse of Port or Madeira wine barrels for the finishing steps (Mosedale, 1995; Russell, 2003). Beer aging also resorts to the reuse of wood from other beverages, as the example of Lambic beers which reuse barrels previously used in wine aging (Coelho, 2020). The previous use changes the composition of extractives by removing compounds but also promoting the sorption of compounds in the wood [namely terpenes, alcohols, esters, aldehydes, norisoprenoids and acids (Coelho, 2020), monomeric anthocyanins, catechins, gallic acid, and *trans*-resveratrol (Barrera-García et al., 2007)], which may then be available for later extraction during a different beverage aging (Coelho, 2020).

14.3.4.4 Barrels reused for other purposes

Barrels can be reused for other purposes, e.g., decoration, whether keeping the barrels intact or deconstruct, using the staves and metal hoops to create something totally new as shown in Fig. 14.5. Furniture and accessories, crafted from used barrels can be structurally sound, aesthetically appealing and highly functional. It allows incorporating salvaged wood with a story, being at the same time a rustic trend.

14.3.5 Innovative technologies for wine spirit's aging

Different approaches have been developed in the searching of alternative technologies for the aging of alcoholic beverages. Several studies are devoted to wood fragments or wood extracts applied to the alcoholic beverages kept in stainless steel tanks, in order to shorten the aging period (economic and social advantage) and also to reduce the quantity of wood used (environmental sustainability). Another research line lies in physical treatments applied in the barrels to accelerate the aging process. Finally, wood fragments combined with the application of physical treatments have also been investigated.

14.3.5.1 Wood fragments

The application of wood fragments has been widely studied in wines (Gómez-Plaza & Bautista-Ortín, 2019), and less often in other alcoholic beverages such as cider brandies (Rodríguez Madrera et al., 2013), apple spirit (Coldea et al., 2020), grape marc spirits (Rodríguez-Solana et al., 2017; Taloumi & Makris, 2017), and sugar cane spirits (Bortoletto & Alcarde, 2015). Regarding the use of wood fragments in the aging of WSs, several works have been performed in the last decade. The effect of wood fragments with different shapes/sizes and from different kinds of wood (oak *vs.* chestnut), in comparison with the aging in barrels, on the physicochemical and sensory properties of the aged WSs was assessed (Caldeira et al., 2010, 2013, 2016, 2017; Canas et al., 2009, 2013, 2016; Cruz et al., 2012). These authors showed that the aged WSs obtained by such alternative technologies had very different chemical composition, resulting from different extraction kinetics during the aging process; the kinetics depended both on the compound nature and on the aging technology. Besides, the results obtained were closely related to the shape/size of fragments and the kinds of wood. In all studies, it was always possible to discriminate chemically the WSs based on the aging technology, although their sensory differentiation was not so evident.

14.3 The aging stage 271

FIGURE 14.5

Examples of barrels reuse. Photos by Despina Galani (a), Mike Bergmann (b), Vamvouras (c), and redcharlie (d) on Unsplash.

In addition, other species like acacia (*Robinia pseudoacacia* L.), mulberry (*Morus alba* L. and *Morus nigra* L.), ash (*Fraxinus* excelsior L.), and cherry (*Prunus avium* L.) are increasingly regarded for wine aging (Fernández de Simón et al., 2009; Tavares et al., 2017), thus being admitted its interest for the WSs aging.

14.3.5.2 Wood fragments combined with micro-oxygenation

The use of wood fragments is often associated with MOX, which consists of directly applying small quantities of oxygen to the beverage, simulating the slow and continuous diffusion of this gas through the barrel (del Álamo-Sanza et al., 2017). Several studies made on red wine showed that the limited amounts of oxygen supplied have beneficial effects on wine quality, and namely on color stabilization (Gómez-Plaza & Bautista-Ortín, 2019).

Regarding the aging of WSs with wood fragments combined with MOX, recent works have been done using chestnut and oak wood (Canas et al., 2019; Granja-Soares et al., 2020); staves combined with MOX were compared with barrels. The WSs aged through the alternative technology presented significantly higher total phenolic content and low molecular compounds contents (phenolic compounds and several odorant compounds) extracted from the wood. These results, which were ascribed to a faster extraction, were more pronounced in the WSs aged with chestnut wood. Accordingly, significantly higher antioxidant activity was found in the aged WSs resulting from alternative technology than from barrels (Nocera et al., 2020). Besides, the WSs resulting from these aging technologies were discriminated by FTIR-ATR spectroscopy combined with FDA and by NIR (Anjos, Caldeira, Roque et al., 2020; Anjos, Martínez Comesaña et al., 2020). From the sensory point of view, for the same aging time, the WSs proceeding from the alternative technology were more evolved and presented significantly higher overall quality (Granja-Soares et al., 2020). In addition, higher amounts of acetaldehyde, 1-butanol, methanol, and alcoholic strength found in WS aged in stainless steel tanks with staves and MOX than in barrels suggest lower evaporation of WS promoted by the alternative technology (Anjos, Caldeira, Pedro, et al., 2020).

More recently, our team started to study the influence of MOX level on the physicochemical and sensory characteristics of the aged WS under the Project POCI-01-0145-FEDER-027819 (https://projects.iniav.pt./oxyrebrand). The first results revealed that the behavior of most of the low molecular weight compounds was significantly influenced by the aging time and the oxygenation level (Canas et al., 2020).

14.3.5.3 Wood extracts

A South African team (van Jaarsveld et al., 2009a,b,c) reported another technological option that consists of obtaining liquid extracts by macerating different kinds of wood chips in water and ethanol solution at different concentrations. After, these macerates, distilled and concentrated by different processes, were mixed with unaged distilled beverages and the chemical composition and sensory acceptability after the aging period was evaluated. They noticed that the best conditions for obtaining the extracts were the use of ethanol at higher concentrations. Influence of oak origin and toasting level on the sensory quality of the beverages was also found. Despite the interesting results achieved, there is a lack of comparison with the aging in barrels.

14.3.5.4 Physical treatments

Given the knowledge about physical and chemical reactions taking place during aging (*vide* 3.1), some physical methods have been explored. Most of them were applied to wine and other alcoholic beverages; little research on WSs has been done.

14.3.5.4.1 Electric field
Zhang, Zeng, Lin, et al. (2013) and Zhang, Zeng, Sun, et al. (2013) studied the application of an electric field (EF) of 1 kV/cm to small barrels of 2 and 5 L filled with the same WS. This treatment increased the extraction of several phenolic compounds from the wood. After the EF treatment for up to 14 months, the amount of phenolic compounds in the WSs was higher than in those aged in barrels without EF treatment.

This kind of physical treatment has also been applied to other alcoholic beverages such as red wine (during aging and in winemaking stages). The main results are summarized by Puértolas et al. (2020).

14.3.5.4.2 Ultrasonic waves combined with wood fragments
Some researcher groups have been applied ultrasounds combined with oak chips to accelerate the aging of "Brandy de Jerez" (Delgado-González et al., 2017; Schwarz et al., 2014, 2020). The best aged spirits were produced in absence of light, in presence of oxygen, at room temperature and with high flow rates (Delgado-González et al., 2017). The authors applied this process to five varietal WSs, whose sensory quality increased with this technical procedure. Recently, it was applied and compared with results of traditional "Brandy de Jerez" aging (Schwarz et al., 2020). The multidimensional analysis of the results showed discrimination between the brandies proceeding from the two aging technologies (barrels *vs.* oak chips and ultrasounds). On the contrary, Schwarz et al. (2014) concluded that the brandies obtained by the accelerated method attained in one month sensory characteristics and acceptability close to those of brandies aged in barrels for 6–18 months.

The influence of the ultrasound application to accelerate the aging process of other alcoholic beverages, such as fermented rice and maize beverages, plum brandies and grape marc spirit, was examined in some works (Balcerek et al., 2017; Chang & Chen, 2002; Taloumi & Makris, 2017); its effect on the beverages' quality was differentiated. Ultrasounds have also been applied in several stages of winemaking, and the main results summarized by García Martín & Sun (2013) showed variable effects on the wine quality.

14.3.5.4.3 Other physical treatments
Other physical treatments, such as gamma irradiation, nanogold photocatalysis, high pressure, and microwave irradiation, applied to some alcoholic beverages are described in the literature (Pielech-Przybylska & Balcerek, 2019), thus admitting the interest of their study in the aging of WSs.

14.4 Concluding remarks
Despite the economic, social, and environmental issues underlying the traditional aging in barrels, many alternatives exist that can concur to ensure the sustainability of WS aging. Indeed, greater achievements can be reached based not only on specific measures related to the WS' aging but rather on those involving the entire value chain (from the forest to the final product), thus implying commitment of all players (forest owners, foresters, coopers, WS producers). The information provided in this chapter points out interesting and promising resources to support such a comprehensive strategy: (i) the use of wood in cooperage from sustainably managed forests, an efficient use of production factors, such as energy and water, and the reuse of wood waste from cooperage in internal and external processes; (ii) a wise and judicious exploring of barrels for WS aging, by extending their useful life, their reuse, and by a proper selection of technological options such as barrel size, different

wood species, and toasting level; (iii) the use of innovative technologies, based on wood fragments (from different wood species and sizes/shapes) combined with MOX, on wood extracts and on physical treatments (EF, ultrasounds, among others).

Further studies on the aging chemistry may provide meaningful clues for the development of new technologies toward a more sustainable wood-based aging of this spirit drink.

Acknowledgments

Funding: The authors thank the financial support of National Funds through FCT—Foundation for Science and Technology under the Project POCI-01-0145-FEDER-027819 (PTDC/OCE-ETA/27819/2017).

References

Abdelaziz, O. Y., Brink, D. P., Prothmann, J., Ravi, K., Sun, M., García-Hidalgo, J., Sandahl, M., Hulteberg, C. P., Turner, C., Lidén, G., & Gorwa-Grauslund, M. F. (2016). Biological valorization of low molecular weight lignin. *Biotechnology Advances, 34*, 1318–1346.

Alañón, M. E., Pérez-Coello, M. S., Díaz-Maroto, I. J., Martín-Alvarez, P. J., Vila-Lameiro, P., & Díaz-Maroto, M. C. (2011). Influence of geographical location, site and silvicultural parameters, on volatile composition of *Quercus pyrenaica* willd. wood used in wine aging. *Forest Ecology and Management, 262*, 124–130.

Anjos, O., Caldeira, I., Pedro, S. I., & Canas, S. (2020). FT-Raman methodology applied to identify different ageing stages of wine spirits. *LWT — Food Science and Technology, 134*, 110179.

Anjos, O., Caldeira, I., Roque, R., Pedro, S. I., Lourenço, S., & Canas, S. (2020). Screening of different ageing technologies of wine spirit by application of near-infrared (NIR) spectroscopy and volatile quantification. *Processes, 8*, 736.

Anjos, O., Carmona, C., Caldeira, I., & Canas, S. (2013). Variation of extractable compounds and lignin contents in wood fragments used in the aging of wine brandies. *BioResources, 8*, 4484–4496.

Anjos, O., Martínez Comesaña, M., Caldeira, I., Pedro, S. I., Eguía Oller, P., & Canas, S. (2020). Application of functional data analysis and FTIR-ATR spectroscopy to discriminate wine spirits ageing technologies. *Mathematics, 8*, 896–916.

Balcerek, M., Pielech-Przybylska, K., Dziekońska-Kubczak, U., Patelski, P., & Strąk, E. (2017). Changes in the chemical composition of plum distillate during maturation with oak chips under different conditions. *Food Technology Biotechnology, 55*, 333–359.

Barrera-García, V. D., Gougeon, R. D., Di Majo, D., De Aguirre, C., Voilley, A., & Chassagne, D. (2007). Different sorption behaviors for wine polyphenols in contact with oak wood. *Journal of Agricultural and Food Chemistry, 55*, 7021–7027.

Belchior, A. P., Canas, S., Caldeira, I., & Carvalho, E. C. (2015). *Aguardentes vinícolas — Tecnologias de produção e envelhecimento. Controlo de qualidade*. Porto: Publindústria, Edições Técnicas.

BNIC. (2009). *L'Encyclopédie du Cognac*. https://www.pediacognac.com (Accessed 14 November 2020).

Bortoletto, A. M., & Alcarde, A. R. (2015). Aging marker profile in cachaça is influenced by toasted oak chips. *Journal of the Institute of Brewing, 121*, 70–77.

Breniaux, M., Renault, P., Meunier, F., & Ghidossi, R. (2019). Study of high power ultrasound for oak wood barrel regeneration: Impact of wood properties and sanitation effect. *Beverages, 5*, 10.

Brunateu, D., Dutel, J.-L., & Trillaud, A. (2012). *Process for regenerating casks or the like and device for the implementation thereof*. US Patent no. 2012/0064805 A1.

Burlacu, E., Nisca, A., & Tanase, C. (2020). A comprehensive review of phytochemistry and biological activities of *Quercus* species. *Forests, 11*, 904.

Caldeira, I., Anjos, O., Belchior, A. P., & Canas, S. (2017). Sensory impact of alternative ageing technology for the producing of wine brandies. *Ciência Técnica Vitivinícola, 32*, 12−22.

Caldeira, I., Anjos, O., Portal, V., Belchior, A. P., & Canas, S. (2010). Sensory and chemical modifications of wine-brandy aged with chestnut and oak wood fragments in comparison to wooden barrels. *Analytica Chimica Acta, 660*, 43−52.

Caldeira, I., Belchior, A. P., & Canas, S. (2013). Effect of alternative ageing systems on the wine brandy sensory profile. *Ciência e Técnica Vivinícola, 28*, 9−18.

Caldeira, I., Mateus, A. M., & Belchior, A. P. (2006). Flavour and odour profile modifications during the first five years of Lourinhã brandy maturation on different wooden barrels. *Analytica Chimica Acta, 563*, 264−273.

Caldeira, I., Santos, R., Ricardo-da-Silva, J. M., Anjos, O., Mira, H., Belchior, A. P., & Canas, S. (2016). Kinetics of odorant compounds in wine brandies aged in different systems. *Food Chemistry, 211*, 937−946.

Canas, S. (2017). Phenolic composition and related properties of aged wine spirits: Influence of barrel characteristics. A review. *Beverages, 3*, 55−76.

Canas, S., Anjos, O., Caldeira, I., & Belchior, A. P. (2019). Phenolic profile and colour acquired by the wine spirit in the beginning of ageing: Alternative technology using micro-oxygenation vs traditional technology. *LWT − Food Science and Technology, 11*, 260−269.

Canas, S., Belchior, A. P., Mateus, A. M., Spranger, M. I., & Bruno de Sousa, R. (2002). Kinetics of impregnation/evaporation and release of phenolic compounds from wood to brandy in experimental model. *Ciência e Técnica Vitivinícola, 17*, 1−14.

Canas, S., Caldeira, I., Anjos, O., Lino, J., Soares, A., & Belchior, A. P. (2016). Physicochemical and sensory evaluation of wine brandies aged using oak and chestnut wood simultaneously in wooden barrels and in stainless steel tanks with staves. *International Journal Food Science Technology, 51*, 2537−2545.

Canas, S., Caldeira, I., & Belchior, A. P. (2009). Comparison of alternative systems for the ageing of wine brandy. Wood shape and wood botanical species effect. *Ciência e Técnica Vitivinícola, 24*, 90−99.

Canas, S., Caldeira, I., & Belchior, A. P. (2013). Extraction/oxidation kinetics of low molecular weight compounds in wine brandy resulting from different ageing technologies. *Food Chemistry, 138*, 2460−2467.

Canas, S., Caldeira, I., Belchior, A. P., Spranger, M. I., Clímaco, M. C., & Bruno-de-Sousa, R. (2018). *Chestnut wooden barrels for the ageing of wine spirits* (p. 16). OIV site (OIV). http://www.oiv.int/en/technical-standards-and-documents/collective-expertise/spirit-beverages.

Canas, S., Danalache, F., Anjos, O., Fernandes, T. A., Caldeira, I., Santos, N., Fargeton, L., Boissier, B., & Catarino, S. (2020). Behaviour of low molecular weight compounds, iron and copper of wine spirit aged with chestnut staves under different levels of micro-oxygenation. *Molecules, 25*, 5266−5291.

Canas, S., Vaz, M., & Belchior, A. P. (2008). Influence de la dimension du fût dans les cinétiques d'extraction/oxydation des composés phénoliques du bois pour les eaux-de-vie Lourinhã. In A. Bertrand (Ed.), *Les eaux-de-vie traditionnelles d'origine viticole* (pp. 143−146). Paris: Lavoisier Tec & Doc.

Cantagrel, R. (1992). *Élaboration et connaissance des spiritueux*. Paris: Lavoisier Tec & Doc.

Cantagrel, R. (2008). La qualité et le renom du Cognac dans le monde, sa place dans l'historie. In A. Bertrand (Ed.), *Les eaux-de-vie traditionnelles d'origine viticole* (pp. 16−38). Paris: Lavoisier Tec & Doc.

Carpena, M., Pereira, A. G., Prieto, M. A., & Simal-Gandara, J. (2020). Wine aging technology: Fundamental role of wood barrels. *Foods, 9*, 1160−1184.

Catarino, S., Pinto, D., & Curvelo-Garcia, A. S. (2003). Validação e comparação de métodos de análise em espectrofotometria de absorção atómica com chama para doseamento de cobre e ferro em vinhos e aguardentes. *Ciência e Técnica Vitivinícola, 18*, 65−76.

Ceccherini, G., Duveiller, G., Grassi, G., Lemoine, G., Avitabile, V., Pilli, R., & Cescatti, A. (2020). Abrupt increase in harvested forest area over Europe after 2015. *Nature, 583*, 72−77.

Cernîsev, S. (2017). Analysis of lignin-derived phenolic compounds and their transformations in aged wine distillates. *Food Control, 73*, 281–290.

CFR. (2020). *Title 27: Alcohol, Tobacco products and Firearms; Part 5 – Labeling and advertising of distilled spirits; Subpart C – Standards of identity for distilled spirits.* https://ecfr.federalregister.gov/current/title-27/chapter-I/subchapter-A/part-5/subpart-C (Accessed 20 November, 2020).

Chang, A. C., & Chen, F. C. (2002). The application of 20 kHz ultrasonicwaves to accelerate the aging of different wines. *Food Chemistry, 79*, 501–506.

Chatonnet, P. (1995). *Influence des procédés de tonnellerie et des conditions d'élevage sur la composition et la qualité des vins élevés en fûts de chêne.* Thèse doctorat de l'Université de Bordeaux II (p. 268). Institut d'Oenologie.

Coelho, E. J. L. (2020). *Valorization of woods from wine ageing and development of methodologies for their reutilization* (Ph.D. thesis) (p. 176). Universidade do Minho, Escola de Engenharia.

Coldea, T. E., Socaciu, C., Mudura, E., Socaci, S. A., Ranga, F., Pop, C. R., Vriesekoop, F., & Pasqualone, A. (2020). Volatile and phenolic profiles of traditional Romanian apple brandy after rapid ageing with different wood chips. *Food Chemistry, 320*, 126643.

Cruz, S., Canas, S., & Belchior, A. P. (2012). Effect of ageing system and time on the quality of wine brandy aged at industrial-scale. *Ciência e Técnica Vitivinícola, 27*, 83–93.

Dec.-Lei No. 323/94. (1994). Estatuto da Região Demarcada das Aguardentes Vínicas da Lourinhã. *Diário da República I Série A, 29*, 7486–7489.

Décret No. 2009-1146. Appellation d'origine contrôlée "Cognac" ou "Eau-de-vie de Cognac" ou "Eau-de-vie des Charentes". *Journal Officiel de la République Française, 221*, 15619–15628.

Décret No. 2009-1285. Appellations d'origine contrôlée "Armagnac", "Blanche Armagnac", "Bas Armagnac", "Haut Armagnac" et "Armagnac-Ténarèze". *Journal Officiel de la République Française, 247*, 17916–17927.

Decreto 521. (2000). *Fija Reglamento de la Denominacion de Origen Pisco.* Chile: Ministerio de Agricultura, 6 pp.

del Álamo-Sanza, M., Cárcel, L. M., & Nevares, I. (2017). Characterization of the oxygen transmission rate of oak wood species used in cooperage. *Journal of Agriculture and Food Chemistry, 65*, 648–655.

Delgado-González, M. J., Sánchez-Guillén, M. M., García-Moreno, M. V., Rodríguez-Dodero, M. C., García-Barroso, C., & Guillén-Sánchez, D. A. (2017). Study of a laboratory-scaled new method for the accelerated continuous ageing of wine spirits by applying ultrasound energy. *Ultrasonics Sonochemistry, 36*, 226–235.

European Commission. (2019). *The European green deal.* Document 52019DC0640.

FDR. (2020). *C.R.C., c. 870. Minister of Justice of Canada* (p. 1121). https://laws-lois.justice.gc.ca/eng/regulations/c.r.c.,_c._870/ (Accessed 20 November 2020).

Fernández de Simón, B., Esteruelas, E., Muñoz, A. M., Cadahía, E., & Sanz, M. (2009). Volatile compounds in acacia, chestnut, cherry, ash, and oak woods, with a view to their use in cooperage. *Journal of Agriculture and Food Chemistry, 57*, 3217–3227.

Flamini, R., Dalla Vedova, A., Cancian, D., Panighel, A., & De Rosso, M. (2007). GC/MS-positive ion chemical ionization and MS/MS study of volatile benzene compounds in five different woods used in barrel making. *Journal of Mass Spectrometry, 42*, 641–646.

Flor, F. J., Leiva, F. J., García, J., Martínez, E., Jiménez, E., & Blanco, J. (2017). Environmental impact of oak barrels production in qualified Designation of Origin of Rioja. *Journal of Cleaner Production, 167*, 208–217.

Flor-Montalvo, F. J., Ledesma, A. S.-T., Cámara, E. M., Jiménez-Macías, E., & Blanco-Fernández, J. (2020). New system to increase the useful life of exhausted barrels in red wine aging. *Foods, 9*, 1686.

FoodDrinkEurope. (2012). *Environmental sustainability vision towards 2030. Achievements, challenges and opportunities* (p. 92). Brussels: Food Drink Europe.

Gallage, N. J., & Moller, B. L. (2018). Vanilla: The most popular flavour. In W. Schwab, B. M. Lange, & M. Wust (Eds.), *Biotechnology of natural products* (pp. 3–24). Cham: Springer International Publishing.

García-Alcaraz, J. L., Montalvo, F. J. F., Cámara, E. M., De La Parte, M. M. P., Jiménez-Macías, E., & Blanco-Fernández, J. (2020). Economic-environmental impact analysis of alternative systems for red wine ageing in re-used barrels. *Journal of Cleaner Production, 244*, 118783.

García Martín, J.-M., & Sun, D.-W. (2013). Ultrasound and electric fields as novel techniques for assisting the wine ageing process: The state-of-the-art research. *Trends in Food Science & Technology, 33*, 40–53.

García-Moreno, M. V., Sánchez-Guillén, M. M., de Mier, M. R., Delgado-González, M. J., Rodríguez-Dodero, M. C., García-Barroso, C., & Guillén-Sánchez, D. A. (2020). Use of alternative wood for the ageing of Brandy de Jerez. *Foods, 9*, 250–268.

Garreau, C. (2008). L'Armagnac. In A. Bertrand (Ed.), *Les eaux-de-vie traditionnelles d'origine viticole* (pp. 39–42). Paris: Lavoisier—Tec & Doc.

Glabasnia, A., & Hofmann, T. (2006). Sensory-directed identification of taste-active ellagitannins in American (*Quercus alba* L.) and European oak wood (*Quercus robur* L.) and quantitative analysis in bourbon whiskey and oak-matured red wines. *Journal of Agricultural and Food Chemistry, 54*, 3380–3390.

Gómez-Plaza, E., & Bautista-Ortín, A. B. (2019). Emerging technologies for aging wines. In A. Morata (Ed.), *Red wine technology* (pp. 149–162). Cambridge, MA: Academic Press.

Granja-Soares, J., Roque, R., Cabrita, M. J., Anjos, O., Belchior, A. P., Caldeira, I., & Canas, S. (2020). Effect of innovative technology using staves and micro-oxygenation on the sensory and odorant profile of aged wine spirit. *Food Chemistry, 333*, 127450.

Grigg, D. (2004). Wine, spirits and beer: World patterns of consumption. *Geography, 89*, 99–110.

Higuchi, T. (1990). Lignin biochemistry: Biosynthesis and biodegradation. *Wood Science and Technology, 24*, 23–63.

IBISWorld Industry Report. (2010). *Global spirits manufacturing: C1122-GL*.

Keller, R. (1987). Différentes variétés de chênes et leur répartition dans le monde. *Connaissance Vigne et Vin, 21*, 191–229.

Krisper, P., Tisler, V., Skubic, V., Rupnik, I., & Kobal, S. (1992). The use of tannins from chestnut (*Castanea sativa*). In R. W. Hemingway, & P. E. Laks (Eds.), *Plant polyphenols* (pp. 1013–1019). New York, NY: Plenum Press.

Léauté, R. (1990). Distillation in alambic. *American Journal of Enology and Viticulture, 41*, 90–103.

Leuschner, C., & Ellenberg, H. (2017). *Ecology of central European forests* (Vol. 1). Cham: Springer.

Liquor Products Act 60. (1989). *South Africa wine industry information and systems, regulations*. http://www.sawis.co.za/winelaw/download/Regulations,_annotated_05_2019.pdf (Accessed 20 November 2020).

López-Sáez, J. A., Glais, A., Robles-López, S., Alba-Sánchez, F., Pérez-Díaz, S., Abel-Schaad, D., & Luelmo-Lautenschlaeger, R. (2017). Unraveling the naturalness of sweet chestnut forests (*Castanea sativa* Mill.) in central Spain. *Vegetation History and Archaeobotany, 26*, 167–182.

Martínez-Gil, A., del Álamo-Sanza, M., Sánchez-Gómez, R., & Nevares, I. (2018). Different woods in cooperage for oenology: A review. *Beverages, 4*, 94–119.

Mitry, D. J., & Smith, D. E. (2009). Convergence in global markets and consumer behaviour. *International Journal of Consumer Studies, 33*, 316–321.

Molder, A., Meyer, P., & Nagel, R.-V. (2019). Integrative management to sustain biodiversity and ecological continuity in Central European temperate oak (*Quercus robur, Q. petraea*) forests: An overview. *Forest Ecology and Management, 437*, 324–339.

Mosedale, J. R. (1995). Effects of oak wood on the maturation of alcoholic beverages with particular reference to whisky. *Forestry: An International Journal of Forest Research, 68*, 203–230.

Nocera, A., Ricardo-da-Silva, J. M., & Canas, S. (2020). Antioxidant activity and phenolic composition of wine spirit resulting from an alternative ageing technology using micro-oxygenation: A preliminary study. *OENO One, 54*, 485−496.

NTP 211.001. (2006). *Norma Técnica Peruana* (p. 16). Lima: Comisión de Reglamentos Técnicos y Comerciales − INDECOPI.

OIV. (2020). *International Code of Oenological Practices, Eco 3/08* (p. 69). Paris: International Organisation of Vine and Wine.

Paluš, H., Parobek, J., Šulek, R., Lichý, J., & Šálka, J. (2018). Understanding sustainable forest management certification in Slovakia: Forest owners' perception of expectations, benefits and problems. *Sustainability, 10*, 2470−2486.

Pettersen, R. C. (1984). The chemical composition of wood. In R. M. Rowell (Ed.), *The chemistry of solid wood, advances in chemistry series* (Vol. 20, pp. 57−126). Washington, DC: ACS.

Pielech-Przybylska, K., & Balcerek, M. (2019). New trends in spirit beverages production. In A. M. Grumezescu, & A. M. Holban (Eds.), *The science of beverages: Vol. 7. Alcoholic beverages* (pp. 65−111). Cambridge, MA: Woodhead Publishing.

Prada, M., Bravo, F., Berdasco, L., Canga, E., & Martínez-Alonso, C. (2016). Carbon sequestration for different management alternatives in sweet chestnut coppice in northern Spain. *Journal of Cleaner Production, 135*, 1161−1169.

Project POCI-01-0145-FEDER-027819. https://projects.iniav.pt./oxyrebrand. (Accessed on 30 December 2020).

Puech, J.-L., Leauté, R., Clot, G., Momdedeu, L., & Mondies, H. (1984). Évolution de divers constituants volatils et phénoliques des eaux-de-vie de Cognac au cours de leur vieillissement. *Sciences des Aliments, 4*, 65−80.

Puech, J.-L., Mourgues, J., Mosedale, J. R., & Léauté, R. (1998). Barrique et vieillissement des eaux-de-vie. In *Œnologie. Fondements scientifiques et technologiques* (pp. 1110−1142). Paris: Lavoisier Tec & Doc.

Puértolas, E., López, N., Condón, S., Alvárez, I., & Raso, J. (2010). Potential applications of PEF to improve red wine quality. *Trends in Food Science & Technology, 21*, 247e255.

Purvis, B., Mao, Y., & Robinson, D. (2019). Three pillars of sustainability: In search of conceptual origins. *Sustainability Science, 14*, 681−695.

Regulation (EC) 110/2008. (2008). Definition, description, presentation, labelling and protection of geographical indications of spirit drinks. *Official Journal of the European Union, L39*, 16−54.

Regulation EU 2019/787. (2019). Definition, description, presentation and labelling of spirit drinks, the use of the names of spirit drinks in the presentation and labelling of other foodstuffs, the protection of geographical indications for spirit drinks, the use of ethyl alcohol and distillates of agricultural origin in alcoholic beverages, and repealing Regulation (EC) No 110/2008. *Official Journal of the European Union, L130*, 1−54.

Rodríguez Madrera, R., Hevia, A. G., & Suárez Valles, B. (2013). Comparative study of two aging systems for cider brandy making. Changes in chemical composition. *LWT - Food Science and Technology, 54*, 513−520.

Rodríguez-Solana, R., Rodríguez-Freigedo, S., Salgado, J. M., Domínguez, J. M., & Cortés-Diéguez, S. (2017). Optimisation of accelerated ageing of grape marc distillate on a micro-scale process using a Box−Benhken design: Influence of oak origin, fragment size and toast level on the composition of the final product. *Australian Journal of Grape and Wine Research, 23*, 5−14.

Roussey, C., Colin, J., Teissier du Cros, R., Casalinho, J., & Perré, P. (2021). *In-situ* monitoring of wine volume, barrel mass, ullage pressure and dissolved oxygen for a better understanding of wine-barrel-cellar interactions. *Journal of Food Engineering, 291*, 110233.

Russell, I. (2003). *Whisky: Technology, production and marketing*. Academic Press.

Schwarz, M., Rodríguez, M. C., Sánchez, M., Guillén, D. A., & Barroso Carmelo, G. (2014). Development of an accelerated aging method for Brandy. *LWT - Food Science and Technology, 59*, 108−114.

Schwarz, M., Rodríguez-Dodero, M. C., Jurado, M. S., Puertas, B. G., Barroso, C., & Guillén, D. A. (2020). Analytical characterization and sensory analysis of distillates of different varieties of grapes aged by an accelerated method. *Foods, 9*, 277.

Smailagić, A., Veljović, S., Gašić, U., Zagorac, D. D., Stanković, M., Radotić, K., & Natić, M. (2019). Phenolic profile, chromatic parameters and fluorescence of different woods used in Balkan cooperage. *Industrial Crops & Products, 132*, 156–167.

spiritsEUROPE. (2020). *100% sustainable from #Farm2Glass. The European spirits sector & the green deal.* Brussels: spiritsEUROPE, 8 pp.

Solomon, A. M. (1983). Pollen morphology and plant taxonomy of white oaks in Eastern North America. *American Journal of Botany, 70*, 481–494.

Sula, R. A., Redka, V. M., Prakh, A. V., Guguchkina, T. I., Ageeva, N. M., Yakuba, & Yu, F. (2011). *Method of the oak barrel regeneration*. RF Patent no. 2428466, Bulletin no. 25.

Taloumi, T., & Makris, D. P. (2017). Accelerated aging of the traditional Greek distillate Tsipouro using wooden chips. Part I: Effect of static maceration vs. ultrasonication on the polyphenol extraction and antioxidant activity. *Beverages, 3*, 5.

Tavares, M., Jordão, A. M., & Ricardo-da-Silva, J. M. (2017). Impact of cherry, acacia and oak chips on red wine phenolic parameters and sensory profile. *OENO One, 51*, 329–342.

Tkach, V., Rumiantsev, M., Kobets, O., Luk'yanets, V., & Musienko, S. (2019). Ukrainian plain oak forests and their natural regeneration. *Forestry Studies, 71*, 17–29.

Trade Map. (2020). *Trade statistics for international business development*. https://www.trademap.org/Country_SelProduct_TS.aspx (Accessed 28 December 2020).

Twede, D. (2005). The cask age: The technology and history of wooden barrels. *Packaging Technology and Science, 18*, 253–264.

van Jaarsveld, F. P., Hattingh, S., & Minnaar, P. (2009a). Rapid induction of ageing character in brandy products — Part I. Effects of extraction media and preparation conditions. *South African Journal of Enology and Viticulture, 30*, 1–15.

van Jaarsveld, F. P., Hattingh, S., & Minnaar, P. (2009b). Rapid induction of ageing character in brandy products — Part II. Influence of type of oak. *South African Journal of Enology and Viticulture, 30*, 16–23.

van Jaarsveld, F. P., Hattingh, S., & Minnaar, P. (2009c). Rapid induction of ageing character in brandy products — Part III. Influence of toasting. *South African Journal of Enology and Viticulture, 30*, 24–37.

Whelan, T., & Kronthal-Sacco, R. (2019). Research: Actually, consumers do buy sustainable product. *Harvard Business Review*. https://hbr.org/2019/06/research-actually-consumers-do-buy-sustainable-products (Accessed 23 December 2020).

Wine Australia Compliance Guide. (2016). Kent Town: Australian Grape and Wine Authority.

Zhang, B., Zeng, X.-A., Lin, W. T., Sun, D.-W., & Cai, J.-L. (2013). Effects of electric field treatments on phenol compounds of brandy aging in oak barrels. *Innovative Food Science & Emerging Technologies, 20*, 106–114.

Zhang, B., Zeng, X. A., Sun, D. W., Yu, S. J., Yang, M. F., & Ma, S. (2013). Effect of electric field treatments on brandy aging in oak barrels. *Food Bioprocess Technology, 6*, 1635–1643.

Further reading

ISC barrels. https://www.iscbarrels.com. (Accessed on 22 December 2020).

Caldeira, I., Vitória, C., Anjos, O., Fernandes, T. A., Gallardo, E., Fargeton, L., Boissier, B., Catarino, S., & Canas, S. (2021). Wine spirit ageing with chestnut staves under different micro-oxygenation strategies: Effects on the volatile compounds and sensory profile. *Applied Sciences, 11*, 3991. https://doi.org/10.3390/app11093991

Canas, S., Anjos, O., Caldeira, I., Fernandes, T. A., Santos, N., Lourenço, S., Granja-Soares, J., Fargeton, L., Boissier, B., & Catarino, S. (2022). Micro-oxygenation level as a key to explain the variation in the colour and chemical composition of wine spirits aged with chestnut wood staves. *LWT — Food Science and Technology, 154*, 112658. https://doi.org/10.1016/j.lwt.2021.112658

Oliveira-Alves, S., Lourenço, S., Anjos, O., Fernandes, T. A., Caldeira, I., Catarino, S., & Canas, S. (2022). Influence of the storage in bottle on the antioxidant activities and related chemical characteristics of wine spirits aged with chestnut staves and micro-oxygenation. *Molecules, 27*, 106. https://doi.org/10.3390/molecules27010106

Schoch, W., Heller, I., Schweingruber, F. H., & Kienast, F. (2004). *Wood anatomy of central European species*. http://www.woodanatomy.ch (Accessed 29 December 2020).

World cooperage. https://www.worldcooperage.com (Accessed 22 December 2020).

CHAPTER 15

Innovative processes for the extraction of bioactive compounds from winery wastes and by-products

Gianpiero Pataro[1], Daniele Carullo[1] and Giovanna Ferrari[1,2]

[1]*Department of Industrial Engineering, University of Salerno, Fisciano, Salerno, Italy;* [2]*ProdAl Scarl — University of Salerno, Fisciano, Salerno, Italy*

15.1 Introduction

Wine production is currently one of the most important and profitable agricultural activities in the world. Grapes of the cultivars *Vitis vinifera* are the basis of the majority of wines produced around the world, with an annual production that, in 2018, has exceeded 70 million tons, of which 37% are produced in Europe, 34% in Asia, and 19% in America (OIV, 2018).

The majority of cropped grapes is processed in wineries (Beres et al., 2017; Hogervorst et al., 2017; Kammerer et al., 2014; Maroun et al., 2017) with a global wine production accounting for 292 MhL in 2018, of which about 64% being produced in Europe (Italy, France, Spain, Germany, and Portugal) (OIV, 2018). The large amount of grapes and wine production is unavoidably associated with the use of a large number of resources, both in terms of water consumption and the use of organic and inorganic products (fertilizers). Moreover, as illustrated in Fig. 15.1, apart from wine product, the winemaking process also generates high loaded wastewater as well as different kind of grape processing wastes and by-products, including grape stalks, grape pomace or marc (skins and seeds), wine lees/sediments, and vine shoots, which account for up to 10%—30% of the processed raw material (Barba et al., 2016; Kammerer et al., 2014; Sirohi et al., 2020). Sustainable wine production should envisage the correct use of the resources, both in the fields and in the winery, as well as the exploitation of grape processing by-products, with a view to a circular economy. These by-products are typically characterized by exceptionally high levels of chemical oxygen demand and high biodegradability (Barba et al., 2016), thus representing a major disposal problem for winery, where they currently find low-added value uses as animal feed or fertilizers (Casazza et al., 2010; Ferrari et al., 2019; Galanakis, 2015; Thirumdas et al., 2020), or to recover renewable energy through anaerobic digestion of semisolid wastes like lees and vinasses (Barba et al., 2016; Moletta, 2005), or sent to the distillery for the production of Grappa from marc distillation (Muhlack et al., 2018).

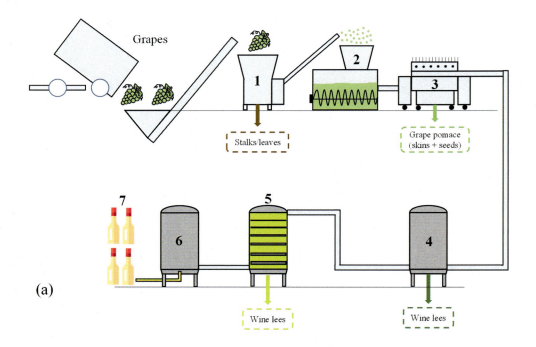
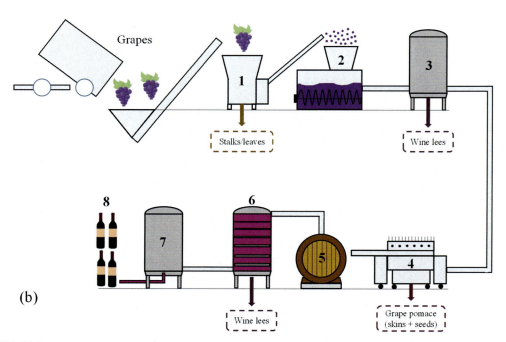

FIGURE 15.1

Simplified schematics of white (A) and red (B) vinification process, highlighting the generation steps of the production of major grape processing by-products. White wine: (1) destemming, (2) crushing, (3) pressing, (4) alcoholic fermentation, (5) filtration, (6) stabilization/storage, (7) bottling. Red wine: (1) destemming, (2) crushing, (3), alcoholic fermentation, (4) pressing, (5) maturation, (6) filtration, (7) stabilization/storage, (8) bottling.

However, all these matrices still represent a reach source of polysaccharides, proteins, lipids, fibers, minerals, and especially of bioactive molecules such as polyphenols, anthocyanins, tannins, and vitamins (Table 15.1), which possess superior antioxidant activity (Andrade et al., 2019), as well as antiinflammatory, antimicrobial, antiviral, and antiaging properties (Sirohi et al., 2020).

Therefore, valorization of by-products derived from the wine industry is a challenge, since it might contribute to lessening the waste disposal problem as well as recovering a gamut of bioactive compounds with functional and health beneficial properties, which may find several applications as natural additives or active ingredients for food, cosmetic, pharmaceutical, and animal feed products (Andrade et al., 2019; Hogervorst et al., 2017; Makris, 2018; Maroun et al., 2017; Nunes et al., 2017; Puertolas & Barba, 2016; Sette et al., 2019), thus highly contributing to a sustainable food chain from an environmental, social, and economical point of view.

The concentration and type of these bioactive compounds may vary greatly depending on several factors, such as geographic origin, climate, time of harvest, and grape varieties, and especially on the winemaking techniques and type of by-product (Table 15.1). Moreover, bioactive compounds are commonly stored within the plant cells, thus their recovery is hindered by the presence of the cell envelope (membranes and wall), which greatly limits their rate of mass transfer during extraction (Agati et al., 2012; Corrales et al., 2008; Donsì, Ferrari, & Pataro, 2010).

Therefore, it is essential that the recovery of bioactive compounds from winery by-products is supported by efficient methods for their selective extraction, optimized for the structure and morphology featuring the plant cells and tissues of the different type of grape processing residues, as well as the different localization of bioactive compounds in the intracellular space.

In this line, nowadays, emerging technologies are attracting increasing interest of researchers and food processors due to their potential to recover effectively and sustainably high-added value compounds with less energy consumed, thus replacing less efficient conventional extraction processes (Barba et al., 2016).

Moreover, their usage is in agreement with the modern concept of green chemistry, which is based on alternative green solvents, in order to minimize the environmental impact and produce extracts with high quality and purity (Ameer et al., 2017; Chemat et al., 2012; Galanakis, 2012; Pataro et al., 2020).

Therefore, this chapter critically analyzes the characteristics of the most promising emerging technologies for the recovery of bioactive compounds from winery by-products, highlighting

Table 15.1 Principal phytochemicals from grape by-products.

Grape by-products	% Grapes (dry basis)	Major phytochemicals
Stalks	2.5–7.5	Fermentable sugars, flavan-3-ols, stilbenes, salified organic acids
Skins	15	Anthocyanins, fibers stilbenes, flavan-3-ols, flavonols, phenolic acids
Seeds	3–6	Flavan-3-ols, stilbenes, fatty acids, phenolic acids
Wine lees	3.5–8.5	Anthocyanins, free aminoacids, mannoproteins, β-1,3-glucans
Vine shoots	5	Phenolic compounds, stilbenes, lignin

Adapted from Sree, V. G., Shree, T. J., Priyanka, A., Sundararajan, R. (2020). Enhanced antioxidant and anticancer properties of pulsed electric field-treated black grape extract. In: K. A. I. Eltris (Eds.), Current topics in medicine and medical research Vol. 8*: Book Publisher International.*

advantages and limitations in view of their future implementation for the sustainable valorization of agro-food wastes and by-products, including those derived from the winery. The discussed innovative techniques are based on electrotechnologies, namely pulsed electric field (PEF) and high−voltage electric discharges (HVED), ultrasound-assisted extraction (UAE), microwave-assisted extraction (MAE), and super/subcritical fluid extraction (SFE, SBFE).

15.2 Extraction technologies for bioactive compounds

The recovery process of bioactive compounds from the intracellular space of plant-based biomasses, such as winery by-products, can be conducted following the dry or wet route and typically involves four main steps (Fig. 15.2): (i) macroscopic matrix pretreatment, (ii) molecules extraction, (iii) separation/purification, and (iv) product formation/stabilization. One of the most important steps of this process is the molecule extraction, in which numerous conventional and nonconventional technologies have been evaluated.

Traditional techniques involve the application of solid-liquid extraction (SLE) simply by means of solvent application and leaching. SLE encompasses conventional methods such as Soxhlet extraction, percolation, and maceration extraction, which are the most widely applied techniques at the industrial scale to extract bioactive compounds from plant matrices (Ameer et al., 2017). During conventional SLE, the biomass, typically in dried form (dry route), is placed in contact with organic or aqueous solvents, depending on the polarity of the target compounds (Corrales et al., 2008; Galanakis, 2012). In particular, for polyphenols recovery from food processing by-products, polar solvents are frequently used, such as aqueous mixtures, preferentially with methanol and ethanol (Frontuto et al., 2019; Sirohi et al., 2020), but also with acetone, or ethyl acetate (Dang et al., 2013; Louli et al., 2004). On the other hand, acetone, dichloromethane, ethyl lactate, dimethyl ether, diethyl ether, ethanol hexane, octane, or their combination have been used in the extraction of different lipophilic compounds (Jessop, 2011).

However, although these methods offer a straightforward approach, they typically suffer from several drawbacks mainly related to the presence of the envelope (membranes and wall) surrounding

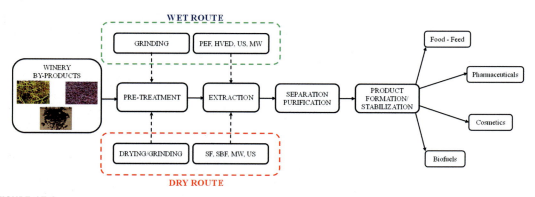

FIGURE 15.2

Schematics of processing steps for the valorization of grape processing by-products through a wet or dry route using either conventional or innovative extraction technologies. *HVED*, high−voltage electric discharges; *MW*, microwave; *PEF*, pulsed electric field; *SBF*, subcritical fluid; *SF*, supercritical fluid; *US*, ultrasound-assisted.

the plant cell and internal organelles (vacuoles), which exert a significant resistance against mass transfer phenomena of solvents and target intracellular compounds, slowing down the solvent extraction process (Donsì, Ferrari, & Pataro, 2010, Donsì, Ferrari, Fruilo, Pataro, 2010).

For these reasons, conventional extraction techniques used to recover valuable compounds from food processing by-products with sufficiently high yield may require excessive consumption of energy, long extraction time, and relatively high temperature, which may cause losses of thermolabile compounds, as well as the use of a large amount of organic solvent to ensure a strong driving force, with negative effects in terms of environmental sustainability and on human health, due to the uncompleted solvent removal from the final product (Barba et al., 2016). Additionally, to enhance the mass transfer rate, conventional solvent extraction methods generally require intensive pretreatments of the raw material, such as comminution, drying, as well as the usage of chemical or enzymatic digestion agents of cell wall/membrane, thus facilitating the penetration of the solvent into the cytoplasm and the diffusion of the solubilized intracellular compounds during the extraction process (Galanakis, 2012, 2015; Hogervorst et al., 2017; Pataro et al., 2020; Romero-Diez et al., 2019). However, these pretreatments are commonly associated with issues such as high energy requirements, degradation of sensitive compounds, especially when high temperatures are involved (e.g., drying), co-extraction of undesirable components, presence of residual chemical agents, hence decreasing the quality of the extracts and complicating the subsequent downstream separation/purification operations.

These underlying drawbacks of the conventional solvent extraction processes have stimulated scientists to explore more selective, cost-effective, and greener techniques for the extraction of bioactive compounds from a wide range of plant-based matrices (Azmir et al., 2013), as will be described in the following paragraph for the valorization of grape processing by-products.

15.3 Innovative extraction methods

To overcome the limitations of conventional extraction methods, several innovative technologies such as electrotechnologies (PEF, HVED)-assisted extraction, UAE, MAE, sub- and supercritical fluids extraction (SBFE, SFE) have been recently proposed by several scientists (Donsì, Ferrari, & Pataro, 2010; Galanakis, 2012, 2015; Hogervorst et al., 2017; Makris, 2018; Maroun et al., 2017, 2018) as an alternative sustainable, efficient, rapid, and cost-effective extraction techniques, based on a green extraction approach, to recover valuable compounds from winery by-products (Fig. 15.2).

In the following paragraphs, the basic principles and recent advances in the application of the above-mentioned innovative technologies for the valorization of winery by-products are deeply discussed and a survey of literature data is reported in Tables 15.2 and 15.3.

15.3.1 Electrotechnologies

Among electrotechnologies, PEF and HVED have been drawing attention as alternative methods for the disruption of plant cells before extraction processes. However, although these technologies are all based on the direct exposure of biomaterial, mainly in wet form, to an external electric field, they differ for the nature of the electric flow (i.e., pulsed or electrical discharge) and the electric field strength (kV/cm) (Barba et al., 2016). These parameters influence not only the design and operational requirements of the corresponding equipment but also the degree of the effects induced in terms of cells damages,

Table 15.2 Summary of innovative extraction technologies (PEF, HVED, UAE, and MAE) to enhance bioactive compounds recovery from grape processing by-products.

Grape by-product	Operating conditions	Extraction conditions	Main findings	Specific energy	Reference
PEF					
Skins	3 kV/cm, 2 Hz, 15 s	50% Ethanol (w/w), 1 h, 70°C, SL ratio = 1:4.5	Enhanced yield of polyphenols (up to 59.5 mg/g DM)	10 kJ/kg of solvent/skins mixture	Corrales et al. (2008)
	0.7–4 kV/cm, 1–200 ms	N/A	Enhanced yield of polyphenols	14.4–111.6 kJ/kg of skins	Cholet et al. (2014)
	9.6–25 kV/cm, 10 Hz, 58.2–151.2 μs	Water, 25°C, SL ratio = 1:4	Enhanced yield of anthocyanins (up to 78 g/L extract) and flavonoids (up to 225 g/L extract)	0.47–3.60 kJ/L of solvent/skins mixture	Medina-Meza and Barbosa-Cánovas (2015)
Seeds	8–20 kV/cm, 0.33 Hz, 0–20 ms	0%–30% Ethanol (w/w), 0–100 min 20–50 °C, SL ratio = 1:5	Enhanced yield of polyphenols (up to 3.4 g/100g DM)	N/A	Boussetta et al. (2012)
Pomace	1.2–3 kV/cm	50% Ethanol (w/w), 7 h 20–35 °C, SL ratio = 1:5	Enhanced yield of polyphenols (up to 2 g/100g DM) and anthocyanins (up to 0.6 g/100g DM)	18 kJ/kg of pomace	Brianceau et al. (2015)
	13.3 kV/cm, 0.5 Hz	Water, 0–2 h 25–30°C, SL ratio = 1:10	Enhanced yield of polyphenols (up to 1.5 mg/g DM) and anthocyanins (up to 0.6 g/100g DM)	0–564 kJ/kg of solvent/pomace mixture	Barba et al. (2015)
Lees	5–25 kV/cm 30–240 μs	N/A	Enhanced yield of mannoproteins	N/A	Martínez et al. (2016)
	5–25 kV/cm 30–105 μs	N/A	Enhanced yield of mannoproteins (up to 240 mg/L of wine)	0.85–72.71 kJ/kg of wine	Martínez et al. (2019)
Vine shoots	13.3 kV/cm, 0.5 Hz 0–1500 pulses	Water, 0–180 min 50°C, SL ratio = 1:20	Enhanced yield of polyphenols (up to 23 mg/g DM)	0.5–800 kJ/kg of solvent/shoots mixture	Rajha et al. (2014)

15.3 Innovative extraction methods

HVED					
Seeds	40 kV/cm, 0–1000 pulses	Water, 50 °C, SL ratio = 1:5	Enhanced yield of polyphenols (up to 8.4 g/100g DM)	N/A	Liu et al. (2011)
	80 kV/cm, 0.33 Hz 1 ms	0%–30% Ethanol (w/w), 0–60 min 50 °C, SL ratio = 1:5	Enhanced yield of polyphenols (up to 8.3 g/100g DM)	N/A	Boussetta et al. (2012)
Pomace	40–133.3 kV/cm, 0.5 Hz	10%–30% Ethanol (w/w), 0–1 h, 20–60 °C, SL ratio = 1:2–1:20	Enhanced yield of polyphenols (up to 2 g/100g DM)	0–800 kJ/kg of solvent/pomace mixture	Boussetta et al. (2011)
	32 kV/cm, 0.5 Hz	Water, 0–2 h 25–30°C, SL ratio = 1:10	Enhanced yield of polyphenols (up to 3 mg/g DM) and anthocyanins (up to 0.4 g/100 g DM)	0–218 kJ/kg of solvent/pomace mixture	Barba et al. (2015)
Stalks	80 kV/cm, 0.5 Hz	0%–30% Ethanol (w/w), pH = 2.5–8.5 SL ratio = 1:5	Enhanced yield of polyphenols (up to 6.6 g/100g FW)	0–188 kJ/kg	Brianceau et al. (2016)
Vine shoots	80 kV/cm, 0.5 Hz	Water, 0–180 min 50°C, SL ratio = 1:20	Enhanced yield of polyphenols (up to 34 mg/g DM)	0.5–260 kJ/kg of solvent/shoots mixture	Rajha et al. (2014)
UAE					
Skins	35 kHz	50% Ethanol (w/w), 1h 70°C, SL ratio = 1:4.5	Enhanced yield of polyphenols (up to 60 mg/g DM)	N/A	Corrales et al. (2008)
	35 kHz	Methanol/HCl (99/1 v/v), 3 sonication cycles (15 + 30 + 15 min), 25°C	Enhanced in yield of polyphenols	N/A	Novak et al. (2008)
	400 W, 24 kHz	Water, 30 min 25–50°C, SL ratio = 1:4	Enhanced yield of anthocyanins (up to 56 g/L extract) and flavonoids (up to 150 g/L extract)	4.1 kJ/cm^2	Medina-Meza and Barbosa-Cánovas (2015)
	20 kHz	8%–92% EtOH (v/v),1–9 min, SL ratio = 1:10	Enhanced in yield of polyphenols (up to 74 mg/g FW)	60–540 kJ/L of solvent/skins mixture	Caldas et al. (2018)

Seeds	200 W, 24 kHz 10%–100% amplitude, 0.1–1 s/s of cycle	Water/Methanol (0%–60% v/v), 1–120 min, 30–70°C SL ratio = 1:25–1:100	Enhanced yield of tartaric acid (up to 200 mg/L extract) and malic acid (up to 50 mg/L extract)	288–2016 kJ/L of solvent/seeds mixture	Palma and Barroso (2002)
	250 W, 40 kHz	40%–60% Ethanol (v/v), 15–30 min, 30–60 °C	Enhanced yield of polyphenols (up to 2.7 mg/g DM) and anthocyanins (up to 1.1 mg/g DM)	2400–5100 kJ/L of solvent/seeds mixture	Ghafoor et al. (2009)
	60 W, 21.5 kHz	Methanol, 60 min 25°C, SL ratio = 1:5	Enhanced yield of polyphenols (up to 56 mg/g DM)	N/A	Casazza et al. (2010)
	50–150 W, 20 kHz 30 min	Methanol, 12 h	Enhanced yield of polyphenols (up to 105 mg/g DM)	13,500 kJ/kg seeds	Da Porto et al. (2013)
Pomace	40–120 kHz	Water, 5–15 min SL ratio = 1:5	Enhanced yield of polyphenols (up to 32.3 mg/100g FW)	15–135 kJ/L of solvent/pomace mixture	González-Centeno et al. (2014)
	25 kHz	50% EtOH, 2.5–80 min 20–50 °C, SL ratio = 1:20	Enhanced yield of polyphenols (up to 25 mg/g DM)	1.02–227.52 kJ/L of solvent/pomace mixture	Tao et al. (2014)
	55 kHz 0.5 s/s per cycle	Water, 2.5–60 min 20–50 °C, SL ratio = 1:20	Enhanced yield of polyphenols (up to 760 mg/100g DM)	65.25–1566 kJ/L of solvent/pomace mixture	González-Centeno et al. (2015)
	300 W, 25 kHz	0%–50% EtOH (v/v), 60 min 20 °C, SL ratio = 1:70	Enhanced yield of polyphenols (up to 439 mg/g dry extract)	N/A	Drosou et al. (2015)
	130 W, 20 kHz pulsed mode	0%–100% Ethanol (v/v) 70% Methanol (v/v), 4–60 min, 20–40°C, SL ratio = 1:8–1:24	Enhanced yield of polyphenols (up to 9.6 mg/g DM)	156–2340 kJ/L of solvent/pomace mixture	Goula et al. (2016)
	160 W, 40 kHz	50% Ethanol (v/v), 5–30 min 25–65 °C, SL ratio = 1:40	Enhanced yield of anthocyanins (up to 18.8 mg/g DM)	720 kJ/L of solvent/ pomace mixture	Bonfigli et al. (2017)
	130 W, 20 kHz pulsed mode	0%–100% Ethanol (v/v), 2–30 min, 20–60 °C, SL ratio = 1:8–1:24	Enhanced yield of polyphenols (up to 48.8 mg/g DM)	78–1170 kJ/L of solvent/pomace mixture	Drevelegka and Goula (2020)

Stalks	600 W, 30 kHz	80% Ethanol (v/v), 0–1380 min, 60 °C, SL ratio = 1:22	Enhanced antioxidant capacity of extract	N/A	Cárcel et al. (2010)
Vine shoots	400 W, 24 kHz	Water, 0–180 min 50°C, SL ratio = 1:20	Enhanced yield of polyphenols (up to 16 mg/g DM)	0.5–3500 kJ/kg of solvent/shoots mixture	Rajha et al. (2014)
MAE Skins	200–600 W, 2450 MHz	0.02–1 mol/L citric acid 0–140 s, SL ratio = 1:5, 1:80	Enhanced yield of anthocyanins (up to 170 mg/100g FW)	N/A	Li et al. (2012)
	2458 MHz	50% EtOH (v/v), 1–30 min SL ratio = 1:10	Enhanced yield of polyphenols (up to 110 mg/g FW)	60–1800 kJ/L of solvent/skins mixture	Caldas et al. (2018)
	N/A	60% Ethanol (v/v), 4 min SL ratio = 1:6.59	Enhanced yield of polyphenols	N/A	Kwiatkowski et al. (2020)
Seeds	650 W, 45 s	32% (w/w) acetone/ 16% (w/w) ammonium citrate	Enhanced yield of polyphenols (up to 110 mg/g DM), flavonoids (up to 52.6 mg/g DM), and proanthocyanidin (up to 31 mg/g DM)	N/A	Dang et al. (2013)
	100–200 W	30%–60% Ethanol (v/v), 2–6 min	Enhanced yield of polyphenols (up to 13 mg/g DM)	120–720 kJ/L of solvent/seeds mixture	Krishnaswamy et al. (2013)
Pomace	200 W	0%–50% EtOH (v/v), 60 min 50°C, SL ratio = 1:50	Enhanced yield of polyphenols	720 kJ/L of solvent/ pomace mixture	Drosou et al. (2015)
	100–600 W	0%–100% Ethanol (v/v), 1–5 min SL ratio = 1:8–1:24	Enhanced yield of polyphenols (up to 36 mg/g DM)	300–9000 kJ/L of solvent/pomace mixture	Drevelegka and Goula (2020)

DM, *dry matter*; FW, *fresh weight*.
N/A = *not available*.
SL ratio: *solid-liquid ratio [g/mL]*.

Table 15.3 Summary of SFE and SBFE processes to enhance bioactive compounds recovery from grape processing by-products.

Grape by-product	Solvent	Pretreatment	P [bar]	T [°C]	t [h]	Main findings	Reference
SFE							
Skins	CO_2 + 7.5% Ethanol	Drying	150	40	0.25	Complete resveratrol recovery	Pascual-Martí et al. (2001)
	CO_2 + 5%–8% Ethanol	Drying + grinding	137/167	37/46	0.5	Enhanced yield of polyphenols (up to 2.5 mg/mL extract)	Ghafoor et al. (2012)
Seeds	CO_2 + Ethanol	Grinding	80/150	35/60	0.5	Enhanced yield of polyphenols (up to 122 mg/kg DM)	Marqués et al. (2013)
Pomace	CO_2 + Ethanol	Drying + grinding	150/300	40	N/A	Enhanced yield of polyphenols (up to 50 mg/g DM) and anthocyanins (up to 1.2 mg/g DM)	Vatai et al. (2009)
	CO_2 + 10% ethanol	Drying + grinding	80	40	7.5	Enhanced yield of polyphenols (up to 3.5 g/100 g DM)	Da Porto et al. (2015)
	CO_2 + 5%–25% Ethanol	Drying + grinding	200	40	N/A	Enhanced yield of polyphenols (up to 3 mg/g DM)	Manna et al. (2015)
SBFE							
Skins	CO_2	Hydroalcoholic extraction	100/130	25/30	N/A	Up to 85% anthocyanins recovery	Bleve et al. (2008)
	Water	Drying + grinding	100	80/120	2	Enhanced yield of polyphenols (up to 77 mg/g DM)	Duba et al. (2015)
Seeds	Water	Drying + grinding	5/15	130/170	0.33/0.5	Enhanced yield of resveratrol (up to 6.9 µg/g DM)	Tian et al. (2017)
Pomace	Water	Drying + grinding	80/150	100/140	1.67	Enhanced yield of polyphenols (up to 31.7 mg/g DM) and flavonoids (up to 15.3 mg/g DM)	Aliakbarian et al. (2012)

N/A = not available.

which may range from mild (PEF) to complete disruption (HVED) of biological cells, thus affecting the extraction efficiency and purity of the extracts. The principles and recent advances in the application of electrotechnologies to extract bioactive compounds from winery by-products are described in the following subsections.

15.3.1.1 Pulsed electric fields

PEF treatment is an innovative nonthermal technology that has been proposed as an alternative to conventional (mechanical, thermal, chemical, and enzymatic) cell disruption methods. PEF processing involves the exposure of plant tissues placed in contact with two electrodes to repetitive short-duration pulses (1 µs −1 ms) of moderate electric field (0.5−10 kV/cm) and relatively low energy input (1−20 kJ/kg), which induces the permeabilization of cell membranes by reversible or irreversible pores formation, known as electroporation or electropermeabilization (Raso et al., 2016). This has been proved to increase the efficiency of the conventional extraction processes of valuable compounds from a wide range of food processing by-products while reducing the energy costs, the solvent consumption, and shortening the treatment time (Donsì, Ferrari, & Pataro, 2010; Frontuto et al., 2019; Pataro et al., 2020; Puertolas & Barba, 2016). The main parameters that affect electroporation are electric field strength and total specific energy input (Raso et al., 2016). In general, increasing the intensity of these parameters enhances electroporation (Raso et al., 2016). For the intensification of mass transfer processes, the setting of these parameters should be such that irreversible electroporation of cell membranes is achieved.

The positive impact of PEF pretreatment on the extraction of bioactive compounds from different winery by-products, such as skins, seeds, pomace, lees, and vine shoots, was observed by several scientists (Table 15.2). For example, the application of a PEF treatment at 3 kV/cm of electric field intensity and relatively low energy input (10 kJ/kg) to skins of the *Dornfelder* variety significantly increased the extractability of phenolic compounds (+68%) and the antioxidant activity (+320%) of water-ethanol extracts, as compared to conventional SLE process (Corrales et al., 2008). Moreover, in the same study, it was found that the antioxidant activity of extracts achieved with PEF was 1.3-fold and 2-fold higher than that achieved upon the application of high hydrostatic pressure (HHP) and ultrasounds (US), respectively. This higher extraction efficiency of PEF as compared with HHP and US was attributed to the capability of the electrical treatment to enhance the inactivation of degrading enzymes. Similar findings were also reported by Barba et al. (2015), who compared the effect of PEF and US pretreatments of aqueous suspensions of grape pomace on the extraction of phenolic compounds and anthocyanins. The authors found that PEF remarkably increased the extraction yield of total phenolic compounds (up to 8%) and especially total anthocyanins (up to 22%) in comparison with US-assisted extractions.

PEF has been also shown to be an effective treatment for the intensification of the polyphenols extraction from grape seeds (Boussetta et al., 2012). Specifically, it was demonstrated that a pre-soaking step of grape seeds from *Pinot Meunier* variety performed in ethanol-water solutions (0%−50%, v/v) prior to the electropermeabilization treatment (8−20 kV/cm, 0−20 ms) yielded a twofold reduction in the diffusion time over conventional extraction. Moreover, when PEF was applied at 50°C to a suspension containing ethanol, the final polyphenols content was about 9 g/100 g dry matter, which was comparable to that achieved upon the application of the HVED-assisted extraction process. In another study, Brianceau et al. (2015) combined PEF treatments of grape pomace from *Dunkelfelder* cultivar (1.2−3 kV/cm; 18 kJ/kg) with a densification step (density = 0.6−1.3 g/cm^3).

The authors observed that the application of a PEF pretreatment at 1.2 kV/cm was enough to intensify the extractability of anthocyanins (+12.8% on average over SLE) during a 7 h conventional maceration process (50% ethanol in water, v/v), independently on the applied extraction temperature.

Recently, it is worth noting that the International Organisation of Vine and Wine (OIV) has approved PEF use in winemaking industries for destemmed and crushed grapes and grape skins (OIV-OENO 634-2020). It has been stated that PEF-induced cell membrane permeabilization of grape skins accelerates extraction, thus resulting in a reduction of maceration time from 10—12 days to 5—6 days, which means the capacity of maceration can be doubled. Additionally, the wine has an improved quality with higher color intensity and higher polyphenol content that makes the wine more suitable for aging (Donsì, Ferrari, Fruilo, & Pataro, 2010; Lopez et al., 2008).

PEF technology can also be applied in the treatment of lees for faster yeast autolysis and to inactivate microorganisms before bottling, resulting in a reduced amount of SO_2 added to the wine to prevent spoilage. For example, the potential of PEF for triggering autolysis and accelerating the release of mannoproteins was evaluated during the aging on lees of *Chardonnay* white wine. These glycoproteins are associated with positive effects such as haze formation reduction, the prevention of tartaric salt precipitation, and the diminution of astringency, along with the improvement of mouthfeel, aroma intensity, and color stability, thereby considerably improving wine quality. The amount of released mannoproteins increased drastically in wines containing PEF-treated yeast and reached its peak after 30 days, while untreated cells required six months. The mannoproteins released in a shorter time from PEF-treated cells featured functional properties similar to those of mannoproteins released during natural autolysis from untreated yeast (Martínez et al., 2019).

As a final point, the application of PEF technology in wineries is particularly interesting due to its potential to improve the quality and productivity of wine, as well as for the sustainable valorization of grape processing by-products through the selective recovery of high-added value compounds. Moreover, PEF can be easily scaled up to process in continuous flow large volumes of either grape or grape processing by-products following a wet route (Fig. 15.2), hence avoiding the need for energy-intensive drying and possibly allowing to reduce the energy demand per unit biomass.

15.3.1.2 High voltage electrical discharges

High voltage electrical discharge (HVED) is a cell disintegration technique of wet biomaterial based on the phenomenon of electrical breakdown of water. During an HVED treatment, high energy is directly released into the aqueous suspension placed in a batch treatment chamber between a high voltage needle electrode and a plated grounded electrode through a plasma channel formed by a short duration (2—5 μs) high current/high voltage electrical discharge (40—60 kV; 10 kA) (Boussetta & Vorobiev, 2014).

Although the mechanisms of HVED are not yet well understood, the combination of electrical breakdown with different secondary phenomena such as high-amplitude pressure shock waves, bubble cavitation, creation of liquid turbulence, and production of active species, occurring during the treatment, are likely the cause of particle fragmentation and cell structural damages, including cell wall disruption, which makes this electrotechnology a more effective cell disintegration technique than PEF. Moreover, air bubbles that are initially present in water or formed due to local heating will likewise be involved in, and accelerate the process (Barba et al., 2016; Boussetta & Vorobiev, 2014; Puertolas & Barba, 2016).

Table 15.2 summarizes the main findings published on the use of HVED technology to intensify the extractability of valuable compounds from winery by-products. In a recent study, it was shown that the application of HVED (32 kV/cm, 0−218 kJ/kg, 0.5 Hz) to an aqueous suspension of grape pomace of *Dunkelfelder* variety markedly increased (up to 116 mg/g FW) the extractability of polyphenols due to the induced product fragmentation, which increased the contact surface between solvent and biomaterial (Barba et al., 2015). Besides, another reason for the improvement of anthocyanins recovery, as compared to the conventional extraction procedure (grinding of grape pomace followed by a diffusion step), was given by the highly turbulent conditions that accelerate the convection of these components from particles to the surrounding medium (Boussetta et al., 2011). In another study, the effect of an HVED treatment, carried out at constant electric field strength (40 kV/cm) and variable number of discharges (1−1000), on the release of polyphenols from previously dried grape seeds in water, was investigated (Liu et al., 2011). It was observed that the extraction yield of phenolic compounds increased with increasing number of delivered discharges up to reach a maximum concentration (8.3 g/100 g dry weight) after 300 discharges, above which a detrimental effect on the polyphenols concentration was detected. This behavior was attributed to the excessive energy load delivered to the samples upon the electrical treatment with a high number of discharges, which might have triggered polyphenols degradation phenomena.

HVED was also found to be a suitable technique for the extraction of antioxidant compounds from vine shoots of grapes of the *Grenache* variety (Rajha et al., 2014). In particular, the improvement of the extraction process was achieved beyond a specific energy threshold, whose minimum value detected in the case of HVED (10 kJ/kg) was 5-fold and 100-fold lower than those needed in the case of PEF and US treatments, respectively. Moreover, HVED produced extracts with the highest polyphenol yield (34.5 mg/g DW), proteins yield (4.5 mg/g DW), and purity (89%).

As per the literature survey, HVED demonstrated a great capability to intensify the extractability of valuable compounds from winery by-products. However, the nonselective extraction and release of all cellular material may create operational problems and increased the subsequent downstream purification costs. Moreover, the technological solutions demand for reliable high pulse power generators along with process limitations (such as operation only in batch mode) might represent the main factors hindering the industrial exploitation of this technique.

15.3.2 Ultrasound-assisted extraction

Ultrasound (US) is a special type of sound wave beyond human hearing in the range between 20 Hz and 1 MHz with intensities greater than 1 W/cm^2, which can induce disruption of plant cells, depending on the frequency utilized (Meullemiestre et al., 2015). In particular, the application of US in the so-called low-frequency range, defined between 18 and 200 kHz, has been extensively explored for the extraction of valuable compounds from biomaterials. During the UAE process, the biomaterial suspended in a liquid medium is typically treated in a batch or continuous flow system, where the energy of acoustic waves initiates a "cavitation" process, which involves the nucleation, growth, and collapse of bubbles, and a propagating shock waveforms jet stream in the surrounding medium causing cell disruption by high shear forces (Luengo et al., 2014; Rabelo et al., 2016; Tao et al., 2014). Moreover, bubble implosion and fragmentation during UAE processes locally produce extreme conditions with estimated temperatures of around 5000 °C and pressures up to 100 MPa, which may further contribute to induce cell damages. Because of this mode of action, US treatment allows to increase the surface contact between solvents and plant tissue, as well as the permeability of cell envelope (membrane, wall), thus allowing to improve the penetration of the solvent in the intracellular space and to accelerate the subsequent diffusion of analytes (Gonzales-Centeno et al., 2015; Rajha et al., 2014).

UAE has been largely explored in the frame of food by-products valorization, especially those deriving from the vinification process, where UAE has successfully employed to extract the major bioactive compounds including phenolics, tartaric acids, esters, and anthocyanins, among others, using water or hydroalcoholic solvents (Table 15.2). However, it has been found that the yield and composition of the extracts are greatly affected by a large number of process parameters namely ultrasonic power, frequency, intensity, duty cycles, geometry of the ultrasonic reactor, solvent type, and temperature (Chemat et al., 2017), which need to be optimized for an efficient recovery of bioactive compounds. In this line, Drevelegka and Goula (2020) using response surface methodology defined the optimal set of values of temperature (56°C), solvent type (53% ethanol-water,v/v), US amplitude (34%), solvent-solid ratio (8 mL/g), and processing time (20 min), which maximize the extraction yield (48.7 mg/g FW) of polyphenolic compounds from the pomace of *Agiorgitiko* grape variety. Similarly, in another study, it was defined the optimal value of different parameters, such as power density (150 W/L), frequency (40 kHz), and extraction time (25 min) to achieve the maximum recovery yields of polyphenols (32.32 mg/100 g FW) and flavonoids (2.04 mg/100g FW) from grape pomace of *Syrah* variety (Gonzales-Centeno et al., 2014).

In several investigations, the extraction efficiency of the UAE was also compared with other conventional and nonconventional extraction technologies. For example, Medina-Meza and Barbosa-Cánovas (2015) carried out a comparative study on the effect of UAE and PEF-assisted extraction of ascorbic acid, polyphenols, flavonoids, and anthocyanins from grape skins (*Vitis Vinifera* L.) using water as an extraction solvent. The authors found that the recovery yields of anthocyanins and flavonoids by UAE were significantly greater than those achieved by the PEF-assisted extraction process, while UAE resulted less efficient in yielding extracts with a high content of total phenolics.

In another study, the capability of UAE to intensify the mass transfer of antioxidant polyphenols from red grape pomace derived from sparkling production was compared with either conventional extraction or MAE (Caldas et al., 2018). It was observed that, within the first 9 min of diffusion, the total phenolic content of extracts obtained by UAE was 2-fold and 1.55-fold higher than those obtained by conventional extraction and MAE, respectively.

Based on the available literature data, it can be stated that UAE offers considerable potential for improving the extraction of intracellular compounds from by-products generated during the vinification process. Nevertheless, the major drawback associated with US processing is the local and overall heat production, which might accelerate degradation reactions of phenolic thermolabile compounds. Therefore, temperature control during treatment is crucial to improve product quality, even though the effectiveness of cell disruption might decrease significantly. Moreover, optimization studies, tailored for each type of by-products investigated, are needed in order to increase the efficiency, decrease the energy demand and improve the extract quality. Finally, trials at larger scales should be carried out to validate the process as well as to assess the economic viability of the UAE process in view of its future exploitation for the valorization of winery by-products.

15.3.3 Microwave-assisted extraction

Microwaves are electromagnetic waves in the frequency range between 300 MHz and 300 GHz, based on the combined effect of two perpendicular oscillating fields, namely electric and magnetic fields

(Angiolillo et al., 2015). When a substance is exposed to microwaves, the latter interact selectively with the dielectric or polar molecules (e.g., water), causing local increases in temperature due to frictional forces from inter-/intramolecular movements.

In the MAE process, microwave irradiation is applied during solvent extraction for quicker and more efficient heating of the extraction medium than through indirect heating, thus contributing to a better and faster dissolution and mass transfer of intracellular compounds. Additionally, microwave irradiation is reported to cause the fast evaporation of water inside the biological cell, with the development of considerable pressures, which contribute destroying the cell wall/membranes and enhance the mass transfer of solvent into the cells and of intracellular compounds into the solvent (Liew et al., 2016). MAE can be performed in open or closed reaction vessels. Open vessels are used for low-temperature extraction at atmospheric pressure whereas closed vessel systems are used for high-temperature extractions (Poojary et al., 2016). MAE may be affected by a large variety of factors, such as microwave power and frequency, treatment duration, moisture content and particle size of the processed matrix, type and concentration of the solvent, solid-to-liquid ratio, extraction temperature, extraction pressure, and number of extraction cycles (Mandal et al., 2007).

Although several studies have been reported on the extraction of valuable compounds from different plant materials (Liew et al., 2016; Tongkham et al., 2017; Yanik, 2017), only limited data are available on the use of MAE for the recovery of valuable compounds from winery by-products (Table 15.2). In recent years, MAE has been successfully employed to extract phenolic, flavonoid, and anthocyanin compounds, from different winery by-products such as grape skins, pomace, and seeds. For example, in the case of MAE of anthocyanins from skins of fresh *Kyoho* grape berries in aqueous citric acid solutions (Li et al., 2012), microwaves contributed to markedly enhance the diffusivity value of the pigments. The higher the power level in the MAE process, the faster the diffusion process. Remarkably, when the solid-to-liquid ratio was increased from 1:20 g/mL up to 1:4 g/mL at a fixed microwave treatment conditions (300 W, 1 min), a further positive effect was observed on the achieved extraction yields. Moreover, a significant interaction between extraction time and microwave power was also detected. In particular, the maximum anthocyanins yield (147.02 mg/100 g FW grape skins) was reached as the power increased from 200 to 600 W, with the extraction time declining from 100 to 40 s (Li et al., 2012).

However, optimization of the MAE processing conditions is crucial to intensify the recovery of high-added value compounds. For instance, the effect of several MAE parameters, such as microwave power, extraction time, and ethanol concentration in the extracting solvent, on total phenols and antioxidant activity of extracts from grape seeds was recently studied using a central composite design (CCD)-based approach (Krishnaswamy et al., 2013). Under the optimized treatment conditions, the predicted extraction yields of polyphenols and antioxidant power were 13 mg and 135 µM ascorbic acid equivalents per gram of raw material, respectively. Moreover, the goodness of fit for the utilized quadratic model was characterized by high values of the determination coefficients ($R^2 = 0.991$ for polyphenols, $R^2 = 0.992$ for antioxidant power).

Overall, results reported in current literature (Table 15.2) highlight that MAE processes have the potential to increase mass transfer rate and extraction yield of valuable compounds from winery by-products, while potentially allowing to decrease the amount of solvent implied during diffusion processes. Nevertheless, this technique is limited to polar solvents and is not suitable for volatile target compounds (Zheng et al., 2011). Moreover, the formation of free radicals, as well as high-temperature increases, may potentially lead to oxidation and degradation of thermolabile compounds during the extraction processes, with a subsequent loss in their functionality. Therefore, optimization and control of processing parameters such as microwave power, processing time, and heating temperature, among others, appears a crucial step before the exploitation of MAE for the valorization of grape processing by-products.

15.3.4 Supercritical fluids extraction

SFE is the most extensively studied emerging extraction technique for the selective recovery of valuable compounds from biomaterials, including suspension of microbial and microalgae cells, and solid matrices (food and food by-product). It has been considered as a sustainable and "green" alternative to the conventional extraction process, which typically follow a dry route and where organic solvents are replaced by supercritical fluids, i.e., fluids with temperature and pressure above or near their critical limit (Da Porto et al., 2015; Galanakis, 2012, 2015). In addition, since supercritical fluids possess low viscosity and high diffusivity, they provide better solvating and transport properties than liquids (Poojary et al., 2016). Among different supercritical fluids, CO_2 is often promoted as environmentally friendly, safe, nontoxic, noncarcinogenic, nonflammable, cheap, and having modest critical conditions (Vatai et al., 2009). Furthermore, it can be applied to a wide range of chemical and biochemical extraction processes, thus replacing conventional extraction techniques with minimized organic solvent consumption (Galanakis, 2012, 2015; Manna et al., 2015). During the supercritical CO_2 extraction process, the biomaterial, typically in dried form, is placed in contact with CO_2 for a certain time (from minutes to hours) in a batch, semibatch, or continuous system, where the solvating power (polarity) of supercritical fluid can be adjusted by manipulating the temperature and pressure of the fluid. Moreover, despite the nonpolar behavior of supercritical CO_2, the selective extraction of a wide range of compounds of different polarity including polyphenols, may occur by adding small amounts of "modifiers" (or co-solvents) such as ethanol, acetone, ethyl acetate, n-hexane, dichloromethane, methanol, and water, able to properly tune CO_2 polarity to that of the target compounds (Ghafoor et al., 2012; Prado et al., 2012; Vatai et al., 2009).

Interestingly, the supercritical CO_2 extraction process does not require complex and costly separation and purification steps of the extract from the organic solvent, since CO_2 is a gas at room temperature and pressure (Molino et al., 2019).

Several investigations have shown the potential of supercritical CO_2 extraction for the recovery of a wide range of bioactive compounds, including stilbenes, anthocyanins, flavonoids, and polyphenols from different winery by-products. In general, it has been observed that the extraction efficiency of valuable compounds from grape skins, seeds, and pomace using supercritical CO_2 extraction increases with CO_2 pressure and temperature up to an optimal level (Ghafoor et al., 2012; Manna et al., 2015; Pascual-Martí et al., 2001). For example, Ghafoor et al. (2012) investigated the effect of extraction temperature (37–46 °C) and pressure (137–167 bar) on the extraction yield of total phenols from seeds of *Vitis labrusca* grapes using a supercritical CO_2 extraction. The optimal extraction conditions were individuated at a temperature ranging between 44 and 46°C, and an intermediate pressure level (153 bar), which yielded 2.41 mg/mL of polyphenols (12.09% recovery). These results may be explained by considering the complex interaction between pressure and temperature occurring during SFE processes. In particular, an increase in pressure up to 153 bar yielded higher solvating power of CO_2 and, hence, an increase in the recovery yields of phenolic compounds. However, further increases in pressure level up to 167 bar did not significantly improve the extraction efficiency, likely due to a reduction of solvent penetration capacity into matrices. Instead, despite an increase in temperature is generally associated with a better solubility of target compounds, it cannot be increased indefinitely, since a structural alteration of phenolics could be induced, thus worsening extract quality. Therefore, the proper tuning of pressure and temperature within SFE processes is of utmost importance to positively impact on high-added value compounds recovery yields.

Several authors have adopted ethanol or hydroalcoholic solutions as a co-solvent with the aim to increase the affinity of supercritical fluids toward target compounds enclosed in cellular tissues of grape pomace (Da Porto et al., 2015; Manna et al., 2015; Vatai et al., 2009), skins (Pascual-Martí et al., 2001), and seeds (Ghafoor et al., 2012). For example, in the work of Vatai et al. (2009), it has been found that the presence of ethanol as a modifier in supercritical CO_2 gave a twofold and a threefold increase in phenolics and anthocyanins recovery, respectively, as when pure CO_2 was employed during SFE process.

Based on the data available in current literature (Table 15.3), the SFE could be claimed as an interesting technique for the selective extraction of phenolics and anthocyanins from winery by-products, despite the difficulty of extracting polar compounds without adding modifiers (Galanakis, 2012). Moreover, preliminary drying or lyophilization processes of biomass aimed at improving the mass transfer phenomena need to be considered, resulting in expensive pretreatments that could limit the industrial application of this technology, together with high equipment costs, which make it difficult to compete with conventional solvent extraction methods.

15.3.5 Subcritical fluids extraction

Similar to SFE, subcritical fluids extraction (SBFE) is a technique that utilizes liquefied fluids slightly below their critical conditions (Aliakbarian et al., 2012). The main advantage over SFE is represented by the possibility to avoid the need for organic solvents as modifiers, due to the large use of water (Tian et al., 2017). In particular, under subcritical conditions, the water dielectric constant may be varied by tuning pressure and temperature conditions, thus making its polarity similar to that of organic solvents, such as ethanol, methanol, and acetone (Duba et al., 2015). Recently, a significant recovery of polyphenols (31.69 mg/g FW) and flavonoids (15.28 mg/g FW) from grape pomace of *Croatina* cultivar using SBFE (116 bar, 140°C) was achieved, when compared to conventional aqueous or ethanolic extraction at room temperature (Aliakbarian et al., 2012). In another study, the main SBFE parameters (pressure, temperature, time, and solid-to-liquid ratio) were optimized for maximizing the extraction of resveratrol from grape seeds, yielding 6.90 µg/g FW, with a 91.98% recovery (Tian et al., 2017). Interestingly, the extraction yield obtained by the SBFE method was significantly higher than those obtained by other techniques (reflux method, UAE, and MAE).

In comparison with the SFE process, SBFE requires more complex downstream separation and purification steps but the final extract purity still appears lower than that obtained by SFE. Therefore, further investigation should be addressed to improve process performances. Moreover, validation of the SBFE process at larger scales is necessary before its exploitation as a valuable method for intensifying the extractability of high-added value compounds from winery by-products.

15.4 Concluding remarks

Winery processing by-products still represent a valuable source of bioactive compounds, of commercial and industrial interest for a more sustainable and bio-based economy. In recent years, different innovative green extraction technologies have emerged, such as the electrotechnologies based on PEF and HVED, UAE and MAE, or pressurized fluid extraction (SFE and SBFE), which ensure higher efficiency and selectivity than conventional processes, as well as a significant reduction in treatment time, solvent and energy consumption.

Among electrotechnologies, PEF can find different applications in winery, both in wine production and valorization of wet grape processing by-products. In this last case, the technology is suitable for selective recovery at room or moderate temperature and with low expenditure of energy of valuable compounds, even though its efficacy strictly depends on raw materials microstructure, hardness, and compactness. HVED, instead, shows a more difficult process control and is a more disruptive technique than PEF, thus leading to higher extraction yield but lower purity, hence implying higher downstream purification costs. MAE and UAE are suitable for enhancing the rate of extraction, either by making the heating process more efficient or disrupting the cell structure, reducing the mass transfer resistances. However, MAE is not desirable for the recovery of thermolabile compounds. Similarly, the occurrence of local excessive increase in temperature during UAE, if not controlled, may also cause degradation of thermolabile compounds. SFE and SBFE are recognized as the most efficient and "green" methods for the recovery of valuable compounds from winery by-products, with the selective behavior imparted by the usage of co-solvents (in the case of SFE). However, although both techniques may provide a significant contribution toward the reduction of solvent consumption, their use is limited by the high capital costs, the high temperatures of operation (SBFE), as well as the need for a highly energy-demanding biomass pretreatment (drying and grinding) prior to the extraction step, which may significantly increase the processing costs.

In the future, additional efforts should be performed in the direction of overcoming the main limitations associated with these techniques by optimizing the processing parameters for improving yields and selectivity with the minimum consumption of energy and solvent, especially from the perspective of industrial scale-up for commercial applications. Moreover, the development of more reliable and affordable processing systems could boost the implementation of these innovative technologies also in agro-food traditional sectors devoted to the processing of seasonal crops, such as winery, thus highly contributing to a sustainable food chain from an environmental and economical point of view.

References

Agati, G., Azzarello, E., Pollastri, S., & Tattini, M. (2012). Flavonoids as antioxidants in plants: Location and functional significance. *Plant Science, 196*, 67−76.

Aliakbarian, B., Fathi, A., Perego, P., & Dehghani, F. (2012). Extraction of antioxidants from winery wastes using subcritical water. *The Journal of Supercritical Fluids, 65*, 18−24.

Ameer, K., Shahbaz, H. M., & Kwon, J.-H. (2017). Green extraction methods for polyphenols from plant matrices and their byproducts: A review. *Comprehensive Reviews in Food Science and Food Safety, 16*, 295−315.

Andrade, M. A., Lima, V., Silva, A. S., Vilarinho, F., Castilho, M. C., Khwaldia, K., & Ramos, F. (2019). Pomegranate and grape by-products and their active compounds: Are they a valuable source for food applications? *Trends in Food Science & Technology, 86*, 68−84.

Angiolillo, L., Del Nobile, M. A., & Conte, A. (2015). The extraction of bioactive compounds from food residues using microwaves. *Current Opinion in Food Science, 5*, 93−98.

Azmir, J., Zaidul, I. S. M., Rahman, M. M., Sharif, K. M., Mohamed, A., Sahena, F., Jahurul, M. H. A., Ghafoor, K., Norulaini, N. A. N., & Omar, A. K. M. (2013). Techniques for extraction of bioactive compounds from plant materials: A review. *Journal of Food Engineering, 117*, 426–436.

Barba, F. J., Brianceau, S., Turk, M., Boussetta, N., & Vorobiev, E. (2015). Effect of alternative physical treatments (Ultrasounds, pulsed electric fields, and high voltage electrical discharges) on selective recovery of biocompounds from fermented grape pomace. *Food and Bioprocess Technology, 8*, 1139–1148.

Barba, F. J., Zhu, Z., Koubaa, M., Sant'Ana, A. S., & Orlien, V. (2016). Green alternative methods for the extraction of antioxidant bioactive compounds from winery wastes and by-products: A review. *Trends in Food Science & Technology, 49*, 96–109.

Beres, C., Costa, G. N. S., Cabezudo, I., da Silva James, N. K., Teles, A. S. C., Cruz, A. P. G., Mellinger-Silva, C., Tonon, R. V., Cabral, L. M. C., & Freitas, S. P. (2017). Towards integral utilization of grape pomace from winemaking process: A review. *Waste Management, 68*, 581–594.

Bleve, M., Ciurlia, L., Erroi, E., Lionetto, G., Longo, L., Rescio, L., Schettino, T., & Vasapollo, G. (2008). An innovative method for the purification of anthocyanins from grape skin extracts by using liquid and sub-critical carbon dioxide. *Separation and Purification Technology, 64*, 192–197.

Bonfigli, M., Godoy, E., Reinheimer, M. A., & Scenna, N. J. (2017). Comparison between conventional and ultrasound-assisted techniques for extraction of anthocyanins from grape pomace. Experimental results and mathematical modeling. *Journal of Food Engineering, 207*, 56–72.

Boussetta, N., & Vorobiev, E. (2014). Extraction of valuable biocompounds assisted by high voltage electrical discharges: A review. *Comptes Rendus Chimie, 17*, 197–203.

Boussetta, N., Vorobiev, E., Deloison, V., Pochez, F., Cordin-Falcimaigne, A., & Lanoisellé, J.-L. (2011). Valorisation of grape pomace by the extraction of phenolic antioxidants: Application of high voltage electrical discharges. *Food Chemistry, 128*, 364–370.

Boussetta, N., Vorobiev, E., Le, L. H., Cordin-Falcimaigne, A., & Lanoisellé, J.-L. (2012). Application of electrical treatments in alcoholic solvent for polyphenols extraction from grape seeds. *Lebensmittel-Wissenschaft und -Technologie- Food Science and Technology, 46*, 127–134.

Brianceau, S., Turk, M., Vitrac, X., & Vorobiev, E. (2015). Combined densification and pulsed electric field treatment for selective polyphenols recovery from fermented grape pomace. *Innovative Food Science & Emerging Technologies, 29*, 2–8.

Brianceau, S., Turk, M., Vitrac, X., & Vorobiev, E. (2016). High voltage electric discharges assisted extraction of phenolic compounds from grape stems: Effect of processing parameters on flavan-3-ols, flavonols and stilbenes recovery. *Innovative Food Science and Emerging Technologies, 35*, 67–74.

Caldas, T. W., Mazza, K. E. L., Teles, A. S. C., Mattos, G. N., Brigida, A. I. S., Conte-Junior, C. A., Borguini, R. G., Godoy, R. L. O., Cabral, L. M. C., & Tonon, R. V. (2018). Phenolic compounds recovery from grape skin using conventional and nonconventional extraction methods. *Industrial Crops and Products, 111*, 86–91.

Cárcel, J. A., García-Péreza, J. V., Mulet, A., Rodríguez, L., & Riera, E. (2010). Ultrasonically assisted antioxidant extraction from grape stalks and olive leaves. *Physics Procedia, 3*, 147–152.

Casazza, A. A., Aliakbarian, B., Mantegna, S., Cravotto, G., & Perego, P. (2010). Extraction of phenolics from *Vitis vinifera* wastes using non-conventional techniques. *Journal of Food Engineering, 100*, 50–55.

Chemat, F., Rombaut, N., Sicaire, A.-G., Meullemiestre, A., Fabiano-Tixier, A.-S., & Abert-Vian, M. (2017). Ultrasound assisted extraction of food and natural products. Mechanisms, techniques, combinations, protocols and applications. A review. *Ultrasonics Sonochemistry, 34*, 540–560.

Chemat, F., Vian, M. A., & Cravotto, G. (2012). Green extraction of natural products: Concept and principles. *International Journal of Molecular Sciences, 13*, 8615–8627.

Cholet, C., Delsart, C., Petrel, M., Gontier, E., Grimi, N., L'Hyvernay, A., Ghidossi, R., Vorobiev, E., Mietton-Peuchot, M., & Gény, L. (2014). Structural and biochemical changes induced by pulsed electric field treatments on cabernet sauvignon grape berry skins: Impact on cell wall total tannins and polysaccharides. *Journal of Agricultural and Food Chemistry, 62*, 2925−2934.

Corrales, M., Toepfl, S., Butz, P., Knorr, D., & Tauscher, B. (2008). Extraction of anthocyanins from grape by-products assisted by ultrasonics, high hydrostatic pressure or pulsed electric fields: A comparison. *Innovative Food Science and Emerging Technologies, 9*, 85−91.

Da Porto, C., Natolino, A., & Decorti, D. (2015). The combined extraction of polyphenols from grape marc: Ultrasound assisted extraction followed by supercritical CO_2 extraction of ultrasound-raffinate. *Lebensmittel-Wissenschaft und -Technologie- Food Science and Technology, 61*, 98−104.

Da Porto, C., Porretto, E., & Decorti, D. (2013). Comparison of ultrasound-assisted extraction with conventional extraction methods of oil and polyphenols from grape (*Vitis vinifera* L.) seeds. *Ultrasonics Sonochemistry, 20*, 1076−1080.

Dang, Y. Y., Zhang, H., & Xiu, Z. K. (2013). Microwave-assisted aqueous two-phase extraction of phenolics from grape (*Vitis vinifera*) seed. *Journal of Chemical Technology & Biotechnology, 89*, 1576−1581.

Donsì, F., Ferrari, G., Fruilo, M., & Pataro, G. (2010). Pulsed electric field-assisted vinification of Aglianico and Piedirosso grapes. *Journal of Agricultural and Food Chemistry, 58*, 11606−11615.

Donsì, F., Ferrari, G., & Pataro, G. (2010). Applications of pulsed electric field treatments for the enhancement of mass transfer from vegetable tissue. *Food Engineering Reviews, 2*, 109−130.

Drevelegka, I., & Goula, A. M. (2020). Recovery of grape pomace phenolic compounds through optimized extraction and adsorption processes. *Chemical Engineering and Processing: Process Intensification, 149*, 107845.

Drosou, C., Kyriakopoulou, K., Bimpilas, A., Tsimogiannis, D., & Krokida, M. (2015). A comparative study on different extraction techniques to recover red grape pomace polyphenols from vinification byproducts. *Industrial Crops and Products, 75*, 141−149.

Duba, K. S., Casazza, A. A., Mohamed, H. B., Perego, P., & Fiori, L. (2015). Extraction of polyphenols from grape skins and defatted grape seeds using subcritical water: Experiments and modeling. *Food and Bioproducts Processing, 94*, 29−38.

Ferrari, V., Taffarel, S. R., Espinosa-Fuentes, E., Oliveira, M. L. S., Saikia, B. K., & Oliveira, L. F. S. (2019). Chemical evaluation of by-products of the grape industry as potential agricultural fertilizers. *Journal of Cleaner Production, 208*, 297−306.

Frontuto, D., Carullo, D., Harrison, S. M., Brunton, N. P., Ferrari, G., Lyng, J. G., & Pataro, G. (2019). Optimization of pulsed electric fields-assisted extraction of polyphenols from potato peels using response surface methodology. *Food and Bioprocess Technology, 2*, 1708−1720.

Galanakis, C. M. (2012). Recovery of high added-value components from food wastes: Conventional, emerging technologies and commercialized applications. *Trends in Food Science & Technology, 26*, 68−87.

Galanakis, C. M. (2015). In C. M. Galanakis (Ed.), *Food waste recovery: Processing technologies and industrial techniques*. ISBN: 978-0-12-800351-0.

Ghafoor, K., Al-Juhaimi, F. Y., & Choi, Y. H. (2012). Supercritical fluid extraction of phenolic compounds and antioxidants from grape (*Vitis labrusca* B.) seeds. *Plant Foods for Human Nutrition, 67*, 407−414.

Ghafoor, K., Choi, Y. H., Jeon, J. Y., & Jo, I. H. (2009). Optimization of ultrasound-assisted extraction of phenolic compounds, antioxidants, and anthocyanins from grape (*Vitis vinifera*) seeds. *Journal of Agricultural and Food Chemistry, 57*, 4988−4994.

González-Centeno, M. R., Comas-Serra, F., Femenia, A., Rosselló, C., & Simal, S. (2015). Effect of power ultrasound application on aqueous extraction of phenolic compounds and antioxidant capacity from grape pomace (*Vitis vinifera* L.): Experimental kinetics and modeling. *Ultrasonics Sonochemistry, 22*, 506−514.

González-Centeno, M. R., Knoerzer, K., Sabarez, H., Simal, S., Rosselló, C., & Femenia, A. (2014). Effect of acoustic frequency and power density on the aqueous ultrasonic-assisted extraction of grape pomace (*Vitis vinifera* L.) − a response surface approach. *Ultrasonics Sonochemistry, 21*, 2176−2184.

Goula, A. M., Thymiatis, K., & Kaderides, K. (2016). Valorization of grape pomace: Drying behavior and ultrasound extraction of phenolics. *Food and Bioproducts Processing, 100*, 132−144.

Hogervorst, J. C., Miljić, U., & Puškaš, V. (2017). Extraction of bioactive compounds from grape processing by-products. In C. M. Galanakis (Ed.), *Handbook of grape processing by-products* (pp. 105−135). Academic Press.

Jessop, P. G. (2011). Searching for green solvents. *Green Chemistry, 13*, 1391−1398.

Kammerer, D. R., Kammerer, J., Valet, R., & Carle, R. (2014). Recovery of polyphenols from the by-products of plant food processing and application as valuable food ingredients. *Food Research International, 65*, 2−12.

Krishnaswamy, K., Orsat, V., Gariépy, Y., & Thangavel, K. (2013). Optimization of microwave-assisted extraction of phenolic antioxidants from grape seeds (*Vitis vinifera*). *Food and Bioprocess Technology, 6*, 441−455.

Kwiatkowski, M., Kravchuk, O., Skouroumounis, G. K., & Taylor, D. K. (2020). Microwave assisted and conventional phenolic and colour extraction from grape skins of commercial white and red cultivars at veraison and harvest. *Journal of Cleaner Production*. https://doi.org/10.1016/j.jclepro.2020.122671

Liew, S. Q., Ngoh, G. C., Yusoff, R., & Teoh, W. H. (2016). Sequential ultrasound-microwave assisted acid extraction (UMAE) of pectin from pomelo peels. *International Journal of Biological Macromolecules, 93*, 426−435.

Liu, D., Vorobiev, E., Savoire, R., & Lanoisellé, J.-L. (2011). Intensification of polyphenols extraction from grape seeds by high voltage electrical discharges and extract concentration by dead-end ultrafiltration. *Separation and Purification Technology, 81*, 134−140.

Li, Y., Xiayang, X., Junhan, W., Zhengfu, W., & Fang, C. (2012). Kinetics and thermodynamics characteristics of microwave assisted extraction of anthocyanins from grape peel. *Transactions of the Chinese Society of Agricultural Engineering, 28*, 326−332.

Lopez, N., Puertolas, E., Condon, S., Alvarez, I., & Raso, J. (2008). Application of pulsed electric fields for improving the maceration process during vinification of red wine: Influence of grape variety. *European Food Research and Technology*. https://doi.org/10.1007/s00217-008-0825-y

Louli, V., Ragoussis, N., & Magoulas, K. G. (2004). Recovery of phenolic antioxidants from wine industry by-products. *Bioresource Technology, 92*, 201−208.

Luengo, E., Condón-Abanto, S., Condón, S., Álvarez, I., & Raso, J. (2014). Improving the extraction of carotenoids from tomato waste by application of ultrasound under pressure. *Separation and Purification Technology, 136*, 130−136.

Makris, D. P. (2018). Green extraction processes for the efficient recovery of bioactive polyphenols from wine industry solid wastes − recent progress. *Current Opinion in Green and Sustainable Chemistry, 13*, 50−55.

Mandal, V., Mohan, Y., & Hemalatha, S. (2007). Microwave assisted extraction - an innovative and promising extraction tool for medicinal plant research. *Pharmacognosy Reviews, 1*, 7−18.

Manna, L., Bugnone, C. A., & Banchero, M. (2015). Valorization of hazelnut, coffee and grape wastes through supercritical fluid extraction of triglycerides and polyphenols. *The Journal of Supercritical Fluids, 104*, 204−211.

Maroun, R. G., Rajha, H. N., El Darra, N., El Kantar, S., Chacar, S., Debs, E., Vorobiev, E., & Louka, N. (2018). Emerging technologies for the extraction of polyphenols from natural sources. In C. M. Galanakis (Ed.), *Polyphenols: Properties, recovery, and applications* (pp. 265−293). Woodhead Publishing.

Maroun, R. G., Rajha, H. N., Vorobiev, E., & Louka, N. (2017). Emerging technologies for the recovery of valuable compounds from grape processing by-products. In C. M. Galanakis (Ed.), *Handbook of grape processing by-products* (pp. 155−181). Academic Press.

Marqués, J. L., Della Porta, G., Reverchon, E., Renuncio, J. A. R., & Mainar, A. M. (2013). Supercritical antisolvent extraction of antioxidants from grape seeds after vinification. *The Journal of Supercritical Fluids, 82*, 238−243.

Martínez, J. M., Cebrián, G., Álvarez, I., & Raso, J. (2016). Release of mannoproteins during *Saccharomyces cerevisiae* autolysis induced by pulsed electric field. *Frontiers in Microbiology, 7*, 1435.

Martínez, J. M., Delso, C., Maza, M. A., Álvarez, I., & Raso, J. (2019). Pulsed electric fields accelerate release of mannoproteins from *Saccharomyces cerevisiae* during aging on the lees of Chardonnay wine. *Food Research International, 116*, 795–801.

Medina-Meza, I. G., & Barbosa-Cánovas, G. V. (2015). Assisted extraction of bioactive compounds from plum and grape peels by ultrasonics and pulsed electric fields. *Journal of Food Engineering, 166*, 268–275.

Meullemiestre, A., Breil, C., Abert-Vian, M., & Chemat, F. (2015). Innovative techniques and alternative solvents for extraction of microbial oils. In A. Meullemiestre, C. Breil, M. Abert-Vian, & F. Chemat (Eds.), *Modern techniques and solvents for the extraction of microbial oils* (pp. 19–42). Springer International Publishing.

Moletta, R. (2005). Winery and distillery wastewater treatment by anaerobic digestion. *Water Science and Technology, 51*, 137–144.

Molino, A., Larocca, V., Di Sanzo, G., Martino, M., Casella, P., Marino, T., Karatza, D., & Musmarra, D. (2019). Extraction of bioactive compounds using supercritical carbon dioxide. *Molecules, 24*, 782.

Muhlack, R. A., Potumarthi, R., & Jeffery, D. W. (2018). Sustainable wineries through waste valorisation: A review of grape marc utilisation for value-added products. *Waste Management, 72*, 99–118.

Novak, I., Janeiro, P., Seruga, M., & Oliveira-Brett, A. M. (2008). Ultrasound extracted flavonoids from four varieties of Portuguese red grape skins determined by reverse-phase high-performance liquid chromatography with electrochemical detection. *Analytica Chimica Acta, 630*, 107–115.

Nunes, M. A., Rodrigues, F., & Oliveira, M. B. P. P. (2017). Grape processing by-products as active ingredients for cosmetic proposes. In C. M. Galanakis (Ed.), *Handbook of grape processing by-products* (pp. 267–292). Academic Press.

OIV (2018). 2018 World Vitiviniculture Situation, OIV Statistical Report on World Vitiviniculture. Retrieved from https://www.oiv.int/public/medias/6371/oiv-statistical-report-on-world-vitiviniculture-2018.pdf.

Palma, M., & Barroso, C. G. (2002). Ultrasound-assisted extraction and determination of tartaric and malic acids from grapes and winemaking by-products. *Analytica Chimica Acta, 458*, 119–130.

Pascual-Martí, M. C., Salvador, A., Chafer, A., & Berna, A. (2001). Supercritical fluid extraction of resveratrol from grape skin of Vitis Vinifera and determination by HPLC. *Talanta, 54*, 735–740.

Pataro, G., Carullo, D., Falcone, M., & Ferrari, G. (2020). Recovery of lycopene from industrially derived tomato processing by-products by pulsed electric fields-assisted extraction. *Innovative Food Science & Emerging Technologies*. https://doi.org/10.1016/j.ifset.2020.102369

Poojary, M. M., Barba, F. J., Aliakbarian, B., Donsì, F., Pataro, G., Dias, D. A., & Juliano, P. (2016). Innovative alternative technologies to extract carotenoids from microalgae and seaweeds. *Marine Drugs, 14*. https://doi.org/10.3390/md14110214

Prado, J. M., Dalmolin, I., Carareto, N. D. D., Basso, R. C., Meirelles, A. J. A., Oliveira, J. V., Batista, E. A. C., & Meireles, M. A. A. (2012). Supercritical fluid extraction of grape seed: Process scale-up, extract chemical composition and economic evaluation. *Journal of Food Engineering, 109*, 249–257.

Puertolas, E., & Barba, F. J. (2016). Electrotechnologies applied to valorization of by-products from food industry: Main findings, energy and economic cost of their industrialization. *Food and Bioproducts Processing, 100*, 172–184.

Rabelo, R. S., Machado, M. T. C., Martínez, J., & Hubinger, M. D. (2016). Ultrasound assisted extraction and nanofiltration of phenolic compounds from artichoke solid wastes. *Journal of Food Engineering, 178*, 170–180.

Rajha, H. N., Boussetta, N., Louka, N., Maroun, R. G., & Vorobiev, E. (2014). A comparative study of physical pretreatments for the extraction of polyphenols and proteins from vine shoots. *Food Research International, 65*, 462–468.

Raso, J., Frey, W., Ferrari, G., Pataro, G., Knorr, D., Teissie, J., & Miklavcic, D. (2016). Recommendations guidelines on the key information to be reported in studies of application of PEF technology in food and biotechnological processes. *Innovative Food Science & Emerging Technologies, 37*, 312–321.

Resolution OIV-OENO 634. (2020). *Treatment of grapes by pulsed electric fields - (PEF)*.

Romero-Díez, R., Matos, M., Rodriguez, L., Bronze, M. R., RodriguezRojo, S., Cocero, M. J., & Matias, A. A. (2019). Microwave and ultrasound pre-treatments to enhance anthocyanins extraction from different wine lees. *Food Chemistry, 272*, 258–266.

Sette, P., Fernandez, A., Soria, J., Rodriguez, R., Salvatori, D., & Mazza, G. (2019). Integral valorization of fruit waste from wine and cider industries. *Journal of Cleaner Production, 242*, 118486.

Sirohi, R., Pandey, J. P., Goel, R., Singh, A., Lohani, U. C., & Kumar, A. (2020). Green processing and biotechnological potential of grape pomace: Current trends and opportunities for sustainable biorefinery. *Bioresource Technology*. https://doi.org/10.1016/j.biortech.2020.123771

Tao, Y., Zhang, Z., & Sun, D.-W. (2014). Kinetic modeling of ultrasound-assisted extraction of phenolic compounds from grape marc: Influence of acoustic energy density and temperature. *Ultrasonics Sonochemistry, 21*, 1461–1469.

Thirumdas, R., Sarangapani, C., & Barba, F. J. (2020). Pulsed electric field applications for the extraction of compounds and fractions (fruit juices, winery, oils, by-products, etc.). In F. J. Barba, O. Parniakov, & A. Wiktor (Eds.), *Pulsed electric fields to obtain healthier and sustainable food for tomorrow* (pp. 227–246). Academic Press.

Tian, J., Wang, Y., Ma, Y., Zhu, P., He, J., & Lei, J. (2017). Optimization of subcritical water extraction of resveratrol from grape seeds by response surface methodology. *Applied Sciences, 7*, 321.

Tongkham, N., Juntasalay, B., Lasunon, P., & Sengkhamparn, N. (2017). Dragon fruit peel pectin: Microwave-assisted extraction and fuzzy assessment. *Agriculture and Natural Resources, 51*, 262–267.

Vatai, T., Škerget, M., & Knez, Z. (2009). Extraction of phenolic compounds from elder berry and different grape marc varieties using organic solvents and/or supercritical carbon dioxide. *Journal of Food Engineering, 90*, 246–254.

Yanik, D. K. (2017). Alternative to traditional olive pomace oil extraction systems: Microwave-assisted solvent extraction of oil from wet olive pomace. *Lebensmittel-Wissenschaft und -Technologie- Food Science and Technology, 77*, 45–51.

Zheng, H., Jin, J., Gao, Z., Huang, H., Ji, X., & Dou, C. (2011). Disruption of chlorella vulgaris cells for the release of biodiesel-producing lipids: A comparison of grinding, ultrasonication, bead milling, enzymatic lysis, and microwaves. *Applied Biochemistry and Biotechnology, 164*, 1215–1224.

CHAPTER 16

The role of pressure-driven membrane processes on the recovery of value-added compounds and valorization of lees and wastewaters in the wine industry

Alexandre Giacobbo[1,2], Andréa Moura Bernardes[1] and Maria Norberta de Pinho[2,3]

[1]*Post-Graduation Program in Mining, Metallurgical and Materials Engineering, (PPGE3M), Federal University of Rio Grande do Sul (UFRGS), Agronomia, Porto Alegre, Rio Grande do Sul, Brazil;* [2]*Center of Physics and Engineering of Advanced Materials, CeFEMA, Instituto Superior Técnico, University of Lisbon, Lisbon, Portugal;* [3]*Chemical Engineering Department, Instituto Superior Técnico, University of Lisbon, Lisbon, Portugal*

16.1 Introduction

Vitiviniculture is one of the most important agro-industrial activities in the world, being even more representative in countries of the Mediterranean region (Ruggieri et al., 2009). According to the *Organization Internationale de la Vigne et du Vin* (OIV), in 2018, the global production of grapes reached 77.8 million tons and 292 million hectoliters of wine were produced (OIV, 2020). Nevertheless, like any industrial activity, along with the economic importance of the wine sector, the environmental problems also arise, which, in turn, are increased in the vintage season due to the large amount of by-products and wastewaters generated in a short time (Ormad et al., 2006). Thus, in just three months, nearby 60%–70% of the wastewaters are generated (Devesa-Rey et al., 2011).

In wineries, the main wastewater sources are related to washing operations, which occur during the crushing and pressing of grapes, as well as in the cleaning of fermentation tanks, barrels, and other equipment and surfaces (Rodrigues et al., 2006). In this way, the wastewaters from the wine sector present large seasonal fluctuations in volume and composition. In terms of volume, a winery generates from 0.3 to 3 L of wastewater per liter of wine produced (Pirra, 2005). With respect to its composition, the wastewater may contain residues of stems, seeds, skins, and lees, as well as musts and wines lost accidentally or during washing operations, and products used for the treatment of wine (coagulants, filtration aids), besides cleaning products, used to wash equipment (Devesa-Rey et al., 2011). The constituents of musts, wines, and lees, namely sugars, ethanol, esters, glycerol, organic acids, phenolic compounds, and yeasts, are also present, in varying proportions, in wastewaters.

In fact, winery wastewater is characterized as highly polluting, with a high organic load and also antibacterial and phytotoxic substances that are resistant to the biological treatment commonly used in this agro-industrial segment (Ormad et al., 2006). On the other hand, many of its constituents are of commercial interest and are liable to recovery, such as phenolic compounds and polysaccharides, for example.

Besides wastewaters, wine lees, as well as other agro-industrial by-products, can be considered a cheap source for the recovery of value-added bioactive compounds (Giacobbo, Bernardes, & de Pinho, 2013; Pinelo et al., 2005). Lees are the residues that accumulate in the tanks containing wine after fermentation, during storage, or after some treatment, as well as those obtained by centrifugation or filtration (Giacobbo et al., 2015). Hence, lees are essentially wine (da Silva, 2003), but saturated with compounds that settle over time, such as tartrates, polysaccharides, phenolic compounds, and microorganism residues (mostly yeasts), among other substances (Pérez-Serradilla & Luque de Castro, 2011). The lees resulted from the alcoholic fermentation are called first racking wine lees, while those derived from the malolactic fermentation are known as second racking wine lees (Giacobbo et al., 2019).

In Fig. 16.1 is displayed an overview of the main by-products generated during wine processing, as well as some recoverable value-added compounds. The yields shown are average values and may vary depending on different aspects such as grape variety and maturation, *terroir*, winemaking technology, among others (da Silva, 2003).

The recovery of substances from wastewaters and/or from by-products is also closely linked to the development of tangible technologies to extract, purify, and concentrate them from such sources. Nevertheless, the process must be carefully conducted since several bioactive compounds may be degraded when exposed to high temperatures, the presence of solvents, or long periods of extraction (Castro-Muñoz et al., 2016).

In this context, due to their inherent characteristics like operating at environmental temperature, separation efficiency, no additives, low energy requirement, and easy scaling up, the use of membrane

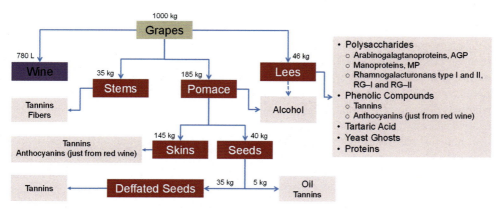

FIGURE 16.1

Overview of the main by-products generated throughout the winemaking process and some of the value-added compounds liable to be recovered. The yields are average values and were taken from da Silva (2003).

separation operations, often integrated with conventional techniques, is a promising alternative for bioactive compounds recovery, especially in steps related to purification and concentration of phenolic compounds and other bioactive phytochemicals present in wastewaters and in extracts from by-products generated in agro-industrial activities. Nevertheless, to assure the success of such an application, some factors are decisive, like the choice of membranes with the appropriate characteristics and the optimization of operating and hydrodynamic conditions.

Fig. 16.1 shows a wide range of materials and substances that may be recovered from winery by-products. Although fibers, seed oil, tartaric acid, and others are also an interesting raw material for different industries, in this chapter the recovery of phenolic compounds and polysaccharides, mainly coming from lees and winery wastewater, will be addressed. Therefore, in this chapter will be presented an overview of conventional methods of extracting phenolic compounds and polysaccharides as well as its purifying, highlighting the role of pressure-driven membrane processes (PDMPs) as integrated processes.

16.2 Value-added compounds found in wastewaters and by-products generated in wine industries

Wastewater from wine industries, as well as other residues and by-products generated in the winemaking process, are important and inexpensive sources of value-added compounds having commercial interest (Pinelo et al., 2005), such as phenolic compounds and polysaccharides, due to their antioxidant properties. In addition, pharmaceutical, cosmetics, and food industries have increased investments in the development and commercialization of natural products, free of chemical additives, bearing in mind the growing demand of the consumer market for this product segment (Bhise et al., 2014). In this way, by-products and wastewaters from agro-industrial activities are emerging as promising alternatives to supply this demand. The average content of bioactive compounds, present in winery wastewater and by-products generated along the winemaking process, has been researched by different authors, who considered different compounds in their analysis, as is displayed in Table 16.1. It is important to highlight that most authors emphasized the total phenolic compounds content and/or the antioxidant activity characteristics, what is associated to the bioactive compounds recovered.

In fact, as can be seen in Table 16.1, the composition and concentration of phytochemicals found in winemaking by-products and wastewater as well as their antioxidant activity varies widely, which can be attributed to several factors such as type of by-product or wastewater, grape variety, grape cultivation method (irrigated or not), harvest, winemaking technology, sample processing, and *terroir* (Alonso et al., 2002; Barcia et al., 2014; Rockenbach et al., 2008; Spranger et al., 2004). Moreover, the adoption of different methodologies to extract and to characterize a particular by-product may also lead to completely different results (Rajha et al., 2014; Tudose-Sandu-Ville et al., 2012). For example, the total phenolic compounds content can be determined by Folin–Ciocalteu method, by high performance liquid chromatography, or by UV-Vis spectrophotometry. Indeed, the discussion of analytical methodologies is not the subject of this chapter, and more details can be found in the literature (Curvelo-Garcia & Barros, 2015; Ignat et al., 2011; Lee et al., 2008; Stefova & Ivanova, 2011).

Chapter 16 Role of PDMPs on recovery of value-added compounds

Table 16.1 Reported composition of wastewaters and by-products from winemaking process (lees and grape pomace).

By-product	Compound	Composition	Ref.
Grape pomace (Merlot)			
	Petunidin-3-glucoside (mg Mv3ge/kg DW)	0.40	Lingua et al. (2016)
	Peonidin-3-glucoside (mg Mv3ge/kg DW)	1.70	
	Malvidin-3-glucoside (mg Mv3ge/kg DW)	96.8	
	Petunidin-3-(6-acetyl)-glucoside (mg Mv3ge/kg DW)	0.40	
	Malvidin-3-(6-acetyl)-glucoside (mg Mv3ge/kg DW)	104	
	Peonidin-3-(6-acetyl)-glucoside (mg Mv3ge/kg DW)	3.30	
	Malvidin-3-(caffeoyl)-glucoside (mg Mv3ge/kg DW)	3.40	
	Delphinidin-3-(6-coumaroyl)-glucoside (mg Mv3ge/kg DW)	7.90	
	Petunidin-3-(6-coumaroyl)-glucoside (mg Mv3ge/kg DW)	24.8	
	Malvidin-3-(6-coumaroyl)-glucoside (mg Mv3ge/kg DW)	143	
	Peonidin-3-(6-coumaroyl)-glucoside (mg Mv3ge/kg DW)	24.6	
	Acetyl pigment A (mg Mv3ge/kg DW)	0.10	
	Coumaroyl vitamin B (mg Mv3ge/kg DW)	0.90	
	Total phenolic compounds (mg GAE/kg DW)	21,220	
Grape pomace (a mixture of 60% cabernet sauvignon, 30% Sangiovese, and 10% Syrah)			
	Total phenolic compounds (mg GAE/kg DW)	2600	Arboleda Mejia et al. (2020)
	Proanthocyanidins (mg CE/kg DW)	490	
	Glucose (mg/kg DW)	460	
	Fructose (mg/kg DW)	4030	
Lees from red wine production (Syrah)			
	Catechin (mg/L)	1.36	Alonso et al. (2002)
	Vanillic acid (mg/L)	1.79–2.93	
	Syringic acid (mg/L)	1.67–3.22	
	Isovanillin (mg/L)	1.00	
	Caftaric acid (mg/L)	4.03–6.23	
	cis-Coutaric acid (mg/L)	0.37	
	trans-Coutaric acid (mg/L)	0.55–1.93	
	Chlorogenic acid (mg/L)	0.59–1.07	
	Caffeic acid (mg/L)	1.39–11.59	

Table 16.1 Reported composition of wastewaters and by-products from winemaking process (lees and grape pomace).—cont'd

By-product	Compound	Composition	Ref.
	cis-p-Coumaric acid (mg/L)	0.38	
	trans-p-Coumaric acid (mg/L)	0.59–2.87	
	Ferulic acid (mg/L)	0.44–2.01	
	Syringaldehyde (mg/L)	0.28–0.54	
	Antioxidant activity (mM TEAC)	18.5–21.0	
Lees from red wine production (Tempranillo)			
	cis-p-Coumaric acid (mg/L)	1.14	Alonso et al. (2002)
	trans-p-Coumaric acid (mg/L)	0.30–0.64	
	Ferulic acid (mg/L)	0.56–0.62	
	Syringaldehyde (mg/L)	0.48–0.64	
	Antioxidant activity (mM TEAC)	7.00–10.0	
Lees from red wine production (Cabernet Sauvignon)			
	Vanillic acid (mg/L)	1.39	Alonso et al. (2002)
	Syringic acid (mg/L)	1.47–2.14	
	Chlorogenic acid (mg/L)	1.94	
	Ferulic acid (mg/L)	0.49–1.10	
	Antioxidant activity (mM TEAC)	4.50–6.50	
Lees from white wine production (Moscatel)			
	p-OH-Phenethyl alcohol (mg/L)	2.09	Alonso et al. (2002)
	Caftaric acid (mg/L)	0.95–4.67	
	cis-Coutaric acid (mg/L)	0.30–0.36	
	Caffeic acid (mg/L)	1.39–3.86	
	Antioxidant activity (mM TEAC)	2.00–4.00	
Lees from white wine production (Palomino Fino)			
	Caftaric acid (mg/L)	4.17	Alonso et al. (2002)
	cis-Coutaric acid (mg/L)	0.18	
	trans-Coutaric acid (mg/L)	0.22	
	Chlorogenic acid (mg/L)	0.80–2.63	
	Antioxidant activity (mM TEAC)	1.00–2.00	
Lees from 1st racking of red wine production (Syrah)			
	Myricetin (mg/kg DE)	4292	Pérez-Serradilla and Luque de Castro (2011)
	Quercetin (mg/kg DE)	13,656	
	Quercetin-3-β-glucoside (mg/kg DE)	8900	
	Caffeic acid (mg/kg DE)	663	
	p-Coumaric acid (mg/kg DE)	2449	

Continued

Table 16.1 Reported composition of wastewaters and by-products from winemaking process (lees and grape pomace).—cont'd

By-product	Compound	Composition	Ref.
	Malvidin-3-glucoside (mg/kg DE)	91.0	
	Malvidin-3-O-(6-p-coumaroyl)-glucoside (mg Mv3ge/kg DE)	11,729	
	Total phenolic compounds (mg GAE/kg DE)	364	
	Antioxidant activity (mM TEAC)	3.93	
Lees from 2nd racking of red wine production (Alicante Bouchet, Syrah, and Tinta Roriz)			
	Total phenolic compounds (mg/L GAE)	1171	Giacobbo, Oliveira, et al. (2013)
	Total polysaccharides (mg/L GE)	4764	
Lees from 2nd racking of red wine production (Merlot)			
	Total phenolic compounds (mg/L GAE)	1065	Giacobbo et al. (2015)
	Total polysaccharides (mg/L GE)	1599	
Lees from red wine production (Syrah)			
	Total phenolic compounds (mg GAE/kg DW)	11,250	Kontogiannopoulos et al. (2017)
	Tartaric acid (mg/kg DW)	575,800	
Lees from Port wine			
	Total phenolic compounds (mg GAE/kg DW)	4200	Romero-Díez et al. (2019)
	Total anthocyanins (mg Mv3ge/kg DW)	6200	
	Antioxidant activity (mM TEAC/kg DW)	402	
Lees from 1st racking of red wine production (Tempranillo)			
	Total phenolic compounds (mg GAE/kg DW)	3703	Romero-Díez et al. (2019)
	Total anthocyanins (mg Mv3ge/kg DW)	4450	
	Antioxidant activity (mM TEAC/kg DW)	0.66	
Lees from 2nd racking of red wine production (Tempranillo)			
	Total phenolic compounds (mg GAE/kg DW)	2342	Romero-Díez et al. (2019)
	Total anthocyanins (mg Mv3ge/kg DW)	2900	
	Antioxidant activity (mM TEAC/kg DW)	0.51	
Lees from 1st racking of red wine production (Merlot)			
	Gallic acid (mg/kg DW)	59.6	Giacobbo et al. (2019)
	Kaempferol-3-O-galactoside (mg/kg DW)	75.8	
	Myricetin-3-O-glucoside (mg/kg DW)	133	

Table 16.1 Reported composition of wastewaters and by-products from winemaking process (lees and grape pomace).—cont'd

By-product	Compound	Composition	Ref.
	Quercetin-3-O-glucuronide (mg/kg DW)	141	
	Myricetin (mg/kg DW)	481	
	Quercetin (mg/kg DW)	713	
	Total phenolic compounds (mg/kg DW)	1603	
	Delphinidin-3-hexoside (mg Cy3ge/kg DW)	3.73	
	Petunidin-3-hexoside (mg Cy3ge/kg DW)	13.4	
	Peonidin-3-hexoside/Pelargonidin-3-hexoside/Malvidin-3-hexoside[a] (mg Cy3ge/kg DW)	193	
	Malvidin-3-hexoside-pyruvate (mg Cy3ge/kg DW)	2.02	
	Malvidin-3-(6-acetyl)-hexoside (mg Cy3ge/kg DW)	26.8	
	Delphinidin-3-(6-coumaroyl)-hexoside (mg Cy3ge/kg DW)	6.75	
	Cyanidin-3-(6-coumaroyl)-hexoside (mg Cy3ge/kg DW)	5.63	
	Malvidin-3-(6-coumaroyl)-hexoside (mg Cy3ge/kg DW)	7.64	
	Peonidin-3-(6-coumaroyl)-hexoside (mg Cy3ge/kg DW)	148	
	Total anthocyanins (mg Cy3ge/kg DW)	407	
Lees from 2nd racking of red wine production (Merlot)			
	Gallic acid (mg/kg DW)	246	Giacobbo et al. (2019)
	Kaempferol-3-O-galactoside (mg/kg DW)	122	
	Myricetin-3-O-glucoside (mg/kg DW)	347	
	Quercetin-3-O-glucuronide (mg/kg DW)	674	
	Myricetin (mg/kg DW)	1159	
	Quercetin (mg/kg DW)	787	
	Total phenolic compounds (mg/kg DW)	3329	
	Delphinidin-3-hexoside (mg Cy3ge/kg DW)	43.7	
	Cyanidin-3-hexoside (mg/kg Cy3ge DW)	10.8	
	Petunidin-3-hexoside (mg Cy3ge/kg DW)	60.3	
	Peonidin-3-hexoside/Pelargonidin-3-hexoside/Malvidin-3-hexoside[a] (mg Cy3ge/kg DW)	206	
	Malvidin-3-hexoside-pyruvate (mg Cy3ge/kg DW)	23.0	
	Petunidin-3-(6-acetyl)-hexoside (mg Cy3ge/kg DW)	28.3	

Table 16.1 Reported composition of wastewaters and by-products from winemaking process (lees and grape pomace).—cont'd

By-product	Compound	Composition	Ref.
	Malvidin-3-(6-acetyl)-hexoside (mg Cy3ge/kg DW)	86.6	
	Delphinidin-3-(6-coumaroyl)-hexoside (mg Cy3ge/kg DW)	46.4	
	Cyanidin-3-(6-coumaroyl)-hexoside (mg Cy3ge/kg DW)	33.7	
	Petunidin-3-(6-coumaroyl)-hexoside (mg Cy3ge/kg DW)	35.5	
	Malvidin-3-(6-coumaroyl)-hexoside (mg Cy3ge/kg DW)	36.8	
	Peonidin-3-(6-coumaroyl)-hexoside (mg Cy3ge/kg DW)	133	
	Total anthocyanins (mg Cy3ge/kg DW)	744	
Lees from red wine production (Nero d'Avola)			
	Total phenolic compounds (mg/L GAE)	1161	Cassano et al. (2019)
	Sugars (mg/L GE)	23.8	
	Anthocyanins (mg Cy3ge/L)	369	
	Resveratrol (mg/L)	0.29	
	Antioxidant activity (mM TEAC)	3.80	
Winery wastewater			
	Total phenolic compounds (mg/L GAE)	98.5	Rodrigues et al. (2020)
	Total polysaccharides (mg/L GE)	248	

[a]*Mix of compounds*; *CE*, (+)-catechin equivalent; *Cy3ge*, cyanidin-3-glucoside equivalent; *DE*, dry extract; *DW*, dry weight; *GAE*, gallic acid equivalent; *GE*, glucose equivalent; *Mv3ge*, Malvidin-3-glucoside equivalent; *TEAC*, Trolox equivalent antioxidant capacity.

16.2.1 Phenolic compounds

Phenolic compounds are secondary metabolites of vegetal origin, which are widely distributed in all higher plants and represent a vast variety of compounds. They are classified into several classes, such as: flavonols, flavanols, flavones, isoflavones, flavanones, anthocyanins, proanthocyanidins, benzoic acids, hydroxycinnamic acids, stilbenes, and lignans (Cassano et al., 2018; Manach et al., Jiménez, 2004). These compounds may have one or more phenolic groups in their structure (Nave et al., 2007), be in some glycosylated form, and/or associated with organic acids and/or molecules with polymerized complexes of high molecular weight (MW), such as tannins. Thus, this wide variety of possibilities results in compounds with different characteristics and a broad distribution of MW, where organic acids with about 120 Da and tannins with a high degree of polymerization (up to 83) are found, which is equivalent to MW of about 25,000 Da (Hanlin et al., 2011; Jackson, 2000; Souquet et al., 1996).

In the last decades, phenolic compounds have been intensively studied, to the point of being considered the "vitamins of the 21st century" (Kammerer et al., 2014). They act as free radical scavengers, as electron or hydrogen donors and as metal chelators (Anastasiadi et al., 2010). In addition, they have high antioxidant capacity, along with antiinflammatory, antiviral, and antimicrobial properties (Ignat et al., 2011). In this way, phenolic compounds act in the prevention of oxidation of nucleic acids, proteins, and lipids (Ky et al., 2014), i.e., they contribute to the prevention of several diseases, such as cancer, diabetes, and osteoporosis, as well as cardiovascular and neurodegenerative diseases (Cassano et al., 2018). Moreover, phenolic compounds have also been highlighted for having excellent properties for preserving food (Selani et al., 2011) and for having an important role in protecting against brain dysfunction (Pinelo et al., 2005).

16.2.2 Polysaccharides

Polysaccharides are biopolymers composed of monosaccharides linked together through glycosidic bonds (Partain, 2000; Zong et al., 2012). These macromolecules are essential to all living organisms, in such a way that they are associated with a variety of vital functions (Srivastava & Kulshreshtha, 1989), being found in greater abundance in algae, fungi, yeasts, bacteria, and plants.

Polysaccharides are the main group of macromolecules in wine, being derived from grapes and microorganisms (Vidal et al., 2003). Arabinogalactanoproteins (AGPs), Arabinogalactans (AGs), Arabinans, Ramnogalacturonan types I (RG–I) and II (RG–II) are derived from the cell walls of grapes, while manoproteins (MPs) are released by yeasts during fermentation (Doco et al., 2003). In a study in which the total fractionation and characterization of the polysaccharides present in red wines was carried out, the authors concluded that 42% are AGPs, 35% are MPs, 19% are RG–II, and 4% are RG–I (Vidal et al., 2003).

In general, AGPs exhibit in their composition about 10% of proteins, arabinose, and galactose (which are linked to the polysaccharide nucleus) and, less frequently, glucuronic acid associated with sugars. AGPs are basically protein-bound AGs and have an MW in the range of 180,000–260,000 Da (Pellerin et al., 1995). Arabinans belong to the AGPs family, are short chain polymers of arabinose and have low MW, with about 6000 Da (Botelho de Sousa et al., 2014).

MPs have a wide range of MW, ranging from 5000 Da to more than 800,000 Da (Saulnier et al., 1991). Its molecular structure is composed of a chain of peptides linked to D-mannose units, where proteins represent about 10% and mannose the remaining 90% (Botelho de Sousa et al., 2014; Doco et al., 2003). Indeed, MPs have aroused special interest in the enological area, mainly due to their positive impact on the sensory quality of wines, such as reducing astringency, improving perception in the mouth, adding complexity and aromatic persistence, increase in sweetness, and roundness (Rigou et al., 2021).

RG–II are polysaccharides of low MW, 5000 to 10,000 Da, which contain 12 glycosidic groups joined by more than 20 different glycosidic bonds. In their composition, RG–II presents sugars such as D–apiosis, L–galactose, 2–O–methyl–L–fucose, 2–O–methyl–D–xylose, among others (Pellerin et al., 1996). In addition, encompassing all the macromolecules present in wine, RG–II are the ones with the highest negative charge density, which can form complexes with divalent and trivalent cations (Doco et al., 2003). In turn, RG–I are alternating polymers of rhamnose and galacturonic acid disaccharides (Oechslin, Lutz, & Amadò, 2003), having an MW between 12,500 and 56,500 Da (Vidal et al., 2003; Yapo et al., 2007).

In recent decades, due to the various applications and characteristics such as biocompatibility, biodegradability, nontoxicity, and therapeutic activities, the interest for using polysaccharides has increased, so that they are emerging as an important class of natural bioactive product (Liu et al., 2015). In addition, many polysaccharides exhibit strong antioxidant properties, resulting in immunomodulatory, antitumor, antiinflammatory, and antifatigue effects, accrediting them as potential new antioxidants (Chen et al., 2012). Furthermore, due to their abilities to interact with tannins, polysaccharides are also associated with many enological phenomena, reducing the astringency of wines (Riou et al., 2002), acting as protective colloids contributing to tartaric stabilization (Gonçalves et al., 2001; Waters et al., 1993), and also as complexing agent of bivalent cations, such as toxic heavy metals (Botelho de Sousa et al., 2014).

16.3 General aspects about the recovery of value-added compounds from agro-industrial by-products and wastewaters

The agro-industrial by-products and wastewaters generated during the processing of fruits and vegetables present a complex composition of bioactive compounds, with vitamins, tocopherols, carotenoids, phenolic compounds, complex carbohydrates, and fibers that have been associated with the supposed beneficial health effects promoted by the consumption of vegetables and fruits (Cassano et al., 2018; Moreira et al., 2017). Consequently, the recovery of these bioactive compounds has been intensely investigated, emerging as an alternative to synthetic substances commonly used in the cosmetic, pharmaceutical, and food industries (Panzella et al., 2020). In fact, the recovery of bioactive compounds is usually associated with a sequence of operations, typically composed by extraction, separation, purification, concentration, and drying.

In general, polysaccharides are recovered by aqueous extraction (Chen et al., 2012; Jia et al., 2014), while phenolic compounds are recovered by extraction with organic solvents such as methanol, ethanol, and acetone, among others (Louli et al., 2004; Makris et al., 2007). However, there are also studies recovering phenolic compounds by extraction with subcritical (Adil et al., 2008) and supercritical (Passos et al., 2010) fluids, microwave (Pérez-Serradilla & Luque de Castro, 2011), high-voltage electrical discharges (Liu et al., 2011), pulsed electric field (Puértolas et al., 2013), and ultrasonication (Bonfigli et al., 2017).

Regardless of the method used for extraction, the resulting extract will contain not only the target compounds, whether polysaccharides or phenolic compounds, but also impurities and other compounds, such as organic acids, sugars, and minerals, in addition to, of course, the solvent used for extraction. Therefore, a separation step is extremely important, both for the removal and recovery of the solvent and for the purification of target compounds. The separation can be performed by chromatographic techniques (Labarbe et al., 1999), adsorption (Soto et al., 2011), capillary electrophoresis (Ignat et al., 2011), among others. However, traditional methods are usually used to purify the target compounds, such as: coagulation and precipitation of impurities (use of chemicals, solvents), adsorption of target compounds in resins, elution of purified target compounds and their concentration by evaporating the solvent (Cassano et al., 2018). Nevertheless, these techniques are inefficient, time-consuming, or use large amounts of chemicals that are often rather toxic, and the development and implementation of more efficient and eco-friendly technologies is of paramount importance both to increase the added value of resources and to minimize impacts on the environment, promoting the circular economy.

16.4 General aspects over pressure-driven membrane processes

Membrane technologies such as reverse osmosis (RO), nanofiltration (NF), ultrafiltration (UF), and microfiltration (MF) are unitary operations of a physical nature that allow the separation of different chemical species from a feed solution by a pressure gradient applied between the two sides of a permselective barrier, the membrane (Fane et al., 2011). As a result, the feed solution is divided in two streams: (i) a permeate stream, containing all the chemical species that permeated the membrane; and (ii) a retentate stream, containing everything that was rejected by it (de Pinho & Minhalma, 2019).

The degree of separation is mainly determined by the membrane's pore-size, which is closely related to its molecular weight cut-off (MWCO), but also, though with less influence, by properties related to solute-membrane and solute-solute interactions, such as surface charge, molecular form, and hydrophobicity (Cassano et al., 2018). MWCO is a parameter frequently used to characterize the selectivity of membranes, and is related to the MW of a reference solute whose rejection is greater than 90% (Baker, 2012). Therefore, a membrane with MWCO of 300 Da is able to reject at least 90% of solutes having 300 Da MW. A schematic representation of the PDMP, namely MF, UF, NF, and RO, is illustrated in Fig. 16.2.

MF membranes have pore sizes from 0.1 to 10 μm and are basically applied to remove suspended solids, bacteria, and colloids (Baker, 2012). UF uses tighter membranes, with pore sizes between 1 and 100 nm (Matsuura, 1993), which corresponds to MWCO from 1000 to 350,000 Da (Cassano et al., 2018), making this unit operation suitable for separation and fractionation of macromolecules. NF, as an intermediate operation between UF and RO, employs membranes with MWCO of 200–1000 Da and is mainly applied to the fractionation of small organic solutes and salts, since it presents rejection coefficients for bivalent salts above 90% and between 40% and 90% for monovalent salts (de Pinho & Minhalma, 2019). Finally, RO membranes have a dense active layer, without pores, presenting high

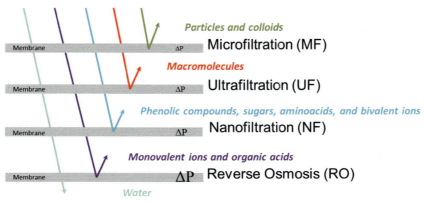

FIGURE 16.2

Schematic description of the retention capacity of the different PDMP.

Adapted from de Pinho, M. N., & Minhalma, M. (2019). Introduction in membrane technologies. In C. M. Galanakis (Ed.), Separation of functional molecules in food by membrane technology (1st ed. pp. 1–29). Chennai: Academic Press. doi:10.1016/B978-0-12-815056-6.00001-2.

selectivity and low productivity when compared to the other PDMP. In other words, they are virtually impermeable to organic solutes and salts and permeable to only water, while they may show rejections of up to 99.9% sodium chloride (de Pinho et al., 2002). In fact, as the pore size of the membranes decreases, greater forces to conduct the operation are required, so that MF, UF, NF, and RO have typical operating pressures in the ranges of 0.1—1 bar, 0.5—8 bar, 5—40 bar, and 20—100 bar, respectively (de Pinho & Minhalma, 2019).

16.5 PDMP in the recovery of polysaccharides and phenolic compounds

In recent years, PDMPs have gained prominence regarding the conventional separation techniques (solvent extraction, precipitation, distillation, centrifugation, etc.) due to their inherent characteristics: (i) no addition of chemicals requirement during separation; (ii) high selectivity; (iii) room temperature operation, enabling the processing of thermolabile substances; (iv) no need of a phase change to effect the separation, saving energy; and (v) modular operation, easing scaling-up (Habert et al., 2006). These characteristics give PDMP applicability in the most different segments of the industry, in separation, purification, and concentration operations, especially in processes related to the production of products containing bioactive and heat-sensitive substances, such as phenolic compounds (Giacobbo, Bernardes, & de Pinho, 2013).

Fig. 16.3 illustrates the conceptual framework of a process based on membrane technologies for the recovery, purification, and concentration of bioactive compounds. In such a process, MF or even open UF may be used to remove suspended solids from wastewaters or extracts from agro-industrial by-products resulting in a clarified extract (permeate stream), and a concentrate stream containing colloids and suspended solids. Subsequently, the substances present in the clarified extract can be separated/purified using tight UF, resulting in a permeate rich in small molecules such as sugars, phenolic compounds, organic acids, and minerals, and a concentrate enriched in polysaccharides and other macromolecules. Finally, NF or RO could be used to concentrate the phenolic compounds and recover the solvent in the permeate stream.

16.5.1 Processing lees and winery wastewater

The study of lees and winery wastewater as raw material to the production of bioactive products has been proposed by different authors.

In previous works, Giacobbo and co-workers (Giacobbo, Bernardes, & de Pinho, 2013; Giacobbo, Oliveira, et al., 2013) proposed an integrated process based on PDMP for the recovery and fractionation of polysaccharides and phenolic compounds present wine lees. In that process, the lees from the second racking were subject to a sedimentation stage after pH adjustment, where 72% of the phenolic compounds and 82% of the polysaccharides were in the settled fraction when the pH was adjusted to 5.4, while at the natural pH of 3.6, only 12% and 15% of polyphenols and polysaccharides were in the settled fraction, respectively. Then, the clarified extract at pH 3.6 was ultrafiltered with a 7600 Da MWCO membrane (characterization described in Giacobbo, Oliveira, et al. (2013)), observing that the permeate flux increased linearly with pressure and tangential velocity. Subsequently, the permeate obtained at the highest operating pressure (4 bar) and the highest

16.5 PDMP in the recovery of polysaccharides and phenolic compounds

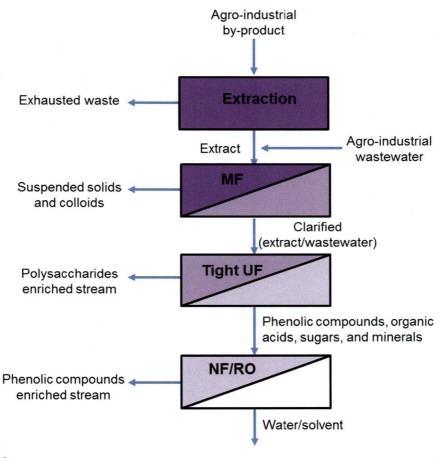

FIGURE 16.3

Conceptual framework of integrated membrane processes for the recovery of bioactive compounds from agro-industrial by-products and wastewater.

tangential velocity (0.87 m/s), containing about 50 mg/L of low MW polysaccharides and 53 mg/L of phenolic compounds, was subjected to fractionation using tight UF and NF membranes characterized in terms of sucrose rejection. UF membranes, characterized as having 8%—17% sucrose rejection, showed rejections to phenolic compounds of 28%—49% and to polysaccharides of 72%—92%, while NF membranes, with 86%—99% sucrose rejection, resulted in rejections to phenolic compounds of 65%—93% and to polysaccharides greater than 99%, demonstrating the applicability of PDMP for fractionation of polysaccharides and phenolic compounds. However, this process showed a low yield (4.5% of phenolic compounds and 1% of low MW polysaccharides recovered), since most of the bioactive compounds were removed by sedimentation or by the 7600 Da UF membrane.

Galanakis et al. (2013) assessed the possibility of using UF to fractionate the phenolic compounds recovered from hydroalcoholic extracts (95/5) of wine lees. Two extracts of lees collected after the first decanting stage of Maratheftiko's red wine production were prepared: one more concentrated, containing about 6450 mg/L of pectin, 3500 mg/L of nonreducing sugars, 1500 mg/L total phenolic compounds, and 265 mg/L of total anthocyanins; and another diluted, containing concentrations about four times lower than the former. These extracts were centrifuged at 4,700g, for 10 min, and then submitted to the fractionation by UF, where three membranes were evaluated: GR40PP, GR70PP, and ETNA01PP, with MWCO of 100,000, 20,000, and 1000 Da, respectively. The authors suggested the sequential use of the membranes ETNA01PP, fluoropolymer and 1000 Da, and GR40PP, polysulfone and 100,000 Da, in order to obtain concentrated stream enriched in hydroxycinnamic acid derivatives in the first stage, subsequently clarifying it from pectin and oligo-galacturonic portions in a second stage.

In a more recent project, Giacobbo et al. (2017, 2015) developed an eco-friendly process involving aqueous extraction and PDMP for the recovery and fractionation of polysaccharides and phenolic compounds from wine lees. In that process, the lees from the second racking underwent aqueous extraction, that is, dilution of 10 times (v/v), to decrease the load of solids and increase the solubility of bioactive compounds (polysaccharides and phenolic compounds). Subsequently, the extract was microfiltered, using membranes with a pore size of 0.2–0.5 μm until a volumetric concentration factor (VCF) of 7–8, representing a recovery of 80%–90% of the initial feed volume as permeate, which contains about 20% of phenolic compounds and 5% of polysaccharides. Then, the polysaccharides and phenolic compounds could be separated by UF, with membranes of 1000–10,000 Da, in which the ETNA10PP membrane (10,000 Da MWCO) showed good fractionation capacity, the largest permeate fluxes, as well as lower incidences of concentration polarization and fouling. Finally, the UF permeate could be concentrated via NF or RO, where an NF membrane, NF270 (400 Da MWCO), showed rejections greater than 92% of phenolic compounds and 99% of polysaccharides, high permeate fluxes, and low concentration polarization incidence. Therefore, this process resulted in four streams: (i) MF concentrate, containing suspended solids and colloids, which must be subjected to another treatment; (ii) UF concentrate, rich in polysaccharides; (iii) NF concentrate, rich in phenolic compounds; and (iv) NF permeate, which can be recycled as reused water or in the aqueous extraction stage.

Later, processing by UF and NF the permeate resulting from the MF of wine lees from a previous study Giacobbo et al. (2015), Giacobbo et al. (2018) demonstrated that by increasing the tangential velocity from 0.48 to 0.96 m/s and decreasing the transmembrane pressure from 15 to 3 bar the mass transfer coefficient at the boundary layer adjacent to the membrane surface rises and, consequently, the concentration polarization and fouling phenomena decrease, and that this also resulted in a slight drop in rejection of phenolic compounds.

Cassano et al. (2019) used an MF hollow fiber polyvinylidenefluoride (PVDF) membrane, with 0.13 μm of pore size, prepared in the laboratory, to clarify an aqueous extract of red wine lees, and three commercial NF membranes to fractionate the bioactive compounds, namely: NP030 and NP010, Microdyn-Nadir polyethersulfone membranes, with MWCO of 300–400 and 1000 Da, respectively, and a 1000 Da MWCO polyamide membrane, MPF36 from Koch Membrane Systems. The aqueous extract was obtained by diluting the lees generated during red wine production (Nero d'Avola variety) with distilled water (25% w/w) and thermostating it in a water bath at 45°C for 1 h, subsequently clarifying it by MF. At the optimized operation conditions, transmembrane pressure of 0.8 bar and feed flow rate of 35 L/h, the MF reached permeate fluxes of about 53 kg/h m^2 at steady state, completely removing the suspended solids from the extract, and resulting in a clarified solution enriched in bioactive compounds, characterized as phenolic compounds, anthocyanins, resveratrol, and sugars. All assayed NF membranes presented high anthocyanin rejection (greater than 93%) and low rejection toward resveratrol and other phenolic compounds. However, among the evaluated membranes, MPF36 exhibited the highest permeate flux (17 kg/h m^2, at 20 bar and VCF of 2) and the lowest rejections to phenolic compounds (26.3%) and resveratrol (10%), emerging thus as the membrane with the best separation factor between these compounds and anthocyanins.

Recently, Rodrigues et al. (2020) assessed the efficiency of micellar enhanced ultrafiltration for the recovery of phenolic compounds and polysaccharides from winery wastewaters (composition displayed in Table 16.1), using Esterquat as a cationic surfactant. The authors inferred that the degree of separation between glucose and phenolic compounds is closely related to the matrix complexity. Thereby, at the optimal operating conditions, rejections of 47.3% for polysaccharides and 78.3% for phenolic compounds were observed, showing that a stream of concentrate rich in phenolic compounds and another of permeate rich in polysaccharides can be obtained.

16.5.2 Processing extracts from other winemaking by-products

The study of other winery by-products, such as grape pomace and seeds, as raw material for the production of bioactive products, has also been carried out.

Santamaría et al. (2002) evaluated different membranes to fractionate phenolic compounds from a hydroalcoholic extract (80/20) of defatted milled grape seeds based on MW. The authors proposed a cascade membrane process to obtain proanthocyanidins-rich streams of different degrees of polymerization. In such a process, the AFC40 membrane, a polyamide NF membrane with 60% CaCl$_2$ rejection, was used to process the extract in diafiltration mode, resulting in a purified permeate containing acids and aldehydes (stream I). The concentrate from the first stage was processed in concentration mode with the PU120 membrane, a polysulfone UF membrane with 20,000 Da MWCO, resulting in a permeate rich in monomers (stream II). The concentrate from the second stage was then processed in diafiltration mode with the FP200 membrane, a PVDF UF membrane with 200,000 Da MWCO, resulting in a concentrate stream enriched in high MW proanthocyanidins, with a degree of polymerization 6.15 (stream III). Finally, the permeate of the third stage was processed in concentration mode with the PU608 membrane, a polysulfone UF membrane with 8000 Da MWCO, resulting in a concentrate stream enriched in high polymerized oligomers with a degree of polymerization 3.37 (stream IV), and a permeate stream enriched in low polymerized oligomers with a degree of polymerization 2.19 (stream V).

Arboleda Mejia et al. (2020) investigated the performance of NF for the recovery of phenolic compounds, with antioxidant activity, from a red grape pomace extract (see the composition in Table 16.1). Three cellulose acetate membranes (CA series: prepared in the laboratory by phase inversion) and one commercial (NF90) were tested. All four membranes, CA400-22, CA316, CA316-70, and NF90, were characterized in terms of reference solutes, displaying sucrose rejections of 16%, 70%, 98%, and 100%, respectively. After ultrasonic-assisted enzymatic extraction, the extract was prefiltered with a nylon filter with a 3 mm diameter orifice, to remove coarse solids, and then processed by NF. Among the tested membranes, CA400-22 presented the higher permeate flux (50 L/h m^2, at 20 bar and 25°C), better performance in terms of separation between sugars and phenolic compounds and low fouling index (of about 23%). During the processing of the extract, this membrane showed low rejections to sugars, 12% and 19% for fructose and glucose, respectively, and high rejections to total phenolic compounds and proanthocyanidins, 73% and 92%, respectively. Therefore, proanthocyanidins and phenolic compounds were recovered in the concentrate stream, while sugars remained in the permeate stream.

Syed et al. (2017) developed a process based on membrane separation aiming at the production of high-value extracts containing a nutraceutical food ingredient, that is, an extract enriched in bioactive monomeric flavan-3-ols obtained from grape pomace. In that process, the grape pomace was first subjected to an extraction step using biocompatible solvents under optimized conditions, i.e., extraction using 40 wt% ethanol/water, at 40°C for 3 days, and an 8:1 mass ratio of the extracting medium to dried and milled grape pomace. This extract was then subject to NF processing with four different NF membranes, in which a 900 Da MWCO membrane (Duramem 900), under the optimal operating conditions of 8 bar pressure and diafiltration volumes of 2, was the membrane that showed the best results, achieving a recovery of about 40% of the monomeric flavan-3-ols in the permeate. This permeate was further concentrated by RO, using the SW30HR membrane, allowing for 90% reuse of the solvent. The authors also stated that the recovery of monomeric flavan-3-ols can be as high as 85%, just increasing the diafiltration volume to 5 in the NF stage.

16.6 Concluding remarks

By-products and residues from the winemaking process, as well as wastewater and extracts from these by-products and residues, contain high levels of bioactive compounds, which have outstanding nutritional and biotechnological importance, resulting in products with high added value, with a wide range of applications, such as in cosmetics, in the pharmaceutical industry, as a food supplement, in enology, etc.

It is important to highlight that these by-products are mostly treated as wastes from the process, consequently, implying additional costs to the company, since they must be properly treated before disposal. On the other hand, the recovery of bioactive compounds would bring economic and environmental benefits, while increasing revenues with the new product (extract obtained from waste) and reducing treatment and disposal expenses, thereby minimizing the associated environmental impacts, promoting the circular economy. Besides, as eco-friendly practices and technologies are adopted in the recovery of these compounds, there is also an increase in environmental gains due to the minimization of impacts. In this context, membrane technologies have demonstrated the ability to achieve both economic and environmental objectives, as they are cleaner technologies and have a great application capacity for the fractionation, purification, and concentration of bioactive compounds from wastewaters and extracts of by-products from agro-industrial activities.

All membrane separation processes evaluated and presented on this chapter show the potential to be used in a process line, where different products may be obtained in different streams. In fact, the successful application of membrane technologies is closely linked to the membrane practitioner's ability to select the best combination of membrane characteristics, solute properties, and hydrodynamic and operating conditions in order to provide high degrees of separation and productivity, which in turn, often require specific case studies. Otherwise, the selection of inadequate membranes, as well as the use of nonoptimized operating conditions, may result in the appearance of undesirable phenomena such as concentration polarization and fouling and, consequently, make the process unfeasible.

References

Adil, İ. H., Yener, M. E., & Bayındırlı, A. (2008). Extraction of total phenolics of sour cherry pomace by high pressure solvent and subcritical fluid and determination of the antioxidant activities of the extracts. *Separation Science and Technology, 43*(5), 1091−1110. https://doi.org/10.1080/01496390801888243

Alonso, Á. M., Guillén, D. A., Barroso, C. G., Puertas, B., & García, A. (2002). Determination of antioxidant activity of wine byproducts and its correlation with polyphenolic content. *Journal of Agricultural and Food Chemistry, 50*(21), 5832−5836. https://doi.org/10.1021/jf025683b

Anastasiadi, M., Pratsinis, H., Kletsas, D., Skaltsounis, A.-L., & Haroutounian, S. A. (2010). Bioactive non-coloured polyphenols content of grapes, wines and vinification by-products: Evaluation of the antioxidant activities of their extracts. *Food Research International, 43*(3), 805−813. https://doi.org/10.1016/j.foodres.2009.11.017

Arboleda Mejia, J. A., Ricci, A., Figueiredo, A. S., Versari, A., Cassano, A., Parpinello, G. P., & de Pinho, M. N. (2020). Recovery of phenolic compounds from red grape pomace extract through nanofiltration membranes. *Foods, 9*(11), 1649. https://doi.org/10.3390/foods9111649

Baker, R. W. (2012). *Membrane technology and applications* (2nd ed.). West Sussex: John Wiley & Sons.

Barcia, M. T., Pertuzatti, P. B., Gómez-Alonso, S., Godoy, H. T., & Hermosín-Gutiérrez, I. (2014). Phenolic composition of grape and winemaking by-products of Brazilian hybrid cultivars BRS Violeta and BRS Lorena. *Food Chemistry, 159*, 95−105. https://doi.org/10.1016/j.foodchem.2014.02.163

Bhise, S., Kaur, A., Gandhi, N., & Gupta, R. (2014). Antioxidant property and health benefits of grape byproducts. *Journal of Postharvest Technology, 2*(1), 1−11.

Bonfigli, M., Godoy, E., Reinheimer, M. A., & Scenna, N. J. (2017). Comparison between conventional and ultrasound-assisted techniques for extraction of anthocyanins from grape pomace. Experimental results and mathematical modeling. *Journal of Food Engineering, 207*, 56−72. https://doi.org/10.1016/j.jfoodeng.2017.03.011

Botelho de Sousa, M., de Pinho, M. N., & Cameira dos Santos, P. (2014). The role of polysaccharides on the grape must ultrafiltration performance. *Ciência e Técnica Vitivinícola, 29*(1), 16−27. https://doi.org/10.1051/ctv/20142901016

Cassano, A., Bentivenga, A., Conidi, C., Galiano, F., Saoncella, O., & Figoli, A. (2019). Membrane-based clarification and fractionation of red wine lees aqueous extracts. *Polymers, 11*(7), 1089. https://doi.org/10.3390/polym11071089

Cassano, A., Conidi, C., Ruby-Figueroa, R., & Castro-Muñoz, R. (2018). Nanofiltration and tight ultrafiltration membranes for the recovery of polyphenols from agro-food by-products. *International Journal of Molecular Sciences, 19*(2), 351. https://doi.org/10.3390/ijms19020351

Castro-Muñoz, R., Yáñez-Fernández, J., & Fíla, V. (2016). Phenolic compounds recovered from agro-food by-products using membrane technologies: An overview. *Food Chemistry, 213*, 753−762. https://doi.org/10.1016/j.foodchem.2016.07.030

Chen, G., Ma, X., Liu, S., Liao, Y., & Zhao, G. (2012). Isolation, purification and antioxidant activities of polysaccharides from *Grifola frondosa*. *Carbohydrate Polymers, 89*(1), 61−66. https://doi.org/10.1016/j.carbpol.2012.02.045

Curvelo-Garcia, A. S., & Barros, P. (Eds.). (2015). *Química Enológica—métodos analíticos avanços recentes no controlo da qualidade de vinhos e de outros produtos vitivinícolas* (1st ed.). Porto: Publindústria, Edições Técnicas.

Devesa-Rey, R., Vecino, X., Varela-Alende, J. L., Barral, M. T., Cruz, J. M., & Moldes, A. B. (2011). Valorization of winery waste vs. the costs of not recycling. *Waste Management, 31*(11), 2327−2335. https://doi.org/10.1016/j.wasman.2011.06.001

Doco, T., Vuchot, P., Cheynier, V., & Moutounet, M. (2003). Structural modification of wine arabinogalactans during aging on lees. *American Journal of Enology and Viticulture, 54*(3), 150−157.

Fane, A. G. T., Wang, R., & Jia, Y. (2011). Membrane technology: Past, present and future. In L. K. Wang, J. P. Chen, Y.-T. Hung, & N. K. Shammas (Eds.), *Membrane and desalination technologies* (pp. 1−45). Totowa: Humana Press.

Galanakis, C. M., Markouli, E., & Gekas, V. (2013). Recovery and fractionation of different phenolic classes from winery sludge using ultrafiltration. *Separation and Purification Technology, 107*, 245−251. https://doi.org/10.1016/j.seppur.2013.01.034

Giacobbo, A., Bernardes, A. M., & de Pinho, M. N. (2013). Nanofiltration for the recovery of low molecular weight polysaccharides and polyphenols from winery effluents. *Separation Science and Technology (Philadelphia), 48*(17). https://doi.org/10.1080/01496395.2013.809762

Giacobbo, A., Bernardes, A. M., & de Pinho, M. N. (2017). Sequential pressure-driven membrane operations to recover and fractionate polyphenols and polysaccharides from second racking wine lees. *Separation and Purification Technology, 173*, 49−54. https://doi.org/10.1016/j.seppur.2016.09.007

Giacobbo, A., Bernardes, A. M., Rosa, M. J., & de Pinho, M. N. (2018). Concentration polarization in ultrafiltration/nanofiltration for the recovery of polyphenols from winery wastewaters. *Membranes, 8*(3), 1−11. https://doi.org/10.3390/membranes8030046

Giacobbo, A., Dias, B. B., Onorevoli, B., Bernardes, A. M., de Pinho, M. N., Caramão, E. B., Rodrigues, E., & Jacques, R. A. (2019). Wine lees from the 1st and 2nd rackings: Valuable by-products. *Journal of Food Science and Technology, 56*(3), 1559−1566. https://doi.org/10.1007/s13197-019-03665-1

Giacobbo, A., do Prado, J. M., Meneguzzi, A., Bernardes, A. M., & de Pinho, M. N. (2015). Microfiltration for the recovery of polyphenols from winery effluents. *Separation and Purification Technology, 143*, 12−18. https://doi.org/10.1016/j.seppur.2015.01.019

Giacobbo, A., Oliveira, M., Duarte, E. C. N. F., Mira, H. M. C., Bernardes, A. M., & de Pinho, M. N. (2013). Ultrafiltration based process for the recovery of polysaccharides and polyphenols from winery effluents. *Separation Science and Technology (Philadelphia), 48*(3), 438−444. https://doi.org/10.1080/01496395.2012.725793

Gonçalves, F., Fernandes, C., & de Pinho, M. N. (2001). White wine clarification by micro/ultrafiltration: Effect of removed colloids in tartaric stability. *Separation and Purification Technology, 22−23*, 423−429. https://doi.org/10.1016/S1383-5866(00)00179-9

Habert, A. C., Borges, C. P., & Nóbrega, R. (2006). *Processos de separação por membranas*. Rio de Janeiro: Editora E-papers.

Hanlin, R. L., Kelm, M. A., Wilkinson, K. L., & Downey, M. O. (2011). Detailed characterization of proanthocyanidins in skin, seeds, and wine of Shiraz and Cabernet Sauvignon wine grapes (*Vitis vinifera*). *Journal of Agricultural and Food Chemistry, 59*(24), 13265−13276. https://doi.org/10.1021/jf203466u

Ignat, I., Volf, I., & Popa, V. I. (2011). A critical review of methods for characterisation of polyphenolic compounds in fruits and vegetables. *Food Chemistry, 126*(4), 1821−1835. https://doi.org/10.1016/j.foodchem.2010.12.026

Jackson, R. S. (2000). Chemical constituents of grapes and wines. In *Wine science* (2nd ed., pp. 232−280). San Diego: Academic Press.

Jia, X., Ding, C., Yuan, S., Zhang, Z., Chen, Y., Du, L., & Yuan, M. (2014). Extraction, purification and characterization of polysaccharides from Hawk tea. *Carbohydrate Polymers, 99*, 319−324. https://doi.org/10.1016/j.carbpol.2013.07.090

Kammerer, D. R., Kammerer, J., Valet, R., & Carle, R. (2014). Recovery of polyphenols from the by-products of plant food processing and application as valuable food ingredients. *Food Research International, 65*, 2−12. https://doi.org/10.1016/j.foodres.2014.06.012

Kontogiannopoulos, K. N., Patsios, S. I., Mitrouli, S. T., & Karabelas, A. J. (2017). Tartaric acid and polyphenols recovery from winery waste lees using membrane separation processes. *Journal of Chemical Technology & Biotechnology, 92*(12), 2934–2943. https://doi.org/10.1002/jctb.5313

Ky, I., Lorrain, B., Kolbas, N., Crozier, A., & Teissedre, P.-L. (2014). Wine by-products: Phenolic characterization and antioxidant activity evaluation of grapes and grape pomaces from six different French grape varieties. *Molecules, 19*(1), 482–506. https://doi.org/10.3390/molecules19010482

Labarbe, B., Cheynier, V., Brossaud, F., Souquet, J.-M., & Moutounet, M. (1999). Quantitative fractionation of grape proanthocyanidins according to their degree of polymerization. *Journal of Agricultural and Food Chemistry, 47*(7), 2719–2723. https://doi.org/10.1021/jf990029q

Lee, J., Rennaker, C., & Wrolstad, R. E. (2008). Correlation of two anthocyanin quantification methods: HPLC and spectrophotometric methods. *Food Chemistry, 110*(3), 782–786. https://doi.org/10.1016/j.foodchem.2008.03.010

Lingua, M. S., Fabani, M. P., Wunderlin, D. A., & Baroni, M. V. (2016). In vivo antioxidant activity of grape, pomace and wine from three red varieties grown in Argentina: Its relationship to phenolic profile. *Journal of Functional Foods, 20*, 332–345. https://doi.org/10.1016/j.jff.2015.10.034

Liu, D., Vorobiev, E., Savoire, R., & Lanoisellé, J.-L. (2011). Intensification of polyphenols extraction from grape seeds by high voltage electrical discharges and extract concentration by dead-end ultrafiltration. *Separation and Purification Technology, 81*(2), 134–140. https://doi.org/10.1016/j.seppur.2011.07.012

Liu, J., Willför, S., & Xu, C. (2015). A review of bioactive plant polysaccharides: Biological activities, functionalization, and biomedical applications. *Bioactive Carbohydrates and Dietary Fibre, 5*(1), 31–61. https://doi.org/10.1016/j.bcdf.2014.12.001

Louli, V., Ragoussis, N., & Magoulas, K. (2004). Recovery of phenolic antioxidants from wine industry by-products. *Bioresource Technology, 92*(2), 201–208. https://doi.org/10.1016/j.biortech.2003.06.002

Makris, D. P., Boskou, G., & Andrikopoulos, N. K. (2007). Recovery of antioxidant phenolics from white vinification solid by-products employing water/ethanol mixtures. *Bioresource Technology, 98*(15), 2963–2967. https://doi.org/10.1016/j.biortech.2006.10.003

Manach, C., Scalbert, A., Morand, C., Rémésy, C., & Jiménez, L. (2004). Polyphenols: Food sources and bioavailability. *The American Journal of Clinical Nutrition, 79*(5), 727–747. https://doi.org/10.1093/ajcn/79.5.727

Matsuura, T. (1993). *Synthetic membranes and membrane separation processes* (1st ed.). Boca Raton: CRC press.

Moreira, M. M., Morais, S., & Delerue-Matos, C. (2017). Environment-friendly techniques for extraction of bioactive compounds from fruits. In A. M. Grumezescu, & A. M. Holban (Eds.), *Soft chemistry and food fermentation — handbook of food bioengineering* (pp. 21–48). London: Academic Press.

Nave, F., Cabrita, M. J., & da Costa, C. T. (2007). Use of solid-supported liquid–liquid extraction in the analysis of polyphenols in wine. *Journal of Chromatography A, 1169*(1), 23–30. https://doi.org/10.1016/j.chroma.2007.08.067

Oechslin, R., Lutz, M. V., & Amadò, R. (2003). Pectic substances isolated from apple cellulosic residue: Structural characterisation of a new type of rhamnogalacturonan I. *Carbohydrate Polymers, 51*(3), 301–310. https://doi.org/10.1016/S0144-8617(02)00214-X

OIV. (2020). *2019 statistical report on world vitiviniculture*. Paris, France: International Organisation of Vine and Wine (OIV). Retrieved from http://oiv.int/public/medias/6782/oiv-2019-statistical-report-on-world-vitiviniculture.pdf.

Ormad, M. P., Mosteo, R., Ibarz, C., & Ovelleiro, J. L. (2006). Multivariate approach to the photo-Fenton process applied to the degradation of winery wastewaters. *Applied Catalysis B: Environmental, 66*(1), 58–63. https://doi.org/10.1016/j.apcatb.2006.02.014

Panzella, L., Moccia, F., Nasti, R., Marzorati, S., Verotta, L., & Napolitano, A. (2020). Bioactive phenolic compounds from agri-food wastes: An update on green and sustainable extraction methodologies. *Frontiers in Nutrition, 7*, 60. https://doi.org/10.3389/fnut.2020.00060

Partain, E. M., III (2000). Industrially important polysaccharides. In C. D. Craver, & C. Carraher (Eds.), *Applied polymer science: 21st century* (1st ed., pp. 303−323). Oxford: Pergamon. https://doi.org/10.1016/B978-008043417-9/50018-0

Passos, C. P., Silva, R. M., Da Silva, F. A., Coimbra, M. A., & Silva, C. M. (2010). Supercritical fluid extraction of grape seed (*Vitis vinifera* L.) oil. Effect of the operating conditions upon oil composition and antioxidant capacity. *Chemical Engineering Journal, 160*(2), 634−640. https://doi.org/10.1016/j.cej.2010.03.087

Pellerin, P., Doco, T., Vida, S., Williams, P., Brillouet, J.-M., & O'Neill, M. A. (1996). Structural characterization of red wine rhamnogalacturonan II. *Carbohydrate Research, 290*(2), 183−197. https://doi.org/10.1016/0008-6215(96)00139-5

Pellerin, P., Vidal, S., Williams, P., & Brillouet, J.-M. (1995). Characterization of five type II arabinogalactan-protein fractions from red wine of increasing uronic acid content. *Carbohydrate Research, 277*(1), 135−143. https://doi.org/10.1016/0008-6215(95)00206-9

Pérez-Serradilla, J. A., & Luque de Castro, M. D. (2011). Microwave-assisted extraction of phenolic compounds from wine lees and spray-drying of the extract. *Food Chemistry, 124*(4), 1652−1659. https://doi.org/10.1016/j.foodchem.2010.07.046

Pinelo, M., Fabbro, P. Del, Manzocco, L., Nuñez, M. J., & Nicoli, M. C. (2005). Optimization of continuous phenol extraction from Vitis vinifera byproducts. *Food Chemistry, 92*(1), 109−117. https://doi.org/10.1016/j.foodchem.2004.07.015

de Pinho, M. N., Geraldes, V., & Minhalma, L. M. (2002). *Integração de operações de membranas em processos químicos: Dimensionamento e Optimização de equipamentos (Integration of operations with membranes in chemical processes: Dimentioning and optimization of equipments)*. Lisbon: IST Press.

de Pinho, M. N., & Minhalma, M. (2019). Introduction in membrane technologies. In C. M. Galanakis (Ed.), *Separation of functional molecules in food by membrane technology* (1st ed., pp. 1−29). Chennai: Academic Press. https://doi.org/10.1016/B978-0-12-815056-6.00001-2

Pirra, A. J. D. (2005). *Characterization and treatment of winery effluents from the Douro Wine Region (Caracterização e tratamento de efluentes vinícolas da Região Demarcada do Douro)*. Vila Real, Portugal: University of Tra s-os-Montes and Alto Douro.

Puértolas, E., Cregenzán, O., Luengo, E., Álvarez, I., & Raso, J. (2013). Pulsed-electric-field-assisted extraction of anthocyanins from purple-fleshed potato. *Food Chemistry, 136*(3), 1330−1336. https://doi.org/10.1016/j.foodchem.2012.09.080

Rajha, H. N., Darra, N. El, Hobaika, Z., Boussetta, N., Vorobiev, E., Maroun, R. G., & Louka, N. (2014). Extraction of total phenolic compounds, flavonoids, anthocyanins and tannins from grape byproducts by response surface methodology. Influence of solid-liquid ratio, particle size, time, temperature and solvent mixtures on the optimization process. *Food and Nutrition Sciences, 05*(04), 397−409. https://doi.org/10.4236/fns.2014.54048

Rigou, P., Mekoue, J., Sieczkowski, N., Doco, T., & Vernhet, A. (2021). Impact of industrial yeast perivative products on the modification of wine aroma compounds and sensorial profile. A review. *Food Chemistry*, 129760. https://doi.org/10.1016/j.foodchem.2021.129760

Riou, V., Vernhet, A., Doco, T., & Moutounet, M. (2002). Aggregation of grape seed tannins in model wine—effect of wine polysaccharides. *Food Hydrocolloids, 16*(1), 17−23. https://doi.org/10.1016/S0268-005X(01)00034-0

Rockenbach, I. I., da Silva, G. L., Rodrigues, E., Kuskoski, E. M., & Fett, R. (2008). Solvent influence on total polyphenol content, anthocyanins, and antioxidant activity of grape (*Vitis vinifera*) bagasse extracts from Tannat and Ancelota - different varieties of *Vitis vinifera* varieties. *Food Science and Technology*. https://doi.org/10.1590/S0101-20612008000500036

Rodrigues, R. P., Gando-Ferreira, L. M., & Quina, M. J. (2020). Micellar enhanced ultrafiltration for the valorization of phenolic compounds and polysaccharides from winery wastewaters. *Journal of Water Process Engineering, 38*, 101565. https://doi.org/10.1016/j.jwpe.2020.101565

Rodrigues, A. C., Oliveira, J. M., Oliveira, J. A., Peixoto, J., Nogueira, R., & Brito, A. G. (2006). Tratamento de efluentes vitivinícolas: uma caso de estudo na região dos vinhos verdes. *Revista Indústria e Ambiente, 40*, 20−25.

Romero-Díez, R., Matos, M., Rodrigues, L., Bronze, M. R., Rodríguez-Rojo, S., Cocero, M. J., & Matias, A. A. (2019). Microwave and ultrasound pre-treatments to enhance anthocyanins extraction from different wine lees. *Food Chemistry, 272*, 258−266. https://doi.org/10.1016/j.foodchem.2018.08.016

Ruggieri, L., Cadena, E., Martínez-Blanco, J., Gasol, C. M., Rieradevall, J., Gabarrell, X., Gea, T., Sort, X., & Sánchez, A. (2009). Recovery of organic wastes in the Spanish wine industry. Technical, economic and environmental analyses of the composting process. *Journal of Cleaner Production, 17*(9), 830−838. https://doi.org/10.1016/j.jclepro.2008.12.005

Santamaría, B., Salazar, G., Beltrán, S., & Cabezas, J. L. (2002). Membrane sequences for fractionation of polyphenolic extracts from defatted milled grape seeds. *Desalination, 148*(1), 103−109. https://doi.org/10.1016/S0011-9164(02)00661-6

Saulnier, L., Mercereau, T., & Vezinhet, F. (1991). Mannoproteins from flocculating and non-flocculating *Saccharomyces cerevisiae* yeasts. *Journal of the Science of Food and Agriculture, 54*(2), 275−286. https://doi.org/10.1002/jsfa.2740540214

Selani, M. M., Contreras-Castillo, C. J., Shirahigue, L. D., Gallo, C. R., Plata-Oviedo, M., & Montes-Villanueva, N. D. (2011). Wine industry residues extracts as natural antioxidants in raw and cooked chicken meat during frozen storage. *Meat Science, 88*(3), 397−403. https://doi.org/10.1016/j.meatsci.2011.01.017

da Silva, L. M. L. R. (2003). Caracterização dos subprodutos da vinificação. *Millenium, 28*, 123−133.

Soto, M. L., Moure, A., Domínguez, H., & Parajó, J. C. (2011). Recovery, concentration and purification of phenolic compounds by adsorption: A review. *Journal of Food Engineering, 105*(1), 1−27. https://doi.org/10.1016/j.jfoodeng.2011.02.010

Souquet, J.-M., Cheynier, V., Brossaud, F., & Moutounet, M. (1996). Polymeric proanthocyanidins from grape skins. *Phytochemistry, 43*(2), 509−512. https://doi.org/10.1016/0031-9422(96)00301-9

Spranger, M. I., Climaco, M. C., Sun, B., Eiriz, N., Fortunato, C., Nunes, A., Leandro, M. C., Avelar, M. L., & Belchior, A. P. (2004). Differentiation of red winemaking technologies by phenolic and volatile composition. *Analytica Chimica Acta, 513*(1), 151−161. https://doi.org/10.1016/j.aca.2004.01.023

Srivastava, R., & Kulshreshtha, D. K. (1989). Bioactive polysaccharides from plants. *Phytochemistry, 28*(11), 2877−2883. https://doi.org/10.1016/0031-9422(89)80245-6

Stefova, M., & Ivanova, V. (2011). Analytical methodology for characterization of grape and wine phenolic bioactives. In C. A. H. I. Özlem Tokusoglu (Ed.), *Fruit and cereal bioactives* (1st ed., pp. 409−427). Boca Raton: CRC Press. https://doi.org/10.1201/b10786

Syed, U. T., Brazinha, C., Crespo, J. G., & Ricardo-da-Silva, J. M. (2017). Valorisation of grape pomace: Fractionation of bioactive flavan-3-ols by membrane processing. *Separation and Purification Technology, 172*, 404−414. https://doi.org/10.1016/j.seppur.2016.07.039

Tudose-Sandu-Ville, Ş., Cotea, V. V., Colibaba, C., Nechita, B., Niculaua, M., & Codreanu, M. (2012). Phenolic compounds in Merlot wines obtained through different technologies in Iaşi vineyard, Romania. *Cercetări Agronomice În Moldova, XLV*(4), 89−98.

Vidal, S., Williams, P., Doco, T., Moutounet, M., & Pellerin, P. (2003). The polysaccharides of red wine: Total fractionation and characterization. *Carbohydrate Polymers, 54*(4), 439−447. https://doi.org/10.1016/S0144-8617(03)00152-8

Waters, E. J., Wallace, W., Tate, M. E., & Williams, P. J. (1993). Isolation and partial characterization of a natural haze protective factor from wine. *Journal of Agricultural and Food Chemistry, 41*(5), 724–730. https://doi.org/10.1021/jf00029a009

Yapo, B. M., Lerouge, P., Thibault, J.-F., & Ralet, M.-C. (2007). Pectins from citrus peel cell walls contain homogalacturonans homogenous with respect to molar mass, rhamnogalacturonan I and rhamnogalacturonan II. *Carbohydrate Polymers, 69*(3), 426–435. https://doi.org/10.1016/j.carbpol.2006.12.024

Zong, A., Cao, H., & Wang, F. (2012). Anticancer polysaccharides from natural resources: A review of recent research. *Carbohydrate Polymers, 90*(4), 1395–1410. https://doi.org/10.1016/j.carbpol.2012.07.026

CHAPTER 17

Sustainable approach to quality control of grape and wine

Piergiorgio Comuzzo, Andrea Natolino and Emilio Celotti
Department of Agricultural Food, Environmental and Animal Science, University of Udine, Udine, Italy

17.1 Introduction and principles of green chemistry

Concern and interest in preserving environment and people's health are constantly increasing in modern society. Chemical industry is known as one of the most relevant and invasive sources of environmental pollutants globally. The concept of Green Chemistry (GC) is based on minimizing or eliminating the use or generation of hazardous substances during synthesis, processing, and application of chemicals, in order to reduce the risks for operators, protecting environment (Wardencki et al., 2005).

The term "Green Chemistry" has been probably used for the first time in 1990, when a paper on the development of chemical industry in Ireland was published (Cathcart, 1990), and only six years later, it began to be interpreted with the concepts and philosophy well known nowadays (Anastas & Williamson, 1996). Before the GC notions became widespread, there was already a well-defined awareness about the need to develop sustainable chemical methodologies, focused on saving solvents and reagents, as well as replacing the most toxic chemicals with other harmless or less harmful compounds (De la Guardia & Armenta, 2011). According to such notions, special attention is paid not only to the risks associated with the management of samples, reagents, and solvents but also to energy consumption and to the waste generated by the numerous steps applied in chemical processes (De la Guardia & Garrigues, 2011).

All these aspects are summarized in the 12 Principles of Green Chemistry, published in 1998 by Anastas and Warner (Anastas & Warner, 1998). Such principles represent the foundations of GC and are based on the following concepts: (i) minimization or nonuse of toxic and hazardous solvents; (ii) reduced generation of residues from chemical processes; (iii) use of renewable and harmless raw materials; (iv) acceleration of chemical reactions by catalysis in order to reduce energy consumption and waste production; (v) increased awareness about chemical products, also after their useful life, to avoid bioaccumulation of harmful degradation products; (vi) maximizing atom and energy economy. According to these aspects, GC is closely linked to the principles of sustainable development and in recent years there is an evident tendency to implement these principles in chemical laboratories and plants.

Therefore, GC deals with the design of chemical products, through their sustainable synthesis, processing, analysis, and destination after use, with the main objective of minimizing the environmental and professional risks of the chemical industry (de Marco et al., 2019). GC principles have been adopted not only at industrial level but also in the other specific areas of chemistry, such as analytical chemistry.

17.2 Green Analytical Chemistry

The concept of Green Analytical Chemistry (GAC) was formulated after several methodological milestones developed to increase the "green character" of analytical protocols. The beginning of GAC dates back to the mid-1970s, when many innovative techniques in the field of sample preparation, measurement, and data handling were introduced, such as flow injection analysis, purge-and-trap techniques, solid phase extraction (SPE), and cloud point extraction. In the subsequent two decades, other advances in analytical technology were proposed, such as microwave-assisted extraction, supercritical fluid extraction, pressurized solvent extraction, and other innovative preparative techniques and injection sample systems.

In 1995, the first statement of principles of what is now known as GAC was published (De la Guardia & Ruzicka, 1995). Three years later, the 12 principles of green chemistry (Anastas & Warner, 1998) were coined for synthetic and industrial chemistry, but only some of them could be adapted to analytical chemistry; for this reason, there was a need to conform the general guidelines of GC to this specific field. Accordingly, in 2013, Gałuszka and co-workers published 12 specific principles for GAC (Gałuszka et al., 2013): (i) if possible, sample treatment should be avoided by using direct methods, (ii) the number and size of the samples should be minimized, (iii) measurements should be performed in situ, (iv) processes and operations should be integrated, (v) if possible, automation and miniaturization of analytical methods should be selected, (vi) derivatization should be avoided, (vii) generation of a large volume of analytical waste should be avoided and proper management of lab waste should be provided, (viii) multianalyte and multiparameter methods should be applied whenever possible, (ix) energy use should be minimal, (x) reagents from renewable sources should be preferred, (xi) toxic reagents and solvents should be eliminated or replaced, (xii) safety of the operators should be improved.

Taking into account these principles and the more general rules of GC, it is possible to highlight four fundamental goals in the greening of analytical methods (Fig. 17.1). Consequently, analysts may act on five main factors for making analytical processes more sustainable: sampling, methods, instrumentation, solvents, and reagents (Gałuszka et al., 2012).

17.3 Greening of analytical procedures
17.3.1 Sampling

Sampling is the first step of each analytical protocol. According to the second principle of GAC, the size and number of samples should be reduced as much as possible (Gałuszka et al., 2013). However, in many cases, this cannot be easily achieved and a sampling procedure is required.

Two different approaches could be adopted for reducing the impact of sampling on analytical processes. First, the use of remote sensing or portable devices for on-site screening; second, the use of

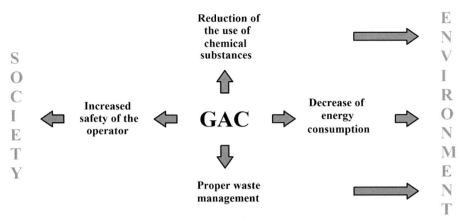

FIGURE 17.1

Key goals of Green Analytical Chemistry (GAC).

statistical methods allowing to properly select the sampling sites, minimizing the number of sampling points (Gałuszka et al., 2013). Furthermore, the application of miniaturized methods and instruments allows to reduce the sample size (Brett, 2007; Gałuszka et al., 2013). Reduction in sample number and size should be done carefully, for ensuring representative and reliable results, taking into account the characteristics of the sample and its heterogeneity.

17.3.2 Analytical methods and instruments

Ideally, sustainable analytical methods should be direct, automatized, miniaturized, and integrated; derivatization procedures, which generally involve significant amounts of toxic solvents and reagents, high levels of exposure for operators and high amounts of waste, should be avoided (De la Guardia & Garrigues, 2012; Płotka-Wasylka et al., 2018).

The integration of analytical processes is important to obtain as much results and information as possible from a single analysis. From this point of view, a perfect chemical instrument should be portable, with a reduced size and weight, miniaturized, automatized, and it should allow the remote sensing (Gałuszka et al., 2013). The use of portable instrumentation drastically reduces sampling time, avoiding sample storage and transport, reducing environmental risks and accelerating decision-making procedures (Chemat et al., 2019). Nowadays, the advancements on informatics and electronics provide high level of portability and miniaturization, ensuring high processing performances and data-saving capacity and accessibility (e.g., on cloud).

17.3.3 Solvents and reagents

Conventional analytical methods and extraction/separation techniques generally require the use of large amount of organic solvents. Most of these solvents are highly volatile, causing air pollution, flammable, and toxic. Several guides report data about the toxicity of the different chemicals and the classification of solvents in different categories (Prat et al., 2014). Thus, whenever possible, these hazardous substances should be replaced with more environmentally friendly compounds, such as water, ethanol, isopropyl alcohol, or other. However, this is often not feasible, since most organic molecules are not soluble in these solvents. A number of studies and researches were focused on green alternatives to conventional organic solvents and four most important categories have been identified: amphiphilic solvents (ASs), ionic liquids (ILs), deep eutectic solvents (DESs), and supercritical fluids (SFs).

Surfactants are the most diffused and successfully used ASs in analytical chemistry, representing the first low-toxicity option to replace conventional solvents (Melnyk et al., 2015). When added to water at concentrations above their critical micelle concentration, they form micellar aggregates with strong solvation properties, interacting with compounds of different polarity (Pacheco-Fernández & Pino, 2019). Surfactants allowed the development of innovative extraction processes and preconcentration schemes, which can be combined with other techniques, such as ultrasounds or microwaves (MWs) (Gürkan et al., 2016; Tang et al., 2017).

ILs are defined as nonmolecular solvents, with melting points below 100°C; they are obtained by the combination of bulky organic cations, with organic and inorganic anions. These solvents are characterized by high thermal and chemical stability, low vapor pressures at ambient temperatures, tunable viscosity and high conductivity (Anderson & Clark, 2018). Several ILs derivatives can be prepared, such as polymeric ILs, ILs-based surfactants, and magnetic ILs (Trujillo-Rodríguez et al., 2019). Several studies have suggested that composition of cation and anion moieties affect the resulting toxicity of ILs (Kudłak et al., 2015) and a more biodegradable and safer natural source should be used for ILs design (Shukla et al., 2018).

DESs are constituted by two components, a hydrogen bond acceptor and hydrogen bond donor, mixed at different ratios. They are a new generation of solvents, with most of the properties featured by ILs, but their preparation is cheaper and they are less toxic than conventional ILs (Cunha & Fernandes, 2018).

An even greener alternative to these organic solvent categories is represented by SFs, especially supercritical carbon dioxide (SC-CO_2). Carbon dioxide easily achieves the supercritical state at pressure and temperature conditions above 73.8 bar and 31.1°C. SC-CO_2 shows several advantages: it is no flammable, no toxic, chemically inert, odorless, cheaper, and easy to recover. Its physical and chemical properties are extremely versatile and can be easily tunable, changing pressure and temperature conditions. SFs are mainly applied in analytical chemistry for sample preparation and SF chromatography (Brunner, 2005).

Basing on these evidences, greening solvents is a fundamental step for reducing or eliminating toxic substances in chemical analysis. This could be achieved also by using more renewable and safer chemicals, e.g., extracted from plants, animals, and microorganism, such as pH and redox indicators, spectrophotometric and fluorimetric reagents, or compounds applied in biosensors (Kradtap Hartwell, 2012).

As discussed, a number of strategies can be adopted to achieve the key goals indicated by GAC principles. It is not possible to formulate guidelines that would be universal for all the potential applications, since analytical chemistry is used in several multidisciplinary fields. In recent decades, agri-food production raised globally and this process is based on increased industrialization aimed at mass production. Grape and wine are an outstanding example because their production achieved a global scale, inducing an exponential increase of its environmental and pollution impact. In view of that, environmental sustainability becomes an essential goal for wine sector, and several strategies could be implemented also in grape and wine analysis (WA).

17.4 Sustainable grape analysis and quality control

Wine quality begins to be managed starting from vineyard. A sustainable wine supply chain adopts techniques able to meet the criteria of environmental and social sustainability, and these concepts may also be applied to grape quality control at all levels, from field to winery. Grape analysis may be managed at three different levels: in lab, after adequate berry sampling, on-field, by using specific portable devices, or when grapes are delivered to the winery, through suitable real-time measurement systems.

17.4.1 Laboratory methods

The traditional protocols used for assessing grape quality and ripeness (e.g., the one developed in 1990 by Yves Glories) Ribéreau-Gayon et al., 2006 are based on appropriate berry sampling and a more or less complex procedure for lab analysis, which become quite tiring when phenolic and cellular ripeness of red grapes are measured, because of the time needed for the extraction of anthocyanins and tannins from the skins Ribéreau-Gayon et al., 2006.

Certain extraction procedures, such as the use of MWs (Celotti, Dell'Oste, et al., 2007), allow to eliminate the long extraction phase required by traditional laboratory methods, facilitating the work of the analyst and becoming a useful alternative, especially in case of high number of samples. MW provides an intensive extraction of the polyphenolic potential of grapes; quantitative information on the amount of polyphenols is retrieved, but no data are obtained on cellular maturity (i.e., their natural extractability during skin maceration), which remains a feature provided by traditional methods Ribéreau-Gayon, Glories, et al., 2006).

Another interesting application for accelerating the evaluation of phenolic ripeness in lab is the determination of the so-called Maturity Trend; the procedure is based on the measurement of the (visible) light reflected by a mash obtained by pulping the berries with a blender; the method provides, in real-time, an index (TM), which is significantly correlated to the phenolic potential of red grapes (Celotti, Della Vedova, et al., 2007).

17.4.2 On-field monitoring

Despite different lab techniques have been proposed for reducing analysis time, the main limitation of the laboratory methods used for grape quality control remains the long time needed for berry sampling in the field. This makes lab approach not very efficient, especially because it does not allow a fast decision-making procedure concerning the definition of harvest date. Moreover, a reliable sampling requires the operator is engaged for several hours per day in collecting berries, frequently in hard working conditions. Sensory evaluation protocols have been suggested as tools to be implemented from this point of view (Rousseau & Delteil, 2000), but the current necessity of reducing analysis time and the need to increase sustainability of vineyard operations can be met by using analysis systems able to supply fast and accurate information about the quality of the grapes directly on-field. Such analytical techniques range from refractometry, to different applications of spectroscopy, including fluorescence and rheological measurements. These measures may have different applications (Fig. 17.2); they allow to objectively define a number of parameters connected to grape quality, by using specific single- or multiparametric, rapid, and nondestructive systems.

- Direct measurement of quality parameters in the vineyard

- Rapid comparison among vineyards, rows, portions of vineyard according to different crop and agronomic variants (irrigation, exposure, fertilization, etc.)

- Definition of ripeness evolution (ripening curves) directly on-field

- Easy integration of different parameters (sugar content, satellite vigor measurements) for the study of the viticultural territory and the management of harvest date

- Elimination of most of the laboratory analysis

- Implementation of viticultural zoning projects

FIGURE 17.2

Possible applications of rapid (real-time) analysis systems in vineyard.

17.4.2.1 Monoparametric systems

Monoparametric portable instruments are available since different years for analyzing grape phenolics in the vineyard. One of these devices has been developed by Celotti and co-workers (Celotti, Carcereri De Prati, Charpentier, & Feuillat, 2008). The instrument is shaped as a clamp (Fig. 17.3) and uses a visible LED light source to measure the light transmitted through a grape skin positioned in a housing within the clamp itself. A portable processing unit, connected with the device, transforms the optical signal, providing an index, named Phenolic Meter Index (PMI), closely related to the polyphenolic

FIGURE 17.3

Portable instrument for measuring Phenolic Meter Index (PMI) on-field; detail of the measuring clamp.

potential of the berry skin (Celotti, Carcereri De Prati, Charpentier, & Feuillat, 2008; Celotti et al., 2009). The instrument allows also the real-time monitoring of red wine color extraction during skin maceration (Celotti, 2012; Celotti et al., 2010).

PMI is a dimensionless quantity and its values are differentials of maturity between samples; conventional concentration data (e.g., in mg/L) are not provided. However, this approach allows to apply the equipment in all situations, supplying useful information on the phenolic potential of the grapes analyzed. The system allows the real-time evaluation of berry samples in the vineyard, providing indications about the phenolic ripeness of the grape, without any calibration need (Celotti, Della Vedova, et al., 2007).

17.4.2.2 Multiparametric systems

Monoparametric systems have been designed in order to evaluate an overall index of polyphenols in the vineyard. Other nondestructive systems have been developed for multiparametric analysis (e.g., sugars, acids, or polyphenols), using different techniques.

One of these systems measures the fluorescence of grape bunches at different spectral zones, when samples are irradiated with different excitation sources, in the UV and VIS field. The instrument has been developed starting from previous chlorophyll analysis systems based on fluorescence sensors. The quality potential of grapes is evaluated directly on-field (Agati et al., 2007; Ben Ghozlen et al., 2010; Ben Ghozlen et al., 2010; Cerovic et al., 2008): when it is placed in a suitable position in front of the bunch, the device provides information on the content of certain analytical parameters, such as flavonoids and anthocyanins, with considerable advantages in terms of immediacy of information and the elimination of long laboratory analyses.

Other systems exploit the potential of other spectroscopic techniques, such as near-infrared (NIR) or mid-infrared spectroscopy (MIR) (Cozzolino, 2015; Fernández-Novales et al., 2019; Ferrer-Gallego et al., 2011; Musingarabwi et al., 2016; Rolle et al., 2012). Analyses are immediate and nondestructive; instruments are positioned in front of the grape bunch, providing multiparametric quantitative data about grape composition. However, as for fluorescence systems, IR spectrometers require consistent calibrations in order to quantify, with suitable accuracy, specific grape analytes.

Magnetic resonance imaging is another interesting application tested for grape analysis on-filed. It is a nondestructive technique allowing to define grape quality parameters (e.g., sugars level and bunch/berry volume) and ripeness, as well as a three-dimensional image reconstruction of bunches (Andaur et al., 2004; Tello et al., 2016).

Hyperspectral imaging has also been proposed for predicting certain grape quality control indices; VIS—NIR hyperspectral cameras can be mounted on all-terrain vehicles, providing quantitative data on different indicators, such as total soluble solids (TSS) and sugar level (Gomes et al., 2017; Gutiérrez et al., 2018), or anthocyanins concentration (Gutiérrez et al., 2018).

Other analytical systems measure the rheological characteristics of the grape, correlating them with certain quality parameters, in particular, berry weight and total flavonoids. The application of these measures in combination with the use of predictive models offers good perspectives for the direct on-field application of such equipments, for evaluating the phenolic quality (PQ) of grapes, defining the harvest date (Brillante et al., 2015; Río Segade et al., 2008; Rolle et al., 2008, 2011).

Many of the portable systems discussed above can be easily integrated with precision viticulture systems (satellite tracking, maps of vigor and GPS) in order to forecast the behavior of a given vineyard and minimize analytical controls during harvest period.

17.4.3 Grape quality control at delivery

In the case of wine cooperatives, where vineyard surface is large and heterogeneous, on-field measurements cannot provide accurate information about the whole grape-growing area. In order to have a complete and objective evaluation of grape quality before vinification, it is necessary to analyze all the samples which are delivered to the winery. Several methodologies for grape quality control have been developed from this point of view. Nowadays, increasingly refined analytical controls are available at grape delivery, with technical solutions able to meet multiple enological needs. Controls may be very accurate, and besides the visual observation and the measurement of sugars and total acidity, other important wine quality parameters may be determined.

Wineries are generally equipped with refractometric stations allowing the real-time measurement of the main quality parameters of grape juice (such as sugars, titratable acidity and pH), as well as the monitoring of must temperature, which is a useful indicator of the conditions of transport/delivery and quality profile of the entering grapes. Coring systems allow the rapid sampling of the grapes directly from trailers and a fast extraction of the juice, which is immediately (real-time) analyzed by such systems. Besides basic analyses (sugars, pH, and titratable acidity), in last years, great attention has been paid to polyphenols.

From this point of view, the most known systems for monitoring grape quality and composition at the delivery are based on infrared (IR) spectroscopy (Cozzolino, 2015; Cozzolino et al., 2006; Versari et al., 2008). Different IR devices have been developed to be used at winery scale. These instruments have the advantage to simultaneously measure several analytical parameters on the basis of a preliminary calibration procedure. IR instruments have high performances and, if adequately calibrated, they allow the punctual monitoring of grape quality, including its initial health conditions. IR measurements are carried out on the grape must obtained by coring; samples must be filtered and this step could negatively affect the accuracy of some analytical indices, such as skin polyphenols. IR devices may be used also for evaluating if the grape is affected by fungal infections (Steel, 2018). This is particularly important when mechanical harvesters are used; in fact, in these conditions, the extent of infection cannot be estimated by visual inspection. IR instruments will be further discussed in the next sections of this chapter, concerning their application to wine analysis.

Other interesting equipments for assessing grape quality are based on reflectance measurements (Celotti & Carcereri De Prati, 1999). These devices provide the simultaneous determination of sugar and phenolic content of red grapes at delivery. The instrument is integrated in a multiparametric monitoring station, equipped with a coring system; it provides a numerical index (named PQ), which correlates well with grape polyphenolic content (Fig. 17.4). PQ is not a concentration value (e.g., in mg/L), but its measure before vinification allows the optimization of the subsequent winemaking procedures, e.g., categorizing the delivered grapes in classes with different PQ, without any calibration need (Celotti et al., 2008a, 2008b; Celotti & Carcereri De Prati, 2000, 2005). In this way, the enological techniques applied on the different categories can be tailored depending on their quality level.

Finally, hyperspectral imaging represents another interesting fast and nondestructive method for assessing the quality of the grapes at delivery (Dambergs et al., 2018, 2020; Lu et al., 2020). By this technique, it is possible to identify if a grape lot is significantly affected by fungal infection or to detect the presence of material other than grape (MOG) (Dambergs et al., 2020). For wine cooperatives, these are useful indicators to define the amount which has to be paid for a given lot of grape, especially in case of mechanical harvest.

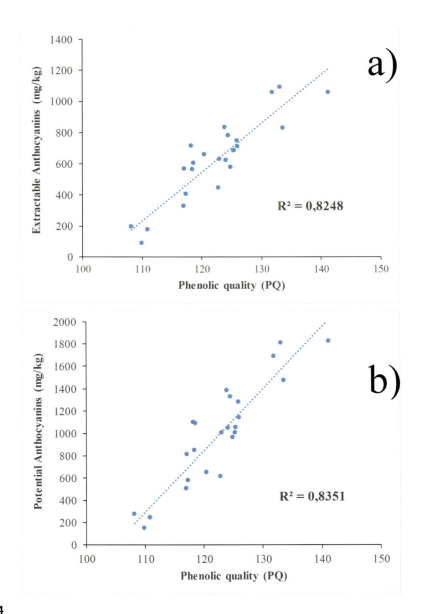

FIGURE 17.4

Correlation between Phenolic Quality (PQ) and extractable **(A)** and potential anthocyanins **(B)** evaluated by the Glories method (cv. Merlot, Israel, September 2003).

The real-time monitoring of grapes during reception provides valuable information for technological purposes and allows to establish an equitable price for grapes depending on its quality. Analytical controls are essential also for the management of the vineyards and the possible implementation of viticultural zoning projects. If control systems are properly integrated, they can provide valuable and complete information about grape quality. Integration improves technical and economical sustainability through the use of rapid, nondestructive, and real-time analysis methods, saving time and reducing the volumes of chemicals and reagents used for grape analysis.

17.5 Sustainable wine analysis and quality control

As mentioned above, GAC aims to make laboratory practices safer to analysts and less energy-demanding (Marcinkowska et al., 2019), reducing the negative impact of chemical analyses on the environment (Gałuszka et al., 2013) and the amounts of waste generated (Armenta et al., 2008). In fact, even if analytical chemistry involves smaller volumes of chemicals compared with industrial processes (Miguel·de la Guardia & Garrigues, 2020), lab activities require the use of many reagents and solvents (Płotka-Wasylka, 2018), producing a nonnegligible impact on environment and human health (Armenta et al., 2008).

Wine analysis (WA) plays a fundamental role in wine production process and trading and involves activities related to process monitoring, quality control, certifications, consumer safety and legislative aspects. The modern approach to WA aims at integrating analytical results in various steps of the supply chain, providing indications useful to properly manage manufacturing, bottling and transport, as well as for dealing the conditions of delivery, selling, and purchasing of bottled and bulk wine. Analytical data are fundamental also for establishing the compliance of a given product and/or manufacturing process with certain contractual agreements or certification schemes (e.g., requirements for protected geographical indication, organic certification or private standards such as BRC-GSFS), as well as with national or international laws, ensuring the quality of wine itself and the protection of the consumer.

The impact of wine laboratories on environment and health of the operators depends on various factors, such as the size of the lab, the number of samples processed, the instrumental facilities available, and the analytical methods used, both concerning the kind of target analytes and the analytical procedures applied.

If one considers the relevant literature related to this field published in the last 20 years, very few scientific papers have been written concerning green procedures in wine analysis (Table 17.1), thus, it might seem that wine sector is not very oriented toward a greening of analytical methods.

However, actually, several efforts have been devoted in last decades for developing rapid and high sample throughput methods (Luque de Castro et al., 2005), and many innovative practices and instrumental facilities have been introduced in wine laboratories for reducing the analysis time minimizing the use of solvents. Therefore, sustainability has often been disguised as speed/simplification of laboratory procedures. Wine labs may be clustered in four main typologies (Table 17.2).

Each of these realities has different needs and different problems regarding sustainability. In the following paragraphs, major critical issues will be discussed, mainly focusing on winery labs, analyzing some of the opportunities available for greening analytical procedures keeping suitable standards of accuracy and reliability.

Table 17.1 Number of papers dealing with green analytical procedures in wine analysis, as reported in Web of Science (WOS) and Scopus databases (update: November 2020).

Matching keywords	Time span	Database	Number of records
"Green analytical chemistry"	2000–2020	WOS[a]	606
		Scopus	545
"Green analytical chemistry" and wine	2000–2020	WOS[a]	21
		Scopus	12
"Analytical method" and green	2000–2020	WOS[a]	4359
		Scopus	1830
"Analytical method" and green and wine	2000–2020	WOS[a]	56
		Scopus	18

[a]*Searched in "all databases."*

Table 17.2 Classification of wine laboratories.

Lab typology	Description
Winery labs	Analytical laboratories that are part of small, medium, or large wineries, operating in the field of quality control and process monitoring
Service labs	Private laboratories, providing analytical services for the wineries, consultancy and certificates of analysis (e.g., for export)
Public laboratories	Analytical laboratories of the control bodies, operating at local, regional, or national level (e.g., customs, agricultural or health ministry), for evaluating authenticity or wholesomeness of the products, as well as to assess their compliance to specific regulations or standards
Research labs	Private or public labs performing wine research (R&D labs of private corporations, universities or other research institutes)

17.5.1 Sustainability issues in winery labs

The capillary introduction of chemical analysis in winemaking has given a significant contribution to the great progress of wine quality observed in the last decades (Linskens & Jackson, 1988). Wine quality control requires the availability of simple and rapid analytical methods, allowing a regular and punctual monitoring of the different production steps and a fast decision-making procedure, able to trigger suitable technological interventions, in case of deviations from the normal winemaking conditions. Sustainability in winery labs involves different aspects, connected with the well-being and health of the operators, as well as with the reduction of waste.

The first point is related to the fact that wine production is mainly concentrated in a short period of the year, during and immediately following harvest (3–4 months generally). In such period, analysts spend frequently many hours a day in lab, standing or generally in uncomfortable positions, with high workloads. In fact, the most common conventional methods used for wine quality control (reducing sugars, total and volatile acidity, free and total sulfur dioxide) consist in traditional analytical techniques, mainly based on titration (Sudario, 1982).

Another issue concerns the production of waste. Generally, the routine quality control in wine labs does not involve reagents characterized by a high toxicity; nevertheless, there are some exceptions.

Reducing sugars are traditionally determined by the Fehling assay, consisting in the titration of a copper (II) sulfate solution with clarified must or wine (Sudario, 1982); the titration is carried out while cupric solution is boiling, so that the method is fairly dangerous and requires attention by the operators. Fehling reagent A contains approximately 69 g/L of copper (II) sulfate pentahydrate; considering that 5 mL of this solution are used for each assay, each must or wine sample consumes approximately 0.35 g of reagent (corresponding to ~90 mg of Cu). This copper is discharged as copper (I) oxide, becoming a nonnegligible source of cumulated waste production, especially if the winery lab processes a large number of samples per day. In fact, the Council Directive 98/83/EC establishes that the maximum copper content in water for human consumption should be 2 mg/L (Council Directive 98/83/EC, 1998) and sugar determination is carried out routinely in wineries, during harvest and fermentation, as well as during the preparation of wines for bottling, when sugar content is generally subjected to corrections.

The Luff-Schoorl assay, that is the method recommended by the International Organization of Vine and Wine (OIV) for measuring sugars and reducing substances in musts and wines (International Organization of Vine and Wine, 2020), provides for a lower concentration of copper sulfate in the copper reagent solution (25 g/L), but the protocol is longer and more complex and this method is generally not used at winery scale.

In addition to the use of copper (II) reagent, the clarification of the samples before Fehling and Luff-Schoorl assay is generally carried out by neutral lead acetate (500 g/L of trihydrate salt) or by zinc (II) ferrocyanide (Carrez reagent) (International Organization of Vine and Wine, 2020; Sudario, 1982), increasing the environmental impact of both these analytical procedures.

To solve these problems, various spectrophotometric methodologies have been developed (Brasil & Reis, 2017), with the aim of making the determination of reducing sugars easier. These methods are mainly based on the oxidation reaction of sugars with copper (II) or with hexacyanoferrate (III) (Başkan et al., 2016; Brasil & Reis, 2017; Friedemann et al., 1962) and they are often coupled with multicommuted flow analysis (FA) systems (Araújo et al., 2000; Brasil & Reis, 2017; Da Silva et al., 2018), with the purpose of reducing reagent consuming, increasing the number of samples analyzed. In addition, techniques such as SPE or in-line dialysis have been suggested for sample clarification (Araújo et al., 2000; Başkan et al., 2016; Da Silva et al., 2018), further reducing the use of toxic reagents and waste generation problems.

Another case of use of harmful reagents in wine laboratories is linked with the determination of free assimilable nitrogen (FAN), an important technological parameter, whose knowledge and management is fundamental for the correct course of alcoholic fermentation. In fact, the addition of nitrogen supplements to grape musts containing insufficient levels of FAN has an essential role for achieving good fermentation kinetics, but the use of excessive amounts of nitrogen may slow down fermentation (Ribéreau-Gayon, Dubourdieu, et al., 2006). The best known method for determining FAN in wineries is the formol index (Sørensen, 1907) which consists in titrating ammonium and the carboxylic groups of amino acids, after blocking the amino functions with an excess of formaldehyde (Sudario, 1982). The method is quite labor intensive, because the pH of the sample must be increased to 8.0–8.3 before titration, for eliminating the influence of titratable acidity (Shively & Henick-Kling, 2001; Sudario, 1982); moreover, formaldehyde is added to the samples in nonnegligible amounts (25 mL of a 37%–40% formalin solution) and due to the toxicity of the reagent and its high volatility (pure formaldehyde boils at $-19°C$, NIST Chemistry Webbook, 2020), the level of exposition of the operators represents a critical point for this lab determination, especially if the number of samples processed per day is high. This determines limitations in the use of formol index in many wineries, especially in small/medium-sized companies (where laboratory fume hoods are generally not present), thus compromising the possibility of using this important processing parameter for optimizing the quality of the wines produced.

The assay proposed by Dukes and Butzke for FAN determination (Dukes & Butzke, 1998) is an interesting example about how a simple analytical protocol may reduce the risk for operators, speeding up lab activities. The test is based on the spectrophotometric determination of primary amino groups, after derivatization with an *o*-phthaldialdehyde/*N*-acetyl-L-cysteine (OPA/CYS) mixture; the assay is not sensitive to proline, with a small response toward ammonium, but the analytical results are well correlated with the high-performance liquid chromatography (HPLC) estimation of amino acids, with good precision and accuracy, low analysis time, limited reagent consumption and toxicity (Dukes & Butzke, 1998). The OPA/CYS derivatization method is also used by different automatic analyzers for primary amino nitrogen determination (see next paragraph).

17.5.2 Automation in winery labs

In last decades, several semiautomatic and automatic solutions have been proposed for accelerating and simplifying wine analysis and quality control.

17.5.2.1 Titrators

Automatic titrators are used since a number of decades for analyzing the most common quality control indices, such as titratable acidity and pH (Budić-Leto et al., 2009), sulfur dioxide (Dorigan Moya, 2019), or reducing sugars (Zironi et al., 1989). Beyond the advantages linked to the reduction of the workload for the operators, the convenience of using automatic titrators is the possibility of reducing the sample and titrant volume and consequently the production of waste. However, titrators are semicontinuous equipments and the intervention of the analyst is required for sample loading, cleaning of the device and changing of reagents when switching to a different analytical method. The main problem connected with the use of titrators is the limited number of samples which can be loaded in a single batch of analysis. Furthermore, titrators generally allow the quantification of a single index at a time (i.e., one parameter for each batch of samples); multiparametric analysis is not possible with this type of instrument, with the exception of titratable acidity and pH which are usually determined simultaneously.

17.5.2.2 Multiparametric and enzymatic analyzers

Continuous analysis systems are useful tools for minimizing analysis time, reducing reagent consuming, and increasing the number of samples processed in routine wine analysis (Luque de Castro et al., 2005). Several types of wine analyzers have been developed for this purpose; just think that, currently, the search for the keywords "wine analyzer" (in English language) on Google allows to identify about 61,000 results (Google, 2021). These equipment allow the multiparametric analysis of wine samples with a limited waste production.

The simplest wine analyzers are small photometers that work by disposable vials containing the reagents needed for performing different determinations. Some of the models available are portable instruments, allowing the on-site quantification of different parameters. Samples and reagents may be dosed by a set of micropipettes and the instrument allows to sequentially determine various parameters on the same sample (e.g., sugars, pH, titratable acidity, malic, lactic and acetic acid, free and total SO_2, anthocyanins and total polyphenols). The advantage of these kind of devices is that analyses become fast and simple being easily manageable by winemakers even without specific training and laboratory skills. This idea well fits with the concept of "equitable analytical chemistry" described by Marcinkowska et al., as one of the social aspects of sustainability in analytical chemistry, that includes analytical methods which are easily available in terms of price and applicability by "everyday users" (Marcinkowska et al., 2019).

When integrated into wine analyzers, FA systems allow to further reduce analysis time, decreasing reagent consumption and waste production. In addition to reducing sugars (see the previous paragraph 17.5.1), different FA methods have been suggested also for the determination of other analytical parameters used in wine quality control, such as sulfur dioxide (Chantipmanee et al., 2017), organic acids (Kritsunankul et al., 2009), titratable acidity (Lima & Reis, 2017), malic acid (de Santana et al., 2014), ethanol (Luque de Castro et al., 2005), total polyphenols (de Andrade et al., 2014; Nalewajko-Sieliwoniuk et al., 2016), iron(II) (Paluch et al., 2020) and total tannins (Infante et al., 2008). According to Luque de Castro and colleagues, the precision of FA methods surpasses or equalizes that of their batch counterparts (Luque de Castro et al., 2005).

Different multiparametric automatic analyzers are commercially available, equipped with trays for the housing of several dozens of sample vials; these instruments allow the dosing of small volumes of samples and reagents (few to some hundreds microliters), the storing of samples and reagents in refrigerated compartments, the stirring of reaction cuvettes and the measurement of multiple parameters simultaneously. Cleaning is also carried out automatically; some of these analyzers are able to process up to 200 samples per hour, with a water consumption of 4 L/h, that is a limited volume, compared with the amount of water consumed for washing glassware after an equal number of traditional determinations.

Many of the parameters acquired by automatic analyzers are determined using enzymatic methods. Enzyme analysis has been proposed since 1960s for beverage analysis (Mayer, 1963). Describing the advantages of applying enzymatic methods to must and wine analysis, Lafon-Lafourcade in 1978 wrote that "if their cost price was not relatively high, their specificity, sensitivity, and rapidity would enable them to compete with the most precise of chemical methods," also suggesting to use them (for wine analysis), only when chemical methods are not specific enough or too laborious (Lafon-Lafourcade, 1978). Nowadays, automatic analyzers allow to significantly increase the number of samples a single commercial enzyme kit may test, by minimizing reagents volume, reducing dramatically the cost of a single determination. The most common parameters determined by commercial enzymatic kits are glucose, fructose, sucrose, glycerol, ethanol, organic acids (ascorbic, gluconic, pyruvic, citric, acetic, tartaric, D- and L-malic, D- and L-lactic), acetaldehyde, L-arginine, urea, ammoniacal nitrogen, and total SO_2.

17.5.2.3 The "IR revolution"
The introduction of IR analyzers represents one of the most recent innovations in wine analysis and in few years it has revolutionized winery lab management practices. According to the OIV, two main families of IR analyzers are available: NIR analyzers and Fourier transform infrared analyzers (FTIR) (International Organization of Vine and Wine, 2010).

The former are scanning spectrometers operating in the NIR region of the spectrum, i.e. at wavelengths between 1100 and 2500 nm (Cozzolino, 2009). The latter (FTIR analyzers) are generally Michelson interferometers (Basalekou et al., 2020; Cozzolino, 2009), operating in the near-middle IR (2000−10000 nm) (International Organization of Vine and Wine, 2010). Differently from traditional (dispersive/scanning) spectrometers, in FTIR spectroscopy, all the light frequencies emitted by the light source pass simultaneously through the sample, without previous selection (Basalekou et al., 2020; International Organization of Vine and Wine, 2010). The result is an interferogram which is mathematically processed by Fourier transform (FT) algorithm for obtaining the IR spectrum (Bauer et al., 2008; International Organization of Vine and Wine, 2010). Compared with scanning devices, FTIR instruments are faster and produce spectra with improved signal-to-noise ratio (Bauer et al., 2008).

Chemometrics plays a fundamental role for the interpretation and analysis of the IR data; due to the complexity of information included in IR spectra, multivariate statistical techniques are required to extract information about the compositional characteristics of a given sample (Cozzolino, 2009). The most common multivariate analysis techniques used for this purpose are principal component analysis, partial least squares, principal component regression, discriminant analysis and artificial neural network (Cozzolino, 2009). The combination of IR methods and chemometric techniques allows the direct and fast detection of the main macroconstituents of musts and wines, comparing the spectrum of an unknown sample with those obtained during the calibration of the instrument (International Organization of Vine and Wine, 2010).

Must and wine samples are processed without any specific preparation procedure, only a preliminary filtration or centrifugation and the elimination of excess carbon dioxide (Dubernet, 2009). All analytical operations are fully automatized (sampling, cleaning), allowing the processing of a large number of samples per day and a limited production of waste, highly reducing the intervention of the operators. However, for avoiding analytical errors, calibration procedures must be carried out correctly.

Calibration of IR analyzers require certain labor; the OIV established detailed guidelines for the management of instrument calibration and the selection of calibration samples (International Organization of Vine and Wine, 2010). Moreover, to periodically test the performances of the equipment, a regular check of calibration is recommended, by comparing the results of the instrument with those obtained from reference analytical methods; this comparison should be carried out daily for the most sensitive parameters (International Organization of Vine and Wine, 2010). The accuracy, repeatability, and detection limit of IR methods depend on different factors, such as the type and number of calibration samples, the parameters to be determined and their concentration, the analytical method used for calibration, as well as any matrix effects due to the interaction of the target molecules with other compounds (Dubernet, 2009).

17.5.3 Sustainability issues in service labs, public labs and research laboratories

In the other lab typologies indicated in Table 17.2, the issues related to improving the sustainability of analytical methods may have different and even more complicated implications.

In service labs, for instance, besides routine wine analysis, more complex analytical procedures are normally managed, such as those aimed at analyzing pesticide residues, toxic contaminants (e.g., ochratoxin A, heavy metals) or unwanted metabolites (e.g., biogenic amines, volatile phenols, or 2,4,6-trichloroanisole). These compounds are often present in fairly low concentrations and their determination requires high accuracy and sensitivity, as well as the use of official or accredited methods.

The same issues can be found in public labs and research laboratories, with other additional problems, such as those connected with the determination of wine authenticity (public labs) or the necessity to change frequently the analytical methods used, depending on the research projects managed at a given time (research labs).

Differently from routine wine analysis, these methods often require complicate preparative procedures (e.g., extraction, preconcentration, derivatization), involving the use of organic solvents and harmful reagents. However, these structures are normally equipped with modern and efficient instrumental facilities and this represents a useful aspect for allowing the greening of lab activities.

In last decades, the development of greener and more sustainable extraction techniques such as solid phase microextraction (Pawliszyn, 2012) or liquid phase microextraction (Sarafraz-Yazdi & Amiri, 2010) and the possibility of using different variants of classic SPE—such as micro-SPE (μ-SPE) (Della Pelle et al., 2016), magnetic-SPE (Ibarra et al., 2015), or dispersive μ-SPE (Xian et al., 2017)—as alternatives to conventional sample preparation procedures (liquid-liquid extraction or SPE), allow reducing the use of solvents and hazardous reagents, minimizing energy consumption, waste production, and increasing the safety for operators and environment (Anthemidis & Samanidou, 2019).

Instruments manufacturers are also oriented toward a more sustainable management of instrumental analyses. For instance, the development of ultra-high performance liquid chromatography and the reduction of the particle size of HPLC columns, allow to significantly reduce the use of solvents and the production of waste, with a positive impact on lab managing costs. Finally, the increased sensitivity and performances of mass spectrometric detectors and the development of new concepts of mass analyzers (hybrid analyzers, Orbitrap mass analyzer) allow to increase method sensitivity, decreasing the detection limit and reducing the need of sample preconcentration.

17.6 Concluding remarks

Many efforts have been carried out in last decades for introducing sustainable laboratory practices in analytical chemistry, as well as for greening analysis methods, reducing the use of solvents and the production of waste. Also in the field of grape and wine quality control, wine companies are moving in this direction, with obvious advantages in terms of environmental impact and wellness of the operators. The consequent reduction of lab managing costs and the perception of consumers that wine is a food product closely linked to a specific territory, will probably determine further developments of this kind of approach in the next years.

If real-time analyses have been successfully tested for monitoring grape quality on-field, on-line measures are still underused in wineries. Despite a number of analytical sensors have been described in scientific literature and could be used for process monitoring at winery scale, only few variables (e.g., temperature, conductivity, volume, pH) are actually exploited through real-time/on-line measurements in enology. Technologies such as smart sensing and Internet of Things will probably have a relevant impact on the changes that will affect wine analysis and on-line monitoring in the future. This will improve the aspects of sustainability related to the concept of "equitable analytical chemistry" described by Marcinkowska and colleagues (Marcinkowska et al., 2019), i.e., the use of simple analytical tools and procedures (e.g., smartphone applications and portable and miniaturized sensors) allowing a given analytical method to become available for a wide number of individuals.

Another relevant aspect, connected to grape and wine analysis, is that an accurate monitoring of the production process, through an efficient application of suitable quality control protocols, may indirectly increase the sustainability of the whole supply chain; just think of how it is possible to reduce the use of certain additives or fining agents, carrying out appropriate wine stability tests.

Therefore, the tools for a sustainable management of grape and wine quality control are already available and well developed and they will be surely further implemented in the future. However, one of the most important factors in making greener wine laboratories is the "operating philosophy" of the lab itself, that is the approach that both management and operators have toward sustainable issues.

References

Agati, G., Meyer, S., Matteini, P., & Cerovic, Z. G. (2007). Assessment of anthocyanins in grape (*Vitis vinifera* L) berries using a noninvasive chlorophyll fluorescence method. *Journal of Agricultural and Food Chemistry, 55*, 1053–1061. https://doi.org/10.1021/jf062956k

Anastas, P. T., & Warner, J. C. (1998). *Green chemistry: Theory and practice*. Oxford University Press.

Anastas, P. T., & Williamson, T. C. (1996). Green chemistry: Designing chemistry for the environment. *Choice Reviews Online, 34*, 1555. https://doi.org/10.5860/choice.34-1555

Andaur, J. E., Guesalaga, A. R., Agosin, E. E., Guarini, M. W., & Irarrázaval, P. (2004). Magnetic resonance imaging for nondestructive analysis of wine grapes. *Journal of Agricultural and Food Chemistry, 52*, 165–170. https://doi.org/10.1021/jf034886c

Anderson, J. L., & Clark, K. D. (2018). Ionic liquids as tunable materials in (bio)analytical chemistry. *Analytical and Bioanalytical Chemistry, 410*, 4565–4566. https://doi.org/10.1007/s00216-018-1125-4

Anthemidis, A., & Samanidou, V. F. (2019). Automation in sample preparation and green analytical perspectives. *Current Analytical Chemistry, 15*, 705. https://doi.org/10.2174/157341101507191015122729

Araújo, A. N., Lima, J. L., Rangel, A. O., & Segundo, M. A. (2000). Sequential injection system for the spectrophotometric determination of reducing sugars in wines. *Talanta, 52*, 59–66. https://doi.org/10.1016/S0039-9140(99)00338-0

Armenta, S., Garrigues, S., & de la Guardia, M. (2008). Green analytical chemistry. *TRAC Trends in Analytical Chemistry, 27*, 497–511. https://doi.org/10.1016/j.trac.2008.05.003

Basalekou, M., Pappas, C., Tarantilis, P. A., & Kallithraka, S. (2020). Wine authenticity and traceability with the use of FT-IR. *Beverages, 6*, 1–13. https://doi.org/10.3390/beverages6020030

Başkan, K. S., Tütem, E., Akyüz, E., Özen, S., & Apak, R. (2016). Spectrophotometric total reducing sugars assay based on cupric reduction. *Talanta, 147*, 162–168. https://doi.org/10.1016/j.talanta.2015.09.049

Bauer, R., Nieuwoudt, H., Bauer, F. F., Kossmann, J., Koch, K. R., & Esbensen, K. H. (2008). FTIR spectroscopy for grape and wine analysis. *Analytical Chemistry, 80*, 1371–1379. https://doi.org/10.1021/ac086051c

Ben Ghozlen, N., Cerovic, Z. G., Germain, C., Toutain, S., & Latouche, G. (2010). Non-destructive optical monitoring of grape maturation by proximal sensing. *Sensors (Switzerland), 10*, 10040–10068. https://doi.org/10.3390/s101110040

Ben Ghozlen, N., Moise, N., Latouch, G., Martinon, V., Mercier, L., Besançon, E., & Cerovic, Z. G. (2010). Assessment of grapevine maturity using a new portable sensor: Non-destructive quantification of anthocyanins. *Journal International Des Sciences de La Vigne et Du Vin, 44*, 1–8.

Brasil, M. A. S., & Reis, B. F. (2017). An automated multicommuted flow analysis procedure for photometric determination of reducing sugars in wine employing a directly heated flow-batch device. *Journal of the Brazilian Chemical Society, 28*, 2013–2020. https://doi.org/10.21577/0103-5053.20170047

Brett, C. M. A. (2007). Novel sensor devices and monitoring strategies for green and sustainable chemistry processes. *Pure and Applied Chemistry, 79*, 1969–1980. https://doi.org/10.1351/pac200779111969

Brillante, L., Tomasi, D., Gaiotti, F., Giacosa, S., Torchio, F., Segade, S. R., Siret, R., Zouid, I., & Rolle, L. (2015). Relationships between skin flavonoid content and berry physical-mechanical properties in four red wine grape cultivars (*Vitis vinifera* L.). *Scientia Horticulturae, 197*, 272–279. https://doi.org/10.1016/j.scienta.2015.09.053z

Brunner, G. (2005). Supercritical fluids: Technology and application to food processing. *Journal of Food Engineering, 67*, 21–33. https://doi.org/10.1016/j.jfoodeng.2004.05.060

Budić-Leto, I., Mešin, N., Gajdoš Kljusurić, J., Pezo, I., & Bralić, M. (2009). Comparative study of the total acidity determination in wine by potentiometric and volumetric titration. *Agriculturae Conspectus Scientificus, 74*, 61–65.

Cathcart, C. (1990). Green chemistry in the emerald isle. *Chemistry and Industry (London), 21*, 684–687.

Celotti, E. (2012). Sistema rapido di cantina per il monitoraggio della macerazione delle uve rosse. *Infowine - Internet Journal of Viticulture and Enology, 7/3*, 1–10.

Celotti, E., & Carcereri De Prati, G. (1999). *Method to evaluate the quality of grapes and relative device (Patent No. EU 1175603)*.

Celotti, E., & Carcereri De Prati, G. (2000). La qualità fenolica delle uve rosse: Valutazione oggettiva mediante misura del colore. *Industrie Delle Bevande, 168*, 378–384.

Celotti, E., & Carcereri De Prati, G. (2005). The phenolic quality of red grapes at delivery: Objective evaluation with colour measurements. *South African Journal of Enology and Viticulture, 26*, 75–82. https://doi.org/10.21548/26-2-2121

Celotti, E., Carcereri De Prati, G., Anaclerio, F., De Luca, E., Scortegagna, E., Ruiz, A., Martinico, S., & Sebastianelli, D. (2009). Innovazione nella misura rapida della qualità fenolica in vigneto. *L'Enologo, 12*, 83–90.

Celotti, E., Carcereri De Prati, G., Charpentier, C., & Feuillat, M. (2008). La mesure de la maturité phénolique directement à la vigne: Expériences en Bourgogne. *Revue Des Œnologues, 127*, 42–45.

Celotti, E., Carcereri De Prati, G., & Fiorini, P. (2008). Modern approach to the red grape quality management. *Infowine - Internet Journal of Viticulture and Enology, 1*, 1–17.

Celotti, E., Carcereri De Prati, G., Sebastianelli, D., & Ruiz, A. E. (2010). Évaluation rapide du potentiel phénolique du raisin rouge et de la couleur du vin rouge. Nouvelle méthode de mesure spectroscopique dans le spectre visible. *Revue Des Œnologues, 136*, 38–42.

Celotti, E., Della Vedova, T., Ferrarini, R., & Martinand, S. (2007). The use of reflectance for monitoring phenolic maturity curves in red grapes. *Italian Journal of Food Science, 19*, 91–100.

Celotti, E., Dell'Oste, S., Fiorini, P., & Carcereri De Prati, G. (2007). Une nouvelle méthode pour l'évaluation des polyphénols des raisins rouges. *Revue Des Œnologues, 125*, 23–27.

Cerovic, Z. G., Moise, N., Agati, G., Latouche, G., Ben Ghozlen, N., & Meyer, S. (2008). New portable optical sensors for the assessment of winegrape phenolic maturity based on berry fluorescence. *Journal of Food Composition and Analysis, 21*, 650–654. https://doi.org/10.1016/j.jfca.2008.03.012

Chantipmanee, N., Alahmad, W., Sonsa-ard, T., Uraisin, K., Ratanawimarnwong, N., Mantim, T., & Nacapricha, D. (2017). Green analytical flow method for the determination of total sulfite in wine using membraneless gas–liquid separation with contactless conductivity detection. *Analytical Methods, 9*, 6107–6116. https://doi.org/10.1039/C7AY01879G

Chemat, F., Garrigues, S., & de la Guardia, M. (2019). Portability in analytical chemistry: A green and democratic way for sustainability. *Current Opinion in Green and Sustainable Chemistry, 19*, 94–98. https://doi.org/10.1016/j.cogsc.2019.07.007

Council Directive 98/83/, E. C. (1998). On the quality of water intended for human consumption - consolidated version (2015). *Official Journal of the European Union, L330*, 32–54.

Cozzolino, D. (2009). Near infrared spectroscopy in natural products analysis. *Planta Medica, 75*, 746–756. https://doi.org/10.1055/s-0028-1112220

Cozzolino, D. (2015). The role of visible and infrared spectroscopy combined with chemometrics to measure phenolic compounds in grape and wine samples. *Molecules, 20*, 726–737. https://doi.org/10.3390/molecules20010726

Cozzolino, D., Dambergs, R. G., Janik, L., Cynkar, W. U., & Gishen, M. (2006). Analysis of grapes and wine by near infrared spectroscopy. *Journal of Near Infrared Spectroscopy, 14*, 279–289. https://doi.org/10.1255/jnirs.679

Cunha, S. C., & Fernandes, J. O. (2018). Extraction techniques with deep eutectic solvents. *TRAC Trends in Analytical Chemistry, 105*, 225–239. https://doi.org/10.1016/j.trac.2018.05.001

Da Silva, P. A. B., De Souza, G. C. S., Paim, A. P. S., & Lavorante, A. F. (2018). Spectrophotometric determination of reducing sugar in wines employing inline dialysis and a multicommuted flow analysis approach. *Journal of the Chilean Chemical Society, 63*, 3918–3924.

Dambergs, B., Jiang, W., Nordestgaard, S., Wilkes, E., & Petrie, P. (2020). *Hyperspectral imaging of Botrytis in grapes.* https://awitc.com.au/wp-content/uploads/2019/07/131-Hyperspectral-imaging-of-Botrytis.pdf. (Accessed 27 December 2020).

Dambergs, B., Nordestgaard, S., Jiang, M., Wilkes, E., & Petrie, P. (2018). Grape and juice handling: Hyperspectral imaging of botrytis in grapes. *Wine & Viticulture Journal, 33*, 22–24.

De Andrade, M. F., de Assis, S. G. F., Paim, A. P. S., & dos Reis, B. F. (2014). Multicommuted flow analysis procedure for total polyphenols determination in wines employing chemiluminescence detection. *Food Analytical Methods, 7*, 967–976. https://doi.org/10.1007/s12161-013-9699-0

De La Guardia, M., & Garrigues, S. (2011). An ethical commitment and an economic opportunity. In M. De La Guardia, & S. Garrigues (Eds.), *Challenges in green analytical chemistry* (pp. 1–12). The Royal Society of Chemistry. https://doi.org/10.1039/9781849732963-00001

De Santana, J. F. S., Belian, M. F., & Lavorante, A. F. (2014). A spectrophotometric procedure for malic acid determination in wines employing a multicommutation approach. *Analytical Sciences, 30*, 657–661. https://doi.org/10.2116/analsci.30.657

De la Guardia, M., & Armenta, S. (2011). Origins of green analytical chemistry. *Comprehensive Analytical Chemistry, 57*, 1–23. https://doi.org/10.1016/B978-0-444-53709-6.00001-X

De la Guardia, M., & Garrigues, S. (2012). The concept of green analytical chemistry. In M. De La Guardia, & S. Garrigues (Eds.), *Handbook of green analytical chemistry* (pp. 1–16). Hoboken (NJ): John Wiley & Sons. https://doi.org/10.1002/9781119940722.ch1

De la Guardia, Miguel, & Garrigues, S. (2020). Past, present and future of green analytical chemistry. In M. De La Guardia, & S. Garrigues (Eds.), *Challenges in green analytical chemistry* (2nd ed., pp. 1–18). The Royal Society of Chemistry. https://doi.org/10.1039/9781788016148-00001

De la Guardia, Miguel, & Ruzicka, J. (1995). Towards environmentally conscientious analytical chemistry through miniaturization, containment and reagent replacement. *Analyst, 120*, 17N. https://doi.org/10.1039/AN995200017N

Della Pelle, F., Di Crescenzo, M. C., Sergi, M., Montesano, C., Di Ottavio, F., Scarpone, R., Scortichini, G., & Compagnone, D. (2016). Micro-solid-phase extraction (μ-SPE) of organophosphorous pesticides from wheat followed by LC-MS/MS determination. *Food Additives & Contaminants Part A, Chemistry, Analysis, Control, Exposure & Risk Assessment, 33*, 291–299. https://doi.org/10.1080/19440049.2015.1123818

Dorigan Moya, H. (2019). Determination of free SO$_2$ in wines using a modified ripper method with potentiometric detection a comparative study with an automatic titrator. *Canadian Journal of Biomedical Research & Technology, 2*, 1–5.

Dubernet, M. (2009). Automatic analysers in oenology. In M. V. Moreno-Arribas, & M. C. Polo (Eds.), *Wine chemistry and biochemistry* (pp. 649–676). New York: Springer. https://doi.org/10.1007/978-0-387-74118-5_28

Dukes, B., & Butzke, C. (1998). Rapid determination of primary amino acids in grape juice using an *o*-phtaldialdehyde/*N*-acetyl-L-cysteine spectrophotometric assay. *American Journal of Enology and Viticulture, 49*, 125–134.

Fernández-Novales, J., Tardáguila, J., Gutiérrez, S., & Paz Diago, M. (2019). On-the-Go VIS + SW - NIR spectroscopy as a reliable monitoring tool for grape composition within the vineyard. *Molecules, 24*, 2795. https://doi.org/10.3390/molecules24152795

Ferrer-Gallego, R., Hernández-Hierro, J. M., Rivas-Gonzalo, J. C., & Escribano-Bailón, M. T. (2011). Determination of phenolic compounds of grape skins during ripening by NIR spectroscopy. *Lebensmittel-Wissenschaft und -Technologie- Food Science and Technology, 44*, 847–853. https://doi.org/10.1016/j.lwt.2010.12.001

Friedemann, T. E., Weber, C. W., & Witt, N. F. (1962). Determination of reducing sugars by oxidation in alkaline ferricyanide solution. *Analytical Biochemistry, 4*, 358–377. https://doi.org/10.1016/0003-2697(62)90137-9

Gałuszka, A., Migaszewski, Z. M., Konieczka, P., & Namieśnik, J. (2012). Analytical eco-scale for assessing the greenness of analytical procedures. *TRAC Trends in Analytical Chemistry, 37*, 61–72. https://doi.org/10.1016/j.trac.2012.03.013

Gałuszka, A., Migaszewski, Z., & Namieśnik, J. (2013). The 12 principles of green analytical chemistry and the SIGNIFICANCE mnemonic of green analytical practices. *TRAC Trends in Analytical Chemistry, 50*, 78–84. https://doi.org/10.1016/j.trac.2013.04.010

Gomes, V. M., Fernandes, A. M., Faia, A., & Melo-Pinto, P. (2017). Comparison of different approaches for the prediction of sugar content in new vintages of whole Port wine grape berries using hyperspectral imaging. *Computers and Electronics in Agriculture, 140*, 244–254. https://doi.org/10.1016/j.compag.2017.06.009

Google. (2021). https://www.google.com/search?ei=rkETYKzwDtuIjLsPqciC6Ao&q=%22wine+analyzer%22&oq=%22wine+analyzer%22&gs_lcp=CgZwc3ktYWIQAzIGCAAQFhAeMgYIABAWEB4yBggAEBYQHjIGCAAQFhAeMgYIABAWEB4yCAgAEBYQChAeMgYIABAWEB4yBggAEBYQHjIGCAAQFhAeMgYIABAWEB46BwgAEECsAM6CAgAEAcQHh. (Accessed 28 January 2021).

Gürkan, R., Korkmaz, S., & Altunay, N. (2016). Preconcentration and determination of vanadium and molybdenum in milk, vegetables and foodstuffs by ultrasonic-thermostatic-assisted cloud point extraction coupled to flame atomic absorption spectrometry. *Talanta, 155*, 38–46. https://doi.org/10.1016/j.talanta.2016.04.012

Gutiérrez, S., Tardaguila, J., Fernández-Novales, J., & Diago, M. P. (2018). On-the-go hyperspectral imaging for the in-field estimation of grape berry soluble solids and anthocyanin concentration. *Australian Journal of Grape and Wine Research, 25*, 127–133. https://doi.org/10.1111/ajgw.12376

Ibarra, I. S., Rodriguez, J. A., Galán-Vidal, C. A., Cepeda, A., & Miranda, J. M. (2015). Magnetic solid phase extraction applied to food analysis. *Journal of Chemistry*. https://doi.org/10.1155/2015/919414. Article ID 919414.

Infante, C. M. C., Soares, V. R. B., Korn, M., & Rocha, F. R. P. (2008). An improved flow-based procedure for microdetermination of total tannins in beverages with minimized reagent consumption. *Microchimica Acta, 161*, 279–283. https://doi.org/10.1007/s00604-007-0875-z

International Organization of Vine and Wine, O. I. V. (2010). *Guidelines on infrared analysers in oenology*. Paris: OIV.

International Organization of Vine and Wine, O. I. V. (2020). Reducing substances. In *Compendium of international methods of analysis* (pp. 348–352). Paris: OIV.

Kradtap Hartwell, S. (2012). Exploring the potential for using inexpensive natural reagents extracted from plants to teach chemical analysis. *Chemistry Education: Research and Practice, 13*, 135−146. https://doi.org/10.1039/c1rp90070f

Kritsunankul, O., Pramote, B., & Jakmunee, J. (2009). Flow injection on-line dialysis coupled to high performance liquid chromatography for the determination of some organic acids in wine. *Talanta, 79*, 1042−1049. https://doi.org/10.1016/j.talanta.2009.03.001

Kudłak, B., Owczarek, K., & Namieśnik, J. (2015). Selected issues related to the toxicity of ionic liquids and deep eutectic solvents—a review. *Environmental Science and Pollution Research, 22*, 11975−11992. https://doi.org/10.1007/s11356-015-4794-y

Lafon-Lafourcade, S. (1978). Application des méthodes enzymatiques à l'analyse des mouts et des vins. *Annales de La Nutrition et de l'alimentation, 32*, 969−974.

Lima, M. J. A., & Reis, B. F. (2017). Fully automated photometric titration procedure employing a multicommuted flow analysis setup for acidity determination in fruit juice, vinegar, and wine. *Microchemical Journal, 135*, 207−212. https://doi.org/10.1016/j.microc.2017.09.016

Linskens, H.-F., & Jackson, J. F. (1988). Wine analysis. In H.-F. Linskens, & J. F. Jackson (Eds.), *Wine analysis* (pp. 1−8). London: Springer.

Luque de Castro, M. D., González-Rodríguez, J., & Pérez-Juan, P. (2005). Analytical methods in wineries: Is it time to change? *Food Reviews International, 21*, 231−265. https://doi.org/10.1081/FRI-200051897

Lu, Y., Saeys, W., Kim, M., Peng, Y., & Lu, R. (2020). Hyperspectral imaging technology for quality and safety evaluation of horticultural products: A review and celebration of the past 20-year progress. *Postharvest Biology and Technology, 170*. https://doi.org/10.1016/j.postharvbio.2020.111318

Marcinkowska, R., Namieśnik, J., & Tobiszewski, M. (2019). Green and equitable analytical chemistry. *Current Opinion in Green and Sustainable Chemistry, 19*, 19−23. https://doi.org/10.1016/j.cogsc.2019.04.003

de Marco, B. A., Rechelo, B. S., Tótoli, E. G., Kogawa, A. C., & Salgado, H. R. N. (2019). Evolution of green chemistry and its multidimensional impacts: A review. *Saudi Pharmaceutical Journal, 27*, 1−8. https://doi.org/10.1016/j.jsps.2018.07.011

Mayer, K. (1963). Über die Anwendung enzymatischer Methoden zur Getränke-analyse. *Mitteilungen Aus Dem Gebiete Der Lebensmittel-Untersuchung Un Hygiene, 54*, 515−519.

Melnyk, A., Namieśnik, J., & Wolska, L. (2015). Theory and recent applications of coacervate-based extraction techniques. *TRAC Trends in Analytical Chemistry, 71*, 282−292. https://doi.org/10.1016/j.trac.2015.03.013

Musingarabwi, D. M., Nieuwoudt, H. H., Young, P. R., Eyéghè-Bickong, H. A., & Vivier, M. A. (2016). A rapid qualitative and quantitative evaluation of grape berries at various stages of development using Fourier-transform infrared spectroscopy and multivariate data analysis. *Food Chemistry, 190*, 253−262. https://doi.org/10.1016/j.foodchem.2015.05.080

Nalewajko-Sieliwoniuk, E., Malejko, J., Pawlukiewicz, A., & Kojło, A. (2016). A novel multicommuted flow method with nanocolloidal manganese(IV)-based chemiluminescence detection for the determination of the total polyphenol index. *Food Analytical Methods, 9*, 991−1001. https://doi.org/10.1007/s12161-015-0274-8

NIST Chemistry Webbook. (2020). *Formaldehyde: Phase change data*. https://webbook.nist.gov/cgi/cbook.cgi?ID=C50000&Mask=4#Thermo-Phase. (Accessed December 2020).

Pacheco-Fernández, I., & Pino, V. (2019). Green solvents in analytical chemistry. *Current Opinion in Green and Sustainable Chemistry, 18*, 42−50. https://doi.org/10.1016/j.cogsc.2018.12.010

Paluch, J., Kościelniak, P., Moleda, I., Machowski, K., Kalinowski, S., Koronkiewicz, S., & Kozak, J. (2020). Novel approach to determination of Fe(II) using a flow system with direct-injection detector. *Monatshefte Für Chemie - Chemical Monthly, 151*, 1305−1310. https://doi.org/10.1007/s00706-020-02649-8

Pawliszyn, J. (2012). 2 - theory of solid-phase microextraction. In J. Pawliszyn (Ed.), *Handbook of solid phase microextraction* (pp. 13−59). Amsterdam: Elsevier. https://doi.org/10.1016/B978-0-12-416017-0.00002-4

Płotka-Wasylka, J. (2018). A new tool for the evaluation of the analytical procedure: Green Analytical Procedure Index. *Talanta, 181*, 204–209. https://doi.org/10.1016/j.talanta.2018.01.013

Płotka-Wasylka, J., Kurowska-Susdorf, A., Sajid, M., de la Guardia, M., Namieśnik, J., & Tobiszewski, M. (2018). Green chemistry in higher education: State of the art, challenges, and future trends. *ChemSusChem, 11*, 2845–2858. https://doi.org/10.1002/cssc.201801109

Prat, D., Hayler, J., & Wells, A. (2014). A survey of solvent selection guides. *Green Chemistry, 16*, 4546–4551. https://doi.org/10.1039/C4GC01149J

Ribéreau-Gayon, P., Dubourdieu, D., Donèche, B., & Lonvaud, A. (2006). Handbook of enology. In *The microbiology of wine and vinifications* (2nd ed., Vol. 1). Hoboken (NJ): John Wiley & Sons.

Ribéreau-Gayon, P., Glories, Y., Maujean, A., & Dubourdieu, D. (2006). Handbook of enology. In *The chemistry of wine stabilization and treatments* (2nd ed., Vol. 2). Hoboken (NJ): John Wiley & Sons.

Río Segade, S., Rolle, L., Gerbi, V., & Orriols, I. (2008). Phenolic ripeness assessment of grape skin by texture analysis. *Journal of Food Composition and Analysis, 21*, 644–649. https://doi.org/10.1016/j.jfca.2008.06.003

Rolle, L., Río Segade, S., Torchio, F., Giacosa, S., Cagnasso, E., Marengo, F., & Gerbi, V. (2011). Influence of grape density and harvest date on changes in phenolic composition, phenol extractability indices, and instrumental texture properties during ripening. *Journal of Agricultural and Food Chemistry, 59*, 8796–8805. https://doi.org/10.1021/jf201318x

Rolle, L., Torchio, F., Lorrain, B., Giacosa, S., Río Segade, S., Cagnasso, E., Gerbi, V., & Teissedre, P. L. (2012). Rapid methods for the evaluation of total phenol content and extractability in intact grape seeds of cabernet-sauvignon: Instrumental mechanical properties and FT-NIR spectrum. *Journal International Des Sciences de La Vigne et Du Vin, 46*, 29–40.

Rolle, L., Torchio, F., Zeppa, G., & Gerbi, V. (2008). Anthocyanin extractability assessment of grape skins by texture analysis. *Journal International Des Sciences de La Vigne et Du Vin, 42*, 157–162.

Rousseau, J., & Delteil, D. (2000). Présentation d'une méthode d'analyse sensorielle des raisins. Principe, méthode et grille d'interprétation. *Revue Francaise d'Œnologie, 183*, 10–13.

Sarafraz-Yazdi, A., & Amiri, A. (2010). Liquid-phase microextraction. *TRAC Trends in Analytical Chemistry, 29*, 1–14. https://doi.org/10.1016/j.trac.2009.10.003

Shively, C. E., & Henick-Kling, T. (2001). Comparison of two procedures for assay of free amino nitrogen. *American Journal of Enology and Viticulture, 52*, 400–401.

Shukla, S. K., Pandey, S., & Pandey, S. (2018). Applications of ionic liquids in biphasic separation: Aqueous biphasic systems and liquid–liquid equilibria. *Journal of Chromatography A, 1559*, 44–61. https://doi.org/10.1016/j.chroma.2017.10.019

Sørensen, S. P. L. (1907). Enzymstudien I: Über die quantitative Messung proteolytischer Spaltungen, Die formoltitrierung. *Biochemische Zeitschrift, 7*, 45–101.

Steel, C. C. (2018). *Determination of thresholds for bunch rot contamination of grapes and techniques to ameliorate associated fungal taints*. https://www.google.com/url?sa=t&rct=j&q=&esrc=s&source=web&cd=&ved=2ahUKEwiGgpjX4rruAhUQ3IUKHTgdBJkQFjAFegQIBhAC&url=https%3A%2F%2Fwww.wineaustralia.com%2Fgetmedia%2Fad6cee75-3330-48ba-b2a9-bcc0b2fb6700%2FCSU-1301-Final-Report&usg=AOvVaw1F5iG-dHq9N4tYaROp. (Accessed 27 January 2021).

Sudario, E. (1982). *L'analisi dei vini e la ricerca delle sofisticazioni* (8th ed.). Italy: F.lli Marescalchi, Casale Monferrato.

Tang, X., Zhu, D., Huai, W., Zhang, W., Fu, C., Xie, X., Quan, S., & Fan, H. (2017). Simultaneous extraction and separation of flavonoids and alkaloids from *Crotalaria sessiliflora* L. by microwave-assisted cloud-point extraction. *Separation and Purification Technology, 175*, 266–273. https://doi.org/10.1016/j.seppur.2016.11.038

Tello, J., Cubero, S., Blasco, J., Tardaguila, J., Aleixos, N., & Ibáñez, J. (2016). Application of 2D and 3D image technologies to characterise morphological attributes of grapevine clusters. *Journal of the Science of Food and Agriculture, 96*, 4575–4583. https://doi.org/10.1002/jsfa.7675

Trujillo-Rodríguez, M. J., Nan, H., Varona, M., Emaus, M. N., Souza, I. D., & Anderson, J. L. (2019). Advances of ionic liquids in analytical chemistry. *Analytical Chemistry, 91*, 505–531. https://doi.org/10.1021/acs.analchem.8b04710

Versari, A., Parpinello, G. P., Mattioli, A. U., & Galassi, S. (2008). Determination of grape quality at harvest using Fourier-transform mid-infrared spectroscopy and multivariate analysis. *American Journal of Enology and Viticulture, 59*, 317–322.

Wardencki, W., Curyło, J., & Namieśnik, J. (2005). Green chemistry - current and future issues. *Polish Journal of Environmental Studies, 14*, 389–395.

Xian, Y., Wu, Y., Dong, H., Guo, X., Wang, B., & Wang, L. (2017). Dispersive micro solid phase extraction (DMSPE) using polymer anion exchange (PAX) as the sorbent followed by UPLC–MS/MS for the rapid determination of four bisphenols in commercial edible oils. *Journal of Chromatography A, 1517*, 35–43. https://doi.org/10.1016/j.chroma.2017.08.067

Zironi, R., Buiatti, S., Dosualdo, D., Baroncini, P., Guidotti, C., & Stefani, R. (1989). Determinazione automatica degli zuccheri riduttori nei mosti e nei vini con elettrodo Pt/redox. *Industrie Delle Bevande, 18*, 513–517.

CHAPTER 18

Life cycle methods and experiences of environmental sustainability assessments in the wine sector

Almudena Hospido[1,a], Beatriz Rivela[2,a] and Cristina Gazulla[3,a]

[1]CRETUS, Department of Chemical Engineering, Universidade de Santiago de Compostela, Santiago de Compostela, Spain; [2]inViable Life Cycle Thinking, Madrid, Spain; [3]Elisava Barcelona School of Design and Engineering, Barcelona, Spain

18.1 The wine supply chain: from land to table

From a life cycle perspective, the assessment of the whole supply is also referred to as *cradle-to-grave* approach, being the cradle the sourcing of the raw materials and the grave the final disposal of the product. This approach is mostly used for business-to-consumers (B2C) products, which encompasses all companies that create products and services that are sold directly to consumers.

If the use and disposal phases are cut out, the analysis is then *cradle-to-gate* and only assesses the product until it leaves the factory gate. Narrowing further the scope, *gate-to-gate* is sometimes used to reduce complexity as only one value-added process in the production chain is assessed, which can later be linked together to complete a larger level of analysis. Those two approaches are mostly used for business-to-business (B2B) products, which encompasses all companies that create products and services geared toward other businesses.

Besides, some intermediate approaches are also possible such as the *gate-to-grave* one that includes distribution, use, and end-of-life stages of a product. Or the *cradle-to-market* which differs from the classical cradle-to-gate perspective in that it also includes the phase of product distribution (Arzoumanidis et al., 2013).

Finally, and connected to the circular economy, a *cradle-to-cradle* has been coined as a variation of the cradle-to-grave by exchanging the waste stage with a recycling process that makes it reusable for another product and then closing the loop.

When referring to the life cycle of wine (Fig. 18.1) and regardless of the designation of origin or geographical indication, the final product is the result of a few stages that can be grouped into the following broad categories:

---

[a]All authors have equally contributed to this chapter

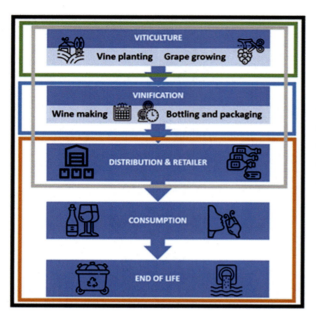

FIGURE 18.1

The life cycle of wine and the different life cycle boundaries.

- **Cradle-to-farm gate** (agricultural phase): The vine is a long-lived shrub (50–70 years in some cases) and every new vineyard planting starts with the physic-chemical analysis of the soil, which together with the climatic condition of a territory, is the information will allow technicians to choose the cultivars best suited to the area. Being a climbing plant, the vine requires a supporting infrastructure (made of wood, concrete, or metal) and, once vine cuttings are planted, it will need two or three years to start producing grapes. During this period and the following productive years, vineyard management is strongly dependent on the microclimate of the area, the characteristics of the soil, the field slope, and grape quality. When the grapes reach optimal maturity (in terms of sugar level, level of acidity, and color), they are collected and conveyed into trailers and transported to the winery for vinification.
- **Gate-to-gate** (vinification phase including packaging): Grapes cannot be stored, so the process of vinification must be initiated immediately.[1] Once the grapes arrive at the winemaking facilities, quality control takes place, each bunch of grapes is deprived of the stem and pressed to promote the fermentation of the entire mass. The must is then pumped into fermenters and yeast is added. The fermentation process is an exothermic reaction that requires temperature control (and this means energy consumption) as it affects the quality of wine (especially white wines). When the entire sugar component has been transformed, the wine is obtained and transferred for aging. When winemakers consider the wine is ready for the market, the wine can either be sold in bulk, or be bottled and packaged.

---

[1] Note that, besides wine, which is the focus here, there are other alcoholic beverages derived from grapes such as cognac, moonshine, or pisco.

- **Gate-to-grave**: Wine can be distributed to retailers or directly to final consumers. Being a worldwide produced and consumed product, the logistics are variable in transport type and average distances covered. Temperature control as well as gentle manipulation are important to guarantee its safe arrival to the final destination. After consumption, waste from primary and secondary packaging is generated and should be properly managed as has to be the human excreta related to a foodstuff as wine is.

18.2 Life cycle—based studies on the wine sector: a review
18.2.1 Lessons learnt from two decades of LCA application

Since the first studies were carried out in 2001, the sustainability assessment of wine production under a life cycle perspective has rising growing interest and is thoroughly examined in several case studies. The growing need to measure and minimize the environmental footprint of the sector has two main driving forces: production is exposed to environmental changes such as global warming and consumers are increasingly aware of the impacts of wine making (Jourdaine et al., 2020).

Life Cycle Assessment (LCA) has been used in the wine sector for different applications: to detect environmental hot spots in supply chains and identify improvement options, for ecolabelling and communication purposes, to compare farming practices, etc. The work by Petti et al. (2015) provided an analysis of the methodologies and standards of the Life Cycle Thinking concept related to wine and reported the results of a comprehensive critical analysis in the domain of the LCA of wine—81 papers published between 2001 and July 2013—to compile a list of scientifically based environmental hot spots and improvements. Focused on the tools created by private and public institutions, Merli et al. (2018) reviewed and analyzed the main environmental impacts of wine production based on LCA studies. More recently, Jourdaine et al. (2020) reviewed 10 LCA papers corresponding to 17 different wine products and identified the Life Cycle Inventory elements that drive the result, which can be useful to simplify the data collection and the comparability of the products in this sector.

The following sections build on previous research on the application of the life cycle perspective to wine chain to provide an overview of key methodological aspects and main scientifically based findings, with the overall goal to synthesize a critical analysis of the sustainability of wine production.

The functional unit: something incomparable cannot be compared

The functional unit (FU) is a central concept in LCA: its specification and the definition of the reference flows determine the materials and processes included in the assessment (Cooper, 2003). The quantified description of the function of the system under analysis is the reference basis for all calculations; thus, among the methodological choices that are made during the execution of an LCA, its definition is essential for modeling and comparative assessments.

However, the FU tends to be defined in a simplistic or insufficient way (Collado-Ruiz & Ostad-Ahmad-Ghorabi, 2010). For the same system, the formulation of the FU can be done using different criteria and frequently only the main functions and constraints are considered, which are not necessarily representative of all the aspects and impacts at stake. This is also the case for the great variety of wines, which are commonly assessed considering a specific amount of product in liters or kilograms as the reference for comparison, without considering the main characteristics of products. Notarnicola et al. (2010) have shown that with more technological production steps, the production of a high-quality wine has a worse environmental performance if the comparison is made based on volume or mass. If a different FU is considered, the results are completely inverted. The variability of results concerning environmental performance of different wines in the world must be analyzed regarding the vinification typology, as shown by Rugani et al. (2013). Comparative generalization of eco-friendly performance between products which are not comparable must be avoided: sensory feedback and other factors such as chloride and sulfate content—which can be characterized according to regulations—should be considered as they potentially influence the impacts associated.

A good example to illustrate the importance of the variety of wines is provided by Petti et al. (2015). When considering the grape varieties of Aglianico for red wine and Chardonnay for white wine in Italy, the main difference is in their maturity stage, which corresponds to the end of August for Chardonnay and the middle of October for Aglianico. Because of this variation, the Chardonnay viticulture needs eight pesticide treatments whereas the Aglianico needs ten of them, with a lower use (aprox. 20%) of pesticides, diesel, and lube oil in the case of Chardonnay. However, the trivial amount of these inputs in the agricultural stage is counterbalanced by the different yields in the two vinifications: according to Notarnicola et al. (2003), one 0.75 L bottle of red wine typically requires from 1.05 to 1.07 kg of grapes, and one of white wine about 1.2 kg.

Moreover, there are technological steps which raise both the quality of the wine and the energy consumption of the process within the same vinification (e.g., storage in barriques and concentration of the must through reverse osmosis).

According to the Environmental Footprint method from the European Commission (EC, 2013), four aspects shall be considered when defining the FU of a Product Environmental Footprint (PEF) study: (i) the function(s)/services(s) provided (what); (ii) the extent of the function or service (how much); (iii) the expected level of quality (how well); and (iv) the duration/lifetime of the product (how long). In the case of wine, the technical secretariat developing the PEF calculation rules for wine agreed on defining moderate consumption of alcoholic beverage as the function provided (what) and 0.75 L of wine as the amount provided (how much), as this is a recognizable unit for both wineries and consumers and 750 mL is a common mandatory nominal quantity for different wine products. Regarding the duration of the product (how long), it is important to remark that wine is exempted by Regulation EU 1168/2011 from a mandatory indication of an expiry date due to its very long shelf life. However, based on the inexistence of objective and commonly accepted criteria or standards to define wine quality, the technical secretariat was not able to define the expected level of quality for this product acknowledging that this limitation requires further developments to improve fair comparisons. The FU is then defined as the consumption of 0.75 L of packaged wine (PEFCR for wine, 2018).

One step further is the comparison between wine and other types of beverages. Mattila et al. (2012) posed the provocative question "beer or wine?" and chose as FU "one restaurant serving corresponding to 1.3 standard drinks (12 g of pure alcohol × 1.3 = 15.6 g)," which is equivalent to 12 cL of wine with 13% alcohol content and 33 cL of beer with 4.72% alcohol content. The study was not oriented to identify the best choice: the overall goal was to illustrate the uncertainty in environmentally conscious decision-making.

Studies at the sectoral and organizational levels can provide relevant insights to industry and policy makers for sustainability improvements. For example, Amienyo (2012) analyzed the UK beverage sector, considering the total annual production and consumption of the beverages in the UK—carbonated soft drinks, beer (lager), wine (red), bottled water, and Scotch whisky—as the FU. From an organizational perspective, Weidema et al. (2016) studied the total environmental impact of the three main product groups distributed by the Nordic Alcohol Monopolies: beer, distilled beverages, and wine. The results of these studies can help governments and beverage manufacturers to formulate appropriate policies and robust strategies, but it is important to remark that their purpose is not the comparison of different types of beverages.

Summing up, the recommendation of the authors of this chapter is to avoid simplification and do not compare incomparable products.

Hot spots

The review of the extensive literature concerning wine LCA studies shows relevant methodological discrepancies but allows identifying those areas of the wine production chain and the key aspects of the life cycle that contribute significantly to the environmental impacts of the overall system. To minimize the risk of leaving out critical points, it is necessary to assess several impact categories instead of using single indicators.

The following aspects are recognized as the hot spots of the wine production chain:

- Viticulture: the production and emissions of synthetic fertilizers and pesticides used for cultivation, as well as the emissions of diesel used in agricultural machineries. Agricultural land occupation, which is mainly driven by the required surface for growing the grapes, shows a high variability related to the yield of the vineyards.
- Vinification/wine making: the electric energy consumption and the emission of volatile organic compounds (VOCs) in the winery.
- Vinification/bottling and packaging: the energy consumption during the production of glass used for bottling.
- Distribution: the fuel consumption in transportation processes. Depending on the type of transport and distance considered, distribution can be the highest contributing phase for climate change (Point et al., 2012).

The following lines present a brief overview of the key elements and methodological aspects of the life cycle of wine, related to the stages described in Section 18.1.

Cradle-to-farm gate—agriculture

The assessment of the agricultural phase has been commonly simplified in previous studies, considering only the year(s) of actual grape production for quantifying input requirements and the release of emissions (Petti et al., 2015). However, the overall life cycle of a vineyard must be considered, including its planting, the first unproductive years, the productive years, and then senescence and disruption (Marenghi, 2005).

The relevance of the timeline perspective has been illustrated by Vazquez-Rowe et al. (2012). Considering a FU of 0.75 L bottle of Ribeiro white wine (Galician, NW Spain appellation), results identified the largest environmental impacts for four different years of production (2007–2010) and showed considerable annual variability in environmental performance. Also focused on Ribeiro appellation, Villanueva-Rey et al. (2015) applied this timeline perspective and presented a temporally based LCA study, to understand how the environmental profile of this specific wine producing area has shifted in the period from 1990 to 2009, analyzing in detail technological improvements and gradual changes in the land use.

The yield of the vineyards cannot be explained by differences in soil quality and climate alone: there are different viticulture regions, with a broad spectrum of grape varieties and diverse viticulture styles. Yield is of paramount importance when it comes to the assessment of the environmental performance of the whole chain and several authors have demonstrated its relevance, especially with regards to GHG emissions per FU (Ponstein et al., 2019; Vazquez-Rowe et al., 2013).

Alternative agricultural techniques such as organic and biodynamic viticulture have been extended in a gradual shift to more sustainable production practices. The results showed by Villanueva-Rey et al. (2014) comparing conventional versus biodynamic viticulture demonstrated that biodynamic production implies the lowest environmental burdens, mainly because an 80% decrease in diesel inputs, due to a lower application of plant protection products and fertilizers, and the introduction of manual work. However, some preliminary assessments suggest that impacts associated with land use and human labor may show different tendencies. More recently, Sinisterra-Solis et al. (2020) assessed the environmental impact of Spanish vineyards in Utiel-Requena protected designations of origin. The results show that, regardless of the grape cultivar, organic systems are more environmentally friendly than the conventional ones.

Gate-to-gate—vinification

Wine production and storage, bottling and packaging activities are the stages considered in a gate-to-gate perspective. The emissions of VOC during the alcoholic fermentation significantly contribute to the impact category of photochemical oxidation. Among the VOC emitted during vinification processes, the most problematic is ethyl alcohol. Other impact categories affected by the vinification processes are those linked to electric energy generation, although technical choices concerning bottling frequently are associated with the highest energy consumption, followed by the refrigeration phase.

Bottling and packaging have been identified as significant contributors to the environmental impacts of the whole chain by many authors. Fusi et al. (2014) assessed the environmental performance of a white wine produced in Sardinia and showed that the environmental performance of the wine chain was mostly determined by the glass bottle production for all impact categories, except for ozone layer depletion. Improvement actions suggested were a lighter glass bottle or the use of polylaminate container instead of the glass bottle. The results of Meneses et al. (2016) for a Catalan wine confirmed the paramount role of bottling: glass production was identified as the major contributor to the environmental load of wine production, being the recycling rate and the bottle weight the key parameters that determine the impact. Authors proposed lightweight bottles and alternative packaging to improve the environmental performance of the wine chain and suggested a glass container deposit legislation to enhance the glass recycling.

Although the traditional bottle is the glass one, there are many alternative packaging systems available in the market. A recent study from Ferrara and De Feo (2020) for the Italian market compared the environmental performance of the traditional single-use glass bottle for wine with four packaging alternatives: aseptic carton, bag-in-box, refillable glass bottle, and multilayer PET bottle. The results showed that the most environmentally sound alternative is the bag-in-box, followed by the aseptic carton, mainly due to the composition of the containers, lower packaging weight relative incidence, and greater palletizing efficiency. Three sensitive parameters were identified: weight of containers, wine distribution distance, and packaging disposal scenario. When decreasing the distribution distance, the environmental performances of refillable glass bottles became comparable to those of aseptic cartons and bag-in-box.

Gate-to-grave

There is an extensive bibliography related to logistics of the food and beverage sector, mainly focused on energy and carbon footprint of product distribution. Distribution over long distances can be very carbon-intensive and generate significant environmental impacts (Christopher, 2011). One of the first studies dealing with logistical options for delivering wine to consumers was published by Cholette and Venkat (2009). Authors used CargoScope, a web-based tool, to estimate energy and carbon emissions of California wines, and found a variability factor of 80 for different supply chain configurations.

There are three principal aspects which must be considered to assess the impact of distribution: distance, cargo mass, and method of carriage. Efficient packaging operations can lead to interesting improvements in the environmental performance of global supply chains (Murphy & Poist, 2003). The bottling either at source or destination is relevant for environmental modeling of wine transportation.

All the literature related to wine distribution confirms the environmental relevance of logistics and shows major variations between the environmental performance of different routing and packaging scenarios (Harris et al., 2018). Thus, special attention must be paid to postproduction stages of the wine supply chain.

Focusing the light: comparison and selection of best practices

LCA has been applied for environmental comparison of alternative technologies, focusing the analysis on specific stages to identify the best choice. Given the importance of aging, some studies have compared alternative systems for keeping organoleptic properties (Garcia Alcaraz, Flor, Martínez Cámara, Pérez de la Parte et al., 2020) or common techniques used for cleaning and disinfection of wine barrels (Garcia Alcaraz, Flor, Martínez Cámara, Sáenz-Diez et al., 2020). The environmental performance of wine fermentation in the two most common types of containers, i.e., steel and concrete tanks, has been analyzed by Flor et al. (2021). New consumer trends such as demand for dealcoholized wines have also received attention and some authors have performed comparative analysis to identify best practices. For example, Margallo et al. (2015) compared three technologies for partial dealcoholisation of wines and found that spinning cone column, one common treatment, showed the highest environmental burdens in all the categories under study.

Integrated approaches toward sustainability of the wine chain

Circular economy, biorefinery, and upcycling have emerged as innovative approaches to face sustainability challenges and identify new opportunities beyond traditional strategies. There is a growing literature on viticulture and winery waste valorization (Cortes et al., 2019, 2020; David et al., 2020). The review of Ahmad et al. (2020) provides an interesting overview of the state-of-the-art of biorefinery opportunities suggesting a promising framework of circular economy for the wine chain.

18.2.2 From a methodological framework point of view

Sustainability is a paradigm that affects many sectors due to an increase of stakeholders' pressure, economic or political reasons. This awareness has led the world of wine to show wide attention to the environment, identifying it as an element of the wine quality itself (Merli et al., 2018). Sustainable vitiviniculture is defined by the Organisation Internationale de la Vigne et du Vin (OIV) as a "global strategy on the scale of the grape production and processing systems, incorporating at the same time the economic sustainability of structures and territories, producing quality products, considering requirements of precision in sustainable viticulture, risks to the environment, products safety and consumer health and valuing of heritage, historical, cultural, ecological and aesthetic aspects" (OIV, 2004). From this definition, it is clear that sustainability in the wine sector is more than just organic, biodynamic, or integrated production, as it incorporates the culture, the landscape, the history, and all the intangible aspects that characterize wine as a product of excellence.

Answering what sustainability in the wine world is requires understanding definitions and principles, but also a review of current practices as done by Flores (2018) in their cross-country analysis about current sustainability assessment frameworks in six countries (South Africa, Australia, New Zealand, US, Chile, and France), which allows a comparison in terms of five different categories:

(1) Structure: self-assessment, footprint, protocol, guide, and certification or label.
(2) Scale: national or regional.
(3) Scope: primary responsibility for planning, conducting, and application policy.
(4) Deepness: how and to what extent the data are processed.
(5) Learning potential: the ability of the proposal to promote learning and improvement in practice.

Even so, the cross-analysis showed a convergence in the main criteria adopted by the frameworks, the local conditions consideration makes hard the proposition of general guidelines to sustainability assessment. In any case, the authors remarked the need to move forward and systematize the current practices and guidelines in a common basis can contribute to the development of sustainability programs, to improve current methods and to encourage the diffusion of sustainability practices to the individuals or in terms of national programs.

Also with comparative and revision purposes, Merli et al. (2018) reviewed the main environmental impacts of wine production, with specific reference to LCA studies, including the main programs from the New World, Europe, and Italy that address sustainability in the wine sector. The authors also reviewed consumers' propensity to buy sustainable wines, finding no conclusive outcomes[2]: even when there is a growing interest regarding environmental sustainability, consumers are not willing to trade-off the organoleptic and sensory quality with product sustainability, and often they are not willing to pay a premium price for a sustainable product because it is hard to perceive its latent characteristics. According to the review, one of the main reasons behind the reticence toward sustainable wines is the information asymmetry and the lack of credibility of the wineries in communicating sustainability initiatives. Merli et al. (2018) compared 30 different sustainability programs (7 of them in Europe and an additional bunch of 15 from Italy)[3] and recognized that this heterogeneity tends to create confusion, with overlapping initiatives, methodologies, and results disorienting the final consumer as well as the companies. So, the real challenge to achieve sustainability in the wine sector consists in arranging programs' features to create a single reference framework in the wine sector and promote scientific consistency, clarity, and transparency for both businesses and consumers.

The in-depth analysis of the different methodologies available at the organization level is beyond the scope of the present chapter, but it is important to mention some of the most applied ones and provide the reader with the links for further information:

---

[2]The reader is referred to their table three for a summary of the key findings of their revision on a selection of studies on sustainable wine and consumers' behavior.
[3]The reader is referred to their tables four to six for a detailed comparison of the programs.

(1) Having noticed the urgent need of developing an international and harmonized system for the calculation of the GHGs emissions specific for the vitivinicultural sector, the OIV has developed the general principles of an international GHG accounting protocol for the vine and wine sector adaptable to each country's situation. The protocol as well as the following supporting documents are available on its website: http://www.oiv.int/en/technical-standards-and-documents/environment-and-vitiviniculture.

(2) The Beverage Industry Environmental Roundtable (BIER) is a technical coalition of leading global beverage companies working together to advance environmental sustainability within the beverage sector. With a broader scope, but also covering wine production, BIER has published a sector guidance that aims to enhance and support the estimation, tracking, and reporting of GHG emissions within the beverage industry. The document can be directly downloaded from BIER website: http://www.bieroundtable.com/wp-content/uploads/Greenhouse-Gas-Emissions-Sector-Guidance.pdf.

(3) Since the 2010s, DG Environment has worked together with the European Commission's Joint Research Center (JRC IES) and other European Commission services toward the development of a harmonized methodology for the calculation of the environmental footprint of products and organizations. This harmonized methodology (EC, 2013) has been tested since then, being wine one of the pilot case studies selected for developing further the PEF method as explained in Section 18.3.3. A description of the process can be found in: https://ec.europa.eu/environment/eussd/smgp/index.htm.

Single indicators and footprints

Available scientific literature on wine production sustainability covers different points of view and different scopes, i.e., system boundaries and sustainability indicators. A predominant focus on single indicators, such as carbon or water footprint, is clear. On the former, the list of references is big. For example, Rugani et al. (2013) critically revised 35 studies with focus on several methodological and conceptual issues behind wine carbon footprinting (CF), such as calculation approaches, labeling, and standardization purposes, while Scrucca et al. (2018) reviewed different experiences regarding CF evaluation in the wine sector focused on nationally/internationally valid protocols and reported the results related to four case studies on Italian wineries. Moving from the product perspective, Sun and Drakerman (2020) identified a gap in the literature as carbon emissions associated with direct-to-consumer sales to wine tourists were not covered neither by the wine tourism not by wine industry CF analyses, so they adapted methods developed in the climate change and sustainable tourism literature specifically for the wine industry and found that cellar door sales associated with both domestic and international wine tourism lead to substantially higher carbon emissions than any other distribution channels.

Focusing on water use, Quinteiro et al. (2014) evaluated the water use associated with a well-known white Portuguese wine using different methods available in LCA to assess the potential environmental harm in ecosystems services derived from freshwater use, being the results and discussion focused on the comparison of the methods which differ significantly concerning the type of freshwater, freshwater scarcity level, and characterization factors considered. Also with the aim of comparing two different approaches when evaluating water use related impacts, Borsato et al. (2019) compared the Water-focused LCA and the Water indicator included in the Italian VIVA certification framework which is based on the Water Footprint Assessment and concluded that the second framework provides more precise recommendations for the optimal management of water use during the vineyard phase, while the LCA approach highlights impact hotspots related to both direct and indirect use of water resources.

Even when moving from the environmental to the social pillar of sustainability the references are scarce, they should not be ignored. Arcese et al. (2017) revised the social LCA approaches available at that time by applying them to the Italian wine sector and pointing out the difficulties associated with finding the right balance between the need of general standards and its application to the complex sector as the wine one. More recently, Martucci et al. (2017) analyzed differences and similarities between the S-LCA methodology and VIVA certification requirements for Italian wine production, evaluating the possibilities for future works to develop the integration of these indicators sets to broader the analyses of the socioeconomic impacts of the wine sector.

To finish, an interesting comparison of different indicators of sustainability is that performed by Amienyo (2012), who ambitiously considered the Life Cycle Sustainability Assessment in the UK beverage sector covering five beverage categories (i.e., carbonated soft drinks, bottled water, beer, red wine, and spirits and liqueurs). The environmental and economic assessments were first carried out at the level of individual supply chains and then extrapolated using a bottom-up approach to first their respective subsectors and then to the UK beverage sector, which was then followed by the social assessment at the sectoral level. According to the author, the findings of this study could help the government and beverage manufacturers to formulate appropriate policies and robust strategies for improving the sustainability in the UK beverage sector, as well as to help consumers to make more informed choices that contribute to sustainable development.

18.3 Environmental product declarations in the wine sector

One of the principal applications of the LCA methodology is to communicate quantitative information about the burdens that products trigger directly or indirectly on the environment.

Like other food industries, market and regulatory forces have increasingly driven the viticulture and wine sector to reduce and communicate their environmental and social performance (Pomarici & Vecchio, 2014). Consequently, a wide landscape of instruments to inform voluntary environmental and/or social information exists worldwide, including that tackling information at the organization level as well as those declaring environmental data at the product level. While other chapters of this book address sustainability certification schemes for viticulture and vinification, this chapter focuses on the declaration of environmental information of the wine's life cycle.

18.3.1 Landscape of environmental labels on wine

Eco-labels signal those products with superior environmental characteristics in comparison to similar nonlabeled products. This voluntary instrument provides easily interpretable information to prompt informed purchasing choices by environmentally responsible consumers (Leire & Thidell, 2005), and therefore eliciting increased demand for products perceived as environmentally favorable (Delmas & Grant, 2014).

In the last decades, the number of eco-labels programs worldwide has grown from a dozen in the 1990s (Delmas & Grant, 2014) to more than 457 programs today (see www.ecolabelindex.com) signaling the increasing use of this communication tool in different product markets, including the wine industry.

ISO distinguishes different types of voluntary environmental labels and declarations all of them tackling the life cycle of the product in a specific way:

- A Type I environmental label is a third-party assessment of a product based on several criteria involved in the environmental impact of a product or material throughout its life cycle (ISO 14024:2018). Type I ecolabel schemes typically apply pass-fail awarding systems and set criteria in a way that only a certain share of the products of a market qualify for the label and, hence, certified products can claim environmental excellence in comparison to noncertified ones (Minkov et al., 2020).
- A Type II claim is a self-declared environmental statement, symbol, and/or graphic made by the manufacturer, importer, distributor, retailer, or anyone else likely to benefit from such claim (ISO 14021:2016). Typical Type II claims are those declaring that a product and/or its packaging are made of recycled material or it is recyclable. If the methodology used to elaborate the self-declared claim is not properly third-party verified, it may be unreliable.
- A Type III environmental declaration presents quantified environmental information of the life cycle of a product to enable comparisons between products fulfilling the same function (ISO 14025:2006). Type III declarations (also known as Environmental Product Declarations, EPDs) are based on independently verified LCA data and are subject to the administration of a program operator. EPDs do not require that environmental thresholds or requirements be fulfilled and, despite allowing comparing products within the same product category, EPDs do not imply a superior environmental performance of the declared product.
- A footprint communicates environmental LCA results addressing an area of concern (e.g., water, climate change, or biodiversity) (ISO 14026:2017).

The implementation of these life cycle communication instruments in the wine sector is uneven. While, at the time of writing (2021.01.25), existing Type I schemes do not include wine as a product category and updated Type III declarations are not available (as discussed later), Type II and carbon and water footprints are more frequent.

It is important to remark that despite organically certified agricultural production covers key environmental issues (e.g., use of pesticides, fertilizers, or additives), it does not tackle the environmental impact of the product over its whole life cycle and it is not considered a Type I ecolabel.

As discussed in Section 18.2.2, CF is one of the most widespread indicators for assessing the environmental effects of food production and consumption (Scrucca et al., 2018). In the last years, some wineries have published the kg of CO_2 equivalent emitted throughout the life cycle of their wine. The release of ISO 14067 in 2013 has standardized the methodology for quantifying the CF of products, including wine. Nevertheless, it is important to remark that CF refers to a single impact category and other relevant indicators must be considered for a more comprehensive analysis of wine production (Scrucca et al., 2018). In that regard, EPDs for wine products are discussed in the next section.

Finally, it is worth noting that while an organization needs to eco-certify its products to label them as such, producers may decide not to inform their customers about it (Delmas & Grant, 2014). Implementing environmental practices considering a life cycle approach should lead to reduce the environmental footprint of wine and even to increase its quality, without the need to inform end consumers. In that respect, effective eco-labels could be those associated with changes in production processes that result in superior products (Delmas & Grant, 2014).

18.3.2 Environmental product declaration and related product category rules

EPDs communicate quantified life cycle—based information which is calculated by applying the LCA methodology following specific calculation rules, i.e., Product Category Rules (PCRs). Among other aspects, PCRs define:

- the product category,
- the unit to which data and results refer to (i.e., FU, or declared unit when the EPD does not cover the full life cycle),
- the system boundary (i.e., which specific processes need to be assessed),
- how to allocate burdens in case several products and co-products are produced at the same time,
- data quality requirements,
- impact categories and impact assessment methodology.

The publication of applicable PCRs is a previous required step to the publication of EPDs for a certain product type. PCRs are published and kept updated by EPD program administrators. Both PCRs and EPDs are valid for a certain period of time to ensure their regular updating. EPDs usually have a temporal validity of five years.

In recent years, the development and use of EPDs has increased for several market sectors, including food and beverages (Del Borghi et al., 2019). However, the number of EPDs of wine published is quite limited. In 2013 in the framework of the LIFE Project HAproWINE, several EPDs of specific wines and a sectorial EPD of the average wine produced in Castilla y León (Spain) were issued. These EPDs were developed according to the PCRs developed by the HAproWINE consortium (available on www.haprowine.eu).

In 2015, three wine producers had EPDs published in the International EPD® System (IES), a global program for type III environmental declarations operating in accordance with ISO 14025. Those EPDs were not updated and no additional ones have been published in the last years. Therefore, at the time of writing, there are no EPDs for wine products registered by IES which online database contains more than 1000 EPDs for different product categories (Del Borghi et al., 2019). However, an increasing interest of wine producers in developing EPDs has led the IES to publish an updated version of the PCRs for wine (IES, 2020).

The PCR for wine is applicable to both still and sparkling wine, either marketed packed or unpacked, and defines key aspects that shall be applied when applying the LCA methodology to this sector, including the following:

- Declared unit: 0.75 L of beverage, including its packaging (primary and secondary). If the product is packaged in different volumes and types, results related to different volumes may also be reported in the EPD.
- System boundary: all cradle-to-grave processes shall be considered, from the production of grapes in the vineyard to the end-of-life processes of packaging waste.
- Allocation rules: as part of grape production, pomace and stems are co-produced together with must, and as part of wine making, lees are also obtained. In both processes, a mass allocation shall be used to allocate production inputs to the different outputs. If appropriate, an economic allocation may also be accepted if representative selling prices of the products (average market price over three years) are considered in a sensitivity analysis.

Table 18.1 Indicators to be used in EPDs of the International EPD System to describe potential environmental impacts of wine.

Parameter	Unit	Upstream (cradle-to-gate)	Core (gate-to-gate)	Downstream (gate-to-grave)	Total
Global warming potential (GWP) fossil	kg CO_2 eq				
GWP biogenic	kg CO_2 eq				
GWP land use and land transformation	kg CO_2 eq				
GWP total	kg CO_2 eq				
Acidification potential (AP)	kg SO_2 eq				
Eutrophication potential (EP)	kg PO_4^{3-} eq				
Formation potential of tropospheric ozone	kg NMVOC eq				
Abiotic depletion potential—elements	kg Sb eq				
Abiotic potential—fossil fuels	MJ, net calorific value				
Water scarcity potential	m^3 eq				

- If site-specific data are not available, direct emissions by the use of fertilizers should refer to default parameters for estimating N_2O (to air), NH_3 (to air), and NO_3 (to water) per kg N synthetic fertilizer and manure applied, and P (to water) per Kg P fertilizer applied.
- Data quality requirements: generally, specific data shall always be used if available.
- Impact category and impact assessment methodology: default impact categories as described in the General Program Instructions of IES (see Table 18.1).

18.3.3 The Product Environmental Footprint process and its implications to the wine sector

With the objective of overcoming "the fragmentation of the internal market as regards different available methods for measuring environmental performance" (EC, 2013), the European Commission adopted in 2013 the Recommendation on the Use of common methods to measure and communicate the life cycle environmental performance of products and organizations. These common methods tackled calculating the environmental footprint of products (PEF method) and organizations (OEF method). Both PEF and OEF methods were developed based on existing, tested and widely used methods and standards including the LCA as defined in ISO 14040-44.

In the following five years, a pilot phase took place during which volunteering organizations tested the method and developed PEFCRs, i.e., Product Environmental Footprint Category Rules, for up to 19 sectors. Since the end of the pilot phase, the EC has established a transition phase to the possible adoption of policies implementing the PEF method. Wine was one of the sectors involved in the PEF pilot testing and PEFCRs for still and sparkling wine were published in April 2018. Part of the rules has been already implemented by the PCRs published by the IES for still and sparkling wine.

While PEF is a LCA-based method to quantify the environmental impacts of products (Zampori & Pant, 2019), PEFCR offers specific guidelines to carry out the LCA study and, therefore, PEFCRs are the PEF equivalent of PCRs (Minkov et al., 2020).

One remarkable difference between EPDs and PEF is the default set of LCIA categories and methods to be used to characterize the environmental life cycle performance of the product declared. While the core PCRs for construction products, i.e., European standard EN 15804:201+A2:2019, has recently aligned to the set of impact categories defined in the PEF method, other sectors including food and beverages, apply a different set of LCIA impact categories and methods. Despite products from different product categories and therefore with different FUs shall not be compared, their environmental results could be combined to assess the impact of complex systems like, for instance, the impact of a city or an average citizen during a year.

Available PEFCRs define environmental benchmarks for the product category, i.e., the average environmental performance of the representative product sold in the EU market (EC, 2013). A representative product may or may not be a real product that one can buy on the EU market, as it can be a virtual product built from the average EU sales-weighted characteristics of all technologies around (Zampori & Pant, 2019). Benchmarks defined in PEFCRs may be used eventually when communicating the environmental performance of a product and comparing the results of a PEF study of one or more products against the benchmark. PEFCRs for wine define the benchmarks for two representative products, i.e., still and sparkling wine. These benchmarks are shown in Table 18.2. Total life cycle results excluding the use phase are provided, whereas results of the use phase are also communicated highlighting the relative importance of this stage directly managed by consumers. While for still wine a representative packaging system was defined considering the different market shares of glass, plastic or bag-in-box containers, in the case of sparkling wine, only glass bottles were considered. Being packaging one of the most relevant aspects of the life cycle of wine, the type of packaging used will determine the distance of a specific product to the PEF benchmark. In fact, the PEFCRs for wine (2018) include an annex where additional benchmarks are presented for wine bottled exclusively in glass as required by regulations of some producing regions. It could be argued if wine bottled in different packaging materials has a comparable quality and therefore if it makes sense to compare an existing wine against the impact of a virtual wine. But on the other hand, the comparison between products may foster environmentally driven innovation.

The virtual product does also takes into account the amount of organic or nonorganic grape used for producing both still and sparkling wine. As this ratio as well as the packaging materials used may change from one year to another, benchmarks should be updated periodically to keep is usefulness.

In contrast to EPDs, PEF requires that, in addition to characterized results, the normalized and weighted results of the representative product are also presented in the PEF study to identify the most relevant impact categories, life cycle stages, processes and elementary flows of a certain product category. Based on normalized and weighted results, PEFCRs for wine identified grape production, wine making, and packaging as the most relevant life cycle stages for both still and sparkling wine (see Section 18.2.1). After the Environmental Footprint pilot phase, a transition phase has been established during which new PEFCRs are being developed while possible policies implementing the PEF and OEF methods are adopted. In that sense, as part of the new Circular Economy Action Plan

Table 18.2 Benchmark values for the life cycle of 0.75 L of still and sparkling wine according to the PEFCR for wine (PEFCRs for still and sparkling PEFCR for wine, 2018).

Impact category	Unit	Still wine Life cycle exc. use stage	Still wine Use stage	Sparkling wine Life cycle exc. use stage	Sparkling wine Use stage
Climate change	kg CO_2 eq	1.5E-00	8.4E-02	2.1E+00	8.1E-02
Climate change—biogenic	kg CO_2 eq	1.1E-01	1.6E-03	9.0E-02	1.5E-03
Climate change-land use and land transformation	kg CO_2 eq	1.4E-03	8.7E-05	1.5E-03	8.8E-05
Ozone depletion	kg CFC-11 eq	3.6E-08	1.2E-10	3.1E-08	1.4E-10
Particulate matter	Disease incidence	8.6E-08	2.5E-09	9.7E-08	2.4E-09
Ionizing radiation, human health	kBq U235 eq	1.3E-01	3.1E-02	1.8E-01	3.12E-02
Photochemical ozone formation, human health	kg NMVOC eq	4.4E-03	1.3E-04	3.6E-03	1.3E-04
Acidification	molt H+ eq	8.5E-03	2.5E-04	8.7E-03	2.5E-04
Eutrophication, terrestrial	mol N eq	2.5E-02	4.6E-04	2.2E-02	5.1E-04
Eutrophication, freshwater	kg P eq	1.6E-04	2.6E-06	1.4E-04	2.7E-06
Eutrophication, marine	kg N eq	4.3E-03	7.3E-05	3.7E-03	7.3E-05
Land use	Dimensionless (pt)	1.6E+02	2.8E-01	1.5E+02	6.6E-01
Water use	m^3 world eq.	8.8E-01	7.0E-02	2.9E+00	7.1E-02
Resource use, minerals and metals	kg Sb eq.	1.5E-05	9.9E-09	1.35E-05	4.0E-08
Resource use, fossils	MJ	1.7E+01	1.33E+00	2.6E+01	1.3E+00

(EC, 2020a), the European Commission proposes that companies substantiate their environmental claims using Environmental Footprint methods. PEF has also a close link to the EU consumer policy revision (New Consumer Agenda) (EU, 2020b) as well as the strategy to promote the global transition to a fair, healthy, and environmentally friendly food system (EU, 2020c), where the EU will promote schemes on environmental footprint calculation to support information to consumers.

At the moment of writing, the communication of quantitative LCA results of wine products in the form of EPD and/or PEF reports is still scarce. However, ongoing political strategies and market initiatives (such as the recently published PCRs in the IES system) combined with the increasing consumer environmental awareness may trigger more companies to declare the environmental information of the wine's life cycle in the upcoming years.

18.4 Sustainability challenges in the wine sector from a life cycle perspective: circularity and methodological developments

The ambitious objectives of the Paris Agreement and of the United Nations Sustainable Development Goals (UN, 2015) require a new way of thinking, a way of taking the whole environmental, social, and economic implications into accounts and Life Cycle Thinking is that way. But even when LCT has permeated the wine sector, the challenge is now to promote a single reference framework to be applied that fosters and guarantees scientific consistency, clarity, and transparency for both businesses and consumers. As it has been described, there is a broad spectrum of programs, methodologies, and indicators that, although do share the life cycle perspective, have been applied with different purposes over not always equivalent systems and arriving in occasions to contradictory results. Methodological challenges, such as the proper definition of the FU, the use of a consequential-LCA perspective,[4] or the assessment of biogenic carbon and temporal dynamics for carbon emission accounting, are currently neglected or only marginally treated. To overcome those, it is necessary not only from a methodological point of view but also for a proper interpretation of results and the more and more important consistent reporting of information related to: (1) the analysis of life cycle hotspots; (2) the determination of weak elements of the methodological approach; and (3) the uncertainty and variability of elementary flows and impact scores.

In the need of redesigning our economy under the umbrella of sustainability, circular economy has emerged as an innovative approach to complement LCT. As described, CE provides a strategic framework for closed-loop material flows and a gateway to LCT, while LCA complements CE by assessing environmental impacts and supporting informed decision-making. But even when CE has permeated the wine sector, a recent systematic literature review (41 selected studies) analyzed how the wine sector has incorporated the premises of CE in its activities and no clear definition regarding CE was found (Calicchio & Maia, 2019).

From the knowledge gained with this chapter, the focus on short-term problems instead of a clear picture of how the wine sector wants to be and wants to be seen in the future together with the lack of environmental awareness by the firms and limited use of EPDs are the main obstacles to overcome those challenges.

References

Ahmad, B., Yadav, V., Yadav, A., Rahman, M. U., Yuan, W. Z., Li, Z., & Wang, X. (2020). Integrated biorefinery approach to valorise winery waste: A review from waste to energy perspectives. *Science of the Total Environment, 719*, 137315. https://doi.org/10.1016/j.scitotenv.2020.137315

Amienyo, D. (2012). *Life cycle sustainability assessment in the UK beverage sector*. PhD thesis, School of Chemical Engineering and Analytical science, University of Manchester, Manchester.

Arcese, G., Lucchetti, M. C., & Massa, I. (2017). Modelling social life cycle assessment framework for the Italian wine sector. *Journal of Cleaner Production, 140*(2), 1027−1036. https://doi.org/10.1016/j.jclepro.2016.06.137

---

[4]Larrea-Gallegos et al. (2019) applied consequential-LCA to grape cultivation although the motivation of the study is pisco instead of wine production.

Arzoumanidis, I., Raggi, A., Petti, L., & Zamagni, A. (2013). The implementation of simplified LCA in agri-food SMEs. In R. Salomone, M. T. Clasadonte, M. Proto, & A. Raggi (Eds.), *Product-Oriented Environmental Management System (POEMS)—improving sustainability and competitiveness in the agri-food chain with innovative environmental management tools* (pp. 151–173). Dordrecht: Springer.

Borsato, E., Giubilato, E., Zabeo, A., Lamastra, L., Criscione, P., Tarolli, P., Marinello, F., & Pizzol, L. (2019). Comparison of water-focused life cycle assessment and water footprint assessment: The case of an Italian wine. *The Science of the Total Environment, 666*, 1220–1231. https://doi.org/10.1016/j.scitotenv.2019.02.331

Calicchio, P., & Maia, J. (2019). How has the wine sector incorporated the premises of circular economy? *Journal of Environmental Science and Engineering B, 8*, 108–117. https://doi.org/10.17265/2162-5263/2019.03.004

Cholette, S., & Venkat, K. (2009). The energy and carbon intensity of wine distribution: A study of logistical options for delivering wine to consumers. *Journal of Cleaner Production, 17*, 1401–1413. https://doi.org/10.1016/j.jclepro.2009.05.011

Christopher, M. (2011). *Logistics and supply chain management*. Harlow: Pearson.

Collado-Ruiz, D., & Ostad-Ahmad-Ghorabi, H. (2010). Fuon theory: Standardizing functional units for product design. *Resources, Conservation and Recycling, 54*, 683–691. https://doi.org/10.1016/j.resconrec.2009.11.009

Cooper, J. S. (2003). Specifying functional units and reference flows for comparable alternatives. *International Journal of Life Cycle Assessment, 8*, 337. https://doi.org/10.1007/BF02978507

Cortés, A., Moreira, M. T., & Feijoo, G. (2019). Integrated evaluation of wine lees valorization to produce value-added products. *Waste Management, 95*, 70–77. https://doi.org/10.1016/j.wasman.2019.05.056

Cortés, A., Oliveira, L. F., Ferrari, V., Taffarel, S. R., Feijoo, G., & Moreira, M. T. (2020). Environmental assessment of viticulture waste valorisation through composting as a biofertilisation strategy for cereal and fruit crops. *Environmental Pollution, 264*, 114794. https://doi.org/10.1016/j.envpol.2020.114794

David, G., Croxatto, G., Sohn, J., Ekman, A., Hélias, A., Gontard, N., & Angellier-Coussy, H. (2020). Using life cycle assessment to quantify the environmental benefit of upcycling vine shoots as fillers in biocomposite packaging materials. *International Journal of Life Cycle Assessment*. https://doi.org/10.1007/s11367-020-01824-7

Del Borghi, A., Moreschi, L., & Gallo, M. (2019). Communication through ecolabels: How discrepancies between the EU PEF and EPD schemes could affect outcome consistency. *International Journal of Life Cycle Assessment, 25*, 905–920. https://doi.org/10.1007/s11367-019-01609-7

Delmas, M. A., & Grant, L. E. (2014). Eco-labeling strategies and price-premium: The wine industry puzzle. *Business & Society, 53*(I), 6–44. https://doi.org/10.1177/0007650310362254

EN 15804:201+A2. (2019). *Sustainability of construction works – environmental product declarations – core rules for the product category of construction products*.

European Commission. (2013). Commission Recommendation 2013/179/EU on the use of common methods to measure and communicate the life cycle environmental performance of products and organisations. *Official Journal of the EU*, 4 May 2013.

European Commission. (2020a). *A new circular economy action plan. For a cleaner and more competitive Europe*. Brussels, 11 March 2020. COM(2020) 98 final.

European Commission. (2020b). *New Consumer Agenda - Strengthening consumer resilience for sustainable recover*. Brussels, 13 November 2020. COM(2020) 696 final.

European Commission. (2020c). *Farm to Fork Strategy - For a fair, healthy and environmentally-friendly food system*.

Ferrara, C., & De Feo, G. (2020). Comparative life cycle assessment of alternative systems for wine packaging in Italy. *Journal of Cleaner Production, 259*, 120888. https://doi.org/10.1016/j.jclepro.2020.120888

Flor, F. J., García-Alcaraz, J. L., Martínez Cámara, E., Jiménez-Macías, E., & Blanco-Fernández, J. (2021). Environmental impact of wine fermentation in steel and concrete tanks. *Journal of Cleaner Production, 278*, 123602. https://doi.org/10.1016/j.jclepro.2020.123602

Flores, S. S. (2018). What is sustainability in the wine world? A cross-country analysis of wine sustainability frameworks. *Journal of Cleaner Production, 172*, 2301−2312. https://doi.org/10.1016/j.jclepro.2017.11.181

Fusi, A., Guidetti, R., & Benedetto, G. (2014, February 15). Delving into the environmental aspect of a Sardinian white wine: From partial to total life cycle assessment. *The Science of the Total Environment, 472*, 989−1000. https://doi.org/10.1016/j.scitotenv.2013.11.148

García-Alcaraz, J. L., Flor, F. J., Martínez Cámara, E., Pérez de la Parte, M. M., Jiménez-Macías, E., & Blanco-Fernández, J. (2020). Economic-environmental impact analysis of alternative systems for red wine ageing in re-used barrels. *Journal of Cleaner Production, 244*, 118783. https://doi.org/10.1016/j.jclepro.2019.118783

García-Alcaraz, J. L., Flor, F. J., Martínez Cámara, E., Sáenz-Diez, J. C., Jiménez-Macías, E., & Blanco-Fernández, J. (2020). Comparative environmental impact analysis of techniques for cleaning wood wine barrels. *Innovative Food Science & Emerging Technologies, 60*, 102301. https://doi.org/10.1016/j.ifset.2020.102301

Harris, I., Rodrigues, V., Pettit, S., Beresford, A., & Liashko, R. (2018). The impact of alternative routeing and packaging scenarios on carbon and sulphate emissions in international wine distribution. *Transportation Research Part D: Transport and Environment, 58*, 261−279. https://doi.org/10.1016/j.trd.2016.08.036

International EPD System (IES). (2020). *Product category rules (PCR) for wine. Product category rules (PCR)* (p. 06). version 1.0 for wine. 10 November 2020.

ISO 14021. (2016). *Environmental labels and declarations - self-declared environmental claims (Type II environmental labelling)*. Geneva: International Organisation for Standardisation.

ISO 14024. (2018). *Environmental labels and declarations − Type I environmental labelling − principles and procedures*. Geneva: International Organisation for Standardisation.

ISO 14025. (2006). *Environmental labels and declarations − Type III environmental declarations − principles and procedures*. Geneva: International Organisation for Standardisation.

ISO 14026. (2017). *Environmental labels and declarations − principles, requirements and guidelines for communication of footprint information*. Geneva: International Organisation for Standardisation.

Jourdaine, M., Loubet, P., Trebucq, S., & Sonnemann, G. (2020). A detailed quantitative comparison of the life cycle assessment of bottled wines using an original harmonization procedure. *Journal of Cleaner Production, 250*, 119472. https://doi.org/10.1016/j.jclepro.2019.119472

Larrea-Gallegos, G., Vázquez-Rowe, I., Wiener, H., & Kahhat, R. (2019). Applying the technology choice model in consequential life cycle assessment: A case study in the Peruvian agricultural sector. *Journal of Industrial Ecology, 23*(3), 601−614. https://doi.org/10.1111/jiec.12812

Leire, C., & Thidell, A. (2005). Product-related environmental information to guide consumer purchasers − A review and analysis of research on perceptions, understanding and use among Nordic consumers. *Journal of Cleaner Production, 13*, 1061−1070. https://doi.org/10.1016/j.jclepro.2004.12.004

Marenghi, M. (2005). *Manuale di viticoltura: Impianto, gestione e difesa del vigneto*. Il Sole 24 Ore Edagricole.

Margallo, M., Aldaco, R., Barceló, A., Diban, N., Ortiz, I., & Irabien, A. (2015). Life cycle assessment of technologies for partial dealcoholisation of wines. *Sustainable Production and Consumption, 2*, 29−39. https://doi.org/10.1016/j.spc.2015.07.007

Martucci, O., Arcese, G., Montauti, Ch, & Acampora, A. (2017). Social aspects in the wine sector: Comparison between social life cycle assessment and VIVA sustainable wine Project indicators. *Resources, 8*(2), 69. https://doi.org/10.3390/resources8020069

Mattila, T., Leskinen, P., Soimakallio, S., & Sironen, S. (2012). Uncertainty in environmentally conscious decision making: Beer or wine? *International Journal of Life Cycle Assessment, 17*, 696−705. https://doi.org/10.1007/s11367-012-0413-z

Meneses, M., Torres, C. M., & Castells, F. (2016). Sensitivity analysis in a life cycle assessment of an aged red wine production from Catalonia, Spain. *The Science of the Total Environment, 562*, 571−579. https://doi.org/10.1016/j.scitotenv.2016.04.083

Merli, R., Preziosi, M., & Acampora, A. (2018). Sustainability experiences in the wine sector: Towards the development of an international indicators system. *Journal of Cleaner Production, 172*, 3791−3805. https://doi.org/10.1016/j.jclepro.2017.06.129

Minkov, N., Lehmann, A., Winter, L., & Finkbeiner, M. (2020). Characterization of environmental labels beyond the criteria of ISO 14020 series. *International Journal of Life Cycle Assessment, 25*, 840−855. https://doi.org/10.1007/s11367-019-01596-9

Murphy, P. R., & Poist, R. F. (2003). Green perspectives and practices: A "comparative logistics" study. *Supply Chain Management: International Journal, 8*(2), 122−131. https://doi.org/10.1108/13598540310468724.

Notarnicola, B., Tassielli, G., & Nicoletti, G. M. (2003). LCA of wine production. In B. Mattsonn, & U. Sonesson (Eds.), *Environmentally-friendly food production*. Cambridge: Woodhead-Publishing.

Notarnicola, B., Tassielli, G., & Settanni, E. (2010). Including more technology in the production of a quality wine: The importance of functional unit. In *Proceedings of VII international conference on life cycle assessment in the agri-food sector* (Vol. II, pp. 235−240). Bari, 22−24th September 2010.

OIV. (2004). *Development of sustainable vitiviniculture*. Resolution CST 1/2004.

PEFCR for wine. (2018). *Product environmental footprint category rules (PEFCR) for still and sparkling wine*. Publication date: 26 April 2018.

Petti, L., Arzoumanidis, I., Benedetto, G., Bosco, S., Cellura, M., De Camillis, C., Fantin, V., Masotti, P., Pattara, C., Raggi, A., Rugani, B., Tassielli, G., & Vale, M. (2015). Life cycle assessment in the wine sector. In B. Notarnicola, R. Salomone, L. Petti, P. A. Renzulli, R. Roma, & A. K. Cerutti (Eds.), *Life cycle assessment in the agri-food sector: Case studies, methodological issues and best practices* (pp. 123−185). Cham: Springer. ISBN 978-3-319-11939-7.

Point, E., Tyedmers, P., & Naugler, C. (2012). Life cycle environmental impacts of wine production and consumption in Nova Scotia, Canada. *Journal of Cleaner Production, 27*, 11−20. https://doi.org/10.1016/j.jclepro.2011.12.035

Pomarici, E., & Vecchio, R. (2014). Millennial generation attitudes to sustainable wine: An exploratory study on Italian consumers. *Journal of Cleaner Production, 66*, 537−545. https://doi.org/10.1016/j.jclepro.2013.10.058

Ponstein, H. J., Ghinoi, S., & Steiner, B. (2019). How to increase sustainability in the Finnish wine supply chain? Insights from a country of origin-based greenhouse gas emissions analysis. *Journal of Cleaner Production, 226*, 768e780. https://doi.org/10.1016/j.jclepro.2019.04.088

Quinteiro, P., Dias, A. C., Pina, L., Neto, B., Ridoutt, B. G., & Arroja, L. (2014). Addressing the freshwater use of a Portuguese wine ('vinho verde') using different LCA methods. *Journal of Cleaner Production, 68*, 46−55. https://doi.org/10.1016/j.jclepro.2014.01.017

Rugani, B., Vázquez-Rowe, I., Benedetto, G., & Benetto, E. (2013). A comprehensive review of carbon footprint analysis as an extended environmental indicator in the wine sector. *Journal of Cleaner Production, 54*, 61−77. https://doi.org/10.1016/j.jclepro.2013.04.036

Scrucca, F., Bonamente, E., & Rinaldi, S. (2018). Carbon footprint in the wine industry. In *Book: Environmental carbon footprints* (pp. 161−196). https://doi.org/10.1016/B978-0-12-812849-7.00007-6

Sinisterra-Solís, N. K., Sanjuán, N., Estruch, V., & Clemente, G. (2020, May 15). Assessing the environmental impact of Spanish vineyards in Utiel-Requena PDO: The influence of farm management and on-field emission modelling. *Journal of Environmental Management, 262*, 110325. https://doi.org/10.1016/j.jenvman.2020.110325

Sun, Y. Y., & Drakerman, D. (2020). Measuring the carbon footprint of wine tourism and cellar door sales. *Journal of Cleaner Production, 266*, 121937. https://doi.org/10.1016/j.jclepro.2020.121937

UN General Assembly. (October, 2015). *Transforming our world: The 2030 Agenda for sustainable development* (Vol. 21). A/RES/70/1.

Vazquez-Rowe, I., Rugani, B., & Benetto, E. (2013). Tapping carbon footprint variations in the European wine sector. *Journal of Cleaner Production, 43*, 146–155. https://doi.org/10.1016/j.jclepro.2012.12.036

Vazquez-Rowe, I., Villanueva-Rey, P., Moreira, M. T., & Feijoo, G. (2012). Environmental analysis of Ribeiro wine from a timeline perspective: Harvest year matters when reporting environmental impacts. *Journal of Environmental Management, 98*, 73–83. https://doi.org/10.1016/j.jenvman.2011.12.009

Villanueva-Rey, P., Vázquez-Rowe, I., Moreira, M. T., & Feijoo, G. (2014). Comparative life cycle assessment in the wine sector: Biodynamic vs. conventional viticulture activities in NW Spain. *Journal of Cleaner Production, 65*, 330–341. https://doi.org/10.1016/j.jclepro.2013.08.026

Villanueva-Rey, P., Vázquez-Rowe, I., Otero, M., Moreira, M. T., & Feijoo, G. (2015). Accounting for time-dependent changes in GHG emissions in the Ribeiro appellation (NW Spain): Are land use changes an important driver? *Environmental Science & Policy, 51*, 215–227. https://doi.org/10.1016/j.envsci.2015.04.001

Weidema, B. P., de Saxcé, M., & Muñoz, I. (2016). *Environmental impacts of alcoholic beverages as distributed by the Nordic Alcohol Monopolies 2014*. Aalborg: 2.-0 LCA consultants.

Zampori, L., & Pant, R. (2019). *Suggestions for updating the product environmental footprint (PEF) method*. EUR 29682 EN. Luxembourg: Publications Office of the European Union. ISBN 978-92-76-00654-1.

CHAPTER 19

Wine packaging and related sustainability issues

Fátima Poças[1,2], José António Couto[1] and Timothy Alun Hogg[1]

[1]*Universidade Católica Portuguesa, CBQF - Centro de Biotecnologia e Química Fina, Laboratório Associado, Escola Superior de Biotecnologia, Rua Diogo Botelho, Porto, Portugal;* [2]*Universidade Católica Portuguesa, CINATE, Escola Superior de Biotecnologia, Porto, Portugal*

19.1 Introduction

This chapter deals with wines as a consumer product, i.e., wines in the form that is presented to the end consumer. These wines are obtained by the consumer either through a retail channel for consumption away from the site of purchase—known as the "off-trade" in wine industry terms, or through food service/HORECA channels—the "on-trade" (Ritchie, 2009). It is recognized that prior to being put into this consumer facing package, a wine might have passed through a number of intermediate storage and transport containers and that these might have a profound effect on the quality and sustainability attributes of a wine. It is likely that most of these containers have been discussed in more detail in other parts of this book and we only present them here when they have a direct impact on the consumer package.

Wine encompasses a broad range of products similar only in that they are made exclusively from grapes and have an alcohol content above 8.5% (v/v) (OIV, 2019), indeed this range is increased even further when we consider that many other products are themselves derived from wine and follow common commercial channels. It is important to state this in a discussion of wine packaging, as the considerations and demands that are placed on a package will be determined by a more restrictive description of the specific wine rather than its nature as a wine per se. As an example, some wines might be put into the final package on the same property as the actual grapes for the wine were grown and will be exported around the world in this form. At the other extreme, some wines are traded internationally in bulk and packaged long distances from where they were originally produced, but consumed close to where they are packaged. Some packaging systems are specifically designed for food service channels, such as large format bag-in-box (BIB), barrels and kegs, while most others are common to retail and foodservice uses. All package types currently employed have a number of attributes that makes them suitable for use with wine. The specific wine type and intended use results in some attributes being given more weight depending on the specific circumstances. These considerations are obviously relevant to the global environmental impact of the product.

For foods generally, it is recognized that the environmental impact of packaging is both highly relevant to the overall environmental burden and also highly variable depending on the food category in question. The wine sector was found to contribute 0.3% to the total of worldwide global annual carbon footprint (CF) of human activities. On average, the CF for a generic bottle of wine is 2.2 ± 1.3 kg CO_2eq from which packaging processes represent 22% (Rugani et al., 2013). In the case of wine, several life cycle assessments (LCAs) studies have concluded that, of all the winemaking chain steps, packaging is that which most contributes to the product's environmental impact, particularly thanks to the large CO_2 emissions required for the glass bottle production (Martins et al., 2018; Rugani et al., 2013; Neto et al., 2013; Point et al., 2012; Sogari et al., 2016). Packaging and bottling operations represent 30% of the energy consumed in EU winemaking sector (Landi et al., 2019).

The importance of packaging in this respect has led to the introduction of the term Packaging Relative Environmental Impact (PREI), which expresses the role that packaging plays in the overall environmental impact of a food product or category in percentage terms (Licciardello, 2017). Importantly a product's PREI should also contemplate the positive sustainability contributions that a packaging system might make, such as in reducing food waste. This measure provides guidance as to where efforts are most needed to reduce the packaging component of the total environmental impact of a particular category of packaged food. PREI depends not only on the packaging system but also on the product itself and a more effective and sustainable packaging is needed for products with a high environmental impact (Boz et al., 2020). The range of products that are sold as wine, the volumes that wines are commercialized under, and the range of materials and formats used will obviously mean that the PREI values for this category will vary greatly. In a review article, Licciardello (2017) gives values for PREI of wines ranging from 34% to 82% indicating that, even in the most modest cases, packaging is a major source of the overall environmental impact. Wine consumers and the distribution chains that serve them are increasingly conscious of the relative role of packaging on environmental impact and pressure to reduce this is continuously increasing.

Environmental impact in the wine sector has been widely studied, but single-issue approaches have been commonly used. A more comprehensive analysis is desirable, since a single indicator does not properly represent the pressure on the environment (Bonamente et al., 2016). This is particularly relevant given the several functions attributed to the wine packaging system: containment and protection from the immediate environment, prevention against oxidation, losses of volatiles and carbon dioxide and microbial contamination; efficiency in distribution and storage, control of volume for commercial, tax, and duty purposes; traceability and authenticity assurance, as well as being a support for the communication of technical and other information to chain stakeholders and to the consumer. Most importantly, wine is not a perishable product, but its organoleptic properties are affected by exposure to oxygen and light. In the case of finished, packaged, still wine, quality is generally diminished by excessive oxygen exposure which should be controlled by the packaging system, but slow and continuous oxygenation may be beneficial for wine aging (Poças et al., 2010). Wine aging, as part of the process to obtain the desired organoleptic properties, and the expected shelf life after packaging, are determinant for the packaging selection. Another aspect that is worthy of consideration is consumer perception and preference. Due to poor or incomplete knowledge, consumers might have, and act on, preferences based on claims for sustainable attributes of packaging that might not actually have any positive effects (Steenis et al., 2017).

Although wine is classically packaged for the consumer in glass bottles closed with a cork stopper, and this is still by far the form under which most wine is sold (OIV, 2020), there is in fact a broad range of packaging systems available and in use. This chapter presents the various consumer packaging systems in use for both the on- and off-trade channels, and describes the major functional attributes of each in addition to an appreciation of their sustainability considerations.

19.2 Packaging systems used for wine

Packaging systems and operations are key factors in the efficient transport of wine and for the environmental impact of wine distribution. Glass bottles sealed with a cork closure are the established, traditional wine packaging system and they are still the most used, both in total volume and in number of units sold. In 2019 the OIV classified 53% of exports as being bottles (OIV, 2020), but the proportion of wine actually sold in this format is likely to be considerably higher as much wine exported in bulk will be packaged close to the consuming market and much of this will be in glass bottles. The OIV figures do not stipulate the type of closure used, but it is expected that the general statement above still holds, indeed sales of natural cork and cork-derived closures have been increasing in recent years (Mullen, 2018). The competition between the various packaging options for wine is certainly great, with a growing range of alternative package sizes and formats becoming available, all possessing attributes that are valued in particular settings, offering greater convenience, economy or, increasingly, environmental credentials (Mordor Intelligence, 2020).

19.2.1 Glass bottle

The glass bottle is typically associated by consumers to be a high quality product and to be a sustainable packaging. Bottles with different shapes such as Bordeaux, Burgundy, Rhône, Porto, and Champagne and different colors and levels of opacity can be found. By far the most common capacity is 750 mL (standard), followed by 375 mL (half bottle) and the 1.5 L (Magnum bottle) although many other capacities are also produced.

Glass is an inert material which does not interact with the wine. The glass surface is hard, smooth, nonporous, and resistant to both high temperature and chemicals and therefore washable, conferring on it a particular facility for reuse. However, a number of issues can be pointed out as potential reasons for reuse of glass bottles not being as high as might be expected. High level of exports and long cellaring are obvious limitations, as well as the high fragmentation of the production and commercialization system. The common argument against reuse is that it costs more to collect and wash the bottles than to produce new ones.

For wines sold for early consumption and for regional distribution, reusable bottles can be considered and growing efforts to reuse glass bottles are reported in France (Provence) and in Australia. An LCA comparing reusing with recycling a standard 750 mL bottle weighing 450 g, within a 50–100 km location in Italy, showed a significant decrease in all environmental impact categories, with the avoided use of virgin glass offsetting the additional resources used for cleaning the reusable bottles (Landi et al., 2019). Removal of labels is a challenging process as nowadays self-adhesive labels are predominant. Bottles designed for reuse would require labels that can be easily washed off, or alternatively those that could resist washing while retaining their quality and this avoiding the need for replacement between uses. There are obvious limitations to this approach but nevertheless it is an avenue worth exploring.

The use of lighter bottles has been suggested to mitigate the environmental impact of glass bottles (Martins et al., 2018). Glass bottles are transported in their final form to the packaging site whereas plastic bottles are most often at least partially formed on or close to this site. Given that over 40% of glass bottles travel more than 300 km to reach the packaging destination (FEVE, 2020), weight reduction is an important measure in the reduction of the environmental impact of both the bottle and the overall product. Lighter bottles are commonly regarded as being less resistant to mechanical stresses and more prone to breakage then heavier ones. However, it is known that other factors also determine mechanical resistance such as shape, distribution of bottle thickness, and potential flaws-especially those located at the external surface. Current production technologies allow for lighter bottles to be manufactured without significantly decreasing their resistance. However restrictions related to specific design requirements for bottles, such as the punt (a marketing requirement) and the dimension profiles of the finish and neck imposed by the standardization of the closure system, limit the application of some technologies or even achieving the light weighing at all in some cases (Soares, 2019).

Glass has been recycled by the wine bottle industry since early 1970s as the melting of cullet (recovered glass) allows for energy savings. As a general rule, energy savings of 2.5%–3% can be achieved for every 10% of cullet added to the input material (Frassine et al., 2016). Although theoretically 100% incorporation of cullet can be achieved, real recycling rates are lower and dependent on the glass color, with transparent uncolored (flint) glass showing much lower recycling rates than brown and green.

19.2.2 Bag-in-box

BIB represented a major change in the packaging of wine when introduced in the 1970s. The concept consists of a flexible, collapsible, four-sided, sealed bag made from several nonlaminated layers, a closure and spout or a tap through which the contents are filled and dispensed, and a rigid outer corrugated board box. Typically bags of two layers are used. The structure of the layers varies, but for wine applications, a common structure consists of an inner layer made of linear, low density polyethylene and an external coextruded layer relying on the barrier properties of ethylene vinyl alcohol copolymer or metalized polyethylene terephthalate (PET). Sometimes polyamide (PA) is included in this layer for increased mechanical resistance. The bag is responsible for the barrier required for preservation and the box provides the mechanical resistance for handling, storage, and transportation. It should be considered that the shelf life of wine packaged in BIB is much shorter than when glass bottle or metal can are used.

The BIBs for wine range from 1 to 5 L capacity for consumer packaging or higher for foodservices. Due to its construction, structure and range of capacities, BIB is a highly efficient system in terms of the ratio of volume delivered to material weight. The boxes can be stacked directly on pallets. This corresponds to lower impacts regarding transportation, both for empty packages from the supplier to the winery and after filling.

All the graphical, legal and consumer relevant information is applied on the box. So the same bag type can be used to fill different wines with similar protection requirements, which is an advantage in managing stocks of unused material.

Recycling of the packaging systems components may integrate into existing recycling waste streams which are implemented for paper and board, but not for the bag because it is considered a plastic mixture, therefore ending up being incinerated or landfilled.

In the case of catering and wine by the glass situations, other configurations exist that employ a flexible pouch to contain the wine. One such configuration consists of a sealable, stainless steel barrel into which the prefilled pouch, or bladder, is placed with the exit pipe protruding out from the top. Compressed air is applied to the space between the inside of the barrel and the pouch expelling the wine on demand. This is essentially a bag in drum configuration in which compressed air rather than gravity is the driving force, but in commercial situations in which this is appropriate, the elimination of the box and tap can be seen as positive environmental contributions.

Flexitank is a similar concept, for much larger dimensions, where the bag, a single-use bladder, is housed in a generic shipping container and used to ship large amounts of wine for packaging closer to market. Flexitanks are filled and emptied by active pumping.

19.2.3 Laminated multimaterial boxes

Since the early 1980s, increasing quantities of popular still wines have been packaged in laminated boxes. This system is produced in paperboard laminated with polyethylene and aluminum foil and is delivered as rolls of the printed packages that are formed-fill-sealed at the packaging site. The packages are formed into brick- or prism-shaped packages with capacities ranging from 200 mL single dose to 1 or 2 L. These systems are highly efficient regarding space usage during transportation and storage of the material and during transportation of both filled and used packages.

Like the BIB, the laminated box system protects the wine from light but the oxygen ingress rate is much higher than in a glass bottle with a cork or aluminum cap, therefore yielding shorter shelf-lives.

Recycling is being promoted by major players but it is recognized that redesign in collection and recycling systems, are still required for an efficient recycling chain. The recovered material is turned into other products, such as roofing tiles, crates, and carton boxes, after blending with other polymers.

19.2.4 Metal (aluminum) can

Wines in aluminum beverage cans are one of the fastest growing sectors in the wine industry, especially in North and South American markets. In the USA, the value of wines packaged in aluminum cans grew ca 30% from 2012 to 2018 (Weed 2019). Main drivers of this growth are (Williams et al., 2018):

- Convenience for consumer: in carrying; due to the light weight, opening; as no corkscrews are required and individual; (187 mL, 250 mL) or double sized portions (375 mL)
- New opportunities of consumption, regarding location and types of events, associated with convenience and portability
- Visual image/branding as cans lend themselves to distinctive designs which promotes consumer appeal.

Like glass, aluminum is infinitely recyclable, and there are well-established domestic recycling streams for it. Reuse is not possible for the common double seamed can.

When comparing to glass bottles, a can's lighter weight during shipping and handling and reduced breakage can account for lower impact, although the overall logistic operations and transport packaging should be considered as these are different. A study commissioned by a can producer indicated that transporting the same total volume of packaged wine in slim 250 mL cans has half the CO_2 emissions of wine transported in glass bottle. Aluminum cans do not break like glass but they dent more or less easily depending of the internal pressure of the wine. For carbonated wines, the internal pressure is sufficient to provide a very good mechanical resistance to the filled, closed can. For still wines, however, the application in the head-space of liquid nitrogen may be required to increase the internal pressure and accordingly the mechanical resistance of the can. Otherwise, the handling operations and transportation packaging must be designed specifically for extra protection.

19.2.5 Plastic bottles

PET bottles have limited use but have found increasing specific applications, such as in airlines, because of the weight reduction when compared to glass bottles, which can reach 35%—45%. Closures are usually plastic or aluminum screw caps. PET bottles can be delivered as injected preforms and be stretch blown into the final bottle near to or at the packaging site and the empty bottle can be transported by neck handling and air conveyers to the filler machine. This brings large savings in transportation.

Today's consumer is sensitive to plastic use and is attracted to "plastic-free" packaging solutions, regardless of the overall impact and true sustainability. In fact it is clear that the CF of PET bottles is lower than that of glass, but it is made from a nonrenewable resource. One important aspect adding to its sustainability credentials is that PET recycling chains are well established with mechanical processes able to give material suited to direct food contact, therefore reducing greatly value losses.

New materials based on sustainable raw materials are being studied for packaging applications, for example, bio-PET and poly lactic acid (PLA). Bio-PET is synthesized partially from bio-derived monomers, with up to 30% renewable materials. One example, PlantBottle, is already on the market. In terms of material characteristics and properties relevant for bottle disposal options, Bio-PET is equal to PET produced 100% from fossil based monomers. Polyethylene furanoate (PEF) is a polyester-like PET, but produced from furandicarboxylic acid and ethylene glycol and both monomers can be bio-based. Better barrier and thermal properties than PET are claimed for PEF, although this polymer does not yet have any commercial applications despite the great expectations for it.

PLA is a polyester synthesized from lactic acid obtained by fermentation of glucose from starch, mainly from corn. Of all bio-based materials PLA has shown the highest commercial potential and is now produced on a comparatively large scale, but one major limitation commonly referred to is the high price and lack of availability, when compared to conventional plastics. Although generally considered a biodegradable material, PLA is actually poorly degradable under simulated ocean and soil conditions, only being compostable at high temperatures. Applications in wine sector are scarce, although development by AIMPLAS (Spain) of a bottle in PLA coated with silicon oxide for gas barrier improvement is worthy of mention.

In general, the low barrier to gases and "scalping" (progressive removal) of wine flavor and aroma compounds, limit plastic bottles usability to short contact times, 3—4 months depending on the specific characteristics of the materials (Pati et al., 2010).

19.2.6 Barrels and kegs

Barrels and kegs are used for food service presentation of wines. While not designed for domestic use, kegs and barrels are used in various types of HORECA situations. The most established configurations are refillable stainless steel barrels of between 20 and 60 L capacity. The wine is in contact directly with the stainless steel and propulsion is by supply of compressed nitrogen or argon, although spritzed or carbonated styles will use carbon dioxide as propellant. Barrels of this type are designed to be professionally cleaned, filled, and distributed full and normally have one way valves for exit of the wine. This obviously requires a considerable logistics support and, as such, has classically been undertaken by those companies with an established business model based on producing and or delivering another beverage—such as beer for service on draught at the point of sale.

As has happened in recent years with beer, the development of single use barrels made from polymer materials has made the keg an option available to a much broader range of wine producers. These can be packaged by a service provider and transported over large distances, as the barrels do not need to be returned for refilling, although they can use the same (or compatible) valve fittings as the stainless steel versions. Some models are designed to have the wine in contact with the semirigid PET housing and propulsion is as for the stainless steel versions. More recent developments also have single-use PET barrels with a laminated bladder inside. In such cases, compressed air can be used to apply pressure to the outside of the bladder and propel the wine through a dispenser. In terms of the plastics use component of the environmental impact, some companies use recyclable plastics for the food contact surfaces and recycled plastics for many other components. As an example, one company making single layer, single use PET kegs, claims that such a solution has a 26% lower LCA score than an equivalent reusable steel keg (Dolium Kegs, 2021).

19.2.7 Paperboard

Paperboard is a major material in the wine packaging systems. It is used as secondary packaging (not in direct contact with the product) as boxes for glass bottles and BIB systems, as trays and wrap-round systems for cans and often as pallet layer separation sheet and corner protection for transportation and box inserts. Typically corrugated board is used with different structures and composing papers characteristics, for fluting and liner layers, according to the mechanical resistance required and marketing considerations that determine printability and the general aspect of the box.

A major positive attribute is that paperboard is made out of a renewable resource. However, the high energy and water demands to produce paper and paperboard are also recognized. Corrugated boxes are widely recycled, but some virgin pulp is always needed because the recycling process shortens the fibers and thus decreases the strength. Recovery of used boxes in (semi) closed-loops is relatively easy and cheap because of their bulk.

The mechanical resistance depends on the grammage of the paper sheets composing the board, the board structure, and also the box construction design. The weight of board used per bottle can range from 10 to 120 g per bottle (Trioli et al., 2015), but the selection depends largely on marketing and commercial aspects.

When used to contain glass and PET bottles, the mechanical strength of the box is of secondary importance, as most of the weight of stacked boxes in a pallet is supported by the bottles themselves. However, for BIB, the resistance of the secondary box is a key attribute.

19.2.8 Cork as closure for wine bottles

Cork is the major material used as closure for wine glass bottles. Given the relevance it represents for wine packaging, some highlights regarding the environmental impact are addressed here.

Wine cork stoppers represent 2/3 of the overall closures in wine sector. Portugal is responsible for 50% of the global raw cork production (APCOR, 2021) and in this country, the cork sector has a high environmental, social, and economic importance. Other closure types are recording a market increase, especially in wines from non-European producing countries (Poças et al., 2010). Cork has, however, shown a resurgence in recent years and this is partly attributable to its intrinsic sustainability credentials. Table 19.1 presents a comparison of emission factors for different wine closures and materials used in labeling and secondary packaging.

An evaluation of the environmental impacts deriving from the production of natural cork stoppers in Portugal showed that the forest management stage, namely pruning and vegetation operations, has the largest contribution in the majority of the impact categories (Demertzi et al., 2016). A more recent industry report indicates, however, the following breakdown through the cork production chain: forest management 0.05, treatment 0.11, production 0.75, and finishing 0.68 kg CO_2eq/1000 stoppers (Amorim Cork, 2019).

Very relevant to the contribution of cork as closure to the global impact of the wine bottle, is the carbon fixing effect of the cork forest. Cork oak absorbs carbon dioxide from the atmosphere and stores it in their perennial tissues and in the soil as organic matter, retaining it for very long periods (Aronson et al., 2009). Cork tree forest (*montado*) can retain ca 6 tonnes of CO_2 per hectare and year (APCOR, 2021). Therefore, the net CF for natural cork is an offset that can be applied to other winery and related activities, such as the production of the glass bottles, which can be particularly relevant for premium wines. International guidelines encourage wineries to include this cork's environmental benefits when calculating the CF of bottled wine (CQC, 2020; OIV, 2011). Industry reports that in a

Table 19.1 Carbon footprint for different type of closures and other packaging components.

Type of closure	gCO_2eq per kg
Natural still wine cork—3.5 g	2310
Agglomerate still wine cork—5.5 g	2200
Screw cap (aluminum 35% recycled + PE seal/tin)—4.8 g	10,600
Screw cap (aluminum75% recycled + PE seal/tin)—4.8 g	7300
Additional tin cap	17,100
Other packaging components	
Paper labels (printed)	2930
Glue (starch)	0550
Plastic film PET (nonrecyclable)	5500
Cardboard	1060

Source: ADEME. (2014). Carbone organique des sols: l'énergie de l'agro-écologie, une solution pour le climat. Agence de l'Environnement et de la Maîtrisse de l'Energie, Référence Ademe 7886, (Angers, France).

scenario based on well-managed cork oak *montado*, there is a forest storage up to −311 kg CO_2/1000 stoppers, through carbon sequestration, and therefore, the carbon balance reaches up to−309 kg CO_2/1000 stoppers, based on a 3.8 g natural cork stopper (Amorim Cork, 2019).

19.3 LCA and environmental assessments for different packaging systems

It is recognized that LCA provides the best framework for assessing the potential environmental impacts of products currently available. In an LCA, the emissions and resources consumed that are linked to a specific product are compiled and documented in a life cycle inventory. An impact assessment is then performed, considering the damage categories: human health, ecosystem quality, and issues related to natural resource use. Impacts considered in a life cycle impact assessment include climate change, ozone depletion, eutrophication, acidification, human toxicity (cancer and noncancer-related), respiratory inorganics, ionizing radiation, ecotoxicity, photochemical ozone formation, land and water use, resource depletion, noise and seabed destruction. The emissions and resources are assigned to each of these impact categories and are then converted into indicators using impact assessment models. Emissions and resources consumed, as well as different product options, can then be cross-compared in terms of the indicators (EU, 2010). The fundamentals of LCA are more deeply developed in another chapter of this book.

LCA is harmonized via international standards (ISO 140404 and 14044) and typically a software package is used that is designed to facilitate in data handling and making environmental comparisons. There is evidence that the choice of software program used can affect the final results and that further consistency and agreement among LCA tools should be a goal of LCA practitioners and users (Speck et al., 2015).

LCA has been used in the field of packaging since the 60s focusing on energy use, resources use, and waste management. A number of references can be mentioned that compared different packaging systems, for foods and beverages, namely beer (Detzel & Monckert, 2009), soft drinks (Falkenstein & Wellenreuther, 2010), milk (Keoleian & Spitzley, 1999). Simon et al. (2016) examined five different packaging materials during their whole life cycle, focusing in detail on the collection of postconsumer bottle systems such as kerbside bin, kerbside bag, deposit-refund, combinations with thermal compression of plastic bottles as well as an attempt made toward examining refill bottles. Following the most recent LCA studies focusing on wine packaging are discussed.

There is a number of publications on LCA of the wine supply chain, most of them including viticulture, winemaking, packaging, and distribution. These works denote that primary and secondary packaging are environmental hotspots of the wine supply chain (Pattara et al., 2012; Point et al., 2012; Mocci 2013).

19.3.1 Carbon footprint and water footprint

CF and water footprint (WF) are widespread indicators obtained by LCA approaches. The assessment of the CF and WF of a red wine bottle (450 g), from the vineyard to the retailer level and including scenarios for the materials' end-of-life, indicates that packaging and bottling phase is responsible for ca 55% of the total CF, while being the third most-impacting phase on the total WF, after distribution and grape production. CF ranges from 1 to 2 kgCO$_2$eq per bottle and total WF is ca 580 L per bottle, depending on the wine type in question (Bonamente et al., 2016). Recycling rates of 77% for the box, 71% for the glass bottle, and 35% for the plastic cap, while 100% of cork landfilling were considered as representative of the Italian scenarios.

Similar results were recently reported for two German wineries: ca 1.8 kg CO$_2$eq/bottle for white and red wine bottles. The WF was between 2.1 and 5.7 L blue water/bottle. The filling, bottling, corking, labeling, and packaging processes, including bottle and package acquisition caused GHG emissions of 505–630 g CO$_2$eq/bottle (Gierling & Blanke, 2021).

Neto et al. (2013) performed an LCA of a white wine, Vinho Verde, produced in Northern Portugal. The largest environmental impact was attributed to viticulture (2 kg CO$_2$eq per bottle—69% of total GHG) followed by bottle production with 0.44 kg CO$_2$eq per bottle.

19.3.2 Comparison between different packages

Alternatives to conventional glass bottles are sometimes marketed as "eco-friendly." However, the role or function and the value they represent to the specific wine product is not the same. There are considerations that need to be addressed when selecting a packaging system, related to the type, quality, and intended shelf-life of the wine, marketing, and commercial issues, as well as consumer preferences. The desired postpackaging storage—cellaring—might be quite extensive even ranging from one or two years to multiple decades. For this type of scenario, only glass presents a suitable barrier to oxygen. Carbonated wines require a barrier to carbon dioxide and a mechanical containment that BIB and the laminated box are not able to provide and it is well known that wine stored in metal can taste different to those stored in glass. The packaging system has a major impact on the wine properties and their evolution over time (Moreira et al., 2016). Therefore, the comparison of different packaging systems in terms of environmental burden would ideally need to incorporate these product—package, functional aspects. Otherwise, comparison is only really valid when focusing on products of equivalent "value" in these terms.

Ferrara and De Feo (2020) compared five alternative systems for wine packaging in Italy, focusing on the importance of secondary and tertiary packaging as well as the efficiency of wine distribution (in terms of palletizing efficiency) in function of the different wine packaging systems considered. These authors compared the traditional single-use glass bottle with laminated multimaterial boxes, BIB, refillable glass bottle, and the multilayer PET bottle. The functional unit of the study was defined as 3 L of wine and distribution by road transportation only in national territory. The optimum palletizing efficiency was considered by maximizing the pallet load patterns using simulation tools. The end of life phase of the packaging systems was modeled considering the disposal scenarios, consisting of incineration, landfilling, and recycling processes, for each component of the systems applicable in the country with rates provided by industry sources. Glass bottles were considered to be 73% recycled, the same scenarios were considered for the bag of the BIB and laminated box (41% landfilling, 32% incineration, and 27% recycling), PET bottles 43% recycled and ca 80% for the corrugated board components of the different systems. For the refillable glass bottles, inputs and emissions from the bottle sanitation process for reuse were also considered (Ferrara et al., 2020).

Consideration of the whole chain and the global packaging system is fundamental to give a full picture of the environmental burden, because often the secondary and tertiary packaging, as well as the warehouse practices, are not the same for different primary packages. For example, most typically glass bottles of wine are distributed in corrugated boxes, while metal cans are in trays and closed with thermo-shrink, plastic wrapping film. Additionally, the selling unit, which is related to the number of bottles or cans to be grouped and respective capacity, is set by regulations, which may also depend on the selling market. Therefore the specific ratio between the distribution packaging materials and the primary packaging needs to be considered and reflected in the functional unit to be selected for the LCA study. Fig. 19.1 shows a conceptual framework for an LCA on wine packaging.

In the study of Ferrara and De Feo (2020), BIB was found to be best performing option, followed by aseptic cartons that demonstrated only a slightly poorer environmental performance. Compared to single use glass bottles, the impacts of BIB were from 60% to 90% lower. The greater sustainability of BIB and aseptic cartons was considered to be due to the lower packaging weight and the higher palletizing efficiency which requires less secondary and tertiary packaging to be produced and transported (Ferrara et al., 2020).

Increasing the share of wine in BIB at the cost of glass bottles was also expected to have a strong potential to decrease value chain−based GHGs in the Finnish wine supply chain. Considering different countries of origin, on average a 0.75 L glass bottle would result in 0.638 kg CO_2eq when produced in non-EU countries, and 0.472 kg CO_2eq when produced within the EU. BIB would cause 0.052 kg, laminated cartons 0.063 kg, and PET bottles 0.182 kg CO_2eq (Ponstein et al., 2019).

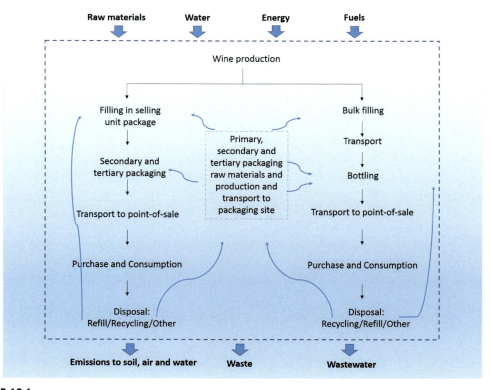

FIGURE 19.1

General framework to assess packaging impact in environment.

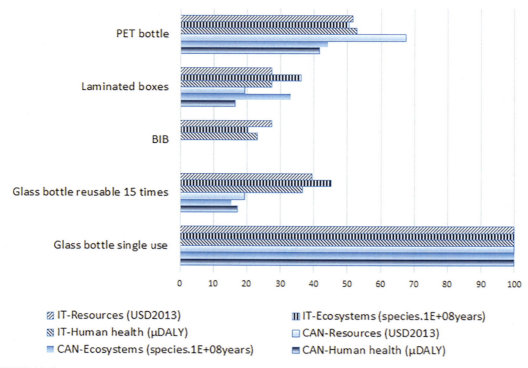

FIGURE 19.2

Impact on damage categories of different packaging systems.

Data from Cleary, J. (2013). Life cycle assessments of wine and spirit packaging at the product and the municipal scale: A toronto, Canada case study. Journal of Cleaner Production, 44, *143–151 and Ferrara, C., Zigarelli, V., & De Feo, G. (2020). Attitudes of a sample of consumers towards more sustainable wine packaging alternatives.* Journal of Cleaner Production, 271, *122581.*

In general for the three categories of damage (see Section 19.3), the Italian and the Canadian studies indicated that PET bottles have an impact of 40%–50% to those corresponding to the glass bottle and laminated box 20%–30% (Cleary, 2013; Ferrara et al., 2020). Fig. 19.2 summarizes the results for the three damage categories obtained in the previous described studies.

19.3.3 Comparison between single use and refillable glass bottle

As discussed above, in spite of the adoption of measures to reduce its impact in the environment, the glass industry is still under considerable pressure in this respect. Returnable bottles were used for decades and may still play a role in specific cases in the near future as most of the LCA studies consider it in the scope of the analyses performed. However, the approaches that the industry have flagged as the way forwards in sustainability are recycling and lightweighting.

The comparison between one-way and refillable glass bottles indicates an impact reduction, as expected, for the refillable bottle. However, this option still demonstrates a worse performance than

BIB because of the secondary and tertiary packaging. The same amount of these were needed as for the single use glass bottles because these materials are not reusable from bottle trip to trip. The environmental performances of refillable glass bottles became comparable to those of aseptic cartons and BIB within a 100 km distribution distance (Ferrara et al., 2020). A similar comparison between single use conventional glass bottle and refiling within a round trip of 200 km for the transport distance of empty bottles between the retailer and the bottle washing/refilling facilities gave results even more favorable to the refilling bottle for 15 uses. Comparing refilling to single use of a common glass bottle (543 g in both cases), this study showed relative percentages for the categories of damage of human health, ecosystems diversity, and resources, respectively 17%, 16%, and 19% (Cleary, 2013).

19.3.4 Impact of recycling of glass bottles

Glass is a permanent material that can be infinitely recycled without losing any of its intrinsic properties, therefore making it an extremely valuable secondary raw material. Recovered glass can effectively replace virgin raw materials in the production processes, ensuring a long-term perspective for resource efficient manufacturing. EU consumption of products packed in glass increased by 39%, while glass recycling increased by 139% (FEVE, 2020). Food contact safety is not an issue in the case of glass recycling because the high temperatures used in the process eliminate any contaminant present. Still the recycling rates correspond to 74% of all postconsumer glass packaging in the EU leaving a large margin for improvement.

Studies reported highlight that the use of virgin glass wine bottles causes a large proportion, up to 30% or even higher, of the overall wine CF. The recycling quota considered in the disposal scenarios has an important effect on the results. Using as reference a recycling quota of 80% and a bottle weight of 365 g, the estimated CF is 0.4–0.5 kg CO_2eq/bottle. The increase of the recycling quota to 85%–95% can reduce the CF by 47%–60% (Gierling & Blanke, 2021).

Glass recycling rates are currently high (up to 80%) in most developed countries, and this scenario is considered a best practice in the context of waste management (Landi et al., 2019). Cullet is the technical term for recovered broken and waste glass that is reprocessed into recycled glass. Incorporation of cullet not only saves virgin materials but also decreases the energy required to melt the furnace input. Nevertheless, high quantities of energy are still required. Landi et al. (2019) concluded that a significant decrease in all the impact categories can be achieved in case of reuse as compared to recycling mainly due to the fact that the production of a new glass bottle from recycled glass always requires the addition of some virgin material that must be mixed with the glass cullet during the melting phase. The authors assumed 42.5% in weight of virgin glass added. However it can be anticipated that the trade-off is largely dependent on the percentage of cullet incorporated.

The major limitations for observing higher recycling rates is the cullet shortage, particularly in wine producing countries that have a high level of exports for the bottled product (Soares, 2019). Glass color is another subject that needs to be addressed regarding recycling. The great majority of wine bottles are flint (uncolored transparent), green and amber, with color choice being made based on marketing and the UV protection required. Color separation is important for color uniformity throughout the production process. The mixing of colors is a problem principally in the production of flint bottles and the percentage of incorporation of recycled cullet in these bottles is typically much lower than in colored ones. These latter have generally standardized tints, like *Georgia green, emerald green, champagne green,* and *dead leaf green* for green bottles (FEVE, 2020). In these dark colors, low percentages of cullet from different colors can be easily diluted in the mass of mainstream glass, thus tolerating higher levels of cullet incorporation.

19.3.5 Impact of lightweighting

Lightweighting, in this context, is the process of decreasing the mass of a glass bottle that holds the same volume of wine while maintaining as much as possible of the bottles functional characteristics. Bottle lightweighting without compromising the mechanical performance has been looked at with great interest and through design and manufacturing improvements and through the introduction of narrow neck press and blow technology - NNPB, the industry has achieved significant results in many markets (for example, soft drinks and beer industries).

A more uniform distribution of the glass thickness throughout the bottle allows an overall reduction of the thickness and thus a lighter bottle is obtained. This can be achieved by a better control of the temperature and viscosity of the molten glass and the reduction in operation time allows (Soares, 2019). The NNPB process also ensures a better control of the glass distribution offering the possibility of obtaining 15% to 30% lighter bottles (Sarwar & Armitage, 2003). However, many wine bottles are not manufactured by this process because it is not compatible with some specific design features of the bottle. The punt at the base of a bottle is often a marketing requirement and dimension profiles of the finish and neck are imposed by the closures standards (CETIE, 2021). The blow and blow process that is used to produce such bottles does not guarantee the same uniform glass distribution as NNPB does, especially at radial level, even despite the above advances in the forming line. This limits the lightweighting that can be achieved in bottles of this type (Soares, 2019). Furthermore, as discussed in a later section, the influence that the weight of the bottle may have on the consumer's perception of quality of the wine itself, must also be taken into account.

The impact of lightweighting single use glass bottles has also been targeted in LCA studies (Cleary, 2013; Ferrara et al., 2020). In the Italian study (Ferrara & De Feo, 2020), the weight considered for a 750 cL glass bottle was 550 g and it was concluded that a 10% decrease in the bottle weight would result in 6% decrease in the global warming potential (GWP). This range of variation of only 10% of the original weight was considered because it could be applicable to all systems considered. In the Canadian study (Cleary, 2013), a 1 L glass bottle weighing 543 g was the standard. A 20% lighter bottle gave values for the three damage categories ca 85% lower than the standard bottle. A value of 10% can be acceptable to describe the variability in weight of some of the lighter package types, but not the range of possible weights in glass bottles derived from different bottle shapes and models as well as production methods. Today, lighter weight bottles may be used, especially for wines segments that might also be packaged in BIB and laminated boxes. As an example, for Bordeaux-shaped bottles on the Portuguese market, an average weight of 475 g was found and specific examples were found as low as 410 g (Soares, 2019).

19.3.6 International distribution

In the previously mentioned studies, transportation was considered only at the regional level of distribution, but the international component must also be considered. In this case, the optimized transport in bulk and carrying out the bottling process on the market country instead of on the sourcing country is of great interest (Point et al., 2012). Harris et al. (2018) studied the impact of alternative routeing and packaging scenarios on carbon and sulfur emissions in international wine distribution. Options studied were freight transport movement including road, rail, or sea, or multimodal combinations. The use of glass bottles, packed in boxes and then stacked onto pallets was compared to wine transportation in flexitanks. The study was conducted for the transportation of wine from Italy and Australia to UK, and showed that the international shipping leg has, in most cases, a much larger CF than the inland transport legs within the UK. Exception is made for the rail scenario using flexitank, where the deep sea shipping and the inland movement yield to similar impact. The use of bulk containers such as flexitanks appears to offer significant advantages both in economic and environmental terms (Harris et al., 2018). Shipping in bulk rather than bottled wine, from Australia to UK, would reduce the GWP by 13%, equivalent to 27,000 t CO_2eq annually (Amienyo et al., 2014). For the Finnish supply chain: a shift in bottling location from the countries of origin to Finland, associated with bulk wine exports to Finland, was concluded to reduce GHG emissions for wine from Chile, Australia, and South Africa, but not for wine from European countries (Ponstein et al., 2019).

It is thus recognized that wine transportation in bulk and filling at the selling market has a potential to reduce the environmental burden. However, given the current European regulations, the majority of wine production must be bottled at source before transportation to the final destinations, particularly for wines that have regional provenance mentioned on the packaging. A revision of the European regulation should be envisaged with a focus on reducing the environmental impact of wine distribution. Interestingly shipping wine in bulk for commercial bottling close to the consuming market was a much used and legally accepted practice in the European wine sector until as late as the 1990s.

19.4 Consumer perceptions of sustainable packaging options for wine

At one level wine is a food product, in that it is eventually ingested and contributes some nutrition to the consumer. However wine is much more than that and is the most studied of all food or beverage categories in terms of how a consumer perceives it (Spence, 2020). Therefore at some price levels, wine is treated as a basic, even commodity product, and in such cases, the sustainability attributes are more likely to be valued by the supply chain and the consumer will be presented with the package that best suits the various demands of by the distribution chains and the legislation.

It is known that the final package factually represents probably the greatest part of the environmental impact of a wine (Licciardello, 2017). However, when it is discussed in the scientific literature in the context of consumer attitudes and perceptions toward sustainability, it is almost always as one of a number of other components (Schaufele & Hamm, 2017).

It is difficult to avoid the conclusion that there is no such thing as a wine consumer, but rather a broad range of consumers that, for convenience, are placed in "segments" that demonstrate similar attitudes and behaviors, including those relevant to sustainability (Capitello & Sirieix 2019). Consumers in different segments have different levels and types of "engagement" with wine per se (Lockshin, 2001). There are segments that are themselves defined by their sustainability concerns (Barber, 2010), but the majority of relevant studies explore the perceptions—and sometimes behaviors—of established segments in relation to sustainability attributes (Capitello & Sirieix, 2019; Ferrara et al., 2020; Sogari et al., 2016; Symoneaux et al., 2019). Such studies lead to a complex picture of how different segments are motivated and respond to different types of sustainability message attributed to a wine.

Sustainability credentials of a wine can be presented in a formal, aggregated way via certification and this itself can take many forms as the most established of types of certification scheme, organic certifications are far the most studied (Symoneaux et al., 2019; Schaufele & Hamm, 2017). However other forms of parallel communication, via advertising or internet based means can also be used to reach the consumer. Whatever the form of presenting the credentials, preferences can be translated into rejection of options that do not comply with their expectations in this respect, or in other manifestations such as their "willingness to pay" a premium for products that satisfy these expectations (Symoneaux et al., 2019; Barber, 2010; Sogari et al., 2016).

As mentioned above, there is very little literature that explores consumer attitudes to sustainability aspects of wine packaging directly and even less that addresses systems other than glass bottles. Ferrara and De Feo (2020) in an exploratory study with Italian consumers, found that those who were less engaged and consumed less, were more likely to accept alternatives such as BIB and cartons. However even in the lower price points, bottles with a pulled cork are sometimes the only form acceptable to more traditional consumers (Ferrara et al., 2020).

Barber (2010) with a study on US consumers asked about their intention to pay (10% or 20 USD per week) more for wines that were packaged in an environmentally friendly way, without stipulating the packaging system. The majority of the approximately 1/3 of correspondents that responded positively were characterized as highly motivated toward environmental issues independent of their engagement with wine.

Concerning the most prevalent package, 750 mL glass bottles, there is a particular consumer-driven challenge to reducing the mass of the glass and thus diminishing the environmental costs outlined above. It is recognized that the format and weight of the bottle are among the most important signifiers of higher quality to many segments of consumers. Sáenz-Navajas et al. (2014) showed that a heavy bottle (>1.3 kg filled) is one of the most positively correlated extrinsic signifiers of high quality among both Spanish and French consumers. When simply considering the price of a wine and the weight of the bottle, a strong positive correlation was found for wines from a number of countries and regions present on the UK market (Piqueras-Fiszman & Spence, 2012). These authors also studied the perceptions of 150 Spanish wine consumers and found a general trend among them to consider a greater weight so signify a higher quality of wine. Again the correspondent population on this study was from a wine culture with a high proportion of sophisticated consumers. In their preliminary study, Ferrara and De Feo (2020) found that the more regular consumers also associated a higher bottle weight with a higher quality of wine. These authors did show evidence that a greater knowledge on the part of the consumer concerning the positive attributes of alternatives might serve to counteract these attitudes.

At the other end of the price spectrum from the commodity wines mentioned above are those that are definitely *premium* products, indeed due to their price and exclusiveness some can even be considered *luxury* items (Velasco & Spence, 2019). The expensive nature of the packaging of these wines, including the weight of the bottle, is intrinsically linked to their nature as premium or luxury products (Velasco & Spence, 2019). Thus producers are reluctant to lightweight their bottles and risk their wines being seen as inferior to competitors.

Wines in such segments have, in this respect, much in common with other high beverages such as spirits and high value liquid food products such as olive oils. In the case of olive oil, "expensive" packaging has been shown to be very important even to the perception of sensory quality (Torres-Ruiz et al., 2013). It is very unlikely that the heavy and excessive packaging for premium and luxury liquid products will remain unaltered. As a nonscientific indicator of this, the major trade fair for luxury packaging, LuxePack, has most of its technical program dedicated to sustainable solutions (LuXepack Monaco, 2021). Specifically for wine, a number of the world's most influential wine journalists are informally, but constantly, campaigning for a reduction in the excessive weight of bottles used by many premium producers (Joseph, 2014).

19.5 Concluding remarks

The package is the first contact that wine, as a physical product, has with the consumer and sustainability issues related to packaging are undeniably high priorities for the wine producer and consumer alike. However, there is a delicate balance that must be achieved between presenting the consumer with efficient and sustainable packages and ensuring that wine's status as an experience good is maintained. Companies are understandably reluctant to potentially decrease the perceived value of a product by presenting it in inappropriate packaging. This is true for the choice of material and format of a package and also for the apparent quality that a particular type of package has within that category.

As the most relevant packaging system, the energy costs and weight mean that glass bottles are consistently penalized in LCAs. However, stoppered glass bottles, particularly with cork as a stopper, have a unique image and association with wine. Together with glass's functional advantages, particularly for wines intended for aging in the package, this image means that it is likely to be maintain its preponderant position and even more so for premium wines. It is also true that through measures such as lightweighting, packaging close to market and improved recycling, the negative sustainability impacts of glass will continue to decrease. Concerning recycling, the continued improvements in sorting will benefit glass as an option as well as many other materials.

References

ADEME. (2014). Carbone organique des sols: l'énergie de l'agro-écologie, une solution pour le climat (Angers, France). In *Agence de l'Environnement et de la Maîtrisse de l'Energie, Référence Ademe 7886*.

Amienyo, D., Camilleri, C., & Azapagic, A. (2014). Environmental impacts of consumption of Australian red wine in the UK. *Journal of Cleaner Production, 72*, 110–119.

Amorim Cork. (2019). *Natural cork stopper carbon footprint, Amorim Cork*. https://www.amorimcork.com/en/sustainability/sustainability-performance-reports-and-studies/. (Accessed December 2020).

APCOR. (2021). *Media center*. www.apcor.pt./en/media-center/. (Accessed February 2021).

Aronson, J., Pereira, J., & Pausas, G. (2009). *Cork oak woodlands on the edge: Ecology, adaptive management and restauration*. Washington D.C., USA: Island Press.

Barber, N. (2010). "Green" wine packaging: Targeting environmental consumers. *International Journal of Wine Business Research, 22*(4), 423–444.

Bonamente, E., Scrucca, F., Rinaldi, S., Merico, M. C., Asdrubali, F., & Lamastra, L. (2016). Environmental impact of an Italian wine bottle: Carbon and water footprint assessment. *The Science of the Total Environment, 560–561*, 274–283.

Boz, Z., Korhonen, V., & Sand, C. K. (2020). Consumer considerations for the implementation of sustainable packaging: A review. *Sustainability, 12*, 2192. https://doi.org/10.3390/su12062192

Capitello, R., & Sirieix, L. (2019). Consumers' perceptions of sustainable wine: An exploratory study in France and Italy. *Economies, 7*, 33.

CETIE. (2021). *International technical center for bottling.* https://www.cetie.org. (Accessed February 2021).

Cleary, J. (2013). Life cycle assessments of wine and spirit packaging at the product and the municipal scale: A toronto, Canada case study. *Journal of Cleaner Production, 44*, 143–151.

CQC. (2020). *Cork quality council.* https://www.corkqc.com/. (Accessed December 2020).

Demertzi, M., Silva, R. P., Neto, B., Dias, A. C., & Arroja, L. (2016). Cork stoppers supply chain: Potential scenarios for environmental impact reduction. *Journal of Cleaner Production, 112*, 1985–1994.

Detzel, A., & Monckert, J. (2009). Environmental evaluation of aluminium cans for beverages in the German context. *International Journal of Life Cycle Assessment, 14*(Suppl. 1), S70–S79.

Dolium Kegs. (2021). https://doliumkegs.com/. (Accessed February 2021).

European Commission - Joint Research. (March 2010). *Centre - Institute for Environment and Sustainability: International Reference Life Cycle Data System (ILCD) Handbook - General guide for Life Cycle Assessment - Detailed guidance.* EUR 24708 EN (First edition). Luxembourg: Publications Office of the European Union, 2010.

Falkenstein, E. V., & Wellenreuther, F. (2010). LCA studies comparing beverage cartons and alternative packaging: Can overall conclusions be drawn? *International Journal of Life Cycle Assessment, 15*(9), 938–945.

Ferrara, C., & De Feo, G. (2020). Comparative life cycle assessment of alternative systems for wine packaging in Italy. *Journal of Cleaner Production, 259*, 120888.

Ferrara, C., Zigarelli, V., & De Feo, G. (2020). Attitudes of a sample of consumers towards more sustainable wine packaging alternatives. *Journal of Cleaner Production, 271*, 122581.

FEVE. (2020). *Facts & products details.* https://feve.org/about-glass/facts-product-details/. (Accessed December 2020).

Frassine, C., Rohde C., & Hirzel, S. (2016). Energy saving options for industrial furnaces – the example of the glass industry. European Council for an Energy Efficient Economy (ECEEE), Industrial Efficiency 2016: Going beyond energy efficiency to deliver savings, competitiveness and a circular economy. Industrial Summer Study Proceedings (pp 467–476). ISBN: 978-91-980482-9-2.

Gierling, F., & Blanke, M. (2021). Carbon reduction strategies for regionally produced and consumed wine: From farm to fork. *Journal of Environmental Management, 278*, 111453.

Harris, I., Sanchez Rodrigues, V., Pettit, S., Beresford, A., & Liashko, R. (2018). The impact of alternative routeing and packaging scenarios on carbon and sulphate emissions in international wine distribution. *Transportation Research Part D, 58*, 261–279.

Joseph, R. (2014). *Jancis [Robinson] hates heavy wine bottles, as I do. Why do winemakers use them?* Tim Atkin. https://timatkin.com/heavy-bottles/. (Accessed February 2021).

Keoleian, A., & Spitzley, D. (1999). Guidance for improving life cycle design and management of milk packaging. *Journal of Industrial Ecology, 3*(1), 111–126.

Landi, D., Germani, M., & Marconi, M. (2019). Analysing the environmental sustainability of glass bottles reuse in an Italian wine consortium. 26th CIRP Life Cycle Engineering Conference. *Procedia CIRP, 80*, 399–404.

Licciardello, F. (2017). Packaging, blessing in disguise. Review on its diverse contribution to food sustainability. *Trends in Food Science & Technology, 65*, 32–39.

Lockshin, L. (2001). Using involvement and brand equity to develop a wine tourism strategy. *International Journal of Wine Marketing, 13*(1), 72–81.

LuXepack Monaco. (2021). *Conference programme.* www.luxepackmonaco.com/en/blog/euroclip/. (Accessed February 2021).

Martins, A. A., Araújo, A. R., Graça, A., Caetano, N. S., & Mata, T. M. (2018). Towards sustainable wine: Comparison of two Portuguese wines. *Journal of Cleaner Production, 183*, 662.

Moccia, L. (2013). Operational research in the wine supply chain. INFOR. *Information Systems and Operational Research, 51*(2), 53–63.

Mordor Intelligence. (2020). *Global wine packaging market - growth, trends, and forecasts (2020–2025)*. India: Industry Reports, Mordor Intelligence.

Moreira, N., Lopes, P., Ferreira, H., Cabral, M., & Guedes de Pinho, P. (2016). Influence of packaging and aging on the red wine volatile composition and sensory attributes. *Food Packaging and Shelf Life, 8*, 14–23.

Mullen, T. (2018). *Why wine corks are on the upswing—Forbes*. https://www.forbes.com/sites/tmullen/2018/02/05/why-wine-corks-are-on-the-upswing/. (Accessed February 2021).

Neto, B., Dias, A. C., & Machado, M. (2013). Life cycle assessment of the supply chain of a Portuguese wine: From viticulture to distribution. *International Journal of Life Cycle Assessment, 18*(3), 590–602.

OIV. (2011). *General principles of the OIV greenhouse gas accounting protocol for the vine and wine sector. RESOLUTION OIV-CST 431-2011*.

OIV. (2019). *International oenological codex 2019 edition*. http://www.oiv.int/en/technical-standards-and-documents/oenological-products/international-oenological-codex. (Accessed December 2020).

OIV. (2020). *OIV-state-of-the-vitivinicultural-sector-in-2019*. http://www.oiv.int/public/medias/7298/oiv-state-of-the-vitivinicultural-sector-in-2019.pdf. (Accessed December 2020).

Pati, S., Mentana, A., La Notte, E., & Del Nobile, M. A. (2010). Biodegradable poly-lactic acid package for the storage of carbonic maceration wine. *LWT-Food Science and Technology, 43*, 1573–1579.

Pattara, C., Raggi, A., & Cichelli, A. (2012). Life cycle assessment and carbon footprint in the wine supply-chain. *Environmental Management, 49*(6), 1247–1258.

Piqueras-Fiszman, B., & Spence, C. (2012). The weight of the bottle as a possible extrinsic cue with which to estimate the price (and quality) of the wine? Observed correlations. *Food Quality and Preference, 25*, 41–45.

Point, E., Tyedmers, P., & Naugler, C. (2012). Life cycle environmental impacts of wine production and consumption in Nova Scotia, Canada. *Journal of Cleaner Production, 27*, 11–20.

Ponstein, H. J., Ghinoi, S., & Steiner, B. (2019). How to increase sustainability in the Finnish wine supply chain? Insights from a country of origin based greenhouse gas emissions analysis. *Journal of Cleaner Production, 226*, 768–780.

Poças, M. F. F., Ferreira, B., Pereira, J., & Hogg, T. (2010). Measurement of oxygen transmission rate through foamed materials for bottle closures. *Packaging Technology and Science, 23*(1), 27–33.

Ritchie, C. (2009). The culture of wine buying in the UK off-trade. *International Journal of Wine Business Research, 21*(3), 194–211.

Rugani, B., Vazquez-Rowe, I., Benedetto, G., & Benetto, E. (2013). A comprehensive review of carbon footprint analysis as an extended environmental indicator in the wine sector. *Journal of Cleaner Production, 54*, 61–77.

Sáenz-Navajas, M.-P., Ballester, J., Peyron, D., & Valentin, D. (2014). Extrinsic attributes responsible for red wine quality perception: A cross-cultural study between France and Spain. *Food Quality and Preference, 35*, 70–85.

Sarwar, M., & Armitage, A. (2003). Tooling requirements for glass container production for the narrow neck press and blow process. *Journal of Materials Processing Technology, 139*, 160–163.

Schaufele, I., & Hamm, U. (2017). Consumers' perceptions, preferences and willingness-to-pay for wine with sustainability characteristics: A review. *Journal of Cleaner Production, 147*, 379–394.

Simon, B., Ben Amour, M., & Foldéniy, R. (2016). Life cycle impact assessment of beverage packaging systems: Focus on the collection of post-consumer bottles. *Journal of Cleaner Production, 112*, 238–248.

Soares, J. (2019). *Is light weighting glass bottles for wine an option? A study on the Portuguese wine market and consumers* (MSc Thesis, Universidade Católica Portuguesa).

Sogari, G., Mora, C., & Menozzi, D. (2016). Factors driving sustainable choice: The case of wine. *British Food Journal, 118*(3), 632–646.

Speck, R., Selke, S., Auras, R., & Fitzsimmons, J. (2015). Choice of Life Cycle Assessment Software Can Impact Packaging System Decisions. *Packaging Technology and Science, 28*, 579–588, 2015.

Spence, C. (2020). Wine psychology: Basic & applied. Cognitive research. *Principles and Implications, 5*, 22.
Steenis, N. D., van Herpen, E., van der Lans, I. A., Ligthart, T. N., & van Trijp, H. C. M. (2017). Consumer response to packaging design: The role of packaging materials and graphics in sustainability perceptions and product evaluations. *Journal of Cleaner Production, 162*, 286–298.
Symoneaux, R., Ugalde, D., & Jourjona, F. (2019). Analysis of the perceptions of wine consumers toward environmental approaches: Support for the management of environmental strategy, 42nd World Congress of Vine and Wine *BIO Web of Conferences, 15*, 03020.
Torres-Ruiz, F. J., Barreda-Tarrazona, R., Vega-Zamora, M., Gutiérrez-Salcedo, M., & Armenteros, E. M. (2013). The perception of quality in the process of olive oil tasting: The effects of packaging attributes. *Journal of Food Agriculture and Environment, 11*(3), 102–105.
Trioli, G., Sacchi, A., Corbo, C., & Trevisan, M. (2015). Environmental impact of vinegrowing and winemaking inputs: An European survey. *Journal of Viticulture and Enology, 7/2*.
Velasco, C., & Spence, C. (2019). Multisensory premiumness. In C. Valasco, & C. Spence (Eds.), *Multisensory packaging*. Palgrave MacMillan.
Weed, A. (2019). Canned wine comes of age. In *Wine spectator*.
Williams, R., Williams, H. A., & Bauman, M. (2018). *Wine-in-a-can marketing implications. Report WICR FR*. WIC Research.

Further reading

Ketelsen, M., Janssen, M., & Hamm, U. (2020). Consumers' response to environmentally-friendly food packaging – a systematic review. *Journal of Cleaner Production, 254*, 120123.

CHAPTER 20

Standards and indicators to assess sustainability: the relevance of metrics and inventories

Ana Marta-Costa[1], Ana Trigo[2], J. Miguel Costa[3] and Rui Fragoso[4]

[1]*CETRAD — Centre for Transdisciplinary Development Studies, University of Trás-os-Montes e Alto Douro (UTAD), Vila Real, Portugal;* [2]*CoLAB Vines&Wines, Association for the Development of Viticulture in the Douro Region, Vila Real, Portugal;* [3]*LEAF — Centre Linking Landscape, Environment, Agriculture and Food Research Center, Instituto Superior de Agronomia, University of Lisbon, Lisbon, Portugal;* [4]*CEFAGE — Center for Advanced Studies in Management and Economics, University of Évora, Évora, Portugal*

20.1 Introduction

The global wine industry has recognized the benefits of embracing a more sustainable approach, and today sustainability is perceived as a competitive and resilience factor for wine stakeholders (Flores, 2018; Keichinger & Thiollet-Scholtus, 2017). Meanwhile climate change is reshaping the wine industry and the need to promote sustainability toward resilience building is crucial for the competitiveness and longevity of the sector (Flores, 2018). If we also account other socioeconomic factors such as rising energy prices, changing market demands, and increasing concerns over chemical exposure, it becomes clear that the global actors of the wine industry must look to implement more sustainable practices (Christ & Burritt, 2013; Gilinsky et al., 2016; Keichinger & Thiollet-Scholtus, 2017; Thiollet-Scholtus & Bockstaller, 2015).

Assessment tools are considered powerful mechanisms to support the transition toward greater sustainability (Ramos, 2019). They allow to strategically tracking production processes to increase efficiency and/or optimize environmental performance (Costa et al., 2020; Merli et al., 2018). However, there are still several limitations concerning the knowledge gap in sustainability as well as methodologies to assess it in agribusiness (Hayati, 2017; Kamali et al., 2017). Currently the predominant focus areas have been either environmental concerns or assessing specific impact categories (e.g., carbon and water footprint). There is an emerging gap relating to social, economic, institutional, political, and ethical factors (Bockstaller, Guichard et al., 2008; Olde et al., 2016).

The rise of sustainability-related awareness among the wine industry resulted in a higher demand of more and better quality data and improved methodologies to assess performance and sustainability credentials during production processes (UNEP, 2014). Nevertheless, there is a lack of general assessment frameworks to evaluate permanent crops such as grapevine (Christ & Burritt, 2013; Flores, 2018; Thiollet-Scholtus & Bockstaller, 2015). It is also perceived that collection of relevant information is seen as a cost rather than as a valuable measurement (Barbosa et al., 2018; Matos & Pirra, 2020; UNEP, 2014). Therefore, there is the need to investigate the long-term viability of the wine sector in all its dimensions.

This work aims to holistically evaluate complex winegrowing systems. Several assessment methodologies were integrated in this study to analyze data gathered from vineyards in the Douro wine region, in northern Portugal, and discuss performance results by benchmarking different assessment approaches. The Douro wine region is one of the world's largest areas of mountain viticulture, and it has the additional challenges of reduced profitability and intensive-labor input (Cichelli et al., 2016; Graça et al., 2017). Therefore, the main goal was to improve the understanding of sustainability issues for this wine region and to provide potential solutions to better assess its sustainability based on environmental, social, and economic credentials.

20.2 Sustainability assessment: major approaches and methodologies
20.2.1 Conceptual theories

The transition toward sustainability has barriers. One major challenge relates to the reduced transparency and lack of standardization and agreement among interested parties (Hayati, 2017) and the fact that there is no universal accepted definition of the concept of sustainability (Keichinger & Thiollet-Scholtus, 2017; Martins et al., 2018; Pomarici & Vecchio, 2019; Velten et al., 2015). The unclear meaning is one of the possible causes for inadequate sustainability assessment research, characterized by biased designs, erroneous interpretations, and incompatibility issues related to difficulties in data comparison (Gafsi & Favreau, 2013; Iyer & Reczek, 2017; Martins et al., 2018; Santiago-Brown et al., 2014).

In order to overcome such conceptual ambiguity, analysts and organizations have been trying to define sustainability in different ways. This attempt to capture higher levels of complexity via new ways of thinking has resulted in a multiplicity of sustainability theoretical models (Hayati, 2017). Currently, the triple bottom line (TBL) theory created by Elkington in 1994 is the most widely known comprehensive model when approaching sustainability issues (Hayati, 2017). Such theory defends "People, Planet, and Profit" as imperative sustainability doctrines, and promotes the idea that sustainable development only occurs when organizations show responsibility toward environmental health, social equity, and economic viability (Graça et al., 2017; Iyer & Reczek, 2017).

New theoretical modeling approaches and the lack of a widely accepted methodology add to the conflicting evaluations of what sustainability performance and associated impacts are. This lack of standardization led to the emergence of a wide variety of distinct and inadequate assessment tools and resulted in bias among sustainability assessments' approaches and erroneous results' interpretations (Dantsis et al., 2010).

20.2.2 Sustainability assessment tools: from simple indicators to complex frameworks

Sustainability assessment tools can be grouped into different groups depending on the level of complexity (Flores, 2018). Sustainability indicators, indicator-set indexes, and assessment frameworks are some of the basic sets of tools that are referred (González-Esquivel et al., 2020; Hayati, 2017).

Contrasting complexity groups mostly differ in the integration methods and/or aggregation procedures used on the assessments' results. While aggregation is the procedure to combine different components or indicators into a single unit, integration is a means by which individual and different indicators are linked to provide a holistic view of the sustainability performance (Bélanger et al., 2015; Van Passel & Meul, 2012).

Sustainability indicators are the most used assessment tools. They can illustrate difficult-to-access information of a particular system, simplify descriptions from complex systems, and provide predictive models or simulations (Bockstaller, Guichard et al., 2008; Parent et al., 2013). These vary from single or isolated indicators to sets of indicators or composite indicators (Bélanger et al., 2015). Composite indicators translate sustainability measurements into a single sustainability index or number by aggregating relevant information from several indicators (González-Esquivel et al., 2020; Marta-Costa & Silva, 2013a).

Single indicators, despite being criticized for providing low quality of prediction, continue to be the most used method to evaluate sustainability of a specific practice (Hayati, 2017). It has been suggested that single indicators should be combined with composite indicators to improve assessment robustness. Additionally, there is a knowledge gap around the efficient integration or aggregation methods when dealing with various factors (Bockstaller, Guichard et al., 2008).

Assessment frameworks, on the other hand, are conceptual and practical efforts designed to guide the entire assessment process - from inception to completion. Having a more complex and rigorous structure based on systematic approaches and disciplinary contents tend to be qualitatively differentiated (Marta-Costa & Silva, 2013b). Five official conceptual assessment frameworks are available (OECD, 2004; Sopilko et al., 2019; UN, 2007):

i. Driving force-state-response frameworks — originally named pressure-state-response model;
ii. Issue-based or theme-based frameworks;
iii. Capital frameworks;
iv. Accounting frameworks;
v. Aggregated indicators.

Other approaches and categories of general assessment frameworks continue to emerge due to the accession of multidisciplinary initiatives among researchers from different fields and disciplines (Abdul et al., 2019; UNEP, 2014).

20.3 Indicators and metrics applied to grapes and wine production

Strategies to measure sustainability more objectively using the three sustainability pillars (environment, society, and economy) are still being asked of the wine industry to improve the use of the sustainability concept, and instead of using it solely for marketing and communication purposes (Merli et al., 2018).

The lack of high-quality data and their availability are causes of concern (Ferrara & Feo, 2018; Vázquez-Rowe et al., 2013). It is, therefore, hard to find an analysis of overall agricultural systems and human-environment interactions from a holistic point of view (Gafsi & Favreau, 2013; Iyer & Reczek, 2017) when trying to define a sustainable grape or wine production system.

Currently the wine industry has a small set of sustainability programs and indicators that have been created or adapted to the sector (Lamastra et al., 2016). The fact that the majority of existing assessment tools are mainly developed for arable crops or livestock production (Christ & Burritt, 2013; Flores, 2018; Hayati, 2017) contributed to neglect crops such as grapevine (Hayati, 2017; Thiollet-Scholtus & Bockstaller, 2015).

Studies using metrics to monitor specific issues in winegrowing systems and the wine supply chain are predominantly based on life cycle assessment (LCA) or on the assessment of greenhouse gases (GHGs) emissions, pesticide impacts and carbon or water footprint (Ferrara & Feo, 2018; Lamastra et al., 2016). However, there are other successful and multidisciplinary initiatives. For example, the holistic French IDEA (*Indicateurs de Durabilité des Exploitations Agricoles or Farm Sustainability Indicators*) method incorporates the three fundamental pillars of sustainability and can be adapted to viticulture (Zahm et al., 2008). However, this method has a poor integration process and does not include data on soil quality and climate conditions (Bockstaller, Wohlfahrt et al., 2008).

In the case of assessments focused on single environmental issues, a connectivity index over 10 existing tools was developed to address the movement of pesticides transfer in water catchments in winegrowing areas (Payraudeau & Gregoire, 2012).

The INDIGO® method was originally developed for arable crops, but was later adapted for viticulture (Thiollet-Scholtus & Bockstaller, 2015). The EIOVI — "Environmental Impact of Organic Viticulture Indicator" was explicitly developed to evaluate the environmental impact of organic viticulture (Fragoulis et al., 2009). In turn, the method of SOECO — "Socioeconomic indicators for viticulture and innovative cultural systems" has been the only model developed specifically for socioeconomic evaluations to characterize the impact of cultural practices at vineyard farm level, on profitability and on human and social capital (Keichinger & Thiollet-Scholtus, 2017).

Other successful case studies tested and validated different assessment methods on vineyards. Certomà and Migliorini (2011) used the MESMIS—"Framework for Assessing the Sustainability of Natural Resource Management Systems" on five organic farms in Tuscany, three of which were vineyards with different management strategies and approaches. More recently, Triviño-Tarradas et al. (2020) validated the INSPIA model — "Initiative for Sustainable Productive Agriculture" originally designed to assess sustainability on annual and permanent crops, on a mixed vineyard and olive-grove farm (Triviño-Tarradas et al., 2020).

The LCA methodology, despite being well recognized by the wine industry for enabling the evaluation of the whole, or part, life cycle, still indicates that there is a wide variability in the winegrowing system's definitions. The lack of original and site-specific inventory data can negatively affect reliability of LCA results and its usefulness to the wine sector (Ferrara & Feo, 2018). Moreover, any sustainability assessment must be linked to the context in which the system being evaluated operates for a matter of clarity (Abdul Murad et al., 2019; Hayati, 2017).

The sustainable wine production model aims to achieve a more rational use of renewable resources and reduce consumption of nonrenewable resources by encouraging local production of agri-food products adapted to the natural and socioeconomic environment. The social aspects of sustainability relate to an equitable income distribution, access to resources and information, and active participation of those involved in research and decision-making processes. It must reflect social values, be in line with traditional institutions and cultures, and generate high levels of autonomy. Economic viability can be achieved by more efficient use of inputs and by adopting the best solutions and practices (González-Esquivel et al., 2020; Marta-Costa, 2010; Marta-Costa & Silva, 2013a).

20.3.1 Environmental dimension and natural resources

Large parts of studies of environmental impacts of the wine industry were carried in Italy and Spain and were focused on a specific wine type attending to the great variety of processes used during the viticulture and winemaking phases (Ferrara & Feo, 2018; Vázquez-Rowe et al., 2013). Among a total of 34 papers focused on the environmental assessment of the wine industry (Ferrara & Feo, 2018), the LCA was the most used. Table 20.1 summarizes the main environmental indicators reported in the literature for evaluation of sustainability in the vineyard and winery systems.

The impact of environmental factors (e.g., soil, climate, and other physical land components) over the production systems is well documented (Costantini & Bucelli, 2014), but their impact on the overall production system and vineyard farm organization of the wine industry has been studied only recently (Costantini et al., 2016).

Wineries also face environmental issues, mainly related to water and wastewater, organic and inorganic solid waste, energy use, GHG emissions, and chemical use (Matos & Pirra 2020). Efficient water use is a key issue, particularly in dry climate regions. Water quality is another issue to have in account as the discharge from wineries can lead to serious contamination of surface and ground water sources (Martins et al., 2018). Organic waste is one of the main discharge elements from wineries and some are already able to treat and reuse it. Management and recycling of inorganic waste (packaging materials, used materials and pallets) has still limitations (Matos & Pirra, 2020).

Wine production requires greats amounts of energy and generates high levels of GHGs. Bottling and distribution logistics are highly carbon intensive and account for around 50% of the CO_2 generated in the entire supply chain (Point et al., 2012). The use of chemicals is also significant in wineries namely for cleaning activities and wine preservation (Costa et al., 2020).

20.3.2 Social dimension and equity

The social dimension of sustainability has often been neglected by researchers compared with the environmental area (Atanda & Öztürk, 2020; Merli et al., 2018; Nilipour 2020; Santos et al., 2019; Trigo et al., 2020). The small number of social indicators reported in literature (Table 20.2) as compared to the existing number of environmental indicators (Table 20.1) supports this supposition. Generally, the vineyard farm labor, quality of life and well-being, and the relationship with the human community are the criteria used to monitor the social dimension of sustainability.

The social dimension has been included in sustainability strategies and assessment tools (Strano et al., 2013). Concepts such as "social responsibility", "socioeconomic equity", and "social enterprise" have emerged and highlight the need to promote both the social and financial motivations of companies (Nilipour, 2020).

According to Luzzani et al. (2020), the impacts of wine industry on stakeholders are in line with the social life cycle assessment (s-LCA) approach, regarding the effects on local community, society, consumers, workers, and value-chain actors. In a framework applied to different Italian wine companies, these authors considered several subthemes to assess social dimension namely the health of community, heritage, cultural and aesthetic aspects, product quality and safety, human resources management, and health of consumers. Arcese et al. (2016), in turn, have studied how s-LCA methodology could be applied to the wine industry. They found that there were limitations due to data collection and procedures used to assess social impacts.

Table 20.1 Selected environmental indicators for the wine supply chain.

Criteria	Indicators	Value-chain position	Sources
Water consumption and efficiency	Water intensity	①②	(1,3)
	Waste water	②	(1)
Soil management	Soil health	①	(4,5)
	Organic matter	①	(3,4,5)
	Soil cover rate	①	(3,4,5)
	Soil tillage index	①	(4,5)
	Soil erosion risk	①	(3,4,5)
	Soil compaction	①	(3)
Air quality	Air quality	①	(3)
	Carbon emissions	②	(1)
	Greenhouse gases balance	①	(4,5)
	Greenhouse per kg	①	(4,5)
Enhancing biodiversity	Natural area	①	(4,5)
	Biodiversity structures	①	(3,4,5)
	Buffers and security areas	①	(4,5)
	Biological controls	①	(3)
Crop management	Crop diversity	①	(4,5)
	Crop rotation	①	(4,5)
	Vine health	①	(3)
	Operational/management continuous improvement	①	(3)
	Certifications	①	(3)
Agrochemicals consumption and efficiency	Fertilizers consumption	①	(2)
	Chemical inputs optimization	①	(3)
	Plant protection products management	①	(2,3,4,5)
	Nitrogen use balance	①	(4,5)
	Nitrogen efficiency	①	(4,5)
	Phosphorus balance	①	(4,5)
	Phosphorus efficiency	①	(4,5)
Energy efficiency	Energy intensity	②	(1)
	Fuel use	①	(3)
Agri-environmental practices	Ratio input/output in farm	①	(3)
	Material intensity	②	(1)
	Solid wastes	②	(1)
	Recycling	①	(3)

Value-chain position: ① *Vineyard;* ② *Winery.*
(1) (Martins et al., 2018); (2) (Santos et al., 2019); (3) (Santiago-Brown et al., 2015); (4) (Triviño-Tarradas et al., 2019); (5) (Triviño-Tarradas et al., 2020).

20.3 Indicators and metrics applied to grapes and wine production

Table 20.2 Social indicators for wine supply chain.

Criteria	Indicators	Value-chain position	Sources
Labor use	Working hours per hectare	①	(2,4,5)
	Workers time in the vineyard farm	①	(3)
	Training/Education	①	(3)
	Staff retention	①	(3)
	Workers' engagement	①	(3)
	Wages paid	①	(2)
	Worker turnover rate	②	(1)
	Beyond legal standards for workers	①	(3)
	Compliance with labor laws	①	(3)
	Workers protective equipment	①	(3)
Quality of life and well-being	Risk of abandonment of agricultural activity	①	(4,5)
	Healthy work environment	①	(3)
	Satisfaction index/happy workers	① ②	(3,4,5)
	Investment in health and safety training	②	(1)
	Social events for workers	①	(3)
	Workers' housing	①	(3)
Interaction with the community	Community benefits	①	(3,6)
	Community health	①	(3,6)
	Happy neighbors	①	(3,6)

Value-chain position: ① Vineyard; ② Winery.
(1) (Martins et al., 2018); (2) (Santos et al., 2019); (3) (Santiago-Brown et al., 2015); (4) (Triviño-Tarradas et al., 2019); (5) (Triviño-Tarradas et al., 2020); (6) (Nilipour, 2020).

A cross-country framework analysis including South Africa, New Zealand, Australia, USA (California), Chile, and France concluded that the social dimension is covered by considering life-quality indicators at work, health, and safety (Flores, 2018). The indicators focused on community involvement were observed only in the frameworks implemented in New Zealand, Australia, and USA.

In a problem of sustainable supply network design in the Portuguese wine industry, Fragoso and Figueira (2021) followed a multiobjective program, where the three dimensions of social, economic, and environmental sustainability were integrated. For social aspects, three indicators were considered: maximization of jobs created, number of infrastructures (existing and created), and number of suppliers.

Varsei and Polyakovskiy (2017) considered in a multiobjective approach, the social impact of suppliers, and location of bottling plants in the supply chain of an Australian wine company. They considered a simplified analytic hierarchy process, which weighted equally indicators of unemployment and regional gross domestic product of each considered location.

Corporate sustainability requires that firms align their environmental and social impacts with their economic objectives (Szolnoki, 2013; Taylor, 2017). Proactive socioenvironmental practices can reduce the negative impacts of wine industry. Such practices can improve employees' health and morale, promote good community relationships, and optimize a company's economic performance and organizational capabilities as well as product innovation and the collaboration with partners or suppliers (Annunziata et al., 2018; Taylor, 2017).

20.3.3 Economic dimension and efficiency

Several economic indicators are described in the literature (Table 20.3). They correspond to the categories of structural conditions of the vineyard farm, profitability, productivity, and efficiency of the systems. Increased efficiency in the use of inputs reduces costs was seen to be related to the economic performance and competitive advantage (Santos et al., 2020).

Table 20.3 Economic indicators for the wine supply chain.

Criteria	Indicators	Value-chain position	Sources
Structural conditions	Current subsidies	①	(2)
	Vineyard farm size	①	(2)
	Succession planning	①	(3)
Income	Yield	① ②	(2,3,4,5)
	Grapes fit for purpose	①	(3)
	Vine health	①	(3)
	Production consistency	①	(3)
	Grapes price	①	(3)
	Wine price	① ②	(3)
	Grape demand	①	(3)
	Fruit quality production	①	(3)
	Brand value	①	(3)
Costs	Production costs	①	(2,4,5)
	Management costs	①	(3)
	Labor costs	①	(3)
	Contracts	①	(3)
	Inputs reduction	①	(3)
	Capital replacement cost	①	(3)
	Land value	①	(3)
	Agricultural capital	①	(2)
	Investment	①	(2)

Table 20.3 Economic indicators for the wine supply chain.—cont'd

Criteria	Indicators	Value-chain position	Sources
Economic efficiency	Profitability	① ②	(3)
	Gross margin	①	(2)
	Net income per hectare	①	(4,5)
	Net income per annual work unit of labor	①	(4,5)
	Labor productivity	①	(3)
	Earning before interests depreciation, taxes and amortizations	②	(1)
	Management continuous assessment	①	(3)
	Ability to benchmark performance	①	(3)
	Return on investment	①	(3)
	Operational efficiency	①	(3)
	Competitiveness indicator	①	(4,5)
	Technical inefficiency	①	(4,5)
Eco-efficiency	Carbon footprint	①	(3)
	Irrigation water application	①	(4,5)
	Water productivity	①	(4,5)
	Energy balance	①	(4,5)
	Energy efficiency	①	(4,5)
	Energy productivity	①	(4,5)
	Nitrogen productivity	①	(4,5)
	Phosphorus productivity	①	(4,5)
	Investments' eco-efficiency related to energy use	②	(6)
	Investments' eco-efficiency related to wastes	②	(6)
	Investments' eco-efficiency related to water use	②	(6)
	Investments' eco-efficiency related to CO_2 emissions	②	(6)
	Investments' eco-efficiency related to chemicals use	②	(6)

Value-chain position: ① *Vineyard;* ② *Winery.*
(1) (Martins et al., 2018); (2) (Santos et al., 2019); (3) (Santiago-Brown et al., 2015); (4) (Triviño-Tarradas et al., 2019); (5) (Triviño-Tarradas et al., 2020); (6) (Olaru et al., 2014).

The assessment of the economic dimension of sustainability includes direct and indirect economic impacts (Corbo et al., 2014). Martins et al. (2018) when comparing the sustainability performance of two Portuguese wines, considered Earning Before Interests Depreciation, Taxes and Amortizations (EBIDTA per functional unit, 0.75 L of wine) to assess the ability of the company to generate revenue to cover operational costs. In an economic performance analysis of the Portuguese wine industry, Faria et al. (2020) used the EBITDA together with the productivity parameter expressed as gross value added/number of employees, financial autonomy ratio, short-term, debt ratio, current ratio (liquidity), total turnover, and net profit. These indicators can be used to assess the economic dimension of sustainability at the different levels of the wine supply chain. Fragoso and Figueira (2021) in their multiobjective model used in their model of economic sustainability the profit and revenue maximization and cost minimization in the wine supply chain. Although unusual, this specification meets the specificities of wine producers to whom profit is not always the single most important economic objective. It has been found that certain companies aimed to produce a top market positioning and differentiated their high-quality wines, while obtaining a premium price (Fragoso and Figueira, 2021).

Wine companies seek to use efficiently their resources and hence cost minimization is a priority in the decision-making process. Varsei and Polyakovskiy (2017) considered in their multiobjective model for the Australian wine industry, minimization of costs in the supply chain as the unique economic objective. Similarly, Fragoso and Figueira (2021) have considered the fixed costs of building and equipping supply chain facilities, and the variable costs with production, transportation, and purchase of materials.

The V.I.V.A. program considered the economic impact of the wine industry on the territory, where suitable business strategies were taken into account (Merli et al., 2018). Corbo et al. (2014) compared several sustainability programmes of the wine industry in Italy, and identified that some of them considered direct, indirect, and local economic impacts. They conclude that a few used life cycle cost methodology.

Luzzani et al. (2020) considered the following subthemes for the economic dimension of sustainability: economic repercussion of structures (vineyards and wineries) and territories, resource depletion, traceability, and procurement practices.

In some sustainability programs, the relationship between supplier and green criteria is used to assess the economic impacts (Flores, 2018) whereas other programs evaluate the economic dimension by considering the wine market share, hiring workers, and salaries (CSWA, 2015; GRI, 2016; Vinos de Chile, 2014).

Efficient vertical coordination between grape production and processing is associated with vineyard farm size and a strong linkage with the territory (Malorgio et al., 2013) and has an important influence on wine quality. Each indicator results from the interaction of the system as a whole and does not rely on just one factor (Santos et al., 2020). The eco-efficiency most accepted definition has been proposed by the World Business Council for Sustainable Development (2000), as the capacity to deliver competitive goods or services that satisfy basic human needs and safeguard ecological sustainability. Nowadays, wine companies explore the role of sustainability in their business models (Brocado & Zicari, 2020) which are built by adaptation to customers' and stakeholders' preferences (McGrath, 2010; Provance et al., 2011). Economic sustainability can work also as a potential source of innovation in business models and including it can create more value to customers (Schaltegger et al., 2016) and can also influence the profit baseline (Broccardo & Zicari, 2020).

20.4 Sustainability assessment essay for winegrowing systems: a case study for the Douro's wine producing region

20.4.1 Context, problem, and aims

The Douro is a world famous wine region due to its port wine production. The Douro is the Portugal's largest grape-growing region (43,863 ha, 22.8% of the total area in 2019) and production volume (1,692,188 hL, 26% in 2019/2020 season) (IVV, 2021). Port wine production makes up about half of the Douro's production (IVV, 2021). The region has multiple microclimates due to its rugged topography and because of that it has been divided into three subregions: Baixo Corgo, Cima Corgo, and Douro Superior. The region is known for its natural and productive heterogeneity (Santos et al., 2020) due to the diverse vineyard landscapes, viticultural systems, and practices (Fig. 20.1).

The Douro's topography explains to a large extent the high labor inputs, which increase its overall costs of production. In addition, labor shortages have become one of the main problems in the sector (Santos et al., 2019). It is a key factor for the Douro's viticulture activities, and as such could undermine future wine production if the high costs are not compensated by increasing growers' fruit prices. Matias et al. (2021) point out that the increase in sales price of grapes by 0.4 €/Kg would prevent foreclosure of about 20% of vineyards in the Douro region, in the next 50 years and considering a fluctuation in labor cost of 1%. In this context, sustainability and competitiveness of the Douro's wine region demands better understanding. Here we consider a TBL perspective, where the end goal should be the well-being of its main actors and the economic, social, and environmental sustainability of the region.

20.4.2 Research design and methodology

Several attempts to measure sustainability of winegrowing systems have been developed (Trigo et al., 2020). This section intends to monitor the degree of sustainability of the winegrowing systems in the Douro, by combining different sustainability assessment methodologies. Data were gathered from a sample of 110 Douro grape producers (average size of 17 ha) by a structured face-to-face survey and using the group discussion developed in 2019. The inquiry was implemented in the season of 2016/2017 (cross-sectional data) and included information about the respondent and the owner or manager, the vineyard and its major inputs and outputs, as well as information on environmental and social issues (e.g., labor use).

The indicators for the sustainability assessment of Douro winegrowing systems were selected using the first steps of the MESMIS evaluation methodology (González-Esquivel et al., 2020). The performance of the Douro winegrowing systems was monitored using 11 criteria, and 27 specific indicators, covering eight sustainability attributes and three areas of sustainable development (see Table 20.4). The 27 indicators were selected on the characteristics of the production units (Cândido et al., 2015). The selected indicators were individually measured for each vineyard and the average value of each was calculated from all the respondents. The data were transformed into a nondimensional value (Singh et al., 2012), and MESMIS (González-Esquivel et al., 2020) was used to analyze the data.

MESMIS can be adapted to different types of information, technical capacity, and local contexts, and can deal with missing data and the rigid structure of many sustainability assessment methods

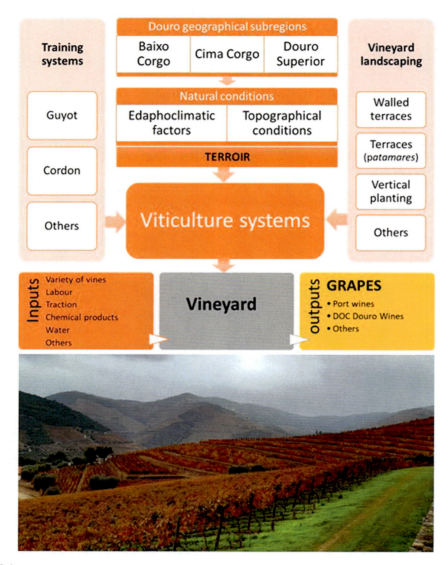

FIGURE 20.1

The heterogeneous vineyard production system of the Douro (top); Cordon trained vineyard with either vertical planting or large walled terraces in autumn (bottom figure).

(Cândido et al., 2015; González-Esquivel et al., 2020). The MESMIS model identifies the most important positive and limiting sustainability elements within the agro-ecosystem grouped in specific diagnostic criteria associated with the attributes of sustainability (productivity, stability, resilience, reliability, adaptability, equity, and self-management). The attribute of productivity gauges efficiency

20.4 Sustainability assessment essay for winegrowing systems

Table 20.4 MESMIS - "Framework for Assessing the Sustainability of Natural Resource Management Systems" indicators applied to the studied 110 vineyard farms from Douro, for the agricultural season 2016/2017.

Criteria	Indicator	Unit	Sustainability area
1. Efficiency	1. Grape yield	Kg/Ha	A
	2. Benefits/costs ratio	—	E
	3. Labor productivity	€/AWU	E
2. Conservation of natural resources	4. Use of fertilizers and chemicals	€/Ha	A
	5. Contribution for physical soil deterioration	Hour/Ha	A
3. System vulnerability	6. Entrepreneur and family income	€/Ha	E
	7. Labor force	Days/Ha	S
	8. Changes of activity in the last 10 years	%	S
	9. Continuity of the activity	%	S
4. Agro-ecological and socioeconomic constraints	10. Land structure	Ha, Number	S
	11. Landscape physiographic quality index	—	A
5. Capacity for change and innovation	12. Adoption of organic farming mode	%	A
	13. Adoption of new production techniques and systems	%	S
	14. Business or investment capacity	€/Ha	E
6. Strengthening of learning and training processes	15. Qualification	%	S
7. Cost and/or benefit distribution	16. Grape sales price	€/Kg	E
	17. Salary	€/Day	S
	18. Subsidies received	€/Ha	A
8. Social participation	19. Participation in sector organizations	%, Number	S
9. Self-sufficiency	20. Degree of dependence on external production factors	€/Ha	E
	21. Debt level	%	E
	22. Grape destination	%	E

Continued

Table 20.4 MESMIS - "Framework for Assessing the Sustainability of Natural Resource Management Systems" indicators applied to the studied 110 vineyard farms from Douro, for the agricultural season 2016/2017.—cont'd

Criteria	Indicator	Unit	Sustainability area
10. Autonomy for decision-making	23. Other sources of income	%	S
	24. Organization of the information	%	E
11. Environmental regulations	25. Waste management	%	A
	26. Soil analysis	%	A
	27. Good agrarian practices	%	A

(A) Environmental; (E) Economic; (S) Social; (Ha) Hectare; (AWU) Annual work unit.

of winegrowing systems and the stability, resilience, and reliability attributes encompass the factors affecting the equilibrium of systems and their surroundings. Adaptability expresses the ability of the system to strike a new equilibrium in its attempts to improve its situation. Equity evaluates how the system is able to distribute by the players and involved stakeholders, in an equitable manner, costs and benefits. Self-management is the ability of a system to control and regulate its interactions with the external world (González-Esquivel et al., 2020).

The best value of each indicator, gathered from the sampled vineyard systems was used as reference value, which assumes an index of 100. This methodology follows González-Esquivel et al. (2020) and is analogous to the economic efficiency measurements in the context of performance assessment (Santos et al., 2020). The data were normalized by transforming the specific indicators into a nondimensional value between 0 and 100.

Integration of multidimensional indicators varies between different evaluation sustainability assessments and it has been supported by aggregating indicators into a composite index (Gómez-Limón & Riesgo, 2009). To evaluate the Douro's sustainability, the final step was to aggregate the indicators into a composite index defined by methods used in the wine sector, in a triple bottom-line perspective just like the MESMIS or INSPIA. Comparison is supported on the previously defined set of indicators.

The MESMIS model aggregates the normalized value of the indicators in its criteria, then in attributes and, finally, in the final composite index through the arithmetic mean. In the INSPIA model, the indicators are weighted and transformed into component scores (aggregated indicators), which were aggregated into a composite score at level 1 (aggregation of basic sustainability indicators), and so on for level 2 (aggregation of aggregated indicators of level 1). The aggregate value of level 1 accounts for profitability, use of inputs, soil management, crop management, enhancing biodiversity, plant protection products (PPPs) management, contribute to GHGs, fertilizer inputs, labor practices, and decent work. The aggregate value of level 2 accounts for profitability, input productivity, farm management, environmental protection, climate change, fertilizer input, welfare, and well-being (Triviño-Tarradas et al., 2019, 2020). INSPIA's composite index is the result of the arithmetic mean of the three aggregated indicators, corresponding to the economic, social, and environmental dimensions (Triviño-Tarradas et al., 2019, 2020). In this work, the INSPIA procedure for aggregating the indicators on levels 1 and 2 followed by the global sustainability index was adapted to the previously defined set of MESMIS indicators.

20.4.3 Results and discussion

The analysis of the 27 MESMIS indicators showed a wide range of normalized values (Fig. 20.2) between 2.7 (indicator 12) and 98.2% (indicator 25) for the Douro's winegrowing systems that could be divided into the three main dimensions of sustainability.

According to the MESMIS assessment model, the indicators were aggregated in their criteria and likewise, into the five aggregated attributes of sustainability (productivity, stability, resilience, reliability, adaptability, equity, and self-management) (Fig. 20.3A).

Results show the low productivity (39.4%) of the Douro winegrowing systems, as well as its stability, resilience, and reliability (48.4%), which is somehow compensated by the equity (62.7%) and self-management (65.8%) with more satisfactory values. The MESMIS method had a composite index of sustainability of 51.4% (Fig. 20.3A).

The results obtained using the INSPIA model to aggregate multiple indicators were similar to the findings from the MESMIS model. The global level of sustainability remains very close to 50 (50.7, Fig. 20.3B). Profitability remained low (18.8), while PPPs management was strong (92.7%) due to the use of integrated production practices.

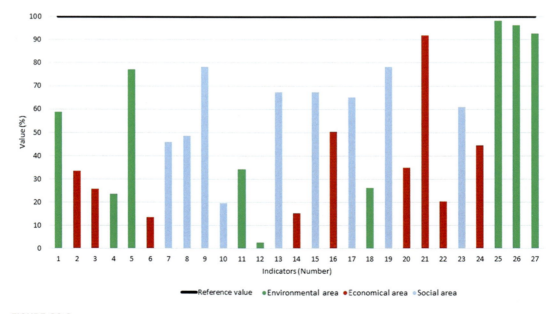

FIGURE 20.2

Normalized indicators of the studied Douro's farms, Portugal, for the agricultural season 2016/2017, by MESMIS method. The reference value is the best value of each indicator, which assumes index 100).

406 Chapter 20 Standards and indicators to assess sustainability

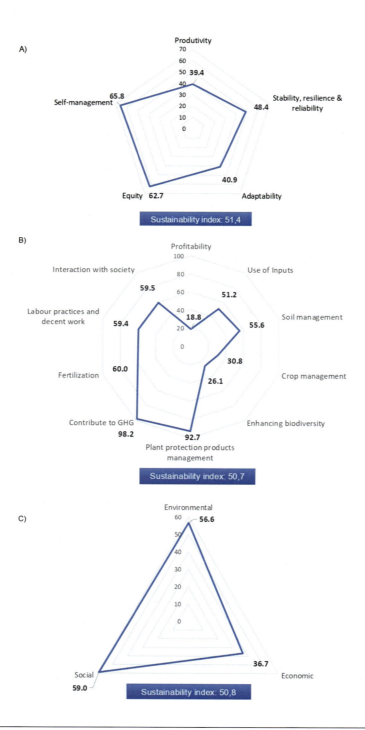

FIGURE 20.3

Aggregated indicators and composite sustainability index of the studied farms, for the agricultural season 2016/2017 by MESMIS (A) and INSPIA (B) models, and by the three sustainability dimensions (C) for the winegrowing systems of the Douro's wine region.

The aggregation of the three main sustainability indicators is shown in Fig. 20.3C. Despite the growing problem of labor shortage, the Douro's wine region reveals a greater sustainability in its social component (59.0), followed by the environmental dimension (56.6). The economic component (36.7) is the weakest for the region in accordance to the data on profitability (Fig. 20.3A and B) and in line with previous reports (Santos et al., 2019, 2020).

20.5 Future trends

The future for the wine industry should involve efforts to implement more standardized indicators and frameworks to assess vineyard, winery, and the other steps of the supply chain such as the transport and logistics (Garcia et al., 2012). This is particularly important for supply chains that span long distances, making transportation and storage of food products very energy-intensive (Taylor, 2017).

Benchmarking reports of logistics should be encouraged to compare stakeholders' performances. These reports would provide comparative data based on the site's history and aggregated data from similar businesses and identify opportunities to improve processes (Garcia et al., 2012). However, effective benchmarking can be limited by insufficient and ambiguous and inconsistent data. Fortunately, the tendency for increasing data availability may help to minimize the problem in the wine supply chain.

In the future, the wine industry will become more technologically and digitally driven. Novel smart vineyard systems provide dashboards and reports based on field sensors providing continuous information on weather, soil, and plant condition to predict vine's responses to stress (Matese and Di Gennaro, 2015). In wineries, more digital-based methods namely in applications to support quality control, made possible by increased offer of smart winery technology which permits to use more efficiently the inputs and reduce costs. In parallel, both winegrowing areas and wineries tend to be served by a high-speed Internet connection which makes possible faster data transfer and also storage. Such large amount of data can be processed and analyzed to support decision-making. Moreover, future technologies will tend to be cheaper, more user-friendly, and more robust (Brunel et al., 2019), which will benefit rural areas.

Precision viticulture and smart farming are ways to improve sustainability (Sarri et al., 2020), via data collection and analysis and more robust decision-making, which enables to reduce inputs and maximize yields and quality (Kitchen et al., 2005; FAO, 2016). Smart farming is based in the use of sophisticated technology, such as navigation satellite systems, soil scanning, data management, and internet of things to optimize processes and add value (Bekmezci, 2015). Gains in production efficiency contribute positively to sustainability of the wine supply chain.

Implementation of lean production strategies should be also considered for the wine sector to meet customer requirements with minimum waste and reduced costs (Banerjee & Ganjeizadeh, 2017; Hill & Hathaway, 2016). However, application of lean production strategies to wine production sector is still limited (Hill & Hathaway, 2016). In the Rioja region (Spain), lean production helped to tackle production problems and improved production and logistics (Jimenez et al., 2012). Other studies point to a reduction in material costs and improved productivity (Hill & Hathaway, 2016; Jasti & Kodali 2016).

Sarri et al. (2020) highlight five major categories of factors to successfully implement lean production in viticulture including knowledge and training, awareness of the operative context, organizational structure, and technology and decision support. Lean production can be applied in different contexts, but it largely depends on the organizational structure of companies (size, layout, and resources available) and the network of suppliers and customers (Longoni et al., 2013).

20.6 Concluding remarks

The sustainability performance assessment done for the Douro's wine region based on the systemic approach of MESMIS evaluation methodology illustrates the sustainability credentials of the region regarding its economic, social, and environmental characteristics. The region showed better performance at social and environmental levels, but it shows a weak performance at economical level due to low productive efficiency, indicating the economic pillar as the weakest of the three sustainability dimensions.

Looking at methodological issues, improved results were obtained when aggregating measured indicators into a composite index by the MESMIS or INSPIA evaluation approaches. In both cases, the low productivity of the Douro winegrowing systems is an important factor and agrees with previous studies based on the intrinsic local/regional difficulties such as its limited accessibility/remoteness and topography, and the higher production costs per hectare (Graça et al., 2017; Santos et al., 2019, 2020).

High-altitude viticulture and grape growers are surviving on the threshold of economic viability. There are also increasing challenges due to climate change (Cichelli et al., 2016; Fraga, 2020), labor shortage, and limitations in the mechanization. The Douro wine region is currently (and will be) in a weak position if the region does not improve its stability, resilience, and adaptability to future threats.

Overall, this work points out the need of more research on sustainability issues and on more effective methods to assess it. Improved evaluation methodologies of sustainability performance can help to improve winegrowing systems and environmental policies to generate a more sustainable development. The wine sector needs more standardized sustainability indicators to support auditing and certification programs and the social component of sustainability must be properly taken into account. Certifications should value the educational component (Moscovici & Reed, 2018) and skills of human resources, which are crucial for socioeconomic performance and innovation (Garcia et al., 2012). Concepts such as "social responsibility", "socioeconomic equity", or "social enterprise" are emerging and translate an increasing social motivation of companies (Nilipour, 2020; Taylor, 2017).

Acknowledgments

This work was supported by the Interreg SUDOE project SOE3/P2/F0917, VINCI - Wine, Innovation, and International Competitiveness funded by FEDER; by the 2019 I&D Research Award from the Fundação Maria Rosa, and through the FCT (Portuguese Foundation for Science and Technology) under the projects UIDB/04011/2020 and UIDB/04007/2020.

References

Abdul Murad, S. M., Hashim, H., Jusoh, M., & Zakaria, Z. Y. (2019). Sustainability assessment framework : A mini review of assessment concept. *Chemical Engineering Transactions, 7*, 379–384.

Annunziata, E., Pussi, T., Frey, M., & Zanni, L. (2018). The role of organizational capabilities in attaining corporate sustainability practices and economic performance: Evidence from Italian wine industry. *Journal of Cleaner Production, 171*, 1300—1311.

Arcese, G., Lucchetti, M. C., & Massa, I. (2016). Modeling social life cycle assessment framework for the Italian wine sector. *Journal of Cleaner Production, 140*, 1027—1036.

Atanda, J. O., & Öztürk, A. (2020). Social criteria of sustainable development in relation to green building assessment tools. *Environment, Development and Sustainability, 22*, 61—87.

Banerjee, A., & Ganjeizadeh, F. (2017). Modeling a leagility index for supply chain sustenance. *Procedia Manufacturing, 11*, 996—1003. https://doi.org/10.1016/j.promfg.2017.07.205

Barbosa, F. S., Scavarda, A. J., Sellitto, M. A., & Lopes Marques, D. I. (2018). Sustainability in the winemaking industry: An analysis of Southern Brazilian companies based on a literature review. *Journal of Cleaner Production, 192*, 80—87.

Bekmezci, M. (2015). Companies' profitable way of fulfilling duties towards humanity and environment by sustainable innovation. *Procedia, Social and Behavioral Sciences, 181*, 228—240.

Bélanger, V., Vanasse, A., Parent, D., Allard, G., & Pellerin, D. (2015). Delta : An integrated indicator-based self-assessment tool for the evaluation of dairy farms sustainability in Quebec, Canada. *Agroecology and Sustainable Food Systems, 39*, 1022—1046.

Bockstaller, C., Guichard, L., Makowski, D., Aveline, A., Girardin, P., & Plantureux, S. (2008a). Agri-environmental indicators to assess cropping and farming systems. A review. *Agronomy for Sustainable Development, 28*(1), 139—149.

Bockstaller, C., Wohlfahrt, J., Hubert, A., Hennebert, P., Zahm, F., Vernier, F., Mazella, N., Keichinger, O., & Girardin, P. (2008b). *Les indicateurs de risque de transfert de produits phytosanitaires et leur validation: exemple de l'indicateur I-Phy* (pp. 103—114). Ingénieries Sciences Eaux & Territoires.

Brocado, L., & Zicari, A. (2020). Sustainability as a driver for value creation: A business model analysis of small and medium entreprises in the Italian wine sector. *Journal of Cleaner Production, 259*, 120852.

Brunel, G., Pichon, L., Taylor, J., & Tisseyre, B. (2019). Easy water stress detection system for vineyard irrigation management. In *12th European conference on precision agriculture, ECPA 2019* (pp. 935—942). July 2019, Montpellier, France.

Cândido, G., Nóbrega, M., Figueiredo, M., & Souto Maior, M. (2015). Sustainability assessment of agroecological production units: A comparative study of idea and MESMIS methods. *Ambiente & Sociedade, 3*, 99—120.

Certomà, C., & Migliorini, P. (2011). The evaluation of sustainability of organic farms in Tuscany. In H. Gökçekus, U. Türker, & J. LaMoreaux (Eds.), *Environmental earth scienceSurvival and sustainability: Environmental concerns in the 21th century* (pp. 165—177). Verlag Berlin Heidelberg: Springer.

Christ, K. L., & Burritt, R. L. (2013). Critical environmental concerns in wine production: An integrative review. *Journal of Cleaner Production, 53*, 232—242.

Cichelli, A., Pattara, C., & Petrella, A. (2016). Sustainability in mountain viticulture. The case of the Valle Peligna. *Agriculture and Agricultural Science Procedia, 8*, 65—72.

Corbo, C., Lamastra, L., & Capri, E. (2014). From environmental to sustainability programs: A review of sustainability initiatives in the Italian wine sector. *Sustainability, 6*, 2133—2159.

Costantini, E. A. C., & Bucelli, P. (2014). Soil and terroir. In S. Kapur, & S. Ersahin (Eds.), *Soil security for ecosystem management* (pp. 97—133). Springer International Publishing.

Costantini, E. A. C., Lorenzetti, R., & Malorgio, G. (2016). A multivariate approach for the study of environmental drivers of wine economic structure. *Land Use Policy, 57*, 53—63.

Costa, J. M., Oliveira, M., Egipto, R., Cid, F., Fragoso, R., Lopes, C. M., & Duarte, E. (2020). Water and wastewater management for sustainable viticulture and oenology in south Portugal — a review. *Ciência Técnica Vitícola, 35*(1), 1—15.

CSWA - California Sustainable Winegrowing Alliance. (2015). *California wine community sustainability.* Report (WWW Document) http://www.sustainablewinegrowing.org/.

Dantsis, T., Douma, C., Giourga, C., Loumou, A., & Polychronaki, E. A. (2010). A methodological approach to assess and compare the sustainability level of agricultural plant production systems. *Ecological Indicators, 10*, 256–263.

Elkington, J. (1994). Towards the sustainable corporation: Win-win-win business strategies for sustainable development. *California Management Review, 36*(2), 90–100.

FAO. (2016). *Citrus fruit fresh and processed.* Ccp.Ci/St/2016. FAO.

Faria, S. S., Lourenço-Gomes, L. S. M., Gouveia, S. H. C., & Rebelo, J. F. (2020). Economic performance of the Portuguese wine industry: A microeconometric analysis. *Journal of Wine Research, 31*(4), 283–300.

Ferrara, C., & Feo, G. (2018). Life cycle assessment application to the wine sector: A critical review. *Sustainability, 10*(395).

Flores, S. S. (2018). What is sustainability in the wine world? A cross-country analysis of wine sustainability frameworks. *Journal of Cleaner Production, 172*, 2301–2312.

Fraga, H. (2020). Climate change: A new challenge for the winemaking sector. *Agronomy, 10*(10), 1465.

Fragoso, R., & Figueira, J. R. (2021). Sustainable supply chain network design: An application to the wine industry in Southern Portugal. *Journal of the Operational Research Society, 72*(6), 1236–1251. https://doi.org/10.1080/01605682.2020.1718015

Fragoulis, G., Trevisan, M., Di Guardo, A., Sorce, A., van der Meer, M., Weibel, F., & Capri, E. (2009). Development of a management tool to indicate the environmental impact of organic viticulture. *Journal of Environmental Quality, 38*, 826–835.

Gafsi, M., & Favreau, J. L. (2013). Indicator-based method for assessing organic farming sustainability. In A. A. Marta-Costa, & E. Silva (Eds.), *Methods and procedures for building sustainable farming systems* (pp. 175–187). Springer Netherlands.

Garcia, F. A., Marchetta, M. G., Camargo, M., Morel, L., & Forradellas, R. (2012). A framework for measuring logistics performance in the wine industry. *International Journal of Production Economics, 135*(1), 284–298.

Gilinsky, A., Newton, S. K., & Vega, R. F. (2016). Sustainability in the global wine industry: Concepts and cases. *Agriculture and Agricultural Science Procedia, 8*, 37–49.

Gómez-Limón, J. A., & Riesgo, L. (2009). Alternative approaches to the construction of a composite indicator of agricultural sustainability: An application to irrigated agriculture in the Duero basin in Spain. *Journal of Environmental Manager, 90*, 3345–3362.

González-Esquivel, C. E., Camacho-Moreno, E., Larrondo-Posadas, L., Sum-Rojas, C., León-Cifuentes, W. E., Vital-Peralta, E., Astier, M., & López-Ridaura, S. (2020). Sustainability of agroecological interventions in small scale farming systems in the Western Highlands of Guatemala. *International Journal of Agricultural Sustainability, 18*(4), 285–299.

Graça, A. R., Simões, L., Freitas, R., Pessanha, M., & Sandeman, G. (2017). Using sustainable development actions to promote the relevance of mountain wines in export markets. *Open Agriculture, 2*, 571–579.

GRI. (2016). *GRI standards.* Retrieved from https://www.globalreporting.org/standards (Accessed on 7 January 2016).

Hayati, D. (2017). *A literature review on frameworks and methods for measuring and monitoring sustainable agriculture.* Technical Report n.22. Global Strategy Technical Report: Rome.

Hill, M., & Hathaway, S. (2016). *Case study: The adoption of lean production in Australian wineries. Using the "market, message and means of communication" framework to design an extension strategy.* Retrieved from https://www.wineaustralia.com/getmedia/c2cc36f6-3186-487f-a63d-9ba873671bf6/Adoption-of-RD-Lean-production-case-study.

IVV. (2021). *Dados estatísticos do setor vitivinícola.* Retrivied from https://www.ivv.gov.pt/np4/estatistica/ (Accessed on 5 February 2021).

Iyer, E. S., & Reczek, R. W. (2017). The intersection of sustainability, marketing, and public policy: Introduction to the special section on sustainability. *Journal of Public Policy and Marketing, 36*(2), 246–254.

Jasti, N. V. K., & Kodali, R. (2016). Development of a framework for lean production system: An integrative approach. *Proceedings of the Institution of Mechanical Engineers, Part B: Journal of Engineering Manufacture, 230*(1), 136–156.

Jimenez, E., Tejeda, A., Perez, M., Blanco, J., & Martinez, E. (2012). Applicability of lean production with VSM to the Rioja wine sector. *International Journal of Production Research, 50*(7), 1890–1904.

Kamali, P. F., Borges, J. A. R., Meuwissen, M. P. M., Boer, I. J. M., & Lansink, A. G. J. M. O. (2017). Sustainability assessment of agricultural systems : The validity of expert opinion and robustness of a multi-criteria analysis. *Agricultural Systems, 157*, 118–128.

Keichinger, O., & Thiollet-Scholtus, M. (2017). Soeco: Indicateurs socio-économiques pour la viticulture et les systèmes de culture innovants. *BIO Web of Conferences 40th World Congress of Vine and Wine, 9*, 04012.

Kitchen, N. R., Sudduth, K. A., Myers, D. B., Drummond, S. T., & Hong, S. Y. (2005). Delineating productivity zones on claypan soil fields using apparent soil electrical conductivity. *Computers and Electronics in Agriculture, 46*, 285–308.

Lamastra, L., Balderacchi, M., Di Guardo, A., Monchiero, M., & Trevisan, M. (2016). A novel fuzzy expert system to assess the sustainability of the viticulture at the wine-estate scale. *The Science of the Total Environment, 572*, 724–733.

Longoni, A., Pagell, M., Johnston, D., & Veltri, A. (2013). When does lean hurt?—an exploration of lean practices and worker health and safety outcomes. *International Journal of Production Research, 51*, 3300–3320.

Luzzani, G., Lamastra, L., Valentino, F., & Capri, E. (2020). Development and implementation of a qualitative framework for the sustainable management of wine companies. *The Science of the Total Environment, 759*(4), 143462. https://doi.org/10.1016/j.scitotenv.2020.143462

Malorgio, G., Grazia, C., Caracciolo, F., & De Rosa, C. (2013). Determinants of wine bottling strategic decisions: Empirical evidence from the Italian wine industry. In E. Giraud-Héraud, & M. Pichery (Eds.), *Wine economics: Quantitative studies and empirical application* (pp. 266–296). New York, NY: Palgrave Macmillan.

Marta-Costa, A. A. (2010). Sustainability study for the rearing of Bovine livestock in mountainous zones. *New Medicine, IX*(1), 4–12.

Marta-Costa, A. A., & Silva, E. (2013a). Approaches for sustainable farming systems assessment. In A. A. Marta-Costa, & E. Silva (Eds.), *Methods and procedures for building sustainable farming systems* (pp. 21–29). Netherlands: Springer.

Marta-Costa, A. A., & Silva, E. (2013b). Methodologies for building sustainable farming systems: The main critical points and questions. In A. A. Marta-Costa, & E. Silva (Eds.), *Methods and procedures for building sustainable farming systems* (pp. 273–277). Netherlands: Springer.

Martins, A. A., Araújo, A. R., Graça, A., Caetano, N. S., & Mata, T. M. (2018). Towards sustainable wine: Comparison of two Portuguese wines. *Journal of Cleaner Production, 183*, 662–676.

Matese, A., & Di Gennaro, S. F. (2015). Technology in precision viticulture: A state of the art review. *International Journal of Wine Research, 7*, 69–81.

Matias, J., Cerveira, A., Santos, C., & Marta-Costa, A. (2021). Influência do Preço da Mão de Obra na Sustentabilidade das Explorações Vitícolas Durienses: Uma Aplicação de ABM. *Revista de Economia e Sociologia Rural, 59*(1), e238886.

Matos, C., & Pirra, A. (2020). Water to wine in wineries in Portugal Douro region: Comparative study between wineries with different sizes. *The Science of the Total Environment, 732*, 139332.

McGrath, R. G. (2010). Business models: A discovery driven approach. *Long Range Planning, 43*(2–3), 247–261.

Merli, R., Preziosi, M., & Acampora, A. (2018). Sustainability experiences in the wine sector: Toward the development of an international indicators system. *Journal of Cleaner Production, 172*, 3791–3805.

Moscovici, D., & Reed, A. (2018). Comparing wine sustainability certifications around the world: History, status and opportunity. *Journal of Wine Research, 29*(1), 1−25.

Nilipour, A. (2020). Introduction to social sustainability. In S. L. Forbes, T. Silva, & A. Gilinsky (Eds.), *Social Sustainability in the global wine industry* (pp. 1−14). Switzerland: Palgrave MacMillan.

OECD. (2004). Statistics, knowledge and policy: Composite indicators of environmental sustainability. In *OECD world forum on key indicators*. Palermo: OECD Publications.

Olaru, O., Galbeaza, M. A., & Bănacu, C. S. (2014). Assessing the sustainability of wine industry in terms of investment. *Procedia Economics and Finance, 15*, 552−559.

Olde, E. M., Oudshoorn, F. W., Sørensen, C. A. G., Bokkers, E. A. M., & Boer, I. J. M. (2016). Assessing sustainability at farm-level: Lessons learned from a comparison of tools in practice. *Ecological Indicators, 66*, 391−404.

Parent, D., Bélanger, V., Vanasse, A., Allard, G., & Pellerin, D. (2013). Method for the evaluation of farm sustainability in Quebec, Canada: The social aspect. In A. A. Marta-Costa, & E. Silva (Eds.), *Methods and procedures for building sustainable farming systems* (pp. 239−250). Netherlands: Springer.

Payraudeau, S., & Gregoire, C. (2012). Modelling pesticides transfer to surface water at the catchment scale: A multi-criteria analysis. *Agronomy for Sustainable Development, 32*(2), 479−500.

Point, E., Tyedmers, P., & Naugler, C. (2012). Life cycle environmental impacts of wine production and consumption in Nova Scotia, Canada. *Journal of Cleaner Production, 27*, 11−20.

Pomarici, E., & Vecchio, R. (2019). Will sustainability shape the future wine market? *Wine Economics and Policy, 8*(1), 1−4.

Provance, M., Donnelly, R. G., & Carayannis, E. G. (2011). Institutional influences on business model choice by new ventures in the microgenerated energy industry. *Energy Policy, 39*(9), 5630−5637.

Ramos, T. B. (2019). Sustainability assessment: Exploring the frontiers and paradigms of indicator approaches. *Sustainability, 11*(824).

Santiago-Brown, I., Metcalfe, A., Jerram, C., & Collins, C. (2014). What does sustainability mean? Knowledge gleaned from applying mixed methods research to wine grape growing. *Journal of Mixed Methods Research, 1*(20), 232−251.

Santiago-Brown, I., Metcalfe, A., Jerram, C., & Collins, C. (2015). Sustainability assessment in wine-grape growing in the new world: Economic, environmental, and social indicators for agricultural businesses. *Sustainability, 7*, 8178−8204.

Santos, M., Galindro, A., Santos, C., Marta-Costa, A., & Martinho, V. (2019). Sustainability evolution of North and Alentejo vineyard regions. *Revista Portuguesa de Estudos Regionais, 50*, 49−63.

Santos, M., Rodríguez, X. A., & Marta-Costa, A. (2020). Efficiency analysis of viticulture systems in the Portuguese Douro region. *International Journal of Wine Business Research, 32*(4), 573−591.

Sarri, D., Lombardo, S., Pagliai, A., Lisci, R., De Pascale, V., Rimediotti, M., Cencini, G., & Vieri, M. (2020). Smart farming introduction in wine farms: A systematic review and a new proposal. *Sustainability, 12*, 7191.

Schaltegger, S., Hansen, E. G., & Lüdeke-Freund, F. (2016). Business models for sustainability: Origins, present research, and future avenues. *Organization & Environment, 29*(1), 3−10.

Singh, R. K., Murty, H. R., Gupta, S. K., & Dikshit, A. K. (2012). An overview of sustainability assessment methodologies. *Ecological Indicators, 15*, 281−299.

Sopilko, N., Myasnikova, O., Shamseev, S., & Kubasova, E. (2019). Modern methods to sustainable development assessment. In M. Markovic, B. Dukanovic, & N. Vukovic (Eds.), *Economy and ecology: Contemporary trends and contradictions* (pp. 209−214) (Moscow, Russia).

Strano, A., De Luca, A. I., Falcone, G., Iofrida, N., Stillitano, T., & Gulisano, G. (2013). Economic and environmental sustainability assessment of wine grape production scenarios in Southern Italy. *Agricultural Sciences, 4*, 12−20.

Szolnoki, G. (2013). A cross-national comparison of sustainability in the wine industry. *Journal of Cleaner Production, 53*, 243−251.
Taylor, S. (2017). *Business of sustainable wine: How to build brand equity in a 21 century wine industry*. Board and Bench Publishing.
Thiollet-Scholtus, M., & Bockstaller, C. (2015). Using indicators to assess the environmental impacts of wine growing activity: The INDIGO® method. *European Journal of Agronomy, 62*, 13−25.
Trigo, A., Marta-Costa, A., & Fragoso, R. (2020). Benchmarking of sustainability assessment tools: Limitations, gaps and potentialities for the agrarian sector. In D. Vrontis, Y. Weber, & E. Tsoukatos (Eds.), *Book proceedings of 13th annual conference of the EuroMed academy of business* (pp. 1145−1164p). EuroMed Press.
Triviño-Tarradas, P., Carranza-Cañadas, P., Mesas-Carrascosa, F. J., & Gonzalez-Sanchez, E. J. (2020). Evaluation of agricultural sustainability on a mixed vineyard and olive-grove farm in southern Spain through the INSPIA model. *Sustainability, 12*(1090), 1−23.
Triviño-Tarradas, P., Gomez-Ariza, M. R., & Basch, G. (2019). Sustainability assessment of annual and permanent crops: The inspia model. *Sustainability, 11*(738), 1−23.
United Nations. (2007). *Indicators of sustainable development: Guidelines and methodologies* (3rd ed.). New York: United Nations Publication, ISBN 978-92-1-104577-2. Retrieved from https://sustainabledevelopment.un.org/content/documents/guidelines.pdf.
United Nations Environment Programme (UNEP. (2014). *Sustainability metrics: Translation and impact on property investment and management*. Retrieved from http://www.unepfi.org/fileadmin/documents/UNEPFI_SustainabilityMetrics_Web.pdf.
Van Passel, S., & Meul, M. (2012). Multilevel and multi-user sustainability assessment of farming systems. *Environmental Impact Assessment Review, 32*, 170−180.
Varsei, M., & Polyakovskiy, S. (2017). Sustainable supply chain network design: A case of the wine industry in Australia. *Omega, 66*, 236−247.
Vázquez-Rowe, I., Rugani, B., & Benetto, E. (2013). Tapping carbon footprint variations in the European wine sector. *Journal of Cleaner Production, 43*, 146−155.
Velten, S., Leventon, J., Jager, N., & Newig, J. (2015). What is sustainable agriculture? A systematic review. *Sustainability, 7*, 7833−7865.
Vinos de Chile. (2014). *Codigo nacional de sustentabilidad de la industria vitivinícola chilena*. Retrieved from http://www.sustentavid.org.
World Business Council for Sustainable Development. (2000). *Eco-efficiency: Creating more with less impact*. World Business Council for Sustainable Development.
Zahm, F., Viaux, P., Vilain, L., Girardin, P., & Mouchet, C. (2008). Assessing farm sustainability with the IDEA method: From the concept of agriculture sustainability to case studies on farms. *Sustainable Development, 16*, 271−281.

Further reading

Baiano, A. (2021). An Overview on Sustainability in the Wine Production Chain. *Beverages, 7*(1), 15. https://doi.org/10.3390/beverages7010015
Corbett, C. J., & Kirsch, D. A. (2001). International diffusion of ISO 14001 certification. *Production and Operations Management, 10*(3), 327−342.
Forbes, S. L., Silva, T.-A., & Gilinsky, A., Jr. (2019). *Social Sustainability in the Global Wine Industry: Concepts and Cases*. Palgrave Pivot: Springer International Publishing. https://doi.org/10.1007/978-3-030-30413-3
Gore, T., Alestig, M., Banerji, S., & Ceccarelli, G. (2021). The Workers Behind Sweden's Italian Wine: An Illustrative Human Rights Impact Assessment of Systembolaget's Italian Wine Supply Chains. *Oxfam Research Report*. https://doi.org/10.21201/2021.7703

Maicas, S., & Mateo, J. J. (2020). Sustainability of Wine Production. *Sustainability, 12*(2), 559. https://doi.org/10.3390/su12020559

Robinson, J. (2004). Squaring the circle? Some thoughts on the idea of sustainable development. *Ecological Economics, 48*(4), 369–384.

Santos, M., Rodríguez, X. A., & Marta-Costa, A. (2021). Productive efficiency of wine grape producers in the North of Portugal. *Wine Economics and Policy, 10*(2), 3–14. https://doi.org/10.36253/wep-8977

Schaltegger, S., & Sturm, A. (1990). Öologische rationalität (German/in English: Environmental rationality). *Unternehmung, 4*, 117–131.

Trigo, A., Marta-Costa, A., & Fragoso, R. (2021). Principles of Sustainable Agriculture: Defining standardised reference points. *Sustainability, 13*(8), 4086. https://doi.org/10.3390/su13084086

van Leeuwen, C., & Seguin, G. (2006). The concept of terroir in viticulture. *Journal of Wine Research, 17*(1), 1–10.

Varas, M., Basso, F., Maturana, S., Pezoa, R., & Weyler, M. (2020). Measuring Efficiency in the Chilean Wine Industry: A Robust DEA Approach. *Applied Economics, 53*(9), 1092–1111. https://doi.org/10.1080/00036846.2020.1826400

CHAPTER 21

The guardianship of Aotearoa, New Zealand's grape and wine industry

Victoria Raw[1], Sophie Badland[2], Meagan Littlejohn[2], Marcus Pickens[3] and Lily Stuart[1]

[1]*The New Zealand Institute for Plant and Food Research Limited, Blenheim, New Zealand;* [2]*New Zealand Winegrowers Inc., Auckland, New Zealand;* [3]*Wine Marlborough, Blenheim, New Zealand*

21.1 Introduction

The concept of sustainability emerged in the 1990s as an important value for businesses, governments, academics, and not least of all for individuals. In 1992, the United Nations (UN) convened the Earth Summit in Rio de Janeiro and produced the UN Framework Convention on Climate Change 'to prevent "dangerous" human interference with the climate system' (United Nations, 2020a). This convention was ratified by 197 countries and was followed by the Kyoto Protocol in 1995, the Paris Agreement in 2015 and the 2019 Climate Action Summit. New Zealand (NZ) is a signatory to these as well as to other international climate change forums, such as the Intergovernmental Panel on Climate Change, the Carbon Neutrality Coalition, and the Pacific Islands Forum (NZ Foreign Affairs & Trade, 2020a).

The term "The Triple Bottom Line" was first used in 1994 (Elkington, 2006) and looked at how environmental, economic, and social values add value to organisations' performances. This has become a widely adopted term. According to Nilipour (2020), sustainability is not only based on economic, social, and environmental developments, but in addition, culture and governance should form part of its framework. The framework should incorporate how culture helps to drive decisions and recognize the importance of regulatory structures, transparency and governance becoming part of business protocols. Nilipour continued by saying that social sustainability is "the least quantifiable … since it cannot be easily measured through metrics" (Nilipour, 2020). In practice, the relationship between a business and society has been referred to as "corporate social responsibility" (CSR), where an organization has economic, legal, ethical, and philanthropic responsibilities toward society (Carroll, 1991).

The UN has created 17 Sustainable Development Goals "to promote prosperity while protecting the planet" by 2030 (UN, 2020b). The goals aim to end poverty by enabling economic and social growth while ensuring that environmental challenges are tackled and the natural environment protected. Although these goals are laudable, they can be achieved only with support from society, government, business, and nongovernmental organizations.

However, long before these sustainability goals and the various international climate forums were established by the UN, Māori (the indigenous peoples of NZ) had long-held sustainability practices and beliefs. Underlying values and principles have helped to guide activities to work in harmony with the environment. Māori people believe that all living things are interconnected through *whakapapa* (genealogy). *Tohu* (indicators), core values and principles determine many cultural practices. The guiding value of *kaitiakitanga* (guardianship) is for the long-term sustainability of the environment whereby a sustainable and balanced ecosystem is one where fauna and flora coexist for the benefit of the people now and for future generations. Māori also believe that, given time, the environment will heal itself.

Care for the environment, the poor and the marginalized are also long-held Judaeo-Christian and Islamic values, among other world faiths, which are extolled in the Old and New Testaments of the Bible and the Quran. More recently Pope Francis (2015), in his encyclical Laudato Si', wrote that "the natural environment is a collective good, the patrimony of all humanity and the responsibility of everyone" (paragraph 95) and that "we should be concerned for future generations" (p. 160).

Sustainability, therefore, is not a quick-fix; rather it is a process that requires a long-term commitment where benefits are not immediate.

The first grapevines in NZ were planted in 1819 in Kerikeri in the Bay of Islands and the first wine was made in the 1830s in nearby Waitangi. With further European settlement came an expansion of wine production in Hawke's Bay in 1851. This was followed by other new migrants who planted vines and made wine in the Auckland region. In time vineyards and wineries were established throughout the country so that by 2020 there were 39,935 ha of vineyards from Northland, at the top of the North Island, to Central Otago, in the south of the South Island (Fig. 21.1), spanning some 1750 km and 717 wineries (New Zealand Wine, 2020a).

The rapid expansion of the wine industry started in earnest in Marlborough in the early 2000s and it became NZ's largest grape growing region. There were 6350 ha in grapes in 2002; by 2006 this had increased to 17,720 ha (Marlborough District Council, 2020); and in 2020 there were 27,808 ha; Hawke's Bay is NZ's second largest grape growing region and from 2002 to 2020 it expanded from 3463 ha to 5034 ha (Zealand Wine, 2020a). By November 2020, NZ's wine industry had reached NZD $2 billion (USD $1.4 billion) in exports (New Zealand Wine, 2020b) (Fig. 21.2).

21.2 NZ's Māori heritage

Aotearoa (NZ) is a bicultural country (Te Ara, 2020). The Treaty of Waitangi, signed on February 6, 1840 between Māori chiefs and the British Crown, is NZ's founding document, although it was never enacted into a statute of law. The Treaty was based on a set of principles that were to form a nation, create a government and British authority. The English and Māori versions of the Treaty differed in their meanings particularly over land and sea ownership, rights and sovereignty. The three articles of the treaty are partnership, protection, and participation. The many breaches of the Treaty saw the creation of the Waitangi Tribunal, in 1975, whereby Māori claims of Treaty violations by the Crown could be redressed and settlements made (NZ History, 2020).

The settlement processes have seen Māori emerge as genuine participants and partners in the wine industry. The compensation packages have included both financial recompense and land repatriation, leading to Māori-owned land to be converted into vineyards for wine production.

21.2 NZ's Māori heritage 417

FIGURE 21.1

Map of New Zealand's wine regions.

Used with permission from New Zealand Winegrowers Incorporated.

21.2.1 Māori values and principles

Māori are streered by core values and principles that enable them to coexist with the environment they interact and live in. These are guided by *tikanga* (customs) and *kawa* (rules); *tikanga* determines what is *tika* (right) and *kāore* (wrong).

For Māori, *kaitiakitanga* is the value of guardianship of the environment. Others include *rangatiratanga* (authority), *whakapapa* (connections), *whānaungatanga* (relationships), *manākitanga* (hospitality), *kotahitanga* (unity), *aroha* (love), and *turangawaewae* (sense of belonging). These values influence the interactions of Māori *kaitiaki* (guardians) with the environment to enhance wellness with other species both physically and spiritually. It is through the values, rules, and the practices of *tikanga* that the guardianship of resources takes place (Harmsworth, 2005). This knowledge and understanding is called *mātauranga* and is dynamic and evolving. *Mātauranga* also shapes language, medicinal practices, weaving, carving, fishing, hunting, and farming practices.

FIGURE 21.2

Vineyards in the Awatere Valley, Marlborough.

Used with permission from the Institute of Plant and Food Research Ltd.

21.2.2 Kaitiakitanga

Kaitiakitanga is the conservation, preservation, and nurturing of the natural world for current and future generations (Haymes, 1998). It has both spiritual and practical meanings that cannot be separated from each other (Harmsworth, 2005; Kawharu, 2000; Royal, 2007). *Kaitiakitanga* links ancestry, environmental issues, and social practices together through responsible management. Resources, therefore, are to be passed on in a better state than when they were received as everything has *mauri* (life) and is *tapu* (sacred). *Kaitiaki* are the guardians that have the knowledge and responsibility to retain, preserve, and maintain the land and all that it contains for their people and future descendants.

21.2.3 Māori and the wine industry

In 2018, a collective, called Tuku, was formed by 5 Māori winemaking families from across *Aotearoa* (NZ). Each family identified with a different *iwi* (tribe) but together shared the values of *kaitiakitanga*, *whakapapa*, *whānaungatanga*, and *manākitanga* in their grape growing, winemaking, and business practices (Tuku Wines, 2019). Tuku is taken from the word *tukutuku* (traditional Māori weaving), synonymous with collaboration, social activity, and bringing people together to create something. This unique collective was formed to support the Māori economy, for the long-term sustainability of the land and for those that work on it.

Another Māori enterprise based in the Nelson and Marlborough regions is Kono, NZ, which produces wine, apples, pears, hops, kiwifruit, and seafood (Kono, 2020). Its principal guide is Te Pae Tawhiti, a 500-year plan for intergenerational success by using sustainable and ethical values; retaining and preserving its cultural heritage; and ensuring enduring support for its products. It has two major wine brands—Kono Wines and Tohu Wines. The latter was the first indigenous wine company to export its wine. *Kaitiakitanga* is one of its core values of protecting, sustaining, and enhancing the natural environment, culture, and people so that Tohu's wine is deeply connected to its origins (Harmsworth, 2005).

Although *kaitiakitanga* is at the fore of these organizations' long-term strategies, the main driver for them is people. Involving their people in all aspects of their business is seen as the enduring answer to the sustainability of their land, environment, business, and people. A 500-year strategy involves using traditional knowledge to develop ecosystems that are less reliant on artificial inputs; it is about integrating a variety of native plants into vineyards; and encouraging the younger generation into all aspects of the business from crop production, winemaking, and marketing.

NZ's Māori heritage places it in a unique position to continue developing its guardianship and stewardship of resources through heritage, hospitality, family ties, love, and physical and cultural practices. In the section on CSR, a range of activities by non-Māori grape growers and winemakers is described. These actions show that they too are guardians demonstrating their desire to enhance the environments in which they live and work.

21.3 New Zealand Winegrowers

Established in 2002, New Zealand Winegrowers Inc. (NZW) is the national organization for NZ's grape and wine industry. It is the "only unified national wine industry body in the world" (New Zealand Wine, 2020c) and has a range of departments that offer services and support for the country's grape growers and winemakers, such as research, advocacy, marketing, and events. Sustainability is a core value for NZW and the greater wine industry; it is a guiding principle for all NZW activities. NZW has identified six key focus areas of sustainability that are aligned with the UN Sustainable Development Goals. Each focus area has an overarching industry goal (New Zealand Wine, 2020d):

- Water: The NZ wine industry will be a world leader in the efficient use and the protection of water quality
- Waste: By 2050 the NZ wine industry will achieve zero waste to landfill
- Pest and disease: To understand, reduce, and mitigate the impacts of existing and potential pests and diseases, as well as being a world leader in sustainable alternatives
- Climate change: By 2050 the NZ wine industry will be carbon neutral
- People: To become an industry of choice for workers
- Soil: To protect and enhance soil health.

21.3.1 Sustainable Winegrowing New Zealand

Sustainable Winegrowing New Zealand (SWNZ) is the sustainability accreditation program owned and operated by NZW. SWNZ was established in 1995 and internationally was one of the first wine industry sustainability accreditation programmes (New Zealand Wine, 2020d). SWNZ was developed during a period of rapid vineyard expansion in NZ to provide grape growers and winemakers with a best practice model for sustainability in vineyards and wineries. The program was designed to ensure better quality assurance throughout the entire production chain (from grapes to glass), to address regulators and consumers' sustainability requirements, verification of claims and the industry's license to operate (New Zealand Wine, 2020d).

SWNZ is recognized as a world-leading program, with 96% of NZ's vineyard producing area being SWNZ-certified (New Zealand Wine, 2020e). The program is also framed around NZW's six focus areas listed previously. SWNZ certification is an annual process (New Zealand Wine, 2020f). Every SWNZ member must submit an annual questionnaire, which contains qualitative and quantitative questions. Members record information about their management practices, sustainability activities, and specific measureable records (e.g., total volume water used). Members are assigned corrective actions if there are any SWNZ standards that are not met. These corrective actions must be completed within an agreed timeframe before SWNZ accreditation can be granted for the next season (New Zealand Wine, 2020d).

All SWNZ-certified vineyards also submit an annual online spray diary. This is a full record of all agrichemical inputs that have been applied to each vineyard block during the past season as well as the harvest date. The compliance processing ensures that only approved products have been applied and that they have been used at the correct application rate and time of the season (New Zealand Wine, 2020g). NZW produces an annual spray schedule that acts as the agrichemical rule-book for SWNZ members. The spray schedule coupled with the annual spray diary submission helps to mitigate residue risks and enhances the traceability of NZ-produced wines (New Zealand Wine, 2020h).

SWNZ is an independently audited certification program. Every member is audited on-site at least every three years to verify the annual information. The independent auditors look at the site setup (e.g., agrichemical storage sheds, irrigation systems installed, winery waste water disposal systems) and request evidence to confirm annual submissions are true and accurate (e.g., copies of resource consents, soil test results, site management plans). Similar to the questionnaire submission, members are assigned corrective actions if there are any requirements that have not been met. The SWNZ accreditation for the upcoming harvest is not granted until the corrective actions assigned have been completed (New Zealand Wine, 2020f).

An analysis of the data collected by SWNZ in 2019 and 2020 found that there was room for improvement to ensure all requirements were aligned with the six focus areas and data were available to track progress towards the industry's sustainability goals. For instance, the review found that there was insufficient evidence to establish an up-to-date product carbon footprint and so targets to reduce greenhouse gas (GHG) emissions had no baseline (Andrews, 2020). Producers did not know how they performed compared with each other or with other industries. As such, SWNZ updated the questionnaire's metric requirements for the 2020/2021 season to ensure progress in all sustainability focus areas could be tracked, including the ability to measure green house gas (GHG) emissions and create GHG baselines. At the end of 2021 all SWNZ vineyards and wineries received their first personalised GHG emissions report, which identified key sources of emissions and benchmarked the site against other SWNZ members regionally and nationally.

21.3.2 Organic viticulture and wineries

Organic Winegrowers New Zealand (OWNZ) is an organic grower-led organization dedicated to supporting organic wine production and its members are also members of NZW. In 2020, 10% of NZ wineries and 2418ha of vineyards were certified as organic (Organic Winegrowers NZ, 2020).

Independent audit verification is essential to be able to claim organic status. Every organic producer goes through an annual audit to ensure their practices meet the required standards. Producers must follow organic cultural practices, such as the use of nonsynthetic herbicides, insecticides, and fungicides in vineyards for three years before attaining full certification through BioGro or AsureQuality (New Zealand Wine, 2020c). For winemaking only certain yeast strains and cleaning products can be used and is dependent on the wine's final export market (BioGro NZ, 2020). This certification is recognized in legislation in a range of international markets. For international consumers it is a key indicator of a product's overall sustainability. BioGro is the leading organic certification business for NZ wine companies and focuses on all organic principles, traceability, animal welfare, biodiversity, and workers' well-being.

In order to become a certified biodynamic producer, through Demeter, a grower must first be certified organic by BioGro (Biodynamics NZ, 2021).

21.3.3 NZW's sustainability policy

Following industry consultation, NZW implemented a new sustainability policy in 2007 for wine entry eligibility to all NZW events, promotions, and competitions. To enter a wine, it must be certified as 100% sustainable through one of the recognized sustainability certification schemes and meet the following criteria, as shown on NZW's website (New Zealand Wine, 2020i):

- 100% of grapes are certified
- 100% of wine processing facilities are certified, as well as brand certification if the brand owner does not own all of the vineyards or wine processing facilities

The policy is further evidence of the NZ wine industry's commitment to sustainability. It ensures that every wine entered in an NZW event, promotion or award has been produced using sustainable grape growing and winemaking methods (New Zealand Wine, 2020d).

21.3.4 Grape and wine research

Through annual grape grower and wine levies, NZW, along with co-investment from Central Government, has funded grape and wine research and development initiatives. In 2017, the Bragato Research Institute (BRI) was established by NZW with Central Government funding to work in collaboration with the industry and other research groups and to help direct future areas of work, conduct commercial trials, and provide extension services (Bragato Research Institute, 2021) (Figs. 21.3 and 21.4).

Large, multiyear and cross-specialization programmes have been established with Crown Research Institutes (CRI) such as Plant and Food Research Ltd. and the National Institute for Water and Atmospheric Research; universities (Lincoln University and University of Auckland); and other research bodies. Areas of specialization from across the different organizations have included winemaking, pathology, soil science, vine physiology, entomology, climate research, bioinformatics, modeling, genomics, and metabolism including flavor and aroma chemistry.

Collaborative programmes such as Lighter Wines looked at producing lower-alcohol Sauvignon blanc wines naturally through vineyard and winery manipulations while still maintaining its characteristic flavors, aromas, and yield; Vineyard Ecosystems was established to understand the links between the environment in, on, above, and under the vine to enhance vineyard resilience and longevity; the Pinot noir Program examined how to produce high-quality Pinot noir while increasing or maintaining yields (Bragato Research Institute, 2021). These shared research programmes have developed a close-knit grape and wine research community.

422 Chapter 21 The guardianship of Aotearoa

FIGURE 21.3
Bragato Research Institute winery in Blenheim.

Used with permission from Bragato Research Institute.

FIGURE 21.4
Bragato Research Institute fermentation vessels.

Used with permission from Bragato Research Institute.

Other funding sources for fundamental science areas, such as PhDs, are available through the universities, Plant and Food Research Ltd., through the Ministry of Business, Innovation and Employment (MBIE) Strategic Science Investment Fund, and other Government-related pools such as the Marsden Fund (administered through the Royal Society of NZ) and the Agricultural and Marketing Research and Development Trust (AGMARDT).

21.3.5 Other NZW initiatives

NZW has actively promoted young talent by initiating programmes to encourage excellence and participation in different aspects of the wine production process (New Zealand Wine, 2020j). It organizes the annual Young Viticulturist and the Young Winemaker of the Year competitions for viticulturists and winemakers under 30 years of age, providing opportunities to upskill, network, and to build strong emerging viticulture and winemaking communities. Each regional competition involves practical, theoretical, and sporting components and the winners go on to compete in a national final. The Women in Wine initiative supports women in the industry and encourages them to take leadership and governance roles. It also encourages the industry to assess their diversity and inclusion culture by conducting diversity workshops and industry surveys, initiating change in businesses to offer more opportunities for women to progress and step up into leadership roles. A mentoring program is also available to all members to develop their careers.

21.3.6 Sustainability guardians

Born out of an industry-led groundswell that some practical sustainability issues were not being fully addressed, NZW launched Sustainability Guardians (New Zealand Wine, 2020k). Through workshop and discussion forums opportunities for peer-to-peer learning, collaboration and innovation were established throughout NZ. The inclusion of other industry personnel (regional councils, suppliers, and waste management companies) provided the chance for wider problem identification, improving best practices and developing strategies for waste management.

In many of NZ's regions, such as Marlborough and Hawke's Bay, distance to and the availability of recovery and recycling services pose a significant problem. This limited infrastructure makes recycling a problem in many rural sectors. Time and cost in dealing with waste products are perennial issues facing all agricultural producers. As such, it was essential to ensure that vineyard and winery by-products could be dealt with effectively and as locally as possible. Identified issues included looking at alternative uses and repurposing waste materials, such as grape marc, lees, filter media, pallets, soft plastics, packaging, bird netting, irrigation line, and vineyard wire (New Zealand Wine, 2020l); reducing initial inputs; increasing the availability of waste recovery services to collect and redistribute; and reducing the cost of transporting waste offsite (New Zealand Wine, 2020l). The working group also established new industry guidelines for the sustainable disposal and storage of CCA-treated vineyard posts (New Zealand Wine, 2020m).

Solutions included wineries actively reducing waste to landfill by incorporating lees in uncovered compost; repurposing of plastics into alternative products (e.g., plastic vineyard posts, plastic pellets); using alternative materials to plastic; and recycling agrichemical containers through national waste recovery schemes (New Zealand Wine, 2020l).

Continuous improvement in seeking practical solutions to waste issues is a positive step in working toward a reduction in inputs and outputs. Further areas, such as worker well-being, must be addressed if the industry is to continue to deliver its sustainability goals.

21.4 Corporate social responsibility

The NZ wine industry has, over the years, taken on a CSR role of the natural environment, people, and culture. Activities by individual businesses, or in collaboration with other organizations, have seen the sponsorship of events and organizations. These philanthropic endeavors demonstrate an engaged commitment to the communities and the social environment in which they operate at national and international levels. They have helped to build a solid base for an innovative and world-leading sustainable wine industry. These diverse initiatives are a vehicle for supporting good causes in the various wine regions across NZ and abroad. Philanthropy and the analysis of what constitutes a social enterprise have been documented (Dressler & Haller, 2020; Fountain & De Silva, 2020), with the type of philanthropic behavior being linked to cultural dimensions. NZ examples of CSR and philanthropic activities by wine businesses are described in the following sections. These examples, and others, are featured on many of the wine businesses' websites under the headings of "sustainability," "our people," "our land," "our story," and "news." This high degree of engagement with consumers through the sharing of these events illustrates the importance that the NZ wine industry associates with CSR and their role as guardians of their environment for future generations.

21.4.1 People, community, and culture

The year 2020 saw 29th annual Hawke's Bay Wine Auction, NZ's most prominent charity wine auction. Initiated by The Hawke's Bay Vintners Group, the auction lots are unique or special wines, most of which are only available at the auction. Proceeds from the annual auction go to Cranford Hospice which provides care for terminally ill patients and their families (Forbes & De Silva, 2020; Hawke's Bay Wine Auction, 2020).

In Marlborough, some local grape growers, winemakers, and associated businesses joined forces to launch Borough Wines in 2020, to support youth in Marlborough by creating a long-term funding stream for the Graeme Dingle Foundation. The Foundation focuses on youth resilience and leadership programmes (Borough Wines, 2020).

In 2018, Craggy Range launched a Children's Christmas appeal to provide educational material and sports equipment for local children in need who might otherwise not have access to such gifts. This initiative brought together many local businesses in Hawke's Bay and Wairarapa including the rural fire service and police and created over 5000 gift packs (Craggy Range, 2020).

Brightwater Vineyards has produced a unique wine label called Heroes of Humanity and proceeds from its sales are donated to the Fifeshire Foundation, a Nelson-based charity that gives assistance to locals in hardship and domestic crisis (Brightwater Vineyards, 2020).

The 27 seconds wine label, in North Canterbury, was set up as a social impact winery in 2017 to raise funds to support Hagar, an international charity that works to combat the global issue of human trafficking, abuse, and slavery (27seconds, 2020). The name refers to the fact that every 27 seconds, a child is sold or trafficked into slavery somewhere in the world. This social enterprise has gained charitable status so that it does not have to pay tax on any profits, increasing its ability to donate to Hagar.

Whitehaven Wines is a sponsor of the annual African Film Festival that showcases the diversity of the cultures and supports film-makers from Africa. In addition, Whitehaven Wines has created a label called Kōparepare, from which proceeds go to LegaSea, which raises public awareness about the management and restoration of marine environments in NZ. It is also the major sponsor of the Whitehaven Graperide, an annual bike race in Marlborough. In 2020 during NZ's lockdown due to COVID-19, Whitehaven Wines purchased restaurant and café vouchers from local businesses. These were given to frontline staff who worked tirelessly throughout this time (Whitehaven Wines, 2020a). These major sponsorship initiatives underpin their efforts to continually invest in the wider community, as well as operating sustainably. Whitehaven Wines' commitment to sustainability is based on its three foundation principles of people, performance, and planet and demonstrates integrity and authenticity (Whitehaven Wines, 2020b).

21.4.2 Research

From 2011 to 2016, the Cresswell Jackson New Zealand Wine Trust provided financial support for NZ grape and wine scientists from universities, CRIs, and polytechnics. The Trust's objectives were to support educational and research projects within the NZ wine industry. The proceeds from an annual wine auction, that were surplus from wine competitions, provided funding for grants (NZ Wine Directory, 2016).

21.4.3 Corporate environmental guardianship

The rapid growth of the grape and wine industry since the early 2000s has led many NZ wine businesses to reassess their impact on the environment. Some have taken on the responsibility to mitigate some of the impacts of what many have seen as a monoculture of grapes, by engaging in activities to regenerate and enhance local ecosystems. These practices are similar to the Māori core value of *kaitiakitanga*.

21.4.3.1 Biodiversity

As a result of Māori and European settlements, which have incorporated farming, forestry, and horticulture as well as urban development, the NZ landscape has been drastically altered. In response to biodiversity losses, there have been initiatives to develop regenerative indigenous vegetation plantings in vineyards and around wetland areas. These schemes have provided financial incentives as well as information on species that are suitable for specific areas.

Yealands Wine Group is a member of International Wineries for Climate Action (IWCA), which seeks to reduce global wine industry carbon emissions. As an expansion of Yealands Wine Group's existing commitment to sustainable winemaking, Yealands Wines Marlborough Sustainability Initiative (YSI) was established in 2020. In association with the Marlborough District Council and the Blenheim Sun newspaper, an annual NZD $46,000 (USD $33,000) has been made available to Marlborough environmental not-for-profit groups to enhance fresh water quality and preserving vulnerable natural areas (Yealands Wines, 2020).

Greening Waipara was an initiative by Lincoln University, Landcare Research and the Hurunui District Council along with the participation of 55 grape growers, the local primary school and other farms. This was a six-year research program that started in 2006 to restore native habitats by planting locally sourced indigenous plants in shelter belts, entranceways, wetland areas, headlands, and in vineyard blocks. The research also looked at using native plant species as hosts for predators and parasitoids by providing nectar sources, further increasing biodiversity. The research investigated the ecosystem services and subsequent marketing opportunities provided by native species when planted in and around vineyards (Shields et al., 2016, pp. 1−22; Tompkins, 2010, pp. 1−301).

The Cawthron Marlborough Environment Awards (Cawthron, 2020) were established in 1997 to encourage local businesses and community groups to undertake environmental projects on sustainable energy and resource usage. The awards have several categories including a wine industry one, sponsored by Wine Marlborough. In 2017 Dog Point Vineyard won the Landscape and Habitat Advancement Award and the Supreme Award for its plantings of native plants, parkland, orchard, woodlands, and organic vineyard across 200 ha. The long-term sustainability of their land through organic farming has enabled Dog Point Vineyard to become a successful business. Benefits to the wider community have included walks through the gardens, barbeques for local charitable fundraising events, and food provision to Crossroads Marlborough Charitable Trust. As part of Dog Point Vineyard's ongoing stewardship of the land, there has been further clearing of exotic willow trees, to make way for native planting regeneration with their neighbors along another shared waterway.

21.4.3.2 Endemic species

The *kārearea* (the NZ falcon) is one of NZ's endangered raptors (Fig. 21.5). The presence of these predators in vineyards deters other birds from damaging grape crops. Since 2010, Brancott Estate has supported the Marlborough Falcon Trust's efforts to restore the *kārearea* population in Marlborough. Funding from Brancott Estate contributed to the building of the rehabilitation and breeding aviary as well as the upkeep and care of the falcons based at one of their vineyards (Marlborough Falcon Trust, 2020).

FIGURE 21.5

The kārearea (NZ Falcon).

Used with permission from Andy Frost Ltd.

In Central Otago, Peregrine Wines' commitment to the natural environment has involved conservation initiatives (Peregrine, 2020). Once found throughout NZ's forests, the ground-feeding *tieke* (the black and red saddleback) were easy prey for introduced pests. Peregrine's staff have helped to relocate *tieke* breeding pairs to new predator-free areas in Fiordland.

The successful regeneration of native bush around one of Ata Rangi's vineyards in Martinborough has occurred through the fencing-off of areas, pest control, and planting of 75,000 trees since 1995. In 2005 Ata Rangi partnered with The Project Crimson Trust and has helped with the propagation of rātā, part of the *Metrosideros* genus (Trees that Count, 2020) that have been planted out across NZ.

These by no means cover all the efforts by different companies within the NZ grape and wine industry. However, they provide a snapshot of the engagement, the extent of the industry's involvement, and the passionate approach toward various causes by those involved. It is the people behind these initiatives who have made them possible. Quoting the words of Clive Paton, the founder of Ata Rangi: "We're not going to see the change in our lifetime, but we're setting it up for our children and grandchildren to keep going" (Trees that Count, 2020).

21.5 Biosecurity

New Zealanders greatly value their country's unique biological heritage; it is intrinsically tied to the sense of national identity, and many have strong spiritual, cultural, or family connections to natural landscapes around the country. NZ is also an agricultural nation and reliant on the success of primary industries for both domestic food production and export revenue. Biosecurity therefore has a twofold importance for NZ; firstly to protect natural heritage and native biodiversity; and secondly to protect primary industries from incursions of unwanted pests and diseases. A national biosecurity engagement campaign, Ko Tātou This is us, encourages all New Zealanders to be *kaitiaki* and take responsibility for keeping NZ safe from the establishment and spread of pests and disease (Ko Tātou This Is Us, 2020).

Thanks to NZ's position as an isolated island nation, the wine industry has developed with relatively little impact from many of the devastating pests and diseases that have affected wine production in other countries. However, huge increases in international travel and trade over the past 2 decades have seen biosecurity risks increase and this has placed strain on border biosecurity systems, inevitably leading to incursions. Recent high-profile examples in other sectors which have led to extensive biosecurity responses include the kiwifruit bacterial pathogen *Pseudomonas syringae* pv. *actinidiae* (2010), the myrtle rust fungus *Austropuccinia psidii* (2017), the cattle pathogen *Mycoplasma bovis* (2017), and the Queensland fruit fly *Bactrocera tryoni* (2019).

In 2016 NZW appointed a full-time biosecurity manager to develop and implement a biosecurity strategy (New Zealand Wine, 2020n) for the NZ wine industry, sending a clear message about the importance of biosecurity to the industry's ongoing sustainability. The overall outcome sought by the strategy was to maximize the protection afforded to grape growers by NZ's biosecurity system. This was to be achieved through three key objectives:

- Maximizing the industry's influence on decision-making and activities in the broader biosecurity system
- Maximizing grape growers' awareness of biosecurity risks and mitigations
- Maximizing grape growers' capacity and capability to participate in biosecurity activities.

In NZ, the Government, via the Ministry for Primary Industries (MPI), and primary production organizations work together to prepare for and respond to biosecurity threats through the Government Industry Agreement (GIA) (GIA, 2020). The majority of NZ's primary production industries are represented through this agreement, enabling collaborative work programmes to be undertaken in readiness for pests and diseases that threaten multiple industry groups, such as *Halyomorpha halys* (the brown marmorated stink bug) and the bacterium *Xylella fastidiosa*. Underpinning the GIA is the premise that industry groups share the costs of readiness work and responses that benefit their sectors, in return for a seat at the governance table and input into biosecurity decision-making affecting their industries. Industries can also develop bilateral agreements with MPI, where they wish to commence readiness work for pests or diseases that are unlikely to be a priority for other GIA partner organizations.

Some of the work NZW has had input into through the GIA includes:

- Development of a readiness work program and response plans for *Halyomorpha halys*
- Development of operational response specifications for *Xylella fastidiosa*
- The development of a nationwide nursery biosecurity scheme.

Outside the GIA, there are less formal collaborations between the biosecurity managers of the horticultural biosecurity groups. They share resources and discuss common issues and approaches to influencing biosecurity best practice and behavioral changes.

NZW promotes biosecurity awareness, particularly with its grape growers and viticulturist members. NZW has worked with Biosecurity New Zealand to encourage people to report any potential biosecurity threats. Posters and car bumper stickers were produced with three key actions: to catch/take a sample, take a photo, and report any unknown pests and diseases that were found (Fig. 21.6).

All grape growers fill in the Biosecurity Vineyard Register annually, which records vineyard location, planted area and which varieties are being grown in each vineyard. In the event of a biosecurity incursion, this enables NZW to easily contact those growers who are at risk of being affected.

Several resources have been developed to assist grape growers with biosecurity preparedness and planning, such as guidelines for best practice, a vineyard biosecurity plan template, a priority pest list called Most Unwanted, and a pest and disease photograph identification guide (New Zealand Wine, 2020n). Biosecurity workshops are held in several wine regions each year to familiarize growers with the available resources and biosecurity presentations are included at large-scale NZW industry events, such as the Bragato Conference and Grape Days.

When developing new vineyards or undertaking replacement planting, NZW members are encouraged to purchase vines that are certified under the Grafted Grapevine Standard (GGS). This was developed in 2006 in conjunction with a technical advisory group which included viticulturists, nursery representatives, and scientists. The original purpose of the GGS was to minimize the risk of vine planting material being infected with *Grapevine leafroll-associated virus-3* (GLRaV-3), one of the most damaging grapevine diseases present in NZ (New Zealand Wine, 2017). Hygiene requirements in the standard ensured that the risk of transmission of other pathogens into newly grafted vines was also greatly reduced. Over time, the GGS has included requirements for vines to be true-to-type by variety and meet certain physical specifications, thus ensuring a high-health product. Nurseries producing GGS-certified vines are independently audited annually to ensure their processes meet the requirements of the standard.

FIGURE 21.6

Snap it, Catch it, Report it poster. Biosecurity campaign by NZW and Biosecurity New Zealand.

Used with permission from New Zealand Winegrowers Incorporated.

The incursion of a new pest or disease is a key risk for the NZ wine industry. Ensuring good biosecurity practices are crucial to maintaining the industry's ongoing economic and environmental sustainability into the future. Partnership and collaboration with the NZ Government and other industry organizations, through the GIA, have improved biosecurity outcomes and ensured that the wine industry has input into and influence on NZ's biosecurity systems. Awareness and participation in biosecurity activities among the grape and wine industry's members and fostering a sense of guardianship are key to identifying and responding to incursions quickly, to protect NZ's wine sector, other primary-sector producers and NZ's biological heritage.

21.6 Natural disaster management

During and after an emergency event, industry bodies, who already advocate for their members across a wide range of functions, are ideally placed to take on key communication roles. They can act as a conduit between their members and the governmental agencies undertaking the response activity. NZW and the regional wine associations have taken on this role in two recent large-scale emergency events—the 2016 Kaikoura earthquake, which affected the Marlborough wine region, and more recently, the COVID-19 pandemic just prior to harvest 2020, which had repercussions across all wine regions. Gathering information from affected industry members to report to local and central government agencies and regional networks, and channeling important information back to industry members, were critical tasks in enabling effective responses to occur.

21.6.1 2016 Kaikoura earthquake

On November 14, 2016, a magnitude 7.8 earthquake struck near Waiau, North Canterbury, 210 km south of Blenheim, Marlborough. This was the most powerful earthquake in the area for over 150 years and ruptures occurred on multiple fault lines (Hamling et al., 2017) causing ground displacement both vertically and horizontally, landslides, and tsunami. In Marlborough, the ground movement caused physical damage to vines, structural damage to buildings, and the collapse of winery infrastructure including tanks and catwalks (Cradock-Henry & Fountain, 2019). The key transport routes of State Highway 1 and the rail line were cut off south of Blenheim.

The wine industry's response was a collaborative effort. A governance group was set up, with managers from NZW and Wine Marlborough acting as controllers with assistance from two MPI response managers, specifically sent to Marlborough to assist the grape and wine industry. The response objectives were to ensure affected members were provided with timely, essential information about health and safety, access to recovery options and overall support.

One of the key actions undertaken as part of the industry response was to hold a public meeting for Marlborough winegrowers in November 2016, soon after the earthquake. This enabled all the relevant information to be provided to affected grape growers and winemakers at once and allowed them to ask questions and share their experiences with others who were facing similar issues. The information provided by affected industry members was then able to be relayed to Government to help with needs assessments and recovery planning.

Since the earthquake, work has been done to upgrade winery infrastructure and tank design to ensure they will not fail should similar events occur in the future. Businesses developed detailed emergency and contingency plans and a local "Primary Industry Adverse Events Network" was convened, which included representatives from all the primary industry groups operating in Marlborough, as well as local and central government agencies (MPI, Civil Defence, Marlborough District Council) and support groups (Federated Farmers, Rural Support Trust). The Kaikoura earthquake experience was also used as a case study, characterizing resilience in NZ's wine industry (Cradock-Henry & Fountain, 2019).

21.6.2 COVID-19

The COVID-19 pandemic affected NZ just prior to the 2020 vintage in some regions, with NZ's first case reported on February 28, 2020. By March 26, 2020, with 283 confirmed or probable cases, the country was in a nationwide "Alert Level 4" lockdown, accompanied by much uncertainty in the wine industry about whether harvest activities would be allowed to continue.

NZW instigated a governance group and activated an industry response with similar objectives to those of the Kaikoura earthquake response—to ensure members were able to access up-to-date information, advice, and support as it became available and to be a conduit for information on industry impact and needs to be passed back to Government. The wine industry was classed as an "essential business" and was given permission to continue essential harvest operations throughout the lockdown period. The NZW advocacy team worked closely with government agencies MPI and MBIE to develop industry-specific guidelines and advice to ensure members could operate safely and that they understood the rules.

The nationwide scale of the pandemic required dedicated communications channels: a hold was placed on all other NZW communication out to membership to ensure important COVID-19-related updates had priority focus. These were sent out regularly as new information came through from Government and industry representatives working with international markets. A dedicated webpage was set up on the nzwine.com website (New Zealand Wine, 2020o). An email inbox and hotline telephone number were established to provide members with a mechanism for requesting and sharing information. A series of regional conference calls was held, with updates provided by senior NZW personnel.

MPI audited several businesses and provided feedback to the industry body. NZW set up a "Safe Harvest Hotline"—this was a number that could be used to report unsafe practices, triggering follow-up to ensure continued social licence to operate as an essential industry.

Regional winegrower associations were important, helping to channel relevant information through to members and making NZW aware of any regionally specific issues. They participated in local forums to ensure that the industry was included in local recovery plans.

Emergency events such as the Kaikoura earthquake and COVID-19 highlight the need for two-way information flow and communication between affected industry members and those who can respond and aid recovery. As an industry body NZW is well-placed to be that conduit and maintains many of the key relationships required as part of its day-to-day business. Having staff who are familiar with NZ's emergency response framework, the Co-ordinated Incident Management System (CIMS), and those who have been exposed to response events is also highly beneficial. These generic skills can be applied across a wide range of events to manage future wine industry risks appropriately. Each adverse event provides unique challenges and lessons learned. NZW has a role in ensuring these lessons are taken on board by members to enable them to prepare, plan ahead to become more resilient to future shocks, and to ensure the long-term sustainability of the grape and wine industry.

21.7 Filling the gap

NZ's horticultural and viticultural industries have seen large changes in the way in which they have recruited their temporary seasonal workforce. Until 2007, seasonal agricultural labor was carried out predominantly by New Zealanders and international backpackers on working holiday visas. A looming labor shortage became increasingly evident owing to the rapid expansion from 2000 to 2006 of plantings of both vineyards and other horticulture crops and low rates of unemployment (The Social Report, 2005). The threat of crops not being picked and not meeting export markets, as well as the long-term viability of these horticultural businesses and the knock-on effects in local communities were of major concern (Corlett, 2019).

To meet this labor shortage, an industry-led initiative, supported by the NZ Government, established the Recognised Seasonal Employer (RSE) Policy in 2007 (Bailey, 2019; Gibson & McKenzie, 2014, pp. 186–210; Hao'uli, 2013). The RSE scheme, in 2020, is administered through different Government ministries to provide a sustainable supply of temporary workers for the horticultural and viticultural industries to pick, pack, and process crops and to raise these industries' economic performance. These workers originally were from Kiribati, Samoa, Tonga, Tuvalu, and Vanuatu, and subsequently it has been extended to Papua New Guinea, Fiji, the Solomon Islands, and Nauru (Bedford et al., 2020a). Employers with existing rights to recruit from overseas can continue to employ workers from other nations. The three main countries sending more than 80% of RSE workers are Samoa, Tonga, and Vanuatu (Bedford, 2020; Nunns et al., 2019).

Another major aim of the RSE scheme was to contribute to economic development and regional growth objectives in the Pacific Island nations (Nunns et al., 2019). NZ has a long-term commitment to sustainable economic and social development in the Pacific region (NZ Foreign Affairs & Trade, 2020b). An annual cap on the number of RSE workers is set and has risen from 8000 places in 2009 to 14,400 in 2019 and 2020 (NZ Immigration, 2020a) following significant growth in the horticulture and viticulture sectors.

To recruit from Pacific Island nations, an RSE employer has to be accredited by Government, comply with tightly defined requirements and to demonstrate to NZ's Ministry of Social Development that there is an insufficient local labor pool of workers before being able to recruit RSE workers. An employer must provide employment agreements to pay workers at least NZ's minimum wage for 30 hours per week; to pay 50% of each worker's international airfare and transport costs; to outline the type of work that will be undertaken; and to provide pastoral care and approved accommodation for workers while they are in NZ (NZ Immigration, 2020b). Likewise an RSE worker must comply with all work and visa conditions. Predeparture inductions and health checks are provided to acquaint workers to life in NZ, covering areas such as climate, clothing, insurance, health services, and taxation.

The detailed requirements set out and the ensuing work outcomes have made the RSE scheme world-leading (Gibson & McKenzie, 2014, pp. 186–210). It has been praised by the International Labour Organisation (Hardie, 2017) and in its Good Practices Database it identified the RSE scheme as having three major themes: fair recruitment, policy coherence, and regional labor mobility (Migrant Labour Migration Branch, 2015).

According to Hao'uli (2013), however, with the degree of requirements imposed on employers and workers, and the financial benefits that have arisen, they have created a "structural imbalance" between groups involved in the scheme and those who are not.

In NZ key impacts of the scheme included providing confidence to the wine industry to implement vineyard growth plans thanks to being able to access a reliable and consistent labor pool. This growth has also increased permanent and seasonal work for New Zealanders (Nunns et al., 2019). Part of this study looked at the scheme's impacts in the different regions. These varied depending on the location but issues such as accommodation were common between them.

In the early years of the scheme competition for accommodation between RSE employers, looking for housing for their workers, and local residents emerged in some rural regions. Local councils, RSE employers, and Government worked to alleviate the housing issues (NZ Immigration, 2020c; Worksafe, 2017). Large purpose-built accommodation facilities, converted backpackers establishments and motels, therefore, became accessible to RSE employees.

With Marlborough's labor shortage, the need for RSE workers has been critical in carrying out the manual winter pruning and summer seasonal work. The increase in planted area has meant that pruning has been brought forward to late April, just after harvest, and carries on until late August. In the early 2000s, pruning would not start until early June. An estimated 2700 RSE workers were working in Marlborough during the winter (Recognised Seasonal Worker Survey, 2020). The contribution that the RSE scheme has made to the viticultural industry is of direct importance and is an essential part of the local industry. Without this labor the manual operations in vineyards would not be completed successfully.

In the Pacific there have been both positive and negative impacts due to the RSE scheme (Bedford, 2019; Bedford et al., 2020b). Some of the positive impacts were increased income to pay for education, living costs, improved housing, and small business start-ups; enabling more women to work; new skill capabilities developed; and the ability to better support communities after natural disasters, particularly hurricane events. Some of the negative outcomes have been the unequal access to the scheme between Pacific Islands and within individual countries; that total wages and remittances have not increased greatly owing to increased costs in NZ; family separation for extended periods of time; change in community activities due to reduced numbers of workers; and changes in traditional values and leadership in local communities. It was identified that in Vanuatu and Tonga, 8% and 12%, respectively, of, mainly, rural-based men aged 20—45 were working either in the RSE or the Australian Seasonal Worker Program, which meant that other family members had to perform their seasonal work in the Islands (Curtain & Howes, 2020).

The RSE scheme has been hailed as a triple-win, as it has contributed social and economic benefits to the local industries in NZ that have grown in size and profitability; the RSE workers' annual incomes have been lifted; and the remittances to the Pacific Islands have enabled improved economic development (Bedford, 2019; Bedford & Bedford, 2017). RSE employers have established a regular, experienced, and therefore productive workforce (Gibson & McKenzie, 2014, pp. 186—210) as people return on an annual basis, reducing the need for training, as over time the workers involved have become skilled and efficient.

The RSE scheme must continue to be fit-for-purpose, to ensure the ongoing sustainability, growth and development of the wine industry for all those involved, including workers' families and communities, while maintaining its status as a world-leading labor mobility system. The RSE scheme is scheduled for a policy review in 2021 given the increase in numbers of workers and issues that have arisen both in NZ and the Pacific Islands, some of which have been addressed here.

21.8 Regional winegrower associations

NZ is fortunate to have a structured support network for the national and regional wine industry which has assisted with the rapid growth and advancement of the industry.

The local bodies that serve the grape growers and winemakers in NZ are referred to as regional winegrower associations. They have developed and grown alongside the needs of the industry. They are located in each of the grape growing regions across the country. There are eight regional associations and a further six subregional winegrower associations, which serve smaller and, or, emerging regions.

Initially there were both regional wine associations and regional grape grower associations. The former followed the establishment and increasing development of grape growing across the country, as wine companies searched beyond their own supply of grapes to increasingly rely on grape growers to supply them. Regional grape grower associations formed to navigate these new relationships. The two types of associations eventually merged into single entities to represent all winemakers and grape growers under a single organization in each region.

These organizations are now funded in whole or in part by a levy collected under the current Wine (Grape Wine Levy) Order 2016 (Ministry of Business Innovation & Employment, 2015), meaning all members contribute financially to the associations of the regions from whence their grapes and wine come.

21.8.1 Role

The original grape grower and wine associations in NZ began life representing the interests of their members. The mandate remains unchanged today. The scale, resources, and focus for each regional association are variable, as the numbers of members and areas of production vary. Regional associations work to develop grape and wine events, promotion, tourism, communication of industry news, and advocacy of their members' interests. Each association is overseen by a board of directors elected from their own local wine industry.

21.8.2 Wine Marlborough

Marlborough is the country's largest wine region in 2020, with 27,808 ha of vines planted; it makes up 70% of NZ's vineyards and 78% of the total vintage in terms of tonnes crushed. Marlborough also has 74% of the country's 694 grape growers (New Zealand Wine, 2020a). In 2020, Sauvignon blanc makes up 80% of the region's planted area, followed by Pinot noir (9%), Pinot gris (4%) and Chardonnay (4%) (New Zealand Wine, 2020p).

In 2020, wine growing and wine production in Marlborough contributed NZD $571 million (USD $401 million) and 18% of GDP to the regional economy, employing (directly and indirectly) 13% of the region's workforce (NZIER, 2020, pp. 1–17). This excluded the temporary seasonal workers such as those under the Recognised Seasonal Employer Scheme.

Marlborough Winegrowers Association Incorporated was established in 1980. In 1992 a limited liability company called Wine Marlborough Limited was established to promote the interests of the Marlborough wine industry and is owned entirely by the Marlborough Winegrowers Association.

21.8.3 Wine Marlborough's activities

Wine Marlborough works on a diverse range of activities: from promotional activities and bespoke events such as hosting domestic and international wine media and wine buyers; to organizing significant events such as the Marlborough Wine & Food Festival (2020), the International Sauvignon Blanc Celebration (Marlborough Sauvignon, 2019), and the Marlborough Wine Show (2020). These events showcase the region's wines. All activities involve members, members' wines, and the sharing of regional information on behalf of all members.

The Wine Marlborough website offers information, education resources, a wine trail map to the region's cellar doors and an industry magazine called *WinePress* (Wine Marlborough, 2020). *WinePress* is printed and distributed monthly and is free to members to keep them informed of grape and wine events and other items of significance.

A significant area of growth has been in the advocacy space, where intervention and interpretation of central and regional government policies are required to enable members to keep pace with changes. Key issues to maintain and enhance Marlborough's environmental credentials are: labor attraction strategies and negotiations with Government; the ongoing rights to take and use water; education around sustainable use of water resources; winery waste minimization; adhering to local authority rules and resource consent issues. These items are able to be addressed through field days, workshops, and seminars for the region's grape growers and winemakers. The importance of such issues highlights the need for continuous adaption to changing needs and circumstances.

The NZ wine industry is connected and collective, in part owing to the presence of regional winegrower associations that work closely with their governing members. By ensuring strong relationships between regional and national wine organizations, they are able to share successes and learn from threats and challenges. Despite there being regional differences in resource availability, there is a strong collective effort to promote all of NZ's wines under the brand of "NZ Wine" as the associations work together alongside NZW to achieve better outcomes for their members.

21.9 Conclusion

Sustainability is a cornerstone of the NZ grape and wine industry. This chapter has identified some of the cultural issues, traditions, and good practices in the wine sector, along with examples of collaborations between the industry and the environments within which it operates. The unified body of NZW and the regional wine associations have enabled the industry to leverage influence and centralize a coordinated response in times of crisis. SWNZ's leading sustainability accreditation program has paved the way for others to follow. NZ's CSR at local, national, and international levels demonstrates that many wine companies deem this to be an important component of their strategic environmental, social, and business management plans. These activities are wide ranging. From environmental, social, cultural, scientific research to sport, these charities and other organizations have been able to find support from benefactors in the wine industry. Biosecurity incursions pose potential large economic, environmental, and social losses to the country. The ongoing work to educate the wine industry highlights the need for vigilance to conserve and protect NZ's flora and fauna for future generations as well as maintaining its primary industries. The RSE scheme's ability to fill the temporary seasonal workforce gap in the horticultural and viticultural industries has been hailed as a success while at the same time helping to lift many Pacific Islanders' annual incomes and their ability to support their community developments. NZ's unique Māori heritage places a strong value on *kaitiakitanga* of all assets that have been handed down and must, therefore, be sustained and enhanced for the future. This and other core values have much to teach about the importance of long-term sustainability. *Whatungarongaro te tangata, toitū te whenua*—the people fade from view but the land remains.

Looking forward, the wine industry must continue to be flexible and resilient in adapting to future sustainability challenges through continued investment in research. How will it fare after COVID-19, given the likely social, environmental, political, and economic global changes? How will the industry maintain NZ's commitment to reduce the impacts of climate change; to be a carbon neutral industry by 2050; to be prepared for new natural disasters and biosecurity incursions; to further reduce agrichemical inputs; and manage labor issues?

Acknowledgments

The authors wish to thank Richard Bedford, Alistair Mitchell, and Nigel Sowman for sharing their insights and knowledge about the RSE scheme and the Cawthron Environment Awards.

References

27 Seconds. (2020). *Your impact*. https://27seconds.co.nz/your-impact/ (Accessed November 2020).

Andrews, J. (2020). *Assessment of land use greenhouse gas (GHG) emissions of the wine sector in New Zealand*. https://www.nzwine.com/media/17918/toitu_report_wineghg-final.pdf.

Bailey, R. (2019). *New Zealand's Recognised Employer Scheme (RSE) 10 year longitudinal study*. Australian National University, Department of Pacific Affairs. http://dpa.bellschool.anu.edu.au/sites/default/files/uploads/2019-06/new_zealands_recognised_employer_scheme_rse_10_year_longitudinal_case_study.pdf.

Bedford, R. (2019). *The 'good house'. Notes from the frield on the contribution the RSE is making to the transformation of housing in five Pacific countries*. https://www.researchgate.net/publication/344036490_The_%27good_house%27.

Bedford, R. (2020). *The RSEs in 2018/19: A profile*. https://www.researchgate.net/publication/344036484_The_RSEs_in_201819_a_profile/citation/download.

Bedford, R., & Bedford, C. (2017). *RSE earnings and remittance surveys Samoans, Tongans and Ni-Vanuatu Employed in Hawke's Bay, the Bay of Plenty and Marlborough for 18–22 weeks, 2014/15 and 2016 Executive Summary*. https://www.researchgate.net/publication/315457808_RSE_EARNINGS_AND_REMITTANCE_SURVEYS_Samoans_Tongans_and_Ni-Vanuatu_Employed_in_Hawke's_Bay_the_Bay_of_Plenty_and_Marlborough_for_18-22_weeks_201415_and_2016_Executive_Summary.

Bedford, C., Nunns, H., & Bedford, R. (2020a). *Participation in the RSE scheme: The myth of opportunity (Blog)*. DevPolicy Blog, Australian National University. https://devpolicy.org/participation-in-the-rse-scheme-the-myth-of-opportunity-20201020-1/.

Bedford, C., Bedford, R., & Nunns, H. (2020b). *RSE impact study: Pacific stream report*. https://www.researchgate.net/publication/343893403_RSE_Impact_Study_Pacific_stream_report/link/5f46d320458515a88b6da915/download.

Biodynamics New Zealand. (2021). *Demeter*. http://biodynamic.org.nz/demeter (Accessed December 2020).

BioGro NZ. (2020). *Organic certification NZ*. https://www.biogro.co.nz/ (Accessed December 2020).

Borough Wines. (2020). *Borough*. https://boroughwine.co.nz/ (Accessed November 2020).

Bragato Research Institute. (2021). *Bragato Research Institute*. https://bri.co.nz/about-us/ (Accessed December 2020).

Brightwater Vineyards. (2020). *Heroes of humanity*. https://www.brightwaterwine.co.nz/heroes-of-humanity/ (Accessed November 2020).

Carroll, A. (1991). The pyramid of corporate social responsibility: Toward the moral management of organizational stakeholders. *Business Horizons, 34*, 39–48. https://doi.org/10.1016/0007-6813(91)90005-G

Cawthron. (2020). *Sound environmental management is good business.* https://www.cmea.org.nz/ (Accessed November 2020).

Corlett, E. (2019). *RSE scheme 'transformed' the New Zealand fruit growing industry.* https://www.rnz.co.nz/national/programmes/eyewitness/audio/2018726025/rse-scheme-transformed-the-new-zealand-fruit-growing-industry (Accessed December 2020).

Cradock-Henry, N., & Fountain, J. (2019). Characterising resilience in the wine industry: Insights and evidence from Marlborough, New Zealand. *Environmental Science & Policy, 94*, 182−190. https://doi.org/10.1016/j.envsci.2019.01.015

Craggy Range. (2020). *A children's Christmas.* https://craggyrange.com/about-us/a-childrens-christmas-foundation/ (Accessed November 2020).

Curtain, R., & Howes, S. (2020). *Managing seasonal labour mobility: Government-central versus government-light.* https://devpolicy.org/managing-seasonal-labour-mobility-government-central-versus-government-light-20201209/ (Accessed December 2020).

Dressler, M., & Haller, C. (2020). Does culture show in philanthropoc engagement? An empirical exploration of German and French wineries. In P. MacMillan (Ed.), *Social sustainability in the global wine industry* (pp. 119−136).

Elkington, J. (2006). Governance for sustainability. *Corporate Governance: An International Review, 14*(6), 522−529. https://doi.org/10.1111/j.1467-8683.2006.00527.x

Forbes, S. L., & De Silva, T.-A. (2020). The Hawke's Bay wine auction: History, motivations and benefits. In P. MacMillan (Ed.), *Social sustainability in the global wine industry* (pp. 75−91).

Fountain, J., & De Silva, T.-A. (2020). 27seconds: A wine brand as a vehicle for social change. In P. MacMillan (Ed.), *Social sustainability in the global wine industry* (pp. 93−105).

Gibson, J., & McKenzie, D. (2014). *Development through seasonal worker programs: The case of New Zealand's RSE program.* https://www.cream-migration.org/publ_uploads/CDP_05_14.pdf.

GIA. (2020). *Government Industry Agreement for Biosecurity Readiness and Response.* https://www.gia.org.nz (Accessed October 2020).

Hamling, I. J., Hreinsdóttir, S., Clark, K., Elliott, J., Liang, C., Fielding, E., & Stirling, M. (2017). Complex multifault rupture during the 2016 M(w) 7.8 Kaikōura earthquake, New Zealand. *Science, 356*(6334). https://doi.org/10.1126/science.aam7194

Hao'uli, E. (2013). Triple wins or trojan horse? Examining the recognised Seasonal,Employer scheme under a TWAIL lens. *New Zealand Yearbook of International Law, 11.* http://www.nzlii.org/nz/journals/NZYbkIntLaw/2013/9.pdf.

Hardie, A. (2017). *RSE a world 'game changer.* Orchardist. https://www.pressreader.com/new-zealand/the-orchardist/20170801/282041917229034.

Harmsworth, G. (2005). *Report on the incorporation of traditional values/tikanga into contemporary Māori business organisation and process.* Landcare Research. https://www.landcareresearch.co.nz/uploads/public/researchpubs/Harmsworth_report_trad_values_tikanga.pdf.

Hawke's Bay Wine Auction. (2020). *Hawke's Bay wine auction.* https://hawkesbaywineauction.co.nz/ (Accessed November 2020).

Haymes, S. (1998). *Defining Kaitiakitanga and the resource management act 1991.* Ko Ngaa take Ture Maori, Auckland University Law Review. http://www.nzlii.org/nz/journals/AukULawRw/1998/11.pdf.

Kawharu, M. (2000). Kaitiakitanga: A Maori anthroploigal persepective of the Maori socio-environmental ethic of resource management. *The Journal of the Polynesian Society, 109*, 349−370.

Ko Tātou This Is Us.(2020). https://www.thisisus.nz/ (Accessed December 2020).

Kono. (2020). *Kono.* https://www.kono.co.nz/about#our-gifts-of-the-land-and-sea (Accessed November 2020).

Marlborough District Council. (2020). *Crop types.* https://www.marlborough.govt.nz/environment/land/land-coverland-use/crop-types (Accessed September 2020).

Marlborough Falcon Trust. (2020). *Brancott estate living land tours.* https://www.mfct.org.nz/brancott-estate-living-land-tours (Accessed November 2020).

Marlborough Sauvignon. (2019). *Marlborough Sauvignon 2019.* https://www.sauvignonnz.com (Accessed December 2020).

Marlborough Wine & Food Festival. (2020). *About the festival.* https://marlboroughwinefestival.com/about (Accessed December 2020).

Marlborough Wine Show. (2020). *The 2020 Marlborough wine show.* https://www.marlboroughwineshow.com (Accessed December 2020).

Migrant Labour Migration Branch. (2015). *The recognized seasonal employers scheme (RSE), New Zealand.* https://www.ilo.org/dyn/migpractice/migmain.showPractice?p_lang=en&p_practice_id=48 (Accessed November 2020).

Ministry of Business Innovation & Employment. (2015). *MBIE's Pacific economic strategy 2015–2021.* https://www.mbie.govt.nz/assets/cf13bcf509/mbie-pacific-economic-strategy-2015-2021.pdf.

New Zealand Foreign Affairs & Trade. (2020a). *Our global agreements.* https://www.mfat.govt.nz/en/environment/climate-change/our-global-agreements/?m=822913#search:b3VyIGdsb2JhbCBhZ3JlZW1lbnRz (Accessed November 2020).

New Zealand Foreign Affairs & Trade. (2020b). *Our approach to aid.* https://www.mfat.govt.nz/en/aid-and-development/our-approach-to-aid (Accessed November 2020).

New Zealand History. (2020). *The Treaty in brief.* https://nzhistory.govt.nz/politics/treaty/the-treaty-in-brief (Accessed November 2020).

New Zealand Immigration. (2020a). *Recognised seasonal employer (RSE) scheme research.* https://www.immigration.govt.nz/about-us/research-and-statistics/research-reports/recognised-seasonal-employer-rse-scheme#:~:text=The%20Recognised%20Seasonal%20Employer%20(RSE,not%20enough%20New%20Zealand%20workers (Accessed October 2020).

New Zealand Immigration. (2020b). *Apply for an agreement to recruit.* https://www.immigration.govt.nz/employ-migrants/hire-a-candidate/employer-criteria/recognised-seasonal-employer/apply-atr (Accessed October 2020).

New Zealand Immigration. (2020c). *Providing accommodation for RSE workers.* https://www.immigration.govt.nz/employ-migrants/explore-your-options/finding-and-hiring-workers-overseas/providing-accommodation-for-rse-workers (Accessed December 2020).

New Zealand Wine. (2020b). *New Zealand wine exports hit $2 billion.* https://www.nzwine.com/media/17686/new-zealand-wine-exports-hit-2-billion.pdf (Accessed December 2020).

New Zealand Wine. (2020c). *Kia ora, welcome to the New Zealand Wine website.* https://www.nzwine.com/ (Accessed November 2020).

New Zealand Wine. (2020d). *NZW sustainability overview.* https://www.nzwine.com/media/16274/nzw-sustainability-overview.pdf.

New Zealand Wine. (2020e). *Sustainable winegrowing NZ.* https://www.nzwine.com/members/sustainability/swnz (Accessed November 2020).

New Zealand Wine. (2020f). *Sustainable winegrowing New Zealand audit procedures for vineyards & wineries.* https://www.nzwine.com/media/3243/audit-procedures-for-vineyards-and-wineries.pdf (Accessed January 2022).

New Zealand Wine. (2020g). *Instructions for updating growth stages & planned applications.* https://www.nzwine.com/media/19140/hd-form-growth-stages-and-planned-applications-instructions.pdf (Accessed January 2022).

New Zealand Wine. (2020h). *New Zealand winegrowers vineyard spray schedule 2021/2022.* https://www.nzwine.com/media/20247/nzw-spray-schedule-2021-hyperlinked-2-september.pdf (Accessed January 2022).

New Zealand Wine. (2020i). *Sustainability policy.* https://www.nzwine.com/members/marketing/user-pays-events-programme/sustainability-policy/ (Accessed December 2020).

New Zealand Wine. (2020j). *Get involved.* https://www.nzwine.com/members/events (Accessed November 2020).

New Zealand Wine. (2020k). *Sustainability guardians*. https://www.nzwine.com/members/sustainability/guardians-programme/ (Accessed December 2020).

New Zealand Wine. (2020l). *Sustainability guardians - waste webinar 1st September*. https://www.nzwine.com/members/events/webinars/sustainability-guardians-waste-webinar/ (Accessed November 2020).

New Zealand Wine. (2020m). *NZW Disposal and storage guidelines for CCA treated vineyard posts*. https://www.nzwine.com/members/sustainability/news/cca-treated-posts/ (Accessed November 2020).

New Zealand Wine. (2020n). *Biosecurity*. https://www.nzwine.com/en/sustainability/biosecurity/ (Accessed November 2020).

New Zealand Wine. (2020o). *COVID-19: NZW response update*. https://www.nzwine.com/en/covid19/ (Accessed November 2020).

New Zealand Wine. (2020p). *Vineyard Report - New Zealand Winegrowers 2020–2023*. https://www.nzwine.com/media/19160/vineyard-report-2021.pdf (Accessed January 2022).

Nilipour, A. (2020). Introduction to social sustainability. In P. MacMillan (Ed.), *Social sustainability in the global wine industry* (pp. 1–14).

Nunns, H., Bedford, C., & Bedford, R. (2019). *RSE impact study: New Zealand stream report*. New Zealand Immigration. https://www.immigration.govt.nz/documents/statistics/rse-impact-study-new-zealand-stream-report.pdf.

NZ Wine. (2017). *Grafted grapevine standard*. https://www.nzwine.com/media/8418/grafted-grapevine-standard-v31-1617-2.pdf (Accessed November 2020).

NZ Wine. (2020a). *Annual report 2020*. https://www.nzwine.com/media/17492/annual-report_2020_final_web.pdf.

NZ Wine Directory. (2016). *Review of the Cresswell Jackson NZ wine trust's activities*. https://nzwinedirectory.co.nz/cresswell-jackson-nz-wine-trust/ (Accessed November 2020).

NZIER. (2020). *Contribution of wine to the Marlborough economy 2020: Wine Marlborough*. https://www.wine-marlborough.co.nz/workspace/uploads/nzier-contribution-of-winegrowing-to-the-marlborough-economy-2020-final_1.pdf.

Organic Winegrowers NZ. (2020). *Organic winegrowers NZ: About*. https://www.organicwinenz.com/about (Accessed November 2020).

Peregrine. (2020). *Conservation*. https://www.peregrinewines.co.nz/story/conservation/ (Accessed December 2020).

Pope Francis. (2015). *Encyclical letter Laudato Si of the holy father Francis on care for our common home*. http://www.vatican.va/content/francesco/en/encyclicals/documents/papa-francesco_20150524_enciclica-laudato-si.html.

Recognised Seasonal Worker Survey. (2020). *Recognised seasonal worker survey*. http://www.nzkgi.org.nz/wp-content/uploads/2020/07/RSE-Doc-June-2020-WEB-FINAL.pdf.

Royal, T. A. C. (2007). Kaitiakitanga – guardianship and conservation. *Te Ara – the Encyclopedia of New Zealand*. https://teara.govt.nz/en/kaitiakitanga-guardianship-and-conservation/print.

Shields, M., Tompkins, J.-M., Saville, D., Meurk, C., & Wratten, S. (2016). *Potential ecosystem service delivery by endemic plants in New Zealand vineyards: Successes and prospects*.

Te Ara. (2020). *Story: Biculturalism*. https://teara.govt.nz/en/biculturalism (Accessed November 2020).

The Social Report. (2005). *Unemployment*. https://www.socialreport.msd.govt.nz/2005/paid-work/unemployment.html (Accessed December 2020).

Tompkins, J.-M. (2010). *Ecosystem services provided by native New Zealand plants in vineyards* (Ph.D. thesis). New Zealand: Lincoln Univeristy.

Trees that Count. (2020). *A great pairing: Forty years of wine and conservation*. https://www.treesthatcount.co.nz/blog/2020/september/a-great-pairing-forty-years-of-wine-and-conservation/ (Accessed November 2020).

Tuku Wines. (2019). *Tuku*. https://tuku.nz/home (Accessed November 2020).

United Nations. (2020a). *Climate change.* https://www.un.org/en/sections/issues-depth/climate-change/ (Accessed November 2020).
United Nations. (2020b). *Sustainable development goals.* https://www.un.org/sustainabledevelopment/sustainable-development-goals/ (Accessed November 2020).
Whitehaven Wines. (2020a). *Whitehaven offers a helping hand.* https://whitehaven.co.nz/blogs/news/whitehaven-offers-a-helping-hand (Accessed December 2020).
Whitehaven Wines. (2020b). *Caring for our land.* https://whitehaven.co.nz/pages/caring-for-our-land (Accessed December 2020).
Wine Marlborough. (2020). https://www.wine-marlborough.co.nz/ (Accessed December 2020).
Worksafe. (2017). *Worker accommodation.* https://www.worksafe.govt.nz/topic-and-industry/worker-accommodation/ (Accessed December 2020).
Yealands Wines. (2020). *Yealands wines Marlborough sustainability initiative.* https://www.yealands.co.nz/sustainability-grant/ (Accessed December 2020).

Further reading

https://www.nzwine.com/media/19522/swnz-vineyard-questionnaire-2021-requirements.pdf, (2021–. (Accessed January 2022).
An unseasonable year: Pacific RSE workers to New Zealand during 2021, (2021–. (Accessed January 2022).
New Zealand's RSE scheme: struggling against COVID, (2021–. (Accessed January 2022).

CHAPTER 22

Sustainable viticulture and behavioral issues: insights from VINOVERT project

Alexandra Seabra Pinto[1], Stéphanie Pérès[2], Yann Raineau[3], Isabel Rodrigo[4] and Eric Giraud-Héraud[5]

[1]*Instituto Nacional de Investigação Agrária e Veterinária, Oeiras, Portugal;* [2]*Univ. Bordeaux, Bordeaux Sciences Agro, BSE, UMR 6060, ISVV, Pessac, France;* [3]*Univ. Bordeaux, Conseil Régional de Nouvelle-Aquitaine, BSE, UMR 6060, ISVV, Pessac, France;* [4]*Instituto Superior de Agronomia, Universidade de Lisboa, Lisboa, Portugal;* [5]*Univ. Bordeaux, INRAE, BSE, UMR CNRS 6060, USC INRAE 1441, ISVV, Villenave d'Ornon, France*

22.1 Introduction

In Europe, the demand for more environmentally, natural and healthy products is rising quickly. Quality standards and social responsibility standards (e.g., ISO 26000, OHSAS 18001, Global Compact, etc.) are developing. For the wine industry, we can also observe these trends: an increase in the consumption of wine certified on environmental requirements (e.g., recently +80% in Sweden for organic wines) (AICEP, 2016), new certifications, increase demands from importers and distribution companies, strong social protest against chemical residues in the environment and in wine, but also against uncontrolled spraying and disrespect for the landscape.

The wine sector in fact uses a number of natural resources (especially water for cleaning operations or irrigation) and has one of the highest levels of use of phytosanitary products in agriculture (Aubert & Enjolras, 2014a). It is also often concerned by the current debates on the labeling of additives, for legitimate reasons (when it concerns recognized allergens) but also by a climate of generalized fears linked to the recent health crises in the agro-food sector. The result is a multitude of regulations and requirements of marketing intermediaries about the conditions of production and labeling, which can be particularly restrictive for businesses. Devices for modifying practices or for promote standardization that reinforce environmental and health performance also have a considerable impact on the entire industrial suppliers in the wine-producing sector (oenological equipment, agricultural supplies, mechanization, packaging, cork industry, etc.) seeking to constantly adapt these intermediate products.

Small and medium size enterprises (SMEs) are clearly in the majority in the European wine sector, particularly in the Southwestern Europe region (CEEV, 2015). If these enterprises wish to maintain their levels of competitiveness, it is necessary to be able to turn to the implementation of new and more sustainable production practices, to anticipate regulations and above all to develop market products corresponding to the new expectations of consumers.

Enterprises face a number of major challenges in adopting practices that are considered more sustainable: the choice of technologies best suited to the business to improve environmental performance, the availability of cost-effective solutions, and behavioral "locks" that inhibit the adoption of practices that are technically feasible and economic profitable. If certain approaches are beginning to be set up with companies and wine producers, they nevertheless remain scattered and are confronted with the problems mentioned below (Giraud-Héraud et al., 2020).

The evolution of international regulations and private standards in the environmental and sanitary field requires a strong capacity of anticipation on the part of companies, both for the adaptation of winegrowers to new practices and new inputs and for the adaptation of processing companies to the requirements of certification and labeling of certain constituents or properties of the wine (allergens, sulfites, calories, etc.). The "green" approach and the "Social and Environmental Responsibility" of companies must integrate as much as possible the entire value chain from vine to glass.

The market potential of sustainability strategies is still difficult to assess and the implementation of adapted marketing positioning for companies requires a better knowledge of the real demand and its characteristics. There are in fact strong contradictions in the behavior of wine consumers, which are more or less noticeably, but heterogeneously, found within national markets: if the organoleptic quality of the products often takes precedence over the other intrinsic characteristics of the products (health and environmental), it is nevertheless necessary to work on products that meet these two requirements and to clarify the trade-offs actually make by consumers.

Behavioral economics, aided by experimental economics, constitutes a turning point in economic science to take better account of the actions actually carried out by agents: understanding the often complex and unstable trade-offs of consumers, but also the behavioral locks of winegrowers (production practices) (Thaler & Sunstein, 2009). The influence of vertical interfirm relations for better marketing can then be studied, as well as the production functions that make it possible to evaluate the technical and economic efficiency of firms and, in fine, to measure costs.

Behavioral economics is also widely used today to understand consumer expectations and reactions to organoleptic information (when consumer tastes the wine) as well as to understand the origin of products, production methods, and particular certifications/claims (Combris et al., 2009; Vecchio, 2013). Experiments are thus being conducted to measure variations in consumers' willingness to pay (WTP) for a wine by fully controlling the information on the characteristics whose effects are measured (Brugarolas et al., 2005; Schäufele and Hamm, 2017).

Bearing in mind the challenges faced by the wine SMEs of the SUDOE region (the Southwest European region), researchers, entrepreneurs, and institutions have established a partnership relationship embodied in a European research project. The VINOVERT project "Wines, Competitiveness, Environmental and Health Policies of Companies in the SUDOE region—support toward the implementation of methodologies" (SOE1/P2/F0246) is an innovative project for the wine sector which took place between 2016 and 2019. This project is part of the specific objective "Improving and increasing the internationalization opportunities of SMEs" of the "Interreg Sudoe Program" under Priority Axis 2—"Competitiveness of SMEs." The project was based on case studies and the research objects were co-constructed with partner companies and institutions.

The VINOVERT project purposes were the development of concrete measures to improve the competitiveness of companies in the SUDOE wine sector by anticipating the environmental and health requirements of markets and marketing channels. The use of behavioral economics has provided a better understanding of the difficulties of modifying viticulture practices and food behavior in favor of wines that perform well from an environmental and health point of view.

This chapter shows how this innovative project has analyzed the agents' behavior (producers and consumers) based on the lessons of behavioral economics. The chapter starts with the presentation of the project framework, from which the objectives and the working groups of the project were defined (part 2). After, two experimental economic works are presented. In the first one, it is evaluated consumers' WTP for wines resulting from sustainable wine-growing practices (part 3). In the second work, the social comparison nudges applied to wine-growing producers is explained (part 4). To finalize, in part 5, it is presented the main project insights.

22.2 VINOVERT—an innovative project

The SUDOE region is diversely equipped to adapt to the reorientation of wine-growing. Producers in the SUDOE region are often at a disadvantage due to the frequently humid climate (particularly in the Atlantic region) which is not conducive to a change in the production system toward more environmentally friendly practices. This climate favors the appearance of fungal diseases and increases the costs of production linked to "good environmental practices."

Spain is the largest organic vineyard in the world (>120,000 ha) (Eurostat, 2021), but consumption is very low. Portugal produces little and consumes almost none, but has developed integrated production. France is the second largest producer and organic wine and sulfite-free wine consumption is increasing. However, competition from organic wine from South America is likely to be exacerbated on the European and American markets. Finally, the solutions are not always viable from a technical and economic point of view and the supply-demand balance often comes up against difficulties in changing behavior or institutional obstacles.

The VINOVERT project (https://www.vinovert.eu) was constructed by a multidisciplinary approach that mobilized work in economics, sociology, agronomy, and oenology. The project was also the result of a research-enterprise partnership that was coordinated by the University of Bordeaux (France). Furthermore, universities and research institutions of France, Spain, and Portugal were working together in association with several wine SMEs of these regions and private and public organizations.

The innovative aspect of the project stems in the introduction of experimental and behavioral economics applied to consumers and to winegrowers. Also, the project has established a connection between agronomic developments (resistant grape varieties, production practices) and winemaking processes for the production of new wines. By integrating cooperatives representing more than 2000 wineries and intermediary companies, the project has guaranteed a wide dissemination and co-construction of the proposed solutions.

The scientific logic of the project was based on two axes: (i) the evaluation of market receptivity (i.e., marketing channels and consumers) for environmental and health guarantees; (ii) the evaluation of the possibilities of reducing inputs on the farm and in the cellar. An analysis of the evolution of standards and regulations was transversal to these two axes. Several partnership research actions were carried out during the course of the work.

The different activities that were defined were focused on:

1. Development of product innovations to meet demand. Production companies, suppliers of oenological equipment and research laboratories worked on the development of experimental wines. Also, sensory analyses and experimental market works, with quality-labeling arbitration and comparison with a North European and intrazone SUDOE experiment were done.

2. Organizational analyses and technical-economic measurement of farm efficiency. Co-construction research-cooperatives on the environmental improvement of wine-growing practices: sociological analyses and incentive procedures (e.g., concrete implementation of "Nudges") to measure the importance of nonmonetary incentives in changing behavior.
3. Measurement of the capacity of farms in the SUDOE region to improve environmental performance and evaluation of technical possibilities for implementing strategies targeted at health and environmental issues; proposal for adapting winemaking methods, sensory analysis, and experimental markets.

The expected results of the VINOVERT project were: (i) A decision-making aid for companies in the sector on the implementation of methodologies to strengthen the environmental and health balance sheet, on the promotion of certifications or claims; (ii) the development of health solutions adapted to the SUDOE climate that are more respectful of the environment while remaining economically profitable; (iii) an assessment of the conversion potential of the SUDOE region on the strategies targeted by the project.

These results would made it possible to develop new approaches to access external markets, but also to assess the conditions for adopting more sustainable production solutions, likely to make the SMEs in the sector viable to face of new competitiveness challenges.

The empirical works that will be presented in the two next points were developed into two different working groups of the VINOVERT project: WP 2—*Conversation capacity/environmental practices;* and WP 4—*Sector organization/institutional environment.*

The WP2 aims was assessing the technical and economic feasibility of pesticide reduction at the vineyard level. Emphasis was placed on the consequences of such a reduction of inputs on the quality of the final product and then on the valorization of these wines by consumers. Field monitoring of plots of land with reduced use of phytosanitary products (including organic farming) were carried out (France and Spain). In Portugal, emphasis was placed on reducing the use of herbicides (and the need to respect the landscape).

The economic feasibility of these schemes in the three countries were assessed. Differentiated winemaking adapted to the specificities of the phytosanitary parameters observed were carried out. The wines were evaluated by panels of consumers in order to reveal their WTP based on organoleptic quality and possible guarantees in terms of health and respect of the landscape.

The WP4 purposes were the identification of institutional obstacles and levers for the adaptation of the wine sector in the SUDOE region. The difficulties in modifying the behavior of companies were analyzed by surveys applied to the key players in the sector and its institutional environment. Surveys at the level of sectors organizations (interprofessional organizations, distributors, etc.) and regulatory bodies (OIV, European Commission, Ministries) to understand future developments in terms of regulation on the part of states or private players (e.g., approval of grape varieties, possible regulations on maximum residue limits for pesticides, regulations on labeling concerning additives, allergens, norms and standards, etc.). Also other surveys were applied to the project's cooperative partners to understand possible barriers to changing practices: evaluation of the effectiveness of nonmonetary incentives through "Nudges."

22.3 Consumers preferences for sustainable practices measured by experimental auctions

VINOVERT has worked on measuring wine consumers' attachment to the reduction of pesticides in viticulture. An extensive literature is available on this market demand. As pointed out by Schäufele and Hamm (2017), these sustainable development considerations are indeed becoming increasingly important for wine consumers in Europe and the United States. And this is gradually being translated into purchasing decisions (Delmas & Grant, 2014; Forbes et al., 2009; Pomarici & Vecchio, 2014).

Much of the work in wine consumption economics has focused on organic wines to assess the premium obtained through this certification in relation to environmental labeling (Brugarolas et al., 2005; Schäufele & Hamm, 2017). More recently, Schäufele and Hamm (2018) proposed to focus on the very large difference between consumer intentions and expectations on the one hand, and actual purchases in a market that often do not confirm declared intentions and expectations (the so-called "attitude-behavior-gap"), on the other hand. The incentive methods found in the literature on experimental economics make it possible to partially resolve this phenomenon by avoiding the "social desirability bias" and by taking into account the income constraints of consumers (Combris et al., 2009; Vecchio, 2013).

Few studies have, however, focused on consumer receptiveness to alternatives to organic certification, either in a less restrictive framework (for winegrowers) of integrated production (for example, for the widely used labels such as "Produção Integrada" in Portugal or "Terra Vitis" in France) or in a more general framework of environmental certification that covers both biodiversity and corporate social responsibility (e.g., "Haute Valeur Environnementale" in France, ISO 26000 at the international level) or, on the contrary, in a more restrictive framework such as the one that prevails with the biodynamic approach (e.g., "Demeter," "Biodyvin" certification, etc.).

However, several authors have shown how difficult it is to measure consumers' real trade-offs between sustainable development attributes on the one hand and organoleptic enhancement on the other, which often remains prominent (Loureiro, 2003; Schmit et al., 2013). In the specific case of organic certification, it is also shown that it could be perceived as a signal of quality in general (Pagliarini et al., 2013).

In VINOVERT project, Fuentès Espinoza et al. (2018) also focused on the (radical) innovation of the "resistant grape variety" solution. These authors showed the difficulty of sensory acceptance by consumers. However, it appears from this study that communication oriented toward environmental and health performance leads to a significant improvement in the position of the resistant variety wine, ultimately placing it at the top of the average qualitative evaluations. On the economic level, the authors showed that this valorization translates into high market shares, gained in the field of conventional wines. Instead, the loss of market share is more limited for premium conventional wine, suggesting that wines of superior quality would be less in direct competition with wines from resistant varieties. These results remain dependent on the specificities of markets and consumers according to their age, gender, or income (Sellers-Rubio & Nicolau-Gonzalbez, 2016; Thomas & Pickering, 2005). This is why it will be necessary to carry out several studies of this type, varying the context, in order to obtain more general lessons. Nevertheless, these results and the methodology used make it possible to better structure the debate, to confirm certain intuitions (e.g., the sensitivity of consumers to environmental performance which would prevent/complete expectations in terms of organoleptic quality) and to invalidate others (the supposed reluctance of consumers to modify grape varieties).

22.3.1 Experimental auction carried out with the partnership of Portuguese wine companies

The experimental market organized in Portugal concerned the question of the elimination of herbicides in vineyards and the valorization of this approach in the markets, both for the quality of the product and also for the improvement of the quality of the landscape. This supposes that the sensitivity of wine buyers has the merit of focusing the VINOVERT project on the purely "altruistic" aspects of wine consumers, insofar as it concerns an attachment to a better preservation (or supposed preservation) of the environment and wine-growing landscapes. Here, too, it is a matter of working on the company's reputation and on the "anchors" that facilitate the long-term appreciation of the products offered on the market.

Two Portuguese partner companies supplied the experience with two wines: a conventional wine (whose grape supply could corresponded to plots using chemical herbicides) and a "Reserva" wine guaranteeing of nonuse of herbicides.

During this Portuguese experiment, two aspects were then specifically studied:

(1) The issue of landscape quality, which is linked to the use or nonuse of herbicides and which is likely to change consumers' WTP in a direct sales context. To this end, it was carried out an "immersion in the landscape" of consumers to compare the WTPs obtained in relation to a control group placed in a traditional laboratory sales situation.
(2) The question of consumer repurchasing beyond short-term considerations. It was asked consumers to elicit a WTP value, not only for a single bottle of wine but also by several bottles in a "linked sale" situation (each time with inseparable batches). The WTP values of "linked sales" for n bottles ($n = 2, 3,$ or 6), always obtained by direct revelation methods, thus made it possible to measure consumer loyalty.

22.3.1.1 Characterization of the consumer sample

The 205 consumers that participated in this experience were all recruited by a specialized company in the Lisbon region. They were compensated of an amount of 30€ as they usually are for participating in marketing works; nevertheless this participation compensation was disconnected with the sale of the wines (i.e., this amount was not considered by the consumers as a "ticket" to buy a bottle of wine, otherwise it created a maximum limit to the declaration of WTP).

The recruitment filters used by the specialized company were not numerous. The main filter was based on the regular consumption and purchase of Portuguese red wine (and therefore implicitly Douro wines, which represents a large part of the supply in Portugal). Another filter was an approximately equal percentage between men and women (in the end it was obtained 53% men and 47% women).

The income level is also monitored in the sample, although this is not a recruitment filter. The majority of the subjects has a net monthly income between €1000 and €1800. In the same way it was obtained the educational level of the sample and more than half of the sample had attended university education.

About the age criterion, 27% of the subjects were under 35 and 24% over 50. The rest (between 35 and 50 years of age) made up 47% of the sample; and the average age was around 42 years. However it was found that these parameters were not really informative to explain the results obtained. The sample was ultimately appeared as a homogeneous population, and the results were explained above all by the nature of the protocol.

The 205 consumers were divided into two groups that were relatively homogeneous with regard to all the sample characteristics mentioned above: Group 1 of 105 consumers and Group 2 of 100 consumers.

22.3.1.2 Experimental protocol

Apart from the specific study on the demand for herbicide reduction in viticulture, the experimental protocol chosen for this experiment was therefore original both in terms of the influence of the "immersion into the landscape" on consumer behavior and in terms of the analysis of the repurchase, or more precisely on the sustainability of their WTP. As already mentioned, these two questions are linked, in that the immersion effect, while interesting in itself, is above all instructive in measuring the anchors they create in the consumer's mind over the long term. Hence the idea of measuring this effect both by a "remove from immersion" of consumers ex-post, i.e., after this immersion in the landscape, while offering consumers the opportunity to buy several bottles of each wine they had previously evaluated in an immersion situation.

To carry out this experimental market in Lisbon, it was used two adjoining rooms of the Hotel and Tourism School of Lisbon (called "room 1" and "room 2," respectively). "Room 1" was a traditional classroom of this school, without any special equipment. "Room 2" was the school's sensory analysis room, equipped with 20 places for wine tastings (isolation of consumers, presence of spittoons, etc.). This room 2, without any modification, was used for the experiment did with the consumers of group 1. On the other hand, the consumers of group 2 benefited from the "immersion into the landscape." This immersion construction was carried out by a specialized company. They made a sound film (atmosphere of naturalness and field of birds in the landscape) with a drone surveying the vineyards of the Douro wines that were used for the experiment. The sound film was projected on a large screen in room 2.

The protocol began in room 1 to explain the direct method of revealing WTPs. The method known as "BDM," by Becker, De Groot, and Marschak (1964), which is widely used in the literature and has the merit of being easily understood by consumers. It was presented with real examples to ensure a good understanding on the part of consumers, who had already received the minimum instructions, via the recruitment company.

Consumers had been warned by the recruitment company that such a procedure would take place, that it would be the subject of an experimental contract in several informational stages. This was repeated in the session, at this stage in room 1. Consumers were further informed that this multistage experimental market would be subject to a random draw at the end of the experiment (choice of stage drawn at random). It was also explained that four wines would be evaluated at each stage, and that one of the wines would also be drawn at random, with a view to selling one bottle (insofar as the declared WTP value would be higher than the price of the wine also drawn at random). This procedure was valid for each consumer.

The Step 1 of the protocol started after these preliminary instructions. The participants were then invited to room 2: the traditional sensory analysis room for group 1 and the landscape immersion room for group 2. However, in this room 2, the consumers of both groups tasted the four wines in the same way and then proposed a WTP value for each of them, with the only information about their origin ("Douro") and their vintage (2017). A WTP registration form dedicated to this purpose was then filled in by the consumers, who then submitted their proposals to the experimenter.

In room 2 (with a constant and controlled temperature of 20°C), the four wines were served previously (during the exhibition time in room 1) and arranged in a different order to limit as much as possible the effects of the order in which the wines were tasted (or in any case, to measure the effects). Consumers formed their evaluations of the wines by tasting them from right to left and without going back and forth.

In Step 2 of the protocol, information on the guarantee of nonuse of herbicides for wines corresponding to the mention "Reserva" (mention not transmitted during the session) was provided to consumers. It was pointed out that for the other two wines, it was not obtained any guarantee from the producers in this respect. Consumers then had to reformulate a new WTP value for each wine and once again wrote their values on a separate sheet which was given to the experimenter. However, it was asked the consumers to report their WTPs for each wine, before informing them that they were going to continue the experiment by returning to room 1.

It is important to note that consumers had not been warned beforehand of the outcome of this experimental market: a step 3 which would not involve the sale of a single bottle, but several in a "linked sale" situation (a term used in industrial economics to specify the sale of batches of products).

At Step 3 of the experiment, the consumers (from group 1 and group 2) therefore returned to room 1, with their WTPs rating dashboard for each wine (WTPs from stage 2 which take into account information on whether or not herbicides are certified for use).

At this stage, the experimenters informed them that they could buy each of the wines on sale by a successive sale procedure: (i) batch of 1 bottle (step 3.1); (ii) batch of 2 bottles (step 3.2); (iii) batch of 3 bottles (step 3.3); (iv) batch of 6 bottles (step 3.4).

Given the lack of prior information (at the time of recruitment) for this procedure, we then proceeded in two stages of declarations:

Question 1: "Yes" or "No" do you accept to buy at a nonzero price a batch of n bottles ($n = 1, 2, 3, 6$) of each of these wines?
Question 2: If you answer "Yes," what is the WTP you propose for each lot?

The BDM procedure was also applied to this part of the experiment. Consumers were warned that if they did not wish to buy the batches of wines, they could always answer "No" each time they were asked the question.

22.3.1.3 The impact of herbicides information on WTP

The first results obtained relate to Steps 1 and 2 of the experiment, which examine the effect of information on herbicides on WTPs, looking at whether this effect is dependent on the organoleptic appreciation of the wine. These results are represented in Fig. 22.1, which provides the WTPs and their heterogeneity (standard deviations) for the entire group of 205 consumers, as well as the evolution of these WTPs between the two stages.

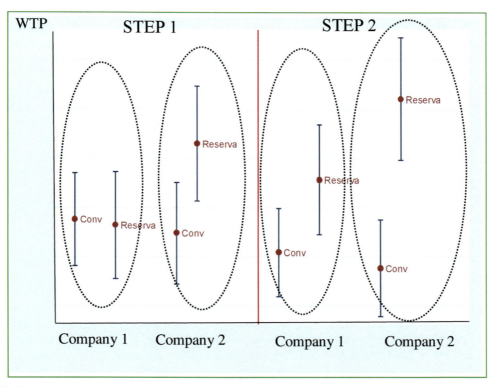

FIGURE 22.1

Evolution of consumers' WTP according to information on herbicide use.

It can be seen that the first three wines are not fundamentally different for consumers. On the other hand, the "Reserva" wine from company 2 is significantly preferred. But what is most important is that the information on pesticides (figure on the right) has a systematically significant effect on the WTPs, and that the percentage increase (+12%) is the same whatever the initial assessment of the wine (the two "Reserva" wines increase their average WTP in the same way).

In addition, it can be seen that on the one hand this information values the two "Reserva" wines and on the other hand it devalues by a comparable level in absolute value (−10%) the wines that cannot make use of this claim. As a result, "Reserva" wines now stand out significantly from conventional wines.

22.3.1.4 The effects of immersion in the landscape

To analyze the effects of the immersion, the data of the two conventional wines that use herbicides and the two "Reserva" wines that do not use them were aggregate, respectively.

The result confirms very well the general trend that was mentioned above. It is even surprising to note to what extent the effect of the information on the use of herbicides is exactly the same from one group to the other when it is observed a very significant translation on the WTPs between the two groups. The landscape immersion greatly increases the WTPs of all the wines, and in a uniform manner. It can therefore be concluded that the effect of the information on herbicides (both valorizing for wines that do not use them and devaluing for those that do) is relatively independent of the qualitative appreciation of the wine and the context in which consumers make this evaluation.

These last results commit us to look at the differences in results obtained between the two groups of consumers (with or without landscape immersion). Table 22.1 shows a very strong influence of landscape immersion, which has the effect of increasing the average WTP of consumers between +36% (for conventional wines) and +45% (for "Reserva" wines). In this table, the results of Group 2 were bolded in order to better highlight the differences between the two groups.

The hierarchy between the wines is the same between the two groups (which again attests to the relevance of comparison by groups). It should also be noted that this trend is repeating itself, even if the number of bottles put on sale is increased. In the immersion group, consumers are much more willing to buy several bottles (2, 3, 6) and this purchase also leads to a higher unit WTP. In other words, landscape immersion leads to a strengthening of consumer loyalty.

22.3.1.5 Main conclusions

This VINOVERT experience has confirmed a growing interest of consumers for environmental certifications and in particular on the specific issue of the use of herbicides. This point is very important because these are pesticides that are hardly ever found in wine (except at extremely high doses that did not concern these wines) and have little potential influence on consumer health. Therefore, the positive reaction (when it is a question of reporting the absence of use) or negative reaction (when a guarantee cannot be given in this respect, whereas it has been given for the "Reserva" wines) is a reaction that can be considered to be mainly oriented toward environmental considerations. Unlike a certification such as Organic, which has this ambiguity, it seems that consumers are really able to value the efforts of producers.

It was initially considered that this appreciation of the efforts of herbicides could be contextualized by the appreciation that consumers have for wine, or even the context in which they appreciate it. However, we have no results to confirm this intuition. Reactions are uniform regardless of quality and context.

Instead, what is particularly interesting in this experience is on the one hand the effect, expected after all, of the context on decisions to buy a wine and the WTPs which were strongly increased in the experience with landscape immersion. Furthermore, this immersion has maintained the level of acceptance of the purchase at a much higher level than the nonimmersion experience and the average WTPs were also much higher.

This observation should lead the entrepreneurial reader to wonder about the influence of marketing channels that are precisely modifying consumers' purchasing contexts. The value of wines is conditioned by the context, with the immersion-landscape context going in the direction of a marketing of the short circuit wine tourism type. It is legitimately to ask if, at a time of (too) strong globalization of wine markets, it would not be time to rethink the reterritorialization of companies. In any case, the attachment to the territories which has made the strength of the European appellations of origin, in order to reinforce the anchoring and loyalty of consumers for the wines of the SUDOE zone.

22.3 Consumers preferences for sustainable practices

Table 22.1 Purchase acceptance and average consumer' WTP for batches of bottles for each wine and group of consumers.

		Company 1				Company 2			
		Conventional		"Reserva"		Conventional		"Reserva"	
		Buyers %	WTP Average unit (WTP≠0) Index	Buyers %	WTP Average unit (WTP≠0) Index	Buyers %	WTP Average unit (WTP≠0) Index	Buyers %	WTP Average unit (WTP≠0) Index
1 bottle	G1	80.90%	79.30	80.95%	88.64	81.90%	69.41	85.71%	100.00
	G2	89.00%	95.05	88.00%	118.50	88.00%	97.80	93.00%	129.49
2 bottles	G1	65.70%	74.91	61.90%	81.14	56.19%	62.82	68.57%	93.04
	G2	79.00%	82.97	79.00%	104.21	75.00%	88.10	83.00%	117.58
3 bottles	G1	56.19%	76.37	55.24%	77.66	49.52%	62.27	62.86%	89.19
	G2	68.00%	79.12	67.00%	100.00	66.00%	82.23	69.00%	107.69
6 bottles	G1	50.00%	69.60	50.00%	64.84	43.81%	55.49	55.24%	79.85
	G2	63.00%	69.23	62.00%	85.71	61.00%	72.34	67.00%	97.07

22.4 The behavioral hypothesis in viticulture validated by nudges

The European wine sector faces major environmental, societal, and economic challenges related to the excessive use of plant protection products. The growing awareness and social questioning about the negative impacts of such use and the growing demand for organic wines, among other aspects, highlight the urgency of such a reduction.

The disparity in treatment levels is high between vineyards, due to local climatic and epidemiological factors (Ambiaud, 2012) but also within vineyards, as shown by the results of the viticulture surveys of the French Ministry of Agriculture. About half of the winegrowers "overdose" their use of phytosanitary products (Aubert & Enjolras, 2014b). In addition, it was observed that the ratio of pesticides remained fairly stable year after year for wine farms despite the climatic conditions of each vintage. The winegrowers who treat the most in a given year generally are those who treated the most the previous year (Aubert & Enjolras, 2014a).

It can therefore be assumed that the use of pesticides is partly a routine activity that has not been questioned, revealing a field of action for levers derived from behavioral economics, such as nudges. Although overdosing is partly linked to behavioral blockages, it is not a question of a knock-on effect or mimicry between winegrowers, but on the contrary of the isolation of certain structures and a lack of meaningful technical references on the issue of treatments adapted to the specific situations of these farms (Raineau, 2018).

Nudges, popularized by Thaler and Sunstein (2009), are now widely used to redirect people's behavior in a desired direction. The processes implemented generally involve two types of intervention: modification of the environment (size and positioning of objects, addition of options or visual elements, decoys, default choices, etc.) and/or provision of information (prosocial messages, provision of norms, direct questioning, etc.).

Among the latter, in the VINOVERT project were experimented the social comparison nudges, developed in line with the pioneering work of Schultz et al. (2007). This type of nudge is based on the provision of information to targeted agents about the decisions made by other agents in order to motivate them to change their choices. The first experiment, conducted in the form of a randomized controlled trial, was carried out within the Gironde cooperative of "Les Vignerons de Tutiac" (France). Experiments were replicated in Spain (with "Martin Codax") and Portugal (with "Adega Cooperativa de Palmela [ACP]"). On each occasion, the experimental project consisted in providing cooperatives members with information on their level of environmental performance (measured here through the Treatment Frequency Index, or TFI) in relation to the performance obtained by other comparable individuals, or groups of individuals, over different timeframes depending on the case. It has been shown that sharing the diversity of performance within a group in this way makes it possible to fill a gap in information in terms of technical references and to initiate a change in practices. In the following points, it will be presented in detail the experimental work carried out in Portugal.

22.4.1 Experience carried out with the partnership of Portuguese wine-growing cooperative

The following case study was carried out in partnership of the "Adega Cooperativa de Palmela" (ACP), located in the metropolitan area of Lisbon. The participation of their technicians and winegrowers was crucial for the nudge application developed during the 2017 and 2018 vineyards' campaigns.

22.4.1.1 The nudge construction

After the 2017 harvest, the field records available from the ACP members were collected. On the basis of the information available there, namely: name, date of application, area treated, dose applied, and name of the disease to be combated by each of the pesticides used in the vineyard, the TFI for plant protection products was calculated for each producer.

The TFI represents the sum of the ratios between the dose applied and the approved dose, for each of the products applied (herbicides, fungicides, insecticides) on a parcel, in proportion to the area actually treated during a crop year (Pingault et al., 2009). Developed in Denmark in the 1980s (Gravesen, 2003), the TFI was integrated into national policies (Germany, Netherlands, France, and UK) to monitor and reduce pesticide use and thus "meet" civil society pressures on the risks of these substances and ensure the competitiveness of their agriculture (Barzman & Dachbrodt-Saaydeh, 2011). In the present study, this calculation related only to fungicides and insecticides.

For the 2017 wine campaign, TFI figures were calculated for 137 wine growers out of a total of 143 "field records" (six of those did not contain the required information or were incorrectly filled in). Of the 137 producers with valid records, 33 produced according the integrated production method. The analysis of the results has shown that the average TFI for the total of records was 11.13, the standard deviation 3.52, and the minimum and maximum TFI was 2.14 and 22.87, respectively. The cumulative percentage of winegrowers, organized by TFI, indicated that only 5% of them had TFI values below 5; and 35% had values below 10, with a greater heterogeneity of values compared with producers with TFI between 10 and 15. Most producers were located in this range: 90% registered TFI below 15, while 55% of producers had TFI values between 10 and 15. It was also found that there were few and fairly spaced TFI records with values between 15 and a maximum of 22.9: only 10% of records.

In view of the nudge application, it was necessary to select two groups of winegrowers with an identical distribution in the statistical variables. Thus, a group with 69 individuals (group 1) elected as a control group, and another with 68 individuals (group 2) elected as the object of that application (nudged group).

Recalling the suggestion that "(…) knowing what other people do—directly or indirectly, through information provided by third parties—has a powerful influence on the behavior of agents" (Schubert, 2017), it was agreed with the ACP that it would be them that would addressed the nudge letters to the 68 members. This decision was more relevant to the objectives of the work, given the function of this entity. The letters were sent just before the start of the phytosanitary treatments of the vineyard, specifically in the second week of March 2018. Since it was found that 1/3 of the total number of 137 wine growers had registered a value of IFT ≤ 10, it was decided to take this threshold as the social reference standard. It was decided that using the mean and not the median would make the values easier to understand.

Based on it, the social universe of 68 producers was broken down into two subuniverses: with IFT equal to or below 10 (24 individuals), and with IFT above 10 (44 individuals). It should be noted that this standard was more "ambitious" compared to that defined in the case studies carried out in France and Spain which opted for the average TFI value. In practical terms, while in Portugal the message to winegrowers (social norm) was "look at what the best are doing" the average IFT shows "what the average is doing."

The nudge letter, an example of which is shown in Fig. 22.2, was sent in line with the ACP. In addition to the descriptive norm included in this letter, an injunctive rule was added, translated into a visual stimulus, in this case a facial expression, or emoji, happy and unhappy. It was thus intended that

> **In 2017, it was possible to obtain a Treatment Frequency Index lower than 10 (*)**
>
> **In 2017, 1 in 3 of the members of Adega Cooperativa de Palmela already reached that index!!!**
>
> **In 2017, your value was 8.46**
>
>
>
> The information contained in this letter is provided to you as an indication and for your private use
>
> (*) The Treatment Frequency Index (**TFI**) measures the total amount of fungicides and insecticides used in your vineyard plot(s), reported at the approved product doses (for the target pests and diseases) and taking into account the total area under your vineyard (on one or more plots).

FIGURE 22.2

Example of a nudge letter with an injunctive standard.

each of the two subgroups should perceive that the quantity of pesticides used in the 2017 agricultural year was socially approved or disapproved, respectively.

22.4.1.2 Injunctive standard effects

It should be noted that while descriptive social standards inform on the prevalence of certain behavior among peers, injunctions directly convey a normative message on what constitutes common (un) approved and potentially sanctioned behavior in their respective sociocultural context (Cialdini & Trost, 1998).

The option to add an injunctive standard was based on two orders of factors. On the one hand, the more people feel that belong to the reference group and the more they feel targeted, the stronger are the positive descriptive encouragements based on descriptive standards; on the other hand, incorporating injunctive standards into descriptive standards helps to contain the "boomerang effect," i.e., "virtuous" people are in line with the average (Schultz et al., 2007).

The "boomerang effect" reported by Thaler and Sunstein (2009) is based on empirical evidence from a study in San Marcos, California. With a view to reducing energy consumption, that study revealed to 300 households not only the amounts of individual energy consumed in the previous weeks, but also the average value of energy consumed by neighboring households. In the following weeks, it was found that users whose consumption was above average had reduced it significantly, in clear contrast to those who had recorded below-average energy consumption which had increased it significantly. According to Thaler and Sunstein, this corresponds to the so-called "boomerang effect" and has produced an important lesson. "If you want to encourage people to adopt socially desirable behavior, never let them know under any circumstances that their present behavior is above average" (Thaler & Sunstein, 2009).

In addition to this, the study revealed another even more interesting conclusion. As those authors reported, half of the households which, in addition to receiving merely descriptive information, had also received a visual stimulus, namely a happy and unhappy facial expression (emoji), showed surprising results. Thus, households whose energy consumption value was above average and who had received an unhappy emoji further reduced their consumption compared to those who had only received verbal information. It was also found that whenever consumers whose energy consumption was lower than the average had received the happy emoji the "boomerang effect" disappeared completely. In other words, when these consumers only received verbal information, they felt that they "had room" for additional consumption, since this was below the average value. However, when verbal information was combined with a visual stimulus, there was no upward adjustment in energy consumption.

22.4.1.3 The results of nudge application

In order to measure the results of the nudge application, after the 2018 harvest, the "field records" of the members of the nudged group (group 2) and the control group (group 1) were collected. The TFI calculations and statistical treatment were then carried out in the same way as in the previous agricultural year. These results were presented at the IX Congress of the Portuguese Association of Agricultural Economics (Rodrigo et al., 2020).

Table 22.2 shows the TFI results for the 102 wine growers for whom it was possible to compare results between 2017 and 2018 and their variation (Δ). To understand these results, two relevant aspects should be mentioned: firstly, of the 137 winegrowers surveyed in 2017, it was only possible to ascertain results for 102 in 2018 (35 "field records" were not delivered to the ACP or were improperly

Table 22.2 TFI values of the two groups of winegrowers (control and nudged).

	Control group (53 winegrowers)			Nudged group (49 winegrowers)		
Year	Average TFI	IFT ≤10 (18)	IFT >10 (35)	Average TFI	IFT ≤10 (20)	IFT >10 (29)
2017	11.22	7.62	13.07	11.16	8.18	13.21
2018	13.07	11.53	13.87	13.21	11.49	14.40
Δ	+1.85	+3.91	+0.80	+2.05	+3.31	+1.19

filled in). Thus, of the 69 individuals in the witness group and the 68 individuals in the nudged group, only 53 and 49, respectively, could be compared. Of the 24 and 44 nudged wine growers who recorded TFI values in 2017 ≤ 10 and >10, only 18 and 29 individuals were able to measure the results. The second aspect to be mentioned concerns the weather conditions experienced by vine growing in the 2018 agricultural year which, prone to mildew and powdery mildew attacks, are in clear contrast to those experienced in the 2017 campaign and thus help to explain the increase in TFI values found.

TFI values increased in 2018 compared to the previous year in both producer groups (control group and nudged group). This increase was slightly higher in the nudged group. The biggest increases were among the winegrowers of both groups who in 2017 had recorded TFI values ≤ 10. The facts stated will certainly not be unrelated to the adverse weather conditions recorded in 2018 for vine growing. If larger quantities of pesticides are identified as a factor in reducing the risks of lower and worse harvests, they will tend to be overestimated when the quantities normally applied are lower. However, comparing the TFI figures of producers who recorded values in 2017 ≤ 10, it appears that the increase in the nudged group is slightly lower compared to the control group.

The results obtained also show that the "boomerang effect" had not disappeared with the presence of the injunctive rule. The TFI values of the nudged group ≤10 increased even in the presence of the happy emoji. Thus, the Portuguese case highlights the persistence of the "boomerang effect" in the application of nudges and, thus, does not confirm the conclusion of the study reported by Thaler and Sunstein (2009). The context in which the experiment took place may have been the explanation for this result. The "boomerang effect" reveals doubts, calling into question its generic solution value.

Finally, it should be noted that the conclusions of this experimental study should take account of its pioneering nature, with only one year of results, and the small size of the total number of ACP members. For example, in the French case, the large number of winegrowers made it possible to identify two groups in addition to the witness group, where different descriptive and injunctive standards were applied.

As nudge application is easy to operate and very inexpensive, its adoption seems to be a useful tool to be adopted either by official bodies or by producers, with a view to reducing the use of pesticides in vine growing. The other note concerns the purposes of applying this nudge in particular. As already mentioned, such aims are not limited to "guiding" future behavior (in this case, reducing the use of pesticides) but also to assessing behavioral "margin." In other words, by making it possible to confront wine growers repeatedly with their practices, this nudge application makes it easier for the "pre-available" to optimize their behavior and consequently to identify ways of reducing the application of pesticides.

22.5 Concluding remarks: VINOVERT project insights

The experimental approaches of an interdisciplinary nature that was carried out in VINOVERT have had a certain success with project partners: trading companies, cooperatives, winegrowers, and producers' organizations. Also, the way of working with them in the co-construction of research objects have increased their scientific relevance as much as their dissemination to professionals.

The VINOVERT has shown and confirmed the interest that companies in the sector could have in considering their Corporate Social Responsibility as a requirement for competitiveness in the medium and long term. It was also shown how wine quality has become multidimensional and that the environmental and health performance of the product is a major issue for future market positioning. Resistant grape varieties, organic certification, and integrated production are possible solutions.

But above all, it has been checked that consumer choices have become complex that they have not given up on organoleptic quality, far from it. It was also argued that the reduction of pesticides in viticulture required a questioning of behaviors as much as technical innovations or monetary or regulatory incentives, and in the end that the ecological transition of the European vineyard also involved a certain organizational revolution.

The fact remains that, as the experimental market did in Portugal (Douro region) has shown, the territorial anchoring of the world of production would once again become the keystone of the reputation system for wines and vineyards, and therefore for companies. On the one hand, because the modes of wine consumption are diversifying more and more and today we no longer speak of "the wine market" but rather "the wine markets": the rapprochement between company and consumer will undoubtedly become, through a sort of reconceptualized short circuit, the anchor point for building buyer loyalty.

On the other hand, because the consumer will no longer be the sole "judge of peace" in a production system: the use of pesticides, the quality of the landscape (in as many different senses as agronomists and society give it), the social links within the company and its terroir, the fair distribution of the value of work, also concern citizens at a local level.

The social responsibility of the wine sector is redefined in these terms. So, it is necessary to transform this intuition into a concrete demonstration, by associating local authorities and citizens to rethink sustainable consumption, the commercial interest of wine tourism and to move toward this still too vague notion of "living well together," which is gradually imposing itself on the within the humanities and social sciences.

References

AICEP. (2016). *Suécia. Vinhos — Breve Apontamento*. AICEP Portugal Global.

Ambiaud, E. (2012). *Pratiques phytosanitaires dans la viticulture en 2010. Fortes disparités de protection contre l'oïdium et le mildiou*. Agreste Primeur, 289. Ministère de l'agriculture, de l'agroalimentaire et de la forêt. http://agreste.agriculture.sg-ppd.maaf.ate.info/IMG/pdf/primeur289.pdf.

Aubert, M., & Enjolras, G. (2014a). The determinants of chemical input use in agriculture: A dynamic analysis of the wine grape-growing sector in France. *Journal of Wine Economics, 9*(1), 1—25.

Aubert, M., & Enjolras, G. (2014b). Between the approved and the actual dose. A diagnosis of pesticide overdosing in French vineyards. *Revue d'Études en Agriculture et Environnement, 95*(3), 327—350.

Barzman, M., & Dachbrodt-Saaydeh, S. (2011). Comparative analysis of pesticide action plans in five European countries. *Pest Management Science, 67*, 1481—1485.

Becker, G. M., DeGroot, M. H., & Marschak, J. (1964). Measuring utility by a single-response sequential method. *Behavorial Science, 9*(3), 226—232.

Brugarolas, M., Martínez-Carrasco, L., Martínez, A., & Rico, M. (2005). Determination of the surplus that consumers are willing to pay for an organic wine. *Spanish Journal of Agricultural Research, 3*(1), 43—51.

CEEV. (2015). *European wine: A solid pillar of the European Union Economy*. Comité Européen des Entreprises Vins.

Cialdini, R., & Trost, M. (1998). Social influence: Social norms, conformity and compliance. In D. T. Gilbert, S. T. Fiske, & G. Lindzey (Eds.), *The handbook of social psychology* (4th ed., pp. 151−192). McGraw-Hill.

Combris, P., Bazoche, P., Giraud-Héraud, E., & Issanchou, S. (2009). Food choices: What do we learn from combining sensory and economic experiments? *Food Quality and Preference, 20*(8), 550−557.

Delmas, M. A., & Grant, L. E. (2014). Eco-labeling strategies and price-premium: The wine industry puzzle. *Business Society Online First, 53*(1), 6−44.

Eurostat. (2021). *Organic farming statistics.* < https://ec.europa.eu/eurostat/databrowser/view/org_cropar/default/table?lang=en|. > Accessed 17.07.21.

Forbes, S. L., Cohen, D. A., Cullen, R., Wratten, S. D., & Fountain, J. (2009). Consumer attitudes regarding environmentally sustainable wine: An exploratory study of the New Zealand marketplace. *Journal of Cleaner Production, 17*(13), 1195−1199.

Fuentes Espinoza, A., Hubert, A., Raineau, Y., Franc, C., & Giraud-Héraud, E. (2018). Resistant grape varieties and market acceptance: An evaluation based on experimental economics. *OENO One, 52*(3). https://oeno-one.eu/.

Giraud-Héraud, E., Lecomte, L., Pérès, S., Raineau, Y., & Seabra Pinto, A. (2020). Quels freins à la transition écologique du vignoble? Enseignements de l'économie. *Revue des Œnologues et des Techniques Vitivinicoles et Œnologiques, 177*, 53−55.

Gravesen, L. (2003). The treatment frequency index: An indicator for pesticide use and dependency as well as overall load on the environment. In *Proceedings of the PAN Europe policy conference* (pp. 28−30). Copenhagen, Denmark.

Loureiro, M. (2003). Rethinking new wines: Implications of local and environmentally friendly labels. *Food Policy, 28*, 547−560.

Pagliarini, E., Laureati, M., & Gaeta, D. (2013). Sensory descriptors, hedonic perception and consumer's attitudes to Sangiovese red wine deriving from organically and conventionally grown grapes. *Frontiers in Psychology, 4*, 1−7.

Pingault, N., Pleyber, E., Champeaux, C., Guichard, L., & Omon, B. (2009). Produits phytosanitaires et protection intégrée des cultures: l'indicateur de fréquence de traitement. *Notes et Etudes Socio-Economiques, 32*, 61−94.

Pomarici, E., & Vecchio, R. (2014). Millennial generation attitudes to sustainable wine: An exploratory study on Italian consumers. *Journal of Cleaner Production, 66*, 537−545.

Raineau, Y. (2018). *Défis environnementaux de la viticulture: Une analyse comportementale des blocages et des leviers d'action.* Thèse de doctorat en sciences économiques, Université de Bordeaux, France.

Rodrigo, I., Seabra Pinto, A., & Giraud-Héraud, E. (2020). Os desafios ambientais da viticultura e a redução do uso de produtos fitofarmacêuticos: Contributo dos nudges. In *Proceedings of IX congress of the Portuguese of agricultural economics* (pp. 1343−1362). Lisbon, Portugal.

Schäufele, I., & Hamm, U. (2017). Consumers' perceptions, preferences and willingness-to-pay for wine with sustainability characteristics. *Journal of Cleaner Production, 147*, 379−394.

Schäufele, I., & Hamm, U. (2018). Organic wine purchase behaviour in Germany: Exploring the attitude-behaviour-gap with data from a household data. *Food Quality and Preference, 63*, 1−11.

Schmit, T. M., Rickard, B. J., & Taber, J. (2013). Consumer valuation of environmentally friendly production practices in wines, considering asymmetric information and sensory effects. *Journal of Agricultural Economics, 64*(2), 483−504.

Schubert, C. (2017). Green nudges: Do they work? Are they ethical? *Ecological Economics, 132*, 329−342.

Schultz, P. W., Nolan, J. M., Cialdini, R. B., Goldstein, N. J., & Griskevicius, V. (2007). The constructive, destructive, and reconstructive power of social norms. *Psychological Science, 18*(5), 429−434.

Sellers-Rubio, R., & Nicolau-Gonzalbez, J. L. (2016). Estimating the willingness to pay for a sustainable wine using a Heckit model. *Wine Economics and Policy, 5*(2), 96–104.

Thaler, R., & Sunstein, C. (2009). *Nudge: Improving decisions about health, wealth and happiness*. Penguin.

Thomas, A., & Pickering, G. (2005). X-it: Gen-X and older wine drinker comparisons in New Zealand. *International Journal of Wine Marketing, 17*(2), 30–48.

Vecchio, R. (2013). Determinants of willingness-to-pay for sustainable wine: Evidence from experimental auctions. *Wine Economics and Policy, 2*(2), 85–92.

CHAPTER 23

Interactive innovation is a key factor influencing the sustainability of value chains in the wine sector

José Muñoz-Rojas[1], María Rivera Méndez[1], José Francisco Ferragolo da Veiga[1], João Luis Barroso[2], Teresa Pinto-Correia[1] and Åke Thidell[3]

[1]MED-Mediterranean Institute for Agriculture, Environment and Development, Universidade de Évora, Évora, Portugal; [2]Programa de Sustentabilidade dos Vinhos do Alentejo, Comissão Vitivinícola Regional Alentejana (CVRA), Évora, Portugal; [3]University of Lund, International Institute for Industrial Environmental Economics (IIIEE), Lund, Sweden

23.1 Introduction

23.1.1 Interactive innovation and food value chain sustainability

The EU Commission (EC) has provided the following description of innovation as part of their European Innovation Partnership for Agricultural productivity and Sustainability (EIP-AGRI): "Innovation is often described as a new idea that proves successful in practice. Innovation may be technological, organizational or social. Innovation may be based on new but also on traditional practices in a new geographical or environmental context. A new idea can consist on a new product, practice, service, production process or a new way of organizing production, etc. Such a new idea turns into an innovation only when it is widely adopted and proves to be useful in practice." (European Council, 2014). Innovation in farming systems thus implies that something new or significantly improved is put in place, but also that such novelties are implemented through successful practices.

Literature on "Innovation" often distinguishes two models of innovation (Knickel et al., 2009; Zahran et al., 2020). Firstly, there is the "linear-type" models of top-down and science-driven innovation, focusing on technologies, aimed at enhancing economic competitiveness and impact in terms of efficiency, effectiveness, and productivity.

On the other hand, one can find the systemic or "alternative" approaches to innovation, which is understood as a social and inherently complex process rather than a singularly technological one (Buller et al., 2019). Indeed, beyond the creation of new knowledge, innovation systems are also inherently social systems (Oliveira et al., 2019). Out of these "alternative" approaches, the most complex one is interactive innovation, which implies adopting a discursive and interactive form of context-specific, multiactor, multiknowledge, and end-user focused engagement that combines social practices with underlying structures (Buller et al., 2019). In such innovative processes, diverse actors and networks mutually, constructively, and intentionally interact across different components and stages of the value chain. Interactive innovation is currently advocated by EU's rural development strategies to enhance sustainability of the agricultural and food sectors.

The European Evaluation Helpdesk (2017) identifies three consecutive requirements to achieve interactive innovation: (1) development of new ideas, (2) consolidation of the processes by which individuals engage with the knowledge innovation system, and (3) enabling a multitude of stakeholders, actors, and related networks acting in institutional and policy environments. To achieve interactive innovation, a social space needs to be created where learning and knowledge sharing are combined via innovation networks, bringing together different actors and stakeholders, with different forms or sources of knowledge (Ingram et al., 2015). Integration of different forms of knowledge represents a challenge for research which thereby evolves much closer to the end-users of the outcomes of research, thus fostering societal processes that otherwise take much longer to kick-off (Nowotny et al., 2004).

Innovation networks can be formal or informal, territorial or sectoral, horizontally or vertically integrated. Moreover, it is now widely acknowledged that innovation networks are dynamic in nature (Klerkx et al., 2010).

Their composition can change over time as the priorities and access to resources by different actors and stakeholders shifts, and they can also vary in shape and size, and in the strength of the relationships between their members. Such networks may well include business clusters, multiple actors and stakeholders, territorial alliances, public-private partnerships and learning agreements (Neumeier, 2012; Zahran et al., 2020).

Overall, interactive innovation seems to call for a shift in the strategic mindset of producers, market agents, and public actors. Also, the value chain needs to move from a competitive toward a cooperative approach to joint strategy making (Ingram et al., 2015; Oliveira et al., 2019; Zahran et al., 2020). This may prove especially relevant for the wine sector, where fierce competition among regional wine trademarks in the globalized market-sphere is at odds with the expectations for an agricultural produce that exceeds the condition of commodity (Merli et al., 2018). Furthermore, according to these same authors, this is also a sector that has at its core territorial and landscape-related concepts such as *terroir*, and thus where distinctiveness and competitive advantages may become more apparent and efficient at above-the-farm scales.

The aim of this chapter is to exemplify how interactive innovation is co-constructed and how it works in the real world, using the Wine Sustainability Program of Alentejo (Portugal) (WASP) as case study. This is a program that is deemed as successful in both enhancing the sustainability standards of the sector (CVRA - Alentejo Regional Winegrowing Commission, 2020b), and that, as we hereby argue, that has achieved this by effectively implementing interactive innovation strategies and approaches.

To test this hypotheses, we assessed how interactive innovation is effectively enacted, ultimately contributing to enhance sustainability of the wine value chains across scales ranging from the farm to the region (e.g., Alentejo wine region).

To do this, we used a mixed method approach (Archibald et al., 2017) that brings together qualitative (document analysis, expert interviews, focus groups, and discussion workshops) along with quantitative (social network analysis [SNA]) methods to elicit the key elements of interactive innovation contributing to the success of the PSVA in achieving its own targets and objectives.

To this date, most wine-chain sustainability studies have focused either on measuring quality standards that can be quantitatively demonstrated, or alternatively on qualitatively eliciting subjective narratives, attitudes, and scenarios helpful to move toward sustainability (Merli et al., 2018). Nonetheless, studies so far have neglected the key role likely played by interactive innovation, a role that is further explored in this chapter.

23.1.2 Eliciting interactive innovations influencing sustainability in the wine sector

The need of the wine sector to enhance its sustainability standards is now inarguable (Gerling, 2015; Forbes et al., 2020; Santini et al., 2013). What remains highly arguable is how sustainability should be enacted, measured, and monitored. Achieving farm level sustainability is a complex venture. As stated by Ohmart (2008): *"agricultural sustainability involves everything you perform on the farm, including economics, environmental impacts of every single agricultural or forestry operation and all aspects of human resources, including not only you and your family but your employees and the surrounding community."*

Therefore, to assess farm level sustainability, a mix-method approach (Archibald et al., 2017) needs to be applied which combines qualitative and quantitative, modeling and participatory, approaches and techniques. As examined by Santini et al. (2013), ample literature exists addressing this issue in the wine sector. According to this author, this is generally performed via indicators that specifically apply either qualitative, quantitative, or mixed methods of research.

Furthermore, studies so far have addressed topics of interest to innovation, such as strategies, entrepreneurial and consumer behavior, supply chain management and certification (Santini et al., 2013). These same authors also identified wine sustainability aspects that have already been addressed in the scientific literature through indicators, including environmental sustainability, organic production models, ecolabelling, greener industrial processes, environmental consumer behavior, sustainable business strategies, and farming-ecology connections. All of these are aspects that can be linked to one, or multiple, aspects of innovation, and that have a strong interactive component.

One interesting example in which indicators for measuring sustainability were adopted exists in Australia (Santiago-Brown et al., 2015). Although such study was grounded on knowledge co-construction, it was somehow restricted to a quantitative and largely biophysical perspective, thus leaving behind key relevant qualitative and social aspects of sustainability that are at the core of interactive forms of innovation (Buller et al., 2019; Cronin et al., 2020). Drawing on all of this previous work, Merli et al. (2018) synthesized the set of existing sustainability initiatives in the wine sector across wine-producing regions and farming systems as diverse as those in Italy, France, Austria, Germany, Spain, California, New Zealand, South Africa, and Chile.

Merli et al. (2018) aimed at developing an international indicator framework, largely drawing on concepts such as Life Cycle Assessment that relies on technological innovation principles. In addition, these authors argued that knowledge co-construction and stakeholder coordination are now inarguably considered as key issues to secure the efficiency and operationality of sustainability assessments in the wine sector, a condition that we hereby advocate can be best accomplished when interactive innovation approaches are adopted.

Another key challenge that underpins all wine sustainability programs worldwide relates to territorial heterogeneities of the wine sector across sociocultural, economic, and ecological conditions that are contingent to each wine-producing region (Merli et al., 2018; Moscovici & Reed, 2018). According to these authors, such heterogeneity demands that both sustainability and related innovation approaches are tackled to safeguard regionally specific social-ecological characteristics of each wine-producing region, synthesized in the concept of "terroir" (for an example of this in Portugal, see Martins et al., 2019).

Furthermore, they also emphasize the need to avoid current application of sustainability standards to justify undesirable green-washing marketing approaches that are increasingly extended in some farming systems, and that are directly contrary to the principles of agroecology (Francis et al., 2003) and transparency. These green-washing strategies too frequently turn consumers skeptical about the environmental performance and benefits of green food and/or products (Goh & Balaji, 2016).

23.1.3 Wine value chains in Portugal and Alentejo

The wine sector is of great importance in Portugal, not only in socioeconomic terms but also from cultural, ecological, and landscape standpoints (Martins et al., 2019). Vineyards and wine have been present in Portugal and the Alentejo since very remote dates, but it was the Romans who firstly developed wine production using a technology based on clay *amphoras* that is still used very similarly in some places in the Alentejo, the so-called "Vinho de Talha" (amphora wine).

The economic dimension of the wine sector in Portugal currently represents 11% of agricultural outputs (meassured in terms of basic prices), with an average annual production value of 817,4 million euros in the period 2015−2019. The Alentejo wine region encompasses 9% of regional agricultural outputs, with an average annual production value of 224,2 million euros for the same period. The Alentejo wine sector currently represents 27.4% of the total economic value of the national wine sector (IVV, 2021).

From the 1980s onwards, the expansion of Portuguese wine quality standards began in Alentejo with the conversion and growth of vineyards and wine production, supported by a fruitful partnership formed by ATEVA—the Technical Association of Viticulturists in Alentejo (created in 1983, by the wine cooperatives and viticulturists), the University of Évora, and the regional and national public bodies CCDRA—Alentejo Regional Coordination and Development Commission, DRAPAL—Regional Directorate of Agriculture and Fisheries of Alentejo, and IVV—Institute of Vine and Wine.

Portugal has one of the worldwide oldest Protected Denomination of Origin (PDO) worldwide, the Douro region (1756). However, in the case of Alentejo, it was only in 1988, after Portugal's accession to the then European Economic Community (then European Union), that the statutes of the first winegrowing zones were approved with a view to improve the production and commercialization of quality wines in specific areas (QWPSR/VQPRD). Currently a Protected Designation of Origin—"PDO?? Alentejo" exists and covers eight subregions (Fig. 23.1), along with a Protected Geographical Indication—"IGP Alentejano" that spans across the NUTs III[1] of Alto (Portalegre), Central (Évora), and Baixo (Beja) Alentejo.

The area currently covered by vineyards in Portugal is 192 401 ha, of which 13%, corresponding to 25 057 ha, are in Alentejo. A slight expansion of 1.4% of its vineyards occurred in Alentejo over the past decade. This is the Portuguese region where most new plantation rights were granted in this period

---

[1] The nomenclature of territorial units for statistics (Nomenclature des Unités territoriales statistiques—NUTS) is a geographical hierarchy, according to which the territory of the European Union is divided into hierarchical levels and related administrative units. The three hierarchical levels are known as NUTS-1, NUTS-2, and NUTS-3 for regions, and LAU-1 and LAU-2 for municipalities. This classification enables cross-border statistical comparisons at various regional levels within all countries of the EU.

23.1 Introduction   465

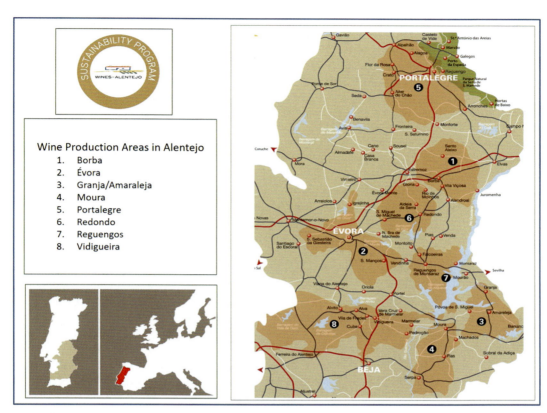

FIGURE 23.1

Map showing the spatial distribution of the eight wine subregions with Protected Denomination of Origin in Alentejo (NUTS-2), comprising the districts (NUTS-3) of Portalegre, Beja, and Évora.

Source: Courtesy of CVRA. (2020a). Comissão Vitivinícola Regional Alentejana. Relatório Anual 2019: Gestão e Contas. https:// www.vinhosdoalentejo.pt./media/Documentos/2019_Relat_rio_de_Actividades_e_Contas_Dem_Financeiras_CLCpdf.pdf. (Accessed on 28 June 2021).

(IVV, 2021). Alentejo is also the Portuguese region where the mean average area of wine producing farms is biggest, 7.58 ha, compared to 1.88 ha nationally, with larger farms representing a greater proportion than in the remaining of the country (GPP, 2020).

Between the agronomic years 2009/10 and 2019/20 national wine production rates shifted between 5.89 million and 6.53 million hl, with a maximum wine production of 7.15 million hl in 2010/11. Portugal currently occupies the 11th position in the world ranking of wine producers. In Alentejo, over the same period, the average yearly production was of 1.05 million hl, shifting between 0.810 million and 0.996 million hl, with a maximum wine production of 1.22 million hl in 2014/15. Alentejo currently represents 16% of national wine production (IVV, 2021). The increase in production and productivity in the region has been largely driven by the expansion of irrigation infrastructures (Costa et al., 2020).

In terms of production of wines suitable for PDO/PGI certification, the yearly average in the last five years was of 5.38 million hl in Portugal, which represents 83% of the total national wine production, corresponding to 3.50 million hl of PDO wines and 1.88 million PGI wines. In Alentejo, the equivalent values are 1.03 million hl, which represents 98% of the total regional wine production, corresponding to 0.56 million hl of PDO wines and 0.47 million PGI wines (IVV, 2021). In Alentejo, the proportion suitable for quality wines is thus remarkably higher than the national average. In this sense, it is also worth noting that 28% of the wine designated as organic at the National level in 2018 was produced in Alentejo.

Wine consumption has steadily increased worldwide over the past decades, moving from 226 million hl of wine in 2000 to 246 million hl in 2018 (IVV, 2019). In Portugal, wine consumption in the same period rose from 4.7 to 5.5 million hl, although with a marked interannual variability. In years 2017 and 2018, there was an increase in consumption justified mainly by the increase in tourism, which has since been practically halted due to the COVID-19 crisis.

The monetary values for both the worldwide exports and imports of wine from Portugal and Alentejo have also steadily increased over the same period, a clear indicator that wine markets are becoming increasingly globalized.

From an overall perspective, Portugal has moved from exporting wine for a value of 519 M Euros in 2000, up to 804 M Euros in 2019. In contrast, imports of wine have remained quite stable (109 M Euros in 2000, and 157 M Euros in 2019), thus shifting positively its national trade balance (IVV, 2019). The main current target countries for exports of Portuguese wines include France (14.4%), the US (10.1%), and the UK (9.4%), the first two of which are direct producing competitors, indicating to the enhanced competitiveness of Portuguese wines. The hierarchy of countries to where wines from Alentejo are now exported is similar, although leaving UK at a much lower position, and incorporating instead Brazil and Angola (CVRA, 2020a).

In relation to sales in the national market, their average numbers in the last four years (2016—2019) indicate to a volume of 2.67 million hl and a monetary value of 951 million euros, with growth rates of 8% and 24%, respectively. Certified wines represented, on average, 43% in volume and 63% in monetary value. The customer share of local bars and restaurants is slightly higher than that of distribution chains, 55% and 45%, respectively. By regions, Alentejo had a market share in 2019 of 37.4% in volume and 39.1% in monetary value, being the leading region in the national market (IVV, 2021).

Once we combine all these data together, a clear impression arises that the wine sector in Alentejo is rapidly gaining economic strength, when compared to the Portuguese and especially, EU trends. Although challenges related to productivity and market competitiveness and differentiation persist in the wine sector of Alentejo, they seem nonetheless to being gradually tackled.

It is important to indicate, at this point, to the legal requirement that is in place for vine farmers, wine producers and also for stock agents, bottlers, exporters, and importers to get inscribed in an official registry, which is regulated at the National level. This secures that a certain capacity exists for the public sector to monitor the trends and changes in the activities of wine farmers and firms.

Overall, it is clear that the Portuguese wine sector is well organized, firstly in terms of producer networking and horizontal cooperation, as shown by wine cooperatives in Portugal coping 37% of the production (IVV, 2021). Another key element of associative capacity for the wine sector in Portugal are the 12 (private law and public utility) Regional Winegrowing Commissions (CVRs) distributed across Portugal to ensure that certification standards are met for Geographical Indications (GIP) and

PDOs. Among these, the CVRA is responsible for the Alentejo. CVRA is actually the key promoter and manager of the WASP, which is the only program for wine sustainability in Portugal.

Despite all these institutional arrangements currently in place, global market drivers are still at the core of the multiple challenges facing the sector (Merli et al., 2018).

We argue that in Alentejo this is effectively tackled through the WASP, which is helping to steer the regional wine sector toward increasing competitiveness by aiming at achieving enhanced sustainability standards that provide the sector with competitive advantages. This program is strongly underpinned by an interactive innovation operational framework. We provide further details in the following section.

23.2 The Wines of Alentejo Sustainability Program: background and implementation

23.2.1 History and outreach

The WASP is an initiative, underpinned by voluntary adherence, promoted by the Alentejo Regional Winegrowing Commission (CVRA) and aiming at engaging not only grape and wine producers in the Alentejo Wine Region but also other stakeholders and actors with a potential role in helping to achieve sustainability goals for the sector.

Collective sustainability plans are common practice in some wine-producing regions across the globe (Merli et al., 2018; Moscovici & Reed, 2018) and have been lately gaining relevance in markets where Alentejo wines are also present. International and domestic wine markets are now starting to demand the application of sustainability principles (Gerling, 2015; Merli et al., 2018). WASP was strongly inspired by several International Organization of Wine and Vine—OIV guidelines (OIV, 2016) alongside with the sustainability program schemes of two major winegrowing regions, California and Chile (Gerling, 2015). Thus, a partnership was formed starting in 2015 with public funding for only two years, originally engaging CVRA and Universidade de Évora along with other nine public and private institutions from the region, which acted as stakeholders (Cronin et al., 2020). This network is now enlarged by over 430 wine producers and vine farmers.

This partnership intends to support improvement of the environmental, social, and economic performance of the Alentejo's wine making activities and promote the recognition of the sustainability standards of the region's wines.

Ultimately, this is expected to help promote and affirm Alentejo's brand in internal and external markets. Overall, WASP was defined by the following specific objectives (CVRA, 2020b):

(i) Articulating the entire wine sector in Alentejo, underpinned by a conceptual and operational framework aimed at increasing social, environmental, and economic well-being at local and regional levels, including the requirement to adopt eco-efficiency principles;
(ii) Characterizing the sustainability performance of producers;
(iii) Contrasting individual results and standards among peers (promoting horizontal cooperation);
(iv) Defining areas for improvement and action plans that will enhance production practices benchmarking for sustainable wine production;

(v) Defining the certification process according to the sustainable production framework that will enable the recognition of the quality of Alentejo Wines based on their sustainability performance;
(vi) Attaining risk minimization;
(vii) Developing applied research projects with participation of companies and fostering cooperation between wine companies and the University of Évora;
(viii) Turning the program outputs and results available, and disseminating them in a timely and efficient manner, including by promotion of best practices/results that can be replicated among producers.

The WASP program was initially (2015−2017) funded with Regional Development Alentejo 2020 funds, but now acts as a bottom-up private self-funded initiative. Over the years is has gathered over 500.000 € of public and private funding. The program started as a regular regional development public-funded program, but upon the success obtained and following request from producers, it has continued ever since, supported through the voluntary financial and logistic support from farmers and other stakeholders, namely cooperatives. It is now managed under a centralized and purposeful central office, which rests in the personal effort, interest, and knowledge basis of two experienced sustainability professionals ("champions" in the jargon of innovation) that were external both to the region and the sector (Cronin et al., 2020).

The WASP originally started as a formal consortium between the CRVA and the University of Évora. This consortium also involved ATEVA, as well as a group of regional winegrowers chosen for their sustainability efforts. This intended to tailor sustainability strategies to the regional contingent characteristics. At this initial stage, some technical reports were produced comprising quantitative sustainability indicators along the whole production chain. This was intended to inform and secure the interest of farmers and other key stakeholders and actors. Nonetheless, upon the end of the publicly funded project, the CVRA decided to extend the program indefinitely upon the interest of participants. The main reason was the added value that participants (voluntary) could achieve for their products by jointly enacting common sustainability standards for their wines (Cronin et al., 2020). In the wine sector, producers are commonly empowered, well-organized, and hold market power (Santini et al., 2013).

In Alentejo, this is underpinned by social and policy support for the wine sector, which is characterized by producers with high education levels and awareness for sustainability and innovation (Martins et al., 2019). Additionally, the fact that WASP is a bottom-up and fully voluntary initiative strengthens its long-term sustainability, due to increased motivation of all participants.

23.2.2 Sustainability quantitative assessments

Since the WASP is oriented at securing minimum sustainability standards by producers in the region, a quantitative assessment methodology becomes especially relevant. The program is highly demanding, as proven by the requirement that, for a producer to become certified, 86% of all sustainability criteria need to score in the highest (so-called "Developed") sustainability level.

To secure that quantitative indicators reflected the problems and challenges faced by farmers and other actors and stakeholders, these were co-constructed by participants in the program. Such indicators have been used to put together a mandatory self-assessment certification tool, which is aligned with the program's objectives, and which is updated yearly by all members of the project.

The matrix of indicators to evaluate sustainability performance permits to measure the changing trends in the overall farming system, jointly including the various stages and actor-networks (Jones, 2009) along the value chain (thus working similarly to a Life Cycle Assessment). The WASP quantitative analysis consists in a self-assessment tool which targets: (1) viticulture; (2) wine cellars; and (3) jointly viticulture and cellars.

For each of these three target groups, and during the first phase of WASP implementation, a set of Primary Intervention Chapters (PICs/CIP) were developed (Fig. 23.2). In a second phase, initiated by the end of 2018, the evaluation methodology was adapted to fit better with the new data and indicators, triggering the implementation of the Secondary Intervention Chapters (SICs/CIS).

Each participant self-evaluates his wine-producing activities according to 11 Intervention Chapters. Combining the results of the self-assessment for each of the 11 PICs, an overall ranking called the "General Sustainability Category" is calculated. This ranking falls within one of four ranges defined (Preinitial, Initial, Intermediate, and Developed). The 11 PICs were developed to include a total of 108 assessment criteria (Fig. 23.3A and B).

According to the CVRA (2020a), the objectives that had been set in the original WASP project are being largely achieved, but not exactly as planned. Due to its links to a Regional Funding Program, Alentejo 2020, along its first two years, the WASP was audited twice by an external and official auditing entity. As a result of these audits, it became evident that the WASP was already reaching far beyond the requirements in the funding program, with the minimum standards required for producers to participate in the WASP being set well above those indicated in the Portuguese legislation for sustainable agriculture.

However, the ultimate goal for every WASP member is to achieve continuous improvement of their sustainability standards. Thus, any arising opportunities for improvement should always be considered. Such opportunities are continuously revised as part of annual action plans, that are periodically reshaped according to any new knowledge acquired, and that aim for each WASP member to achieve the status of "Developed." Once each WASP member has reached this status, seven SICs, including a total of 63 sustainability assessment criteria are defined as new targets to be achieved individually, thus permitting tailored plans to be implemented for each individual producer. Currently, the WASP includes a total of 18 Intervention Chapters with 171 evaluation criteria in total (CVRA, 2020b).

Once the category to which each WASP belongs at any given moment has been clearly established, the SICs become available for them to apply. A process of validation and recognition will only proceed when the WASP member achieves the Final Category of "Developed" in the SIC (Fig. 23.4). This audit is performed under the ISO/IEC 17021 (certification of management systems), thus establishing a direct link with standard quality systems (CVRA, 2020c).

In order to ensure the credibility and reliability of the program, the results of the self-evaluation for each WASP member are subjected to an internal audit by the Alentejo Regional Wine Growing Commission, which is performed according to the abovementioned ISO/IEC 17021 standards. The self-evaluation serves as an annual individual diagnosis on the sustainable practices of WASP members (CVRA, 2020c).

Planning for the WASP started in 2013 and the Program was officially launched to producers in May 2015.

PRIMARY INTERVENTION CHAPTERS

- Grape Production
- Soil Management
- Water Management in the Vineyard
- Disease & Pests Management in the Vineyard
- Energy Management in the Vineyard
- Waste Management in Vineyard
- Water Conservation in the Cellar
- Energy Management in the Cellar
- Waste Management in the Cellar
- Disease & Pest Management in the Cellar
- Human Resources

SECONDARY INTERVENTION CHAPTERS

- Wine Quality
- Sustainable Management of Ecosystems
- Air Quality
- Management & Handling of Materials in Production
- Packaging options & Components
- Community & Skateholders
- Socio-Economics & Regional Development

FIGURE 23.2

List of the productive aspects measured by each of the WASP voluntary adherents during the primary and secondary intervention chapters of the program.

Source: Courtesy of CVRA. (2020b). Wines of Alentejo Sustainability Program: Intervention chapters, sustainability indicators, thresholds and results. http://sustentabilidade.vinhosdoalentejo.pt./en/wines-of-alentejo-sustainability-programme. (Accessed on 29 June 202)1.

FIGURE 23.3A

Examples of various aspects measured by each WASP voluntary adherents during the primary and secondary intervention chapters of the program, in this case including key agro-ecological assessment criteria adopted by the program.

Source: Courtesy of CVRA. (2020b). Wines of Alentejo Sustainability Program: Intervention chapters, sustainability indicators, thresholds and results. *http://sustentabilidade.vinhosdoalentejo.pt./en/wines-cf-alentejo-sustainability-programme. (Accessed on 29 June 2021).*

472 Chapter 23 Interactive innovation is a key factor

FIGURE 23.3B

Examples of various aspects measured by each WASP voluntary adherents during the primary and secondary intervention chapters of the program, in this case including key socioeconomic assessment criteria adopted by the program.

Source: Courtesy of CVRA. (2020b). Wines of Alentejo Sustainability Program: Intervention chapters, sustainability indicators, thresholds and results. *http://sustentabilidade.vinhosdoalentejo.pt./en/wines-of-alentejo-sustainability-programme. (Accessed on 29 June 2021).*

With a strategy based on actions requiring low financial investment leading to fast returning benefits (considered by the project leader and manager as "low hanging fruits"), the 11 PICs were released and made available to the wider public. This was done to demonstrate the clear benefits for Alentejo's grape and wine producers of joining and implementing the WASP.

This strategy had almost immediate results, with the Program gaining 93 members by the end of 2015, a number that has increased every year. By the end of 2018, seven new intervention chapters

MINIMUM VALUE REQUIRED FOR DESIGNATION OF THE GENERAL SUSTAINABILITY CATEGORY OF WASP

GENERAL SUSTAINABILITY CATEGORY	VALUE	PERCENTAGE
PI — Pre-Initial	≤ 1,8	≤ 45%
I — Initial	≤ 2,4	≤ 60%
Int — Intermediate	≤ 3,4	≤ 85%
D — Developed	≥ 3,5	≥ 86%

FIGURE 23.4

Thresholds of values required for the achievement of the various categories of Sustainability status for producers, ending with the "Developed" one toward which all participants must continuously work.

Source: Courtesy of CVRA. (2020b). Wines of Alentejo Sustainability Program: Intervention chapters, sustainability indicators, thresholds and results. http://sustentabilidade.vinhosdoalentejo.pt./en/wines-of-alentejo-sustainability-programme. (Accessed on 29 June 2021).

were opened—SICs—to members who had already reached the level required by implementing the 11 PICs. The highly heterogeneous group of members (e.g., wide range of property sizes, economic and financial capacities, team sizes, objectives, or strategies), the approach for continuous improvement adopted by the WASP, allows members to develop and implement the Program at their own speed. Currently the Program has 445 voluntary members and growing.

23.2.3 How have sustainability assessment results evolved?

The CVRA has so far conducted two assessments of the regional implementation of the WASP. The CVRA firstly assessed the implementation and continuous improvement of the PICs on a regional basis up until 2018 (Fig. 23.5A). This evaluation was later expanded to include the SICs in 2019 (Fig. 23.5B).

When analyzing the PIC assessments synthesized as spider graphs (Fig. 23.5A and B), it becomes self-evident that WASP participants extend and compress their different sustainability impacts according to an apparently random pattern. This is due to two characteristics of the WASP: the continuous improvement model which assigns values for each member's evaluation trend, thus continuously "expanding" the graph, and the constant increase in new members, which decreases the

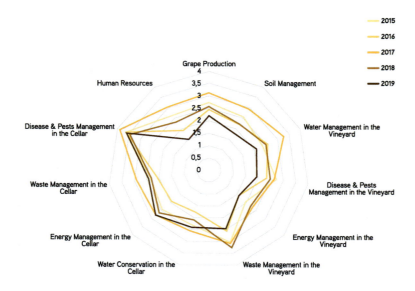

FIGURE 23.5A

Graphical representation of the results of the primary intervention chapters of the WASP program for the years 2015–2019.

Source: Courtesy of CVRA. (2020b). Wines of Alentejo Sustainability Program: Intervention chapters, sustainability indicators, thresholds and results. http://sustentabilidade.vinhosdoalentejo.pt./en/wines-of-alentejo-sustainability-programme. (Accessed on 29 June 2021).

average levels of sustainability implemented. Such trends are influenced by the time-lapse between sustainability action requirement and implementation, resulting in a graphical contraction of results (Fig. 23.5A and B).

Interestingly, the shape of the spider graph on PICs remains very similar as years go by, with waste management and water management in the wine cellar, and disease and pest management in the vineyard ranking top, and human resources and energy management in the vineyard ranking last (Fig. 23.5A). Ranking at the top are PICs that are not strictly part of the sustainability program, and that are triggered by the irregularity of water availability in the Alentejo region and by the fear of pests by most wine producers in the region. Some other PICs are alien to most farmers, because from an economic point of view, their importance might not be so obvious, i.e., human resources, which could explain their lower ranking marks. Regarding the SICs, which are only ones valued ahead of the PICs, one finds wine quality, socioeconomic and regional development and community, actors and stakeholders with top marks, while the sustainable management of ecosystems ranks last (Fig. 23.5B). Thus, it is obvious why social objectives are reached sooner by most farmers and wine producers that the environmental ones, which require higher financial investment and therefore imply higher levels of risk. However, as years go by and more producers reach the SICs, we might see changes in these outputs and indicators.

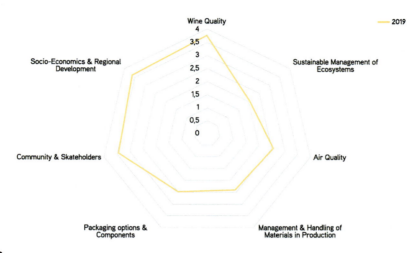

FIGURE 23.5B
Graphical representation of the results of the secondary intervention chapters of the WASP program for the year 2019.

Source: Courtesy of CVRA. (2020b). Wines of Alentejo Sustainability Program: Intervention chapters, sustainability indicators, thresholds and results. http://sustentabilidade.vinhosdoalentejo.pt./en/wines-of-alentejo-sustainability-programme. (Accessed on 29 June 2021).

23.3 Assessing WASP's interactive innovation toward enhanced sustainability

23.3.1 Methods and stages

This section describes how interactive innovation actions with an impact in sustainability standards can be identified and disentangled across its multiple facets and resulting complexity. This was done by combining qualitative and quantitative methods, using the WASP as a case study example.

The methodological approach that we defined to elicit interactive innovation in the WASP includes collecting data on five types of project interactions (Buller et al., 2019; Cronin et al., 2020). For each step, it becomes necessary to identify and characterize stakeholders and actor-networks (Jones, 2009), structures, functions, institutions, and good practices involved. These five interactions are the following: (1) Interactions with funding mechanisms and institutions; (2) Internal project interactions; (3) External project interactions; (4) Interactions with the policy environment; and (5) Interactive societal challenges. To gather all this information, desk-based research, in-depth interviews, and a participatory (online) workshop with key stakeholders were implemented as follows:

(1) **Desk-based research**: This first step was intended: (i) to collate information about the WASP project implementation in general and its diverse stages, funding, and outreach; (ii) to gather available information about the program including technical reports and publications by the Universidade de Évora over the initial stage of the WASP and about other similar activities by the coordinator and each of the partners; (iii) to identify the legal and economic status of WASP partners, the geographical scope where they perform their activities and also about legal and economic relationships between them; (iv) to gain an initial overview of the social, cultural, economic, and policy context on which the WASP is framed. The main information sources used for this first evaluation stage included the WASP Website program information, along with project management documents and deliverables provided by: (a) Partners (CVRA, Universidade de Évora and ATEVA); (b) Social media: CVRA, ATEVA, Individual Producers; (c) Newspapers articles; (d) Scientific papers: Some links provided by the University of Évora team that worked on the project; (e) Others: Periodic statistical datasets and periodic reports provided by CVRA (Cronin et al., 2020).

(2) **In depth interviews** with 15 experts (including producers, academics, managers, representatives of associations, and technicians) identified as key stakeholders of relevance to the five interactions. To select relevant stakeholders and good informants, four clusters were preliminary identified: (1) Wine producers' associations in the region, including ATEVA and CVRA; (2) Academic entities involved in the WASP, especially the U. Évora (involved in the initial and publicly funded stage) and Lund University (involved later on along the implementation stage, via annual cooperation on specific sustainability issues identified in WASP implementation), and also others indirectly engaged; (3) Innovative farmers, firms and cooperatives from Alentejo engaged in the program and in charge of both or either wine production, and vine cultivation; (4) Public entities involved with the WASP throughout the initial public-funded and the follow-up self-funded phases, and that ought to be considered as external stakeholders.

The individual interviews were conducted online (except for the three initial ones, which were performed live) due to the COVID-19—related restrictions. Interviews were recorded and each interviewee was furthermore asked to comment on our interpretation of the findings. Following a first preliminary live interview with the WASP project leader, different models of interview templates were designed and delivered to all engaged project partners, beneficiaries/participants, external stakeholders, and funding bodies. Additional interviews with experts were arranged with these and other relevant stakeholders. In addition, 25 further individual enquiries were performed with actors and stakeholders (if possible, including the same ones responding to in-depth interviews) including representatives of diverse sectors engaged in different functions of the WASP, and also those engaged along the various stages in the WASP. These extra interviews were intended to perform a SNA (Prell et al., 2009). These SNA enquiries were aimed at eliciting the degree of intensity and information exchanges in the mutual interactions among different stakeholders in the WASP project, and how these shifted along the stages of project preparation (2015), public funding (2015—2017) and self-funding (2017—2021) stages. This allowed us to identify and evaluate mutual relationships and linkages among the different stakeholders and institutions along the diverse stages of the project.

(3) **Participatory (online) workshop with key stakeholders** (15) was conducted, online, on the July 22, 2020 to jointly evaluate and refine findings, and to identify and disentangle future options to improve interactive innovation in the WASP. Stakeholders involved in the workshop

include the main stakeholders already interviewed plus a selection of others involved in the SNA, including academics and technicians, wine producers, vine farmers, representatives of associations and the WASP, managers of cooperatives and public officers. The main sources of data and informing the workshops were the key points identified upon the in-depth interviews and SNA, which were presented and discussed among all participants during half a day of open debate.

23.3.2 Results of WASP's interactive innovation assessment

Reflecting upon outputs and responses obtained through the joint qualitative and quantitative assessment that we performed over the WASP allowed us identify certain interactive innovation aspects and factors that have been, so-far, largely neglected in the key role they play in driving sustainability standards of the wine sector, in these case by facilitating the successful implementation of the WASP.

These aspects include the key role of individual champions in steering more dynamic strategies, leading by example, of open-ended actor-networks (Jones, 2009) for knowledge and action co-construction, the sheer importance of shared values, and lastly the relevance of horizontal cooperation (among farmers, thus mainly through peer to peer learning) and vertical coordination (along the value chain) for sustainability purposes.

23.3.2.1 Dynamic leadership

Dynamic leadership can act as a crucial driver to enact interactive innovation (Buller et al., 2019). According to our results in the analysis of interactive innovation in the WASP, the presence of a leader or champion (in this case, the WASP manager) since the beginning of the project has proven essential to achieve the following targets: (1) identify the institutional support required for success, (2) generate trust and boost motivation by engaging key stakeholders in the decision-making process and in the development of novel products and practices, (3) expand and diversify the network of participants, (4) disseminate and embed innovation across a variety of contexts and realms, and (5) advance the approach beyond "project-term" funding toward longer-term programs.

This "champion" has had close working relationships with the CVRA's management allowing him to influence decision-making processes, while remaining autonomous enough to establish contacts and alliances with partners, stakeholders, and external actors, and coordinate the project as well as to organize knowledge and information sharing.

Communication within the project has been largely valued by participants as very good. The role and competence of the WASP coordinator is a determinant factor and he acts as catalyst to facilitate moving forward.

According to the WASP managing body, constant work is required for the revitalization of what has already been committed and has not yet been fulfilled, or what has been done and should continue because it has been proven a success, and lastly, also of what has not yet been done, but could and even should potentially be implemented.

It seems apparent that the "champion" could not be easily replaced because he acts as change agent, rather than manager, so his loss before the WASP enters a routine could lead to a loss of enthusiasm from participants, and undermine the success of the project. The ability of a person to become a champion is mainly drawn from their soft skills: their charisma and the form in which they engage and inspire others, rather than by the content of what they communicate (Block, 2019).

23.3.2.2 Shared values

Arising from our analysis, it has become evident that defining a vision based on shared values is an important step for interactive innovation projects, and that ultimately, shared values trigger enhanced sustainability actions and foster more effective participation (EU Commission, 2021).

A first argument is that sharing values provides a common perspective. In the case of the WASP, shared values have acted as a framework underpinning common objectives, actions, and activities. Additionally, they enhance internal relationships, fostering trust and driving behaviors, routines and preferences.

We found that shared values may lead farmers to tackle risks by adapting their practices to new challenges arising. Regarding trust-building within the project, something that is quite unique to the WASP is that distrust was never a problem, as noticed by the program managers. The CVRA has been established and working in the region for many years, and therefore it has created strong trust bonds across the wine sector key stakeholders and actors. When the program was created, already existing trust bonds, plus the shared values, conformed a solid base in which to start working, sharing knowledge, activities, and practices, while securing their accountability: "the whole system is based on trust," according to the CVRA leader (Cronin et al., 2020).

How were shared values created in this case then? A first set of values was established when the project was originally devised. Initially sharing underpinning values turns the execution of projects simpler, faster, and more consistent. Additionally, we found that some values arise along a project, via active listening across stakeholders and actors, cooperating and sharing knowledge, although this may result in tensions.

23.3.2.3 Stakeholder engagement: network evolution process over time

The key barrier encountered by the WASP has been some cultural and mindset aspects of local farmers and wine makers. These aspects are clearly reflected in regional farming discourses, which are largely underpinned by conservatism and a related reticence toward changes and innovations (Pinto-Correia et al., 2019). In Portugal this aspect may be a deciding factor compared to the US or Australia, where other similar sustainability programs are also in place for the wine sector (Merli et al., 2018).

Alentejo, despite being a region with an urgent need to innovate, is still characterized by a farming community largely afraid of the unknown, despite many of the larger-scale farmers and land managers being highly educated (Pinto-Correia et al., 2019). Therefore, constant effort is needed to raise awareness across producers about the added value of innovative projects, including this one.

This currently being a voluntary and free project, certain barriers exist that need overcoming. Nevertheless, and in order to achieve its goals, those invited to speak at workshops or other events tend to be largely the farmers and winemakers who are prepared to present their experiences and results, problems, and solutions. In fact, the active commitment of stakeholders in the WASP network is a key factor for its continuity, deepening the dissemination of results and fostering new innovations.

One of the key roles that WASP project managers have performed is to monitor the network and to identify who are the most active and critical stakeholders and those who are not. Carrying out this exercise in successive stages along the program shows how the initial network evolved. Firstly, networks have possibilities to expand their results if the initial design relies on a central stakeholder that has a reputation, an established structure of relationships in the territory, and is independent of the project's resources.

The business orientation of the key stakeholders and actors involved in the WASP is the second factor to consider. These stakeholders act by aligning the whole process toward applicable results, getting involved in the dissemination and recruitment of new participants, thus enhancing the possibilities for the continuity of the program. On the contrary, if a given stakeholder or actor has a strong project orientation, the focus is on the expected results and the effort slows down once the funding ends.

Finally, it is likely that some more peripheral stakeholders with the capacity to promote, finance, and disseminate, can join actively when projects are focused on mature sectors. For this reason, the participation of such stakeholders and actors is even more important when we are dealing with pioneering activities, and/or scenarios with low population densities, such as Alentejo.

23.4 Final reflections and conclusions

Sustainability programs for the wine sector have been in place worldwide for over 30 years and are rapidly spreading (Merli et al., 2018) under a paradigm shift where co-constructing innovation strategies is key for gaining market power, and thus also for achieving financial sustainability, which is especially relevant in a rapidly globalized world. Such programs are, so far, largely focused on quantitatively monitoring the sustainability of the diverse productive functions performed by voluntary producers and other stakeholders along the value chain, by using quantitative indicators and standards that allow for a certain trademark to be collectively secured incurring in competitive advantages.

This is a key aspect to consider when aiming for a clear shift to happen in a sector, farming, that is largely conservative and reticent to forms of innovation other than technological (Pinto-Correia et al., 2019). It is important, that on the mid-to longer-term timescales, other more mixed quantitative-qualitative and interactive aspects be also considered. In the WASP, such aspects include issues related to leadership, sharing values, and enhancing networking and stakeholder and actor engagement.

Within our own case study, all of these aspects seem to have been successfully fostered mainly via the incentives for participation and continuation that, throughout the various phases of the program, have been put in place by the WASP managers. Another key factor contributing for the success of the WASP relates to the open-ended and flexible nature of the program.

Nonetheless, it must also be said that the next big step is to reach enough critical mass, which is considered by the WASP managers as being more than 60% of the wine sector economic agents of the Alentejo, a target needed to turn this free and voluntary project into a mandatory and sector-wide project. In the short and medium term, the goal of the current managers of the WASP is to become certified by a third party, that is, from an outsider entity that can state independently and unbiasedly whether the project is worthy and deserves to be acknowledged (CVRA, 2020c).

The CVRA has now produced a referential internal document defining the process to evaluate how the WASP is being implemented into the management system of each program member, so that they can all achieve their own individual WASP certification. Ultimately, four certifying bodies entities accredited by the IPAC (Portuguese Institute of Accreditation) with ISO/IEC 17021 (certification of Management Systems) will perform the certification audits (CVRA, 2020c).

Based on our findings, some questions arise regarding the current impact of interactive innovations on the long term, including:

Will the innovative actions undertaken so far be enough to reach the threshold of engagement and commitment required to move the whole sector toward enhanced levels of sustainability?

How are the preconditions for participants to succeed influenced by the contingent sociocultural conditions that define each regional wine sector and related actor-networks?

In the case some of these conditions fail, how could the current levels of success in Alentejo be maintained? and

Could this last target perhaps be potentially achieved by establishing a trademark, thus securing competitive marketing advantages across regional to global levels?

The success of the WASP program for the wine sector can also inspire and motivate other farming sectors of the Alentejo region where, similarly to most other Mediterranean regions, key climate and sustainability challenges are threatening the immediate survival of essential farming and food systems and related ecosystems (Prosperi et al., 2014; Muñoz-Rojas et al., 2019). Such challenges include climate change, biodiversity and water scarcity, soil erosion and fertility loss, local labor shortages and population aging, and the permanence of largely conservative mindsets that hamper innovation.

All of this, along with the lack of a clear strategy to cope more efficiently with the globalization of the sector, puts in risk multifunctional vineyard landscapes and "terroirs" that over the past couple of millennia have been co-produced by multiple stakeholders and actors mutually aligned and coordinated. Nonetheless, further opportunities for sustainability-oriented interactive innovation may arise, such as those linked to digital networks (Fielke et al., 2020), resulting in alternative pathways for enhanced sustainability, and that will surely become central in the short- to middle-term timescales at which the most influential policies and farming strategies are defined. Whether the wine sector embraces them or not is yet to be seen. Its longer-term sustainability will ultimately depend on this.

Acknowledgments

The authors would like to thank the European Commission, which through its H2020-LIAISON project (Grant agreement ID: 773418), has funded the interactive innovation research aspects condensed in this chapter. Also the CVRA, which through its WASP program, funded by the Alentejo 2020 RD Program, conducted the sustainability assessment hereby examined. Additionally, authors from the University of Évora were also funded by the FCT—Foundation for Science and Technology (Portugal), under the Project UIDB/05183/2020. Last, we would like to thank Prof. Miguel Costa, from the ISA-Lisbon, for inviting us to publish our research in this book, and to the reviewers who allowed us to improve our text and figures, to make them publishable.

References

Archibald, M. N., Radil, A. I., Zhang, X., & Hanson, W. E. (2017). Current mixed methods practices in qualitative research: A content analysis of leading journals. *International Journal of Qualitative Methods, 2015*, 5–33. https://doi.org/10.1177/160940691501400205

Block, J. H. (2019). Rural entrepreneurship; between start-ups, hidden champions and family businesses. *The Journal of Business, Economics, Sustainability, Leadership and Innovation, 3*. https://doi.org/10.37659/2663-5070-2019-3-14-19

Buller, H., van Dijk, L., Fieldsend, A., & Varga, E. (2019). *D.1.1. Moving innovation from the sporadic to systemic: Innovation governance strategy, approaches and practices and role of different stakeholders and governance bodies on Global and International Scale*. H2020 LIAISON Project. GA no. 77341 Accessed on 28 March 2021 https://liaison2020.eu/your-material/?language=English.

Costa, J. M., Oliveira, M., Egipto, R., Fragoso, R., Lopes, C. M., & Duarte, E. (2020). Water and wastewater management for sustainable wine production in dry Mediterranean regions. *Ciência e Técnica Vitivinícola, 35*(1), 1–15. https://doi.org/10.1051/ctv/20203501001

Cronin, E., Foselle, S., Fieldsend, A., Menet, A., Couzy, C., Rogge, E. L., & Von Münchaussen, S. (2020). *Deliverable 4.1. 32 case study portraits, with practice abstracts, of interactive Innovation projects and initiatives across Europe*. LIAISON-H2020 Project Accessed on 28 March 2021 https://liaison2020.eu/wp-content/uploads/2021/01/Deliverable-4.1-Case-Study-Portraits.pdf.

CVRA. (2020a). *Comissão Vitivinícola Regional Alentejana. Relatório Anual 2019: Gestão e Contas* Accessed on 28 June 2021 https://www.vinhosdoalentejo.pt/media/Documentos/2019_Relat_rio_de_Actividades_e_Contas_Dem_Financeiras_CLCpdf.pdf.

CVRA. (2020b). Wines of Alentejo Sustainability Program: Intervention chapters, sustainability indicators, thresholds and results. http://sustentabilidade.vinhosdoalentejo.pt/en/wines-of-alentejo-sustainability-programme. (Accessed 29 June 2021).

CVRA. (2020c). *Wines of Alentejo Sustainability Program:* Certification Accessed on 29 June 2021 http://sustentabilidade.vinhosdoalentejo.pt/en/wasp-certificationast.

EU Commission. (2021). *The EIP-AGRI concept* Accessed on 28 March 2021 https://ec.europa.eu/eip/agriculture/en/eip-agri-concept.

European Council. (2014). *Communication from the commission to the European Parliament, the Council, the European economic and social committee and the committee of the regions. Research and innovation as sources of renewed growth*. Brussels, 10 June 2014. 339 final version Accessed on 28 March 2021 https://eur-lex.europa.eu/procedure/EN/1042071.

European Evaluation Helpdesk for Rural Development. (2017). *Evaluation of innovation in rural development programs 2014-2020* Accessed on 28 March 2021 https://enrd.ec.europa.eu/evaluation/publications/evaluation-innovation-rural-development-programmes-2014-2020_en.

Fielke, S., Taylor, B., & Jakku, E. (2020). Digitalisation of agricultural knowledge and advice networks: A state-of-the-art review. *Agricultural Systems, 180*, 102763. https://doi.org/10.1016/j.agsy.2019.102763

Forbes, S. L., De Silva, T.-A., & Gilinsky, A., Jr. (2020). *Social sustainability in the global wine industry. Concepts and cases* (p. 204). Springer Nature, ISBN 978-3-030-30413-3.

Francis, C., Lieblein, G., Gliessman, S., Breland, T. A., Creamer, N., Harwood, R., Salomonsson, L., Helenius, J., Rickerl, D., Salvador, R., Wiedenhoeft, M., Simmons, S., Allen, P., Altieri, M., Flora, C., & Poincelot, R. (2003). Agroecology: The ecology of food systems. *Journal of Sustainable Agriculture, 22*(3), 99–118. https://doi.org/10.1300/J064v22n03_10

Gerling, C. (2015). *Environmentally sustainable viticulture practices and practicality*. Apple Academic Press, ISBN 978-1-77188-112-8. eBook ISBN: 978-1-4987-2229-2.

Goh, S. K., & Balaji, M. S. (2016). Linking green skepticism to green purchase behavior. *Journal of Cleaner Production, 131*, 629–638. https://doi.org/10.1016/j.jclepro.2016.04.122

GPP. (2020). *Gabinete de Planeamento, Políticas e Administração Geral*. Estatisticas Agrícolas INE 2019 Accessed on 28 March 2021 https://www.gpp.pt/index.php/estatistica-agricolas-estruturais-e-de-producao/estatisticas-agricolas-estruturais-e-de-producao.

Ingram, J., Maye, D., Kirwan, J., Curry, N., & Kubinakova, K. (2015). Interactions between niche and regime: An analysis of learning and innovation networks for sustainable agriculture across Europe. *The Journal of Agricultural Education and Extension, 21*(1), 55–71. https://doi.org/10.1080/1389224X.2014.991114

IVV. (2019). *Instituto da Vinha e do Vinho*. Reforma da PAC 2021–2027 plano estratégico da PAC análise sectorial: Vinho. Contributo IVV, IP Accessed on 28 March 2021.

IVV. (2021). *Instituto da Vinha e do Vinho*. Dados Estatisticos do setor vitivinícola-2021 Accessed on 29 March 2021 https://www.ivv.gov.pt/np4/estatistica/.

Jones, O. (2009). Actor-network theory: A world of networks. In O. Jones (Ed.), *Nature-culture* (pp. 309–323). International Encyclopedia of Human Geography. https://doi.org/10.1016/B978-008044910-4.00716-1

Klerkx, L., Aarts, N., & Leeuwis, C. (2010). Adaptive management in agricultural innovation systems: The interactions between innovation networks and their environment. *Agricultural Systems, 103*(6), 390–400. https://doi.org/10.1016/j.agsy.2010.03.012

Knickel, K., Brunori, G., Rand, S., & Proost, J. (2009). Towards a better conceptual framework for innovation processes in agriculture and rural development: From linear models to systemic approaches. *Journal of Agricultural Education and Extension, 15*(2), 131–146. https://doi.org/10.1080/13892240902909064

Martins, A. A., Costa, M. C., Araújo, A. R., Morgado, A., Pereira, J. M., Fontes, N., Graça, A., Caetano, N. S., & Mata, T. M. (2019). Sustainability evaluation of a Portuguese "terroir" wine. In *BIO web of conferences* (Vol. 12, p. 03017). EDP Sciences.

Merli, R., Preziosi, M., & Acampora, A. (2018). Sustainability experiences in the wine sector: Toward the development of an international indicators system. *Journal of Cleaner Production, 172*, 3791–3805. https://doi.org/10.1016/j.jclepro.2017.06.129

Moscovici, D., & Reed, A. (2018). Comparing wine sustainability certifications around the world: History, status and opportunity. *Journal of Wine Research, 29*(1), 1–25. https://doi.org/10.1080/09571264.2018.1433138

Muñoz-Rojas, J., Pinto-Correia, T., & Napoleone, C. (2019). Farm and land system dynamics in the Mediterranean: Integrating different spatial-temporal scales and management approaches. *Land Use Policy, 88*(2019), 104082. https://doi.org/10.1016/j.landusepol.2019.104082

Neumeier, S. (2012). Why do social innovations in rural development matter and should they be considered more seriously in rural development research? — Proposal for a stronger focus on social innovations in rural development research. *Sociologia Ruralis, 52*(1), 48–69. https://doi.org/10.1111/j.1467-9523.2011.00553.x

Nowotny, H., Scott, P., & Gibbons, M. (2004). *Re-thinking science. Knowledge and the public in a age of uncertainty* (p. 288). Polity Press.

Ohmart, C. (2008). Innovative outreach increases adoption of sustainable winegrowing practices in Lodi region. *California Agriculture, 62*(4), 142–147. https://doi.org/10.3733/ca.v062n04p142

OIV. (2016). *OIV general principles of sustainable Vitiviniculture-environmental-social-economic and cultural aspects OIVCST518-2016* Accessed on 23 June 2021 https://www.oiv.int/js/lib/pdfjs/web/viewer.html?file=/public/medias/4943/oiv-cst-518-2016-en.pdf.

Oliveira, M. D. F., Gomes da Silva, F., Ferreira, S., Teixeira, M., Damásio, H., Dinis Ferreira, A., & Gonçalves, J. M. (2019). Innovations in sustainable agriculture: Case study of Lis Valley irrigation district, Portugal. *Sustainability, 11*, 331. https://doi.org/10.3390/su11020331

Pinto-Correia, T., Muñoz-Rojas, J., Thorsoe, M., & Noe, E. (2019). Governance discourses reflecting tensions in a multifunctional land use system in decay; tradition versus modernity in the Portuguese Montado. *Sustainability, 11*(12), 3363. https://doi.org/10.3390/su11123363

Prell, C., Hubacek, K., & Reed, M. (2009). Stakeholder analysis and social network analysis in natural resource management. *Society & Natural Resources, 22*, 501–518. https://doi.org/10.1080/08941920802199202

Prosperi, P., Allen, T., Padilla, M., Peri, I., & Cogill, B. (2014). Sustainability and food & nutrition security: A vulnerability assessment framework for the Mediterranean region. *Sage Open, 4*(2). https://doi.org/10.1177/2158244014539169, 2158244014539169.

Santiago-Brown, I., Metcalfe, A., Jerram, C., & Collins, C. (2015). Sustainability assessment in wine-grape growing in the new world: Economic, environmental, and social indicators for agricultural businesses. *Sustainability, 7*(7), 8178–8204. https://doi.org/10.3390/su7078178

Santini, C., Cavicchi, A., & Casini, L. (2013). Sustainability in the wine industry: Key questions and research trends. *Agricultural and Food Economics, 1*(1), 1−14. https://doi.org/10.1186/2193-7532-1-9

Zahran, Y., Kassem, H. S., Naba, S. M., & Alotaibi, B. A. (2020). Shifting from fragmentation to integration: A proposed framework for strengthening agricultural knowledge and innovation system in Egypt. *Sustainability, 12*(12), 5131. https://doi.org/10.3390/su12125131

Further reading

Bonjean, I., & Mathijs, E. (2016). How transaction costs shape market power: Conceptualization and policy implications. In *12th European IFSA symposium, programme and book of abstracts, Social and technological transformation of farming systems: Diverging and converging pathways* (p. 107). IFSA. https://www.harper-adams.ac.uk/events/ifsa/papers/5/5.4%20Bonjean%20and%20Mathijs.pdf. (Accessed 28 March 2021).

CHAPTER 24

European wine policy framework—The path toward sustainability*

João Onofre

Head of Unit for Wines, Spirits and Horticultural Crops, European Commission, Brussels, Belgium

24.1 Introduction

The European wine sector is a story of success. The European Union (EU) is the world's leading producer of wine, accounting for 44% of world wine-growing areas and 65% of production (OIV, 2021).

The wine sector accounts for three million direct jobs in the EU. Moreover, European wine is an iconic sector symbolizing quality, know-how, culture, and tradition. Today, over 70% of the EU wine is under a certified quality seal, with a virtually endless variety of regional types, of which 1141 *Protected Designations of Origin* (PDOs) and 438 *Protected Geographical Indications* (PGIs).

In spite of its enormous variability, all wine regions in the EU work under an extensive and unique regulatory set of rules that guarantee respect of the single market and fair competition within the Union. They are compiled in the Common Market Organisation regulation (EU) No 1308/2013 mainly in its articles 39 to 54, 62 to 72, 80 to 88, and 92 to 123 (European Union, 2013). They contain, among others, definitions; requirements for operators, registers, and controls; rules for planting authorizations (wine is the only sector having production limitations under the Common Agricultural Policy); rules on oenological practices, rules on labeling, presentation, PDOs/PGIs and traditional terms and a market support system aiming at long-term market stability (European Union, 2013).

Following the implementation of the 1999 and 2008 wine reforms, the EU wine sector has fully addressed the challenge of competitiveness. The former introduced first restructuring measures and the latter enhanced the adaptation to the market with additional measures through dedicated national support programmes, tailor-made for EU wine-producing Member States and regions. Major elements of these programmes include restructuring and reconversion of vineyards, investments in wine cellars, and promotion in third countries. This resulted in a quick turnaround from wine surpluses of the past, toward higher quality (70% of EU wines are marketed under quality labels), with a constant average unit value increase since 2008.

---

*Disclaimer: The information and views set out in this chapter are those of the author and do not necessarily reflect the official opinion of the European Commission.

The overall value of EU exports attained 11.9 billion € in 2019, compared to 5.6 billion in 2008 (European Commission, 2019). More importantly, quality and reputation of EU wines have become a door opener for all EU quality agricultural products in recent international trade agreements (Japan, Canada, and Mexico).

Sustainability and addressing citizens' concerns about environment, food, and health are the major challenges for the European wine sector for the next 20 years.

In 2020, the COVID-19 pandemic strongly impacted competitiveness of the EU wine sector. First and foremost, it led to a major drop of the EU's wine consumption, following the lockdown of the restaurants, specialized wine shops, and tourism in general. Social aspects of wine consumption are also relevant in this context, as virtual elimination of social contacts took a strong toll on wine consumption. Secondly, the export market slowed down significantly as business in the two main export markets has been hampered by both tariffs (e.g., USA) and economic slowdown (e.g., China). This forced the EU Commission to adopt a specific set of exceptional measures to address surplus wine stocks (namely via crisis distillation and crisis storage). At the time of writing (Spring 2020), following the impact of the COVID-19 second wave, market unbalance will be subsisted during the first half of 2021. Spring frosts affecting major wine areas in France and Italy and easing of national lockdowns following the roll-out of the vaccination process will probably rebalance the wine market at the later part of 2021.

24.2 Environmental aspects of wine production

Since the last quarter of the 19th century, copper has been the main active ingredient used against widespread downy mildew and against other less common vine diseases (e.g., black rot, anthracnose, and bacterial diseases). For more than 150 years, European vineyards have been subjected to unrestricted use of copper products, in some years to the tone of 70–80 kg/ha. Copper is a highly toxic metal that can have very damaging impacts on environment, pollution of groundwater, biodiversity, and human health (EFSA, 2018; EIP-Agri, 2019). In particular, the EFSA peer review concluded that there is a risk to operators applying copper-based formulations on grapes and other crops (e.g., tomatoes), justifying the recommendation to wear adequate personal protective equipment. The EFSA's report also identified a high risk to birds and mammals, to aquatic organisms, including sediment dwellers, and to soil macroorganisms, including earthworms, for all the representative uses (grapes, tomatoes, and cucurbits). Following this risk assessment, and in accordance to the European regulation No 2018/1981, since January 1, 2019 were imposed strong limitations to the use of copper in agriculture, also applicable for organic agriculture (total application of maximum 28 kg of copper per hectare over a period of seven years) (European Union, 2018).

Fig. 24.1 displays a map elaborated by the European Commission's Joint Research Center, showing copper concentration in European soils. It shows a striking correlation between major wine areas in France, Italy, Spain, or Portugal and copper concentration in soils as already pointed out in recent literature (Lamichhane et al., 2018; EIP-Agri, 2019; Karimi et al., 2021).

This map reflects a historical situation of accumulation of copper in soils since the early days of its use at the end of the 19th century. Today, copper remains one of the essential tools to control important diseases in organic viticulture, but it is no longer a main active ingredient in conventional viticulture since other more effective ingredients have been developed by the agrochemical industry.

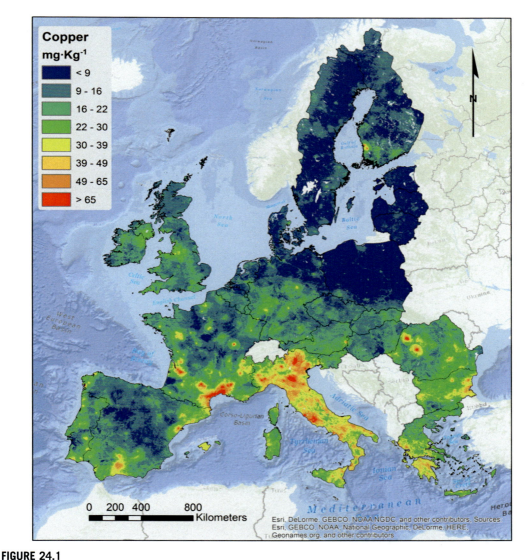

FIGURE 24.1

Concentration of copper in EU soils (European Commission Joint Research Center, 2019).

In this respect, European vineyards use 1.8% of the total used agricultural land area, but they account for 20% of the total consumption of plant protection products in EU. Although highly variable according to different regions and weather during the growth season, typical European vineyards, under medium disease pressure, apply an average of 12—15 fungicide treatments (Pedneault & Provost, 2016; Pertot et al., 2017; Teissedre et al., 2017), but the number can be higher in some situations (e.g., adverse climate conditions), not counting the use of insecticides, acaricides, and herbicides (EIP-Agri, 2019). This has a huge cost for EU wine growers (estimated at over one billion €/year), not always compensated by market receipts. More importantly, this comes at a high environmental cost.

Other than the direct environmental impact, current levels of pesticide use in wine growing also harm the reputation of the sector. Indeed, there are regular and constant press stories about residues in wine and health effects of abundant pesticide application in wine regions. Moreover, citizens' concerns about health and the environment have led to action and regulation.

In 2009, the EU Member States approved Directive 2009/128/EC on the Sustainable Use of Pesticides (European Commission, 2009). This Directive states that pesticide use in areas that could cause contamination (along roads, railway lines, close to surface water or sewage systems) should be eliminated or reduced as far as possible. The Directive also calls for a reduction in pesticide use or risks in areas used by the general public, such as public parks and gardens, sports and recreation grounds, school grounds and children's playgrounds, and in the close vicinity of healthcare facilities as well as in recently treated areas used by or accessible to agricultural workers. As a result of this Directive, many cities decided to go pesticide free. For example, Germany prohibited pesticide use on nonagricultural land, Italy and Slovenia banned glyphosate use in public areas, and France and Luxembourg banned all pesticide use in public areas. In addition, in France, in December 2019, the Ministry of Agriculture published a ruling whereby minimal distances for treatment free zones have to be respected: five meters for low crops such as vegetables and cereals and ten meters for high crops: fruit trees and vines. Also 60 municipalities in France imposed stricter limits with distances ranging up to 150 m for treatment free zones around housing, schools, public areas, and along waterways. The proximity of vineyards and residential areas in many wine-growing regions cause vineyards to be greatly affected by such restrictive measures. Wine growers will need to adapt to these new conditions and take up the challenge of responding to society's demands for improved sustainability. In addition, new private "sustainable" labels for EU wines are being promoted privately or seeking official certification.

24.3 Technical solutions to the challenges

The EU wine legislation is focused mostly on *Vitis vinifera*. According to Regulation (EU) No 1308/2013, interspecific hybrids are not allowed within PDOs and some specific hybrid varieties are banned in all the EU (*Noah, Othello, Isabelle, Jacquez, Clinton, and Herbemont*) (European Union, 2013). In the 1930s, many European countries began to restrict the use of highly productive American direct producer vines and their hybrids, in a context of economic crisis, overproduction, and underconsumption. Some of the arguments put forward at that time for this prohibition included the bad quality and taste as well as the high content in methanol of such varieties which has not been scientifically confirmed yet. This ban was enshrined in European law in 1970, upheld since that time and included in the current CMO under Article 81(2) of Regulation (EU) No 1308/2013 (European Union, 2013). However, even if today's hybrid varieties are totally different from the ones forbidden in the past, the main arguments for opposing hybrid varieties lie on quality concerns and risks of overproduction. Hybrids are also highly resistant to diseases, and require fewer pesticides (Merdinoglu et al., 2018; González-Centeno et al., 2019). This would benefit both wine growers, who would be less exposed to chemicals, and consumers who request more "natural" wines, as shown by the boom in the market of organic wines or the so-called "natural wines."

In addition, resistant varieties are a crucial tool for wine growers especially when important/useful biocides are withdrawn from the market or their use is significantly restricted. Using less fungicides is less costly for farmers, and leads to a lower pressure on the target fungi, and therefore to a lower risk that they will develop a resistance thereto. As previously stated, under medium disease pressure, at least 12 fungicide treatments per season are needed for traditional *V. vinifera* varieties grown under conventional management (EIP-Agri, 2019). In a study including 183 fungus resistant grape (FRG) varieties grown in six different EU countries, the number of fungicide treatments was reduced by 73% and 82% in organic vineyards with low and medium disease pressure, respectively (Pedneault & Provost, 2016). In a survey involving 65 German vineyards under organic management, growers reported having to spray FRG varieties 3.8 times per season on average (Pedneault & Provost, 2016). Estimations are that growing FRG varieties could cut production costs by two in French vineyards (Pedneault & Provost, 2016).

Other studies confirm the high amount of pesticides used on vines every year. A recent study calculated that the total treatment frequency index (TFI) of a standard conventional wine reached 16.9 and 12.7 for a premium conventional wine, while the total TFIs of an organic wine and of a resistant variety wine stood at 2 for both (Fuentes Espinoza et al., 2018). Another research on resistant varieties expected to reduce the average TFI from 12, currently observed for the traditional varieties, to 2 in the future vine-growing systems (Merdinoglu et al., 2018). Similar findings were highlighted in a communication from the University of Udine at the sixth international symposium of the Oenoviti International network, i.e., there is a potential reduction of fungicide treatments from 15—18 to 2—3 per season when using resistant varieties (Teissedre et al., 2017). Resistant varieties allow to produce wines with virtually zero treatments. A producer using the variety *Solaris* has stated that in nine years he applied only 2 treatments against diseases. Hybrids can reduce the need for pesticides in wine growing by 86% (Cailliatte, 2018), but more studies are still needed to assess berry and wine quality issues in such varieties.

24.4 Wine production and climate change

Wine production in Europe is based on the concept of "terroir," which is a French term that intends to encompass the variety of agricultural and environmental influences determining the particular characteristics of wine. Europe has built its reputation for high quality wines on the immense variability and richness of its "terroirs." However, this may also prove to be a major vulnerability in times of accelerating climate change. The EU wine sector is possibly one of the most exposed to the challenges imposed by climate change. Contrary to other products, the EU wine sector cannot simply transpose its "terroirs" and replicate wines with the same characteristics in other areas of the EU. Also, in some areas, in particular with steep slopes, low soil fertility or very rocky terrain, wine is the only possible and economically viable agricultural production.

There are multiple evidences of climate changes affecting EU vineyards, namely increasing market volatility. In fact, in the last five years, the EU production attained both record high and low harvests. In Southern Europe, where most of Europe's landmark wine areas are located, the wine cultural cycle shortened significantly: harvest is starting two months earlier than 20 years ago (Van Leeuwen & Darriet, 2016). It is difficult to maintain maximum alcohol levels of 15% vol. Drought and high temperatures are affecting negatively yields and quality of the grapes (Van Leeuwen & Darriet, 2016).

Wine aromas develop in the final part of the cycle and with shorter life cycles wine regions face difficulties in respecting typical patterns for their key varieties and final products (Van Leeuwen et al., 2020). On the other hand, warmer weather favors expansion of viticulture toward Northern, Central, and Eastern Europe, and reduces the need in these areas for chaptalization. We have been witnessing expansion of vines in areas considered until now unfit for vine cultivation, such as Belgium, The Netherlands, Denmark, Sweden, Poland, and the Baltic States. The same trend can be observed for the UK.

24.5 Markets and consumers expectations

Wine consumption in the EU has been significantly decreasing in the last decades. While we see (moderate) increases in consumption in nonproducing or low-producing countries, the overall trend is marked by steep declines in wine consumption in major wine production countries (France, Italy, and Spain). For example, alcohol consumption in France has declined steadily for 40 years, mainly due to a significant drop in wine consumption. Between 1961 and 2003, alcohol consumption fell by 47.5%, from 17.7 to 9.3 L of pure alcohol per year per capita. At the same time, wine consumption fell by more than half (61.5%), from 126.1 to 48.5 L of wine per year and per capita.

According to the *Institut National de la Statistique et des Études Économiques* (INSEE), in France, this decrease concerns wines of current consumption (2.4% decrease per year on average since 1960), while quality wines have seen their consumption increase at a rate of 2.7% per year (Planetoscope, 2021). Moreover, EU's quality policy for the wine sector managed to increase the share of the marketed wines with a geographical indication, achieving thus a welcome effect of "less but better" wine consumption. In 2020, the EU consumed an estimated volume of wine of about 112 million hL, a value in line with 2019 (OIV, 2021). This is the result of the counterbalancing among countries with opposite trends. If, on the one hand, there are countries like Italy (24.5 million hL, +10.0% with respect to its five-years average) and Germany (19.8 million hL, +0.2%) that increased their consumption with respect to 2019, on the other hand other EU Member States, including for example Spain (9.6 million hL, −6.8% compared to 2019), Portugal (4.6 million hL, −0.6%), Romania (3.8 million hL, −1.9%), and Belgium (2.6 million hL, −3.1%), show a negative trend with respect to 2019 (OIV, 2021). France level of consumption remained unchanged as compared to 2019. The UK, in Europe, but outside the UE, has an estimated consumption of 13.3 million hL (+2.2%/2019).

Regarding the consumption in third countries, different trends were observed. The USA reached 33.0 million hL, in line with 2019, proving to be a resilient market. Concerning China, a 17.4% drop with respect to 2019 was observed, while in South America, overall wine consumption increased in 2020 in comparison with 2019. These consumption variations can be associated to COVID-19 sanitary crisis, lockdown measures, disruption of Horeca channel, and decrease of tourism. Fig. 24.2 illustrates wine consumptions by countries in 2020 and evolution of consumption in major countries (OIV, 2021).

The EU wine industry has been facing these challenges of consumption by pushing toward quality and increase in value added, as well as investing in promotion efforts in third countries, with support from the EU budget. To be noted that wine promotion in third countries is specifically aimed at addressing direct competitors (namely New World wines) and not at stimulating wine consumption itself.

24.5 Markets and consumers expectations 491

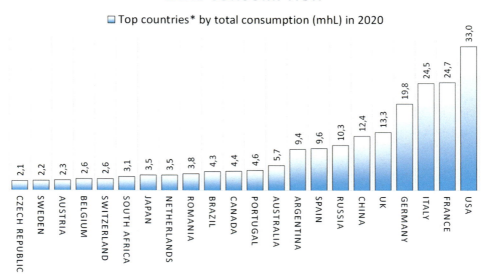

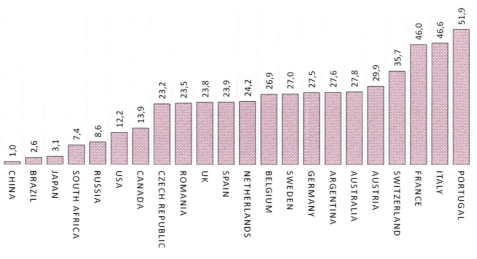

FIGURE 24.2

Wine consumption per capita and per country in 2020 (OIV, 2021).
* Countries with consumption above 2 mhL.

Several reasons can explain the consistent decline in the EU's wine consumption, mostly linked to urbanization of society and increasing awareness of the dangers of excessive alcohol consumption. This has followed the evolution of society in wine-producing countries, where wine moved from being a "basic foodstuff" to a conviviality drink. Such reduction of consumption was partly compensated by an increase in the average quality, leading to a significant unit price rise for wines.

Consumers of the future and millennials are very susceptible to trends and fashions: rosés, sparkling wines, dealcoholized wines, wine bars are expanding. On the other side, EU support to the sector is being challenged and wine is even being compared to tobacco because of the consequences of irresponsible drinking.

In order to address health and responsible drinking, there is an increasing demand for low alcohol beverages (Bucher et al., 2018). While alcohol-free beer is currently marketed in the EU and it is a very successful product, the EU's legislation today does not allow low or zero alcohol wines to be marketed under the "wine" product designation. In addition, there is also a major concern by EU consumers in relation to the long-term exemption for alcoholic drinks in providing complete information on nutritional value and ingredients. If applicable for all food products (i.e., yoghurts, cookies, olive oil, and processed meats), why should the wine sector be exempted from providing such information? For this reason, and following the Commission report on this exemption dated from 2017 (European Commission, 2017a), and which concluded that the status quo is no longer warranted, the alcoholic beverage industry has come up with self-regulatory proposals to label at least the energy and to provide ingredients lists, possibly online. Such self-regulatory efforts have led the EU legislator to make legislative proposals for the energy and ingredients labeling of wine and aromatized wine products, which are in the pipeline and are likely to enter into force in the next months.

24.6 EU policy framework toward increased sustainability of the wine sector

24.6.1 CAP reform proposals

On June 1, 2018, the European Commission adopted COM (2018)394 final (CAP reform proposals) (European Commission, 2018), which includes a "wine package" containing the following elements:

- *Wine planting authorizations*—The proposal does not touch the current baseline policy principle that maximum expansion of the wine area is limited to 1% of the existing wine area, i.e., possibility to expand production potential by taking account of a factor of "natural" decline of wine area in most Member States.
- *Elimination of all existing bans on hybrid varieties*—The proposal opens the possibility to the Member States and the wine sector to use: (a) varieties belonging to the species *Vitis labrusca*; (b) varieties that were once forbidden for reasons not scientifically/clearly justified at the time when they were introduced and in any case no longer valid; (c) interspecific hybrid varieties of *V. vinifera* with other *Vitis* species in PDOs—as is already the case for PGIs.
The proposal does not, however, require Member States and geographical indications specifications to authorize such varieties. In case the Member States consider that such varieties should not be grown on their territory, they can simply not include them in their list of authorized vine varieties. Similarly, if wine growers do not see any advantages in using such varieties, they

can choose not to cultivate them. It aims actually at eliminating all existing exclusions on hybrids thus paving the way for new hybrid varieties with improved pest and disease resistance and better adapted to climate change. While ancient varieties have their faithful consumers, but a limited expansion potential, new hybrid varieties have a huge potential to reduce pesticide use (Cailliatte, 2018). Finally, it allows unlocking of R&D on wine hybrid varieties, allowing in the future development of hybrids to address effects of climate change (i.e., drought resistance).

- *Expansion of wine definition to cover low alcohol products*—Current rules exclude wine products not covered by quality labels with alcohol content lower than 8.5% vol. to be marketed as "wine." The above-mentioned regulation proposal opens up the wine definition to cover zero and low alcohol wines under the definition of "wine." This proposal aims at capturing new wine trends within the "wine family" and address societal expectations. It also aims at putting EU producers on an equal footing with some external competitors that are by law allowed to produce and export their wines to the World market.
- *Obligation to provide consumer information on nutrition/ingredients*—Formally, such proposal was not made by the Commission, but has it has been introduced in the interinstitutional legislative process by both Member States in the Council and the European Parliament, and probably will remain in the final legal text still to be adopted.

According to the provisions of the EU Regulation 1169/2011 (Food Information for the Consumer—FIC Regulation), alcoholic drinks with more than 1.2% alcohol by volume are exempted from the obligations imposed on all other food products to supply information for consumers on nutritional values and ingredients (European Union, 2011). As part of the Common Agricultural Policy reform process, an amendment to the FIC Regulation has been introduced in the draft regulations in order to make sure such information will be mandatory for all wines in the future. Such an amendment is crucial to address the societal concerns related to the nutritional content of wine.

After more than 40 years of consumer information rules at EU level, it goes against the long-term interests of the wine sector to continue being perceived as unwilling to show the same level of consumer information as compared to all other food products. Integrating the wine sector in the regular obligation of consumer information actually protects the wine sector from proliferation of national laws that have the potential to fragment the internal market. For this reason, the wine sector has welcomed such legislative proposals, which will preserve the unity of the Single Market.

At this moment, these proposals, that are an integral part of the overall CAP reform package, have not yet been adopted by the co-legislators. This is due mostly to delays in the adoption of the MFF 2021−2027 (Multiannual Financial Framework). As the CAP is one of the major policies in spending terms of EU's budget, both Member States (Council of Ministers) and the European Parliament have been reluctant to accept policy proposals without the accompanying the proper financial background. In view of the positions taken by both co-legislators, it can, however, be estimated that the bulk of the "wine package" will eventually pass into European law, if and when the political conditions will converge toward an agreement.

24.6.2 Green deal/farm to fork strategy

On May 20, 2020, the EU Commission adopted a communication (COM (2020)381 final) on a *Farm to Fork Strategy* (F2F) for a fair, healthy, and environmentally friendly food system

(European Commission, 2020). The F2F is at the heart of the *European Green Deal*, an innovative, sustainable, and inclusive growth strategy to boost the economy, improve citizen's health, and protect the environment. The F2F represents a quantum leap for the EU's food system. For more than 20 years, the EU has successfully built a food policy framework based on food safety. The F2F implies a move to the next level: sustainability. This is a fundamental change of paradigm, expanding the focus from "product safety" toward economic, social, and environmental process. This was largely anticipated by the June 2018 CAP reform proposals for wine that focused almost exclusively on environmental and societal sustainability. The adoption of the F2F strategy significantly reinforces the case for an early adoption of the 2018 package. Furthermore, the F2F offers an important set of opportunities and challenges for the wine sector that will need to be carefully considered. In particular:

- *Targets on chemical pesticides*—The Commission will take action to reduce the overall use and risk of chemical pesticides by 50% and the use of more hazardous pesticides by 50% in 2030. This will be achieved by revising the Sustainable Use of Pesticides Directive, enhancing provisions on integrated pest management and promoting greater use of safe alternative ways to protect harvests from pests and diseases.

  While vine hybrids have a significant potential in reducing application of hazardous pesticides, the wine sector will need to undertake more ambitions and urgent actions to tackle harmful effects of pesticides. It is understood that the process of vineyard restructuring toward hybrids will start only in some years, and it would be a slow process, yielding results only in the medium term. More importantly, the wine sector is relatively "conservative" toward innovation, preferring in general to rely on the typical varietal mix in existing PDO's. However, climate change and reduction of availability of chemical molecules will necessarily force the introduction of new varieties and new oenological practices.

- *Targets on nutrient excess*—The Commission will act to reduce nutrient losses (especially nitrogen and phosphorus) by at least 50%, while ensuring there is no deterioration in soil fertility.

- *Targets and action plan on organics*—F2F and the biodiversity strategy aim at supporting the achievement of at least 25% of the EU's agricultural land under organic farming in 2030. Since the widening of the scope of the EU's organic label to the wine sector in 2012, the organic wine market developed significantly. In 2017, within the EU, Spain, Italy, and France were the main producing countries for organic grapes. Spain dominated the market with 107, 000 ha of certified grapes, Italy was in second position with 105,000 ha followed by France with some 78,000 ha. According to the projections provided by the IWSR Drinks Market Analysis company, commissioned by the interprofessional organization SudVinBio (MON-Viti, 2021), the European supply should increase by 2022. Growth in the three leading countries will follow the same trend and in 2022, Spain is predicted to have 160,000 ha of organic vineyards, Italy 120,000 ha, and France 100,000 ha (MON-Viti, 2021).

However, current estimated share of organic grapes wine production in EU wine production corresponds only to 10.3% of the total areas under wine in 2017 (EIP-Agri, 2019) and it faces two major limitations toward further expansion:

- *Marketing*—For the most prestigious wines, organic is not a major marketing strategy. There are claims that some wine regions actually produce under organic conditions, but are not certified as such. While there is no factual evidence of this, it is clear that other marketing elements besides the organic label prevail.

- *Production*—Organic wines are more dependent on copper use than traditional wines, simply because there are no other available alternatives under organic production mode. Until the moment that such an alternative could be found, any expansion of organic wines risks increasing the use of copper in viticulture. Table 24.1 shows the area of organic or under conversion to organic grapes per Member State in the EU and predicted trends.
- *Promotion of a sustainable food consumption and facilitating shift toward healthy diets*—Under the F2F communication, the EU Commission aims mostly at reversing the rise in overweight and obesity rates by 2030. Alcohol is not listed among the foodstuffs whose promotion and consumption is to be regulated in view of a more healthy and sustainable diet (contrary to red/processed meat and other foods high in fat, sugars, and salt). However, there is always a general sensitiveness at the EU level concerning alcohol consumption. The wine sector has an interest in anticipating this trend and address moderate alcohol consumption upfront, as well as participate on ongoing/future initiatives on food information for consumers and sustainability.

Table 24.1 Area (ha) of organic or under conversion to organic, of grapes per EU Member State (European Commission, 2017b).

	Total area viticulture	Area of wine production	Area of table grapes and raisins	Area of organic grapes and under conversion	Percentage organic area	Trends[a] 2021–2022
Austria	48,050	48,050	0	5663	11.8%	↗
Belgium	34,110	31,980	2130	4092	12%	↗
Czech Republic	15,810	15,810	0	788	5%	↗
Croatia	21,900	21,600	300	1010	4.6%	↗
Cyprus	5930	5310	620	264	4.5%	↗
France	750,460	745,230	5230	78,502	10.5%	↗
Germany	100,260	100,260	0	7201	7.2%	↗
Greece	101,750	61,870	39,960	4424	4.3%	↗
Hungary	67,080	63,850	3230	1716	2.6%	↗
Italy	675,260	634,120	43,140	105,000	15.5%	↗
Luxemburg	1260	1260	0	123	9.8%	↗
Malta	680	640	50	9	1.9%	↗
Portugal	178,840	176,810	2040	3504	2.0%	↗
Romania	175,320	168,450	6870	2169	1.2%	↗
Slovenia	15,860	15,840	20	561	3.5%	↗
Slovakia	8470	8280	190	124	1.5%	↗
Spain	937,760	921,650	16,120	106.897	11.4%	↗
Total	3,138,640	3,020,920	119,910	322,047	10.3%	↗

[a]Predicted trends (MON-Viti, 2021).

- *Geographical indication (GI) sustainability initiative*—As part of the F2F, the Commission is launching the GI sustainability initiative. As wines PDO's/PGI's are the backbone of the EU GI system, it is clear that the wine sector will have to be a major actor driving sustainability criteria on the GI protection system.
- *Enhanced import provisions*—F2F stipulates that imported food must comply with relevant EU regulations and standards. While as such this provision poses no problems for wine, the wine sector is highly dependent on the export market. The wine sector has a strong interest to support and defend the Commission's "Green diplomacy" and remain as the flagship of EU's agricultural exports.

24.6.3 Research and innovation in the wine sector

The EU clearly recognizes the role that research and innovation play to address the challenges of the agricultural sector, including the wine sector, whether it means increasing productivity sustainably, ensuring a better economic viability, tackling climate change or using resources more efficiently. Certainly, research is crucial as it provides the necessary knowledge base. Yet, innovation is also important. In the past, innovation was often considered a consequence of the one-way transfer of scientific results to the agricultural practice. However, innovation also takes place on the ground, when new ideas are generated by groups of people gathering together to discuss practical problems and to develop solutions. These groups can include farmers, researchers, advisors, entrepreneurs, experts from nongovernmental organizations, to mention just a few examples.

The EU finances both research and innovation, mainly through the EU research framework program (currently HORIZON 2020) and the EU rural development policy. Other EU funding sources include COST, LIFE, and INTERREG. HORIZON 2020 showed a strong priority toward research in agriculture by earmarking EUR 4.5 billion Euros for research and innovation on food security, sustainable agriculture, and the bioeconomy during the period 2014–2020. In that programming period, the EU financed several dozen research and innovation projects that relate in one way or another to viticulture and winemaking practices.

Research topics include inter alia vine diseases (e.g., WINETWORK), climate change mitigation (e.g., CLIM4VITIS), reduction of use and impact of pesticides (e.g., NOVATERRA), genetic study of vine (e.g., GRAPEINNOVATION), substitution of sulfites in wine (e.g., 4SURE), eco-innovative maceration (e.g., ULTRAWINE), yeast biodiversity (e.g., MITOGRESSION), etc. A more complete overview of funded projects can be found in the report of the EIP-AGRI Focus Group "Diseases and pests in viticulture" (EIP-Agri, 2019) and in the IRIS database (Marsden et al., 2016).

Regarding rural development, innovation and knowledge transfer is one of six strategic priorities that must be addressed by each program. H2020 also supported projects related to knowledge transfer and that promote innovation via demonstration (e.g., NEFERTITI). Just like for other priorities and focus areas, programming authorities are free to select those measures that are most likely to deliver the best outcomes taking into account the specificity and needs of a respective program region. For innovation, co-operation, knowledge transfer and information actions, advisory services, investments in physical assets, farm and business development are the most relevant actions.

The co-operation measure plays a central role in the implementation of the European Innovation Partnership (EIP) for agriculture. Under this measure, support can be given among others for the establishment and operation of operational groups of the EIP. The rural development regulation also establishes the EIP network at EU level. This network facilitates the exchange of information and knowledge sharing by interacting directly with the operational groups in Europe. It provides a help desk function, animates discussions, screens and reports on research results, etc.

A so-called EIP focus group, gathering researchers, farmers, and advisors, was established in 2017 to specifically discuss the issue of pests and diseases in viticulture. These results can be found in a final report available on the web. In addition, a number of EIP operational groups deal with wine and viticulture-related innovation. The link below leads to a list of over 40 innovation projects in the EIP database (https://ec.europa.eu/eip/agriculture/en/focus-groups/diseases-and-pests-viticulture) (EIP-Agri, 2021).

Regarding higher education and training financed by EU, the Vinifera Euromaster program should be highlighted as an example of success story and good practice for the EU Commission (https://ec.europa.eu/programmes/erasmus-plus/projects/eplus-project-details/#project/2010-0139).

24.7 Concluding remarks

The EU wine sector became in the last 10 years one of the most successful sectors in the EU's agricultural economy. While maintaining its competitiveness, increasing sustainability and addressing public concerns on food and health will be the major challenges for the European wine sector for the next 20 years. The EU Commission will continue in the future to support development of the EU wine sector by providing an updated regulatory framework, flexible and tailor-made internal support policies, and market access in third countries via international agreements. But continued success in addressing the future challenges will depend mainly on the EU's wine industry capacity to understand these challenges that require proactivity and capacity to continuously innovate. Reduction of pesticide use, adaptation to climate change and mitigating its effects, promoting knowledge transfer and addressing consumers concerns are major issues that need to be addressed upfront by the EU wine industry, in order to maintain its cutting-edge advantage, its competitiveness, and its reputation in the medium and long term at a global scale.

References

Bucher, T., Deroover, K., & Stockley, C. (2018). Low-alcohol wine: A narrative review on consumer perception and behaviour. *Beverages, 4*, 82. https://doi.org/10.3390/beverages4040082

Cailliatte, R. (October 2018). What kind of wine for tomorrow? Presentation on diseases resistant vines: Towards the vineyards of the future. In *INRA, plant breeding and biology division*. Event at the European Parliament.

EFSA. (2018). Conclusion on pesticides peer review. Peer review of the pesticide risk assessment of the active substance copper compounds copper(I), copper(II) variants namely copper hydroxide, copper oxychloride, tribasic copper sulfate, copper(I) oxide, Bordeaux mixture. *European Food Safety Authority (EFSA) Journal, 16*(1), 5152. https://efsa.onlinelibrary.wiley.com/doi/10.2903/j.efsa.2018.5152.

EIP-Agri. (2019). *Diseases and pests in viticulture*. Final Report https://ec.europa.eu/eip/agriculture/en/focus-groups/diseases-and-pests-viticulture.

EIP-Agri. (2021). https://ec.europa.eu/eip/agriculture/en/eip-agri-projects.

European Commission. (2009). Directive 2009/128/EC of the European Parliament and of the Council of 21 October 2009 establishing a framework for Community action to achieve the sustainable use of pesticides. *Official Journal of the European Union*. L 309/71.

European Commission. (2017a). *Report from the Commission to the European Parliament and the Council regarding the mandatory labelling of the list of ingredients and the nutrition declaration of alcoholic beverages.* COM(2017)58 final.

European Commission. (2017b). *DG agriculture and rural development.* Unit B4 - Organics. Direct communication.

European Commission. (2018). *Proposal for a Regulation of the European Parliament and of the Council amending Regulation (EU) No 1308/2013 establishing a common organisation of the markets in agricultural products, (EU) no 1151/2012 on quality schemes for agricultural products and foodstuffs, (EU) No 215/2014 on the definition, description, presentation, labelling and the protection of geographical indications of aromatised wine products, (EU) No 228/2013 laying down specific measures for agriculture in the outermost regions of the Union and (EU) No 229/2018 laying down specific measures for agriculture in favour of the smaller Aegean islands.* COM(2018)394 final.

European Commission. (2019). *DG agriculture and rural development.* https://agridata.ec.europa.eu/extensions/DashboardWine/WineTrade.html. (Accessed January 2021).

European Commission. (2020). *Communication from the commission to the European parliament, the Council, the European economic and social committee and the committee of the regions - a farm to Fork strategy for a fair, healthy and environmentally-friendly food system.* COM(2020)381 final.

European Union. (22 November 2011). Regulation (EU) No 1169/2011 of the European parliament and of the Council of 25 October 2011 on the provision of food information to consumers, amending regulations (EC) No 1924/2006 and (EC) No 1925/2006 of the European Parliament and of the Council, and repealing commission directive 87/250/EEC, Council directive 90/496/EEC, commission directive 1999/10/EC, directive 2000/13/EC of the European Parliament and of the Council, commission Directives 2002/67/EC and 2008/5/EC and commission regulation (EC) No 608/2004. *Official Journal of the European Union, L 304*, 18.

European Union. (20 December 2013). Regulation (EU) No 1308/2013 of the European Parliament and of the Council. establishing a common organisation of the markets in agricultural products and repealing Council Regulations (EEC) No 922/72, (EEC) No 234/79, (EC) No 1037/2001 and (EC) No 1234/2007. Article 81(2)(b). *Official Journal of the European Union, L347*, 671.

European Union. (14 December 2018). Commission Implementing Regulation (EU) 2018/1981 of 13 December 2018 renewing the approval of the active substances copper compounds, as candidates for substitution, in accordance with Regulation (EC) No 1107/2009 of the European Parliament and of the Council concerning the placing of plant protection products on the market, and amending the Annex to Commission Implementing Regulation (EU) No 540/2011. *Official Journal of the European Union L, 317*, 16.

Fuentes Espinoza, A., Hubert, A., Raineau, Y., Franc, C., & Giraud-Héraud, É. (2018). Resistant grape varieties and market acceptance: An evaluation based on experimental economics. *OENO One, 52*(3). https://doi.org/10.20870/oeno-one.2018.52.3.2316

González-Centeno, R., Chira, K., Miramont, C., Escudier, J.-L., Samson, A., Salmon, J.-M., Ojeda, H., & Teissedre, P. L. (2019). Disease resistant bouquet vine varieties: Assessment of the phenolic, aromatic, and sensory potential of their wines. *Biomolecules, 9*(12), 793. https://doi.org/10.3390/biom9120793

Karimi, B., Masson, V., Guilland, C., Leroy, E., Pellegrinelli, S., Giboulot, E., Maron, P.-A., & Ranjard, L. (2021). Ecotoxicity of copper input and accumulation for soil biodiversity in vineyards. *Environmental Chemistry Letters, 19*, 2013−2030. https://doi.org/10.1007/s10311-020-01155-x

Lamichhane, J. R., Osdaghi, E., Behlau, F., Kohl, J., Jones, J. B., & Aubertot, J.-N. (2018). Thirteen decades of antimicrobial copper compounds applied in agriculture. A review. *Agronomy for Sustainable Development, 38*, 28. https://doi.org/10.1007/s13593-018-0503-9

Marsden, E., Mackey, A., & Plonsky, L. (2016). The IRIS Repository: Advancing research practice and methodology. In A. Mackey, & E. Marsden (Eds.), *Advancing methodology and practice: The IRIS repository of instruments for research into second languages* (pp. 1−21). New York, NY: Routledge.

Merdinoglu, D., Schneider, C., Prado, E., Wiedemann-Merdinoglu, S., & Mestre, P. (2018). Breeding for durable resistance to downy and powdery mildew in grapevine. *OENO One, 52*(3), 203–209.

MON-Viti. (2021). *La production mondiale de vin bio augmentera dans tous les pays d'ici 2022*. https://www.mon-viti.com/articles/international/la-production-mondiale-de-vin-bio-augmentera-dans-tous-les-pays-dici-2022. (Accessed January 2021).

OIV. (2021). *State of the vitivinicultural world in 2020*. International Organisation of Vine and Wine. https://www.oiv.int/public/medias/7909/oiv-state-of-the-world-vitivinicultural-sector-in-2020.pdf.

Pedneault, K., & Provost, C. (2016). Fungus resistant grape varieties as a suitable alternative for organic wine production: Benefits, limits, and challenges. *Scientia Horticulturae, 208*, 54–77.

Pertot, I., Caffi, T., Rossi, V., Mugnai, L., Hoffmann, C., Grando, M. S., Gary, C., Lafond, D., Duso, C., Thiery, D., Mazzoni, V., & Anfora, G. (2017). A critical review of plant protection tools for reducing pesticide use on grapevine and new perspectives for the implementation of IPM in viticulture. *Crop Protection, 97*, 70–84. https://doi.org/10.1016/j.cropro.2016.11.025

Planetoscope. (2021). https://www.planetoscope.com/Le-Vin/872-consommation-mondiale-de-vin.html.

Teissedre, P. L., Riesen, R., & Rienth, M. (2017). New resistant grape varieties and alternatives to pesticides in viticulture for quality wine production. In *Proceedings of the 6th Œnoviti international symposium*. Œnoviti International Network. Changins, Suisse. https://en.calameo.com/read/005916410da44419d517a.

Van Leeuwen, C., Barbe, J.-C., Darriet, P., Geffroy, O., Gomès, E., Guillaumie, S., Helwi, P., Laboyrie, J., Lytra, G., Le Menn, N., Marchand, S., Picard, M., Pons, A., Schüttler, A., & Thibon, C. (2020). Recent advancements in understanding the terroir effect on aromas in grapes and wines. *Oeno One, 54*(4). https://doi.org/10.20870/oeno-one.2020.54.4.3983

Van Leeuwen, C., & Darriet, F. (2016). The impact of climate change on viticulture and wine quality. *Journal of Wine Economics, 11*(1), 150–167.

Index

Note: 'Page numbers followed by "f" indicate figures and "t" indicate tables.'

A

Abscisic acid (ABA), 29
Actual grapevine diversity, 26–28
Aged wine spirit, 259, 262f
    technological process of, 261–262
Agricultural sustainability, 194
Agro-industrial by-products and wastewaters
    value-added compounds from, 314
Agronomical response, 26
Alcoholic fermentation (AF), 239
Alternative agricultural techniques, 356
Amplified fragment length polymorphism (AFLP), 46
Anaerobic digestion, 228–229
Analytical methods and instruments, 329
Arabinogalactanoproteins (AGPs), 313
Artificial neural networks (ANNs), 125
Association for the Development of Viticulture in Douro Region (ADVID), 152

B

Bag-in-box, 374–375
Barena method, 269
Barrels
    kegs, 376–377
    regeneration, 269
    reuse for aging, 270
Behavioral economics, 442, 452
Behavioral hypothesis in viticulture, 452–456
Best linear unbiased predictors (BLUPs), 75
Beverage Industry Environmental Roundtable (BIER), 359
Bioactive compounds, 283
    extraction technologies for, 284–285
Biodiversity, 425–426
    conservation, 92–93, 194
Biomass production, 94–95
Biosecurity, 427–429
Boomerang effect, 455–456
Botanical species in cooperage, 265–266
Bragato Research Institute fermentation vessels, 422f
Bragato Research Institute winery, 422f
Brightwater Vineyards, 424
By-products generated, wineindustries, 307–314

C

CAP reform proposals, 492–493

Carbon dioxide reuse, 189–191, 190t
Carbon footprint, 379–380
Cawthron Marlborough Environment Awards, 426
$\delta^{13}C$ genetic variability, 67–68, 77f
Chemical pesticides, 494
Climate change effects
    leaf metabolites, 167
    phenology, 166
    physiology, 166–167
    yield and berry quality attributes, 168
Climate regulation, 95–96
Climate variables, 69
Clonal selection
    multienvironmental trials for, 51, 52f
    programs, 30
Closed systems, 10
Coefficients of variation (CV), 55, 57t–58t
Commission Work Program 2020, 13
Consumer acceptability, 137–138
Contrasted soil water scenarios, 69–71
Cooperage process and sustainability, 267
Co-ordinatedIncident Management System (CIMS), 431
Cork as closure for wine bottles, 377–379
Cork stoppers, 10
Corporate environmental guardianship, 425–427
Corporate social responsibility (CSR), 415, 424–427
COVID-19 pandemic, 5, 431
Cradle-to-farm gate, 352, 355–356
Cradle-to-market, 351
CRISPR/Cas9, 244–245
Crop stress coefficients, reproductive cycle, 111–112
Cultivated grapevine, 26–28

D

Data access, 139
Decision support systems (DSSs), 152
Deep learning, 133–139
Designations of origin (DOs), 262
Desk-based research, 476
Digital tools, 135–137
Dimethyl dicarbonate (DMDC), 11
Direct root-zone irrigation (DRZ), 107
Disease management, 150–153, 151f
Dog Point Vineyard, 426
Douro Demarcated Region (DDR), 150–151
Douro'swine producing region, 401–407

501

Drought response variability, 29
Drought tolerance phenotyping, grapevine populations
    soil heterogeneity, 66–68
        best linear unbiased predictors (BLUPs), 75
        $\delta^{13}C$ genetic variability, 67–68, 77f
        climate variables, 69
        contrasted soil water scenarios, 69–71
        experimental setup, 68–69
        high-throughput measurements, 67–68
        plant performance phenotyping, 67–68
        QTL detection genetic variability, 67–68, 77f
        quantitative trait locus (QTL) studies, 66
        soil characteristics hinder drought tolerance studies variations, 66–67
        spatial distribution of predawn leaf water potential within the field, 71–72, 72f, 73t
        spatial variations statistical methods, 67
        water deficit, 67–68
Dynamic leadership, 477

E

Earning Before InterestsDepreciation, Taxes and Amortizations (EBIDTA), 400
Ecoagriculture landscapes, 194
Ecological-based pest management strategies, 155–156
E-commerce, 11
Economic dimension and efficiency, 398–400
Ecosystem service (ES), 85, 86f, 88–96
Electrical resistivity measurements (ERa), 66–67
Electric field, 273
Electrotechnologies, 285–294
Empirical best linear unbiased predictors (EBLUPs), 50
Endemic species, 426–427
Energy audit in wineries, 222–223
Energy consumption in the winery, 223–226, 224f
Energy demand reduction, 226–228
Energy reduction, 9
Energy use
    anaerobic digestion, 228–229
    energy audit in wineries, 222–223
    energy consumption in the winery, 223–226, 224f
    energy demand reduction, 226–228
    Italian winery, energy audit of, 231–232
    renewable energy utilization, 228–231
    solar systems, 230–231
    TESLA research project, 233
    thermochemical conversion processes, 229–230
    wineries located in Veneto (Italy), energy assessment related to, 233–236
Environmental dimension and natural resources, 395
Environmental Footprint method, 354
Environmental labels on wine, landscape of, 360–361

Enzymatic processes, 190
Erosion control, 90
EU Project TESLA—Transferring energy, 9
European wine policy framework, 485–486
    environmental aspects of wine production, 486–488
    markets and consumers expectations, 490–492
    sustainability of wine sector, 492–497
    technical solutions to challenges, 488–489
    wine production and climate change, 489–490
Evapotranspiration (ET), 106–107
Experimental protocol, 447–448

F

Forecasting systems, 152
Forest ecosystems, 266–267
"Framework for Assessing the Sustainability of Natural Resource Management Systems", 403t–404t
France
    cases studies characterization, 211
    electrical energy consumption rate, 212
    water usage, 211
Free assimilable nitrogen (FAN), 338–339
Functional biodiversity, 193–194
Functional unit (FU), 353

G

G20 Action Plan, 12
Gate-to-gate-vinification, 356
Gate-to-grave, 357
G×E interaction, 51
Genetically modified organisms (GMOs), 87
Genetic diversity, 27
Genetic intravarietal grapevine diversity, 53–59, 54t–55t
Genome editing, 244–245
Genome Wide Association Studies (GWAS), 66
Genomics, 240–244
Genotype-by-environment (G×E) interaction effect, 50–51
Geographical indication (GI) sustainability initiative, 496
Glass bottle, 373–374
Glass bottles recycling, 383–384
Glass recycling rates, 383
Global greenhouse gas (GHG), 187–188
Global warming potential (GWP), 384
Global wine industry, 391
    Douro'swine producing region, 401–407
    grapes and wine production, 393–400
    sustainability assessment, 392–393
Good agricultural practices (GAPs), 86–87, 166
"Good environmental practices", 443
Government-Industry Agreement (GIA), 428
Grafted Grapevine Standard (GGS), 428
Grape by-products, 283t

Grape composition, 115−117
Grape microbiome, 246
Grape processing by-products, 290t−291t
Grapevine, 45, 47−51
    conservation, 47−49
    disease management, 150−153, 151f
    ecological-based pest management strategies, 155−156
    integrated pest management (IPM), 148, 148t, 149f
    mating disruption (MD) technique, 154−155
    noncrop habitats (NCHs), 155−156
    pest management, 153−159, 154t
    polyclonal selection, 49−50
    quality control of, 393−400
        Green Analytical Chemistry (GAC), 328
        greening of analytical procedures, 328−330
        principles of green chemistry, 327−328
        research, 421−423
        sustainable grape analysis and quality control, 331−336
        sustainable wine analysis and quality control, 336−342
    representative sampling of, 47
    Scaphoideus titanus Ball, 158−159
    Sustainable Use of Pesticide (SUP) Directive (Directive 2009/128/EC), 148
    vibrational mating disruption (VMD), 158−159
Grapevine cultivar, 31
Grapevine water status assessment, 108−111
Gravity-flow wineries, 187−188, 192−193
Green Analytical Chemistry (GAC), 328
Green chemistry, 327−328
Green deal/farm to fork strategy, 493−496
Greening of analytical procedures, 328−330
Greening Waipara, 425−426
Green methods, 299

H

Habitat provision contamination, 91−92
Heat stress physiology, 6
Heterogeneous vineyard production system, 402f
High-throughput measurements, 67−68
High voltage electrical discharges (HVED), 293−294
Hot spots, 355
Hygienic question, 193

I

In depth interviews, 476
INDIGO® method, 394
Injunctive standard effects, 454−455
Innovative extraction methods, 285−298, 286t−289t
    electrotechnologies, 285−294
    microwave-assisted extraction, 295−297
    subcritical fluids extraction (SBFE), 298
    supercritical fluids extraction (SFE), 297−298
    ultrasound-assisted extraction, 294−295
Integrated pest management (IPM), 148, 148t, 149f
Interactive innovations
    and food value chain sustainability, 461−463
    influencing sustainability in, 463−464
International distribution, 384−385
International EPD® System (IES), 362
International Organisation of Vine and Wine (OIV), 260, 293
International Sauvignon Blanc Celebration, 434−435
International Wineries for Climate Action (IWCA), 425
Intravarietal grapevine diversity, conservation and evaluation of
    clonal selection, multienvironmental trials for, 51, 52f
    coefficient of variation (CV), 55
    genetic intravarietal grapevine diversity, 53−59, 54t−55t
    grapevine, 45, 47−51
        conservation, 47−49
        polyclonal selection, 49−50
        representative sampling of, 47
    intravarietal genetic variability, 56−59
    molecular understanding, 46
    Portugal practical applications, 59−60, 60f
    relative bias (RB), 55−56
    relative mean square error (RMSE), 55−56
IR revolution, 340−341
Irrigation nonconventional water, 115−117
Irrigation scheduling, 113−115
Italy
    case study characterization, 212
    electrical energy consumption rate, 213
    energy audit of, 231−232
    renewable resources, 213
    UV-C plant sanitization, 213
    water usage, 213

K

Kaikoura earthquake 2016, 430
Kaitiakitanga, 418
Kaolin characterization, 168−169

L

Laboratory methods, 331
Labor mobility system, 433
Laminated multimaterial boxes, 375
Leaf metabolism, 176−177
Life cycle assessment (LCA), 12, 353, 379−385, 394
Living Building Challenge (LBC) sustainability certification, 188
Luff-Schoorl assay, 338

M

Machine learning, 133–139
Magnetic resonance imaging, 333
Maori
    heritage, 416–419
    values and principles, 417
    and the wine industry, 418–419
Markets and consumers expectations, 490–492
Marlborough Winegrowers Association, 434
Mating disruption (MD) technique, 154–155
Mediterranean vineyards, environmental sustainability of
    abscisic acid (ABA), 29
    actual grapevine diversity, 26–28
    agronomical response, 26
    cultivated grapevine, 26–28
    drought response variability, 29
    genetic diversity, 27
    genomics tools, 34–35
    molecular response, 26
    new breeding technologies, 34–35
    physiological response, 26
        changing environments, intracultivar variability in, 30–31, 32f
        water stress, intercultivar variability in, 28–30
    semiarid conditions, rootstocks selection for, 31–34, 33f
    stomatal regulation, 29
    water use efficiency (WUE), 26
    wild grapevines, 27
Metabolomics, 240–244
Metagenomics, 240–244
Metal (aluminum), 375–376
Metaproteomics, 243
Metatranscriptomics, 240–244
Microbiological control of wine production
    alcoholic fermentation (AF), 239
    CRISPR/Cas9, 244–245
    genome editing, 244–245
    genomics, 240–244
    grape microbiome, 246
    metabolomics, 240–244
    metagenomics, 240–244
    metaproteomics, 243
    metatranscriptomics, 240–244
    new strains, 250–251
    omics technologies, 240–244
    proteomics, 240–244
    SO_2 reduction, 246–248
    spontaneous *vs.* inoculated fermentations, 248–250, 249f
    transcriptomics, 240–244
Microwave-assisted extraction, 294–295
Molecular response, 26
Molecular understanding, 46
Monoparametric systems, 332–333
Multiparametric and enzymatic analyzers, 339–340

N

Natural disaster management, 430–431
New Zealand's grape and wine industry, 415–416
    biosecurity, 427–429
    corporate social responsibility, 424–427
    Maori heritage, 416–419
    natural disaster management, 430–431
    regional winegrower associations, 433–435
    winegrowers, 419–423
Noncrop habitats (NCHs), 155–156
Normalized different vegetation index (NDVI), 112
Nudge application, 455–456
Nudge construction, 453–454
Nutrient excess, 494
Nutrients from wastewater, 207

O

Oenology, 192–193
    by-products, 193–194
    management, 193–194
Omics technologies, 240–244
Online marketing, 11
Organic viticulture and wineries, 420–421
Organisation Internationale de la Vigne et du Vin (OIV), 357–358
Organization for Economic Co-operation and Development (OECD), 12
Organ temperature reduction, 169–171
Ownership, 139

P

Packaging Relative Environmental Impact (PREI), 372
Paperboard, 377
Participatory (online) workshop with key stakeholders, 475–477
PDMP, 316–320
Pest management, 153–159, 154t
Phenolic compounds, 312–313, 316–320
Photosynthetic activity, 171–176
Photosynthetically active radiation (PAR), 165
Physical treatments, 272–273
Physiological responses
    changing environments, intracultivar variability in, 30–31, 32f
    water stress, intercultivar variability in, 28–30
Plant-based applications, 136–137
Plant performance phenotyping, 67–68
Plastic bottles, 376

Polysaccharides, 313–314, 316–320
Polyvinylidenefluoride (PVDF) membrane, 318–319
Portugal
    case study characterization, 208–210
    electrical energy consumption rate, 210–211
    practical applications, 59–60, 60f
    water use, 210
    wine companies, 446–450
    wine-growing cooperative, 452–456
Portuguese Association for Grapevine Diversity (PORVID), 59
Precision agriculture (PA), 125–126
Precision and digital viticulture
    artificial intelligence (AI), 125, 132–139
    artificial neural networks (ANNs), 125
    consumer acceptability, 137–138
    data access, 139
    deep learning, 133–139
    digital tools, 135–137
    machine learning, 133–139
    ownership, 139
    plant-based applications, 136–137
    precision agriculture (PA), 125–126
    remote sensing, 132–139
    security using AI, 139
    soil-based applications, 135–136
    vineyard management, remote sensing for, 127–131
    winemaking, 137–138
Predawn leaf water, spatial distribution of, 71–72, 72f, 73t
Pressure-driven membrane processes, 315–316
Proactive socioenvironmental practices, 398
Processed kaolin particle film (PKPF)
    berries and wine, 177
    climate change effects
        leaf metabolites, 167
        phenology, 166
        physiology, 166–167
        yield and berry quality attributes, 168
    costs, 178
    environment, 178
    good agricultural practices (GAPs), 166
    Kaolin characterization, 168–169
    leaf metabolism, 176–177
    organ temperature reduction, 169–171
    photosynthetic activity, 171–176
    photosynthetically active radiation (PAR), 165
    pros and cons, 178
    radiation excess reflection, 169–171
    short-term adaptation strategy, 168–177
    vine water status, 171–176
Processing lees and winery wastewater, 316–319
Product Category Rules (PCRs), 362
Product Environmental Footprint (PEF), 354
Product environmental footprint process, 363–365
Project Crimson Trust, 427
Protected Denomination of Origin (PDO), 464
Proteomics, 240–244
Pulsed electric fields (PEF), 292–293

Q

Quantitative trait locus (QTL) studies, 66
    detection genetic variability, 67–68, 77f

R

Radiation excess reflection, 169–171
Refillable glass bottle, 382–383
Reform and Opening up policy, 4
Regenerative wineries water management
    environmental impacts, 202–203
    France
        cases studies characterization, 211
        electrical energy consumption rate, 212
        water usage, 211
    future challenges, 213–214
    Italy
        case study characterization, 212
        electrical energy consumption rate, 213
        renewable resources, 213
        UV-C plant sanitization, 213
        water usage, 213
    Portugal
        case study characterization, 208–210
        electrical energy consumption rate, 210–211
        water use, 210
    strategies, 206–208
        nutrients from wastewater, 207
        resources efficient use, 206–207
        solid waste streams, recovering bioenergy from, 208
        solid waste, value-added products from, 207–208
        water recovering, 207
    water cycle, 203–206
        wastewater treatment technologies, 204–206, 204t
        water use efficiency, 203–204
Regional winegrower associations, 431, 433–435
Relative bias (RB), 55–56
Relative mean square error (RMSE), 55–56, 57t–58t
Remote sensing, 132–139
Renewable energy, 192, 228–231
Residues valorization, 195
"Resistant grape variety" solution, 445–446
Resources efficient use, 206–207
Rhizosphere interactions, 32
RNAseq analysis, 34

S

Scaphoideus titanus Ball, 158–159
Scarcer water resources, 6
Second racking wine lees, 306
Security using AI, 139
Selective amplification of microsatellite polymorphic loci (SAMPL), 46
Semiarid conditions, rootstocks selection for, 31–34, 33f
Shared values, 478
Short-term adaptation strategy, 168–177
Single indicators and footprints, 359–360
Single leaf, 105–107
Small and medium size enterprises (SMEs), 441
"Social and Environmental Responsibility", 442
Social dimension and equity, 395–398
Social life cycle assessment (s-LCA), 8, 395
Socioenvironmentally sound, 13
Sodium adsorption ratio (SAR) criteria, 116
Soil and water management, legislation issues for, 12–13
Soil-based applications, 135–136
Soil characteristics hinder drought tolerance studies variations, 66–67
Soil compaction avoidance, 88–89, 89f
Soil degradation, 7
Soil health and conservation, 7
Soil heterogeneity, 66–68
    best linear unbiased predictors (BLUPs), 75
    $\delta^{13}C$ genetic variability, 67–68, 77f
    climate variables, 69
    contrasted soil water scenarios, 69–71
    experimental setup, 68–69
    high-throughput measurements, 67–68
    plant performance phenotyping, 67–68
    predawn leaf water potential within the field, spatial distribution of, 71–72, 72f, 73t
    QTL detection genetic variability, 67–68, 77f
    soil characteristics hinder drought tolerance studies variations, 66–67
    spatial variations statistical methods, 67
    water deficit, 67–68
Soil organic matter (SOM) mineralization, 87
Soil-plant-atmosphere water status, 109–111, 110f
Soil threats, 88–96
Soil water content (SWC), 108–109
Solar systems, 230–231
Solid-liquid extraction (SLE), 284
Solid waste streams, recovering bioenergy from, 208
Solid waste, value-added products from, 207–208
Solvents and reagents, 329–330
SO_2 reduction, 246–248
Spontaneous *vs.* inoculated fermentations, 248–250, 249f
Staff training, 7
Stakeholder engagement, 478–479
Stomatal regulation, 29
Subcritical fluids extraction (SFE), 290t–291t, 297–298
Supercritical fluids extraction (SBFE), 290t–291t, 298
Supply chain issues
    general aspects, 11
    metrics and analytical tools, 12
Sustainability guardians, 423
Sustainability policy, 421
Sustainability quantitative assessments, 468–473
Sustainability-related awareness, 391–392
Sustainability wine industry
    concept and issues, 1–3
    COVID-19, 5
    dimethyl dicarbonate (DMDC), 11
    future prospects, 14–16
    origin, 13
    quality, 13
    R&D, 16
    risks and concerns of, 5
    socioenvironmentally sound, 13
    soil and water management, legislation issues for, 12–13
    supply chain issues
        general aspects, 11
        metrics and analytical tools, 12
    vineyard issues, 6–8
    winemaking issues, 8–11
    wine quality certification issues, 13
    worldwide, 3–5
Sustainable grape analysis and quality control, 331–336
Sustainable packaging options for wine, 385–387
Sustainable Use of Pesticide (SUP) Directive (Directive 2009/128/EC), 148
Sustainable viticultural systems, soil management in
    biodiversity conservation, 92–93
    biomass production, 94–95
    climate regulation, 95–96
    ecosystem service (ES), 85, 86f, 88–96
    erosion control, 90
    genetically modified organisms (GMOs), 87
    good agricultural practices (GAPs), 86–87
    habitat provision contamination, 91–92
    soil compaction avoidance, 88–89, 89f
    soil organic matter (SOM) mineralization, 87
    soil threats, 88–96
    vineyards soil management, 96–97
    water quality and supply, 90–91
Sustainable wine analysis and quality control, 336–342
Sustainable Winegrowing New Zealand (SWNZ), 419–420
Sustainable wine production, 281

T

TESLA research project, 233
Thermochemical conversion processes, 229–230
The Triple Bottom Line, 415
Titrators, 339
Total treatment frequency index(TFI), 489
Transcriptomics, 240–244
Triple bottom line (TBL) theory, 392
Type I environmental label, 361
Type II claim, 361
Type III environmental declaration, 361

U

Ultrasonic waves combined with wood fragments, 273
Ultrasound-assisted extraction, 292–293
UV-C plant sanitization, 213

V

Value-added compounds in wastewaters, 307–314
Vapor pressure deficit (VPD), 108
Vibrational mating disruption (VMD), 158–159
Vine performance, 115–117
Vine water status, 171–176
Vineyard issues, 6–8
Vineyard management, remote sensing for, 127–131
Vineyards soil management, 96–97
Vineyard water balance
    components of, 106f
    crop stress coefficients, reproductive cycle, 111–112
    direct root-zone irrigation (DRZ), 107
    evapotranspiration (ET), 106–107
    grape composition, 115–117
    grapevine water status assessment, 108–111
    irrigation nonconventional water, 115–117
    irrigation scheduling, 113–115
    normalized different vegetation index (NDVI), 112
    single leaf to, 105–107
    sodium adsorption ratio (SAR) criteria, 116
    soil-plant-atmosphere water status, 109–111, 110f
    soil water content (SWC), 108–109
    vapor pressure deficit (VPD), 108
    vine performance, 115–117
    wastewater and saline water, 115–117
    water-saving strategies, 113–115
Vinho de Talha, 464
VINOVERT project, 441–444, 456–457
    behavioral hypothesis in viticulture, 452–456
    sustainable practices measured by experimentalauctions, 445–450
Viticulture, behavioral hypothesis in, 452–456
Vitiviniculture, 305
V.I.V.A. program, 400
Volumetric concentration factor (VCF), 318

W

WASP
    interactive innovation assessment, 477–479
    interactive innovation toward enhanced sustainability, 475–479
Wastewaters
    and saline water, 115–117
    value-added compounds in, 307–314
Water
    deficit, 67–68
    footprint, 379–380
    management and saving, 191–192
    preservation, 192
    quality/supply, 90–91
    recovering, 207
Water cycle, 203–206
    wastewater treatment technologies, 204–206, 204t
    water use efficiency, 203–204
Water-saving strategies, 113–115
Water usage, 213
Water use efficiency (WUE), 26, 67–68
Whitehaven Wines, 424–425
Wild grapevines, 27
Wine chain
    integrated approaches toward sustainability of, 357
Wine consumption, 466
Winegrowers, 419–423
Wineindustries, by-products generated in, 307–314
Winemaking, 137–138, 308t–312t
    by-products, 319–320
    issues, 8–11
    process, 192–193
Wine Marlborough, 434
Wine packaging and related sustainability issues, 371–379
    bag-in-box, 374–375
    barrels and kegs, 376–377
    cork as closure for wine bottles, 377–379
    glass bottle, 373–374
    laminated multimaterial boxes, 375
    LCA and environmental assessments for different packaging systems, 379–385
    metal (aluminum), 375–376
    paperboard, 377
    plastic bottles, 376
    sustainable packaging options for wine, 385–387
Wine planting authorizations, 492
Wine production, 281
    carbon dioxide reuse, 189–191, 190t
    certification issues, 13

Wine production (*Continued*)
   and climate change, 489–490
   environmental aspects of, 486–488
   functional biodiversity, 193–194
   oenology, 192–193
      by-products, 193–194
      management, 193–194
   renewable energy, 192
   water management and saving, 191–192
   winemaking process, 192–193
   winery level, systems and initiatives at, 188–189
Wineries located in Veneto (Italy), energy assessment related to, 233–236
Winery labs, automation in, 339–341
Winery level, systems and initiatives at, 188–189
Winery wastewater, 306
Wine sector
   environmental product declarations in, 360–365
   interactive innovation and food value chain sustainability, 461–463
   interactive innovations influencing sustainability in, 463–464
   life cycle-based studies on, 353–360
   research and innovation in, 496–497
   sustainability challenges in, 366
   sustainability of, 492–497
   WASP interactive innovation toward enhanced sustainability, 475–479
   Wines of Alentejo Sustainability Program, 467–474
   wine supply chain, 351–353
   wine value chains in Portugal and Alentejo, 464–467
Wines of Alentejo Sustainability Program, 467–474
Wine spirit production, 259–260
   aged wine spirit and its production process, 260–262
   aging stage, 264–273
Wine supply chain, 351–353, 396t–397t
Wine value chains in Portugal and Alentejo, 464–467
Wooden barrels, 268–270
Wood extracts, 272
Wood fragments, 270–271
   combined with microoxygenation, 272
Wood reuse to more sustainable process, 268

Y

Yealands Wine Group, 425

Printed in the United States
by Baker & Taylor Publisher Services